biology

today and tomorrow
with physiology

biology

today and tomorrow

Cecie Starr

Christine A. Evers

Lisa Starr

BROOKS/COLE
CENGAGE Learning

Australia • Brazil • Japan • Korea • Mexico • Singapore • Spain • United Kingdom • United States

BROOKS/COLE
CENGAGE Learning™

Biology: Today and Tomorrow With Physiology, International Edition
Cecie Starr, Christine A. Evers, Lisa Starr

Senior Acquisitions Editor, Life Sciences: Peggy Williams

Publisher, Life Sciences: Yolanda Cossio

Assistant Editor: Shannon Holt

Editorial Assistant: Sean Cronin

Media Editor: Lauren Oliveira

Marketing Manager: Tom Ziolkowski

Marketing Coordinator: Lianne Ramsauer

Marketing Communications Manager: Darlene Macanan

Content Project Manager: Hal Humphrey

Art Director: John Walker

Manufacturing Planner: Karen Hunt

Rights Acquisitions Specialist: Roberta Broyer

Production Service: Grace Davidson & Associates, Inc.

Photo Researcher: Paul Forkner, Myrna Engler Photo Research Inc.

Text Researcher: Pablo D'Stair

Copy Editor: Anita Wagner

Illustrators: ScEYEnce Studios, Lisa Starr, Precision Graphics

Title Page Image: pharaoh eagle-owl; Biosphoto / Xavier Eichaker

Cover Image: atlas moth newly hatched caterpillar; Minden Pictures / Ch'ien Lee

Compositor: Lachina Publishing Services

Library of Congress Control Number: 2012933002

For permission to use material from this text or product, submit all requests online at **www.cengage.com/permissions**. Further permissions questions can be e-mailed to **permissionrequest@cengage.com**.

International Edition:

ISBN-13: 978-1-133-59055-2

ISBN-10: 1-133-59055-1

Cengage Learning International Offices

Asia
www.cengageasia.com
tel: (65) 6410 1200

Australia/New Zealand
www.cengage.com.au
tel: (61) 3 9685 4111

Brazil
www.cengage.com.br
tel: (55) 11 3665 9900

India
www.cengage.co.in
tel: (91) 11 4364 1111

Latin America
www.cengage.com.mx
tel: (52) 55 1500 6000

UK/Europe/Middle East/Africa
www.cengage.co.uk
tel: (44) 0 1264 332 424

Represented in Canada by
Nelson Education, Ltd.
www.nelson.com
tel: (416) 752 9100 / (800) 668 0671

Cengage Learning is a leading provider of customized learning solutions with office locations around the globe, including Singapore, the United Kingdom, Australia, Mexico, Brazil, and Japan. Locate your local office at: **www.cengage.com/global**

For product information and free companion resources:
www.cengage.com/international
Visit your local office: **www.cengage.com/global**
Visit our corporate website: **www.cengage.com**

Printed in the United States of America
2 3 4 5 15 14 13 12

Contents in Brief

Contents

UNIT FOUR ECOLOGY

UNIT FIVE HOW ANIMALS WORK

Preface

It's an exciting time to be a biologist, with a wealth of new discoveries every day, and biology-related issues such as climate change and stem cell research often making headlines. However, this avalanche of information can intimidate non-scientists. This book was designed and written specifically for students who most likely will not become biologists and may never again take another science course. It is an accessible and engaging introduction to biology that provides future decision-makers with an understanding of basic biology and the process of science.

Accessible Text Understanding stems mainly from making connections between concepts and details, so a text with too little detail reads as a series of facts that beg to be memorized. However, excessive detail can overwhelm the introductory student. Thus, we constantly strive to strike the perfect balance between level of detail and accessibility. We once again revised the text to eliminate details that do not contribute to a basic understanding of essential concepts. We also know that English is a second language for many students, so we avoid idioms and aim for a clear, straightforward style.

This edition includes many analogies to familiar objects and phenomena that will help students understand abstract concepts. For example, in the discussion of transpiration in Chapter 27 (Plant Form and Function), we explain that a column of water is drawn upward through xylem as a drinker draws fluid up through a straw. A graphic of a drink with a straw (*right*) reinforces the explanation in the text.

A Wealth of Timely Applications This book is packed with everyday applications of biological processes. At every opportunity, we enliven discussions of biological processes with references to their effects on human health and the environment. (In the index, you'll find health-related applications denoted by red squares and environmental applications by green squares.) Each chapter begins with a section that explores a current event or controversy directly related to the chapter's content. For example, a discussion of binge drinking on college campuses introduces the concept of metabolism in Chapter 4.

The introductory topic is integrated into the chapter material and the final **REVISITED** section of the chapter. This final section uses chapter content to reinforce and expand the topic presented in the opening essay. In chapter 4, for

example, this section presents a metabolic pathway that breaks down alcohol, linking the function of enzymes in this pathway to hangovers, the alcohol flushing reaction, alcoholism, and cirrhosis. The section is illustrated with a photo of a cirrhotic human liver and a very ill Gary Reinbach, who died at age 22 of alcoholic liver disease.

Emphasis on the Process of Science To strengthen a student's analytical skills and offer insight into contemporary research, each chapter includes an exercise called **DIGGING INTO DATA**. The exercise consists of a short text passage—usually about a published scientific experiment—and a table, chart, or other graphic that presents experimental data. A student can use information in the text and graphic to answer a series of questions. For example, the exercise in Chapter 2 asks students to interpret results of a study that examined the effect of dietary fat intake on "good" and "bad" cholesterol levels.

Throughout the book, descriptions of current research and photos of the scientists who do it drive home the message that science is an ongoing endeavor carried out by a diverse community of people. For example, Chapter 14 (Plants and Fungi) describes an international team of plant scientists (*right*) now seeking ways to protect the world's wheat supply form a newly evolved strain of fungal disease. We discuss not only what researchers have already discovered, but also how discoveries were made, how our understanding has changed over time, and what remains under investigation.

In-Text Learning Tools To emphasize connections between biological topics, each chapter begins with **WHERE YOU HAVE BEEN**, a brief list of concepts that were introduced earlier and will reappear in the chapter. A similar **WHERE YOU ARE GOING** at the close of the chapter text points toward later discussions that will be based on the current chapter's material.

The chapter itself consists of several numbered sections that contain a manageable chunk of information. Every section ends with a boxed **TAKE-HOME MESSAGE** in which we pose a question that reflects the critical content of the section, and then answer the question in bulleted list format. Every chapter has at least one **FIGURE IT OUT QUESTION** with an answer immediately following. These questions allow students to quickly check their understanding as they read. Mastering scientific vocabulary challenges many students, so we have included an **ON-PAGE GLOSSARY** of key terms introduced in each two-page spread, in addition to a complete glossary at the book's end. The end-of-chapter material features a **VISUAL SUMMARY** that reinforces each chapter's key concepts. A **SELF-QUIZ** poses multiple choice and other short answer questions for self-assessment (answers are in Appendix I). A set of more challenging **CRITICAL THINKING QUESTIONS** provides thought-provoking exercises for the motivated student.

Design and Content Revisions The following list highlights some of the revisions to each chapter. A section-by-section guide to revised content and figures is available upon request from your sales representative.

Introduction - Chapter 1 has an expanded discussion of science and many new photos. Material on taxonomy was moved to this chapter for an early introduction.

Unit I How Cells Work
• Chapter 2 has an expanded discussion of hydrogen bonding and its importance in liquid water. Molecular models were updated. A new photo of PET scans illustrates the use of tracers.
• Chapter 3 includes a section that considers how we define life.
• Chapter 4 has new photos illustrating energy. Enzyme discussion has updated graphics that include a model of an active site.
• Chapter 5 introduction now includes material on anthropogenic carbon dioxide emissions. A new overview diagram illustrates aerobic respiration in a mitochondrion.

Unit II Genetics
• Chapter 6 has a new introductory essay about cloned 9/11 rescue dog Trackr.
• Chapter 7 now includes epigenetics and related discoveries.
• Chapter 8 reorganized; expanded discussion of sexual vs. asexual reproduction now includes the Red Queen hypothesis.
• Chapter 9 has many new photos and includes a discussion of Tay-Sachs. Genetic screening section has been heavily revised.
• Chapter 10 opens and concludes with sections about SNP testing and the implications of information it provides. DNA profiling updates the material on DNA fingerprinting.

Unit III Evolution and Diversity
• Chapter 11 fossil section is expanded with many new photos.
• Chapter 12 has much expanded coverage of cladistics, including examples of practical applications. Behavioral isolation now illustrated with peacock spiders; adaptive radiation, with honeycreepers; coevolution, with large blue butterflies and ants.
• Chapter 13 now opens with a section about how understanding the evolutionary history of pathogens such as HIV can help us fight disease. New art depicts viral reassortment.
• Chapter 14 now opens with a section about how wheat stem rust fungus threatens the global food supply.

• Chapter 15 has heavily revised coverage of primate and human evolution, and coverage of Neanderthals has expanded.

Unit IV Ecology
• Chapter 16 has a new Revisited section discussing geese that caused a plane crash. Ecological footprints added.
• Chapter 17 includes a new subsection about plant-herbivore interactions.
• Chapter 18 has a reorganized discussion of biomes.

Unit V How Animals Work
• Chapter 19 has updated information about stem cell research and improved coverage of thermoregulatory strategies.
• Chapter 20 now discusses the negative health effects of sitting.
• Chapter 21 has a new opening section about sudden cardiac arrest, automatic external defibrillators, and CPR.
• Chapter 22 has a new section consolidating material on antigen receptors and processing; many new photos added.
• Chapter 23 has updated coverage of first aid for choking and USDA nutrition recommendations.
• Chapter 24 has improved coverage of vision.
• Chapter 25 has updated coverage of endocrine disruptors.
• Chapter 26 has updated coverage of *in-vitro* fertilization and other assisted reproductive technologies.

Unit VI How Plants Work
• Chapter 27 has many new photos and updated artwork; added material on stem and leaf specializations.
• Chapter 28 has been reorganized and expanded. It includes updated art and many new photos; updated information about colony collapse disorder; and new material on diversity and function of flower structure as related to pollination vectors. The plant hormone sections have been heavily revised to reflect recent research breakthroughs and the resulting paradigm shifts.

Acknowledgments

No list can convey our thanks to the team of dedicated people who made this book happen, including our academic advisors listed on the following page. We could not have done this book without Grace Davidson, who coordinated the flow of files, photos, and art, and whose gentle and patient prodding kept us all on schedule. Paul Forkner tenaciously tracked down the many exceptional photos that make this book unique. Copyeditor Anita Wagner and proofreader Diane Miller ensured our text is clear and concise. At Cengage Learning, Yolanda Cossio and Peggy Williams unwaveringly supported us and our ideals.

LISA STARR, CHRISTINE EVERS, AND CECIE STARR   *March, 2012*

Academic Advisors of Recent Editions

We owe a special debt to the members of our advisory board, listed below. They helped us shape the book's design and to choose appropriate content. We appreciate their guidance.

Andrew Baldwin, *Mesa Community College*
Charlotte Borgeson, *University of Nevada, Reno*
Gregory A. Dahlem, *Northern Kentucky University*
Gregory Forbes, *Grand Rapids Community College*
Hinrich Kaiser, *Victor Valley Community College*
Lyn Koller, *Pierce College*
Terry Richardson, *University of North Alabama*

We also wish to thank the reviewers listed below.

Idris Abdi, *Lane College*
Meghan Andrikanich, *Lorain County Community College*
Lena Ballard, *Rock Valley College*
Barbara D. Boss, *Keiser University, Sarasota*
Susan L. Bower, *Pasadena City College*
James R. Bray Jr., *Blackburn College*
Mimi Bres, *Prince George's Community College*
Randy Brewton, *University of Tennessee*
Evelyn K. Bruce, *University of North Alabama*
Steven G. Brumbaugh, *Green River Community College*
Chantae M. Calhoun, *Lawson State Community College*
Thomas F. Chubb, *Villanova University*
Julie A. Clements, *Keiser University, Melbourne*
Francisco Delgado, *Pima Community College*
Elizabeth A. Desy, *Southwest Minnesota State University*
Brian Dingmann, *University of Minnesota, Crookston*
Josh Dobkins, *Keiser University, online*
Hartmut Doebel, *The George Washington University*
Pamela K. Elf, *University of Minnesota, Crookston*
Johnny El-Rady, *University of South Florida*
Patrick James Enderle, *East Carolina University*
Jean Engohang-Ndong, *BYU Hawaii*
Ted W. Fleming, *Bradley University*
Edison R. Fowlks, *Hampton University*
Martin Jose Garcia Ramos, *Los Angeles City College*
J. Phil Gibson, *University of Oklahoma*
Judith A. Guinan, *Radford University*
Carla Guthridge, *Cameron University*
Laura A. Houston, *Northeast Lakeview–Alamo College*
Robert H. Inan, *Inver Hills Community College*
Dianne Jennings, *Virginia Commonwealth University*
Ross S. Johnson, *Chicago State University*
Susannah B. Johnson Fulton, *Shasta College*
Paul Kaseloo, *Virginia State University*
Ronald R. Keiper, *Valencia Community College West*
Dawn G. Keller, *Hawkeye Community College*
Ruhul H. Kuddus, *Utah Valley State College*
Dr. Kim Lackey, *University of Alabama*
Vic Landrum, *Washburn University*
Lisa Maranto, *Prince George's Community College*
Catarina Mata, *Borough of Manhattan Community College*
Kevin C. McGarry, *Keiser University, Melbourne*
Timothy Metz, *Campbell University*
Ann J. Murkowski, *North Seattle Community College*
Alexander E. Olvido, *John Tyler Community College*
Joshua M. Parke, *Community College of Southern Nevada*
Elena Pravosudova, *Sierra College*
Nathan S. Reyna, *Howard Payne University*
Carol Rhodes, *Cañada College*
Todd A. Rimkus, *Marymount University*
Laura H. Ritt, *Burlington County College*
Lynette Rushton, *South Puget Sound Community College*
Erik P. Scully, *Towson University*

Marilyn Shopper, *Johnson County Community College*
Jennifer J. Skillen, *Community College of Southern Nevada*
Jim Stegge, *Rochester Community and Technical College*
Lisa M. Strain, *Northeast Lakeview College*
Jo Ann Wilson, *Florida Gulf Coast University*

We were also fortunate to have conversations with the following workshop attendees. The insights they shared proved invaluable.

Robert Bailey, *Central Michigan University*
Brian J. Baumgartner, *Trinity Valley Community College*
Michael Bell, *Richland College*
Lois Borek, *Georgia State University*
Heidi Borgeas, *University of Tampa*
Charlotte Borgenson, *University of Nevada*
Denise Chung, *Long Island University*
Sehoya Cotner, *University of Minnesota*
Heather Collins, *Greenville Technical College*
Joe Conner, *Pasadena Community College*
Gregory A. Dahlem, *Northern Kentucky University*
Juville Dario-Becker, *Central Virginia Community College*
Jean DeSaix, *University of North Carolina*
Carolyn Dodson, *Chattanooga State Technical Community College*
Kathleen Duncan, *Foothill College, California*
Dave Eakin, *Eastern Kentucky University*
Lee Edwards, *Greenville Technical College*
Linda Fergusson-Kolmes, *Portland Community College*
Kathy Ferrell, *Greenville Technical College*
April Ann Fong, *Portland Community College*
Kendra Hill, *South Dakota State University*
Adam W. Hrincevich, *Louisiana State University*
David Huffman, *Texas State University, San Marcos*
Peter Ingmire, *San Francisco State*
Ross S. Johnson, *Chicago State University*
Rose Jones, *NW-Shoals Community College*
Thomas Justice, *McLennan Community College*
Jerome Krueger, *South Dakota State University*
Dean Kruse, *Portland Community College*
Dale Lambert, *Tarrant County College*
Debabrata Majumdar, *Norfolk State University*
Vicki Martin, *Appalachian State University*
Mary Mayhew, *Gainesville State College*
Roy Mason, *Mt. San Jacinto College*
Alexie McNerthney, *Portland Community College*
Brenda Moore, *Truman State University*
Alex Olvido, *John Tyler Community College*
Molly Perry, *Keiser University*
Michael Plotkin, *Mt. San Jacinto College*
Amanda Poffinbarger, *Eastern Illinois University*
Johanna Porter-Kelley, *Winston-Salem State University*
Sarah Pugh, *Shelton State Community College*
Larry A. Reichard, *Metropolitan Community College*
Darryl Ritter, *Okaloosa-Walton College*
Sharon Rogers, *University of Las Vegas*
Lori Rose, *Sam Houston State University*
Matthew Rowe, *Sam Houston State University*
Cara Shillington, *Eastern Michigan University*
Denise Signorelli, *Community College of Southern Nevada*
Jennifer Skillen, *Community College of Southern Nevada*
Jim Stegge, *Rochester Community and Technical College*
Andrew Swanson, *Manatee Community College*
Megan Thomas, *University of Las Vegas*
Kip Thompson, *Ozarks Technical Community College*
Steve White, *Ozarks Technical Community College*
Virginia White, *Riverside Community College*
Lawrence Williams, *University of Houston*
Michael L. Womack, *Macon State College*

Invitation to Biology

1

Whether or not you have studied biology, you already have an intuitive understanding of life on Earth because you are part of it. Every one of your experiences with the natural world—from the warmth of the sun on your skin to the love of your pet—contributes to that understanding.

1.1 The Secret Life of Earth

In this era of detailed satellite imagery and cell phone global positioning systems, could there possibly be any places left on Earth that humans have not yet explored? Actually, there are plenty of them. For example, a team of scientists was recently dropped by helicopter into the middle of a vast and otherwise inaccessible cloud forest on top of New Guinea's Foja Mountains. Within a few minutes, the explorers realized that their landing site, a dripping, moss-covered swamp, had been untouched by humans. Team member Bruce Beehler remarked, "Everywhere we looked, we saw amazing things we had never seen before. I was shouting. This trip was a once-in-a-lifetime series of shouting experiences."

How did the explorers know they had landed in uncharted territory? For one thing, the forest was filled with plants and animals previously unknown even to native peoples that have long inhabited other parts of the region. During the next month, the team members discovered many new species, including a rhododendron plant with flowers the size of plates and a frog the size of a pea. They also came across hundreds of species that are on the brink of extinction in other parts of the world, and some that supposedly had been extinct for decades. The animals had never learned to be afraid of humans, so they could easily be approached. A few were discovered as they casually wandered through camps (Figure 1.1).

New species are discovered all the time, often in places much more mundane than Indonesian cloud forests. How do we know what species a particular organism belongs to? What is a species, anyway, and why should discovering a new one matter to anyone other than a scientist? You will find the answers to such questions in this book. They are part of the scientific study of life, **biology**, which is one of many ways we humans try to make sense of the world around us.

Trying to understand the immense scope of life on Earth gives us some perspective on where we fit into it. For example, the current rate of extinctions is about 1,000 times faster than normal, and human activities are responsible for the acceleration. At this rate, we will never know about most of the species that are alive on Earth today. Does that matter? Biologists think so. Whether or not we are aware of it, humans are intimately connected with the world around us. Our activities are profoundly changing the entire fabric of life on Earth. The changes are, in turn, affecting us in ways we are only beginning to understand.

Ironically, the more we learn about the natural world, the more we realize we have yet to learn. But don't take our word for it. Find out what biologists know, and what they do not, and you will have a solid foundation on which to base your own opinions about the human connection—your connection—with all life on Earth.

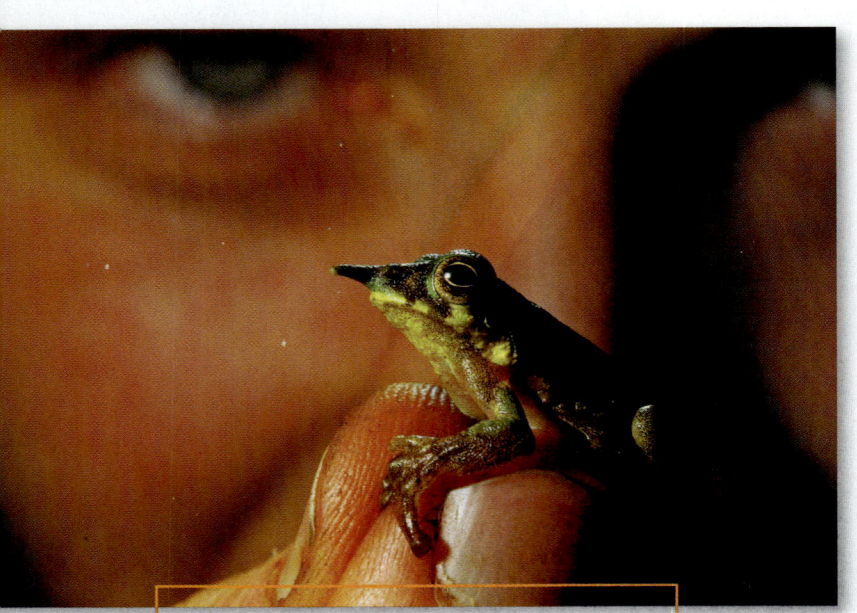

FIGURE 1.1 Paul Oliver discovered this tiny tree frog perched on a sack of rice during a particularly rainy campsite lunch. The explorers dubbed the new species "Pinocchio frog" after the Disney character because the male frog's long nose inflates and points upward during times of excitement.

Tim Laman/ National Geographic Stock.

1.2 Life Is More Than the Sum of Its Parts

Biologists study all aspects of life, past and present. What, exactly, is the property we call "life"? We may never come up with a good definition, because living things are too diverse, and they consist of the same basic components as nonliving things. When we try to define life, we end up only identifying properties that differentiate living from nonliving things. Even so, we do understand that life is organized in successive levels, with new properties emerging at each level (Figure 1.2). Complex properties, including life, often emerge from the interactions of much simpler parts. For an example, in the drawings *below*, a property called "roundness" emerges

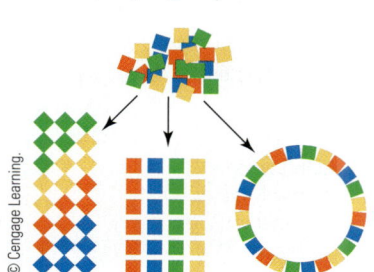

when the parts are organized one way, but not other ways. The idea that different structures can be assembled from the same basic building blocks is a common theme in our world, and also in biology. Consider **atoms**, which are fundamental units of matter—the building blocks of all substances ❶. Atoms bond together as **molecules** ❷. There are no atoms unique to living things, but there are unique molecules. In today's world, only living things make the "molecules of life," which are lipids, proteins, DNA, RNA, and complex carbohydrates. The property of "life" appears at the next level, when many molecules of life become organized as a cell. A **cell** is the smallest unit of life ❸. Some cells live and reproduce independently; others do so as part of a multicelled organism. An **organism** is an individual that consists of one or more cells ❹. Cells of multicelled organisms are typically organized as tissues, organs, and organ systems that interact to keep the individual's body working properly. A **population** is a group of interbreeding individuals of the same species living in a given area ❺. At the next level, a **community** consists of all populations of all species in a given area ❻. Communities may be large or small, depending on the area defined. The next level of organization is the **ecosystem**, which is a community interacting with its physical and chemical environment ❼. The most inclusive level, the **biosphere**, encompasses all regions of Earth's crust, waters, and atmosphere in which organisms live ❽.

FIGURE 1.2 Animated! Levels of organization in nature, from simpler to more complex.

❶ Atoms are fundamental units of matter.

❷ Molecules consist of atoms.

❸ Cells consist of molecules.

❹ Organisms consist of cells.

❺ Populations consist of organisms.

❻ Communities consist of populations.

❼ Ecosystems consist of communities interacting with their environment.

❽ The biosphere consists of all ecosystems on Earth.

© Cengage Learning 2012.

Take-Home Message

How do living things differ from nonliving things?

■ All things, living or not, consist of the same building blocks: atoms. Atoms join as molecules.

■ The unique properties of life emerge as certain kinds of molecules become organized into cells.

■ Higher levels of life's organization include multicelled organisms, populations, communities, ecosystems, and the biosphere.

atom Fundamental building block of all matter.
biology The scientific study of life.
biosphere All regions of Earth where organisms live.
cell Smallest unit of life.
community All populations of all species in a given area.
ecosystem A community interacting with its environment.
molecule Two or more atoms bonded together.
organism Individual that consists of one or more cells.
population Group of interbreeding individuals of the same species that live in a given area.

A Producers harvest energy from the environment. Some of that energy flows from producers to consumers.

sunlight energy

Producers
plants and other self-feeding organisms

B Nutrients that become incorporated into the cells of producers and consumers are eventually released back into the environment (by decomposition, for example). Producers take up some of the released nutrients.

Consumers
animals, most fungi, many protists, bacteria

C All of the energy that enters the world of life eventually flows out of it, mainly as heat released back to the environment.

FIGURE 1.3 Animated!
The one-way flow of energy and the cycling of materials in the world of life. The photo *above* shows a producer acquiring energy and nutrients from the environment, and consumers acquiring energy and nutrients by eating the producer.

Credits: left, © Victoria Pinder, www.flickr.com/photos/vixstarplus; right, © Cengage Learning.

1.3 How Living Things Are Alike

A set of key features distinguishes living organisms from nonliving things. All living organisms require ongoing inputs of energy and raw materials; all sense and respond to change; and all have DNA that guides their functioning.

❭ Organisms Require Energy and Nutrients Not all living things eat, but all require energy and nutrients on an ongoing basis. A **nutrient** is a substance that an organism needs for growth and survival but cannot make for itself. Organisms spend a lot of time acquiring energy and nutrients because both are essential to maintain life's organization and functioning. However, what type of energy and nutrients they acquire varies considerably depending on the type of organism. The differences allow us to classify all living things into two categories: producers and consumers. **Producers** make their own food using energy and simple raw materials they get from non-biological sources. Plants are producers that use the energy of sunlight to make sugars from water and carbon dioxide (a gas in air), a process called **photosynthesis**. By contrast, **consumers** cannot make their own food. They get energy and nutrients by feeding on other organisms. Animals are consumers. So are decomposers, which feed on the wastes or remains of other organisms. The wastes and remains of consumers end up in the environment, where they serve as nutrients for producers. Said another way, nutrients cycle between producers and consumers.

FIGURE 1.4 Organisms sense and respond to conditions inside and outside of themselves. This baby orangutan is laughing in response to being tickled. Apes and humans make different sounds when being tickled, but the airflow patterns are so similar that we can say apes really do laugh.

© Dr. Marina Davila Ross, University of Portsmouth.

Energy, however, is not cycled. It flows from the environment, through organisms, and back to the environment. This flow maintains the organization of every living cell and body, and it also influences how individuals interact with one another and their environment. The energy flow is one-way, because with each transfer, some energy escapes as heat, and cells cannot use heat as an energy source. Thus, energy that enters the world of life eventually leaves it (Figure 1.3). We return to this topic in Chapter 5.

> **Organisms Sense and Respond to Change** Living things detect and respond to conditions both inside and outside of themselves (Figure 1.4). As an example, after you eat, the sugars from your meal enter your bloodstream. The added sugars set in motion a series of events that causes cells throughout the body to take up sugar faster, so the sugar level in your blood quickly falls. This response keeps your blood sugar level within a certain range, which in turn helps keep your cells alive and your body functioning.

The fluid portion of your blood is a component of your internal environment, which is all of the body fluids outside of cells. Unless that internal environment is kept within certain ranges of composition, temperature, and other conditions, your body cells will die. By sensing and adjusting to change, you and all other organisms keep conditions in the internal environment within a range that favors survival. **Homeostasis** is the name for this process, and it is a defining feature of life.

> **Organisms Grow and Reproduce** With little variation, the same types of molecules perform the same basic functions in every organism. For example, information encoded in an organism's **DNA** (deoxyribonucleic acid) guides the ongoing metabolic activities that sustain the individual through its lifetime. Such activities include **development**: the process by which the first cell of a new individual becomes a multicelled adult; **growth**: increases in cell number, size, and volume; and **reproduction**: processes by which individuals produce offspring.

Individuals of every natural population are alike in certain aspects of their body form and behavior because their DNA is very similar: Orangutans look like orangutans and not like caterpillars because they inherited orangutan DNA,

Energy flow through the world of life maintains the organization of every living cell and body.

consumer Organism that gets energy and nutrients by feeding on tissues, wastes, or remains of other organisms.

development Process by which the first cell of a new individual becomes a multicelled adult.

DNA Deoxyribonucleic acid; carries hereditary information that guides development and functioning.

growth In multicelled species, an increase in the number, size, and volume of cells.

homeostasis Process in which an organism keeps its internal conditions within tolerable ranges by sensing and responding to change.

nutrient Substance that an organism needs for growth and survival but cannot make for itself.

photosynthesis Process by which producers use light energy to make sugars from carbon dioxide and water.

producer Organism that makes its own food using energy and nonbiological raw materials from the environment.

reproduction Process by which parents produce offspring.

which differs from caterpillar DNA in the information it carries. **Inheritance** refers to the transmission of DNA to offspring. All organisms receive their DNA from one or more parents.

DNA is the basis of similarities in form and function among organisms. However, the details of DNA molecules differ, and herein lies the source of life's diversity. Small variations in the details of DNA's structure give rise to differences among individuals, and also among types of organisms. As you will see in later chapters, these differences are the raw material of evolutionary processes.

DNA

Take-Home Message

How are all living things alike?

- A one-way flow of energy and a cycling of nutrients sustain life's organization.
- Organisms sense and respond to conditions inside and outside themselves. They make adjustments that keep conditions in their internal environment within a range that favors cell survival, a process called homeostasis.
- Organisms develop and function based on information encoded in their DNA, which they inherit from their parents. DNA is the basis of similarities and differences in form and function.

1.4 How Living Things Differ

Living things differ tremendously in their observable characteristics, or traits. Various classification schemes help us organize what we understand about the scope of this variation, which we call Earth's **biodiversity**.

For example, organisms can be grouped on the basis of whether they have a **nucleus**, which is a sac with two membranes that encloses and protects a cell's DNA. **Bacteria** (singular, bacterium) and **archaea** (singular, archaeon) are two types of organisms whose DNA is *not* contained within a nucleus. All bacteria and archaea are single-celled, which means individual organisms consist of one cell (Figure 1.5). Collectively, they are the most diverse representatives of life. Different kinds are producers or consumers in nearly all parts of the biosphere. Some inhabit such extreme environments as frozen desert rocks, boiling sulfurous lakes, and nuclear reactor waste. The first cells on Earth may have faced similarly hostile environments.

Traditionally, organisms without a nucleus have been called **prokaryotes**, but this designation is an informal one. Despite their similar appearance, bacteria and archaea are less related to one another than we had once thought. Archaea are actually more closely related to **eukaryotes**, organisms whose DNA is contained within a nucleus. Some eukaryotes live as individual cells; others are multicelled (Figure 1.6). Eukaryotic cells are typically larger and more complex than bacteria or archaea.

Structurally, **protists** are the simplest eukaryotes. As a group they vary a great deal, from single-celled consumers to giant, multicelled producers.

Fungi (singular, fungus) are eukaryotic consumers that secrete substances to break down food externally. Their cells then absorb nutrients released from the

A Bacteria are the most numerous organisms on Earth. *Left*, a bacterium with a row of iron crystals that acts like a tiny compass; *right*, spiral cyanobacteria.

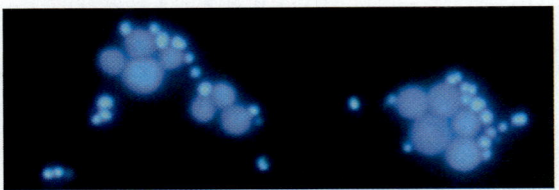

B Archaea may resemble bacteria, but they are more closely related to eukaryotes. These are two types of archaea from a hydrothermal vent on the seafloor.

FIGURE 1.5 Animated! A few representative bacteria and archaea.

Credits: (a) left, Dr. Richard Frankel; right, © Susan Barnes; (b) © Dr. Harald Huber, Dr. Michael Hohn, Prof. Dr. K.O. Stetter, University of Regensburg, Germany.

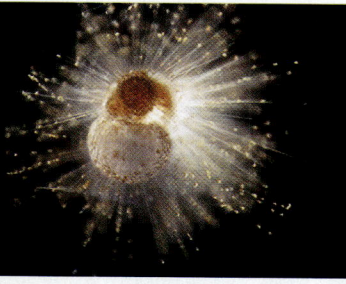

Protists are a group of extremely diverse eukaryotes that range from giant multicelled seaweeds to microscopic single cells. Biologists now view "protists" as a collection of major groups, some only distantly related to others.

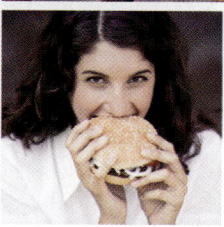

Animals are multicelled eukaryotes that ingest tissues or juices of other organisms. All actively move about during at least part of their life.

Fungi are eukaryotic consumers that secrete substances to break down food externally. Most are multicelled.

Plants are multicelled eukaryotes, most of which are photosynthetic. Nearly all have roots, stems, and leaves. Plants are the primary producers in land ecosystems.

> **FIGURE 1.6 Animated!** A few representative eukaryotes.
>
> Credits: clockwise from top left, © Lewis Trusty/ Animals Animals; Courtesy of Allen W. H. Bé andDavid A. Caron; © Tom & Pat Leeson, Ardea London Ltd.; © Martin Zimmerman, *Science*, 1961, 133:73-79, © AAAS.; © Pixtal/ SuperStock; Lady Bird Johnson Wildflower Center; © John Lotter Gurling/ Tom Stack & Associates; © Dr. Dennis Kunkel/ Visuals Unlimited; JupiterImages Corporation.

breakdown. Many fungi are decomposers. Individuals of most types, including those that form mushrooms, are multicellular. Yeasts and some other fungi are single-celled.

Plants are multicelled eukaryotes that live mainly on land. Nearly all are photosynthetic producers. Besides feeding themselves, plants and other photosynthesizers serve as food for most of the other organisms in the biosphere.

Animals are multicelled consumers that ingest tissues or juices of other organisms. Herbivores graze, carnivores eat meat, scavengers eat remains of other organisms, parasites withdraw nutrients from the tissues of a host, and so on. Animals develop through a series of stages that lead to the adult form. All kinds actively move about during at least part of their lives.

Take-Home Message

How do living things differ from one another?

- Organisms differ in their details; they show tremendous variation in observable characteristics, or traits.

- We can divide Earth's biodiversity into broad groups based on traits such as having a nucleus or being multicellular.

animal Multicelled eukaryotic consumer that develops through a series of stages and moves about during part or all of its life.

archaeon Member of a group of single-celled organisms that lack a nucleus but are more closely related to eukaryotes than to bacteria.

bacterium Member of the most diverse and well-known group of single-celled organisms that lack a nucleus.

biodiversity Scope of variation among living organisms.

eukaryote Organism whose cells characteristically have a nucleus.

fungus Single-celled or multicelled eukaryotic consumer that breaks down material outside itself, then absorbs nutrients released from the breakdown.

inheritance Transmission of DNA to offspring.

nucleus Double-membraned sac that encloses a cell's DNA.

plant A multicelled, typically photosynthetic eukaryote.

prokaryote Single-celled organism without a nucleus.

protist Member of a diverse group of simple eukaryotes.

domain	Eukarya	Eukarya	Eukarya	Eukarya	Eukarya
kingdom	Plantae	Plantae	Plantae	Plantae	Plantae
phylum	Magnoliophyta	Magnoliophyta	Magnoliophyta	Magnoliophyta	Magnoliophyta
class	Magnoliopsida	Magnoliopsida	Magnoliopsida	Magnoliopsida	Magnoliopsida
order	Apiales	Rosales	Rosales	Rosales	Rosales
family	Apiaceae	Cannabaceae	Rosaceae	Rosaceae	Rosaceae
genus	*Daucus*	*Cannabis*	*Malus*	*Rosa*	*Rosa*
species	*carota*	*sativa*	*domestica*	*acicularis*	*canina*
common name	wild carrot	marijuana	apple	prickly rose	dog rose

FIGURE 1.7 Taxonomic classification of five species that are related at different levels. Each species has been assigned to ever more inclusive groups, or taxa: in this case, from genus to domain.

Figure It Out: Which of the plants shown here are in the same order?

Answer: Marijuana, apple, prickly rose, and dog rose

1.5 What Is a "Species"?

Each time we discover a new **species**, or unique kind of organism, we name it. **Taxonomy**, a system of naming and classifying species, began thousands of years ago, but naming species in a consistent way did not become a priority until the eighteenth century. At the time, European explorers who were just discovering the scope of life's diversity started having more and more trouble communicating with one another because species often had multiple names. For example, the dog rose (a plant native to Europe, Africa, and Asia) was alternately known as briar rose, witch's briar, herb patience, sweet briar, wild briar, dog briar, dog berry, briar hip, eglantine gall, hep tree, hip fruit, hip rose, hip tree, hop fruit, and hogseed—and those are only the English names! Species often had multiple scientific names too, in Latin that was descriptive but often cumbersome. The scientific name of the dog rose was *Rosa sylvestris inodora seu canina* (odorless woodland dog rose), and also *Rosa sylvestris alba cum rubore, folio glabro* (pinkish white woodland rose with smooth leaves).

An eighteenth-century naturalist, Carolus Linnaeus, standardized a two-part naming system that we still use. By the Linnaean system, every species is given a unique two-part scientific name. The first part is the name of the **genus** (plural, genera), a group of species that share a unique set of features. The second part is the specific epithet. Together, the genus name and the specific epithet designate one species. Thus, the dog rose now has one official name, *Rosa canina*, that is recognized worldwide.

Genus and species names are always italicized. For example, *Panthera* is a genus of big cats. Lions belong to the species *Panthera leo*. Tigers belong to a different species in the same genus (*Panthera tigris*), and so do leopards (*P. pardus*). Note how the genus name may be abbreviated after it has been spelled out once.

〉 A Rose by Any Other Name The individuals of a species share a unique set of inherited traits. For example, giraffes normally have very long necks, brown spots on white coats, and so on. These are morphological (structural) traits. Individuals of a species also share biochemical traits (they make and

A six-kingdom classification system in which all eukaryotes have been sorted into one of four kingdoms: protists, plants, fungi, and animals. The protist kingdom includes the most ancient multi-celled and all single-celled eukaryotes.

FIGURE 1.8 Animated!
Two ways to see the big picture of life. Lines in such diagrams indicate evolutionary connections.
© Cengage Learning.

A three-domain system sorts all life into three domains: Bacteria, Archaea, and Eukarya. The Eukarya domain includes all eukaryotes.

use the same molecules) and behavioral traits (they respond the same way to certain stimuli, as when hungry giraffes feed on tree leaves).

We can rank species into ever more inclusive categories based on some subset of traits it shares with other species. Each rank, or **taxon** (plural, taxa), is a group of organisms that share a unique set of traits. Each category above species—genus, family, order, class, phylum (plural, phyla), kingdom, and domain—consists of a group of the next lower taxon (Figure 1.7). Using this system, we can sort all life into a few categories (Figure 1.8).

It is easy to tell that orangutans and tigers are different species because they look very different. Distinguishing species that share a more recent ancestor may be much more challenging (Figure 1.9). In addition, traits shared by members of a species often vary a bit among individuals, such as eye color does among people. How do we decide if similar-looking organisms belong to different species or not? The short answer is that we rely on whatever information we have. Early naturalists studied anatomy and distribution—essentially the only methods available at the time—so species were named and classified according to what they looked like and where they lived. Today's biologists are able to compare traits that the early naturalists did not even know about, including biochemical ones such as DNA sequence. For example, Linnaeus grouped plants by the number and arrangement of reproductive parts, a scheme that resulted in odd pairings such as castor-oil plants with pine trees. Having more information today, we place these plants in separate phyla.

Evolutionary biologist Ernst Mayr defined a species as one or more groups of individuals that potentially can interbreed, produce fertile offspring, and do not interbreed with other groups. This "biological species concept" is useful in many cases, but it is not universally applicable. For example, we may never know whether separate populations could interbreed even if they did get together. As another example, populations often continue to interbreed even as they diverge, so the exact moment at which two populations become two species is often impossible to pinpoint. We return to speciation and how it occurs in Chapter 12, but for now it is important to remember that a "species" is a convenient but artificial construct of the human mind.

FIGURE 1.9 Four butterflies, two species: Which are which?

The top row shows two forms of the species *Heliconius melpomene*; the bottom row, two forms of *Heliconius erato*. *H. melpomene* and *H. erato* never cross-breed. Their alternate but similar patterns of coloration evolved as a shared warning signal to predatory birds that these butterflies taste terrible.

© 2006 Axel Meyer, "Repeating Patterns of Mimicry." *PLoS Biology* Vol. 4, No. 10, e341 doi:10.1371/journal.pbio.0040341. Used with Permission.

Take-Home Message

How do we keep track of all the species we know about?

- Each species has a unique, two-part scientific name.
- Classification systems group species on the basis of shared traits.

genus A group of species that share a unique set of traits.
species Unique type of organism.
taxon Group of organisms that share a unique set of traits.
taxonomy The science of naming and classifying species.

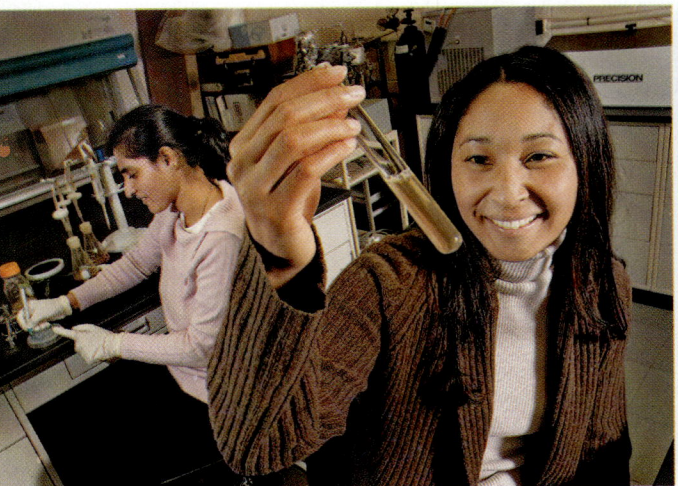

A Improving the efficiency of biofuel production from agricultural wastes.

B Studying benefits of weedy buffer zones on farms.

C Discovering medically active natural products in new species of marine animals.

FIGURE 1.10 Examples of research in the field of biology.

Credits: (a) © Roger W. Winstead, NC State University; (b) Photo by Scott Bauer, USDA/ ARS; (c) Courtesy of Susanna López-Legentil; (d) Volker Steger/ Photo Researchers, Inc.; (e) Cape Verde National Institute of Meteorology and Geophysics and the U.S. Geological Survey; (f) National Cancer Institute.

How do my own biases affect what I'm learning?

1.6 The Science of Nature

Most of us assume that we do our own thinking—but do we, really? You might be surprised to find out just how often we let others think for us. Consider how a school's job, which is to impart as much information as possible to students, meshes perfectly with a student's job, which is to acquire as much knowledge as possible. In the resulting rapid-fire transfer of information, it can be easy to forget about the quality of what is being transferred. Any time you accept information without question, you allow someone else to think for you.

❯ Thinking About Thinking **Critical thinking** is the deliberate process of judging the quality of information before accepting it. "Critical" comes from the Greek *kriticos* (discerning judgment). When you think this way, you move beyond the content of information to consider supporting evidence, bias, and alternative interpretations. How does the busy student manage it? Critical thinking does not necessarily require extra time, just a bit of extra awareness. There are many ways to do it. For example, you might ask yourself some of the following questions while you are learning something new:

What message am I being asked to accept?

Is the message based on facts or opinion?

Is there a different way to interpret the facts?

What biases might the presenter have?

How do my own biases affect what I'm learning?

Asking yourself questions like these is a way of being conscious about learning. It can help you decide whether to allow new information to guide your beliefs and actions.

❯ How Science Works Critical thinking is a big part of **science**, the systematic study of the observable world and how it works. A scientific line of inquiry usually begins with curiosity about something observable, such as a noticeable decrease in the number of birds in a particular area. Typically, a scientist will read about what others have discovered before making a **hypothesis**, which is a testable explanation for a natural phenomenon. An example of a

D Sequencing the human genome. **E** Looking for fungi traveling in atmospheric dust. **F** Devising a vaccine for cancer.

hypothesis would be, "In my neighborhood, the number of birds is decreasing because the number of cats is increasing." A **prediction**, or statement of some condition that should exist if the hypothesis is correct, comes next. Making predictions is called the if–then process, in which the "if" part is the hypothesis, and the "then" part is the prediction.

Next, a scientist will devise ways to test a prediction. Tests may be performed on a **model**, or analogous system, if working with an object or event directly is not possible. For example, animal diseases are often used as models of similar human diseases. Careful observations are one way to test predictions that flow from a hypothesis. So are **experiments**: tests designed to support or falsify a prediction. A typical experiment explores a cause-and-effect relationship.

Researchers use variables to investigate cause-and-effect relationships. **Variables** are characteristics or events that differ among individuals or over time. Biological systems are complex, which means they involve many variables that are difficult to study separately. Thus, biology researchers often test two groups of individuals simultaneously. An **experimental group** is a set of individuals that have a certain characteristic or receive a certain treatment. This group is tested side by side with a **control group**, which is identical to the experimental group except for one variable: the characteristic or the treatment being tested. Any differences in experimental results between the two groups should be an effect of changing the variable.

Test results—**data**—that are consistent with the prediction are evidence in support of the hypothesis. Data inconsistent with the prediction are evidence that the hypothesis is flawed and should be revised.

A necessary part of science is reporting one's results and conclusions in a standard way, such as in a peer-reviewed journal article. The communication gives other scientists an opportunity to evaluate the information for themselves, both by checking the conclusions drawn and by repeating the experiments.

Forming a hypothesis based on observation, and then systematically testing it, evaluating it, and sharing the results are collectively called the **scientific method** (Table 1.1).

There are many different ways to do research, particularly in biology (Figure 1.10). Some biologists do surveys; they observe without making hypotheses. Some make hypotheses and leave the experimentation to others. However,

Table 1.1	The Scientific Method
1. Observe some aspect of nature.	
2. Think of an explanation for your observation (in other words, form a hypothesis).	

3. Test the hypothesis.
    a. Make a prediction based on the hypothesis.
    b. Test the prediction using experiments or surveys.
    c. Analyze the results of the tests (data).	

4. Decide whether the results of the tests support your hypothesis or not (form a conclusion).	
5. Report your results to the scientific community.	

control group In an experiment, a group of individuals who are not exposed to the variable being tested.

critical thinking Judging information before accepting it.

data Values or other factual information obtained from experiments or surveys.

experiment A test designed to support or falsify a prediction.

experimental group In an experiment, a group of individuals who are exposed to a variable.

hypothesis Testable explanation of a natural phenomenon.

model Analogous system used for testing hypotheses.

prediction Statement, based on a hypothesis, about a condition that should exist if the hypothesis is correct.

science Systematic study of the observable world.

scientific method Systematically making, testing, and evaluating hypotheses.

variable A characteristic or event that differs among individuals or over time.

despite a broad range of subject matter, scientific experiments are typically designed in a consistent way, so the effects of changing one variable at a time can be measured. To give you a sense of how biology experiments work, we summarize two published studies here.

In 1996 the U.S. Food and Drug Administration (the FDA) approved Olestra®, a fat replacement manufactured from sugar and vegetable oil, as a food additive. Potato chips were the first Olestra-containing food product on the market in the United States. Controversy about the food additive soon raged. Many people complained of intestinal problems after eating the chips and thought that the Olestra was at fault. Two years later, researchers at Johns Hopkins University School of Medicine designed an experiment to test the hypothesis that this food additive causes cramps.

The researchers predicted *if* Olestra causes cramps, *then* people who eat Olestra will be more likely to get cramps than people who do not. To test their prediction, they used a Chicago theater as a "laboratory." They asked 1,100 people between the ages of thirteen and thirty-eight to watch a movie and eat potato chips. Each person got an unmarked bag that contained 13 ounces of chips. In this experiment, individuals who ate Olestra-containing potato chips constituted the experimental group, and individuals who ate regular chips were the control group. The variable was the presence or absence of Olestra in the chips.

A few days after the experiment was finished, the researchers contacted all of the people and collected any reports of post-movie gastrointestinal problems. Of 563 people making up the experimental group, 89 (15.8 percent) complained about cramps. However, so did 93 of the 529 people (17.6 percent) making up the control group—who had eaten the regular chips. In this experiment, people were about as likely to get cramps whether or not they ate chips made with Olestra. These results did not support the prediction, so the researchers concluded that eating Olestra does not cause cramps (Figure 1.11).

A different experiment that took place in 2005 investigated whether certain behaviors of peacock butterflies defend these insects from predation by birds. The researchers performing this experiment began with two observations. First, when a peacock butterfly rests, it folds its wings, so only the dark underside shows (Figure 1.12**A**). Second, when a butterfly sees a predator approaching, it repeatedly flicks its wings open, while also moving them in a way that produces a hissing sound and a series of clicks (Figure 1.12**B**).

The researchers were curious about why the peacock butterfly flicks its wings. After they reviewed earlier studies, they came up with two hypotheses that might explain the wing-flicking behavior:

1. Although wing-flicking probably attracts predatory birds, it also exposes brilliant spots that resemble owl eyes. Anything that looks like owl eyes is known to startle small, butterfly-eating birds, so exposing the wing spots might scare off predators.

2. The hissing and clicking sounds produced when the peacock butterfly moves its wings may be an additional defense that deters predatory birds.

The researchers then used their hypotheses to make the following predictions:

1. *If* peacock butterflies startle predatory birds by exposing their brilliant wing spots, *then* individuals with wing spots will be less likely to get eaten by predatory birds than those without wing spots.

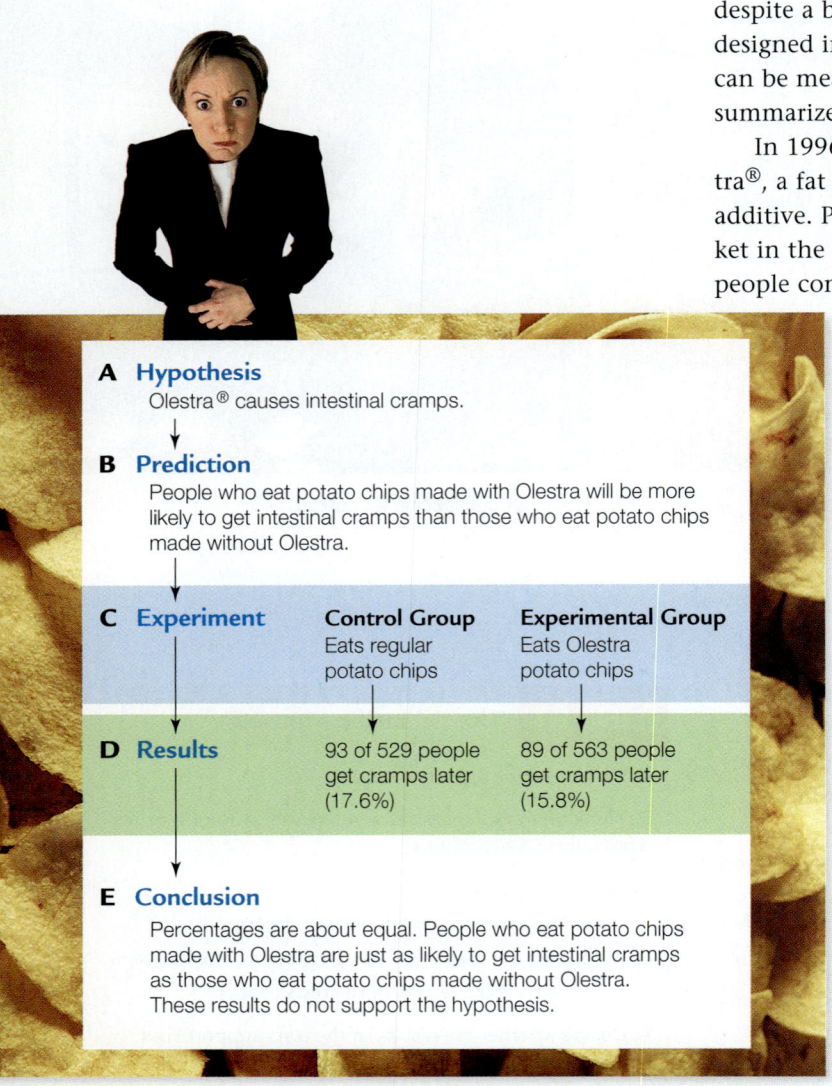

A Hypothesis
Olestra® causes intestinal cramps.

B Prediction
People who eat potato chips made with Olestra will be more likely to get intestinal cramps than those who eat potato chips made without Olestra.

C Experiment

	Control Group	Experimental Group
	Eats regular potato chips	Eats Olestra potato chips

D Results

	93 of 529 people get cramps later (17.6%)	89 of 563 people get cramps later (15.8%)

E Conclusion
Percentages are about equal. People who eat potato chips made with Olestra are just as likely to get intestinal cramps as those who eat potato chips made without Olestra. These results do not support the hypothesis.

FIGURE 1.11 The steps in a scientific experiment to determine if Olestra causes cramps. A report of this study was published in the *Journal of the American Medical Association* in January of 1998.

A With wings folded, a resting peacock butterfly looks a bit like a dead leaf.

B When a bird approaches, a butterfly repeatedly flicks its wings open. This behavior exposes brilliant spots and also produces hissing and clicking sounds.

C Researchers tested whether peacock butterfly wing flicking and hissing reduce predation by blue tits.

FIGURE 1.12 Testing peacock butterfly defenses. Researchers painted out the spots of some butterflies, cut the sound-making part of the wings on others, and did both to a third group; then exposed each butterfly to a hungry blue tit. Results are listed *right*, in Table 1.2.

Figure It Out: What percentage of butterflies with no spots and no sound survived the test?
Answer: 20 percent

Credits: (a) © Matt Rowlings, www.eurobutterflies.com; (b) © Adrian Vallin; (c) © Antje Schulte.

Table 1.2	Results of Peacock Butterfly Experiment*			
Wing Spots	Wing Sound	Total Number of Butterflies	Number Eaten	Number Survived
Spots	Sound	9	0	9 (100%)
No spots	Sound	10	5	5 (50%)
Spots	No sound	8	0	8 (100%)
No spots	No sound	10	8	2 (20%)

* Proceedings of the Royal Society of London, Series B (2005) 272: 1203–1207.

2. *If* peacock butterfly sounds deter predatory birds, *then* sound-producing individuals will be less likely to get eaten by predatory birds than silent individuals.

The next step was the experiment. The researchers used a marker to paint the wing spots of some butterflies, and scissors to cut off the sound-making part of the wings of others. A third group had both treatments: their wing spots were painted and cut. The researchers then put each butterfly into a large cage with a hungry blue tit (Figure 1.12**C**) and watched the pair for thirty minutes.

Table 1.2 lists the results of the experiment. All of the butterflies with unmodified wing spots survived, regardless of whether they made sounds. By contrast, only half of the butterflies that had spots painted out but could make sounds survived. Most of the silenced butterflies with painted-out spots were eaten quickly. The test results confirmed both predictions, so they support the hypotheses. Predatory birds are indeed deterred by peacock butterfly sounds, and even more so by wing spots.

Take-Home Message

What is science?

- The scientific method consists of making, testing, and evaluating hypotheses, and sharing results. It is a way of critical thinking, or systematically judging the quality of information before allowing it to guide one's beliefs and actions.

- Experiments systematically measure the effect on a natural system of changing a variable.

- Natural processes are often influenced by many interacting variables. Experiments help researchers unravel causes of complex natural processes by focusing on the effects of changing a single variable.

1.7 Analyzing Experimental Results

〉Sampling Error Researchers can rarely observe all individuals of a group. For example, the explorers you read about in Section 1.1 did not—and could not—survey every uninhabited part of the Foja Mountains. The cloud forest alone cloaks more than 2 million acres (Figure 1.13**A**), so surveying all of it would take unrealistic amounts of time and effort. Besides, tromping about even in a small area can damage delicate forest ecosystems.

Given such limitations, researchers often look at subsets of an area, a population, or an event. They test or survey the subset, then use the results to make generalizations. However, generalizing from a subset is risky because the subset may not be representative of the whole. Consider the golden-mantled tree kangaroo, an animal that was first discovered in 1993 on a single forested mountaintop in New Guinea. For more than a decade, the species was never seen outside of that habitat, which is getting smaller every year because of human activities. Thus, the golden-mantled tree kangaroo was considered to be one of the most endangered animals on the planet. Then, in 2005, the New Guinea explorers discovered that this kangaroo species is fairly common in the Foja Mountain cloud forest (Figure 1.13**B**). As a result, biologists now believe its future is secure, at least for the time being.

Sampling error is a difference between results obtained from testing a subset of a group, and results from testing the whole group. Sampling error may be unavoidable, as illustrated by the example of the golden-mantled tree kangaroo. However, knowing how it can occur helps researchers design their experiments to minimize it. For example, sampling error can be a substantial problem with a small subset, so experimenters try to start with a relatively large sample, and they typically repeat their experiments (Figure 1.14).

To understand why such practices reduce the risk of sampling error, think about what happens each time you flip a coin. There are two possible outcomes: The coin lands heads up, or it lands tails up. Thus, the chance that the coin will land heads up is one in two (1/2), which is a proportion of 50 percent. However, when you flip a coin repeatedly, it often lands heads up, or tails up, several times in a row. With just 3 flips, the proportion of times that heads actually land up may not even be close to 50 percent. With 1,000 flips, the proportion of times that the coin lands heads up is likely to be near 50 percent.

In cases like flipping a coin, it is possible to calculate probability. **Probability** is the measure, expressed as a percentage, of the chance that a particular outcome will occur. That chance depends on the total number of possible outcomes. For instance, imagine that 10 million people enter a random drawing to win a car. There are 10 million possible outcomes, so each person has the same probability of winning the item: 1 in 10 million, or (a very improbable) 0.00001 percent.

Analysis of experimental data often includes calculations of probability. If a result is very unlikely to have occurred by chance alone, it is said to be **statistically significant**. In this context, the word "significant" does not refer to the result's importance. It means that a statistical analysis shows the result has a very low probability (typically, less than a 5 percent chance) of being skewed by sampling error. As you will see in the next section, every scientific result—even a statistically significant one—has a probability of being incorrect.

Variation in a set of data is typically shown as error bars on a graph. Error bars that measure variation around an average indicate precision—the data's closeness in values (Figure 1.15).

A The cloud forest that covers about 2 million acres of New Guinea's Foja Mountains is extremely remote and difficult to access, even for natives of the region. The first major survey of this forest occurred in 2005.

B Kris Helgen holds a golden-mantled tree kangaroo he found during the 2005 Foja Mountains survey. This kangaroo species is extremely rare in other areas, so it was thought to be critically endangered prior to the expedition.

FIGURE 1.13 Example of how generalizing from a subset can lead to an incorrect conclusion. It is also an example of how science is self-correcting.

Credits: (a) Tim Laman/ National Geographic Stock; (b) © Bruce Beehler/ Conservation International.

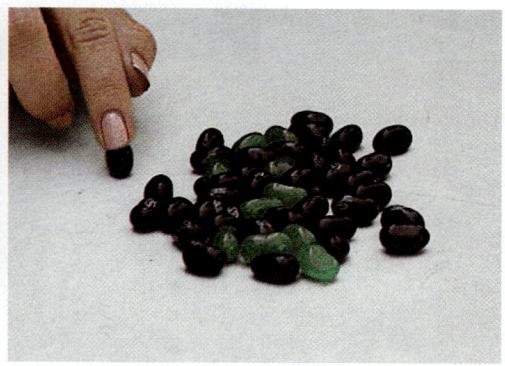

A Natalie, blindfolded, randomly plucks a jelly bean from a jar. The jar contains 120 green and 280 black jelly beans, so 30 percent of the jelly beans in the jar are green, and 70 percent are black.

B The jar is hidden from Natalie's view before she removes her blindfold. She sees one green jelly bean in her hand and assumes that the jar must hold only green jelly beans.

C Still blindfolded, Natalie randomly picks out 50 jelly beans from the jar. She ends up picking out 10 green and 40 black ones.

D The larger sample leads Natalie to assume that one-fifth of the jar's jelly beans are green (20 percent) and four-fifths are black (80 percent). This sample more closely approximates the jar's actual green-to-black ratio of 30 percent to 70 percent. The more times Natalie repeats the sampling, the greater the chance she has of guessing the actual ratio.

FIGURE 1.14 How sample size affects sampling error.

© Gary Head.

> Bias in Interpreting Results Experimenting with a single variable apart from all others is not often possible, particularly when studying humans. For example, remember that the people who participated in the Olestra experiment were chosen randomly, which means the study was not controlled for gender, age, weight, medications taken, and so on. Such variables may well have influenced the experiment's results.

Humans are by nature subjective, and scientists are no exception. Researchers risk interpreting their results in terms of what they want to find out. That is why they typically design experiments that will yield quantitative results, which are counts or some other data that can be measured or gathered objectively. Quantitative results minimize the potential for bias, and also give other scientists an opportunity to repeat the experiments and check the conclusions drawn from them. This last point gets us back to the role of critical thinking in science. Scientists expect one another to recognize and put aside bias in order to test hypotheses in ways that may prove them wrong. If a scientist does not, then others will, because exposing errors is just as useful as applauding insights. The scientific community consists of critically thinking people trying to poke holes in one another's ideas. Ideally, their collective efforts make science a self-correcting endeavor.

Take-Home Message

How do scientists avoid potential pitfalls of sampling error and bias when doing research?

- Researchers minimize sampling error by using large sample sizes and by repeating their experiments.
- A statistical analysis can show the probability that a result has occurred by chance alone.
- Science is a self-correcting process because it is carried out by a community of individuals who continually retest and recheck one another's ideas.

FIGURE 1.15 Example of error bars in a graph. This graph was adapted from the peacock butterfly research described in Section 1.6.

The researchers recorded the number of times each butterfly flicked its wings in response to an attack by a bird. The dots represent average frequency of wing flicking for each sample set of butterflies. The error bars that extend above and below the dots indicate the range of values—the sampling error.

Figure It Out: What was the fastest rate at which a butterfly with no spots or sound flicked its wings?

Answer: 22 times per minute

© Cengage Learning.

probability The chance that a particular outcome of an event will occur; depends on the total number of outcomes possible.

sampling error Difference between results derived from testing an entire group of events or individuals, and results derived from testing a subset of the group.

statistically significant Refers to a result that is statistically unlikely to have occurred by chance.

Table 1.3	Examples of Scientific Theories
Theory	**Main Premises**
Atomic theory	All substances consist of atoms.
Big bang	The universe originated with an explosion and continues to expand.
Cell theory	All organisms consist of one or more cells, the cell is the basic unit of life, and all cells arise from existing cells.
Evolution	Change occurs in the inherited traits of a population over generations.
Global warming	Human activities are causing Earth's average temperature to increase.
Plate tectonics	Earth's crust is cracked into pieces that move in relation to one another.

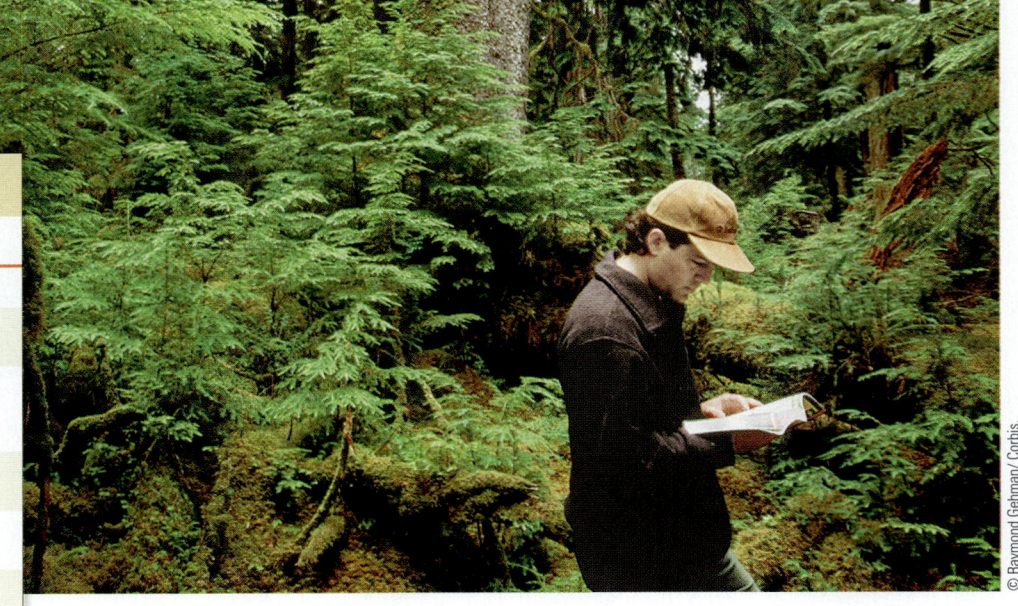

© Raymond Gehman/ Corbis.

1.8 The Nature of Science

Suppose a hypothesis stands even after years of tests. It is consistent with all data ever gathered, and it has helped us make successful predictions about other phenomena. When a hypothesis meets these criteria, it is considered to be a **scientific theory** (Table 1.3). To give an example, all observations to date have been consistent with the hypothesis that matter consists of atoms. Scientists no longer spend time testing this hypothesis for the compelling reason that, since we started looking 200 years ago, no one has discovered matter that consists of anything else. Thus, scientists use the hypothesis, now called atomic theory, to make other hypotheses about matter.

Scientific theories are our best objective descriptions of the natural world. However, they can never be proven absolutely, because to do so would necessitate testing under every possible circumstance. For example, in order to prove atomic theory, the atomic composition of all matter in the universe would have to be checked—an impossible task even if someone wanted to try.

Like all hypotheses, a scientific theory can be disproven by a single observation or result that is inconsistent with it. For example, if someone discovers a form of matter that does not consist of atoms, atomic theory would have to be revised. The potentially falsifiable nature of scientific theories means that science has a built-in system of checks and balances. A theory is revised until no one can prove it to be incorrect. The theory of evolution, which states that change occurs in a line of descent over time, still holds after a century of observations and testing. As with all other scientific theories, no one can be absolutely sure that it will hold under all possible conditions, but it has a very high probability of not being wrong. Few other theories have withstood as much scrutiny.

You may hear people apply the word "theory" to a speculative idea, as in the phrase "It's just a theory." This everyday usage of the word differs from the way it is used in science. Speculation is an opinion, belief, or personal conviction that is not necessarily supported by evidence. A scientific theory is different. By definition, a scientific theory is supported by a large body of evidence, and it is consistent with all known facts.

A scientific theory also differs from a **law of nature**, which describes a phenomenon that has been observed to occur in every circumstance without fail, but for which we do not have a complete scientific explanation. The laws of thermodynamics, which describe energy, are examples. We understand *how* energy behaves, but not exactly *why* it behaves the way it does.

law of nature Generalization that describes a consistent natural phenomenon that has an incomplete scientific explanation.

scientific theory Hypothesis that has not been disproven after many years of rigorous testing.

›The Limits of Science Science helps us be objective about our observations in part because of its limitations. For example, science does not address many questions, such as "Why do I exist?" Answers to such questions can only come from within as an integration of the personal experiences and mental connections that shape our consciousness. This is not to say subjective answers have no value, because no human society can function for long unless its individuals share standards for making judgments, even if they are subjective. Moral, aesthetic, and philosophical standards vary from one society to the next, but all help people decide what is important and good. All give meaning to our lives.

Neither does science address the supernatural, or anything that is "beyond nature." Science neither assumes nor denies that supernatural phenomena occur, but scientists often cause controversy when they discover a natural explanation for something that was thought to have none. Such controversy often arises when a society's moral standards are interwoven with its understanding of nature. For example, Nicolaus Copernicus proposed in 1540 that Earth orbits the sun. Today that idea is generally accepted, but the prevailing belief system had Earth as the immovable center of the universe. In 1610, astronomer Galileo Galilei published evidence for the Copernican model of the solar system, an act that resulted in his imprisonment. He was publicly forced to recant his work, spent the rest of his life under house arrest, and was never allowed to publish again.

As Galileo's story illustrates, exploring a traditional view of the natural world from a scientific perspective is often misinterpreted as a violation of morality. As a group, scientists are no less moral than anyone else, but they follow a particular set of rules that do not necessarily apply to others: Their work concerns only the natural world, and their ideas must be testable by other scientists.

Science helps us communicate our experiences without bias. As such, it may be as close as we can get to a universal language. We are fairly sure, for example, that the laws of gravity apply everywhere in the universe. Intelligent beings on a distant planet would likely understand the concept of gravity. We might well use gravity or another scientific concept to communicate with them, or anyone, anywhere. The point of science, however, is not to communicate with aliens. It is to find common ground here on Earth.

Take-Home Message

What is a scientific theory?

- A scientific theory is a time-tested hypothesis that is consistent with all known facts. It is our most objective way of describing the natural world.

1.9 The Secret Life of Earth (revisited)

 Earth hosts at least 100 million species. That number is only an estimate because we are still discovering them. For example, a mouse-sized opossum and a cat-sized rat turned up on a return trip to the Foja Mountains. Other recently discovered species include a leopard in Borneo; a wolf in Egypt; a dolphin in Australia; spiders in California (Figure 1.16); a giant crayfish in Tennessee; a rat-eating plant in the Philippines; and carnivorous sponges near Antarctica. Each new species discovered is a reminder that we do not yet know all of the organisms living on our own planet. We don't even know how many to look for. You can find information about the 1.8 million species we do know about in the Encyclopedia of Life, an online reference maintained by collaborative effort (www.eol.org).

Science helps us communicate our experiences without bias.

FIGURE 1.16 Dr. Jason Bond holds a new species of trapdoor spider he discovered living in sand dunes along California's central coast. Bond named the spider *Aptostichus stephencolberti*, after TV personality Stephen Colbert.

Courtesy East Carolina University.

WHERE YOU ARE GOING . . .

This book parallels nature's levels of organization, from atoms to the biosphere. Learning about the structure and function of atoms and molecules will prime you to understand how living cells work. Learning about processes that keep a single cell alive can help you understand how multicelled organisms survive. Knowing what it takes for organisms to survive can help you see why and how they interact with one another and their environment.

Summary

Section 1.1 **Biology** is the systematic study of life. We have encountered only a fraction of the organisms that live on Earth, in part because we have explored only a fraction of its inhabited regions.

Section 1.2 Biologists think about life at different levels of organization, with different properties emerging at successively higher levels. Life emerges at the level of the **cell**.

All matter consists of **atoms**, which combine as **molecules**. An **organism** consists of one or more cells. A **population** is a group of individuals of a species in a given area; a **community** is all populations of all species in a given area. An **ecosystem** is a community interacting with its environment. The **biosphere** includes all regions of Earth that hold life.

Section 1.3 Life has underlying unity in that all living things have similar characteristics:

(1) All organisms require energy and **nutrients** to sustain themselves. **Producers** harvest energy from the environment to make their own food by processes such as **photosynthesis**; **consumers** eat other organisms, or their wastes and remains.

(2) Organisms keep the conditions in their internal environment within ranges that their cells are able to tolerate—a process called **homeostasis**.

(3) **DNA** contains information that guides an organism's form and function, which include **development**, **growth**, and **reproduction**. The passage of DNA from parents to offspring is called **inheritance**.

Section 1.4 The many types of organisms that currently exist on Earth differ greatly in details of body form and function. **Biodiversity** is the sum of differences among living things. **Bacteria** and **archaea** are **prokaryotes**, single-celled organisms whose DNA is not contained within a **nucleus**. The DNA of single-celled or multi-celled **eukaryotes** (**protists**, **plants**, **fungi**, and **animals**) is contained within a nucleus.

Section 1.5 Each **species** has a two-part name. The first part is the **genus** name. When combined with the specific epithet, it designates the particular species. With **taxonomy**, species are ranked into ever more inclusive **taxa** on the basis of shared traits.

Section 1.6 **Critical thinking**, the self-directed act of judging the quality of information as one learns, is an important part of **science**. Generally, a researcher observes something in nature, forms a **hypothesis** (testable explanation) for it, then makes a **prediction** about what might occur if the hypothesis is correct. Predictions are tested with observations, **experiments**, or

both. Experiments typically are performed on an **experimental group** as compared with a **control group**, and sometimes on **models**. Conclusions are drawn from experimental results, or **data**. A hypothesis that is not consistent with data is modified or discarded. The **scientific method** consists of making, testing, and evaluating hypotheses, and sharing results.

Biological systems are usually influenced by many interacting **variables**. Research approaches differ, but experiments are typically designed in a consistent way. Experiments test whether and how changing a variable influences a system. A researcher changes a variable, then observes the effects of the change. This practice allows the researcher to study a cause-and-effect relationship in a complex natural system.

Section 1.7 A small sample size increases the potential for **sampling error** in experimental results. In such cases, a subset may be tested that is not representative of the whole. Researchers design experiments carefully to minimize sampling error and bias, and they use **probability** rules to check the **statistical significance** of their results. Scientists check and test one another's work, so science is a self-correcting process.

Section 1.8 Science helps us be objective about our observations because it is concerned only with testable ideas about observable aspects of nature. Opinion and belief have value in human culture, but they are not addressed by science. A **scientific theory** is a long-standing hypothesis that is useful for making predictions about other phenomena. It is our best way of describing reality. A **law of nature** describes something that occurs without fail, but our scientific explanation of why it occurs is incomplete.

Self-Quiz   Answers in Appendix I

1. _____ are fundamental building blocks of all matter.
   - a. Cells
   - b. Atoms
   - c. Organisms
   - d. Molecules

2. The smallest unit of life is the _____ .
   - a. atom
   - b. molecule
   - c. cell
   - d. organism

3. _____ is the transmission of DNA to offspring.
   - a. Reproduction
   - b. Development
   - c. Homeostasis
   - d. Inheritance

4. A process by which an organism produces offspring is called _____ .
   - a. reproduction
   - b. inheritance
   - c. development
   - d. homeostasis

5. Organisms require _____ and _____ to maintain themselves, grow, and reproduce.
   - a. sunlight; energy
   - b. cells; raw materials
   - c. nutrients; energy
   - d. DNA; cells

6. _____ move around for at least part of their life.
   a. Organisms
   b. Plants
   c. Animals
   d. Prokaryotes

7. By sensing and responding to change, organisms keep conditions in the internal environment within ranges that cells can tolerate. This process is called _____ .
   a. sampling error
   b. development
   c. homeostasis
   d. critical thinking

8. DNA _____ .
   a. guides functioning and development
   b. is the basis of traits
   c. is transmitted from parents to offspring
   d. all of the above

9. A butterfly is a(n) _____ (choose all that apply).
   a. organism
   b. domain
   c. species
   d. eukaryote
   e. consumer
   f. producer
   g. prokaryote
   h. trait

10. A bacterium is _____ (choose all that apply).
    a. an organism
    b. single-celled
    c. an animal
    d. a eukaryote

11. Bacteria, Archaea, and Eukarya are three _____ .
    a. organisms
    b. domains
    c. consumers
    d. producers

12. A control group is _____ .
    a. a set of individuals that have a characteristic under study or receive an experimental treatment
    b. the standard against which an experimental group is compared
    c. the experiment that gives conclusive results

13. Science addresses only that which is _____ .
    a. alive
    b. observable
    c. variable
    d. indisputable

14. Fifteen randomly selected students are found to be taller than 6 feet. The researchers concluded that the average height of a student is greater than 6 feet. This is an example of _____ .
    a. experimental error
    b. sampling error
    c. a subjective opinion
    d. experimental bias

15. Match the terms with the most suitable description.
    _____ life
    _____ probability
    _____ species
    _____ scientific theory
    _____ hypothesis
    _____ prediction
    _____ producer

    a. statement of what you expect to see if the hypothesis is correct
    b. unique type of organism
    c. property that emerges at the level of the cell
    d. time-tested hypothesis
    e. testable explanation
    f. measure of chance
    g. makes its own food

Critical Thinking

1. A person is declared to be dead upon the irreversible cessation of spontaneous body functions: brain activity, or blood circulation and respiration. However, only about 1% of a person's cells have to die in order for all of these things to happen. How can someone be dead when 99% of his or her cells are still alive?

2. Explain the difference between a one-celled organism and a single cell of a multicelled organism.

Digging Into Data

Peacock Butterfly Predator Defenses

The photographs *below* represent experimental and control groups used in the peacock butterfly experiment that was discussed in Section 1.7. See if you can identify the experimental groups, and match them up with the relevant control group(s). *Hint:* Identify which variable is being tested in each group (each variable has a control).

 A Wing spots painted out

 D Wings cut but not silenced

 B Wing spots visible; wings silenced

 E Wings painted but spots visible; wings cut but not silenced

 C Wing spots painted out; wings silenced

 F Wings painted but spots visible

Scientific Paper; Adrian Vallin, Sven Jakobsson, Johan Lind and Christer Wiklund, *Proc. R. Soc. B* (2005 272, 1203, 1207). Used with permission of The Royal Society and the author.

3. Why would you think twice about ordering from a restaurant menu that lists only the second part of the species name (not the genus) of its offerings? *Hint:* Look up *Ursus americanus, Ceanothus americanus, Bufo americanus, Homarus americanus, Lepus americanus,* and *Nicrophorus americanus.*

4. Once there was a highly intelligent turkey that had nothing to do but reflect on the world's regularities. Morning always started out with the sky turning light, followed by the master's footsteps, which were always followed by the appearance of food. Other things varied, but food always followed footsteps. The sequence of events was so predictable that it eventually became the basis of the turkey's theory about the goodness of the world. One morning, after more than 100 confirmations of the goodness theory, the turkey listened for the master's footsteps, heard them, and had its head chopped off.

   Any scientific theory is modified or discarded upon discovery of contradictory evidence. The absence of absolute certainty has led some people to conclude that "facts are irrelevant because they can change." If that is so, should we stop doing scientific research? Why or why not?

5. In 2005, researcher Woo-suk Hwang reported that he had made immortal stem cells from human patients. His research was hailed as a breakthrough for people affected by degenerative diseases, because stem cells may be used to repair a person's own damaged tissues. Hwang published his results in a peer-reviewed journal. In 2006, the journal retracted his paper after other scientists discovered that Hwang's group had faked their data.

   Does the incident show that results of scientific studies cannot be trusted? Or does it confirm the usefulness of a scientific approach, because other scientists discovered and exposed the fraud?

2 Molecules of Life

This chapter explores the first level of life's organization—atoms—and provides an example of how different arrangements of the same building blocks can form different products (Section 1.2). You will begin to explore mechanisms of homeostasis, the process by which organisms keep themselves in a state that favors cell survival (Section 1.3).

2.1 Fear of Frying

The human body requires only about a tablespoon of fat each day to stay healthy, but most people in developed countries eat far more than that. The average American eats about 70 pounds of fat per year, which may be part of the reason why the average American is overweight. Being overweight increases one's risk for many chronic illnesses. However, the total quantity of fat in one's diet may be less important to health than the types of fats. Fats are more than inert molecules that accumulate in strategic areas of our bodies. As major components of cell membranes, fats have powerful effects on cell function.

The typical fat molecule has three fatty acids, each with a long chain of carbon atoms that can vary a bit in structure. We call fats with a certain arrangement of hydrogen atoms around those carbon chains *trans* fats. Small amounts of *trans* fats occur naturally in red meat and dairy products, but the main source of *trans* fats in the American diet is an artificial food product called partially hydrogenated vegetable oil.

Hydrogenation is a manufacturing process that adds hydrogen atoms to vegetable oils in order to change them into solid fats. In 1908, Procter & Gamble Co. developed partially hydrogenated soybean oil as a substitute for the more expensive solid animal fats they had been using to make candles. However, the demand for candles began to wane as more households in the United States became wired for electricity, and P & G began to look for another way to sell its proprietary fat. Partially hydrogenated vegetable oil looks a lot like lard, so in 1911 the company began aggressively marketing it as a revolutionary new food: a solid cooking fat with a long shelf life, mild flavor, and lower cost than lard or butter. By the mid-1950s, hydrogenated vegetable oil had become a major part of the American diet. It was (and still is) found in a tremendous number of manufactured and fast foods (Figure 2.1).

For decades, hydrogenated vegetable oil was considered healthier than animal fats because it was made from plants, but we now know otherwise. *Trans* fats raise the level of cholesterol in our blood more than any other fat, and they directly alter the function of our arteries and veins. The effects of such changes are quite serious. Eating as little as 2 grams a day (about 0.4 teaspoon) of hydrogenated vegetable oil measurably increases a person's risk of atherosclerosis (hardening of the arteries), heart attack, and diabetes. A small serving of french fries made with hydrogenated vegetable oil contains about 5 grams of *trans* fat.

All organisms consist of the same kinds of molecules, but small differences in the way those molecules are put together can have big effects in a living organism. With this concept, we introduce you to the chemistry of life. This is your chemistry, and it makes you far more than the sum of your body's molecules.

FIGURE 2.1 *Trans* fats, an unhealthy food, are abundant in hydrogenated oils commonly used to make manufactured and fast foods.

© GraÃŞa Victoria/ Shutterstock

an atom

proton
neutron
electron

FIGURE 2.2 Animated! Atoms consist of electrons moving around a core, or nucleus, of protons and neutrons. Models such as this diagram cannot show what atoms really look like. Electrons zoom around in fuzzy, three-dimensional spaces about 10,000 times bigger than the nucleus. © Cengage Learning 2012.

2.2 Start With Atoms

Life's unique characteristics start with the properties of different atoms. Even though atoms are about 20 million times smaller than a grain of sand, they consist of even smaller subatomic particles. Positively charged **protons** (p+) and uncharged **neutrons** occur in an atom's core, or **nucleus**. Negatively charged **electrons** (e⁻) move around the nucleus (Figure 2.2). **Charge** is an electrical property: Opposite charges attract, and like charges repel.

A typical atom has about the same number of electrons and protons. The negative charge of an electron is the same magnitude as the positive charge of a proton, so the two charges cancel one another. Thus, an atom with exactly the same number of electrons and protons carries no net charge.

The number of protons in the nucleus is called the **atomic number**, and it determines the type of atom, or element. **Elements** are pure substances, each consisting only of atoms with the same number of protons in their nucleus (Figure 2.3). For example, the atomic number of carbon is 6, so all atoms with six protons in their nucleus are carbon atoms, no matter how many electrons or neutrons they have. Carbon, the substance, consists only of carbon atoms, and all of those atoms have six protons. Each of the 118 known elements has a symbol that is typically an abbreviation of its Latin or Greek name. Carbon's symbol, C, is from *carbo*, the Latin word for coal. Coal is mostly carbon. Appendix VI shows a periodic table of the elements.

Carbon and all other elements occur in different forms, or **isotopes**, that differ in their number of neutrons. We refer to an isotope by its total number of protons and neutrons, which is the isotope's **mass number**. Mass number is shown as a superscript to the left of an element's symbol. For example, atoms of the most common carbon isotope, ^{12}C, have six protons and six neutrons; those of ^{14}C have six protons and eight neutrons (6 + 8 = 14).

^{14}C (carbon 14) is a radioactive isotope, or **radioisotope**. Atoms of a radioisotope have an unstable nucleus that breaks down spontaneously. As a nucleus breaks down, it emits radiation—subatomic particles, energy, or both—a process called **radioactive decay**. Each radioisotope decays at a predictable rate into predictable products. For example, when carbon 14 decays, one of its neutrons splits into a proton and an electron. The nucleus emits the electron as radiation. Thus, a carbon atom with eight neutrons and six protons (^{14}C) becomes a nitrogen atom, with seven neutrons and seven protons (^{14}N).

Each radioisotope decays at a fixed rate, regardless of temperature, pressure, or whether the atoms are part of molecules. The rate is so predictable that we can say with certainty that about half of the atoms in any sample of ^{14}C will be

mass number → 12
element symbol → C
atomic number → 6
element name → carbon

elemental substance

FIGURE 2.3 Example of an element: carbon.

Credits: left; Theodore Gray/ Visuals Unlimited, Inc.; right, © Cengage Learning.

atomic number Number of protons in the atomic nucleus; determines the element.

charge Electrical property; opposite charges attract, and like charges repel.

electron Negatively charged subatomic particle.

element A pure substance that consists only of atoms with the same number of protons.

isotopes Forms of an element that differ in the number of neutrons their atoms carry.

mass number Of an isotope, the total number of protons and neutrons in the atomic nucleus.

neutron Uncharged subatomic particle in the atomic nucleus.

nucleus Core of an atom; occupied by protons and neutrons.

proton Positively charged subatomic particle that occurs in the nucleus of all atoms.

radioactive decay Process by which atoms of a radioisotope emit energy and subatomic particles when their nucleus spontaneously disintegrates.

radioisotope Isotope with an unstable nucleus.

FIGURE 2.4 PET scans use radioactive tracers to form a digital image of a process in the body's interior. These two PET scans allow us to compare the activity of a molecule called MAO-B in the body of a non-smoker (*left*) and a smoker (*right*). The activity is color-coded from *red* (highest) to *purple* (lowest). Low MAO-B activity is associated with violence, impulsiveness, and other behavioral problems.

Brookhaven National Laboratory.

^{14}N atoms after 5,730 years. The predictable rate of radioactive decay makes it possible for scientists to estimate the age of a rock or fossil by measuring its isotope content (we return to this topic in Section 11.4).

Radioisotopes are often used in **tracers**, which are substances with a detectable component attached. Typically, a radioactive tracer is a molecule in which an atom (such as ^{12}C) has been replaced with a radioisotope (such as ^{14}C). A radioactive tracer can be delivered into a biological system such as a cell, body, or ecosystem, and then followed as it moves through the system with instruments that detect radiation. For example, PET (short for positron-emission tomography) helps us "see" a functional process inside the body. By this procedure, a radioactive sugar or other tracer is injected into a patient. Inside the patient's body, cells with differing rates of activity take up the tracer at different rates. A scanner detects radioactive decay wherever the tracer is, then translates that data into an image (Figure 2.4).

〉 Why Electrons Matter Electrons are really, really small. How small are they? If they were as big as apples, you would be about 3.5 times taller than our solar system is wide. Simple physics explains the motion of, say, an apple falling from a tree, but electrons are so tiny that such everyday physics does not explain their behavior. For example, electrons carry energy, but only in incremental amounts. An electron gains energy only by absorbing the exact amount needed to boost it to the next energy level. Likewise, it loses energy only by emitting the exact difference between two energy levels. This concept will be important to remember when you learn how cells harvest and release energy.

Imagine that an atom is a multilevel apartment building with a nucleus in the basement. Each "floor" of the building corresponds to a certain energy level,

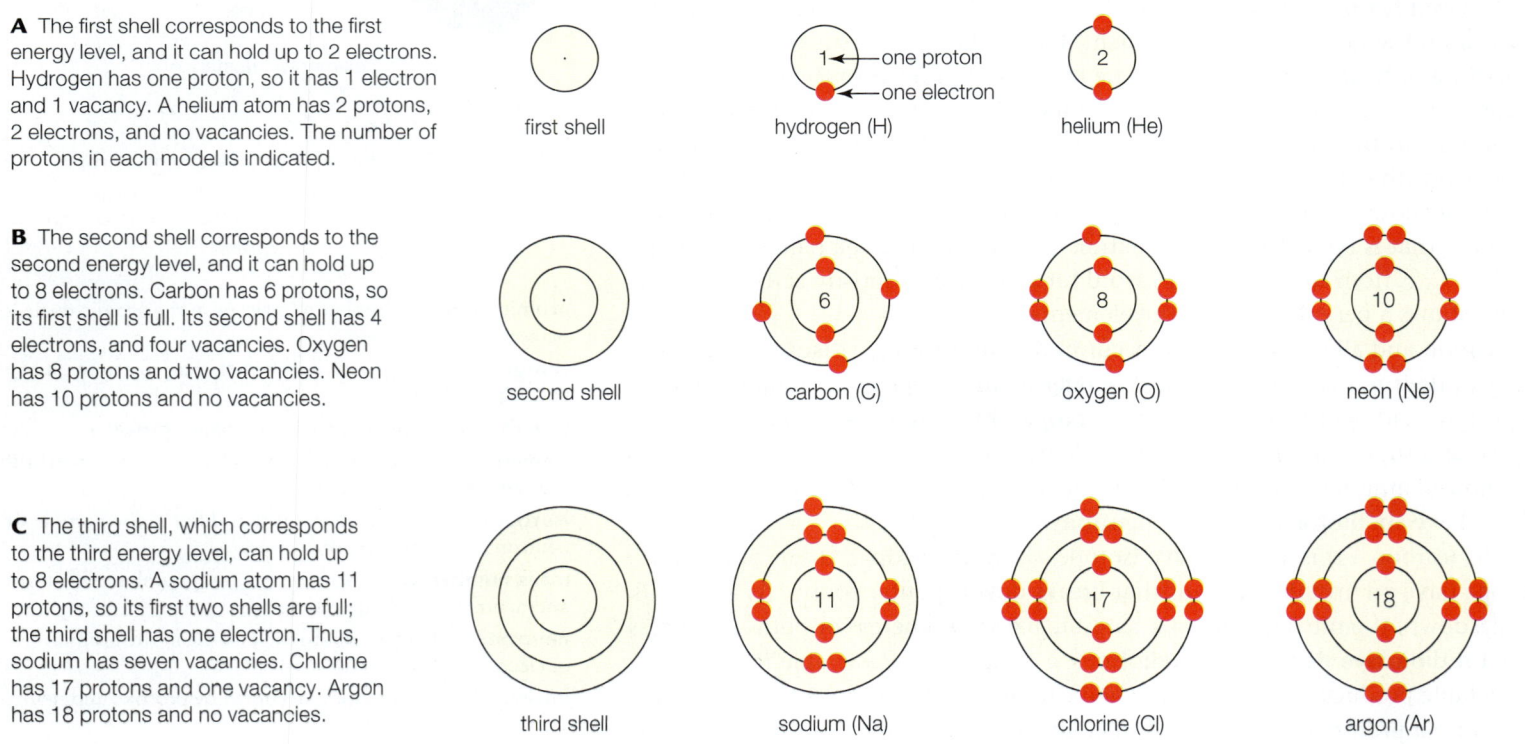

A The first shell corresponds to the first energy level, and it can hold up to 2 electrons. Hydrogen has one proton, so it has 1 electron and 1 vacancy. A helium atom has 2 protons, 2 electrons, and no vacancies. The number of protons in each model is indicated.

first shell

1 ←— one proton ←— one electron
hydrogen (H)

2
helium (He)

B The second shell corresponds to the second energy level, and it can hold up to 8 electrons. Carbon has 6 protons, so its first shell is full. Its second shell has 4 electrons, and four vacancies. Oxygen has 8 protons and two vacancies. Neon has 10 protons and no vacancies.

second shell

6
carbon (C)

8
oxygen (O)

10
neon (Ne)

C The third shell, which corresponds to the third energy level, can hold up to 8 electrons. A sodium atom has 11 protons, so its first two shells are full; the third shell has one electron. Thus, sodium has seven vacancies. Chlorine has 17 protons and one vacancy. Argon has 18 protons and no vacancies.

third shell

11
sodium (Na)

17
chlorine (Cl)

18
argon (Ar)

FIGURE 2.5 Animated! Shell models. Each circle (shell) represents one energy level. We fill the shells with electrons from the innermost shell out, until there are as many electrons as the atom has protons. **Figure It Out:** Which of these models have unpaired electrons in their outer shell?

Credits: © Cengage Learning.

Answer: Hydrogen, carbon, oxygen, sodium, and chlorine

and each has a certain number of "rooms" available for rent. Electrons like to occupy those rooms in pairs. **Shell models** (Figure 2.5) help us visualize how electrons populate atoms. In this model, nested "shells" correspond to successively higher energy levels. Thus, each shell includes all rooms on one floor of our atomic apartment building. We draw a shell model of an atom by filling shells with electrons (represented as balls or dots), from the innermost shell out, until there are as many electrons as the atom has protons. There is only one room on the first floor, the lowest energy level (Figure 2.5**A**). It fills up first. In hydrogen, the simplest atom, one electron occupies that room. Helium, with two protons, has two electrons that fill the room—and the first shell. In larger atoms, more electrons rent the second-floor rooms (Figure 2.5**B**), then the third-floor rooms (Figure 2.5**C**), and so on.

When an atom's outermost shell is filled with electrons, we say that it has no vacancies. Any atom is in its most stable state when it has no vacancies. By contrast, when an atom's outermost shell can accommodate another electron, it has a vacancy. Atoms with vacancies tend to eliminate them by interacting with other atoms; in other words, they are chemically active. For example, the sodium atom (Na) in Figure 2.5**C** has one electron in its outer (third) shell, which can hold eight. With seven vacancies, we can predict that this atom is chemically active.

In fact, this particular sodium atom is not just active, it is extremely so. Why? The shell model shows that a sodium atom has an unpaired electron, but in the real world, electrons really like to be in pairs when they occupy atoms. Solitary atoms that have unpaired electrons are called **free radicals**. With some exceptions, free radicals have a very strong tendency to interact with other atoms, and such interactions make them dangerous to life. A free radical sodium atom can easily evict its one unpaired electron, so that its second shell—which is full of electrons—becomes its outermost, and no vacancies remain. This is the atom's most stable state, and in fact the vast majority of sodium atoms on Earth are like this one, with 11 protons and 10 electrons.

Atoms with an unequal number of protons and electron are called **ions**. Ions carry a net (or overall) charge. Sodium ions (Na⁺) offer an example of how atoms gain a positive charge by losing an electron (Figure 2.6**A**). Other atoms gain a negative charge by accepting an electron. For example, an uncharged chlorine atom has 17 protons and 17 electrons. Its outer shell can hold eight electrons, but it has only seven. This atom has one vacancy and one unpaired electron, so we can predict—correctly—that it is chemically very active. An uncharged chlorine atom easily fills its third shell by accepting an electron. When that happens, the atom becomes a chloride ion (Cl⁻) with 17 protons, 18 electrons, and a net negative charge (Figure 2.6**B**).

Atoms of elements such as helium, neon, and argon have no vacancies when they have as many electrons as protons. These elements occur most frequently on Earth as solitary, uncharged atoms.

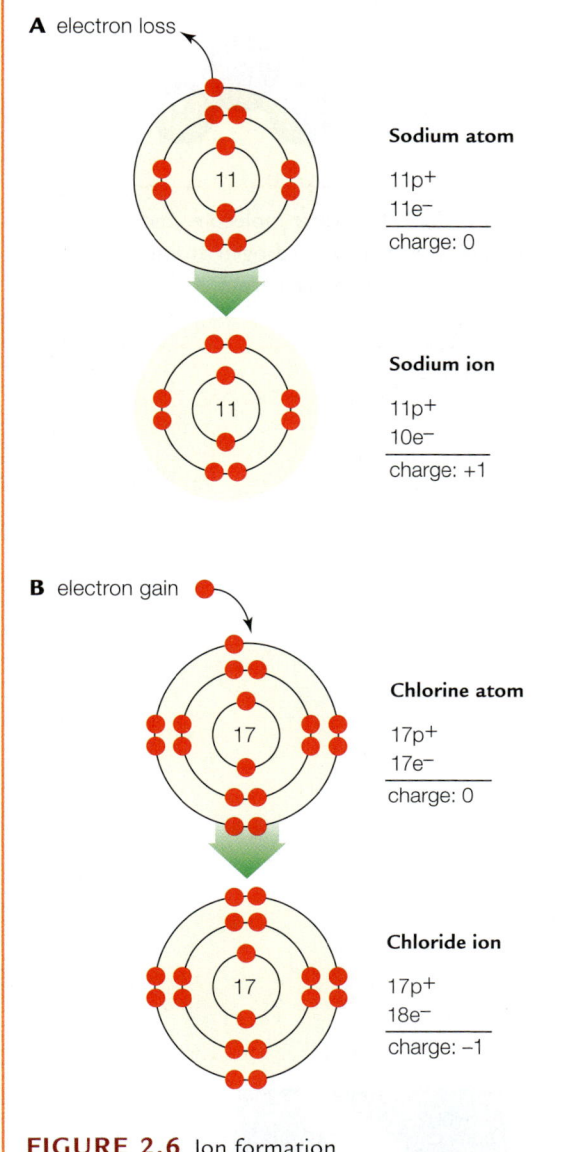

FIGURE 2.6 Ion formation.

A A sodium atom (Na) becomes a positively charged sodium ion (Na⁺) when it loses the electron in its third shell. The atom's full second shell is now its outermost, so it has no vacancies.

B A chlorine atom (Cl) becomes a negatively charged chloride ion (Cl⁻) when it gains an electron and fills the vacancy in its third, outermost shell.

© Cengage Learning.

Take-Home Message

What are atoms?

- Atoms consist of electrons moving around a nucleus of protons and neutrons. The number of protons determines the element.

- Atoms tend to get rid of vacancies, for example by gaining or losing electrons (thereby becoming ions).

free radical Atom with an unpaired electron.
ion Atom that carries a charge because it has an unequal number of protons and electrons.
shell model Model of electron distribution in an atom.
tracer A molecule with a detectable component.

one oxygen atom —
two hydrogen atoms — H   H

FIGURE 2.7 The water molecule. Each water molecule has two hydrogen atoms bonded to the same oxygen atom.

© Cengage Learning.

2.3 From Atoms to Molecules

An atom can get rid of vacancies by participating in a chemical bond with another atom. A **chemical bond** is an attractive force that arises between two atoms when their electrons interact. Chemical bonds link atoms into molecules. In other words, each molecule consists of atoms held together in a particular number and arrangement by chemical bonds. For example, a water molecule consists of three atoms: two hydrogen atoms bonded to the same oxygen atom (Figure 2.7). A water molecule is also a **compound**, which means it has atoms of two or more elements. Other molecules, including molecular oxygen (a gas in air), have atoms of one element only.

The term "bond" applies to a continuous range of atomic interactions. However, we can categorize most bonds into distinct types based on their properties. Which type forms depends on the atoms taking part in the molecule.

〉 Ionic Bonds Two ions may stay together by the mutual attraction of their opposite charges, an association called an **ionic bond**. Ionic bonds can be quite strong. Ionically bonded sodium and chloride ions make sodium chloride (NaCl), which we know as common table salt. A crystal of this substance consists of a lattice of sodium and chloride ions interacting in ionic bonds (Figure 2.8**A**). Ions retain their respective charges when participating in an ionic bond (Figure 2.8**B**). Thus, one "end" of an ionically bonded molecule has a positive charge, and the other "end" has a negative charge. Any such separation of charge into distinct positive and negative regions is called **polarity** (Figure 2.8**C**).

© Cengage Learning.

〉 Covalent Bonds In a **covalent bond**, two atoms share a pair of electrons, so each atom's vacancy becomes partially filled (Figure 2.9). Sharing electrons links the two atoms, just as sharing a pair of earphones links two friends (*inset*). Covalent bonds can be stronger than ionic bonds, but they are not always so.

Table 2.1 shows different ways of representing covalent bonds. In structural formulas, a line between two atoms represents a single covalent bond, in which two atoms share one pair of electrons. For example, molecular

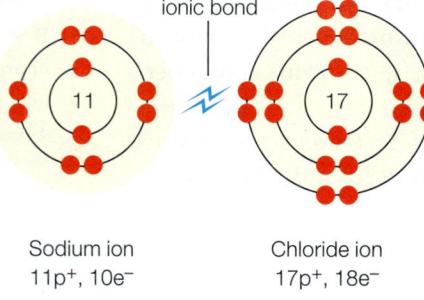

Na⁺ Cl⁻

ionic bond

Sodium ion
11p⁺, 10e⁻

Chloride ion
17p⁺, 18e⁻

+ Na⁺   Cl⁻ −

positive ←——→ negative
charge   charge

A Each crystal of table salt consists of many sodium and chloride ions locked together in a cubic lattice by ionic bonds.

B The strong mutual attraction of opposite charges holds a sodium ion and a chloride ion together in an ionic bond.

C Ions taking part in an ionic bond retain their charge, so the molecule itself is polar. One side is positively charged (represented by a *blue* overlay); the other side is negatively charged (*red* overlay).

FIGURE 2.8 Animated! An example of ionic bonding: table salt, or NaCl.

Credits: (a) top, Gary Head; bottom left, © Bill Beatty/ Visuals Limited; (a right, b,c) © Cengage Learning.

Table 2.1 Representing Covalent Bonds in Molecules

Common name	Water	Familiar term.
Chemical name	Dihydrogen monoxide	Describes elemental composition.
Chemical formula	H_2O	Indicates unvarying proportions of elements. Subscripts show number of atoms of an element per molecule. The absence of a subscript means one atom.
Structural formula	H—O—H	Represents each covalent bond as a single line between atoms.
Structural model		Shows relative sizes and positions of atoms in three dimensions.
Shell model		Shows how pairs of electrons are shared in covalent bonds.

hydrogen (H_2) has one covalent bond between hydrogen atoms (H—H). Two, three, or even four covalent bonds may form between atoms when they share multiple pairs of electrons. For example, two atoms sharing two pairs of electrons are connected by two covalent bonds. Such double bonds are represented by a double line between the atoms. A double bond links the two oxygen atoms in molecular oxygen (O=O). Three lines indicate a triple bond, in which two atoms share three pairs of electrons. A triple covalent bond links the two nitrogen atoms in molecular nitrogen (N≡N).

Double and triple bonds are not distinguished from single bonds in structural models, which show positions and relative sizes of the atoms in three dimensions. The bonds are shown as one stick connecting two balls, which represent atoms. Elements are usually coded by color:

carbon   hydrogen   oxygen   nitrogen   phosphorus

Atoms share electrons unequally in a polar covalent bond. A bond between an oxygen atom and a hydrogen atom in a water molecule is an example. In this case, one atom (the oxygen) pulls the electrons a little more toward its side of the bond, so that atom bears a slight negative charge. The atom at the other end of the bond (the hydrogen, in this case) bears a slight positive charge. Covalent bonds in compounds are usually polar. By contrast, atoms participating in a nonpolar covalent bond share electrons equally. There is no difference in charge between the two ends of such bonds. Molecular hydrogen (H_2), oxygen (O_2), and nitrogen (N_2) are examples.

Molecular hydrogen

(H—H)

Two hydrogen atoms, each with one proton, share two electrons in a nonpolar covalent bond.

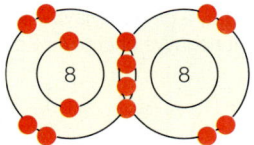

Molecular oxygen

(O=O)

Two oxygen atoms, each with eight protons, share four electrons in a double covalent bond.

Water

(H—O—H)

Two hydrogen atoms share electrons with an oxygen atom in two covalent bonds. The bonds are polar because the oxygen exerts a greater pull on the shared electrons than the hydrogens do.

FIGURE 2.9 Animated! Covalent bonds, in which atoms fill vacancies by sharing electrons. Two electrons are shared in each covalent bond. When sharing is equal, the bond is nonpolar. When one atom exerts a greater pull on the electrons, the bond is polar. The number of protons in each nucleus is shown.

© Cengage Learning.

Take-Home Message

How do atoms interact in chemical bonds?

- A chemical bond forms between atoms when their electrons interact. Depending on the atoms taking part in it, the bond may be ionic or covalent.

- An ionic bond is a strong mutual attraction between ions of opposite charge.

- Atoms share a pair of electrons in a covalent bond. When the atoms share electrons unequally, the bond is polar.

chemical bond An attractive force that arises between two atoms when their electrons interact.

compound Molecule that has atoms of more than one element.

covalent bond Type of chemical bond in which two atoms share a pair of electrons.

ionic bond Type of chemical bond in which a strong mutual attraction links ions of opposite charge.

polarity Any separation of charge into distinct positive and negative regions.

A Polarity of the water molecule. Each of the hydrogen atoms in a water molecule bears a slight positive charge (represented by a *blue* overlay). The oxygen atom carries a slight negative charge (*red* overlay).

negative charge

positive charge

a hydrogen bond

B A hydrogen bond is an attraction between a hydrogen atom and another atom taking part in a separate polar covalent bond.

C The many hydrogen bonds that form among water molecules impart special properties to liquid water.

FIGURE 2.10 Animated! Hydrogen bonding in water. © Cengage Learning.

2.4 Hydrogen Bonds and Water

Life evolved in water. All living organisms are mostly water, many of them still live in it, and all of the chemical reactions of life are carried out in water-based fluids. What makes water so fundamentally important for life?

Water has unique properties that arise from the two polar covalent bonds in each water molecule. Overall, the molecule has no charge, but the oxygen atom carries a slight negative charge, and the two hydrogen atoms carry a slight positive charge. Thus, the molecule itself is polar (Figure 2.10**A**).

The polarity of individual water molecules attracts them to one another. The slight positive charge of a hydrogen atom in one water molecule is drawn to the slight negative charge of an oxygen atom in another. This type of interaction is called a hydrogen bond. A **hydrogen bond** is an attraction between a covalently bonded hydrogen atom and another atom taking part in a separate polar covalent bond (Figure 2.10**B**). Like ionic bonds, hydrogen bonds form by the mutual attraction of opposite charges. However, unlike ionic bonds, hydrogen bonds do not make molecules out of atoms, so they are not chemical bonds.

Hydrogen bonds lie on the weaker end of the spectrum of atomic interactions; they form and break much more easily than covalent or ionic bonds. Even so, many of them form, and collectively they are quite strong. As you will see, hydrogen bonds stabilize the characteristic structures of biological molecules such as DNA and proteins. They also form in tremendous numbers among water molecules (Figure 2.10**C**). The extensive hydrogen bonding among water molecules gives special properties to liquid water that make life possible.

〉 Water Is an Excellent Solvent The polarity of the water molecule and its ability to form hydrogen bonds make water an excellent **solvent**, which means that many other substances easily dissolve in it. Substances that dissolve easily in water are **hydrophilic** (water-loving). Ionic solids such as sodium chloride (NaCl) dissolve in water because the slight positive charge on each hydrogen atom in a water molecule attracts negatively charged ions (Cl^-), and the slight negative charge on the oxygen atom attracts positively charged ions

(Na^+). Hydrogen bonds among many water molecules are collectively stronger than an ionic bond between two ions, so the solid dissolves as water molecules tug the ions apart and surround each one (*inset*).

Sodium chloride is called a **salt** because it releases ions other than H^+ and OH^- when it dissolves in water (more about these ions in the next section). When a substance such as NaCl dissolves, its component ions disperse uniformly among the molecules of liquid, and it becomes a **solute**. A uniform mixture such as salt dissolved in water is called a **solution**. Chemical bonds do not form between molecules of solute and solvent, so the proportions of the two substances in a solution can vary.

Nonionic solids such as sugars dissolve easily in water because their molecules can form hydrogen bonds with water molecules. Hydrogen bonding with water does not break the covalent bonds of such molecules; rather, it dissolves the substance by pulling individual molecules away from one another and keeping them apart.

Water does not interact with **hydrophobic** (water-dreading) substances such as oils. Oils consist of nonpolar molecules, and hydrogen bonds do not form between nonpolar molecules and water. When you mix oil and water, the water breaks into small droplets, but quickly begins to cluster into larger drops as new hydrogen bonds form among its molecules. The bonding excludes molecules of oil and pushes them together into drops that rise to the surface of the water. The same interactions occur at the thin, oily membrane that separates the watery fluid inside cells from the watery fluid outside of them. As you will see in Chapter 3, such interactions give rise to the structure of cell membranes.

> Water Stabilizes Temperature **Temperature** is a way to measure the energy of molecular motion. All molecules jiggle nonstop, and they jiggle faster as they absorb heat. Hydrogen bonding keeps water molecules from jiggling as much as they would otherwise, so it takes more heat to raise the temperature of water compared with other liquids. Temperature stability is an important part of homeostasis, because most of the molecules of life function properly only within a certain range of temperature.

Below 0°C (32°F), water molecules do not jiggle enough to break hydrogen bonds, and they become locked in the rigid, lattice-like bonding pattern of ice (Figure 2.11). Individual water molecules pack less densely in ice than they do in water, which is why ice floats on water. Sheets of ice that form on the surface of ponds, lakes, and streams can insulate the water under them from subfreezing air temperatures. Such "ice blankets" protect aquatic organisms during cold winters.

> Water Has Cohesion Molecules of some substances resist separating from one another, and the resistance gives rise to a property called **cohesion**. Water has cohesion because hydrogen bonds collectively exert a continuous pull on its individual molecules. You can see cohesion in water as surface tension, which means that the surface of liquid water behaves a bit like a sheet of elastic (*inset*).

Cohesion is a part of many processes that sustain multicelled bodies. As one example, water molecules constantly escape from the surface of liquid water as vapor, a process called **evaporation**. Evaporation is resisted by hydrogen bonding among water molecules. In other words, overcoming water's cohesion takes energy. Thus, evaporation sucks energy (in the form of heat) from liquid water, and this lowers the water's surface temperature. Evaporative water loss helps you and some other mammals cool off when you sweat in hot, dry weather. Sweat, which is about 99 percent water, cools the skin as it evaporates.

Cohesion works inside organisms, too. Consider how plants absorb water from soil as they grow. Water molecules evaporate from leaves, and replacements are pulled upward from roots. Cohesion makes it possible for columns of liquid water to rise from roots to leaves inside narrow pipelines of vascular tissue. In some trees, these pipelines extend hundreds of feet above the soil (Section 27.6 returns to this topic).

Take-Home Message

What gives water the special properties that make life possible?

- Extensive hydrogen bonding among water molecules, which arises from the polarity of the individual molecules, gives water special properties.

- Liquid water is an excellent solvent. Hydrophilic substances such as salts and sugars dissolve easily in water to form solutions. Hydrophobic substances do not dissolve in water.

- Water also stabilizes temperature, and it has cohesion.

FIGURE 2.11 Ice. *Above*, hydrogen bonds lock water molecules in a rigid lattice in ice. Molecules in this lattice pack less densely than in liquid water (compare Figure 2.10**C**), so ice floats on water. *Below*, a covering of ice can insulate water underneath it, thus keeping aquatic organisms from freezing during harsh winters.

Credits: top, © Cengage Learning; bottom, www.flickr.com/photots/roseo-fredrock.

cohesion Property of a substance that arises from the tendency of its molecules to resist separating from one another.
evaporation Transition of a liquid to a vapor.
hydrogen bond Attraction between a covalently bonded hydrogen atom and another atom taking part in a separate covalent bond.
hydrophilic Describes a substance that dissolves easily in water.
hydrophobic Describes a substance that resists dissolving in water.
salt Ionic compound that releases ions other than H^+ and OH^- when it dissolves in water.
solute A dissolved substance.
solution Homogeneous mixture of solute and solvent.
solvent Liquid that can dissolve other substances.
temperature Measure of molecular motion.

2.5 Acids and Bases

Concentration refers to the amount of a particular solute that is dissolved in a given volume of fluid. Hydrogen ion (H⁺) concentration is a special case. We measure the amount of hydrogen ions in a solution using a value called **pH**. In liquid water, some of the water molecules spontaneously separate into hydrogen ions and hydroxide ions (OH⁻). These ions combine again to form water. When the number of H⁺ ions equals the number of OH⁻ ions in the liquid, the pH is 7, or neutral. A one-unit decrease in pH corresponds to a tenfold increase in the number of H⁺ ions, and a one-unit increase corresponds to a tenfold decrease in the number of H⁺ ions (Figure 2.12). One way to get a sense of the scale is to taste dissolved baking soda (pH 9), pure water (pH 7), and lemon juice (pH 2).

Substances called **bases** accept hydrogen ions, so they can raise the pH of fluids and make them basic, or alkaline (above pH 7). **Acids** give up hydrogen ions when they dissolve in water, so they lower the pH of fluids and make them acidic (below pH 7).

Nearly all of life's chemistry occurs near pH 7. Under normal circumstances, fluids inside cells and bodies stay within a consistent range of pH because they are buffered. A **buffer** is a set of chemicals that can keep pH stable by alternately donating and accepting ions that affect pH. For example, two chemicals, carbonic acid and bicarbonate, are part of a homeostatic mechanism that normally keeps your blood pH between 7.35 and 7.45. Carbonic acid forms when carbon dioxide gas dissolves in the fluid portion of blood. It can dissociate into a hydrogen ion and a bicarbonate ion, which in turn recombine to form carbonic acid:

$$H_2CO_3 \longrightarrow H^+ + HCO_3^- \longrightarrow H_2CO_3$$

carbonic acid   bicarbonate   carbonic acid

An excess of OH⁻ ions in blood causes the carbonic acid in it to release H⁺ ions. These combine with the excess OH⁻ ions to form water, which does not affect pH. Excess H⁺ in blood combines with the bicarbonate, so it does not affect pH.

Any buffer can neutralize only so many ions. Even slightly more than that limit and the pH of the fluid will change dramatically. Buffer failure can be catastrophic in a biological system because most biological molecules can function properly only within a narrow range of pH. For instance, when breathing is impaired suddenly, carbon dioxide gas that accumulates in tissues ends up as excess carbonic acid in blood. The resulting decline in blood pH may cause the person to enter a dangerous level of unconsciousness called a coma.

Burning fossil fuels such as coal releases sulfur and nitrogen compounds that affect the pH of rain and other forms of precipitation. Rainwater is not buffered, so the addition of acids or bases has a dramatic effect. In places with a lot of fossil fuel emissions, the rain and fog can be more acidic than vinegar. The corrosive effects of this acid rain are visible in urban areas (*inset*). Acid rain also changes the pH of water in soil, lakes, and streams (we return to the topic of acid rain in Section 18.5).

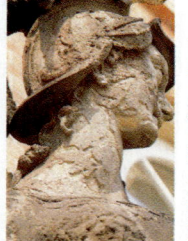

W. K. Fletcher/ Photo Researchers, Inc.

pH	Substance
14	drain cleaner
13	oven cleaner
12	bleach / hair remover
11	household ammonia
10	milk of magnesia / hand soap / toothpaste
9	Tums / detergents / baking soda
8	seawater / egg white / blood, tears
7	pure water
6	milk / butter / corn
5	urine, tea, typical rain / black coffee / bread / beer
4	bananas / tomatoes, wine / orange juice
3	vinegar / cola / lemon juice / acid rain
2	
1	gastric fluid
0	battery acid

more basic ↑
more acidic ↓

FIGURE 2.12 A pH scale. Here, *red* dots signify hydrogen ions (H⁺) and *blue* dots signify hydroxyl ions (OH⁻). Also shown are the approximate pH values for some common solutions.

This pH scale ranges from 0 (most acidic) to 14 (most basic). A change of one unit on the scale corresponds to a tenfold change in the amount of H⁺ ions.

Figure It Out: What is the approximate pH of cola?

Answer: 2.5

Art: © Cengage Learning; photos: © JupiterImages Corporation.

Take-Home Message

Why are hydrogen ions important in biological systems?

- pH reflects the number of hydrogen ions in a fluid. Most biological systems function properly only within a narrow range of pH.
- Acids release hydrogen ions in water; bases accept them.
- Buffers help keep pH stable. Inside organisms, they are part of homeostasis.

2.6 Organic Molecules

The same elements that make up a living body also occur in nonliving things, but their proportions differ. For example, compared to sand or seawater, a human body contains a much larger proportion of carbon atoms. Why? Unlike sand or seawater, a body consists of a very high proportion of the molecules of life—complex carbohydrates, lipids, proteins, and nucleic acids—which in turn consist of a high proportion of carbon atoms. Molecules that have primarily hydrogen and carbon atoms are said to be **organic**. The term is a holdover from a time when such molecules were thought to be made only by living things, as opposed to the "inorganic" molecules that formed by nonliving processes.

Carbon's importance to life arises from its versatile bonding behavior. Carbon can bond to many different elements, and each carbon atom can form four covalent bonds. Many organic compounds have a backbone—a chain of carbon atoms—to which other atoms attach. The ends of a backbone may join so that the carbon chain forms one or more ring structures (Figure 2.13). The versatility means that carbon atoms can be assembled into a variety of organic molecules.

As you will see in the next few sections, the function of an organic molecule depends on its structure. The structure of many organic molecules is quite complex, so structural formulas representing them may not show some bonds, for clarity (Figure 2.14**A**). Hydrogen atoms bonded to a carbon backbone may also be omitted, and other atoms as well. Carbon ring structures are often represented as polygons (Figure 2.14**B**). If no atom is shown at a corner or at the end of a bond, a carbon is implied there. Ball-and-stick models are useful for representing smaller organic compounds (Figure 2.14**C**). Space-filling models can show the overall shape of large molecules (Figure 2.14**D**). Proteins and nucleic acids are often represented as ribbon structures, which, as you will see in Section 2.9, show how the molecule folds and twists.

〉 From Structure to Function All biological systems are based on the same organic molecules, but the details of those molecules differ among organisms. Just as atoms bonded in different numbers and arrangements form different molecules, simple organic building blocks bonded in different numbers and arrangements form different versions of the molecules of life.

Cells assemble complex carbohydrates, lipids, proteins, and nucleic acids from small organic molecules. These small organic molecules—simple sugars, fatty acids, amino acids, and nucleotides—are called **monomers** when they

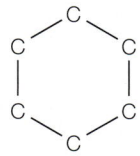

A Carbon's versatile bonding behavior allows it to form a variety of structures, including rings.

B Carbon rings form the framework of many sugars, starches, and fats, such as those found in doughnuts.

FIGURE 2.13 Carbon rings.

Credits: (a) © Cengage Learning; (b) © JupiterImages Corporation..

acid Substance that releases hydrogen ions in water.
base Substance that accepts hydrogen ions in water.
buffer Set of chemicals that can keep the pH of a solution stable by alternately donating and accepting ions that contribute to pH.
concentration The amount of molecules or ions per unit volume of a solution.
monomer Molecule that is a subunit of polymers.
organic Describes a molecule that consists mainly of carbon and hydrogen atoms.
pH Measure of the amount of hydrogen ions in a fluid.

glucose

A A structural formula for an organic molecule can be very complicated.

glucose

B Structural formulas of organic molecules are often simplified by using polygons as symbols for rings, and omitting the labels for some or all of the atoms.

glucose

C A ball-and-stick model shows the arrangement of atoms in three dimensions.

glucose

D A space-filling model can show a molecule's overall shape.

FIGURE 2.14 Modeling an organic molecule. © Cengage Learning.

A Condensation. Cells build a large molecule from smaller ones by this reaction. An enzyme removes a hydroxyl group from one molecule and a hydrogen atom from another. A covalent bond forms between the two molecules; water also forms.

B Hydrolysis. Cells split a large molecule into smaller ones by this water-requiring reaction. An enzyme attaches a hydroxyl group and a hydrogen atom (both from water) at the cleavage site.

FIGURE 2.15 Animated!
Metabolism: two common reactions by which cells build and break down organic molecules. © Cengage Learning.

© JupiterImages Corporation.

are used as subunits of larger molecules. Molecules that consist of multiple monomers are called **polymers**. Cells build polymers from monomers, and break down polymers to release monomers. Such processes, in which molecules change, are called **reactions**.

Reactions that run constantly inside cells help them stay alive, grow, and reproduce. **Metabolism** refers to reactions and all other activities by which cells acquire and use energy as they make and break apart organic compounds. Metabolism also requires **enzymes**, which are organic molecules that make reactions proceed faster than they would on their own. For example, in a common metabolic reaction called condensation, an enzyme covalently bonds two monomers together (Figure 2.15**A**). In hydrolysis, the reverse of condensation, an enzyme splits an organic polymer into its component monomers (Figure 2.15**B**).

Take-Home Message

How are all of the molecules of life alike?

■ The molecules of life (carbohydrates, lipids, proteins, and nucleic acids) are organic, which means they consist mainly of carbon and hydrogen atoms.

■ The structure of an organic molecule starts with its carbon backbone, a chain of carbon atoms that may form a ring.

■ Cells assemble large polymers from smaller monomers of simple sugars, fatty acids, amino acids, and nucleotides. They also break apart polymers into their component monomers.

2.7 Carbohydrates

Carbohydrates consist of carbon, hydrogen, and oxygen atoms in a 1:2:1 ratio. The simplest carbohydrates are sugars, or monosaccharides. "Saccharide" is from a Greek word that means sugar. Common sugars have a backbone of five or six carbon atoms. Most are water soluble, so they are easily transported through the water-based internal fluids of all organisms. Monosaccharides that are components of the nucleic acids DNA and RNA have five carbon atoms. Glucose, which has six carbon atoms, can be used as a fuel to drive cellular processes, or as a structural material to build larger molecules. Disaccharides are polymers of two sugar units. Sucrose, which is our table sugar, is a disaccharide, with one unit of glucose and one of fructose (another monosaccharide).

"Complex" carbohydrates, or polysaccharides, are straight or branched chains of hundreds or thousands of sugar monomers. There may be one type or many types of monomers in a polysaccharide. The most common polysaccharides are cellulose, starch, and glycogen. All consist only of glucose monomers, but as substances their properties are very different. Why? The answer begins with differences in patterns of covalent bonding that link their glucose units.

Cellulose, the major structural material of plants, is the most abundant organic molecule in the biosphere. It consists of long, straight chains of glucose monomers locked into tight, sturdy bundles by hydrogen bonds (Figure 2.16**A**). These tough cellulose fibers act like reinforcing rods that help stems resist wind and other forms of mechanical stress.

Cellulose does not dissolve in water, and it is not easily broken down. Some bacteria and fungi make enzymes that can break it apart into its component sugars, but humans and other mammals do not. Dietary fiber, or "roughage," usually refers to the cellulose and other indigestible polysaccharides in our

A Cellulose, a structural component of plants. Chains of glucose units stretch side by side and hydrogen-bond at many —OH groups. The hydrogen bonds stabilize the chains in tight bundles that form long fibers. Very few types of organisms can digest this tough, insoluble material.

B In starch, a series of glucose units form a chain that coils. Starch is the main energy reserve in plants, which store it in their roots, stems, leaves, seeds, and fruits.

C Glycogen. In animals, this poly-saccharide functions as an energy reservoir. It is especially abundant in the liver and muscles of active animals, including people.

FIGURE 2.16 Animated! Three of the most common complex carbohydrates and their locations in a few organisms. Each polysaccharide consists only of glucose units, but different bonding patterns that link the subunits result in substances with very different properties.

Credits: (a–c), © Cengage Learning; middle, © JupiterImages Corporation.

vegetable foods. Bacteria that live in the guts of termites and grazers such as cattle and sheep help these animals digest the cellulose in plants.

In starch, a different covalent bonding pattern between glucose monomers makes a chain that coils up into a spiral (Figure 2.16**B**). Like cellulose, starch does not dissolve easily in water, but it is not as stable as cellulose. These properties make starch ideal for storing chemical energy in the watery, enzyme-filled interior of plant cells. Most plants make much more glucose than they can use. The excess is stored as starch inside cells that make up roots, stems, and leaves. However, because it is insoluble, starch cannot be transported out of the cells and distributed to other parts of the plant. When sugars are in short supply, hydrolysis enzymes break the bonds between starch's glucose monomers. The released glucose, which is soluble, can be transported out of the cells. Humans also have enzymes that hydrolyze starch, so this carbohydrate is an important component of our food.

The covalent bonding pattern in glycogen forms highly branched chains of glucose monomers (Figure 2.16**C**). In animals, glycogen is the sugar-storage equivalent of starch in plants. Muscle and liver cells store it to meet a sudden need for glucose. These cells break down glycogen to release its glucose subunits.

Take-Home Message

What are carbohydrates?

- Subunits of simple carbohydrates (sugars), bonded in different ways, form various types of complex carbohydrates.

- Cells use carbohydrates for energy or as structural materials.

carbohydrate Molecule that consists primarily of carbon, hydrogen, and oxygen atoms in a 1:2:1 ratio.

cellulose Tough, insoluble carbohydrate that is the major structural material in plants.

enzyme Organic molecule that speeds up a reaction without being changed by it.

metabolism All the enzyme-mediated chemical reactions by which cells acquire and use energy as they build and break down organic molecules.

polymer Molecule that consists of multiple monomers.

reaction Process of molecular change.

2.8 Lipids

Lipids are fatty, oily, or waxy organic compounds. Many lipids incorporate **fatty acids**, which are small organic molecules that consist of a carbon chain "tail" topped with an acidic "head." A fatty acid's long tail is hydrophobic (hence the name "fatty"), but its head is a hydrophilic carboxyl group (Figure 2.17). You are already familiar with the properties of fatty acids because these molecules are the main component of soap. The hydrophobic tails of the fatty acids in soap attract oily dirt, and the hydrophilic heads dissolve the dirt in water.

Saturated fatty acids have only single bonds in their tails. In other words, their carbon chains are fully saturated with hydrogen atoms (Figure 2.17**A**). Saturated fatty acid tails are flexible and they wiggle freely. The tails of unsaturated fatty acids have one or more double bonds that limit their flexibility (Figure 2.17**B,C**). These bonds are *cis* or *trans*, depending on the way the hydrogens are arranged around them (Figure 2.17**D,E**).

> **Fats** The carboxyl group head of a fatty acid can easily form a covalent bond with another molecule. When it bonds to a glycerol, a type of alcohol, it loses its hydrophilic character and makes a fat. **Fats** are lipids with one, two, or three fatty acids bonded to the same glycerol. A fat with three fatty acid tails is called a **triglyceride**. Triglycerides are entirely hydrophobic, so they do not dissolve in water. Most "neutral" fats, such as butter and vegetable oils, are examples. Triglycerides are the most abundant and richest energy source in vertebrate bodies. They are concentrated in tissue that insulates and cushions body parts.

Butter, cream, and other high-fat animal products have a high proportion of **saturated fats**, which means they consist mainly of triglycerides

"head" (hydrophilic carboxyl group)

hydrophobic "tail"

A stearic acid (saturated) **B** linoleic acid (omega-6) **C** linolenic acid (omega-3) **D** oleic acid (cis) **E** elaidic acid (trans)

FIGURE 2.17 Animated! Fatty acids. Double bonds in the tails are colored *red*.

A The tail of stearic acid is fully saturated with hydrogen atoms. **B** Linoleic acid, with two double bonds, is unsaturated. The first double bond occurs at the sixth carbon from the end of the tail, so linoleic acid is called an omega-6 fatty acid. Omega-6 and **C** omega-3 fatty acids are "essential fatty acids," which means your body does not make them and they must come from food.

D The hydrogen atoms (in *blue*) around the double bond in oleic acid are on the same side of the tail. Most other naturally occurring unsaturated fatty acids have these *cis* bonds. **E** Hydrogenation creates abundant *trans* bonds, with hydrogen atoms on opposite sides of the tail (in *blue*).

Figure It Out: Are the double bonds in linolenic acid *cis* or *trans*?

Answer: cis

© Cengage Learning.

with three saturated fatty acid tails. Saturated fats tend to be solid at room temperature because their floppy saturated tails can pack tightly together. Most vegetable oils are **unsaturated fats**, which means they consist mainly of triglycerides with one or more unsaturated fatty acid tails. Each double bond in a fatty acid tail makes a rigid kink. Kinky tails do not pack tightly, so unsaturated fats are typically liquid at room temperature. The partially hydrogenated vegetable oils that you learned about in Section 2.1 are an exception. They are solid at room temperature. The special *trans* double bond keeps fatty acid tails straight, allowing them to pack tightly just like saturated fats do.

hydrophilic head

one layer of lipids

one layer of lipids

two hydrophobic tails

A A phospholipid molecule   **B** A lipid bilayer

FIGURE 2.18 Phospholipids as components of cell membranes. A double layer of phospholipids—the lipid bilayer—is the structural foundation of all cell membranes.
© Cengage Learning.

❯ **Phospholipids** A **phospholipid** has two fatty acid tails and a head that contains phosphate (Figure 2.18**A**). The tails are hydrophobic, but the phosphate is highly polar and it makes the head very hydrophilic. The opposing properties of a phospholipid molecule give rise to cell membrane structure. Phospholipids are the most abundant lipids in cell membranes, which have two layers of lipids (Figure 2.18**B**). The heads of one layer are dissolved in the cell's watery interior, and the heads of the other layer are dissolved in the cell's fluid surroundings. In such **lipid bilayers**, all of the hydrophobic tails are sandwiched between the hydrophilic heads. You will read more about the structure of cell membranes in Chapter 3.

❯ **Waxes** A **wax** is a complex, varying mixture of lipids with long fatty acid tails bonded to long-chain alcohols or carbon rings. The molecules pack tightly, so the resulting substance is firm and water-repellent. Plant leaves secrete a layer of waxes that helps restrict water loss and repel parasites. Secreted waxes also protect, lubricate, and soften our skin and hair. Waxes, together with fats and fatty acids, make feathers waterproof. Bees store honey and raise new generations of bees inside honeycomb made from wax that they secrete.

❯ **Steroids** All eukaryotic cell membranes contain **steroids**, which are lipids with a rigid backbone of four carbon rings and no fatty acid tails. Cholesterol is the most common steroid in animal cell membranes. The animal body also remodels cholesterol into other compounds, including bile salts that help digest fats; vitamin D that keeps teeth and bones strong; and steroid hormones such as estrogens and testosterone, which govern the development of sexual traits and other processes of sexual reproduction (Figure 2.19).

an estrogen   testosterone

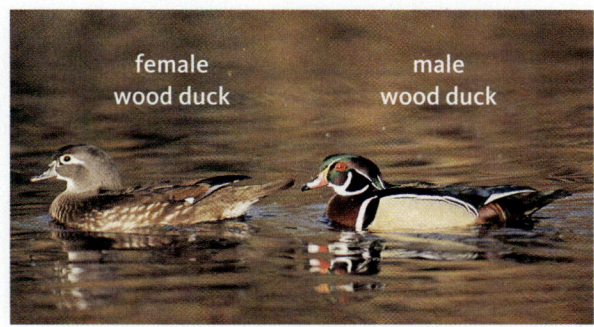

female wood duck   male wood duck

FIGURE 2.19 Estrogen and testosterone, steroid hormones that cause different traits to arise in males and females of many species such as wood ducks.

Credits: Top, © Cengage Learning; bottom, Tim Davis/ Photo Researchers, Inc.

Take-Home Message

What are lipids?

- Lipids are fatty, waxy, or oily organic compounds. Common types include fats, phospholipids, waxes, and steroids.
- Triglycerides are fats that serve as energy reservoirs in vertebrate animals.
- Phospholipids are the main component of cell membranes.
- Waxes are components of water-repelling and lubricating secretions.
- Steroids occur in eukaryotic cell membranes. Some are remodeled into other molecules.

fat Lipid that consists of a glycerol molecule with one, two, or three fatty acid tails.

fatty acid Organic compound that consists of a chain of carbon atoms with an acidic carboxyl group at one end.

lipid Fatty, oily, or waxy organic compound.

lipid bilayer Double layer of lipids arranged tail-to-tail; structural foundation of all cell membranes.

phospholipid A lipid with a phosphate in its hydrophilic head, and two nonpolar fatty acid tails.

saturated fat Fat that consists mainly of triglycerides with three saturated fatty acid tails.

steroid A type of lipid with four carbon rings and no fatty acid tails.

triglyceride A fat that has three fatty acid tails.

unsaturated fat Fat that consists mainly of triglycerides with one or more unsaturated fatty acid tails.

wax Water-repellent mixture of lipids with long fatty acid tails bonded to long-chain alcohols or carbon rings.

methionine

valine

methionine — valine

❶ A peptide bond forms between two amino acids (here, methionine and valine).

| methionine | valine | histidine | leucine | threo |

❷ More amino acids become linked into a polypeptide chain. The sequence of amino acids in the chain is the protein's primary structure. Each type of protein has a unique primary structure.

FIGURE 2.20 **Animated!** How protein structure arises. Chapter 7 returns to protein synthesis.

Credits: (3–5) 1BBB, A third quaternary structure of human hemoglobin A at 1.7-A resolution. Silva, M.M., Rogers, P.H., Arnone, A., Journal: (1992) *J.Biol.Chem.* 267: 17248-17256; (1,2,6) © Cengage Learning.

2.9 Proteins

Of all biological molecules, proteins are the most diverse in both structure and function. Structural proteins support cell parts and, as part of tissues, multicelled bodies. Feathers, hooves, and hair, as well as tendons and other body parts, consist mainly of structural proteins. A tremendous number of different proteins, including some structural types, are active participants in all processes that sustain life. Most enzymes that drive metabolic reactions are proteins. Proteins move substances, help cells communicate, and defend the body.

Proteins are polymers, and cells make all of the thousands of different kinds they need from only twenty kinds of monomers called amino acids. An **amino acid** is a small organic compound with an amine group ($-NH_2$), a carboxyl group (the acid), and a side chain called an "R group" specific to the type of amino acid. In most amino acids, all three groups are attached to the same carbon atom:

amine group

carboxyl group

R group

A protein's biological activity arises from and depends on its shape.

Protein synthesis involves covalently bonding amino acids into a chain (Figure 2.20). The bond that forms between the two amino acids is called a **peptide bond** ❶. The process is repeated hundreds or thousands of times, so a long chain of amino acids called a polypeptide forms ❷. The sequence of amino acids in the polypeptide is called the protein's primary structure.

One of the fundamental ideas in biology is that structure dictates function. This idea is particularly appropriate as applied to proteins, because a protein's biological activity arises from and depends on its shape. There are several levels of protein structure beyond amino acid sequence. Even before a polypeptide has been completely synthesized, it begins to twist and fold as hydrogen bonds form among the amino acids of the chain. This hydrogen bonding may cause parts of the polypeptide to form coils or flat sheets, and these patterns constitute the protein's secondary structure ❸. The primary structure of each type of protein is unique, but most proteins have coils and sheets.

© Cengage Learning.

❸ Secondary structure arises as the polypeptide twists into a coil or sheet held in place by hydrogen bonds. Most proteins have coils and sheets.

❹ Tertiary structure occurs when the coils and sheets fold up into a domain. In this example, the coils of a globin chain form a pocket.

❺ Some proteins have two or more polypeptide chains. Hemoglobin, shown here, consists of four globin chains (*green* and *blue*). Each globin pocket now holds a heme group (*red*).

❻ Fibrous proteins aggregate by the many thousands into much larger structures. The filaments that make up hair are examples; they consist of tightly bundled keratin proteins.

From "Structure of the Rotor of the V-Type Na+-ATPase from Enterococcus hirae" by Murata, et all. *Science* 29 April 2005: 654-659. DOI1.1126/ science.1110064. Used with permission.

Much as an overly twisted rubber band coils back upon itself, hydrogen bonding between nonadjacent regions of the protein makes its coils and sheets fold up into even more compact domains. Such domains are part of a protein's overall three-dimensional shape, or tertiary structure ❹. Tertiary structure is what makes a protein a working molecule. For example, sheets or coils of some proteins curl up into a barrel shape (*inset*). A barrel domain often functions as a tunnel for small molecules, allowing them to pass, for example, through a cell membrane. Globular domains of enzymes form chemically active pockets that can make or break bonds of other molecules.

Many proteins also have quaternary structure, which means they consist of two or more polypeptide chains that are closely associated or covalently bonded together ❺. Most enzymes are like this, with multiple polypeptide chains that collectively form a roughly spherical shape.

Fibrous proteins aggregate by many thousands into much larger structures, with their polypeptide chains organized into strands or sheets. The keratin in your hair is an example ❻. Some fibrous proteins contribute to the structure and organization of cells and tissues. Others, such as the protein filaments in muscle cells, help cells, cell parts, and multicelled bodies move.

Proteins with attached sugars or lipids are glycoproteins and lipoproteins, respectively. Some lipoproteins form when enzymes covalently bond lipids to a protein. Other lipoproteins are aggregate structures that consist of variable amounts and types of proteins and lipids.

> **The Importance of Protein Structure** Protein shape depends on hydrogen bonds and other interactions that heat, some salts, shifts in pH, or detergents can disrupt. Such disruption causes proteins to **denature**, which means they unwind and otherwise lose their three-dimensional shape. Once a protein's shape unravels, so does its function.

You can see denaturation in action when you cook an egg. A protein called albumin is a major component of egg white. Cooking does not break the covalent bonds of albumin's primary structure, but it does disrupt the hydrogen bonds that maintain the protein's shape. When a translucent egg white turns opaque, the albumin has been denatured. For a very few proteins, denaturation is reversible if normal conditions return, but albumin is not one of them. There is no way to uncook an egg.

Prion diseases are the dire aftermath of proteins that change shape. Examples include mad cow disease (bovine spongiform encephalitis, or BSE), Creutzfeldt–

amino acid Small organic compound that is a subunit of proteins. Consists of a carboxyl group, an amine group, and a characteristic side group (R), all typically bonded to the same carbon atom.

denature Regarding a biological molecule, to become so altered in shape that some or all function is lost.

peptide bond A bond between the amine group of one amino acid and the carboxyl group of another. Joins amino acids in proteins.

protein Organic compound that consists of one or more chains of amino acids (polypeptides).

A Charlene Singh, here being cared for by her mother, was one of the three people who developed symptoms of vCJD disease while living in the United States. She died in 2004.

B Slice of brain tissue from a person with vCJD. Characteristic holes and prion protein fibers radiating from several deposits are visible.

FIGURE 2.21 Variant Creutzfeldt–Jakob disease (vCJD). Two hundred people have died from vCJD since 1990.

ATP Adenosine triphosphate. Nucleotide that consists of an adenine base, a ribose sugar, three phosphate groups.

DNA Deoxyribonucleic acid. Nucleic acid that carries hereditary information about traits; consists of two nucleotide chains twisted in a double helix.

nucleic acid Single- or double-stranded chain of nucleotides joined by sugar–phosphate bonds; DNA or RNA.

nucleotide Small organic compound that consists of a five-carbon sugar, a nitrogen-containing base, and one or more phosphate groups.

prion Infectious protein.

RNA Ribonucleic acid. Some types have roles in protein synthesis.

Jakob disease in humans, and scrapie in sheep. These infectious diseases may be inherited, but more often they arise spontaneously. All are characterized by relentless deterioration of mental and physical abilities and eventual death (Figure 2.21**A**).

All prion diseases begin with a protein that occurs normally in mammals. One such protein, PrPC, is found in cell membranes throughout the body, but we still know very little about what it does. Very rarely, a PrPC protein spontaneously misfolds. In itself, a single misfolded protein molecule would not pose much of a threat. However, when this particular protein misfolds, it becomes a **prion**, or infectious protein. The altered shape of a misfolded PrPC protein causes normally folded PrPC proteins to misfold too. Because each protein that misfolds becomes infectious, the number of prions increases exponentially.

The shape of misfolded PrPC proteins allows them to align tightly into long fibers. The fibers begin to accumulate in the brain in water-repellent patches that disrupt brain cell function, causing symptoms such as confusion, memory loss, and lack of coordination. Tiny holes form in the brain as its cells die (Figure 2.21**B**). Eventually, the brain becomes so riddled with holes that it looks like a sponge.

In the mid-1980s, an epidemic of mad cow disease in Britain was followed by an outbreak of a new variant of Creutzfeldt–Jakob disease (vCJD) in humans. The cattle became infected by the prion after eating feed prepared from the remains of scrapie-infected sheep, and people became infected by eating beef from infected cattle (prions are not denatured by cooking or typical treatments that inactivate other infectious agents). The use of animal parts in livestock feed is now banned in many countries, and the number of cases of BSE and vCJD has since declined. Cattle with BSE still turn up, but so rarely that they pose little threat to human populations.

Take-Home Message

Why is protein structure important?

■ A protein's function depends on its structure, which consists of chains of amino acids that twist and fold into functional domains.

■ Changes in a protein's structure may also alter its function.

2.10 Nucleic Acids

Nucleotides are small organic molecules that function as energy carriers, enzyme helpers, chemical messengers, and subunits of DNA and RNA. Each consists of a sugar with a five-carbon ring, bonded to a nitrogen-containing base and one or more phosphate groups. The nucleotide **ATP** (adenosine triphosphate) has a row of three phosphate groups attached to its ribose sugar (Figure 2.22**A**). When the outer phosphate group of an ATP is transferred to another molecule, energy is transferred along with it. You will read more about phosphate groups and their important metabolic role in Chapters 4 and 5.

Nucleic acids are polymers, chains of nucleotides in which the sugar of one nucleotide is joined to the phosphate group of the next (Figure 2.22**B**). An example is **RNA**, or ribonucleic acid, named after the ribose sugar of its component nucleotides. RNA consists of four kinds of nucleotide monomers, one of which is ATP. RNA molecules carry out protein synthesis, which we discuss in detail in Chapter 7.

DNA, or deoxyribonucleic acid, is a nucleic acid named after the deoxyribose sugar of its component nucleotides. A DNA molecule consists of two chains of nucleotides twisted into a double helix (Figure 2.22**C**). Hydrogen bonds between the nucleotides hold the two chains together.

Each cell starts life with DNA inherited from a parent cell. That DNA contains all information necessary to build a new cell and, in the case of multicelled organisms, an entire individual. The cell uses the order of nucleotide bases in DNA—the DNA sequence—to guide production of RNA and proteins. Parts of the DNA sequence are identical or nearly so in all organisms. Different species (and individual organisms) have unique DNA sequences (Chapter 6 returns to DNA structure and function).

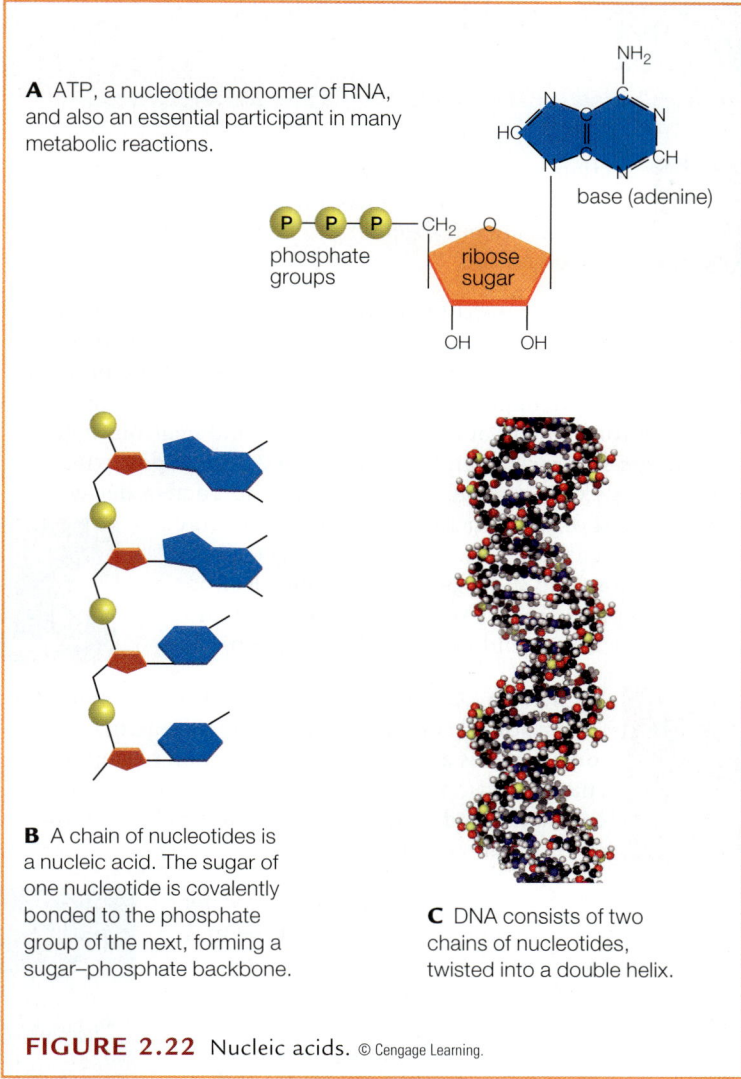

A ATP, a nucleotide monomer of RNA, and also an essential participant in many metabolic reactions.

base (adenine)

phosphate groups

ribose sugar

B A chain of nucleotides is a nucleic acid. The sugar of one nucleotide is covalently bonded to the phosphate group of the next, forming a sugar–phosphate backbone.

C DNA consists of two chains of nucleotides, twisted into a double helix.

FIGURE 2.22 Nucleic acids. © Cengage Learning.

Take-Home Message

What are nucleotides and nucleic acids?

■ Nucleotides are monomers of nucleic acids. They also have important metabolic roles, for example as energy carriers.

■ DNA encodes information that guides the production of RNA and proteins.

■ RNA carries out protein synthesis.

2.11 Fear of Frying (revisited)

JupiterImages Corporation.

Trans fatty acids are relatively rare in unprocessed foods, so it makes sense from an evolutionary standpoint that our bodies may not have enzymes to deal with them efficiently. The enzymes that hydrolyze *cis* fatty acids have difficulty breaking down *trans* fatty acids, a problem that may be a factor in the ill effects of *trans* fats. All prepackaged foods in the United States are now required to list *trans* fat content, but may be marked "zero grams of *trans* fats" even when a single serving contains up to half a gram.

WHERE YOU ARE GOING . . .

In Chapter 3, you will read more about lipids and proteins in cell membranes. Electrons, carbohydrates, enzymes, and nucleotides will come up again as you learn how energy drives metabolism (Chapters 4 and 5). Chapter 6 revisits DNA structure and function, and Chapter 7, protein synthesis. Section 5.6 returns to the dangers of free radicals. Section 11.4 explains how radioisotopes can be used to date rocks and fossils. Section 18.5 reconsiders acid rain. Section 27.6 covers water's cohesion in the context of water transport in plants.

Summary

Section 2.1 All organisms consist of the same kinds of molecules. Seemingly small differences in the way those molecules are put together can have big effects inside a living organism.

Section 2.2 Atoms consist of **electrons**, which carry a negative **charge**, moving about a **nucleus** of positively charged **protons** and uncharged **neutrons**. The number of protons (the **atomic number**) determines the **element**; **isotopes** of an element differ in the number of neutrons. The total number of protons and neutrons is the **mass number**. **Tracers** can be made with **radioisotopes**, which spontaneously emit particles and energy by **radioactive decay**. A **shell model** of an atom represents electron energy levels as concentric circles. Atoms tend to get rid of vacancies. Many do so by gaining or losing electrons, thereby becoming **ions**. **Free radicals** (atoms with unpaired electrons) tend to be very active chemically.

Section 2.3 **Chemical bonds** join atoms as molecules. A **compound** consists of two or more elements. An **ionic bond** is a strong mutual attraction of oppositely charged ions. Atoms share a pair of electrons in a **covalent bond**. Covalent bonds have **polarity** if the electrons are not shared equally.

Section 2.4 The two polar covalent bonds in water molecules give rise to an overall polarity of the molecule. **Hydrogen bonds** that form among water molecules in tremendous numbers are the basis of water's unique properties: a capacity to act as a **solvent** for **salts** and other polar **solutes**; resistance to **temperature** changes; and **cohesion**. Hydrophilic substances dissolve easily in water to form **solutions**; hydrophobic substances do not. **Evaporation** is the transition of liquid to vapor.

Section 2.5 A solute's **concentration** refers to the amount of solute in a given volume of fluid; **pH** reflects the number of hydrogen ions (H^+). At neutral pH (7), the amounts of H^+ and OH^- ions are equal. **Acids** release hydrogen ions in water; **bases** accept them. A **buffer** can keep a solution within a consistent range of pH. Most cell and body fluids are buffered because most molecules of life work only within a narrow range of pH.

Section 2.6 The molecules of life are **organic**, so they consist mainly of carbon and hydrogen atoms. Chains or rings of carbon atoms form the backbones of these molecules. Cells build complex carbohydrates, lipids, proteins, and nucleic acids from **monomers** of simple sugars, fatty acids, amino acids, and nucleotides, respectively. Reactions that assemble and break down these **polymers** require **enzymes** and are part of **metabolism.**

glucose

Section 2.7 Enzymes assemble complex **carbohydrates** such as **cellulose**, glycogen, and starch from simple carbohydrate (sugar) subunits. Cells use carbohydrates for energy, and as structural materials.

Section 2.8 Cells use **lipids** for energy and as structural materials. **Fats** have **fatty acid** tails; **triglycerides** have three. **Saturated fats** are mainly triglycerides with three saturated fatty acid tails (only single bonds link their carbons). **Unsaturated fats** are mainly triglycerides with one or more unsaturated fatty acids. A **lipid bilayer** (of mainly **phospholipids**) is the structural foundation of all cell membranes. **Waxes** are part of water-repellent and lubricating secretions. **Steroids** occur in cell membranes; some are remodeled into other molecules.

Section 2.9 The shape of a **protein** is the source of its function. Protein structure begins as a linear sequence of **amino acids** linked by **peptide bonds** into a polypeptide. Polypeptides twist into loops, sheets, and coils that can pack further into functional domains. Many proteins, including most enzymes, consist of two or more polypeptides. Fibrous proteins aggregate into much larger structures. A protein's shape may be disrupted by shifts in pH or temperature, or exposure to detergent or some salts. If that happens, the protein unravels, or **denatures**, and it loses its function. **Prion** diseases are a consequence of misfolded proteins.

Section 2.10 **Nucleotides** consist of a sugar, a phosphate group, and a nitrogen-containing base. Nucleotides are monomers of **DNA** and **RNA**, which are **nucleic acids**. Some nucleotides have additional functions. For example, **ATP** energizes many kinds of molecules by phosphate-group transfers. DNA encodes heritable information that guides the synthesis of RNA and proteins. RNA molecules carry out protein synthesis.

Self-Quiz Answers in Appendix I

1. Which of the following statements is incorrect?
   a. Isotopes have the same atomic number and different mass numbers.
   b. Atoms have about the same number of electrons as protons.
   c. All ions are atoms.
   d. Free radicals are dangerous because they emit energy.
   e. A carbon atom can share electrons with up to 4 other atoms.

2. What is the name of an atom that has one proton and no electrons?

3. The mutual attraction of opposite charges holds atoms together as molecules in a(n) _____ bond.
   a. ionic   c. polar covalent
   b. hydrogen   d. nonpolar covalent

4. Rank the following types of bonds by polarity, with 1 being the least polar, and 3 being the most polar:
   _____ 1   a. ionic
   _____ 2   b. polar covalent
   _____ 3   c. nonpolar covalent

5. A(n) _____ substance repels water.
   a. acidic   c. hydrophobic
   b. basic   d. polar

6. When dissolved in water, a(n) _____ donates H^+ and a(n) _____ accepts H^+.
   a. acid; base   c. buffer; solute
   b. base; acid   d. base; buffer

Digging Into Data

Effects of Dietary Fats on Lipoprotein Levels

Cholesterol that is made by the liver or that enters the body from food does not dissolve in blood, so it is carried through the bloodstream by lipoproteins. Low-density lipoprotein (LDL) carries cholesterol to body tissues such as artery walls, where it can form deposits associated with cardiovascular disease. Thus, LDL is often called "bad" cholesterol. High-density lipoprotein (HDL) carries cholesterol away from tissues to the liver for disposal, so HDL is often called "good" cholesterol. In 1990, Ronald Mensink and Martijn Katan published a study that tested the effects of different dietary fats on blood lipoprotein levels. Their results are shown in Figure 2.23.

1. In which group was the level of LDL ("bad" cholesterol) highest?

2. In which group was the level of HDL ("good" cholesterol) lowest?

3. An elevated risk of heart disease has been correlated with increasing LDL-to-HDL ratios. Rank the three diets according to their predicted effect on cardiovascular health.

protein   lipid

an HDL particle

| | Main Dietary Fats | | | |
	cis fatty acids	trans fatty acids	saturated fats	optimal level
LDL	103	117	121	<100
HDL	55	48	55	>40
ratio	1.87	2.44	2.2	<2

FIGURE 2.23 Effect of diet on lipoprotein levels. Researchers placed 59 men and women on a diet in which 10 percent of their daily energy intake consisted of *cis* fatty acids, *trans* fatty acids, or saturated fats. Blood LDL and HDL levels were measured after three weeks on the diet; averaged results are shown in mg/dL (milligrams per deciliter of blood). All subjects were tested on each of the diets. The ratio of LDL to HDL is also shown.

This lipoprotein image was made by Amy Shih and John Stone using VMD and is owned by the Theoretical and Computational Biophysics Group, NIH Resource for Macromolecular Modeling and Bioinformatics, at the Beckman Institute, University of Illinois at Urbana-Champaign. Labels added to the original image by book author.

7. _____ is a simple sugar (a monosaccharide).
   a. Glucose   c. Ribose   e. a and c
   b. Sucrose   d. Starch   f. a, b, and c

8. Unlike saturated fats, the fatty acid tails of unsaturated fats incorporate one or more _____ .
   a. phosphate groups   c. double bonds
   b. glycerols   d. single bonds

9. Which of the following is a class of molecules that encompasses all of the other molecules listed?
   a. triglycerides   c. waxes   e. lipids
   b. fatty acids   d. steroids f. phospholipids

10. _____ are to proteins as _____ are to nucleic acids.
   a. Sugars; lipids   c. Amino acids; hydrogen bonds
   b. Sugars; proteins   d. Amino acids; nucleotides

11. A denatured protein has lost its _____ .
   a. hydrogen bonds   c. function
   b. shape   d. all of the above

12. Which of the following are not found in DNA?
   a. amino acids   c. nucleotides
   b. sugars   d. phosphate groups

13. Match the terms with their most suitable description.
   _____ hydrophilic   a. protons > electrons
   _____ atomic number   b. number of protons in nucleus
   _____ hydrogen bonds   c. polar; dissolves easily in water
   _____ positive charge   d. collectively strong
   _____ temperature   e. protons < electrons
   _____ negative charge   f. measure of molecular motion

14. Match each molecule with its most suitable description.
   _____ wax   a. protein primary structure
   _____ starch   b. an energy carrier
   _____ triglyceride   c. water-repellent secretions
   _____ DNA   d. carries heritable information
   _____ polypeptide   e. sugar storage in plants
   _____ ATP   f. richest energy source in animals

15. Match each molecule with its component(s).
   _____ protein   a. glycerol, fatty acids, phosphate
   _____ phospholipid   b. glycerol, fatty acids
   _____ fat   c. nucleotide monomers
   _____ nucleic acid   d. glucose monomers
   _____ cellulose   e. sugar, phosphate, base
   _____ nucleotide   f. amino acid monomers
   _____ wax   g. glucose, fructose
   _____ sucrose   h. fatty acids, carbon rings

Critical Thinking

1. Alchemists were the forerunners of modern-day chemists. Many of these medieval scholars and philosophers spent their lives trying to transform lead (atomic number 82) into gold (atomic number 79). Explain why they never did succeed in that endeavor.

2. Draw a shell model of a lithium atom (Li), which has 3 protons, then predict whether the majority of lithium atoms on Earth are uncharged, positively charged, or negatively charged.

3. Polonium is a rare element with 33 radioisotopes. The most common one, ^{210}Po, has 82 protons and 128 neutrons. When ^{210}Po decays, it emits an alpha particle, which is a helium nucleus (2 protons and 2 neutrons). ^{210}Po decay is tricky to detect because alpha particles do not carry very much energy compared to other forms of radiation. They can be stopped by, for example, a sheet of paper or a few inches of air. That is one reason that authorities failed to discover toxic amounts of ^{210}Po in the body of former KGB agent Alexander Litvinenko until after he died suddenly and mysteriously in 2006. What element does an atom of ^{210}Po change into after it emits an alpha particle?

4. In the following list, identify the carbohydrate, the fatty acid, the amino acid, and the polypeptide:
   a. NH_2—CHR—COOH   c. (methionine)$_{20}$
   b. $C_6H_{12}O_6$ d. $CH_3(CH_2)_{16}COOH$

Cell Structure 3

...Where You Have Been

Reflect on the Section 1.2 overview of life's levels of organization. You will now begin to see how the nonliving molecules of life—carbohydrates (2.7), lipids (2.8), proteins (2.9), and nucleic acids (2.10)—form cells (1.2) and carry out functions that define life (1.3). You will also see an application of tracers (2.2), and revisit the philosophy of science (1.9).

3.1 Food for Thought

We find bacteria at the bottom of the ocean, high up in the atmosphere, miles underground—essentially anywhere we look. Mammalian intestines typically harbor fantastic numbers of them, but bacteria are not just stowaways there. Intestinal bacteria make vitamins that mammals cannot, and they crowd out more dangerous germs. Cell for cell, bacteria that live in and on a human body outnumber the person's own cells by about ten to one.

Escherichia coli is one of the most common intestinal bacteria of warm-blooded animals. Most of the hundreds of types, or strains, of *E. coli* are harmless. A few of the known harmful strains make a toxic protein that can severely damage the lining of the human intestine. After ingesting as few as ten cells, a person may become ill with severe cramps and bloody diarrhea that lasts up to ten days. In some people, complications of infection result in kidney failure, blindness, paralysis, and death. Each year, about 265,000 people in the United States become infected with toxin–producing *E. coli* (Figure 3.1).

Toxic strains of *E. coli* live in the intestines of other animals—mainly cattle, deer, goats, and sheep—apparently without sickening them. Humans are exposed to the bacteria when they come into contact with feces of animals that harbor it, for example, by eating contaminated ground beef. During slaughter, meat can come into contact with feces. Bacteria in the feces stick to the meat, then get thoroughly mixed into it during the grinding process. Unless contaminated meat is cooked to at least 71°C (160°F), live bacteria will enter the digestive tract of whoever eats it.

People also become infected with toxic *E. coli* by eating fresh fruits and vegetables that have come into contact with animal feces. Washing produce with water does not remove the bacteria because they are sticky. In June 2011, more than 4,000 people in Germany and France were sickened after eating sprouts, and 49 of them died. The outbreak was traced to a single shipment of contaminated fenugreek seeds from Egypt.

The impact of such outbreaks, which occur with disturbing regularity, extends beyond the casualties. The contaminated sprouts cost growers in the European Union at least $600 million in lost sales. In 2011 alone, the United States Department of Agriculture (USDA) recalled 36.7 million pounds of ground meat products contaminated with toxic bacteria, at a cost in the billions of dollars. Such costs are eventually passed along to taxpayers and consumers.

Food growers and processors are now using procedures that they hope will reduce the number and scope of these outbreaks. Meat and produce are being tested for some bacteria before sale, and improved documentation should allow a source of contamination to be pinpointed more quickly.

FIGURE 3.1 Toxin-producing bacteria can contaminate foods. *Top*, cells of *E. coli* strain O157:H7 (*red*) on intestinal cells of a small child. This type of bacteria can cause a serious intestinal illness in people who eat foods contaminated with it, such as ground beef or fresh produce (*bottom*).

Top, © Stephanie Schuller/ Photo Researchers, Inc.; bottom, © JupiterImages Corporation.

3.2 What, Exactly, Is a Cell?

You learned in Section 1.2 that the cell is the smallest unit with the properties of life. Let's now turn to the details of cell structure and function.

❯ The Cell Theory Hundreds of years of observations led to the way we now answer the question, What is a cell? Today we know that a cell carries out metabolism and homeostasis, and reproduces either on its own or as part of a larger organism. By this definition, each cell is alive even if it is part of a multi-celled body, and all living organisms consist of one or more cells. We also know that cells reproduce themselves by dividing, so it follows that all existing cells must have arisen by division of other cells. Later chapters discuss the processes by which cells divide, but for now all you need to know is that a cell passes its hereditary material—its DNA—to offspring during those processes. Taken together, these four generalizations constitute the **cell theory**, a foundation of modern biology (Table 3.1).

❯ Components of All Cells Cells vary in shape and in what they do, but all share certain organizational and functional features: a plasma membrane, cytoplasm, and DNA (Figure 3.2).

Every cell has a **plasma membrane**, an outer membrane that separates the cell's contents from its external environment. A plasma membrane is selectively permeable, which means it allows only certain materials to cross, so it controls exchanges between the cell and its environment. Phospholipids (Section 2.8) are the most abundant type of lipid in cell membranes. As you will see in Section 3.3, many different proteins embedded in a lipid bilayer or attached to one of its surfaces carry out membrane functions.

The plasma membrane encloses a jellylike mixture of water, sugars, ions, and proteins called **cytoplasm**. Some or all of a cell's metabolism occurs in the cytoplasm, and the cell's internal components, including organelles, are suspended in it. **Organelles** are structures that carry out special metabolic functions inside a cell. Membrane-enclosed organelles compartmentalize tasks such as building, modifying, and storing substances.

All cells start out life with DNA, though a few types of cells lose it as they mature. Only eukaryotic cells have a **nucleus** (plural, nuclei), an organelle with a double membrane that contains the cell's DNA. A few bacteria have lipid bilayers enclosing their DNA, but the structure of this membrane differs from a nuclear membrane. In most bacteria and archaea, the DNA is suspended directly in cytoplasm.

❯ Cell Size Almost all cells are too small to see with the naked eye. Why? The answer begins with the processes that keep a cell alive. A living cell must exchange substances with its environment at a rate that keeps pace with its metabolism. These exchanges occur across the plasma membrane, which can handle only so many exchanges at a time. The rate of exchange across a plasma membrane depends on its surface area: the bigger it is, the more substances can cross the membrane during a given interval. Thus, cell size is limited by a physical relationship called the **surface-to-volume ratio**. By this ratio, an object's volume increases with the cube of its diameter, but its surface area increases only with the square.

Table 3.1	The Cell Theory

1. Every living organism consists of one or more cells.

2. The cell is the structural and functional unit of all organisms. A cell is the smallest unit of life, individually alive even as part of a multicelled organism.

3. All living cells arise by division of preexisting cells.

4. Cells contain hereditary material, which they pass to their offspring when they divide.

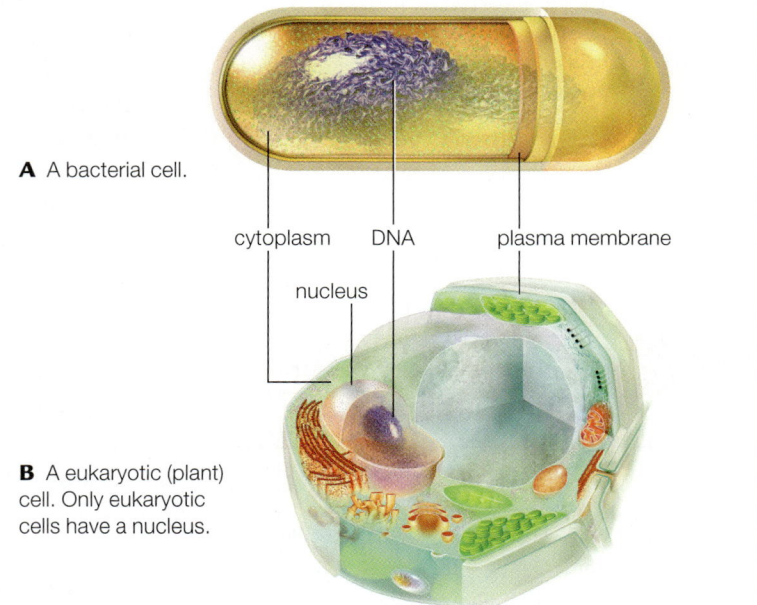

A A bacterial cell.

cytoplasm   DNA   plasma membrane

nucleus

B A eukaryotic (plant) cell. Only eukaryotic cells have a nucleus.

FIGURE 3.2 Animated! Cells. Archaea are similar to bacteria in overall structure; both are generally much smaller than eukaryotic cells. If the two cells depicted here had been drawn to the same scale, the bacterium would be about this big:

© Cengage Learning.

cell theory Theory that all organisms consist of one or more cells, which are the basic unit of life; all cells come from division of preexisting cells; and all cells pass hereditary material to offspring.

cytoplasm Semifluid substance enclosed by a cell's plasma membrane.

nucleus Of a cell, an organelle with two membranes that holds the cell's DNA.

organelle Structure that carries out a special metabolic function inside a cell.

plasma membrane A cell's outermost membrane.

surface-to-volume ratio A relationship in which the volume of an object increases with the cube of the diameter, and the surface area increases with the square.

Diameter (cm)	2	3	6
Surface area (cm²)	12.6	28.2	113
Volume (cm³)	4.2	14.1	113
Surface-to-volume ratio	3:1	2:1	1:1

FIGURE 3.3 Examples of surface-to-volume ratio. This physical relationship between increases in volume and surface area limits the size and influences the shape of cells.

© Cengage Learning.

Table 3.2	Common Units of Length		
Unit		**Equivalent**	
		Meter	Inch
centimeter	cm	1/100	0.394
millimeter	mm	1/1000	0.0394
micrometer	μm	1/1,000,000	0.0000394
nanometer	nm	1/1,000,000,000	0.0000000394
meter	m	100 cm	39.4
		1,000 mm	
		1,000,000 μm	
		1,000,000,000 nm	

Apply the surface-to-volume ratio to a round cell. As Figure 3.3 shows, when a cell expands in diameter, its volume increases faster than its surface area does. Imagine that a round cell expands until it is four times its original diameter. The volume of the cell has increased 64 times (4^3), but its surface area has increased only 16 times (4^2). Each unit of plasma membrane must now handle exchanges for four times as much cytoplasm ($64 \div 16 = 4$). If the cell gets too big, the inward flow of nutrients and the outward flow of wastes across that membrane will not be fast enough to keep the cell alive.

Surface-to-volume limits also affect cell shape. For example, muscle cells in your thighs are as long as the muscle in which they occur, but each is thin, so it exchanges substances efficiently with fluids in the tissue surrounding it.

❯ How Do We See Cells? Most cells are 10–20 micrometers in diameter, about fifty times smaller than the unaided human eye can perceive (Figure 3.4), so no one even knew cells existed until well after the first microscopes were invented. Like their early predecessors, many modern microscopes rely on visible light to illuminate objects. As you will learn in Chapter 5, all light travels in waves. This property makes light bend when it passes through curved glass lenses. Inside a light microscope, such lenses focus light that passes through a specimen, or bounces off of one, into a magnified image. Photographs of images enlarged with a microscope are called micrographs.

Phase-contrast microscopes shine light through specimens. Most cells are nearly transparent, so their internal details may not be visible unless they are first stained, or exposed to dyes that only some cell parts soak up. Parts that absorb the most dye appear darkest. Staining results in an increase in contrast (the difference between light and dark) that allows us to see a greater range of detail (Figure 3.5**A**). Surface details can be revealed by reflected light (Figure 3.5**B**).

With a fluorescence microscope, a cell or a molecule serves as the light source; it fluoresces, or emits energy in the form of light, when a laser beam is focused on it. Some molecules fluoresce naturally (Figure 3.5**C**). More typically, researchers attach a light-emitting tracer (Section 2.2) to the cell or molecule of interest.

FIGURE 3.4 Relative sizes. *Below,* the diameter of most cells is between 1 and 100 micrometers. Table 3.2 (*left*) shows conversions among units of length; also see Units of Measure, Appendix V.

Credits: from left, © Cengage Learning; Virus, CDC; Mitochondria, Chloroplast, Bacteria, Eukaryotic cells, © Cengage Learning; Louse, Edward S. Ross; Frog egg, © Cengage Learning; Ant, Frog, © A Cotton Photo/ Shutterstock; Rat, © Pakhnyushcha/ Shutterstock; Goose, © Vasyl Helevachuk/ Shutterstock; Boy, © Piotr Marcinski/ Shutterstock; Giraffe, © Valerie Kalyuznnyy/ Photos.com; Whale, © Dorling Kindersley/ the Agency Collection/ Getty Images; Tree, © Cengage Learning.

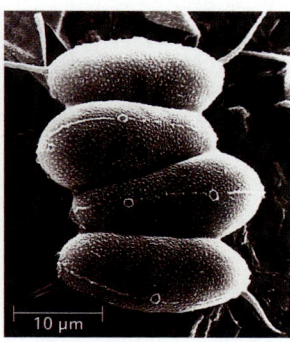

A Phase-contrast light microscopes yield high-contrast images of transparent specimens. Dark areas have taken up dye.

B Reflected light microscopes capture light reflected from the surface of specimens.

C This fluorescence micrograph shows fluorescent light emitted by chlorophyll molecules in the cells.

D Transmission electron micrographs reveal fantastically detailed images of internal structures.

E Scanning electron micrographs show surface details. SEMs may be artificially colored to highlight specific details.

FIGURE 3.5 Different microscopes reveal different characteristics of the same organism, a green alga (*Scenedesmus*).

Figure It Out: About how big are these cells?

Answer: About 20 μm in length and 8 μm in width

Credits: (a,b; d,e) Jeremy Pickett-Heaps, School of Botany, University of Melbourne; (c) © Prof. Franco Baldi.

Other microscopes can reveal finer details. For example, electron microscopes use magnetic fields to focus a beam of electrons onto a sample. Electrons travel in wavelengths much shorter than those of visible light, so these microscopes resolve details thousands of times smaller than light microscopes do. Transmission electron microscopes direct electrons through a thin specimen, and the specimen's internal details appear as shadows in the resulting image (Figure 3.5**D**). Scanning electron microscopes direct a beam of electrons across the surface of a specimen that has been coated with a thin layer of metal. The irradiated metal emits electrons and x-rays, which are converted into an image of the surface (Figure 3.5**E**).

Take-Home Message

How are all cells alike?

- The cell is the fundamental unit of all life.

- All cells start life with a plasma membrane, cytoplasm, and a region of DNA, which, in eukaryotic cells only, is enclosed by a nucleus. The surface-to-volume ratio limits cell size and influences cell shape.

- Different types of microscopes reveal different aspects of cell structure.

human eye (no microscope)

small animals

largest organisms

frog eggs

| 100 μm | 1 mm | 1 cm | 10 cm | 1 m | 10 m | 100 m |

A Phospholipids are the most abundant component of eukaryotic cell membranes. Each phospholipid molecule has a hydrophilic head and two hydrophobic tails.

phosphate group

hydrophilic head

two hydrophobic tails

B In a watery fluid, phospholipids spontaneously line up into two layers: the hydrophobic tails cluster together, and the hydrophilic heads face outward, toward the fluid. This lipid bilayer forms the framework of all cell membranes. Many types of proteins intermingle among the lipids—a few that are typical of plasma membranes are shown opposite.

one layer of lipids

one layer of lipids

FIGURE 3.6 Animated! Cell membrane structure. **A–C** Organization of lipids in cell membranes. **D–G** Examples of membrane proteins.
© Cengage Learning.

A cell's basic structure is essentially a lipid bilayer bubble filled with fluid.

fluid

© Cengage Learning.

3.3 The Structure of Cell Membranes

In all organisms, cell membranes are lipid bilayers that consist mainly of phospholipids (Figure 3.6**A**). The polar head of a phospholipid interacts with water molecules; the nonpolar fatty acid tails do not. As a result of these properties, phospholipids swirled into water spontaneously organize themselves into a lipid bilayer sheet or bubble (Figure 3.6**B**). A cell's basic structure is essentially a lipid bilayer bubble filled with fluid.

Other molecules, including cholesterol and proteins, are embedded in or attached to the lipid bilayer of every cell membrane. Many of these molecules move around the membrane more or less freely. A cell membrane behaves like a two-dimensional liquid of mixed composition, so we describe it as a **fluid mosaic**. The "mosaic" part of this phrase comes from a membrane's mixed composition of lipids and proteins. The fluidity occurs because the phospholipids in a typical cell membrane are not bonded to one another. They stay organized as a bilayer as a result of collective hydrophobic and hydrophilic attractions, which, on an individual basis, are relatively weak. Thus, phospholipids in a bilayer can drift sideways and spin around their long axis, and their tails are free to wiggle.

Different kinds of cells may use different kinds of lipids in their membranes. Archaea do not even build their phospholipids with fatty acids. Instead, they use molecules that have reactive side chains, so the tails of archaeal phospholipids form covalent bonds with one another. As a result of this rigid crosslinking, archaeal phospholipids do not drift, spin, or wiggle in a bilayer. Thus, the membranes of archaea are more rigid than those of bacteria or eukaryotes, a characteristic that may help these cells survive in extreme habitats.

❭ **Membrane Proteins** A cell membrane physically separates an external environment from an internal one, but that is not its only task. Many types of

C Recognition proteins such as this MHC molecule tag a cell as belonging to one's own body.

D Receptor proteins bind substances outside of the cell. This one is a B cell receptor. B cell receptors help the body eliminate toxins and infectious agents such as bacteria.

E Transport proteins bind to molecules on one side of the membrane, and release them on the other side. This one transports glucose.

F This transport protein, an ATP synthase, makes ATP when hydrogen ions flow through its interior.

Extracellular Fluid

Lipid
Bilayer

Cytoplasm

proteins are associated with a cell membrane, and each type adds a specific function to it. Thus, a cell membrane can have different characteristics depending on the proteins in it. For example, the plasma membrane has certain proteins that no internal cell membrane has. **Adhesion proteins** in the plasma membrane fasten cells together in animal tissues. **Recognition proteins** function as unique identity tags for each individual or species (Figure 3.6**C**). Being able to recognize "self" imparts the potential ability to distinguish nonself (foreign) cells or particles. **Receptor proteins** bind to specific extracellular substances such as hormones or toxins, or to molecules on another cell's plasma membrane (Figure 3.6**D**). Binding triggers a change in the cell's activities that may involve metabolism, movement, division, or even cell death. Different receptor proteins occur on different cells, but all are critical for homeostasis.

Additional proteins, including enzymes, occur on all cell membranes. **Transport proteins** move specific substances across a membrane, typically by forming a channel through it (Figure 3.6**E,F**). These proteins are important because lipid bilayers are impermeable to most substances, including ions and polar molecules. Some transport proteins are open channels through which a substance moves on its own across a membrane. Others use energy to actively pump a substance across. We return to the topic of transport across membranes in the next chapter.

Take-Home Message

What is a cell membrane?

- The foundation of all cell membranes is the lipid bilayer: two layers of phospholipids, tails sandwiched between heads.
- Proteins that associate with lipid bilayers add various functions to a membrane.

adhesion protein Plasma membrane protein that helps cells stick together.

fluid mosaic Model of a cell membrane as a two-dimensional fluid of mixed composition.

receptor protein Plasma membrane protein that binds to a particular substance outside of the cell.

recognition protein Plasma membrane protein that tags a cell as belonging to self (one's own body or species).

transport protein Protein that passively or actively assists specific ions or molecules across a membrane.

FIGURE 3.7 Examples of bacteria (*this page*) and archaea (*facing page*).

Credits: (a) Rocky Mountain Laboratories, NIAID, NIH; (b) © R. Calentine/ Visuals Unlimited; (c) Cryo-EM image of *Haloquadratum walsbyi*, isolated from Australia. Courtesy of Zhuo Li (City of Hope, Duarte, California, USA), Mike L. Dyall-Smith (Charles Sturt University, Australia), and Grant J. Jensem (California Institute of Technology, Pasadena, California, USA); (d,e) © K.O. Stetter & R. Rachel, Univ. Regensburg.

A Protein filaments, or pili, anchor bacterial cells to one another and to surfaces. Here, *Salmonella typhimurium* cells (*red*) use their pili to invade human cells.

B Ball-shaped *Nostoc* cells are a type of freshwater photosynthetic bacteria. The cells in each strand stick together in a sheath of their own jellylike secretions.

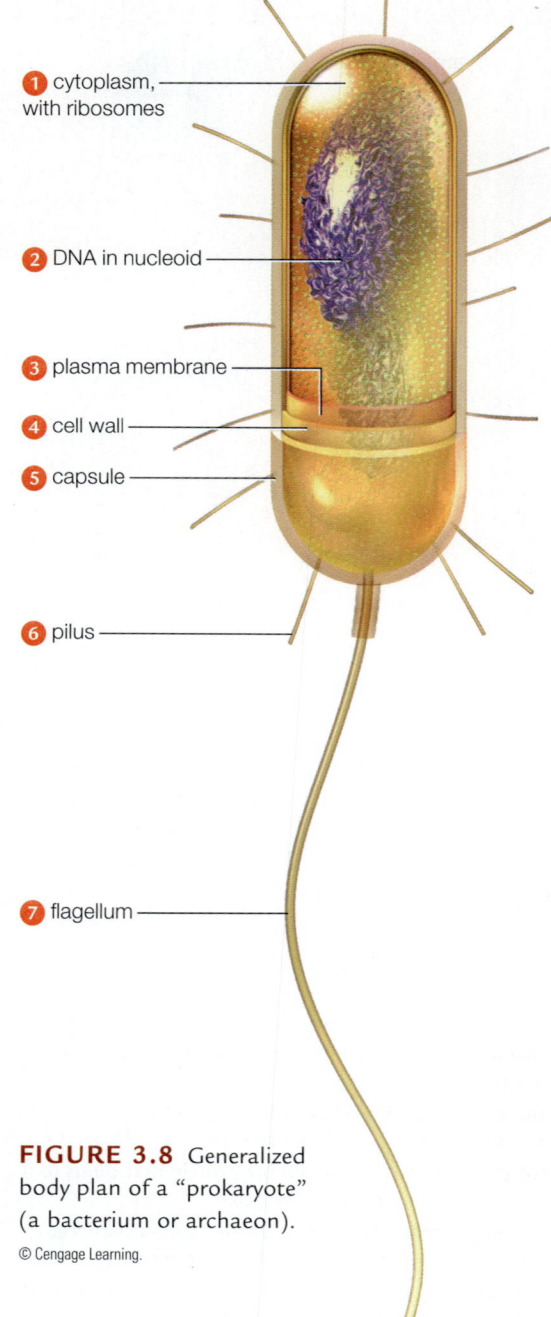

1. cytoplasm, with ribosomes
2. DNA in nucleoid
3. plasma membrane
4. cell wall
5. capsule
6. pilus
7. flagellum

FIGURE 3.8 Generalized body plan of a "prokaryote" (a bacterium or archaeon).

© Cengage Learning.

3.4 Introducing Prokaryotic Cells

All bacteria and archaea are single-celled, and none have a nucleus. Outwardly, they appear so similar that archaea were once thought to be an unusual group of bacteria. Both were classified as prokaryotes, a word that means "before the nucleus." By 1977, it had become clear that archaea are more closely related to eukaryotes than to bacteria, so they were given their own separate domain. The term "prokaryote" is now considered an informal designation.

As a group, bacteria and archaea are the smallest and most metabolically diverse forms of life that we know about. They inhabit nearly all of Earth's environments, including some very hostile places. The two kinds of cells differ in structure and metabolism. Chapter 13 revisits them in more detail; here, we present an overview of their structure (Figures 3.7 and 3.8).

Most bacteria and archaea are not much bigger than a few micrometers. None have a complex internal framework, but protein filaments under the plasma membrane reinforce the cell's shape. Such filaments also act as scaffolding for internal structures. The cytoplasm of these cells ❶ contains many **ribosomes** (organelles upon which polypeptides are assembled), and in some species, additional organelles. The cell's genes typically occur on one large circular molecule of DNA located in an irregularly shaped region of cytoplasm called the nucleoid ❷. In some species, the nucleoid is enclosed by a membrane.

The plasma membrane ❸ of all bacteria and archaea selectively controls which substances move into and out of the cell, as it does for eukaryotic cells. The plasma membrane bristles with proteins that carry out important metabolic processes. For example, part of the plasma membrane of cyanobacteria (some are shown in Figure 3.7**B**) folds into the cytoplasm. Molecules that carry out photosynthesis are embedded in this membrane.

A durable **cell wall** ❹ that surrounds the plasma membrane of nearly all bacteria and archaea imparts shape to the cell. A cell wall is permeable to water, so dissolved substances easily cross it on the way to and from the plasma membrane. Sticky polysaccharides form a slime layer or capsule around the wall of many types of bacteria ❺. These sticky structures help the cells adhere to many

C The square archaea *Haloquadratum walsbyi* thrives in brine pools saltier than soy sauce. Gas-filled organelles (white structures) buoy these highly motile cells, which can aggregate into flat sheets reminiscent of tile floors.

`0.5 μm`

D *Ferroglobus placidus* prefers superheated water spewing from the ocean floor. The durable composition of archaeal lipid bilayers (note the gridlike texture) keeps their membranes intact at extreme heat and pH.

`0.7 μm`

E *Metallosphaera prunae*, an archaeon discovered in a smoking pile of ore at a uranium mine, prefers high temperatures and low pH. (*White* shadows are an artifact of electron microscopy.)

`1 μm`

types of surfaces (such as fresh produce and ground meat), and they also offer protection against some predators and toxins.

Some bacteria and archaea have protein filaments called **pili** (singular, pilus) projecting from their surface ❻. Pili help cells cling to or move across surfaces (some are visible in Figure 3.7**A**). Many prokaryotes also have one or more **flagella** (singular, flagellum), which are are long, slender cellular structures used for motion ❼. A bacterial flagellum rotates like a propeller that drives the cell through fluid habitats.

❯ Biofilms Bacteria often live so close together that an entire community shares a layer of secreted slime. A communal living arrangement in which single-celled organisms live in a shared mass of slime is called a **biofilm** (Figure 3.9). In nature, a biofilm usually consists of multiple species, all entangled in their own mingled secretions. It may include bacteria, algae, fungi, protists, and archaea. Participating in a biofilm allows the cells to linger in a favorable spot rather than be swept away by fluid currents, and to reap the benefits of living communally. For example, rigid or netlike secretions of some species serve as permanent scaffolding for others; species that break down toxic chemicals allow more sensitive ones to thrive in polluted habitats that they could not withstand on their own; and waste products of some serve as raw materials for others. Later chapters discuss medical implications of biofilms, including dental plaque.

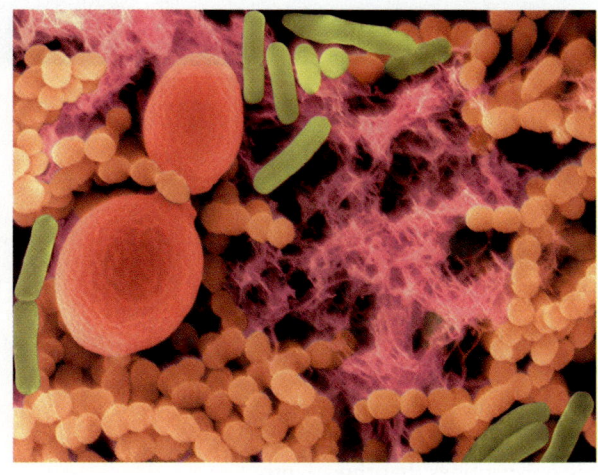

FIGURE 3.9 A biofilm: Oral bacteria in dental plaque. Three species of bacteria (*tan*, *green*) and a yeast (*red*) stick to one another and to teeth via a gluelike mass of shared, secreted polysaccharides (*pink*). Other secretions of these organisms cause cavities and periodontal disease.

© Dennis Kinkel Microscopy, Inc./ Phototake.

Take-Home Message

How are bacteria and archaea alike?

■ Bacteria and archaea do not have a nucleus. Most kinds have a cell wall around their plasma membrane. The permeable wall reinforces and imparts shape to the cell body.

■ The structure of bacteria and archaea is relatively simple, but as a group these organisms are the most diverse forms of life.

biofilm Community of microorganisms living within a shared mass of slime.

cell wall Semirigid but permeable structure that surrounds the plasma membrane of some cells.

flagellum Long, slender cellular structure used for movement.

pilus A protein filament that projects from the surface of some bacterial cells.

ribosome Organelle of protein synthesis.

FIGURE 3.10 Animated!

Common components of eukaryotic cells. This is an animal cell.

1 A plasma membrane controls the kinds and amounts of substances that move into and out of a cell.

2 The nucleus contains, protects, and controls access to DNA.

3 Ribosomes attached to rough endoplasmic reticulum (ER) assemble polypeptides that thread into the ER's interior.

4 Enzymes inside smooth ER make lipids and break down toxins, fatty acids, and carbohydrates.

5 Vesicles transport, store, or digest substances.

6 Golgi bodies finish, sort, and ship lipids and proteins.

7 Mitochondria make ATP.

8 Cytoskeletal elements provide structural support; move cell parts or the whole cell.

9 Centrioles produce and organize microtubules.

© Cengage Learning 2010.

3.5 A Peek Inside a Eukaryotic Cell

Protists, fungi, plants, and animals are eukaryotes. Some of these organisms are independent, free-living cells; others consist of many cells working together as a body. All eukaryotic cells start out life with a nucleus, ribosomes, and other organelles (Figures 3.10 and 3.11). Like the plasma membrane around a cell **1**, a membrane around an organelle controls the types and amounts of substances that enter and exit it. Such control maintains a special internal environment that allows the organelle to carry out its particular function. That function may be isolating toxic or sensitive substances from the rest of the cell, transporting substances through cytoplasm, or providing a favorable environment for a specific metabolic reaction or other process, to give some examples.

❯ The Nucleus A nucleus serves two important functions. First, it keeps the cell's genetic material—its one and only copy of DNA—away from metabolic processes that might damage it. Isolated in its own compartment, the cell's DNA stays separated from the bustling activity of the cytoplasm **2**.

The second function of a nucleus is to control the passage of certain molecules between the nucleus and the cytoplasm. The nuclear membrane, or **nuclear envelope**, carries out this function. It consists of two lipid bilayers folded together as a single membrane. Membrane proteins embedded in the two lipid bilayers aggregate into thousands of tiny pores that span the nuclear envelope. Some bacteria have membranes around their DNA, but we do not consider the bacteria to have nuclei because there are no pores in these membranes.

Large molecules, including RNA and proteins, cannot cross a lipid bilayer on their own. Nuclear pores function as gateways for these molecules to enter and exit a nucleus. Protein synthesis offers an example of why this movement is important. Protein synthesis occurs in cytoplasm, and it requires the participation of many molecules of RNA. RNA is produced in the nucleus. Thus, RNA molecules must move from nucleus to cytoplasm, and they do so through nuclear pores. Proteins must move through the pores in the other direction, because RNA synthesis occurs in the nucleus, and it requires the participation of many proteins produced in the cytoplasm.

› The Endomembrane System The endomembrane system is a series of interacting organelles between the nucleus and the plasma membrane. Its main function is to make lipids, enzymes, and other proteins destined for secretion, or for insertion into cell membranes. It also destroys toxins, recycles wastes, and has other special functions. The system's components vary among different types of cells, but here we present the most common ones.

Part of the endomembrane system is an extension of the nuclear envelope called **endoplasmic reticulum**, or **ER**. ER forms a continuous compartment that folds into flattened sacs and tubes. Two kinds of ER, rough and smooth, are named for their appearance in electron micrographs. Thousands of ribosomes attached to the outer surface of rough ER give this organelle its "rough" appearance ❸. These ribosomes make proteins that thread into the interior of the ER as they are assembled. Inside the ER, the proteins take on their tertiary structure (Section 2.9). Some of them become part of the ER membrane itself. Others end up as enzymes in the smooth ER ❹. Smooth ER has no ribosomes on its surface, so it does not make protein. Smooth ER enzymes make most of the lipids that form the cell's membranes. They also break down carbohydrates, fatty acids, and some drugs and poisons.

Small, membrane-enclosed **vesicles** form by budding from other organelles or from the plasma membrane ❺. Many types transport substances from one organelle to another, or to and from the plasma membrane. **Vacuoles**, which appear empty under a microscope, have various functions depending on cell type. Many isolate or dispose of waste, debris, and toxins. A large, fluid-filled central vacuole keeps plant cells plump (a central vacuole is illustrated in Figure 3.2**B**). **Lysosomes** in animal cells contain powerful enzymes that break down wastes, ingested cells, and cellular debris delivered by other vesicles. Enzymes in **peroxisomes** break down fatty acids, amino acids, and toxins such as alcohol.

Some vesicles fuse with and empty their contents into a **Golgi body**. This organelle has a folded membrane that often looks like a stack of pancakes ❻. Enzymes in a Golgi body put finishing touches on proteins and lipids that have been delivered from the ER. They attach phosphate groups or sugars, and cut certain polypeptides. The finished products (membrane proteins, proteins for secretion, and enzymes) are sorted and packaged into new vesicles that carry them to the plasma membrane or to lysosomes.

› Mitochondria The **mitochondrion** (plural, mitochondria) ❼ is a eukaryotic organelle that specializes in making ATP by aerobic respiration (Chapter 5 details this metabolic pathway). A mitochondrion has two membranes, one highly folded inside the other, that form its ATP-making machinery. Nearly all

nuclear envelope　mitochondrion　DNA in nucleus　nuclear pore　rough ER　0.5 µm

FIGURE 3.11 The nucleus of a cell taken from a mouse's pancreas.
© Kenneth Bart.

endoplasmic reticulum (ER) Organelle that is a continuous system of sacs and tubes extending from the nuclear envelope. Smooth ER makes lipids and breaks down carbohydrates and fatty acids; ribosomes on the surface of rough ER synthesize polypeptides.

Golgi body Organelle that modifies polypeptides and lipids, then packages the finished products into vesicles.

lysosome Enzyme-filled vesicle that breaks down cellular wastes and debris.

mitochondrion Eukaryotic organelle that produces ATP by aerobic respiration.

nuclear envelope A double membrane that constitutes the outer boundary of the nucleus. Pores in the membrane control which substances can cross.

peroxisome Enzyme-filled vesicle that breaks down amino acids, fatty acids, and toxic substances.

vacuole A fluid-filled, empty-looking organelle that isolates or disposes of waste, debris, or toxic materials.

vesicle Small, membrane-enclosed organelle; different kinds store, transport, or break down their contents.

outer membrane

outer compartment

inner compartment

inner membrane

0.5 µm

A Sketch and transmission electron micrograph of a mitochondrion. This organelle specializes in producing large quantities of ATP.

two outer membranes

stroma

inner membrane

1 µm

B Each chloroplast is enclosed by two outer membranes. Photosynthesis occurs at a much-folded inner membrane. The transmission electron micrograph shows a chloroplast from a tobacco leaf; lighter patches are nucleoids where DNA is stored.

FIGURE 3.12 Animated! Bacteria-like organelles.

Figure It Out: What organelle is visible to the upper right in the micrograph of the mitochondrion?

Answer: Rough ER

Credits: (a,b) left, © Cengage Learning; (a) Micrograph, Keith R. Porter; (b) Dr. Jeremy Burgess/ Photo Researchers, Inc.

eukaryotic cells have mitochondria, which resemble bacteria in size, form, and biochemistry (Figure 3.12**A**). They have their own DNA and ribosomes, and they divide independently of the cell. Such clues led to a theory that mitochondria evolved from aerobic bacteria that took up permanent residence inside a host cell (we return to this topic in Section 13.3).

> **Chloroplasts** Photosynthetic cells of plants and many protists contain **chloroplasts**, which are organelles specialized for photosynthesis. Most chloroplasts are oval or disk-shaped (Figure 3.12**B**). Each has two outer membranes enclosing a semifluid interior, the stroma, that contains enzymes and the chloroplast's own DNA. A third, highly folded membrane forms a single, continuous compartment inside the stroma. Photosynthesis occurs at this third membrane. In many ways, chloroplasts resemble photosynthetic bacteria, and they are thought to have evolved from them.

> **The Cytoskeleton** Between the nucleus and plasma membrane of all eukaryotic cells is a system of interconnected protein filaments collectively called the **cytoskeleton**. Elements of the cytoskeleton reinforce, organize, and move cell structures, and often the whole cell (see Figure 3.9 ❽). Some of these elements are always present. Others form only at certain times.

Microtubules are long, hollow cylinders that consist of subunits of the protein tubulin (Figure 3.13**A**). They grow from and are organized by barrel-shaped organelles called centrioles (see Figure 3.9 ❾). Microtubules form a dynamic scaffolding for many cellular processes, rapidly assembling when they are needed and then disassembling when they are not. For example, before a eukaryotic cell divides, microtubules assemble, separate the cell's duplicated chromosomes, then disassemble. As another example, microtubules that form in the growing end of a young nerve cell support and guide its lengthening in a particular direction.

Microfilaments are fibers that consist primarily of subunits of the protein actin (Figure 3.13**B**). They strengthen or change the shape of eukaryotic cells. Crosslinked, bundled, or gel-like arrays of them make up the cell cortex, a reinforcing mesh under the plasma membrane. Actin microfilaments that form at the edge of a cell drag or extend it in a certain direction (Figure 3.13**C**). In muscle cells, microfilaments interact to bring about contraction.

Different types of **intermediate filaments** support cells and tissues, and they are the most stable parts of a cell's cytoskeleton. These filaments form a framework that lends structure and resilience to cells and tissues. Some kinds underlie and reinforce membranes, including the nuclear envelope.

Among many accessory molecules associated with cytoskeletal elements are **motor proteins**, which move cell parts when energized by a phosphate-group transfer from ATP. A cell is like a bustling train station, with molecules and structures being moved continuously throughout its interior. Motor proteins are like freight trains, dragging their cellular cargo along tracks of dynamically assembled microtubules and microfilaments (Figure 3.14).

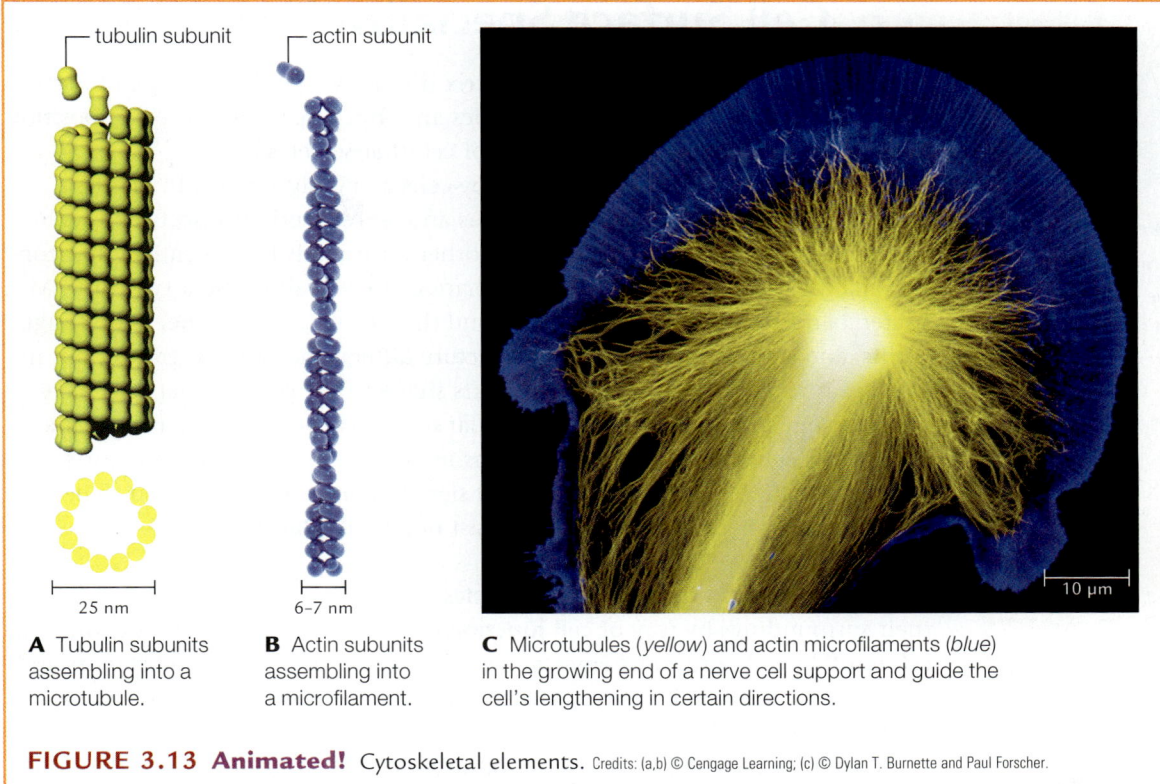

A Tubulin subunits assembling into a microtubule.

B Actin subunits assembling into a microfilament.

C Microtubules (*yellow*) and actin microfilaments (*blue*) in the growing end of a nerve cell support and guide the cell's lengthening in certain directions.

tubulin subunit

actin subunit

25 nm

6–7 nm

10 μm

FIGURE 3.13 Animated! Cytoskeletal elements. Credits: (a,b) © Cengage Learning; (c) © Dylan T. Burnette and Paul Forscher.

⟩ Cilia, Flagella, and False Feet Motor proteins also move external structures such as flagella and cilia. Eukaryotic flagella are structures that whip back and forth to propel cells such as sperm (*right*) through fluid. They have a different structure and type of motion than prokaryotic flagella.

© Cengage Learning.

Cilia (singular, cilium) are short, hairlike structures that project from the surface of some cells. Cilia are usually more profuse than flagella. The coordinated waving of many cilia propels cells through fluid, and stirs fluid around stationary cells. For example, the cilia on thousands of cells lining your airways sweep inhaled particles away from your lungs.

Astrid Hanns-FriederMichler/ Photo Researchers, Inc.

Amoebas (*left*) and other types of eukaryotic cells form **pseudopods**, or "false feet." As these temporary, irregular lobes bulge outward, they move the cell and can engulf a target such as prey. Elongating microfilaments force the lobe to advance in a steady direction. Motor proteins that are attached to the microfilaments drag the plasma membrane along with them.

Take-Home Message

What do all eukaryotic cells have in common?

- All eukaryotic cells start life with a nucleus, ribosomes, and other organelles.

- An organelle's membrane maintains an internal environment that allows it to carry out a special function. The nucleus protects and controls access to a cell's DNA. ER, vesicles, and Golgi bodies interact to make and modify proteins and lipids. Mitochondria produce ATP. Chloroplasts specialize in photosynthesis.

- A cytoskeleton of protein filaments is the basis of eukaryotic cell shape, internal structure, and movement.

chloroplast Organelle of photosynthesis in the cells of plants and many protists.

cilia Short, movable structures that project from the plasma membrane of some eukaryotic cells.

cytoskeleton Network of interconnected protein filaments that support, organize, and move eukaryotic cells and their internal structures.

intermediate filament Stable cytoskeletal element that structurally supports cells and tissues.

microfilament Cytoskeletal element that reinforces cell membranes; fiber of actin subunits.

microtubule Cytoskeletal element involved in movement; hollow filament of tubulin subunits.

motor protein Type of energy-using protein that interacts with cytoskeletal elements to move the cell's parts or the whole cell.

pseudopod A temporary protrusion that helps some eukaryotic cells move and engulf prey.

FIGURE 3.14 Animated!

A motor protein (*tan*) drags cellular freight (here, a *pink* vesicle) as it inches along a microtubule. © Cengage Learning.

cuticle

outer cell of leaf

photosynthetic cell inside leaf

FIGURE 3.15 A plant ECM. Section through a plant leaf showing cuticle, a protective covering of deposits secreted by living cells. George S. Ellmore.

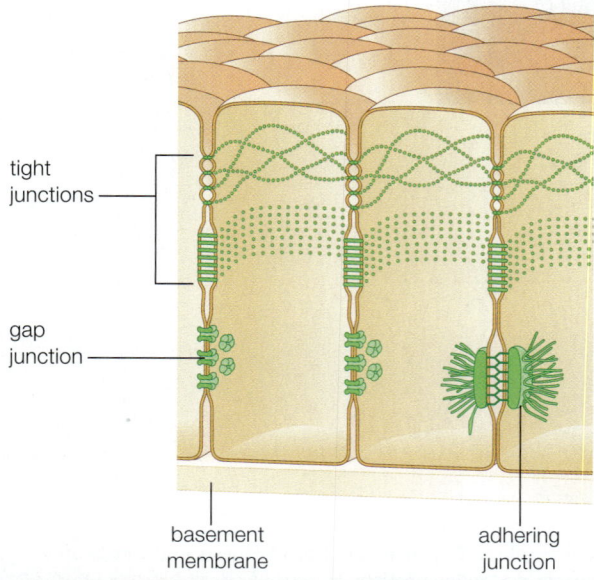

tight junctions

gap junction

basement membrane

adhering junction

FIGURE 3.16 Animated! Cell junctions.

Top, cell junctions common in animal tissues: tight junctions, gap junctions, and adhering junctions.

Bottom, the micrograph shows how a profusion of tight junctions (*green*) seals abutting surfaces of kidney cell membranes to form a waterproof tissue. The DNA in each cell nucleus appears *red*.

Top, © Cengage Learning; bottom, © ADVANCELL (Advanced In Vitro Cell Technologies; S.L.) www.advancell.com.

cell junction Structure that connects a cell to another cell or to extracellular matrix.

cuticle Secreted covering at a body surface.

extracellular matrix (**ECM**) Complex mixture of cell secretions; its composition and function vary by cell type.

3.6 Cell Surface Specializations

Many cells secrete an **extracellular matrix** (**ECM**), a complex mixture of molecules that often includes polysaccharides and fibrous proteins. The composition and function of ECM vary by the type of cell that secretes it.

A **cuticle** is a type of ECM secreted by cells at a body surface. In plants, a cuticle of waxes and proteins helps stems and leaves fend off insects and retain water (Figure 3.15). Crabs, spiders, and other arthropods have a cuticle that consists mainly of chitin, a tough polysaccharide. A cell wall is also a type of ECM. Bacteria and archaea secrete a wall around their plasma membrane, as do fungi, plants, and some protists. Cell wall structure differs among these groups, but in all cases it protects, supports, and imparts shape to the cell. Animal cells have no walls, but some secrete an extracellular matrix called basement membrane. Basement membrane is a sheet of fibrous material that structurally supports and organizes tissues, and it has roles in cell signaling. It is not considered to be a cell membrane because it does not consist of a lipid bilayer.

❭ Cell Junctions In multicelled species, cells interact with one another and their surroundings by way of **cell junctions**, which are structures that connect a cell to other cells and to its environment. Cells send and receive substances and signals through some junctions. Other kinds help cells recognize and stick to each other and to ECM.

Three types of cell junctions are common in animal tissues (Figure 3.16). In tissues that line body surfaces and internal cavities, rows of proteins form tight junctions between plasma membranes of adjacent cells. These junctions prevent body fluids from seeping between the cells. For example, tight junctions sealing cells in the lining of the stomach normally keep acidic fluid from leaking out. If a bacterial infection damages this lining, acid and enzymes can erode the underlying layers. The result is a painful peptic ulcer.

Strong adhering junctions, which are composed of adhesion proteins, snap cells to one another. They also connect microfilaments and intermediate filaments inside the cell to ECM outside the cell. Skin and other tissues subject to abrasion or stretching have a lot of adhering junctions. These cell junctions also strengthen contractile tissues such as heart muscle.

Gap junctions are channels that connect the cytoplasm of adjoining animal cells, thus permitting water, ions, and small molecules to pass directly from the cytoplasm of one cell to another. These channels allow entire regions of cells to respond to a single stimulus. Heart muscle and other tissues in which the cells perform a coordinated action have many gap junctions. In plants, open channels called plasmodesmata extend across plant cell walls to connect the cytoplasm of adjacent cells. Like gap junctions, plasmodesmata also allow substances to flow quickly from cell to cell.

Take-Home Message

What structures form on the outside of eukaryotic cells?

- Many cells secrete an extracellular matrix (ECM). ECM varies in composition and function depending on the cell type.
- Plant cells, fungi, and some protists have a porous wall around their plasma membrane. Animal cells do not have walls.
- A secreted waxy cuticle helps protect the exposed surfaces of soft plant parts.
- Cell junctions structurally and functionally connect cells in tissues. In animal tissues, cell junctions also connect cells with basement membrane.

3.7 The Nature of Life

What exactly makes a cell, or an organism that consists of them, alive? Living things have a high proportion of the organic molecules of life, but so do the remains of dead organisms in seams of coal. Living things also use energy to reproduce themselves, but computer viruses, which are arguably not alive, can do that too. So how do biologists, who study life as a profession, describe life? Their best description consists of a list of properties associated with things we know to be alive. You have already learned about two of these properties:

1. They make and use the organic molecules of life.

2. They consist of one or more cells.

The remainder of this book details the other properties:

3. They engage in self-sustaining biological processes such as metabolism.

4. They change over their lifetime, by developing, maturing, and aging.

5. They use DNA as their hereditary material when they reproduce.

6. They have the collective capacity to change over successive generations.

© R. Llewellyn/ Superstock, Inc.

Together, these six properties characterize living things as different from nonliving things.

Atoms in organic molecules are the stuff of you, and us, and all of life. Yet it takes far more than organic molecules to complete the picture. Life continues only as long as an ongoing flow of energy sustains its organization. With energy and the hereditary codes of DNA, matter becomes organized, generation after generation. Even with the death of individuals, life elsewhere is prolonged: With each death, molecules released are taken up as raw materials by new generations.

Take-Home Message

What is life?

■ Biologists describe the characteristic of "life" in terms of a set of properties unique to living things.

■ In living things, the molecules of life are organized as one or more cells that engage in self-sustaining biological processes, use DNA as their hereditary material, and change over lifetimes and generations.

3.8 Food for Thought (revisited)

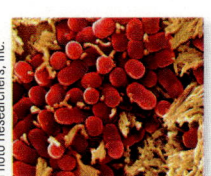

© Stephanie Schuller/ Photo Researchers, Inc.

Some think we should prevent our food from getting contaminated by enacting stricter laws governing farming practices and food preparation. Others think the safest way to protect consumers from food poisoning is by sterilizing food to kill toxic *E. coli* and other bacteria that may be in it. For example, recalled, contaminated ground beef is typically cooked or otherwise sterilized, then processed into ready-to-eat foods. Raw beef trimmings are effectively sterilized when sprayed with ammonia and ground to a paste. The resulting meat product is routinely added as a filler to hamburger patties, fresh ground beef, hot dogs, lunch meats, sausages, frozen entrees, canned foods, and other items sold to quick service restaurants, hotel and restaurant chains, institutions, and school lunch programs.

WHERE YOU ARE GOING . . .

Chapter 4 explores cell function; Chapter 5, photosynthesis and aerobic respiration. We revisit protein synthesis and control over it in Chapter 7. Some cellular structures are required for cell division (Chapter 8). Chapter 13 details the structures, evolution, and metabolism of bacteria, archaea, and protists. Cell structures introduced in this chapter return in the context of the physiology of animals (Chapters 19–26) and plants (Chapters 27 and 28). In Chapter 22, you will see how bacteria can be helpful or harmful to one's health.

Summary

Section 3.1 Bacteria are found in all parts of the biosphere, including the human body. Huge numbers inhabit our intestines, but most of these are beneficial. A few can cause disease. Contamination of food with disease-causing bacteria can result in food poisoning that is sometimes fatal.

Section 3.2 Cells differ in size, shape, and function, but all start out life with a **plasma membrane**, **cytoplasm**, and a region of DNA. Most cells have additional components.

In eukaryotic cells, DNA is contained within a **nucleus**, which is a membrane-enclosed **organelle**. All cell membranes, including the plasma membrane and organelle membranes, are selectively permeable and consist mainly of phospholipids organized as a lipid bilayer. The **surface-to-volume ratio** limits cell size.

By the **cell theory**, all organisms consist of one or more cells; the cell is the smallest unit of life; each new cell arises from another, preexisting cell; and a cell passes hereditary material to its offspring.

Section 3.3 A cell membrane can be described as a **fluid mosaic**, which means it behaves like a two-dimensional liquid of mixed composition—lipids (mainly phospholipids) and proteins. The lipids are organized as a double layer in which the nonpolar fatty acid tails of both layers are sandwiched between the polar heads.

All cell membranes may have enzymes and **transport proteins**. Plasma membranes can also incorporate **receptor proteins**, **adhesion proteins**, and **recognition proteins**.

Section 3.4 Bacteria and archaea, informally grouped as "prokaryotes," are the most diverse forms of life. These single-celled organisms have no nucleus, but all have DNA and **ribosomes**. Many have a permeable but protective **cell wall** and a sticky capsule, as well as motile structures (**flagella**) and other projections (**pili**). Bacteria and other microbial organisms often share living arrangements in **biofilms**.

Section 3.5 All eukaryotic cells start out life with a nucleus and other organelles. The nucleus protects and controls access to the cell's DNA. Membrane proteins form pores in the **nuclear envelope** that control the movement of molecules into and out of the nucleus.

Endoplasmic reticulum (**ER**) is a continuous system of sacs and tubes extending from the nuclear envelope. Ribosome-studded rough ER makes proteins; smooth ER makes lipids and breaks down carbohydrates, fatty acids, and some toxins. **Golgi bodies** modify proteins and lipids before sorting them into vesicles. Different types of **vesicles** store, degrade, or transport substances through the cell. Enzymes in **peroxisomes** break down substances such as amino acids, fatty acids, and toxins. **Lysosomes** contain enzymes that break down wastes and cellular debris for

recycling. Fluid-filled **vacuoles** have various functions, including storage and disposal of wastes and toxins. A large central vacuole keeps plant cells plump.

Other eukaryotic organelles include **mitochondria** (which produce ATP by aerobic respiration) and **chloroplasts** (which specialize in photosynthesis).

A **cytoskeleton** organizes a eukaryotic cell's interior, reinforces its shape, and helps move its parts. Interactions between ATP-driven **motor proteins** and hollow, dynamically assembled **microtubules** bring about movement of cells and cell parts such as eukaryotic flagella and **cilia**. A **microfilament** mesh reinforces plasma membranes. Elongating microfilaments help bring about movement of **pseudopods**. **Intermediate filaments** support cells and tissues.

Section 3.6 Many cells secrete an **extracellular matrix** (**ECM**) that has different functions depending on the cell type. In animals, a secreted basement membrane supports and organizes cells in tissues. Plant cells, fungi, and many protists secrete a wall around the plasma membrane. A **cuticle** is an ECM secreted by cells at a body surface.

Cell junctions connect cells to one another and to their environment. Plasmodesmata connect the cytoplasm of adjacent plant cells. In animals, gap junctions form open channels between adjacent cells; adhering junctions anchor cells to one another and to basement membrane; and tight junctions form a waterproof seal between cells in some tissues.

Section 3.7 All living things make and use the molecules of life; consist of one or more cells that engage in self-sustaining biological processes; change over their lifetime; and pass their DNA to offspring that can change over generations.

Self-Quiz   Answers in Appendix I

1. Despite the diversity of cell type and function, all cells have these three things in common:
   a. cytoplasm, DNA, and organelles with membranes
   b. a plasma membrane, DNA, and a nuclear envelope
   c. cytoplasm, DNA, and a plasma membrane
   d. a cell wall, cytoplasm, and DNA

2. Every cell is descended from another cell. This idea is part of _____ .
   a. evolution   c. the cell theory
   b. the theory of heredity d. cell biology

3. The surface-to-volume ratio _____ .
   a. does not apply to prokaryotic cells   c. constrains cell size
   b. is part of the cell theory   d. b and c

4. True or false? Some protists start out life with no nucleus.

5. Unlike eukaryotic cells, prokaryotic cells _____ .
   a. have no plasma membrane   c. have no nucleus
   b. have RNA but not DNA   d. a and d

6. Cell membranes consist mainly of _____ and _____ .
   a. lipids; carbohydrates   c. lipids; carbohydrates
   b. phospholipids; proteins   d. phospholipids; ECM

7. Most membrane functions are carried out by _____ .
   a. proteins
   b. phospholipids
   c. nucleic acids
   d. hormones

8. Which of the following statements is correct?
   a. Ribosomes are only found in bacteria and archaea.
   b. Some animal cells are prokaryotic.
   c. Only eukaryotic cells have mitochondria.
   d. The plasma membrane is the outermost boundary of all cells.

9. In a lipid bilayer, the _____ of all the lipid molecules are sandwiched between all of the _____ .
   a. hydrophilic tails; hydrophobic heads
   b. hydrophilic heads; hydrophilic tails
   c. hydrophobic tails; hydrophilic heads
   d. hydrophobic heads; hydrophilic tails

10. The main function of the endomembrane system is _____ .
    a. building and modifying proteins and lipids
    b. isolating DNA from toxic substances
    c. secreting extracellular matrix onto the cell surface
    d. producing ATP by aerobic respiration

11. Enzymes contained in _____ break down worn-out organelles, bacteria, and other particles.
    a. lysosomes
    b. mitochondria
    c. endoplasmic reticulum
    d. peroxisomes

12. Put the following structures in order according to the pathway of a secreted protein:
    a. plasma membrane
    b. Golgi bodies
    c. endoplasmic reticulum
    d. post-Golgi vesicles

13. No animal cell has a _____ .
    a. plasma membrane
    b. flagellum
    c. lysosome
    d. cell wall

14. _____ connect the cytoplasm of plant cells.
    a. Plasmodesmata
    b. Adhering junctions
    c. Tight junctions
    d. Adhesion proteins

15. Match each cell part with its main function.
    _____ mitochondrion   a. connects cells
    _____ chloroplast   b. protective covering
    _____ ribosome   c. ATP production
    _____ nucleus   d. protects DNA
    _____ cell junction   e. protein synthesis
    _____ flagellum   f. maintains internal environment
    _____ cell membrane   g. photosynthesis
    _____ cuticle   h. movement

Critical Thinking

1. In a classic episode of *Star Trek*, a gigantic amoeba engulfs an entire starship. Spock blows the cell to bits before it has a chance to reproduce. Think of at least one problem a biologist would have with this particular scenario.

2. In plants, the cell wall forms as a young plant cell secretes polysaccharides onto the outer surface of its plasma membrane. Being thin and pliable, this primary wall allows the cell to enlarge and change shape. At maturity, cells in some plant tissues deposit material onto the primary wall's inner surface. Why doesn't this secondary wall form on the outer surface of the primary wall?

Digging Into Data

Organelles and Cystic Fibrosis

CFTR is a transport protein in the plasma membrane of cells lining cavities and ducts of the lungs, liver, pancreas, intestines, and reproductive system. The transporter moves chloride ions out of the cells. Water that follows the ions creates a thin film that allows mucus to slide easily through these structures.

People with cystic fibrosis (CF) have too few copies of the CFTR protein in the plasma membranes of their cells. Not enough chloride ions leave the cells, and so not enough water leaves them either. The result is thick, dry mucus that clogs the airways to the lungs and other passages. Symptoms include difficulty breathing and chronic lung infections.

In most people with CF, one amino acid of the CFTR protein is missing. A protein with this change is made correctly, and it can transport ions correctly, but it never reaches the plasma membrane to do its job. In 2000, researchers investigated the cellular location of the defective protein (Figure 3.17).

1. Which organelle contains the least amount of CFTR protein in normal cells? In cells with the deletion?

2. In which organelle is the amount of CFTR protein most similar in both types of cells?

3. Where is the CFTR protein with the deletion getting held up?

FIGURE 3.17 Amounts of CFTR protein associated with endoplasmic reticulum, vesicles traveling from ER to Golgi, and Golgi bodies in CF cells and normal cells. © Cengage Learning.

P.L. Walne and J. H. Arnott, Planta, 77:325–354, 1967.

3. A student is examining different samples with a microscope. She discovers the single-celled organism above swimming in water from a freshwater pond. What kind of microscope is she using?

4. Which structures can you identify in the organism above? Is it a prokaryotic or eukaryotic cell?

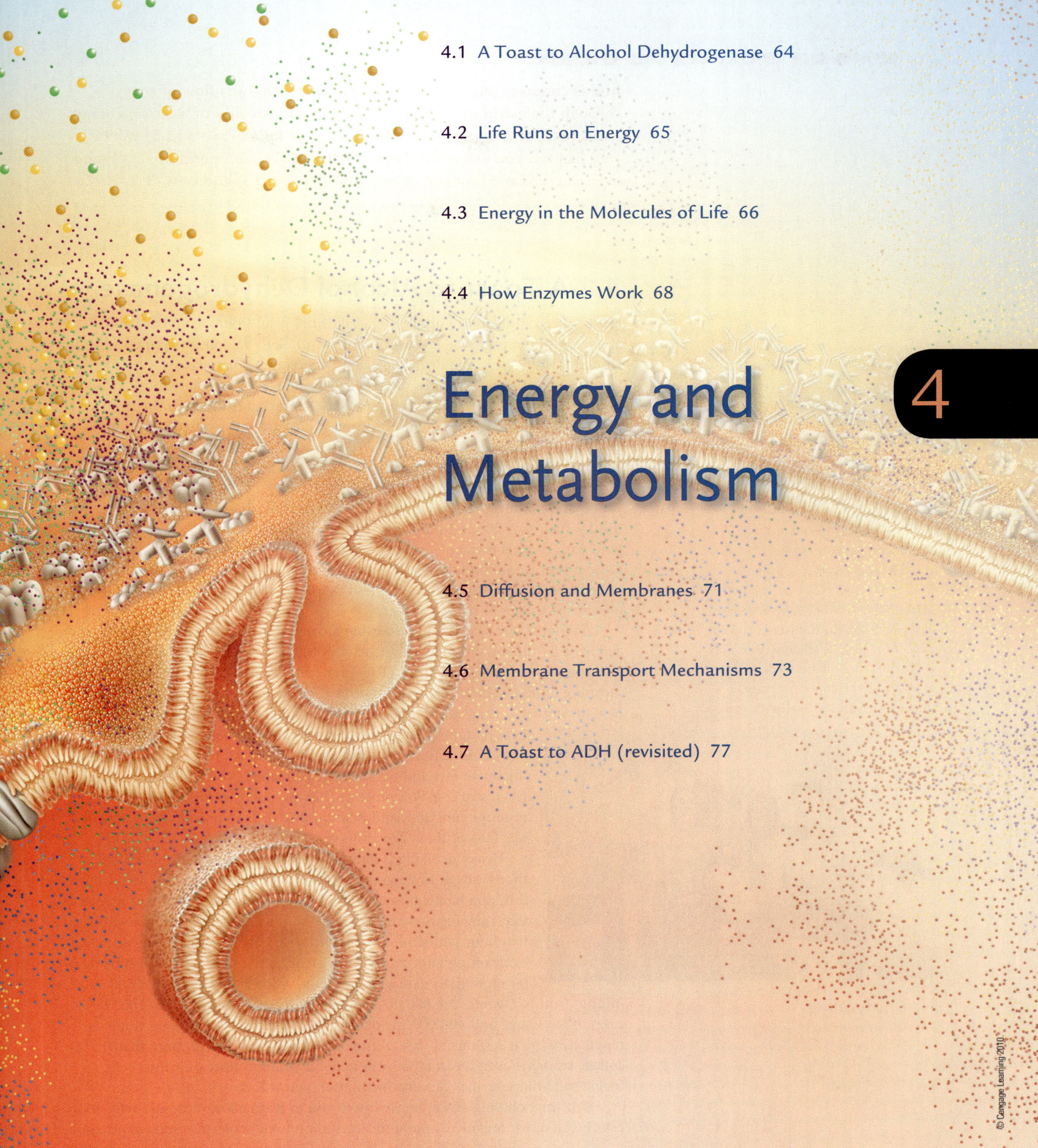

Energy and Metabolism

4

In this chapter, you will gain insight into the one-way flow of energy through the world of life (Sections 1.2, 1.3). A review of electron energy levels (2.2) and the selective permeability of lipid bilayers (3.3) will be very helpful. You will revisit laws of nature (1.9), temperature (2.4), and the structure and function of molecules (2.6–2.10) and cells (3.3–3.6).

4.1 A Toast to Alcohol Dehydrogenase

Most college students are under the legal drinking age, but alcohol use and abuse continues to be the most serious drug problem on college campuses in the United States. Before you drink, consider what you are consuming. A bottle of beer, a glass of wine, or a shot of vodka all contain the same amount of alcohol or, more precisely, ethanol. Ethanol molecules move quickly from the stomach and small intestine into the bloodstream. Almost all of the ethanol ends up in the liver, a large organ in the abdomen. Liver cells have impressive numbers of enzymes (Section 2.6). One enzyme, alcohol dehydrogenase (ADH), helps break down ethanol and other toxic compounds (Figure 4.1).

Ethanol and its breakdown products damage liver cells, so the more a person drinks, the fewer liver cells are left to do the breaking down. Ethanol also interferes with normal metabolic processes. For example, in the presence of ethanol, oxygen that would ordinarily take part in breaking down fatty acids is diverted to breaking down the ethanol. As a result, fats tend to accumulate as large globules in the tissues of heavy drinkers.

Long-term, heavy drinking can cause alcoholic hepatitis, a disease characterized by inflammation and destruction of liver tissue. It can also lead to cirrhosis, or scarring of the liver. (The term cirrhosis is from the Greek *kirros*, meaning orange-colored, after the abnormal skin color of people with the disease.) Eventually, the liver of a heavy drinker may just quit working, with dire health consequences. The liver is the largest gland in the human body, and it has many important functions. In addition to breaking down fats and toxins, it helps regulate the body's blood sugar level, and it makes proteins that are essential for blood clotting, immune function, and maintaining the solute balance of body fluids. Loss of these functions can be deadly.

Heavy drinking is dangerous in the short term too. Tens of thousands of undergraduate students have been polled about their drinking habits in recent surveys. More than half of them reported that they regularly drink five or more alcoholic beverages within a two-hour period—a self-destructive behavior called binge drinking. Consuming large amounts of alcohol in a brief period of time does far more than damage one's liver. Aside from the related 500,000 injuries from accidents, the 600,000 assaults by intoxicated students, 100,000 cases of date rape, and 400,000 incidences of unprotected sex among students, binge drinking is responsible for killing or causing the death of more than 1,700 college students every year. Ethanol is toxic: If you put more of it into your body than your enzymes can deal with, then you will die.

With this sobering example, we invite you to learn about how and why your cells break down organic compounds, including toxic molecules such as ethanol.

alcohol dehydrogenase

FIGURE 4.1 Alcohol dehydrogenase. This enzyme helps the body break down toxic alcohols such as ethanol, thus making it possible for humans to drink beer, wine, and other alcoholic beverages.

Top, © Cengage Learning; bottom, © wavebreakmedia ltd/ Shutterstock.

4.2 Life Runs on Energy

Energy is formally defined as the capacity to do work, but this definition is not very satisfying. Even the brilliant physicists who study it cannot say what energy is, exactly. However, even without a perfect definition, we have an intuitive understanding of energy just by thinking about familiar forms of it, such as light, heat, electricity, and motion. We also understand intuitively that one form of energy can be converted to another. Think about how a lightbulb changes electricity into light, or how an automobile changes gasoline into the energy of motion.

Thermodynamics is the study of heat and other forms of energy (*therm* is a Greek word for heat; *dynam* means energy). Because our understanding of energy is incomplete, our best descriptions of how it behaves are laws of nature rather than theories (Section 1.9). For example, we know that the total amount of energy before and after every conversion is always the same. In other words, energy cannot be created or destroyed—a phenomenon called the **first law of thermodynamics**. We also know that energy also tends to spread out, or disperse, until no part of a system holds more than another part. Think about how heat flows from a hot frying pan to cool air in a kitchen until the temperature of both is the same. We never see cool air raising the temperature of a hot pan. Again, this phenomenon is always true, but we do not know exactly why. Thus, the tendency of energy to disperse is called the **second law of thermodynamics**.

Work occurs as a result of energy transfers. For example, a plant cell powers glucose synthesis by absorbing light energy from the sun. This particular energy transfer involves the conversion of one form of energy (light) to another (chemical energy). Most other types of cellular work occur by the transfer of chemical energy from one molecule to another.

As you learn about such processes, remember that every time energy is transferred, a bit of it disperses. The energy lost from the transfer is usually in the form of heat. As a simple example, a typical incandescent lightbulb converts about 5 percent of the energy of electricity into light. The remaining 95 percent of the energy ends up as heat that radiates from the bulb.

Dispersed heat is not very useful for doing work, and it is not easily converted to a more useful form of energy. Because some of the energy in every transfer disperses as heat, and heat is not useful for doing work, we can say that the total amount of energy available for doing work in the universe is always decreasing.

Is life an exception to this inevitable flow? An organized body is hardly dispersed. Energy becomes concentrated in each new organism as the molecules of life organize into cells. Even so, living things constantly use energy—to grow, to move, to acquire nutrients, to reproduce, and so on—and some energy is lost in every one of these processes. Unless those losses are replenished with energy from another source, the complex organization of life will end.

Most of the energy that fuels life on Earth comes from the sun. In our world, energy flows from the sun, through producers, then consumers (Figure 4.2). During this journey, energy is transferred many times. With each transfer, some energy escapes as heat until, eventually, all of it is permanently dispersed. However, the second law of thermodynamics does not say how quickly the dispersal has to happen. Energy's spontaneous dispersal is resisted by chemical bonds.

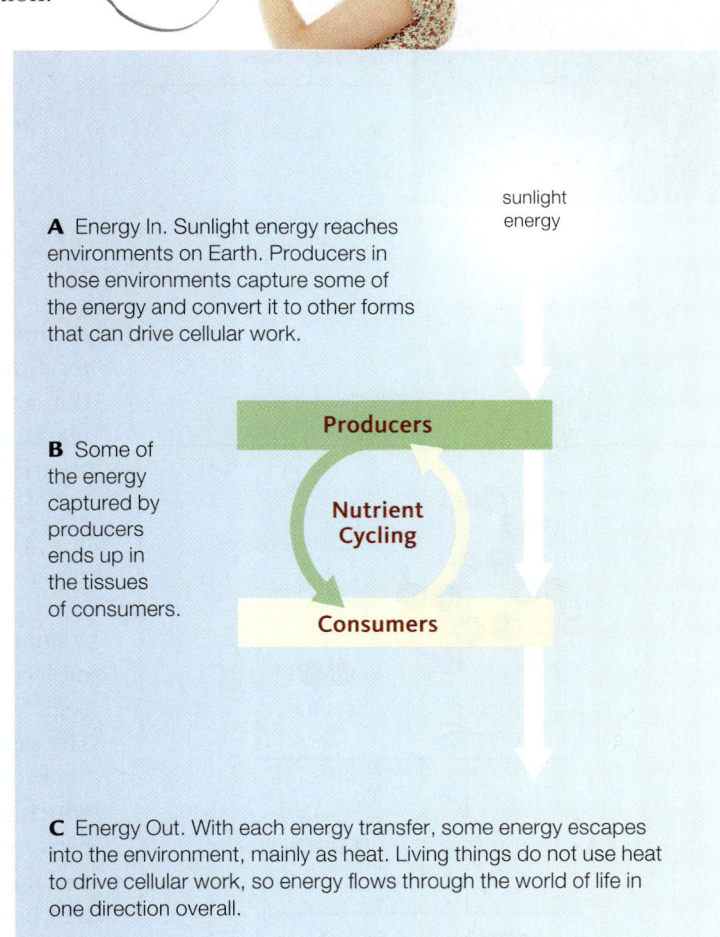

A Energy In. Sunlight energy reaches environments on Earth. Producers in those environments capture some of the energy and convert it to other forms that can drive cellular work.

sunlight energy

B Some of the energy captured by producers ends up in the tissues of consumers.

Producers

Nutrient Cycling

Consumers

C Energy Out. With each energy transfer, some energy escapes into the environment, mainly as heat. Living things do not use heat to drive cellular work, so energy flows through the world of life in one direction overall.

FIGURE 4.2 Animated! A one-way flow of energy into living organisms compensates for a one-way flow of energy out of them. Energy inputs drive a cycling of materials among producers and consumers.

Credits: top left, bottom, © Cengage Learning; top right, © Piotr Marcinski/ Shutterstock.

energy The capacity to do work.
first law of thermodynamics Energy cannot be created or destroyed.
second law of thermodynamics Energy tends to disperse spontaneously.

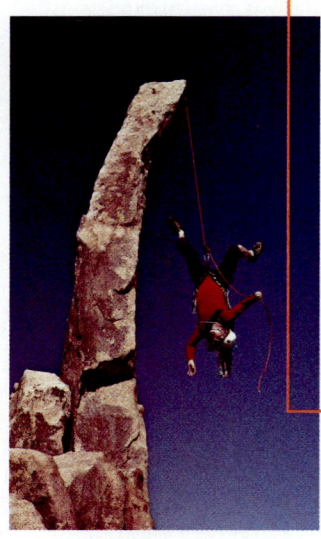

FIGURE 4.3
Illustration of potential energy. By opposing the downward pull of gravity, the rope attached to the rock prevents the man from falling. Similarly, a chemical bond keeps two atoms from moving apart.
© Greg Epperson/ Shutterstock.

The energy in chemical bonds is a type of potential (stored) energy (Figure 4.3). Think of all the bonds in the countless molecules that make up your skin, heart, liver, fluids, and other body parts. Those bonds hold the molecules, and you, together—at least for the time being.

Take-Home Message

What is energy?

- Energy is the capacity to do work. It can be converted from one form to another, but it cannot be created or destroyed. It disperses spontaneously.

- Some energy is lost during every transfer or conversion. Thus, organisms can maintain their complex organization only as long as they replenish themselves with energy they harvest from someplace else.

4.3 Energy in the Molecules of Life

During a chemical reaction, one or more **reactants** (molecules that enter the reaction and become changed by it) become one or more **products** (molecules that are produced by the reaction). Intermediate molecules may form between reactants and products. We show a chemical reaction as an equation in which an arrow points from reactants to products:

$$6CO_2 \ + \ 6H_2O \ \xrightarrow{\text{light energy}} \ C_6H_{12}O_6 \ + \ 6O_2$$

carbon dioxide   water   glucose   oxygen

A number before a chemical formula in such equations indicates the number of molecules; a subscript indicates the number of atoms of that element per molecule. Note that atoms shuffle around in a reaction, but they never disappear: The same number of atoms that enter a reaction remain at the reaction's end.

Every chemical bond holds energy, and the amount of energy depends on which elements are taking part in the bond. For example, the covalent bond between an oxygen and a hydrogen atom in any water molecule holds a certain amount of energy, which differs from the amount of energy held by a covalent bond between two oxygen atoms in molecular oxygen (O_2). Thus, in most reactions, the energy of the reactants differs from the energy of the products. Reactions in which the reactants have less energy than the products require a net (or overall) energy input to proceed (Figure 4.4**A**). In other reactions, the reactants have greater energy than the products. Such reactions release energy (Figure 4.4**B**).

> Why Earth Does Not Go Up in Flames The molecules of life release energy when they combine with oxygen. For example, think of how a spark ignites wood. Wood is mostly cellulose, which consists of long chains of repeating glucose monomers (Section 2.7). A spark starts a reaction that converts the cellulose in wood and oxygen (in air) to water and carbon dioxide. This reaction releases enough energy to start the same reaction with other cellulose and oxygen molecules. That is why wood keeps burning after it has been lit.

Earth is rich in oxygen—and in potential energy-releasing reactions. Why doesn't it burst into flames? Luckily, chemical bonds do not break without at least a small input of energy, even in an energy-releasing reaction. We call this input activation energy. **Activation energy** is the minimum energy input

glucose
$C_6H_{12}O_2$

6 oxygen
O_2

A energy in

B energy out

6 carbon dioxide
CO_2

6 water
H_2O

Energy

FIGURE 4.4 Energy inputs and outputs in chemical reactions.

A Some reactions convert molecules with lower energy to molecules with higher energy, so they require a net energy input to proceed.

B Other reactions convert molecules with higher energy to molecules with lower energy, so they end with a net energy output.

Figure It Out: Which law of thermodynamics explains energy inputs and outputs in chemical reactions?

Answer: The first law

© Cengage Learning.

© Cengage Learning.

FIGURE 4.5 Animated! Activation energy. Most reactions, including energy-releasing ones such as burning wood cellulose, will not begin without at least a small input of energy. This activation energy is shown in the graph above as a bump in an energy hill. Reactants in this example have more energy than the products. Activation energy keeps this and other energy-releasing reactions from starting spontaneously.

Credits: left, © Westend61/ Superstock; right, © Cengage Learning.

required to get a chemical reaction started. It is a bit like a hill that reactants must climb before they can coast down the other side to products (Figure 4.5).

Both energy-requiring and energy-releasing reactions have activation energy, but the amount varies with the reaction. Consider guncotton (nitrocellulose), a highly explosive derivative of cellulose. Christian Schönbein accidentally discovered a way to manufacture it when he used his wife's cotton apron to wipe up a nitric acid spill on his kitchen table, then hung it up to dry next to the oven. The apron exploded. Being a chemist in the 1800s, Schönbein immediately tried marketing guncotton as a firearm explosive, but it proved to be too unstable to manufacture. So little activation energy is needed to make guncotton react with oxygen that it tends to explode unexpectedly. Several manufacturing plants burned to the ground before guncotton was abandoned for use as a firearm explosive. The substitute? Gunpowder, which has a higher activation energy for a reaction with oxygen.

> **Energy In, Energy Out** Cells store energy by running energy-requiring reactions that build organic compounds (Figure 4.6A). For example, light energy drives the overall reactions of photosynthesis, which produce glucose from carbon dioxide and water. Unlike light, glucose can be stored in a cell. Cells harvest energy by running energy-releasing reactions that break the bonds of organic compounds (Figure 4.6B). Most cells do this when they carry out the overall reactions of aerobic respiration, which releases the energy of glucose by breaking the bonds between its carbon atoms. You will see in the next sections how cells use energy released from some reactions to drive others (we return to the reactions of photosynthesis and aerobic respiration in Chapter 5).

small molecules (e.g., carbon dioxide, water) → energy-requiring reactions → organic compounds (carbohydrates, fats, proteins)

A Cells store energy in the chemical bonds of organic compounds.

organic compounds (carbohydrates, fats, proteins) → energy-releasing reactions → small molecules (e.g., carbon dioxide, water)

B Cells retrieve energy stored in the chemical bonds of organic compounds.

FIGURE 4.6 Cells store and retrieve energy in the chemical bonds of organic molecules.

© Cengage Learning.

Take-Home Message

How do cells use energy?

- Cells store and retrieve energy by making and breaking chemical bonds.
- Some reactions require a net input of energy. Others end with a net release of energy.
- Most chemical reactions require an input of activation energy to begin.

activation energy Minimum amount of energy required to start a reaction.
product A molecule that is produced by a reaction.
reactant A molecule that enters a reaction and is changed by participating in it.

4.4 How Enzymes Work

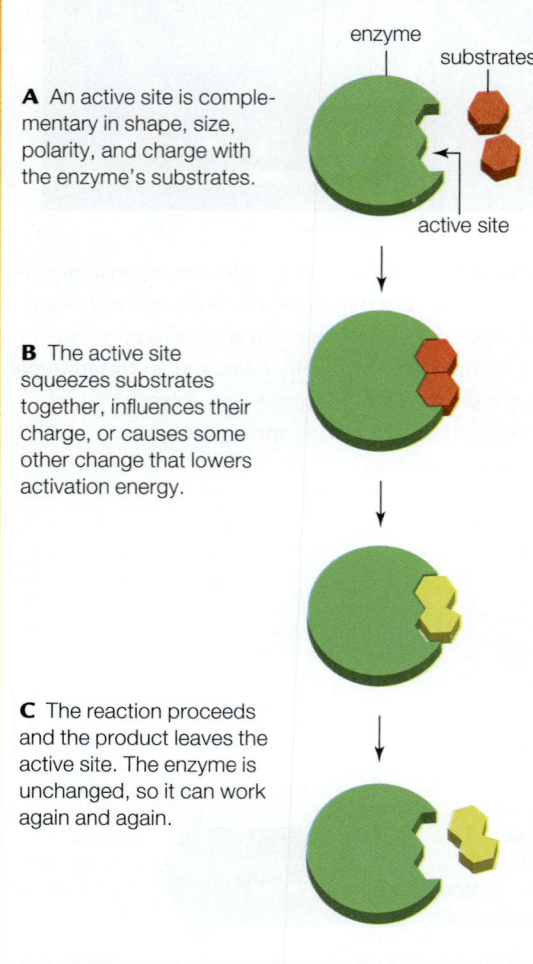

A An active site is complementary in shape, size, polarity, and charge with the enzyme's substrates.

B The active site squeezes substrates together, influences their charge, or causes some other change that lowers activation energy.

C The reaction proceeds and the product leaves the active site. The enzyme is unchanged, so it can work again and again.

D For simplicity, enzymes and active sites are often depicted as blobs or geometric shapes. This model shows the actual contours of an active site in an enzyme (hexokinase) that adds a phosphate group to six-carbon sugars. A phosphate group is meeting up with a glucose molecule in the active site.

FIGURE 4.7 How an active site works.

Credits: (a–c) © Cengage Learning; (d) PDB ID: 1GZX; Paoli, M., Liddington, R., Tame, J., Wilinson, A., Dodson, G.; Crystal Structure of T state hemoglobin with oxygen bound at all four haems. *J. Mol.Bio.*, v256, pp. 775–792, 1996.

> The Need for Speed In cells, the making and breaking of chemical bonds requires enzymes. Why? Consider that centuries might pass before a sugar molecule would break down to CO_2 and water on its own, yet the same process takes less than a millisecond inside your cells. Enzymes make the difference. Remember from Section 2.6 that an enzyme makes a reaction run much faster than it would on its own. Enzymes are unchanged by participating in a reaction, so they can work again and again.

Some enzymes are RNAs, but most are proteins. Each enzyme recognizes specific reactants, or **substrates**, and alters them in a specific way. Such specificity occurs because an enzyme's polypeptide (or nucleotide) chains fold up into a pocket called an **active site**, where substrates bind and where reactions proceed (Figure 4.7). An active site is complementary in shape, size, polarity, and charge to the enzyme's substrate. This fit is the reason why each enzyme acts only on specific reactants. An enzyme's active site squeezes reactants together, redistributes their charge, or causes some other change. The change reduces activation energy, so it lowers the barrier that prevents the reaction from proceeding.

> Factors That Influence Enzyme Activity An enzyme's function can be enhanced or inhibited by other molecules. Some of these regulatory molecules exert their effects by binding directly to an active site; others bind elsewhere on the enzyme. In the latter case, binding of the regulatory molecule alters the overall shape of the enzyme (Figure 4.8).

The pH of most human body fluids is between 6 and 8 (Section 2.5). Most of our enzymes work best in this range, but some function at much higher or lower pH (Figure 4.9**A**). Pepsin, a digestive enzyme, works best between pH 1 and 2 **❶**. Pepsin begins the process of protein digestion in the stomach, which is very acidic (pH 2). During digestion, the contents of the stomach pass into the small intestine, where the pH rises to about 9. Pepsin denatures above pH 5.5, so this enzyme becomes inactivated in the small intestine. Here, protein digestion continues with the assistance of trypsin, an enzyme that tolerates the higher pH **❷**.

Adding heat boosts the energy of a system, which is why molecular motion increases with temperature (Section 2.4). The greater the energy of reactants, the closer a reaction is to its activation energy. Thus, the rate of an enzymatic reaction typically increases with temperature, but only up to a point (Figure 4.9**B**). An enzyme denatures above a characteristic temperature. Then, the reac-

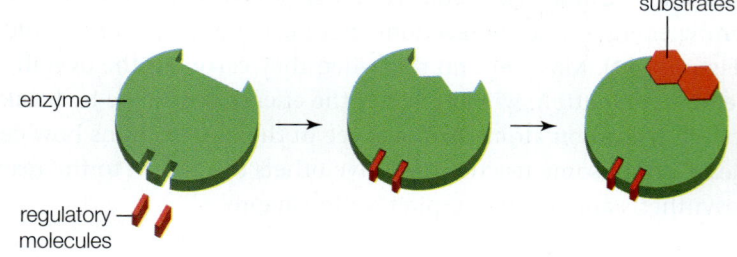

FIGURE 4.8 Regulatory molecule binding to enzymes. Some types of regulatory molecules (*red*) bind to an enzyme in a place other than the active site. This binding changes the shape of the enzyme in a way that enhances or inhibits its function.

Figure It Out: In this example, how does the binding of the regulatory molecules help or hinder the enzyme's function?

Answer: It allows the enzyme to bind its substrates.

© Cengage Learning.

tion rate falls sharply as the shape of the enzyme changes and it stops functioning (Section 2.9). Body temperatures above 42°C (107.6°F) adversely affect the function of many of your enzymes, which is why severe fevers are dangerous.

The activity of many enzymes is also influenced by the amount of salt in the surrounding fluid. Too little salt, and polar parts of the enzyme attract one another so strongly that the enzyme's shape changes. Too much salt interferes with the hydrogen bonds that hold the enzyme in its characteristic shape, and the enzyme denatures.

> **Cofactors** Most enzymes can function properly only with assistance from metal ions or small organic molecules. Such enzyme helpers are called **cofactors**. Many dietary vitamins and minerals are essential because they are cofactors or are precursors for them. **Coenzymes** are organic cofactors. Coenzymes carry chemical groups, atoms, or electrons from one reaction to another, and often into or out of organelles. In some reactions, they stay tightly bound to the enzyme. In others, they participate as separate molecules.

Unlike enzymes, many coenzymes are modified by taking part in a reaction, then become regenerated in other reactions. For example, the coenzyme NAD+ (nicotinamide adenine dinucleotide) can accept electrons and hydrogen atoms, thus becoming NADH. NAD+ forms again when electrons and hydrogen atoms are removed from the NADH:

NAD+ + electrons + H+ ⟶ **NADH** ⟶ NAD+ + electrons + H+

In cells, the nucleotide ATP (Section 2.10) functions as a coenzyme in many reactions. ATP has three phosphate groups, and the bonds between these groups hold a lot of energy. When a phosphate group is transferred to or from a nucleotide, energy is transferred along with it. Thus, the nucleotide can receive energy from an energy-releasing reaction, and it can also donate energy that contributes to the "energy in"

part of an energy-requiring reaction. ATP is such an important currency in a cell's energy economy that we use a cartoon coin (*inset*) to symbolize it.

A reaction in which a phosphate group is transferred from one molecule to another is called a **phosphorylation**. ADP (adenosine diphosphate) forms when an enzyme transfers a phosphate group from ATP to another molecule. Cells constantly run this reaction in order to drive a variety of reactions. Thus, they constantly have to replenish their stockpile of ATP—by running energy-requiring reactions that phosphorylate ADP:

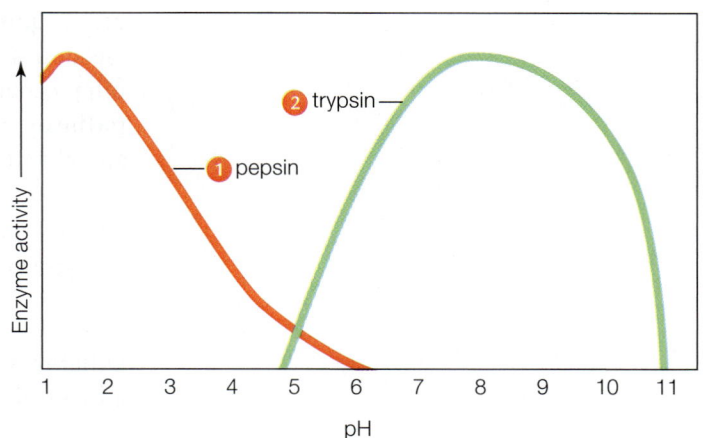

A Each enzyme functions best within a characteristic range of pH. On either side of that range, the reaction rate falls sharply because the enzyme denatures.

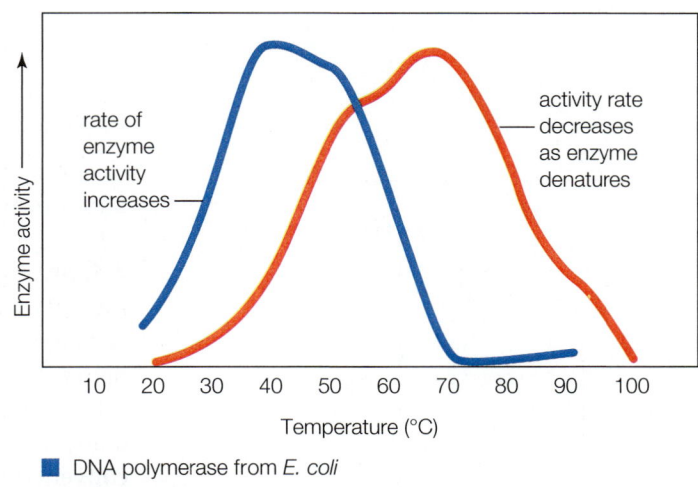

■ DNA polymerase from *E. coli*
■ DNA polymerase from *Thermus aquaticus*

B Enzyme activity varies with temperature. The rate of an enzymatic reaction typically increases with temperature, up to a point. After that point, the rate falls sharply as the enzyme denatures. *E. coli* inhabits the gut (normally 37°C); *T. aquaticus* prefers geothermal hot springs.

FIGURE 4.9 Enzymes, temperature, and pH.

Figure It Out: What is the optimal pH for trypsin?   *Answer: About 8*

© Cengage Learning.

active site Pocket in an enzyme where substrates bind and a reaction occurs.

coenzyme An organic cofactor.

cofactor A metal ion or organic molecule that associates with an enzyme and is necessary for its function.

phosphorylation A phosphate-group transfer.

substrate Of an enzyme, a reactant that is specifically acted upon by the enzyme.

> **Metabolic Pathways** Metabolism, remember, refers to activities by which cells acquire and use energy as they build and break down organic molecules (Section 2.6). Building, rearranging, or breaking down an organic substance often occurs stepwise, in a series of enzymatic reactions called a **metabolic pathway**. Some metabolic pathways are linear, meaning that the reactions run straight from reactant to product:

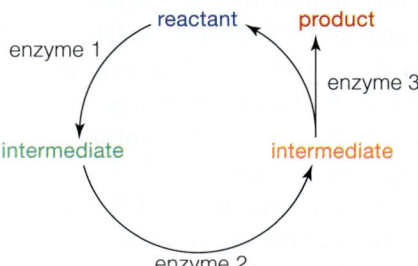

Other metabolic pathways are cyclic. In a cyclic pathway, the last step regenerates a reactant for the first step:

Both linear and cyclic pathways are common in cells.

> **Controlling Metabolism** Cells conserve energy and resources by making only what they need at any given moment—no more, no less. How does a cell adjust the types and amounts of molecules it produces? Several mechanisms help a cell maintain, raise, or lower its production of thousands of different substances. For example, reactions do not only run from reactants to products. Many also run in reverse at the same time, with some of the products being converted back to reactants. The rates of the forward and reverse reactions often depend on the concentrations of reactants and products: A high concentration of reactants pushes the reaction in the forward direction, and a high concentration of products pushes it in the reverse direction.

Other mechanisms more actively regulate pathways. Certain molecules in a cell govern how fast enzyme molecules are made, or influence the activity of enzymes that have already been built. For example, the end product of a series of enzymatic reactions may inhibit the activity of one of the enzymes in the series, an effect called **feedback inhibition** (Figure 4.10).

> **Electron Transfers** The bonds of organic molecules hold a lot of energy that can be released in a reaction with oxygen. One such reaction—burning—releases the energy of organic molecules all at once, explosively (Figure 4.11**A**). Cells use oxygen to break the bonds of organic molecules, but they have no way to harvest the explosive burst of energy that occurs during burning. Instead, they break the molecules apart in several steps that release the energy in small, manageable increments. Most of these steps are electron transfers, in which one molecule accepts electrons from another.

An **electron transfer chain** is an organized series of reaction steps in which membrane-bound enzymes and other molecules give up and accept electrons in turn. Electrons are at a higher energy level when they enter a chain than when they leave. Electron transfer chains can harvest the energy given off by an elec-

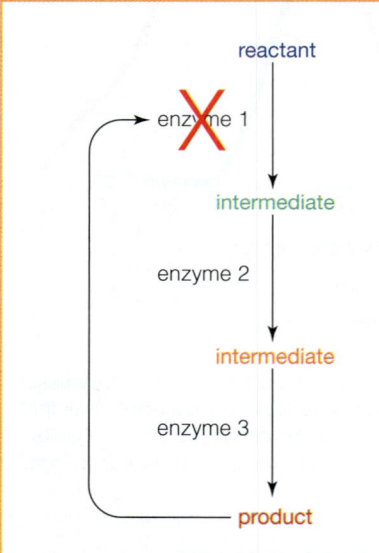

FIGURE 4.10 Animated!
Feedback inhibition. In this example, three kinds of enzymes act in sequence to convert a substrate to a product, which inhibits the activity of the first enzyme.

Figure It Out: Is this an example of a cyclic or a linear pathway?

Answer: This is a linear pathway.
© Cengage Learning.

© Cengage Learning.

glucose
+
oxygen

carbon
dioxide
+
water

glucose **1**
+
oxygen H^+ e^- **2**

carbon
dioxide
+
water e^- **3**

1 An input of activation energy splits glucose into carbon dioxide, electrons, and hydrogen ions (H^+).

2 Electrons lose energy as they move through an electron transfer chain. Energy released by electrons is harnessed for cellular work.

3 Electrons, hydrogen ions, and oxygen combine to form water.

A Glucose and oxygen react (burn) when exposed to a spark. Energy is released all at once as light and heat when carbon dioxide and water form.

B In cells, the same overall reaction occurs in a stepwise fashion that involves an electron transfer chain. Energy is released in amounts that cells are able to harness for cellular work.

FIGURE 4.11 Comparing uncontrolled and controlled energy release.

Credits: (a) © JupiterImages Corporation; (b) © Cengage Learning.

tron as it drops to a lower energy level (Section 2.2 and Figure 4.11**B**). In the next chapter, you will learn about how coenzymes deliver electrons to electron transfer chains in photosynthesis and aerobic respiration. Energy released at certain steps in those chains helps drive the synthesis of ATP.

Take-Home Message

What is a metabolic pathway?

- A metabolic pathway is a sequence of enzyme-mediated reactions that builds, breaks down, or remodels an organic molecule. Some pathways involve electron transfer chains.

- Enzymes greatly enhance the rate of specific reactions. Each works best within a characteristic range of temperature, pH, and salt concentration. Many require the assistance of cofactors.

- ATP couples reactions that release energy with reactions that require energy.

- Cells conserve energy and resources by producing only what they require at a given time. This metabolic control arises from regulatory molecules and other mechanisms that influence metabolic pathways and individual enzymes.

4.5 Diffusion and Membranes

Andrew Lambert Photography/ Photo Researchers, Inc.

Diffusion is the spontaneous spreading of molecules or ions (*left*), and it is an essential way in which substances move into, through, and out of cells. Atoms are always jiggling, and this internal movement causes molecules to randomly collide with one another millions of times each second. Rebounds from the collisions propel them through a liquid or gas,

diffusion Spontaneous spreading of molecules or atoms through a liquid or gas.

electron transfer chain Array of enzymes and other molecules in a cell membrane that accept and give up electrons in sequence, thus releasing the energy of the electrons in steps.

feedback inhibition Mechanism by which a change that results from some activity decreases or stops the activity.

metabolic pathway Series of enzyme-mediated reactions by which cells build, remodel, or break down an organic molecule.

resulting in a gradual and complete mixing. How fast the mixing occurs depends on five factors:

1. Size. It takes more energy to move a large molecule than it does to move a small one. Thus, smaller molecules diffuse more quickly than larger ones.

2. Temperature. Molecules move faster at higher temperature, so they collide more often. Thus, the higher the temperature, the faster the rate of diffusion.

3. Concentration. A difference in solute concentration between adjacent regions of a solution is called a concentration gradient. Solutes tend to diffuse "down" their concentration gradient, from a region of higher concentration to one of lower concentration. Why? When molecules are more crowded, they collide more often. Thus, during a given interval, more molecules get bumped out of a region of higher concentration than get bumped into it. Any substance tends to diffuse in a direction set by its own concentration gradient, not by the gradients of any other solutes that may be sharing the same space in the solution.

4. Charge. Each ion or charged molecule in a fluid contributes to the fluid's overall electric charge. A difference in charge between two regions of the fluid can affect the rate and direction of diffusion between them. For example, positively charged substances (such as sodium ions) will tend to diffuse toward a region with an overall negative charge.

5. Pressure. Diffusion may be affected by a difference in pressure between two adjoining regions. Pressure squeezes molecules together, and molecules that are more crowded collide and rebound more frequently. Thus, diffusion occurs faster at higher pressures.

❯ Osmosis and Tonicity Remember from Section 3.3 that lipid bilayers are selectively permeable: Water can cross them, but ions and most polar molecules cannot (Figure 4.12). When the fluids on either side of a selectively permeable membrane differ in solute concentration, water will diffuse across the membrane in a direction that depends on tonicity. Tonicity refers to the relative concentration of solutes in fluids separated by a selectively permeable membrane. Fluids that are **isotonic** have the same overall solute concentration. When the overall solute concentrations of the two fluids differ, the fluid with the lower concentration of solutes is said to be **hypotonic** (*hypo–*, under). The other one, with the higher solute concentration, is **hypertonic** (*hyper–*, over).

When a selectively permeable membrane separates two fluids that are not isotonic, water will diffuse from the hypotonic fluid into the hypertonic one (Figure 4.13). The diffusion will continue until the two fluids are isotonic, or until some pressure against the hypertonic fluid counters it. The diffusion of water across a membrane is so important in biology that it is given a special name: **osmosis**.

If a cell's cytoplasm is hypertonic with respect to the fluid outside of its plasma membrane, water diffuses into it. If the cytoplasm is hypotonic with respect to the fluid on the outside, water diffuses out. In either case, the solute concentration of the cytoplasm changes. If it changes enough, the cell's enzymes will stop working, with potentially lethal results. Many cells have built-in mechanisms that compensate for differences in tonicity between cytoplasm and external fluid. In cells with no such mechanism, the volume—and solute concentration—of cytoplasm will change as water diffuses into or out of the cell (Figure 4.14**A–C**).

The rigid cell walls of plants and many protists, fungi, and bacteria can resist an increase in the volume of cytoplasm even in hypotonic environments. In the case of plant cells, cytoplasm usually contains more solutes than soil water does.

FIGURE 4.12 Animated! Selective permeability of lipid bilayers. Hydrophobic molecules, gases, and water molecules can cross a lipid bilayer on their own. Ions in particular and most polar organic molecules such as glucose cannot.

© Cengage Learning.

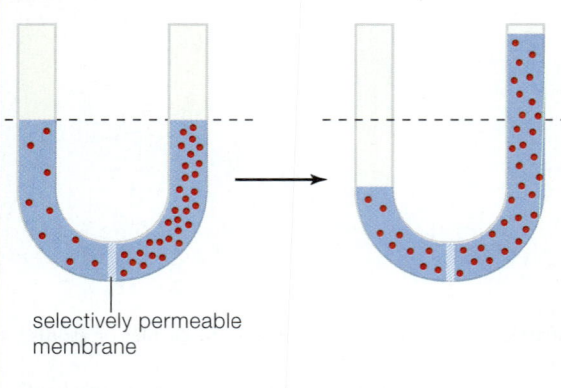

FIGURE 4.13 Animated! Osmosis. Water moves across a selectively permeable membrane that separates two fluids of differing solute concentration. The fluid volume changes in the two compartments as water diffuses across the membrane. © Cengage Learning.

A Red blood cells immersed in an isotonic solution do not change in volume. The fluid portion of blood is normally isotonic with cytoplasm.

B Red blood cells immersed in a hypertonic solution shrivel up as water diffuses out of them.

C Red blood cells immersed in a hypotonic solution swell up as water diffuses into them.

D Osmotic pressure keeps plant parts erect. These cells in an iris petal are plump with cytoplasm.

E Cells from a wilted iris petal. The cytoplasm shrank, and the plasma membrane has pulled away from the cell wall.

FIGURE 4.14 Animated! Effects of tonicity in human red blood cells (**A–C**) and iris petal cells (**D,E**).

Credits: (a–c) M. Sheetz, R. Painter, and S. Singer, *Journal of Cell Biology*, 70:193 (1976) by permission of The Rockefeller University Press; (d–e) Claude Nuridsany & Marie Perennou/Photo Researchers, Inc.

Thus, water usually diffuses from soil into a plant—but only up to a point. Cell walls keep plant cells from expanding very much, so an influx of water causes pressure to build up inside these cells. Pressure that a fluid exerts against a structure that contains it is called **turgor**. When enough pressure builds up inside a plant cell, water stops diffusing into its cytoplasm. The amount of turgor that is enough to stop osmosis is called osmotic pressure.

Osmotic pressure keeps walled cells plump, just as high air pressure inside a tire keeps it inflated. A young land plant can resist gravity to stay erect because its cells are plump with cytoplasm (Figure 4.14**D**). When soil dries out, it loses water but not solutes, so the concentration of solutes increases in soil water. If the soil water becomes hypertonic with respect to cytoplasm, water will start diffusing out of the plant's cells, so their cytoplasm shrinks (Figure 4.14**E**). As turgor inside the cells decreases, the plant wilts.

Take-Home Message

What influences the movement of ions and molecules?

■ Molecules or ions tend to diffuse into an adjoining region of fluid in which they are not as concentrated.

■ The steepness of a concentration gradient as well as temperature, molecular size, charge, and pressure affect the rate of diffusion.

■ Osmosis is a net diffusion of water between two fluids that differ in solute concentration and are separated by a selectively permeable membrane.

4.6 Membrane Transport Mechanisms

A cell's survival often depends on its ability to increase, decrease, and maintain concentrations of specific solutes in its internal fluids. Every cell requires raw materials from its environment, and each must also dispose of wastes. Those that live in salty, acidic environments must also keep their internal composition stable even when that composition differs—often dramatically—from that of the external environment. This control is vital, because if a cell's internal fluids

hypertonic Describes a fluid that has a high overall solute concentration relative to another fluid.

hypotonic Describes a fluid that has a low overall solute concentration relative to another fluid.

isotonic Describes two fluids with identical solute concentrations.

osmosis Diffusion of water across a selectively permeable membrane; occurs when the fluids on either side of the membrane are not isotonic.

turgor Pressure that a fluid exerts against a membrane, wall, or other structure that contains it.

A A glucose molecule (here, in extracellular fluid) binds to a glucose transporter (*gray*) in the plasma membrane.

B Binding causes the transport protein to change shape.

C The glucose molecule detaches from the transport protein on the other side of the membrane (here, in cytoplasm), and the protein resumes its original shape.

FIGURE 4.15 Animated! Passive transport of glucose. **Figure It Out:** In this example, which fluid is hypotonic: extracellular fluid or cytoplasm?

Answer: Cytoplasm

PDB files from NYU Scientific Visualization Lab.

varied a lot, say in pH or the amount of salt, enzymes in the fluids would stop working, and so would the cell.

Ions and most polar molecules (with the notable exception of water) can cross lipid bilayers only with the help of transport proteins (Section 3.3). Each type of transport protein moves a specific ion or molecule; glucose transporters only transport glucose, calcium pumps only pump calcium, and so on. The specificity of transport proteins means that the amounts and types of many substances that cross a cell membrane depend on which transport proteins are embedded in it.

Transport protein specificity allows cells and membrane-enclosed organelles to control the volume and composition of their fluid interior by moving particular solutes one way or the other across the membrane. For example, a glucose transporter can move glucose across a plasma membrane, but it cannot move phosphorylated glucose. Enzymes in cytoplasm phosphorylate glucose as soon as it enters the cell, thus preventing the molecule from moving back through the glucose transporter and leaving the cell.

> **Passive Transport** In **passive transport**, the movement of a solute (and the direction of the movement) through a transport protein is driven entirely by the solute's concentration gradient. For this reason, passive transport is also called facilitated diffusion. The solute simply binds to the passive transport protein, and the protein releases it to the other side of the membrane (Figure 4.15).

A glucose transporter is an example of a passive transport protein. This protein changes shape when it binds to a molecule of glucose. The shape change moves the solute to the opposite side of the membrane, where it detaches. Then, the transporter reverts to its original shape. Some passive transporters do not change shape; they form permanently open channels through a membrane. Others are gated, which means they open and close in response to a stimulus such as binding to a signaling molecule or a shift in electric charge.

> **Active Transport** Maintaining a particular solute's concentration at a certain level often means transporting the solute against its gradient, to the side of a membrane where it is more concentrated. Transporting a solute against its concentration gradient requires energy. In **active transport**, a transport protein uses energy to pump a solute against its gradient across a cell membrane. After a solute binds to an active transporter, an energy input (for example, in the form of a phosphate-group transfer from ATP) changes the shape of the protein. The change causes the transport protein to release the solute to the other side of the membrane.

A calcium pump is an example of an active transport protein. This protein moves calcium ions across cell membranes (Figure 4.16). Calcium ions act as potent messengers inside cells, and they affect the activity of many enzymes. Thus, the calcium ion concentration in cytoplasm is very tightly regulated. Calcium pumps in the plasma membrane of all eukaryotic cells can

A Two calcium ions (*pink* balls) bind to the transport protein (*gray*).

B Energy in the form of a phosphate group is transferred from ATP to the protein. The transfer causes the protein to change shape so that it ejects the calcium ions to the opposite side of the membrane.

C After it loses the calcium ions, the transport protein resumes its original shape.

FIGURE 4.16 Animated!
Active transport of calcium ions.

© Cengage Learning.

keep the concentration of calcium in cytoplasm thousands of times lower than it is in extracellular fluid.

Nearly all of the cells in your body have active transport proteins called sodium–potassium pumps (Figure 4.17). Sodium ions in cytoplasm diffuse into the pump's open channel and bind to its interior. A phosphate-group transfer from ATP causes the pump to change shape so that its channel opens to extracellular fluid, where it releases the sodium ions. Then, potassium ions from extracellular fluid diffuse into the channel and bind to its interior. The transporter releases the phosphate group and reverts to its original shape. The channel opens to the cytoplasm, where it releases the potassium ions.

Bear in mind that the membranes of all cells, not just those of animals, have active transport proteins. In plants, for example, active transport proteins in the plasma membranes of leaf cells pump sugars into tubes that thread throughout the plant body.

> Membrane Trafficking When a lipid bilayer is disrupted, it seals itself, because the disruption exposes the fatty acid tails of the phospholipids to their watery surroundings. In watery fluids, lipids spontaneously rearrange themselves so their nonpolar tails are together (Section 3.3). A vesicle forms when a patch of membrane bulges into the cytoplasm because the hydrophobic tails of the lipids in the bilayer are repelled by water on both sides. The water "pushes" the

active transport Energy-requiring mechanism in which a transport protein pumps a solute across a cell membrane against its concentration gradient.

passive transport Mechanism by which a concentration gradient drives the movement of a solute across a cell membrane through a transport protein. Requires no energy input.

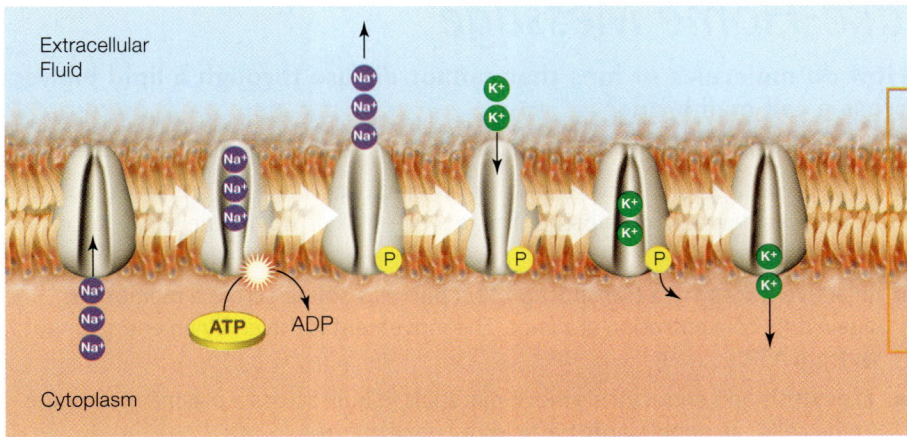

FIGURE 4.17 The sodium–potassium pump. This active transport protein (*gray*) transports sodium ions (Na^+) from cytoplasm to extracellular fluid, and potassium ions (K^+) in the other direction. The transfer of a phosphate group (P) from ATP provides energy required for transporting the ions against their concentration gradient.

© Cengage Learning.

FIGURE 4.18 Animated! Membrane crossings. A plasma membrane is a hub of activity: Molecules and ions (colored balls) are constantly flowing into and out of a cell via transport proteins embedded in its plasma membrane. Vesicles are also taking in or expelling bulk amounts of solutes and much larger particles.

❶ Endocytosis begins as a small patch of plasma membrane sinks inward.

❷ As the membrane balloons into the cell, it wraps itself around a small volume of extracellular fluid, along with the solutes or particles it contains.

❸ The balloon pinches off inside the cell as a vesicle, which may deliver its contents to an organelle.

© Cengage Learning 2013.

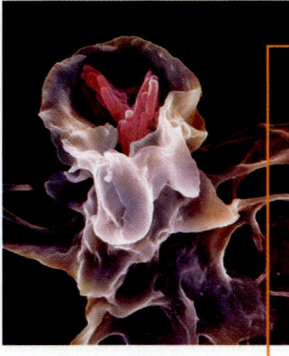

FIGURE 4.19 Animated!
Phagocytosis. This micrograph shows a phagocytic white blood cell engulfing several *Tuberculosis* bacteria (*red*).

Photo Researchers, Inc.

tails together, which helps round off the bulge as a vesicle, and also seals the rupture in the membrane.

As part of vesicles, patches of membrane constantly move to and from the cell surface (Figure 4.18). In a process called **endocytosis**, a cell takes up substances in bulk near its outer surface. A small patch of plasma membrane balloons inward ❶, taking extracellular fluid with it. The balloon sinks farther into the cytoplasm ❷, and then pinches off as a vesicle ❸. The vesicle delivers its contents to an organelle or stores them in a cytoplasmic region. In **exocytosis**, a vesicle moves to the cell's surface, and the protein-studded lipid bilayer of its membrane fuses with the plasma membrane. As the exocytic vesicle loses its identity, its contents are released to the surroundings.

Phagocytosis (which literally means "cell eating") is an endocytic pathway in which cells such as amoebas engulf microorganisms, cellular debris, or other large particles. Animals have phagocytic white blood cells that can engulf viruses and bacteria, cancerous body cells, and other threats (Figure 4.19). Temporary projections from the cell's surface engulf a target and fuse around it. The engulfed target ends up inside a vesicle in the cell's cytoplasm.

Take-Home Message

How do molecules or ions that cannot diffuse through a lipid bilayer cross a cell membrane?

- Transport proteins help specific molecules or ions to cross cell membranes.

- In passive transport, a solute binds to a transport protein that releases it on the opposite side of the membrane. The movement is driven by the solute's concentration gradient.

- In active transport, a transport protein pumps a solute across a membrane against its concentration gradient. The movement is driven by an energy input, as from ATP.

- Exocytosis and endocytosis move materials in bulk across a plasma membrane. Some cells can engulf large particles by phagocytosis.

endocytosis Process by which a cell takes in extracellular substances in bulk; a portion of the cell's plasma membrane balloons inward and pinches off as a vesicle.
exocytosis Process by which a cell expels a vesicle's contents to extracellular fluid.
phagocytosis Endocytic pathway by which a cell engulfs particles such as microbes or cellular debris.

4.7 A Toast to ADH (revisited)

In humans and other animals, alcohol dehydrogenase (ADH) is part of a pathway that detoxifies alcohols made by gut-inhabiting bacteria, and those in foods such as ripe fruit. This pathway also breaks down the tiny quantities of alcohols that form as by-products of other metabolic pathways.

ADH converts ethanol to acetaldehyde, an organic molecule even more toxic than ethanol and the most likely source of various hangover symptoms. A different enzyme, ALDH, very quickly converts acetaldehyde to nontoxic acetate. Both enzymes use the coenzyme NAD^+ to accept electrons and hydrogen atoms. Thus, the overall pathway of ethanol metabolism in humans is:

$$\text{ethanol} \xrightarrow[\text{NAD}^+ \quad \text{NADH}]{\text{ADH}} \text{acetaldehyde} \xrightarrow[\text{NAD}^+ \quad \text{NADH}]{\text{ALDH}} \text{acetate}$$

In the average adult human body, this metabolic pathway can detoxify between 7 and 14 grams of ethanol per hour. The average alcoholic beverage contains between 10 and 20 grams of ethanol, which is why having more than one drink in any two-hour interval may result in a hangover.

Defects in ADH or ALDH can affect alcohol metabolism. For example, if a person's ADH is overactive, acetaldehyde accumulates faster than ALDH can detoxify it:

$$\text{ethanol} \xrightarrow{\text{ADH}} \begin{array}{l} \text{acetaldehyde} \\ \text{acetaldehyde} \\ \text{acetaldehyde} \end{array} \xrightarrow{\text{ALDH}} \text{acetate}$$

People with an overactive form of ADH become flushed and feel ill after drinking even a small amount of alcohol. The unpleasant experience may be part of the reason that these people are less likely to become alcoholic than others.

Having underactive ALDH also causes acetaldehyde to accumulate:

$$\text{ethanol} \xrightarrow{\text{ADH}} \begin{array}{l} \text{acetaldehyde} \\ \text{acetaldehyde} \\ \text{acetaldehyde} \end{array} \xrightarrow{\textbf{\textcolor{red}{X}}} \text{acetate}$$

Underactive ALDH is associated with the same effect—and the same protection from alcoholism—as overactive ADH. Both types of variant enzymes are common in people of Asian descent. For this reason, the alcohol flushing reaction is often called "Asian flush."

Having an underactive ADH enzyme has the opposite effect. It slows alcohol metabolism, so people with low ADH activity may not feel the ill effects of drinking alcoholic beverages as much as other people. When these people drink alcohol, they have a tendency to become alcoholics. Compulsive, uncontrolled drinking damages an alcoholic's health and social relationships. The study mentioned in Section 4.1 showed that one-quarter of the undergraduate students who binged also had other signs of alcoholism.

Alcoholics will continue to drink despite the knowledge that doing so has tremendous negative consequences. In the United States, alcohol abuse is the leading cause of cirrhosis of the liver, a condition in which the liver becomes so scarred, hardened, and filled with fat that it loses its function (Figure 4.20**A**,**B**). It stops making the protein albumin, so the solute balance of body fluids is disrupted, and the legs and abdomen swell with watery fluid. It cannot remove drugs and other toxins from the blood, so they accumulate in the brain—which impairs mental functioning and alters personality. Restricted blood flow through the liver causes veins to enlarge and rupture, so internal bleeding is a risk. The damage to the body results in a heightened susceptibility to diabetes and liver cancer. Once cirrhosis has been diagnosed, a person has about a 50 percent chance of dying within 10 years (Figure 4.20**C**).

© Cengage Learning.

A Normal, healthy human liver.

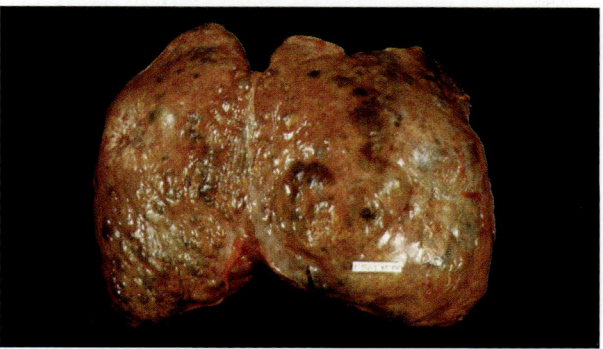

B A cirrhotic human liver.

C Gary Reinbach, who died at the age of 22 from alcoholic liver disease shortly after this photograph was taken, in 2009. The odd color of his skin is a symptom of cirrhosis.

FIGURE 4.20 Alcoholic liver disease.

Credits: (a) Southern Illinois University/ Photo Researchers, Inc.; (b) Martin M. Rotker/ Photo Researchers, Inc.; (c) Stuart Clark/ The Sunday Times/ nisyndication.

WHERE YOU ARE GOING . . .

Chapter 5 details metabolic pathways by which cells store and retrieve energy in chemical bonds. Enzymes are critical for DNA replication (6.4), gene expression (7.2–7.5), biotechnology (10.2, 10.3), and digestion (23.3). Membrane transport of ions is discussed in the context of muscle contraction (20.4), nervous system signaling (24.2), stomata function (27.6), and plant cell signaling (28.7, 28.8). Section 27.6 returns to sugar transport in plants.

Summary

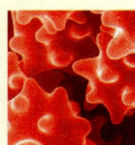

(**hypotonic**) toward the region with a higher solute concentration (**hypertonic**). There is no net movement of water between **isotonic** solutions. Osmotic pressure is the amount of **turgor** (fluid pressure against a cell membrane or wall) that stops osmosis.

Section 4.1   Currently the most serious drug problem on college campuses is binge drinking, which is associated with alcoholism. Drinking more alcohol than the body's enzymes can detoxify can be lethal in both the short term and the long term.

Section 4.2   **Energy** is the capacity to do work. Energy cannot be created or destroyed (**first law of thermodynamics**), but it can be converted from one form to another and transferred between objects or systems. Energy tends to disperse spontaneously (**second law of thermodynamics**). A bit disperses at each energy transfer, usually in the form of heat.

Living things can maintain their organization only as long as they harvest energy from someplace else. Energy flows in one direction through the biosphere, starting mainly from the sun, then into and out of ecosystems. Producers and then consumers use the energy to assemble, rearrange, and break down organic molecules that cycle among organisms throughout ecosystems.

Section 4.3   Cells store and retrieve energy by making and breaking chemical bonds in chemical reactions, in which **reactants** are converted to **products**. Some reactions require a net energy input; others end with a net energy release. **Activation energy** is the minimum energy input required to start a reaction.

Section 4.4   Enzymes greatly enhance the rate of a chemical reaction. Each has an **active site** and works on a particular **substrate** within a characteristic range of temperature, salt concentration, and pH. Many enzymes require assistance from **coenzymes** or other **cofactors**.

ATP functions as an energy carrier between reaction sites in cells. It has three phosphate groups; when one of them is transferred to another molecule, energy is transferred along with it. Phosphate-group transfers (**phosphorylations**) to and from ATP couple reactions that release energy with reactions that require energy.

Cells build, convert, and break down substances in enzyme-mediated reaction sequences called **metabolic pathways**. Controls over enzymes and metabolic pathways allow cells to conserve energy and resources by producing only what they require. **Feedback inhibition** is an example of metabolic control. **Electron transfer chains** in some pathways harvest electron energy in small, manageable increments.

Section 4.5   Molecules or ions tend to spread spontaneously (**diffuse**), with the eventual result being a gradual and complete mixing. A concentration gradient is a difference in the concentration of a solute between adjoining regions of solution. The steepness of the gradient, temperature, solute size, charge, and pressure influence the diffusion rate.

Osmosis is the diffusion of water across a selectively permeable membrane, from the region with a lower solute concentration

Section 4.6   Gases, water, and small nonpolar molecules can diffuse across a lipid bilayer. Most other molecules, and ions in particular, cross only with the help of transport proteins.

Transport proteins allow a cell or membrane-enclosed organelle to control which substances enter and exit. The types of transport proteins in a membrane determine which substances can cross it. Calcium pumps and other **active transport** proteins use energy, such as a phosphate transfer from ATP, to pump a solute against its concentration gradient. **Passive transport** proteins work without an energy input; a solute's movement is driven by its concentration gradient.

Substances in bulk and large particles are moved across plasma membranes by processes of **exocytosis** and **endocytosis**. In exocytosis, a cytoplasmic vesicle fuses with the plasma membrane, and its contents are released to the outside of the cell. The vesicle's membrane lipids and proteins become part of the plasma membrane. In endocytosis, a patch of plasma membrane balloons into the cell, and forms a vesicle that sinks into the cytoplasm. Some cells engulf large particles such as prey or cell debris by the endocytic pathway of **phagocytosis**.

Self-Quiz   Answers in Appendix I

1. _____ is life's primary source of energy.
   a. Food   b. Water   c. Sunlight   d. ATP

2. Which of the following statements is not correct?
   a. Energy cannot be created or destroyed.
   b. Energy cannot change from one form to another.
   c. Energy tends to disperse spontaneously.

3. If we liken a chemical reaction to an energy hill, then an _____ reaction is an uphill run.
   a. energy-requiring   c. ATP-assisted
   b. energy-releasing   d. both a and c

4. _____ are always changed by participating in a reaction. (Choose all that are correct.)
   a. Enzymes   c. Reactants
   b. Cofactors   d. Active sites

5. Enzymes _____ .
   a. are proteins, except for a few RNAs
   b. lower the activation energy of a reaction
   c. are changed by participating in a reaction
   d. a and b

6. One environmental factor that influences enzyme function is _____ .
   a. temperature   c. light
   b. wind   d. radioactivity

7. A metabolic pathway _____ .
   a. may build or break down molecules
   b. generates heat
   c. can include an electron transfer chain
   d. all of the above

8. Which of the following statements is incorrect?
   a. Some metabolic pathways are cyclic.
   b. Glucose can diffuse through a lipid bilayer.
   c. Feedback inhibition controls some metabolic pathways.
   d. All coenzymes are cofactors.
   e. Osmosis is a case of diffusion.

9. Ions or molecules tend to diffuse from a region where they are _____ (more/less) concentrated to another where they are _____ (more/less) concentrated.

10. _____ cannot easily diffuse across a lipid bilayer.
    a. Water   c. Ions
    b. Gases   d. all of the above

11. Transporters that require an energy boost help sodium ions across a cell membrane. This is a case of _____ .
    a. passive transport   c. facilitated diffusion
    b. active transport   d. a and c

12. If you immerse a red blood cell in a hypotonic solution, water will _____ .
    a. diffuse into the cell   c. show no net movement
    b. diffuse out of the cell d. move in by endocytosis

13. Fluid pressure against a wall or cell membrane is called _____ .
    a. osmosis   c. diffusion
    b. turgor   d. osmotic pressure

14. Vesicles form in _____ .
    a. endocytosis   c. phagocytosis
    b. exocytosis   d. a and c

15. Match each term with its most suitable description.
    ____ reactant   a. assists some enzymes
    ____ phagocytosis   b. forms at reaction's end
    ____ first law   c. enters a reaction
         of thermodynamics   d. requires energy boost
    ____ product   e. one cell engulfs another
    ____ cofactor   f. energy cannot be created
    ____ diffusion   or destroyed
    ____ passive transport   g. faster with a gradient
    ____ active transport   h. no energy boost required
    ____ ATP   i. currency in an energy economy

Critical Thinking

1. Beginning physics students are often taught the basic concepts of thermodynamics with two phrases: First, you can't win. Second, you can't break even. Explain.

2. Water molecules tend to diffuse in response to their own concentration gradient. How can water be more or less concentrated?

3. Dixie Bee wanted to make JELL-O shots for her next party, but felt guilty about encouraging her guests to consume alcohol. She tried to compensate for the toxicity of the alcohol by adding pieces of healthy fresh pineapple to the shots, but when she did, the JELL-O never solidified. What happened? Hint: JELL-O is mainly sugar and a gelatinous mixture of proteins.

4. The enzyme trypsin is sold as a dietary enzyme supplement. Explain what happens to trypsin taken with food.

5. The enzyme catalase combines two hydrogen peroxide molecules ($H_2O_2 + H_2O_2$) to make two molecules of water ($2H_2O$). A gas also forms. What is the gas?

Digging Into Data

One Tough Bug

Ferroplasma acidarmanus is a species of archaea discovered in an abandoned California copper mine (Figure 4.21**A**). These cells use an energy-harvesting pathway that combines oxygen with iron–sulfur compounds in minerals such as pyrite. The reaction dissolves the minerals, so groundwater that seeps into the mine ends up accumulating high concentrations of metal ions such as copper, zinc, cadmium, and arsenic. Another reaction product, sulfuric acid, lowers the pH of the resulting solution to zero.

Unwalled *F. acidarmanus* cells maintain their internal pH at a cozy 5.0 despite living in an environment with a composition essentially the same as hot battery acid. Thus, researchers investigating *Ferroplasma* metabolic enzymes were surprised to discover that most of the cells' enzymes function best at very low pH (Figure 4.21**B**).

1. What does the dashed line in the graph signify?

2. Of the four enzymes profiled in the graph, how many function optimally at a pH lower than 5? How many retain significant function at pH 5?

3. What is the optimal pH for *Ferroplasma* carboxylesterase?

A Deep inside one of the most toxic sites in the United States: Iron Mountain Mine, in California. The water in this stream, which is about one meter (3 feet) wide, is hot (around 40°C, or 104°F), heavily laden with arsenic and other toxic metals, and has a pH of zero. The slime streamers growing in it are a biofilm dominated by a species of archaea, *Ferroplasma acidarmanus*.

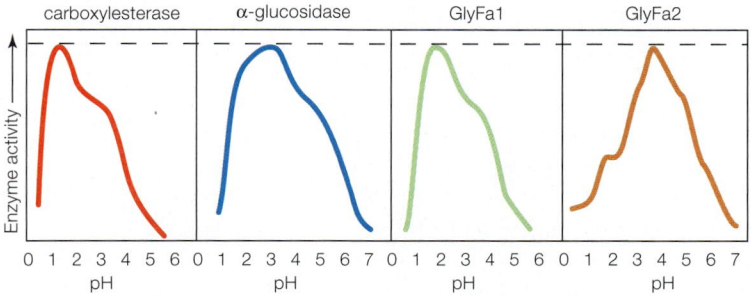

B pH profiles of four enzymes isolated from *Ferroplasma*. Researchers had expected these enzymes to function best at the cells' cytoplasmic pH (5.0).

FIGURE 4.21 pH anomaly of *Ferroplasma* enzymes.

(a) Courtesy of Dr. Katrina J. Edwards; (b) From Golyshina et al., *Environmental Microbiology*, 8(3): 416–425. © 2006 John Wiley and Sons. Used with permission of the publisher.

5 Capturing and Releasing Energy

This chapter explores the main metabolic pathways (Section 4.4) by which organisms harvest energy from the sun (4.2, 4.3). We also revisit producers and consumers (1.3), electrons and energy levels (2.2), chemical bonds (2.3), organic compounds (2.6–2.9), chloroplasts and mitochondria (3.5), electron transfers (4.4), gradients (4.5), and membrane transport (4.6).

5.1 A Burning Concern

Your body is about 9.5 percent carbon by weight, which means that you contain an enormous number of carbon atoms. Where did they all come from? Those atoms may have passed through other consumers before you ate them, but at some point they were components of photosynthetic organisms—mainly plants. Plants get their carbon atoms from carbon dioxide (CO_2), a gas in air. Your carbon atoms—and those of most other organisms on land—were recently part of Earth's atmosphere, in molecules of CO_2.

Humans and other animals get our energy and carbon by breaking down organic molecules that have been assembled by other organisms. By contrast, plants and other photosynthetic producers get energy directly from the environment, and carbon atoms from inorganic molecules such as CO_2. **Photosynthesis** is a metabolic pathway in which sunlight energy is harnessed to build glucose molecules from CO_2 and water.

Photosynthesizers remove CO_2 from the atmosphere, and lock its carbon atoms in organic compounds that make up their tissues. When they and other organisms break down organic compounds for energy, carbon atoms are released in the form of CO_2, which then reenters the atmosphere. Since photosynthesis evolved, these two processes have constituted a balanced cycle of the biosphere: The amount of CO_2 that photosynthesis removes from the atmosphere is roughly the same amount that organisms put back into it—at least it was, until humans came along.

As early as 8,000 years ago, humans began burning forests to clear land for agriculture. When trees and other plants burn, most of the carbon in their tissues is released into the atmosphere as CO_2. Today, we are burning a lot more than our ancestors ever did. We are also burning fossil fuels—coal, petroleum, and natural gas—to satisfy our greater and greater demands for energy (Figure 5.1). Fossil fuels are the organic remains of ancient photosynthetic organisms. When we burn these fuels, we release the carbon they have stored for hundreds of millions of years back into the atmosphere, mainly as CO_2.

Our activities have put Earth's atmospheric cycle of carbon dioxide far out of balance, because we are adding far more CO_2 to the atmosphere than photosynthetic organisms are removing from it. In 2010 alone, we released 10 billion tons of CO_2 into the atmosphere, most of it from burning fossil fuels—an increase of 5.9 percent over 2009, and 49 percent over 1990. Such alarming statistics are why researchers are scrambling to find a way to make cost-effective biofuels. Biofuels are oils, gases, or alcohols made from organic matter (mainly plant material) that is not fossilized. Using biofuels does not add more CO_2 to the atmosphere; rather, growing plant matter for fuel recycles CO_2 that is already in it.

FIGURE 5.1 Products of fossil fuel combustion in the sky over Los Angeles on a sunny day. The color of smog comes from one of its components, a brown toxic gas called nitric oxide. Invisible components include other nitrogen and sulfur compounds, carbon dioxide, organic molecules, and mercury and other heavy metals.

© Brooke Schreier Ganz.

FIGURE 5.2 Properties of light. Visible light makes up a very small part of a large spectrum of electromagnetic energy radiating from the sun. This energy undulates across space as waves that we measure in nanometers (nm). About 25 million nanometers are equal to 1 inch. Raindrops or a prism can separate white light into its different wavelengths, which we see as different colors.

Credits: left, © Photodisc/ Getty Images; right, © Cengage Learning.

5.2 Capturing Rainbows

Energy flow through nearly all ecosystems on Earth begins when photosynthesizers intercept sunlight. Harnessing the energy of sunlight for work is a complicated business, or we would have figured out how to do it in an economically sustainable way by now. Plants do it by converting light energy to chemical energy, which they and most other organisms use to drive cellular work. In order to understand how that happens, you have to understand a little about the nature of light.

Light is a form of electromagnetic energy, which means it travels through space in waves, a bit like waves move across an ocean. The distance between the crests of two successive waves of light is a **wavelength**, which we measure in nanometers (nm). Visible light travels in wavelengths between 380 and 750 nm (Figure 5.2). We see all of these wavelengths combined as white light, and particular wavelengths in this range as different colors. When white light passes through a prism (or raindrops that act as tiny prisms), it separates into its component colors and makes a rainbow.

Photosynthesizers use pigments to capture visible light. A **pigment** is an organic molecule that selectively absorbs light of certain wavelengths. Wavelengths of light that are not absorbed are reflected, and that reflected light gives each pigment its characteristic color. For example, a pigment that absorbs violet, blue, and green light reflects the rest of the visible light spectrum—yellow, orange, and red light. This pigment would appear orange to us.

Chlorophyll *a* is the most common photosynthetic pigment in plants, and also in photosynthetic protists and bacteria. Chlorophyll *a* absorbs violet, red, and orange light, so it appears green to us. Accessory pigments, including other chlorophylls, absorb additional wavelengths of visible light for photosynthesis. Accessory pigments are multipurpose molecules. Their antioxidant properties help protect plants and other organisms from the damaging effects of ultraviolet (UV) light in the sun's rays; their appealing colors attract animals to ripening

chlorophyll *a* Main photosynthetic pigment in plants.
pigment An organic molecule that selectively absorbs light of certain wavelengths.
photosynthesis Metabolic pathway in which light energy is harnessed to drive glucose synthesis from CO_2 and water.
wavelength Distance between the crests of two successive waves of light.

FIGURE 5.3 Animated! A few photosynthetic pigments.

The curves in this graph show the efficiency at which each pigment absorbs the different wavelengths of visible light. Line color indicates the pigment's characteristic color. Using a combination of pigments allows photosynthetic organisms to maximize the range of wavelengths they can capture for photosynthesis.

Credits: top, © Photobac/ Shutterstock; bottom, © Cengage Learning.

fruit or pollinators to flowers. You may already be familiar with some of these pigments. For example, carrots are orange because they contain a lot of beta-carotene (which is often abbreviated as β-carotene). Roses are red and violets are blue because of their anthocyanin content.

Most photosynthetic organisms maximize the range of wavelengths they can capture for photosynthesis by using a combination of pigments (Figure 5.3). In plants, chlorophylls are usually so abundant that they mask the colors of other pigments, which is why typical leaves are green. The green leaves of many plants change color during autumn because they stop making pigments in preparation for a period of dormancy. Chlorophyll breaks down faster than the other pigments, so the leaves turn red, orange, yellow, or violet as their chlorophyll content declines and their accessory pigments become visible.

Pigment molecules are a bit like antennas specialized for receiving light energy of only certain wavelengths. Absorbing energy excites a pigment's electrons. Remember, an energy input can boost an electron to a higher energy level (Section 2.2). The excited electron returns quickly to a lower energy level by emitting the extra energy. As you will see, photosynthetic cells can capture energy emitted from an electron returning to a lower energy level. Arrays of chlorophylls and other photosynthetic pigments in these cells hold on to the energy by passing it back and forth. When the energy reaches a special pair of chlorophylls, the reactions of photosynthesis begin.

All life is sustained by inputs of energy, but not all forms of energy can sustain life.

Take-Home Message

How do photosynthesizers absorb light?

- Energy radiating from the sun travels through space in waves.
- Humans perceive different wavelengths of visible light as different colors. The shorter the wavelength, the greater the energy.
- Pigments absorb light at specific wavelengths. Photosynthetic species use pigments such as chlorophyll *a* to harvest the energy of light for photosynthesis.

5.3 Storing Energy in Sugar

All life is sustained by inputs of energy, but not all forms of energy can sustain life. Sunlight, for example, is abundant here on Earth, but it cannot be used to directly power protein synthesis or other energy-requiring reactions that all organisms must run in order to stay alive. Photosynthesis converts the energy of light into the energy of chemical bonds. Unlike light, chemical energy can power the reactions of life, and it can be stored for use at a later time.

In plants and many protists, photosynthesis occurs in certain types of cells (Figure 5.4). These cells contain chloroplasts (Section 3.5), which have many specializations for photosynthesis. Plant chloroplasts have two outer membranes, and they are filled with a thick, cytoplasm-like fluid called **stroma**. Suspended in the stroma are the chloroplast's own DNA, some ribosomes, and an inner, much-folded **thylakoid membrane**. The folds of a thylakoid membrane typically form stacks of interconnected disks called thylakoids. The space enclosed by the thylakoid membrane is a single, continuous compartment.

Photosynthesis is often summarized this way:

$$6CO_2 \; + \; 6H_2O \; \xrightarrow{\text{light energy}} \; C_6H_{12}O_6 \; + \; 6O_2$$

carbon   water   glucose   oxygen
dioxide

The equation means that photosynthesis converts carbon dioxide and water to glucose and oxygen. However, keep in mind that photosynthesis is not a single reaction. Rather, it is a metabolic pathway, a series of many reactions (Section 4.4). These reactions occur in two stages. Molecules embedded in the thylakoid membrane carry out the reactions of the first stage, which are driven by light and thus called the light-dependent reactions. The "photo" in photosynthesis means light, and it refers to the conversion of light energy to the chemical bond energy of ATP during this stage. Water molecules are converted to oxygen gas (O_2), and released hydrogen ions and electrons end up in the coenzyme NADPH (Figure 5.5**A**).

The "synthesis" part of photosynthesis refers to the reactions of the second stage, which build glucose molecules from carbon dioxide and water. These reaction run in the stroma. They are collectively called the light-independent reactions because light energy does not power them. Instead, they run on energy delivered by the coenzymes that formed in the first stage (Figure 5.5**B**).

FIGURE 5.4 Animated! Zooming in on the site of photosynthesis in cells of typical leafy plants. The micrograph shows chloroplast-stuffed cells in a leaf of *Plagiomnium ellipticum*, a type of moss.

Credits: top, © Dr. Ralf Wagner, www.dr-ralf-wagner.de.; bottom left and right, © Cengage Learning.

two outer membranes of chloroplast

stroma

part of thylakoid membrane system:

thylakoid compartment, cutaway view

50 µm

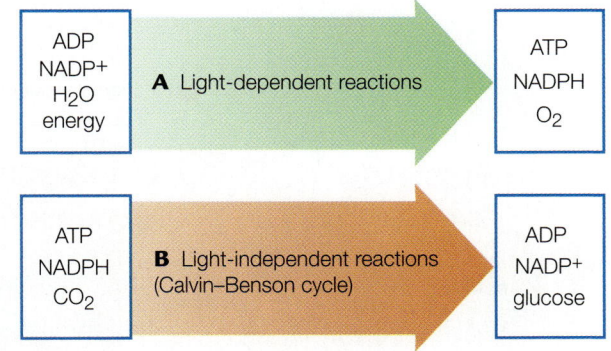

FIGURE 5.5 Inputs and outputs of the two stages of photosynthesis.

© Cengage Learning.

Take-Home Message

Where in a eukaryotic cell does photosynthesis take place?

- In the first stage of photosynthesis, light energy drives the formation of oxygen gas, ATP, and other coenzymes. These light-dependent reactions occur at the thylakoid membrane of chloroplasts.

- The second stage of photosynthesis, the light-independent reactions, occurs in the stroma of chloroplasts. ATP and other coenzymes drive the synthesis of glucose from water and carbon dioxide.

stroma The cytoplasm-like fluid between the thylakoid membrane and the two outer membranes of a chloroplast.

thylakoid membrane A chloroplast's highly folded inner membrane system; forms a continuous compartment in the stroma.

5.4 The Light-Dependent Reactions

When a chlorophyll or other photosynthetic pigment absorbs light, one of its electrons jumps to a higher energy level (shell). The electron quickly drops back down to a lower shell by emitting its extra energy. In the thylakoid membrane, energy emitted by an electron is not lost to the environment. In this special membrane, photosynthetic pigments occur in clusters held together by proteins. The clusters can hold onto energy by passing it back and forth.

The reactions of photosynthesis begin when energy being passed around the thylakoid membrane reaches a photosystem (Figure 5.6). Photosystems are groups of hundreds of chlorophylls, accessory pigments, and other molecules. When a photosystem absorbs energy ❶, it ejects electrons.

A photosystem can lose only a few electrons before it must be restocked with more. Where do replacements come from? This photosystem gets more electrons by pulling them off of water molecules in the thylakoid compartment. It is really difficult to pull electrons off of water molecules; a photosystem is the only biological system that can do it. The reaction is so strong that it causes the water molecules to break apart into hydrogen ions and oxygen atoms ❷. The oxygen atoms combine and diffuse out of the cell as oxygen gas (O_2).

The electrons ejected from the photosystem immediately enter an electron transfer chain in the thylakoid membrane ❸. Remember that electron transfer chains can harvest the energy of electrons in small, usable increments (Section 4.4). Electrons ejected from the photosystem pass from one molecule of the electron transfer chain to the next, and they release a bit of energy at each step. The molecules of the chain use the released energy to actively transport hydrogen ions (H^+) across the membrane, from the stroma into the thylakoid compartment ❹. Thus, the flow of electrons through electron transfer chains sets up and maintains a hydrogen ion gradient across the thylakoid membrane.

FIGURE 5.6 Animated! Light-dependent reactions of photosynthesis in the thylakoid membrane. Electrons that travel through two different electron transfer chains end up in NADPH, which delivers them to sugar-building reactions in the stroma.

© Cengage Learning.

❶ Light energy ejects electrons from a photosystem.

❷ The photosystem pulls replacement electrons from water molecules, which break apart into oxygen and hydrogen ions. The oxygen leaves the cell as O_2.

❸ The electrons enter an electron transfer chain in the thylakoid membrane.

❹ Energy lost by the electrons as they move through the chain is used to actively transport hydrogen ions from the stroma into the thylakoid compartment. A hydrogen ion gradient forms across the thylakoid membrane.

❺ Light energy ejects electrons out of another photosystem. Replacement electrons come from an electron transfer chain.

❻ The ejected electrons move through a second electron transfer chain, then combine with $NADP^+$ and H^+, so NADPH forms.

❼ Hydrogen ions in the thylakoid compartment are propelled through the interior of ATP synthases by their gradient across the thylakoid membrane.

❽ Hydrogen ion flow causes ATP synthases to attach phosphate to ADP, so ATP forms in the stroma.

After the electrons have moved through the first electron transfer chain, they are accepted by another photosystem. This photosystem absorbs light energy, which causes it to eject electrons ❺. These electrons immediately enter a second, different electron transfer chain. At the end of this chain, the coenzyme $NADP^+$ accepts the electrons along with H^+, so NADPH forms ❻:

$$NADP^+ + 2e^- + H^+ \longrightarrow \boxed{NADPH}$$

Hydrogen ions in the thylakoid compartment want to follow their concentration gradient by moving back into the stroma. However, ions cannot diffuse through lipid bilayers (Section 4.5). H^+ leaves the thylakoid compartment only by flowing through transport proteins called ATP synthases, which are embedded in the thylakoid membrane ❼. Hydrogen ion flow through an ATP synthase causes this protein to attach a phosphate group to ADP ❽, so ATP forms in the stroma. The process by which the flow of electrons through electron transfer chains drives ATP formation is called **electron transfer phosphorylation**.

Take-Home Message

What happens during the first stage of photosynthesis?

- In the light-dependent reactions of photosynthesis, chlorophylls and other pigments in the thylakoid membrane transfer the energy of light to photosystems.
- Absorbing energy causes electrons to leave photosystems and enter electron transfer chains in the membrane. The flow of electrons through the transfer chains sets up hydrogen ion gradients that drive ATP formation.
- Oxygen is released, and electrons end up in NADPH.

5.5 The Light-Independent Reactions

The enzyme-mediated reactions of the **Calvin–Benson cycle** build glucose in the stroma of chloroplasts (Figure 5.7). These reactions are light-independent because light energy does not power them. Instead, they run on ATP and NADPH that formed in the light-dependent reactions.

Light-independent reactions use carbon atoms from CO_2 to make glucose. The process of incorporating carbon atoms from an inorganic source into an organic molecule is called **carbon fixation**. In most plants, photosynthetic protists, and some bacteria, the enzyme rubisco fixes carbon by attaching CO_2 to a five-carbon compound known as RuBP. The resulting six-carbon molecule splits at once into two 3-carbon intermediates called PGA, which continue in the cycle. It takes six cycles of Calvin–Benson reactions to fix the six carbon atoms necessary to make one glucose molecule (*left*).

glucose

Plants can use the glucose they make in the light-independent reactions as building blocks for other organic molecules, or they can break it down to access the energy held in its bonds. However, most of the glucose is converted at once to sucrose or starch by other pathways that conclude the light-independent reactions. When glucose is needed in other parts of the plant, the starch is broken down to its glucose monomers and exported from the cell.

> **Adaptations to Hot and Dry Climates** Several adaptations allow plants to live in regions where sunlight is intense and water is scarce or sporadically available. For example, a thin, waterproof coating called a cuticle limits

FIGURE 5.7 Animated! The Calvin–Benson cycle. This sketch shows a cross-section of a chloroplast with the light-independent reactions cycling in the stroma. The steps shown are a summary of six cycles of the Calvin–Benson reactions and their product, one glucose molecule. © Cengage Learning.

Calvin–Benson cycle Light-independent reactions of photosynthesis; cyclic carbon-fixing pathway that forms sugars from CO_2.

carbon fixation Process by which carbon from an inorganic source such as carbon dioxide gets incorporated into an organic molecule.

electron transfer phosphorylation Process in which electron flow through electron transfer chains sets up a hydrogen ion gradient that drives ATP formation.

A Tiny pores called stomata are visible in this close-up of a leaf. Stomata close to conserve water on hot, dry days, and this causes oxygen to accumulate inside the plant's tissues. The buildup makes sugar production inefficient in C3 plants.

B Crabgrass "weeds" overgrowing a lawn. Crabgrasses, which are C4 plants, thrive in hot, dry summers, when they easily outcompete Kentucky bluegrass and other fine-leaved C3 grasses commonly planted in residential lawns.

C The jade plant, *Crassula argentea*, and other CAM plants survive in hot deserts by opening stomata to fix carbon only at night. They run the Calvin–Benson cycle during the day, when stomata are closed.

FIGURE 5.8 Adaptations to hot, dry weather.

Credits: (a) © D. Kucharski & K. Kucharska/ Shutterstock; (b) Image courtesy msuturfweeds.net; (c) © Tamara Kulikova/ Shutterstock.

---

water loss by evaporation from aboveground plant parts. However, a cuticle also prevents gases from entering and exiting a plant by diffusing through cells at the surfaces of leaves and stems. Gases play a critical role in photosynthesis, so photosynthetic parts are often studded with tiny, closable gaps called **stomata** (singular, stoma; Figure 5.8**A**). When stomata are open, oxygen produced by the light-dependent reactions can diffuse from photosynthetic cells into the air, and carbon dioxide for the light-independent reactions can diffuse from air into the plant's photosynthetic tissues.

Plants that use only the Calvin–Benson cycle are called **C3 plants**, because a three-carbon molecule (PGA) is the first stable intermediate to form in the light-independent reactions. C3 plants typically conserve water on hot, dry days by closing their stomata. However, when stomata are closed, oxygen produced by the light-dependent reactions cannot escape from the plant, and it builds up in the plant's tissues. When oxygen is present in large amounts, it can bind to rubisco's active site in place of carbon dioxide. The result is that the plant loses carbon instead of fixing it, so sugar production slows. C3 plants compensate for this alternate pathway by making a lot of rubisco: It is the most abundant protein on Earth.

Over the past 50 to 60 million years, an additional set of reactions that compensates for rubisco's inefficiency evolved independently in many plant lineages. Plants that use the additional reactions also close stomata on dry days, but their sugar production does not decline. Examples are corn, switchgrass, and bamboo. We call these plants **C4 plants** because the first stable intermediate to form in their carbon-fixing reactions is a four-carbon compound. Such plants fix carbon twice, in two kinds of cells. In the first cell, carbon is fixed by an enzyme that does not use oxygen even when the oxygen level is high. The resulting intermediate is transported to a second cell and converted to carbon dioxide, where rubisco fixes carbon for a second time as the CO_2 enters the Calvin–Benson cycle. The extra reactions keep the CO_2 level high near rubisco, thus maintaining sugar production even during hot, dry conditions (Figure 5.8**B**).

Succulents, cacti, and other **CAM plants** have a carbon-fixing pathway that allows them to conserve water even in desert regions with extremely high daytime temperatures. CAM stands for crassulacean acid metabolism, after the Crassulaceae family of plants in which this pathway was first studied (Figure 5.8**C**). Like C4 plants, CAM plants fix carbon twice, but the reactions occur at different times rather than in different cells. Stomata on a CAM plant open at night, when lower temperatures minimize evaporative water loss. Then, carbon is fixed for the first time from CO_2 in the air. The product of the cycle, a four-carbon acid, is stored in the cell's central vacuole. When the stomata close the next day, the acid moves out of the vacuole and becomes broken down to CO_2, which is fixed for the second time when it enters the Calvin–Benson cycle.

Take-Home Message

What happens during the second stage of photosynthesis?

- During the light-independent reactions (the second stage of photosynthesis), ATP and NADPH that formed in the light-dependent reactions drive the synthesis of glucose from CO_2.

- Stomata close to conserve water when conditions are hot and dry, so oxygen builds up in the plant's tissues. In C3 plants, this buildup interferes with carbon fixation, thus slowing sugar production.

- Additional carbon-fixing pathways allow C4 and CAM plants to maintain high sugar production on hot and dry days.

5.6 Photosynthesis and Aerobic Respiration: A Global Connection

The first cells on Earth did not tap into sunlight. Like some modern archaea, these ancient organisms extracted energy and carbon from simple molecules such as methane and hydrogen sulfide—gases that were plentiful in the nasty brew that constituted Earth's early atmosphere (Figure 5.9 ❶). The first photosynthetic cells evolved about 3.2 billion years ago, probably in shallow ocean waters. Sunlight offered these organisms an essentially unlimited supply of energy, and they were very successful. Oxygen gas (O_2) released from uncountable numbers of water molecules began seeping out of photosynthetic cells, and it accumulated in the ocean and the atmosphere. From that time on, the world of life would never be the same ❷.

Molecular oxygen had been a very small component of Earth's early atmosphere before photosynthesis evolved. The new abundance of atmospheric oxygen exerted tremendous selection pressure on all organisms. Why? Oxygen gas reacts easily with metals such as enzyme cofactors, and free radicals (Section 2.2) form during those reactions. Free radicals damage biological molecules, so they are dangerous to life. The ancient cells had no way to detoxify free radicals, and most of them quickly died out. Only a few lineages persisted in deep water, muddy sediments, and other **anaerobic** (oxygen-free) habitats. New metabolic pathways that detoxified the free radicals evolved in the survivors. Cells with these pathways were the first **aerobic** organisms—they could live in the presence of oxygen. As you will see, one of the pathways, aerobic respiration, put oxygen's reactive properties to good use. **Aerobic respiration** is one of several pathways by which organisms use the energy stored in carbohydrates to make ATP. This pathway requires oxygen, and it produces carbon dioxide and water—the raw materials from which photosynthesizers make sugars. With this connection, the cycling of carbon, hydrogen, and oxygen through living things came full circle (*left*).

energy

Photosynthesis

CO₂
H₂O

glucose
O₂

Aerobic Respiration

energy

FIGURE 5.9 Then ❶ and now ❷, a view of how Earth's atmosphere was permanently altered by the evolution of photosynthesis. Photosynthesis is now the main pathway by which energy and carbon enter the web of life.
© Cengage Learning 2010.

> **Extracting Energy From Carbohydrates** Photosynthesizers capture energy from the sun, and store it in the form of carbohydrates. They and most other organisms use energy stored in carbohydrates to run the diverse reactions that sustain life. However, carbohydrates rarely participate in such reactions, so how do cells harness their energy? In order to use the energy stored in carbohydrates, cells must first transfer it to molecules such as ATP, which does participate in many of the energy-requiring reactions that a cell runs. The energy transfer occurs when cells break the bonds of a carbohydrate's carbon backbone. Energy released as those bonds are broken drives ATP synthesis.

There are a few different kinds of carbohydrate-breakdown pathways, but aerobic respiration is the one that typical eukaryotic cells use at least most of the time. Aerobic respiration yields much more ATP than other carbohydrate breakdown pathways. You and other multicelled organisms could not live without its higher yield.

aerobic Involving oxygen or occurring in its presence.
aerobic respiration Oxygen-requiring metabolic pathway that breaks down carbohydrates to produce ATP.
anaerobic Occurring in the absence of oxygen.
C3 plant Type of plant that uses only the Calvin–Benson cycle to fix carbon.
C4 plant Type of plant that minimizes photorespiration by fixing carbon twice, in two cell types.
CAM plant Type of plant that minimizes photorespiration by fixing carbon twice, at different times of day.
stomata Gaps that open on plant surfaces; allow water vapor and gases to diffuse into and out of plant tissues.

glucose

Glycolysis

2 ATP
2 NAD+

4 ATP (2 net)

2 NADH   2 pyruvate

2 NADH

2 NADH   2 acetyl–CoA   2 CO$_2$

Krebs
Cycle

4 CO$_2$

6 NADH
2 FADH$_2$

2 ATP

ATP
ATP
ATP

Electron Transfer
Phosphorylation

32 ATP

oxygen   H$_2$O

Cytoplasm

A The first stage, glycolysis, occurs in the cell's cytoplasm. Enzymes convert a glucose molecule to 2 pyruvate for a net yield of 2 ATP. 2 NAD+ combine with electrons and hydrogen ions during the reactions, so 2 NADH also form.

Mitochondrion

B The second stage occurs in mitochondria. The 2 pyruvate are converted to acetyl–CoA, which enters the Krebs cycle. CO$_2$ forms and leaves the cell. 2 ATP, 8 NADH, and 2 FADH$_2$ form during the reactions.

C The third and final stage, electron transfer phosphorylation, occurs inside mitochondria. 10 NADH and 2 FADH$_2$ donate electrons and hydrogen ions to electron transfer chains. Electron flow through the chains sets up hydrogen ion gradients that drive ATP formation. Oxygen accepts electrons at the end of the chains.

FIGURE 5.10 Animated! Aerobic respiration. The reactions start in the cytoplasm and end in mitochondria. © Cengage Learning.

> **Aerobic Respiration** Aerobic respiration begins in cytoplasm (Figure 5.10**A**). A set of reactions, which are collectively called **glycolysis**, convert one six-carbon molecule of glucose into two three-carbon molecules of pyruvate. Two ATP are invested to begin the reactions, but four ATP form by the end. Thus, we say that the net (overall) yield is two ATP per glucose. Glycolysis also produces two NADH. After glycolysis, aerobic respiration continues in two more stages that run inside mitochondria (Section 3.5).

The second stage of aerobic respiration breaks down pyruvate. The reactions begin when the pyruvate enters the inner compartment of a mitochondrion, where it becomes converted to carbon dioxide and an intermediate, acetyl–CoA. A cyclic pathway called the **Krebs cycle** then breaks down the acetyl–CoA to carbon dioxide (Figure 5.10**B**). At the end of the second stage of aerobic respiration, two molecules of pyruvate have been converted to six molecules of carbon dioxide. Two ATP have formed. Also, 8 NAD+ and 2 FAD (another coenzyme) combined with electrons and hydrogen ions, so 8 NADH and 2 FADH$_2$ formed.

At this point, one glucose molecule has been broken down completely; its six carbon atoms have exited the cell, in six CO_2:

Four ATP also formed, but almost all of the energy released from the breakdown of the glucose molecule is now carried by twelve coenzymes (eight NADH and two $FADH_2$) that accepted electrons during the reactions. These coenzymes now power the third and final stage of aerobic respiration, electron transfer phosphorylation (Figure 5.10C).

The third-stage reactions begin when the coenzymes deliver their cargo of electrons and hydrogen ions to electron transfer chains in the inner mitochondrial membrane (Figure 5.11 ❶). As the electrons pass through the electron transfer chains, they give up energy bit by bit. Molecules in the chains use that energy to actively transport hydrogen ions from the inner mitochondrial compartment to the outer one ❷. At the end of the chains, oxygen molecules accept electrons and combine with hydrogen ions, so water forms ❸. Aerobic respiration, which literally means "taking a breath of air," refers to oxygen as the final electron acceptor in this pathway. Every breath you take provides your aerobically respiring cells with a fresh supply of oxygen.

As hydrogen ions are pumped from the inner to the outer compartment of the mitochondrion, a hydrogen ion concentration gradient forms across the inner mitochondrial membrane.

The gradient attracts the hydrogen ions back toward the inner mitochondrial compartment. The ions cross the inner membrane by flowing through the interior of ATP synthases ❹. The flow causes these transport proteins to attach a phosphate group to ADP, so ATP forms ❺.

Thirty-two ATP typically form in the third stage of aerobic respiration. Add four ATP from the first and second stages, and the overall yield from the breakdown of one glucose molecule is thirty-six ATP.

Take-Home Message

How do cells access the chemical energy stored in carbohydrates?

- Eukaryotic cells typically convert the chemical energy of carbohydrates to the chemical energy of ATP by the oxygen-requiring pathway of aerobic respiration.
- Aerobic respiration begins in cytoplasm with glycolysis, and ends in the mitochondrion with electron transfer phosphorylation.
- A typical net yield of aerobic respiration is thirty-six ATP per glucose. Carbon dioxide and water also form.

FIGURE 5.11 The final stage of aerobic respiration: electron transfer phosphorylation.

❶ Coenzymes deliver electrons to electron transfer chains in the inner mitochondrial membrane.

❷ Energy lost by the electrons as they move through the transfer chain causes hydrogen ions to be pumped from the inner to the outer compartment. A hydrogen ion gradient forms across the inner mitochondrial membrane.

❸ Oxygen (O_2) accepts electrons and hydrogen ions at the end of mitochondrial electron transfer chains, so water forms.

❹ Hydrogen ions flow back to the inner compartment through ATP synthases. The flow drives the formation of ATP from ADP and phosphate ❺.

Figure It Out: Which other pathway discussed in this chapter involves electron transfer phosphorylation?

Answer: The light-dependent reactions of photosynthesis

© Cengage Learning.

Aerobic respiration literally means "taking a breath of air."

glycolysis Set of reactions in which glucose is broken down to two pyruvate for a net yield of two ATP.

Krebs cycle Cyclic pathway that helps break down pyruvate to carbon dioxide during aerobic respiration.

5.7 Fermentation

Most types of eukaryotic cells use aerobic respiration exclusively, or they use it most of the time. Many bacteria, archaea, protists, and some eukaryotic cells can harvest energy from carbohydrates by anaerobic pathways of **fermentation**. Fermentation and aerobic respiration begin with the same reactions—glycolysis—in the cytoplasm. Unlike aerobic respiration, fermentation pathways also come to an end in cytoplasm. These pathways convert pyruvate to other molecules, but do not fully break it down to carbon dioxide and water as occurs in aerobic respiration. Electrons do not flow through electron transfer chains, so no more ATP forms. However, electrons are removed from NADH, so NAD$^+$ is regenerated. Regenerating this coenzyme allows glycolysis—along with the small ATP yield it offers—to continue. Thus, the net ATP yield of fermentation consists of the two ATP that form in glycolysis (see Figure 5.10**A**). Electrons are accepted by organic molecules (not oxygen) at the end of the reactions, so fermentation pathways do not require oxygen in order to proceed.

Fermentation helps cells of aerobic species produce ATP under anaerobic conditions. It also provides enough energy to sustain many anaerobic species, including bacteria, fungi, and single-celled protists that inhabit sea sediments, animal guts, improperly canned food, sewage treatment ponds, or deep mud. Some of these organisms, including the bacteria that cause botulism, cannot tolerate aerobic conditions, and will die when exposed to oxygen.

> **Alcoholic Fermentation** In **alcoholic fermentation**, the pyruvate from glycolysis is converted to ethyl alcohol, or ethanol. First, 3-carbon pyruvate is split into carbon dioxide and 2-carbon acetaldehyde. Then, electrons and hydrogen are transferred from NADH to the acetaldehyde, forming NAD$^+$ and ethanol (Figure 5.12**A**). Alcoholic fermentation in a fungus, *Saccharomyces cerevisiae*, sustains these yeast cells as they grow and reproduce. It also helps us produce beer, wine, and bread (Figure 5.12**B**).

Beer brewers typically use germinated, roasted, and crushed barley as a carbohydrate source for *Saccharomyces* fermentation. Ethanol produced by the fermenting yeast cells makes the beer alcoholic, and CO_2 makes it bubbly. Hops flowers add flavor and help preserve the finished product. Winemakers start with crushed grapes for *Saccharomyces* fermentation. The yeast cells convert sugars in the grape juice to ethanol.

Bakers take advantage of alcoholic fermentation by *Saccharomyces* cells to make bread from flour, which contains starches and a protein called gluten. When flour is kneaded with water, the gluten polymerizes in long, interconnected strands that make the resulting dough stretchy and resilient. Yeast cells in the dough produce CO_2 as they ferment the starches. The gas accumulates in bubbles that are trapped by the gluten mesh. As the bubbles expand, they cause the dough to rise. The ethanol produced by the fermentation reactions evaporates during baking.

> **Lactate Fermentation** In **lactate fermentation**, the electrons and hydrogen ions carried by NADH are transferred directly to pyruvate. This reac-

NADH   NAD$^+$

pyruvate   carbon dioxide   acetaldehyde   ethanol

A The last stages of alcoholic fermentation produce CO_2, ethanol, and NAD$^+$.

B One product of alcoholic fermentation in *Saccharomyces* cells (ethanol) makes beer alcoholic; another (CO_2) makes it bubbly. Holes in bread are pockets where CO_2 released by fermenting *Saccharomyces* cells accumulated in the dough. The micrograph shows budding *Saccharomyces* cells.

FIGURE 5.12 Animated!
Alcoholic fermentation.

Credits: (a) © Cengage Learning; (b) top left, © Elena Boshkovska/ Shutterstock; top right, © optimarc/ Shutterstock; bottom, Dr. Dennis Kunkel/ Visuals Unlimited.

JupiterImages Corporation.

tion converts pyruvate to 3-carbon lactate and also converts NADH to NAD+ (Figure 5.13**A**).

Some organisms that carry out lactate fermentation spoil food, but we use others to preserve it. For instance, *Lactobacillus* bacteria break down lactose in milk by fermentation. We use this bacteria to produce dairy products such as buttermilk, cheese, and yogurt. Yeast species ferment and preserve pickles, corned beef, sauerkraut, and kimchi.

Animal skeletal muscles, which move bones, consist of cells fused as long fibers. The fibers differ in how they make ATP. Red fibers have many mitochondria and produce ATP by aerobic respiration. These fibers sustain prolonged activity such as marathon runs. They are red because they contain an abundance of myoglobin, a protein that stores oxygen for aerobic respiration in muscle tissue (Figure 5.13**B**). White muscle fibers contain few mitochondria and no myoglobin, so they do not carry out a lot of aerobic respiration. Instead, they make most of their ATP by lactate fermentation. This pathway makes ATP quickly but not for long, so it is useful for quick, strenuous activities such as weight lifting or sprinting (Figure 5.13**C**). The low ATP yield does not support prolonged activity. That is one reason why chickens cannot fly very far: Their flight muscles consist mostly of white fibers (thus, the "white" breast meat). Chickens fly only in short bursts. More often, they walk or run. Their leg muscles consist mostly of red muscle fibers, the "dark meat." Most human muscles consist of a mixture of white and red fibers, but the proportions vary among muscles and among individuals. Great sprinters tend to have more white fibers. Great marathon runners tend to have more red fibers. Section 20.4 offers a closer look at skeletal muscle fibers and how they work.

Take-Home Message

What is fermentation?

- ATP can form by carbohydrate breakdown in anaerobic fermentation pathways.
- Fermentation pathways begin with glycolysis in cytoplasm, and they also end in cytoplasm.
- The end product of alcoholic fermentation is ethanol. The end product of lactate fermentation is lactate. Both pathways have a net yield of two ATP per glucose molecule. The ATP forms during glycolysis.
- Fermentation reactions regenerate the coenzyme NAD+, without which glycolysis (and ATP production) would stop.

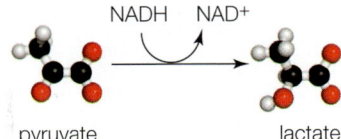

$$\text{NADH} \quad \text{NAD}^+$$

pyruvate → lactate

A The last stage of lactate fermentation produces lactate and NAD+.

B Lactate fermentation occurs in white muscle fibers, visible in this cross-section of human thigh muscle. The red fibers, which make ATP by aerobic respiration, sustain endurance activities.

C Intense activity such as sprinting quickly depletes oxygen in muscles. Under anaerobic conditions, ATP is produced mainly by lactate fermentation in white muscle fibers. Fermentation does not make enough ATP to sustain this type of activity for long.

FIGURE 5.13 Lactate fermentation.

Credits: (a) © Cengage Learning; (b) © William MacDonald, M.D.; (c) © sportgraphic/Shutterstock.

5.8 Alternative Energy Sources in Food

Glycolysis converts glucose to pyruvate, and the Krebs cycle reactions transfer electrons from the pyruvate to coenzymes. Removing electrons from a molecule is called oxidation. Oxidizing an organic molecule can break the covalent bonds of its carbon backbone. Aerobic respiration fully oxidizes glucose, completely dismantling it carbon by carbon.

Cells also dismantle other organic molecules by oxidizing them. Complex carbohydrates, fats, and proteins in food can be converted to molecules that enter glycolysis or the Krebs cycle. As in glucose metabolism, many coenzymes accept electrons, and the energy of the electrons they carry ultimately drives the synthesis of ATP in electron transfer phosphorylation.

alcoholic fermentation Anaerobic carbohydrate breakdown pathway that produces ATP, CO_2, and ethanol.
fermentation An anaerobic pathway by which cells harvest energy from carbohydrates.
lactate fermentation Anaerobic carbohydrate breakdown pathway that produces ATP and lactate.

› Complex Carbohydrates The digestive system breaks down starch and other complex carbohydrates to monosaccharide subunits (Figure 5.14**A**). These sugars are quickly taken up by cells for glycolysis ❶. Unless ATP is being used quickly, its concentration rises in the cytoplasm. A high concentration of ATP causes sugars to be diverted away from glycolysis and into a pathway that forms glycogen (Section 2.7). Liver and muscle cells especially favor the conversion of glucose to glycogen, and these cells contain the body's largest stores of it. Between meals, the liver maintains the glucose level in blood by converting stored glycogen to glucose.

What happens if you eat too many carbohydrates? When the blood level of glucose gets too high, acetyl–CoA is diverted away from the Krebs cycle and into a pathway that makes fatty acids. That is why excess dietary carbohydrate ends up as fat.

› Fats Remember from Section 2.8 that a fat molecule has a glycerol head and one, two, or three fatty acid tails. Cells dismantle fat molecules by first breaking the bonds that connect the glycerol with the fatty acids

starch (a complex carbohydrate)　　　　glucose

A Complex carbohydrates are broken down to their monosaccharide subunits, which can enter glycolysis ❶.

a fat

glycerol head

fatty acid tails

B Fats are broken down by separating the glycerol head from the fatty acid tails. The fatty acids are converted to acetyl–CoA ❷, and the glycerol is converted to PGAL ❸.

alanine (an amino acid)　　　　pyruvate

C Amino acids are converted to acetyl–CoA, pyruvate, or an intermediate of the Krebs cycle ❹.

FIGURE 5.14 Animated! A variety of organic compounds from food can enter the reactions of aerobic respiration.

© Cengage Learning.

Food

| Fats | Complex Carbohydrates | Proteins |

fatty acids　　glycerol　　glucose, other simple sugars　　amino acids

❷　　❸　　❶　　❹

acetyl–CoA　　PGAL　　　　acetyl–CoA

Glycolysis

NADH　　pyruvate

intermediate of Krebs cycle

Krebs Cycle

NADH, FADH$_2$

ATP

ATP

ATP

Electron Transfer Phosphorylation

(Figure 5.14**B**). Nearly all cells in the body oxidize free fatty acids by splitting their long backbones into two-carbon fragments. These fragments are converted to acetyl–CoA, which can enter the Krebs cycle ❷. The glycerol is converted to an intermediate of glycolysis ❸.

On a per carbon basis, fats are a richer source of energy than carbohydrates. Carbohydrate backbones have many oxygen atoms. A fat's long fatty acid tails are carbon chains that typically have no oxygen atoms bonded to them, so they have a longer way to go to become oxidized—more reactions are required to fully break them down. Coenzymes accept electrons in these reactions. The more coenzymes that accept electrons, the more electrons can be delivered to the ATP-forming machinery of electron transfer phosphorylation.

© shabaneiro/ Shutterstock.

> **Proteins** Some enzymes in your digestive system split dietary proteins into their amino acid subunits, which are absorbed into the bloodstream. Cells can use amino acids to build proteins or other molecules, but when you eat more protein than your body needs, amino acids become broken down further. The amino group is removed, and it becomes ammonia (NH_3), a waste product that the body eliminates in urine. The carbon backbone is split, and acetyl–CoA, pyruvate, or an intermediate of the Krebs cycle forms, depending on the amino acid (Figure 5.14**C**). Cells can divert these organic molecules into the Krebs cycle ❹.

Take-Home Message

Can the body use organic molecules other than glucose for energy?

- Complex carbohydrates, fats, and proteins can be broken down to yield ATP.
- First the digestive system and then individual cells convert molecules in food into intermediates of glycolysis or the Krebs cycle.

5.9 A Burning Concern (revisited)

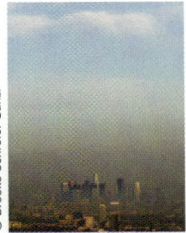
© Brooke Schreier Ganz.

Tiny pockets of Earth's ancient atmosphere remain in Antarctica, preserved in snow and ice that have been accumulating in layers, year after year, for millions of years (Figure 5.15). Air and dust trapped in each layer reveal the composition of the atmosphere that prevailed when the layer formed. These layers tell us that the atmospheric CO_2 level was relatively stable for about 10,000 years before the industrial revolution began in the mid-1800s. Since then, the CO_2 level has been steadily rising. Today, the atmospheric CO_2 level is higher than it has been for *15 million years*.

Atmospheric carbon dioxide affects Earth's climate, so this increase in CO_2 is contributing to global climate change. We are seeing a warming trend that mirrors the increase in CO_2 levels: Earth is now the warmest it has been for 12,000 years. The trend is affecting biological systems everywhere. Life cycles are changing: Birds are laying eggs earlier; plants are flowering earlier than usual; mammals are hibernating for shorter periods. Migration patterns and habitats are also changing. These changes may be too fast for many species, and the rate of extinctions is rising.

Under normal circumstances, extra carbon dioxide stimulates photosynthesis, which means extra carbon dioxide uptake. However, changes in temperature and moisture patterns as a result of global warming are offsetting this benefit because they are proving harmful to plants and other photosynthetic organisms.

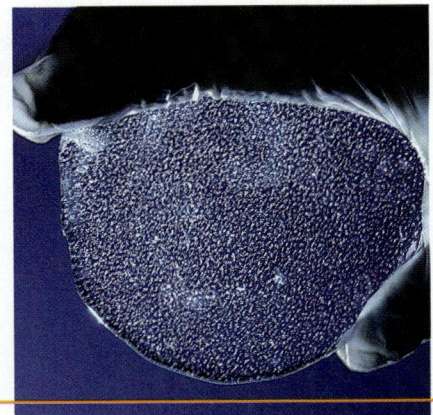
FIGURE 5.15 A slice of ancient history. Air bubbles trapped in Antarctic ice core slices such as this one are samples of Earth's atmosphere as it was when the ice formed. The deeper the slice, the older the air bubbles.

www.photo.antarctica.ac.uk.

WHERE YOU ARE GOING . . .

Carbon dating is covered in Section 11.4; carbon- and energy-harvesting strategies, in Section 13.5; the carbon cycle, in Section 17.6. Chapter 14 discusses evolutionary adaptations of plants; Chapters 27 and 28, plant structure and function. In Chapter 18, you will see more human impacts on the biosphere. Muscle function returns in Chapter 20; how the body acquires oxygen for respiration in Chapter 21; and digestion and nutrition in Chapter 23.

Summary

Section 5.1 By the metabolic pathway of **photosynthesis**, the energy of light is used to build glucose from water and carbon dioxide. Plants and other photosynthetic producers remove CO_2 from the atmosphere; aerobic respiration by producers and consumers puts it back. Humans disrupt the balance of these two processes by burning fossil fuels, which adds extra CO_2 to the atmosphere. The resulting imbalance is contributing to global warming.

Section 5.2 Photosynthetic **pigments** absorb visible light of particular **wavelengths** for photosynthesis. Light that is not absorbed is reflected as a pigment's characteristic color. The main photosynthetic pigment, **chlorophyll _a_**, absorbs violet and red light, so it appears green. Accessory pigments absorb additional wavelengths.

Section 5.3 In chloroplasts, the light-dependent reactions of photosynthesis occur at a much-folded **thylakoid membrane**. The membrane forms a continuous compartment in the chloroplast's **stroma**, where the light-independent reactions occur.

The following diagram summarizes the two stages of photosynthesis:

Section 5.4 Photosynthetic pigments in the thylakoid membrane absorb light energy and pass it to photosystems, which then release electrons. The electrons flow through electron transfer chains in the thylakoid membrane, and end up in NADPH. Molecules of the electron transfer chain use energy released by the electrons to set up a hydrogen ion gradient across the thylakoid membrane. The ions flow back across the membrane through ATP synthases, causing these proteins to produce ATP (a process called **electron transfer phosphorylation**). Photosynthesis releases oxygen because a photosystem replaces lost electrons by pulling them from water molecules, which break apart as a result.

Section 5.5 The ATP and NADPH that form in the light-dependent reactions power the light-independent reactions of the **Calvin–Benson cycle**, which builds glucose from CO_2. Closing **stomata** allows a plant to conserve water, but it also limits gas exchange. Oxygen buildup in plant tissues reduces the efficiency of glucose production in **C3 plants**. Additional **carbon fixation** reactions in **C4 plants** and **CAM plants** make sugar production more efficient on hot, dry days.

Section 5.6 Photosynthesis changed the composition of Earth's early atmosphere, with profound effects on life's evolution. Organisms that could not tolerate the increased oxygen content persisted only in **anaerobic** habitats. Oxygen-detoxifying pathways evolved, allowing organisms to thrive in **aerobic** conditions.

Most modern organisms convert the chemical energy of carbohydrates to the chemical energy of ATP by oxygen-requiring **aerobic respiration**. In eukaryotes, this pathway starts with **glycolysis** in cytoplasm, and ends in mitochondria. Coenzymes pick up electrons in aerobic respiration's first two stages, glycolysis and the **Krebs cycle**. The energy of those electrons drives ATP synthesis in the third stage, electron transfer phosphorylation. At the end of the electron transfer chains, oxygen accepts electrons and hydrogen ions, so water forms.

Section 5.7 Anaerobic **fermentation** pathways begin with glycolysis and finish in the cytoplasm. A molecule other than oxygen accepts electrons at the end of these reactions. The end product of **alcoholic fermentation** is ethyl alcohol, or ethanol. The end product of **lactate fermentation** is lactate. The final steps of fermentation serve to regenerate NAD^+, which is required for glycolysis to continue, but they produce no ATP. Thus, the breakdown of one glucose molecule in either alcoholic or lactate fermentation yields only the two ATP that form in glycolysis.

Section 5.8 In humans and other organisms, the simple sugars from carbohydrate breakdown, the glycerol and fatty acids from fat breakdown, and the carbon backbones of amino acids from protein breakdown may enter aerobic respiration at various reaction steps.

Self-Quiz   Answers in Appendix I

1. In a land plant, most of the carbon dioxide used in photosynthesis comes from _____ .
   - a. glucose
   - b. the atmosphere
   - c. water
   - d. soil

2. _____ is/are the main energy source that drives photosynthesis.
   - a. Sunlight
   - b. Hydrogen ions
   - c. Oxygen
   - d. Carbon dioxide

3. In the light-dependent reactions, _____ .
   - a. carbon dioxide is fixed
   - b. ATP forms
   - c. CO_2 accepts electrons
   - d. sugars form

4. When a photosystem absorbs light, _____ .
   - a. sugar phosphates are produced
   - b. electrons are transferred to ATP
   - c. RuBP accepts electrons
   - d. it ejects electrons

5. The atoms in the oxygen molecules released during photosynthesis come from _____ .
   - a. glucose
   - b. carbon dioxide
   - c. water
   - d. hydrogen ions

6. Is the following statement true or false? Plants make all of their ATP by photosynthesis.

Digging Into Data

Energy Efficiency of Biofuel Production From Corn, Soy, and Prairie Grasses

Most biomass currently used for biofuel production in the United States consists of food crops—mainly corn, soybeans, and sugarcane. In 2006, David Tilman and his colleagues published the results of a 10-year study comparing the net energy output of various biofuels. The researchers grew a mixture of native perennial grasses without irrigation, fertilizer, pesticides, or herbicides, in sandy soil that was so depleted by intensive agriculture that it had been abandoned. They measured the usable energy in biofuels made from the grasses, from corn, and from soy. They also measured the energy it took to grow and produce each kind of biofuel (Figure 5.16).

1. About how much energy did ethanol produced from one hectare of corn yield? How much energy did it take to grow and produce that ethanol?

2. Which biofuel tested had the highest ratio of energy output to energy input?

3. Which of the three crops would require the least amount of land to produce a given amount of biofuel energy?

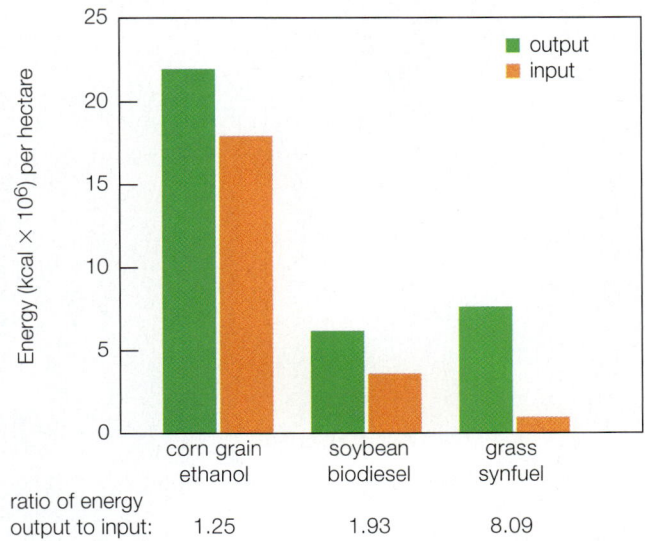

ratio of energy output to input: 1.25   1.93   8.09

FIGURE 5.16 Energy inputs and outputs of biofuels from corn and soy grown on fertile farmland, and grassland plants grown in infertile soil. One hectare is about 2.5 acres.

© Cengage Learning.

7. After photosynthesis evolved, its by-product, _____ , accumulated and changed the atmosphere.

8. Glycolysis starts and ends in the _____ .
   a. nucleus
   b. mitochondrion
   c. plasma membrane
   d. cytoplasm

9. The Calvin–Benson cycle starts when _____ .
   a. light is available
   b. carbon dioxide is attached to RuBP
   c. electrons leave a photosystem

10. In eukaryotes, aerobic respiration is completed in the _____ .
    a. nucleus
    b. mitochondrion
    c. plasma membrane
    d. cytoplasm

11. In eukaryotes, fermentation is completed in the _____ .
    a. nucleus
    b. mitochondrion
    c. plasma membrane
    d. cytoplasm

12. In the third stage of aerobic respiration, _____ is the final acceptor of electrons.
    a. water   b. hydrogen   c. oxygen   d. NADH

13. Which of the following is *not* produced by an animal muscle cell operating under anaerobic conditions?
    a. heat
    b. lactate
    c. ATP
    d. NAD$^+$
    e. pyruvate
    f. all are produced

14. Hydrogen ion flow drives ATP formation during _____ .
    a. photosynthesis
    b. aerobic respiration
    c. fermentation
    d. the Calvin–Benson cycle
    e. both a and b
    f. all of the above

15. Your body cells can use _____ as an alternative energy source when glucose is in short supply.
    a. fatty acids
    b. glycerol
    c. amino acids
    d. all of the above

16. Match the term with the best description.
    ____ pyruvate
    ____ fermentation
    ____ mitochondrion
    ____ pigment
    ____ carbon dioxide
    ____ rubisco
    ____ chloroplast

    a. no oxygen required
    b. site of photosynthesis
    c. product of glycolysis
    d. aerobic respiration ends here
    e. carbon-fixing enzyme
    f. like an antenna
    g. big in the atmosphere

Critical Thinking

1. About 200 years ago, Jan Baptista van Helmont wanted to know where growing plants get the materials necessary for increases in size. He planted a tree seedling weighing 5 pounds in a barrel filled with 200 pounds of soil and then watered the tree regularly. After five years, the tree weighed 169 pounds, 3 ounces, and the soil weighed 199 pounds, 14 ounces. Because the tree had gained so much weight and the soil had lost so little, he concluded that the tree had gained all of its additional weight by absorbing the water he had added to the barrel, but of course he was incorrect. What really happened?

2. As you learned, membranes impermeable to hydrogen ions are required for electron transfer phosphorylation. Membranes in mitochondria serve this purpose in eukaryotes. Prokaryotic cells do not have this organelle, but they do make ATP by electron transfer phosphorylation. How do you think they do it, given that they have no mitochondria?

3. While looking into an aquarium, you see bubbles coming from an aquatic plant (*right*). What are the bubbles?

© E. R. Degginger.

DNA Structure and Function

6

6.1 A Hero Dog's Golden Clones

On September 11, 2001, an off-duty Canadian police officer, Constable James Symington, drove his search dog Trakr from Nova Scotia to Manhattan. Within hours of arriving, the dog led rescuers to the area where the final survivor of the World Trade Center attacks was buried. She had been clinging to life, pinned under rubble from the building where she had worked. Symington and Trakr helped with the search and rescue efforts for three days nonstop, until Trakr collapsed from smoke and chemical inhalation, burns, and exhaustion (Figure 6.1).

Trakr survived the ordeal, but later lost the use of his limbs from a degenerative neurological disease probably linked to toxic smoke exposure at Ground Zero. The hero dog died in April 2009, but his DNA lives on in his genetic copies—his **clones**. Symington's essay about Trakr's superior nature and abilities as a search and rescue dog won the Golden Clone Giveaway, a contest to find the world's most clone-worthy dog. Trakr's DNA was shipped to Korea, where it was inserted into donor dog eggs, which were then implanted into surrogate mother dogs. Five puppies, all clones of Trakr, were delivered to Symington in July 2009.

Many adult animals have been cloned besides Trakr, but cloning mammals is still unpredictable and far from routine. Typically, less than 2 percent of the implanted embryos result in a live birth. Of the clones that survive, many have serious health problems.

Why the difficulty? Even though all cells of an individual inherit the same DNA, an adult cell uses only a fraction of it compared with an embryonic cell. To make a clone from an adult cell, researchers must reprogram its DNA to function like the DNA of an egg. Even though we are getting better at doing that, we still have a lot to learn.

FIGURE 6.1 James Symington and his dog Trakr at Ground Zero, September 2001.
© James Symington.

So why do we keep trying? The potential benefits are enormous. Already, cells of cloned human embryos are helping researchers unravel the molecular mechanisms of human genetic diseases. Such cells may one day be induced to form replacement tissues or organs for people with incurable diseases. Endangered animals might be saved from extinction; extinct animals may be brought back. Livestock and pets are already being cloned commercially.

Perfecting methods for cloning animals brings us closer to the possibility of cloning humans, both technically and ethically. For example, if cloning a lost cat for a grieving pet owner is acceptable, why would it not be acceptable to clone a lost child for a grieving parent? Different people have very different answers to such questions, so controversy over cloning continues to rage even as techniques improve. Understanding the basis of heredity—what DNA is and how it works—will inform your own opinions about cloning.

A Proteins structurally organize DNA into chromosomes.

B A human chromosome.

C DNA, the substance, extracted from human cells.

6.2 Chromosomes

Stretched out end to end, the DNA in a single human cell would be about 2 meters (6.5 feet) long. How can that much DNA pack into a nucleus that is less than 10 micrometers in diameter? Such tight packing is possible because proteins associate with the DNA and help keep it organized. In cells, DNA molecules and their associated proteins form structures called **chromosomes** (Figure 6.2). Each DNA molecule consists of two strands twisted into a double helix ➊ (more about the double helix in the next section). At the first level of chromosomal organization, the DNA wraps twice at regular intervals around "spools" of proteins called **histones** ➋. These DNA–histone spools appear in micrographs as beads on a string. Interactions among histones and other proteins twist the spooled DNA into a tight fiber ➌. This fiber coils, and then it coils again into a hollow cylinder that looks a bit like an old-style telephone cord ➍.

During most of the cell's life, each chromosome consists of one DNA molecule. When the cell prepares to divide, it duplicates all of its chromosomes, so that both of its offspring will get a full set. After duplication, each chromosome consists of two DNA molecules, or **sister chromatids**, attached to one another at a constricted region called the **centromere**:

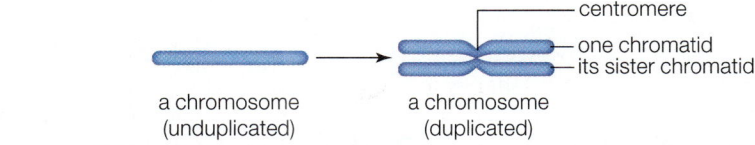

a chromosome
(unduplicated)

centromere
one chromatid
its sister chromatid

a chromosome
(duplicated)

As you will see in Chapter 8, the cell's duplicated chromosomes condense into their familiar "X" shapes ➎ just before the process of division begins.

❭ Chromosome Number The DNA of a eukaryotic cell is divided up among several chromosomes that differ in length and shape. The total number of chromosomes—the **chromosome number**—is a characteristic of the species. For example, the chromosome number of oak trees is 12, so the nucleus of a cell from an oak tree has 12 chromosomes. The chromosome number of human body cells is 46, so human body cells have 46 chromosomes.

FIGURE 6.2 **Animated!** Chromosome structure.

➊ The DNA molecule itself has two strands twisted into a double helix.

➋ At regular intervals, the DNA molecule (*blue*) wraps around a core of histone proteins (*purple*).

➌ The DNA and proteins associated with it twist tightly into a fiber.

➍ The fiber coils and then coils again to form a hollow cylinder.

➎ At its most condensed, a duplicated chromosome has an X shape.

Figure It Out: What is the yellow structure in the upper *left* corner of (A)?   *Answer: A cell nucleus*

Credits: (a) © Cengage Learning 2010; (b) Andrew Syred/ Photo Researchers, Inc.; (c) Patrick Landmann/ Photo Researchers, Inc.

centromere Of a duplicated eukaryotic chromosome, constricted region where sister chromatids attach to each other.

chromosome A structure that consists of DNA and associated proteins; carries part or all of a cell's genetic information.

chromosome number The sum of all chromosomes in a cell of a given species.

clone Genetically identical copy of an organism.

histone Type of protein that structurally organizes eukaryotic chromosomes.

sister chromatid One of two attached members of a duplicated eukaryotic chromosome.

© Cengage Learning.

Actually, human body cells have two sets of 23 chromosomes. Having two sets of chromosomes means these cells are **diploid**, or *2n*. A technique called karyotyping reveals an individual's diploid complement of chromosomes. With this technique, cells taken from the individual are treated to make the chromosomes condense, and then stained so the chromosomes can be distinguished under a microscope. A micrograph of a single cell is digitally rearranged so the images of the chromosomes are lined up by centromere location, and arranged according to size, shape, and length. The finished array constitutes the individual's **karyotype** (Figure 6.3**A**). A karyotype shows how many chromosomes are in the individual's cells, and can also reveal major structural abnormalities.

> **Types of Chromosomes** All except one pair of a diploid cell's chromosomes are **autosomes**, which are the same in both females and males. The two autosomes of each pair have the same length, shape, and centromere location. They also hold information about the same traits. Think of them as two sets of books on how to build a house. Your father gave you one set. Your mother had her own ideas about wiring, plumbing, and so on. She gave you an alternate set that says slightly different things about many of those tasks.

Members of a pair of **sex chromosomes** differ between females and males. The differences determine an individual's sex. The sex chromosomes of humans are called X and Y. The body cells of typical human females have two X chromosomes (XX); those of typical human males have one X and one Y chromosome (XY). XX females and XY males are the rule among fruit flies, mammals, and many other animals, but there are other patterns. Female butterflies, moths, birds, and certain fishes have two nonidentical sex chromosomes (Figure 6.3**B**); the two sex chromosomes of males are identical. Environmental factors (not sex chromosomes) determine sex in some species of invertebrates, turtles, and frogs. As an example, the temperature of the sand in which sea turtle eggs are buried determines the sex of the hatchlings.

A Karyotype of a female human, with identical sex chromosomes (XX).

B Karyotype of a female chicken, with nonidentical sex chromosomes (ZW).

FIGURE 6.3 Animated! Karyotypes.

Credits: (a) © University of Washington Department of Pathology; (b) With kind permission from Springer Science+Business Media: *Chromosome Research*, Volume 17, Number 1, 99 113, DOI: 10.1007/s10577-009-9021-6; *Avian comparative genomics: reciprocal chromosome painting between domestic chicken (Gallus gallus) and the stone curlew (Burhinus oedicnemus, Charadriiformes)—An atypical species with low diploid number*; Wenhui Nie, Patricia C. M. O'Brien, Bee L. Ng, Beiyuan Fu, Vitaly Volobouev, Nigel P. Carter, Malcolm A. Ferguson-Smith and Fengtang Yang; fig 2a.

Take-Home Message

What are chromosomes?

- In cells, DNA and associated proteins are organized as chromosomes.
- A eukaryotic cell's DNA is divided among some characteristic number of chromosomes, which differ in length and shape.
- Members of a pair of sex chromosomes differ between males and females. Chromosomes that are the same in males and females are called autosomes.

6.3 Fame, Glory, and DNA Structure

A strand of DNA is a polymer of nucleotides that have been linked into a chain (Section 2.10). Only four types of nucleotides make up DNA: Each consists of a five-carbon sugar, three phosphate groups, and a nitrogen-containing base after which it is named (Figure 6.4). Just how those four nucleotides—adenine (A), thymine (T), cytosine (C), and guanine (G)—are arranged in DNA was a puzzle that took over 50 years to solve. As molecules go, DNA is gigantic, and chromosomal DNA has a complex structural organization. Both factors made the molecule difficult to work with given the laboratory methods at the time.

base

three phosphate groups

adenine (A)
deoxyadenosine triphosphate

thymine (T)
deoxythymidine triphosphate

cytosine (C)
deoxycytidine triphosphate

guanine (G)
deoxyguanosine triphosphate

FIGURE 6.4 Animated! The four nucleotides that make up DNA. Each kind has three phosphate groups, a deoxyribose sugar (*orange*), and a nitrogen-containing base (*blue*) after which it is named.

© Cengage Learning.

Clues about DNA's structure started coming together around 1950, when Erwin Chargaff, one of many researchers trying to determine the structure of DNA, made two important discoveries about the molecule. First, the amounts of adenine and thymine are identical, as are the amounts of cytosine and guanine (A = T and C = G). We call this discovery Chargaff's first rule. Chargaff's second discovery, or rule, is that the DNA of different species differs in its proportions of adenine and guanine.

Meanwhile, American biologist James Watson and British biophysicist Francis Crick had been sharing ideas about the structure of DNA. The helical pattern of secondary structure that occurs in many proteins (Section 2.9) had just been discovered, and Watson and Crick suspected that the DNA molecule was also a helix. The two spent many hours arguing about the size, shape, and bonding requirements of the four kinds of nucleotides that make up DNA. They pestered chemists to help them identify bonds they might have overlooked, fiddled with cardboard cutouts, and made models from scraps of metal connected by suitably angled "bonds" of wire.

Biochemist Rosalind Franklin had also been working on the structure of DNA. Like Crick, Franklin specialized in x-ray crystallography, a technique in which x-rays are directed through a purified and crystallized substance. Atoms in the substance's molecules scatter the x-rays in a pattern that can be captured as an image. Researchers can use the pattern to calculate the size, shape, and spacing between any repeating elements of the molecules—all of which are details of molecular structure.

Franklin had already used x-ray crystallography to solve the complex and unorganized structure of coal. She had been told she would be the only one in her department working on the structure of DNA, so she did not know that Maurice Wilkins was already doing the same thing just down the hall.

Wilkins and Franklin had been given identical samples of carefully prepared DNA. Franklin's meticulous work with her sample yielded the first clear x-ray diffraction image of DNA as it occurs inside cells (Figure 6.5), and she gave a presentation on this work in 1952. DNA, she said, had two chains twisted into a double helix, with a backbone of phosphate groups on the outside, and bases arranged in an unknown way on the inside. She had calculated DNA's diameter, the distance between its chains and between its bases, the angle of the helix,

FIGURE 6.5 Rosalind Franklin and her x-ray diffraction image of DNA. This image was the final link in a long chain of clues that led to the discovery of DNA's structure.

NLM

autosome Any chromosome other than a sex chromosome.
diploid Having two of each type of chromosome characteristic of the species (2n).
karyotype Image of an individual's complement of chromosomes arranged by size, length, shape, and centromere location.
sex chromosome Member of a pair of chromosomes that differs between males and females.

and the number of bases in each coil. Crick, with his crystallography background, would have recognized the significance of the work—if he had been there. Watson was in the audience but he was not a crystallographer, and he did not understand the implications of Franklin's x-ray diffraction image or her calculations.

Franklin started to write a research paper on her findings. Meanwhile, and perhaps without her knowledge, Watson reviewed Franklin's x-ray diffraction image with Wilkins, and Watson and Crick read a report detailing Franklin's unpublished data. Franklin's data provided Watson and Crick with the last piece of the DNA puzzle. In 1953, they put together all of the clues that had been accumulating for the last fifty years and built the first accurate model of the DNA molecule. On April 25, 1953, Rosalind Franklin's work appeared third in a series of articles about the structure of DNA in the journal *Nature*. Wilkins's research paper was the second article in the series. The work of Franklin and Wilkins supported with experimental evidence Watson and Crick's theoretical model, which was presented in the first article.

> **The Double Helix** Watson and Crick proposed that a DNA molecule consists of two chains (or strands) of nucleotides, running in opposite directions and coiled into a double helix (Figure 6.6**A**). Covalent bonds between the sugar of one nucleotide and the phosphate of the next form the sugar–phosphate backbone of each chain. Hydrogen bonds between the internally positioned bases hold the two strands together. Only two kinds of base pairings form: A to T, and G to C, which explains the first of Chargaff's rules.

Most scientists had assumed (incorrectly) that the bases had to be on the outside of the helix, because they would be more accessible to DNA-copying enzymes that way. You will see in the next section how these enzymes access the nucleotide bases on the inside of DNA's double helix.

Dozens of scientists contributed to the discovery of DNA's structure, but only three received recognition from the general public for their work. Rosalind Franklin died in 1958. Because the Nobel Prize is not given posthumously, she did not share in the 1962 honor that went to Watson, Crick, and Wilkins (Figure 6.6**B**,**C**) for the discovery of the structure of DNA.

← sugar–phosphate backbone

← hydrogen bonds link internally positioned nucleotide bases

A Structure of DNA, as illustrated by a composite of three different models. The two sugar–phosphate backbones coil in a helix around internally positioned bases.

FIGURE 6.6 Animated! DNA structure.
Figure It Out: What do the yellow balls represent?
Answer: Phosphate groups

Credits: (a) PDB ID: 1BBB; Silva, M.M., Rogers, P.H., Arnone, A.: A Third Quaternary Structure of Human Hemoglobin at 1.7-Å resolution. *J. Biol. Chem.* 276 pp. 17248. © 1992 American Society for Biochemistry and Molecular Biology. Used with permission. (b) A. C. Barrington Brown, 1968 J. D. Watson; (c) nobelprize.org.

B Watson (*left*) and Crick (*right*) with their model of DNA.

C Maurice Wilkins.

> DNA's Base-Pair Sequence Just two kinds of base pairings give rise to the incredible diversity of traits we see among living things. How? Even though DNA is composed of only four nucleotides, the *order* in which one nucleotide follows the next—the **DNA sequence**—can vary tremendously (which explains Chargaff's second rule). For example, a small piece of DNA from any organism might be:

© Cengage Learning.

Note how the two strands match. They are complementary, which means each base on one strand pairs with a suitable partner base on the other. This base-pairing pattern (A to T, G to C) is the same in all molecules of DNA. However, the DNA sequence differs among species, and even among individuals of the same species. The information encoded by that sequence is the basis of traits that define species and distinguish individuals. Thus DNA, the molecule of inheritance in every cell, is the basis of life's unity. Variations in its nucleotide sequence are the foundation of life's diversity.

Take-Home Message

What is the structure of DNA?

- A DNA molecule consists of two nucleotide chains (strands) running in opposite directions and coiled into a double helix. Internally positioned nucleotide bases hydrogen-bond between the two strands. A pairs with T, and C with G.

- The sequence of nucleotides in a DNA strand—the DNA sequence—is the basis of traits that define species and distinguish individuals.

- DNA sequences vary among species and among individuals. This variation is the basis of life's diversity.

> Variations in the nucleotide sequence of DNA are the foundation of life's diversity.

6.4 DNA Replication and Repair

Before a cell reproduces, it must copy its DNA so that each of its future offspring will inherit a full complement of chromosomes. **DNA replication** is the process by which a cell copies its DNA. During this energy-intensive process, enzymes and other molecules first open the double helix to expose the internally positioned bases, and then link nucleotides into new strands of DNA based on the sequence of those bases.

A cell's genetic information consists of the order of nucleotides in the DNA of its chromosomes. Descendant cells must get an exact copy of that information, or inheritance will go awry. Thus, each chromosome is copied in entirety, and the two chromosomes that result from replication are duplicates of the parent molecule.

In eukaryotes, DNA replication necessarily takes place inside the nucleus, which houses the chromosomes. An enzyme called **DNA polymerase** is central to the process. There are several types of DNA polymerases, but all can assemble a new strand of DNA from nucleotides. Each type requires one strand of DNA that serves as a template, or guide, for the synthesis of another. Each polymerase also requires a primer in order to initiate DNA synthesis. A **primer** is a short, single strand of DNA or RNA that is complementary to a targeted DNA sequence.

DNA polymerase DNA replication enzyme; uses a DNA template to assemble a complementary strand of DNA.

DNA replication Energy-requiring process by which a cell duplicates its DNA before it divides.

DNA sequence The order of nucleotides in a strand of DNA.

primer Short, single strand of DNA or RNA that base-pairs with a targeted DNA sequence.

enzymes —

primer —

DNA polymerase —

DNA ligase —

1 As replication begins, enzymes begin to unwind and separate the two strands of DNA.

2 Primers base-pair with the exposed single DNA strands.

3 Starting at primers, DNA polymerases (*green* boxes) assemble new strands of DNA from nucleotides, using the parent strands as templates.

4 DNA ligase seals any gaps that remain between bases of the "new" DNA, so a continuous strand forms.

5 Each parental DNA strand (*blue*) serves as a template for assembly of a new strand of DNA (*magenta*). Both strands of the double helix serve as templates, so two double-stranded DNA molecules result.

FIGURE 6.7 Animated! DNA replication, in which a double-stranded molecule of DNA is copied in entirety. Two double-stranded DNA molecules form; one strand of each is parental (old), and the other is new, so DNA replication is said to be semiconservative.

© Cengage Learning.

Before DNA replication, a chromosome consists of one molecule of DNA—one double helix (Figure 6.7). As replication begins, enzymes break the hydrogen bonds that hold the double helix together, so the two DNA strands unwind and separate **1**. Another enzyme constructs short primers that base-pair with the single-stranded areas of the DNA **2**. The primers serve as attachment points for DNA polymerases, which begin assembling new strands of DNA on each of the two parent strands. As a DNA polymerase moves along a strand of DNA, it uses the sequence of bases as a template to assemble a new strand of DNA from nucleotides **3**.

The polymerase follows base-pairing rules: It adds a T to the end of the new DNA strand when it reaches an A in the template DNA strand; it adds a G when it reaches a C; and so on. Thus, the base sequence of each new strand of DNA is complementary to its template (parental) strand. The enzyme DNA ligase seals any gaps, so the new DNA strands are continuous **4**.

Each nucleotide provides energy for its own attachment to the end of a growing strand of DNA. Remember from Section 4.4 that the bonds between phosphate groups hold a lot of energy. Two of a nucleotide's three phosphate groups are removed when it is added to a DNA strand. Breaking those bonds releases enough energy to drive the attachment.

As each new DNA strand lengthens, it winds up with its template strand into a double helix. So, after replication, two double-stranded molecules of DNA have formed **5**. One strand of each molecule is parental (old), and the other is new; hence the name of the process, **semiconservative replication**. Each new strand of DNA is complementary in sequence to one of the two parent strands, so both double-stranded molecules produced by DNA replication are duplicates of the parent molecule.

❯ How Mutations Arise Sometimes the wrong base is added to a growing DNA strand during replication; at other times, a nucleotide gets lost, or an extra one slips in. Either way, the new DNA strand will no longer be complementary to the parent strand. Most of these replication errors occur simply because DNA polymerases work very fast, copying about 50 nucleotides per second in eukaryotes, and up to 1,000 per second in bacteria. Mistakes are inevitable, and some types of DNA polymerases make a lot of them. Luckily, most DNA polymerases proofread their work. They can correct a mismatch by reversing the synthesis reaction to remove the mispaired nucleotide, then resuming synthesis in the forward direction.

Replication errors also occur after the cell's DNA gets broken or otherwise damaged, because DNA polymerases do not copy damaged DNA very well. In most cases, repair enzymes and other proteins remove and replace damaged or mismatched bases in DNA before replication begins.

When these proofreading and repair mechanisms fail, an error becomes a **mutation**, a permanent change in a cell's DNA sequence. Repair enzymes cannot recognize a mutation after the DNA has been replicated, because each new DNA strand base-pairs properly with its parent strand. Thus, the mutation is passed to the cell's descendants, their descendants, and so on.

Mutations alter DNA's genetic instructions, so they may have a harmful outcome. For example, cancer begins with mutations. Rosalind Franklin died at the age of 37, of ovarian cancer probably caused by extensive exposure to x-rays dur-

A Major breaks (*red* arrows) in chromosomes of a human white blood cell after exposure to ionizing radiation. Pieces of broken chromosomes often become lost during DNA replication.

thymine dimer

B Thymine dimer. This type of DNA damage is caused by exposure to UV light between 230 and 380 nanometers in wavelength.

FIGURE 6.8 Two types of DNA damage that can lead to mutations.

Credits: (a) O. Shovman, A. C. Riches, D. Adamson, and P. E. Bryant, "An improved assay for radiation-induced chromatid breaks using a colcemid block and calyculin-induced PCC combination," *Mutagenesis* (2008) 23(4): 267–270 first published online March 6, 2008 doi:10.1093/mutage/geno09, by permission of Oxford University Press; (b) © Cengage Learning.

ing her work. At the time, the link between x-rays, mutations, and cancer was not understood. We now know that electromagnetic energy with a wavelength shorter than 320 nanometers, including x-rays, most ultraviolet (UV) light, and gamma rays, can knock electrons out of atoms. Such ionizing radiation damages DNA, breaking it into pieces that get lost during replication (Figure 6.8**A**).

UV light in the range of 320–380 nanometers does not have enough energy to knock electrons out of an atom. However, it can cause a covalent bond to form between adjacent thymine or cytidine bases. The result is a nucleotide dimer that kinks the DNA strand (Figure 6.8**B**). DNA polymerase tends to copy the kinked part incorrectly during replication, and mutations are the outcome. Exposing unprotected skin to sunlight increases the risk of cancer because the UV wavelengths cause nucleotide dimers to form. For every second a skin cell spends in the sun, 50–100 of these dimers form in its DNA.

Some natural or synthetic chemicals cause mutations. For instance, some of the chemicals in tobacco smoke transfer methyl groups (—CH_3) to nucleotide bases in DNA. Nucleotides altered in this way do not base-pair correctly. Other chemicals in the smoke are converted by the body to compounds that are easier to excrete, and the breakdown products bind irreversibly to DNA. Replication errors are the outcome in both cases, and these can lead to mutation.

We return to the topic of cancer-causing mutations in Section 8.5. Keep in mind that not all mutations are dangerous. As you will see in later chapters, mutations give rise to the variation in traits that is the raw material of evolution.

Take-Home Message

What happens during DNA replication?

- When a cell copies its DNA, each strand of the double helix serves as a template for synthesis of a new, complementary strand of DNA. Two double helices result.

- Replication errors can give rise to mutations, which is why DNA damage by environmental agents such as UV light and chemicals can cause cancer.

- Proofreading and repair mechanisms usually maintain the integrity of a cell's genetic information by correcting mispaired bases and fixing damaged DNA.

mutation Permanent change in DNA sequence.

semiconservative replication Describes the process of DNA replication, which produces two copies of a DNA molecule: one strand of each copy is new, and the other is a strand of the original DNA.

A A cow egg is held in place by suction through a hollow glass tube called a micropipette. DNA is identified by a *purple* stain.

B Another micropipette punctures the egg and sucks out the DNA. All that remains inside the egg's plasma membrane is cytoplasm.

C A new micropipette prepares to enter the egg at the puncture site. The pipette contains a cell grown from the skin of a donor animal.

D The micropipette enters the egg and delivers the skin cell to a region between the cytoplasm and the plasma membrane.

E After the pipette is withdrawn, the donor's skin cell is visible next to the cytoplasm of the egg. The transfer is now complete.

F An electric current causes the foreign cell to fuse with and empty its nucleus into the cytoplasm of the egg. The egg begins to divide, and an embryo forms.

FIGURE 6.9 Animated! Somatic cell nuclear transfer, using cattle cells. This series of micrographs was taken by scientists at Cyagra, a company that specializes in cloning livestock.

Courtesy of Cyagra, Inc., www.cyagra.com.

differentiation Process by which cells become specialized during development; occurs as different cells in an embryo begin to use different subsets of their DNA.

reproductive cloning Technology that produces genetically identical individuals.

somatic cell nuclear transfer (**SCNT**) Reproductive cloning method in which the DNA from a body cell is transferred into an unfertilized egg.

6.5 Cloning Adult Animals

The word "cloning" means making an identical copy of something, and it can refer to deliberate interventions in reproduction that produce an exact genetic copy of an organism. Genetically identical organisms occur all the time in nature, arising most often by the process of asexual reproduction (which we discuss in Chapter 8). Embryo splitting, another natural process, results in identical twins. The first few divisions of a fertilized egg form a ball of cells that sometimes splits spontaneously. If both halves of the ball continue to develop independently, identical twins result. Artificial embryo splitting has been routine in research and animal husbandry for decades. With this technique, a ball of cells is grown from a fertilized egg in a laboratory. The ball is teased apart into two halves, each of which goes on to develop as a separate embryo. The embryos are implanted in surrogate mothers, who give birth to identical twins. Artificial twinning and any other technology that yields genetically identical individuals is called **reproductive cloning**.

Twins get their DNA from two parents that typically differ in their DNA sequence. Thus, although twins produced by embryo splitting are identical to one another, they are not identical to either parent. Animal breeders who want an exact copy of a specific individual may turn to a cloning method that starts with a single cell taken from an adult organism. Such procedures present more of a technical challenge than embryo splitting. Unlike a fertilized egg, a body cell from an adult will not automatically start dividing. It must first be tricked into rewinding its developmental clock.

All cells descended from a fertilized egg inherit the same DNA. Thus, the DNA in each living cell of an individual is like a master blueprint that contains enough information to build an entirely new individual. As different cells in a developing embryo start using different subsets of their DNA, they become different in form and function, a process called **differentiation**. Differentiation is usually a one-way path in animal cells. Once a cell specializes, all of its descendant cells will be specialized the same way. By the time a liver cell, muscle cell, or other specialized cell forms, most of its DNA has been turned off, and is no longer used.

To clone an adult, scientists must first transform one of its differentiated cells into an undifferentiated cell by turning its unused DNA back on. In **somatic cell nuclear transfer** (**SCNT**), a researcher removes the nucleus from an unfertilized egg, then inserts into the egg a nucleus from an adult animal cell (Figure 6.9). A somatic cell is a body cell, as opposed to a reproductive cell (*soma* is a Greek word for body). If all goes well, the egg's cytoplasm reprograms the transplanted DNA to direct the development of an embryo, which is then implanted into a surrogate mother. The animal that is born to the surrogate is genetically identical with the donor of the nucleus.

SCNT is now a common practice among people who breed prized livestock. Among other benefits, many more offspring can be produced in a given time frame by cloning than by traditional breeding methods. Cloned animals have the same championship features as their DNA donors (Figure 6.10). Offspring can also be produced from a donor animal that is castrated or even dead.

As the techniques become routine, cloning a human is no longer only within the realm of science fiction. SCNT is already being used to produce human embryos for research. Researchers harvest undifferentiated (stem) cells from the cloned human embryos, then use them to study human processes. For example,

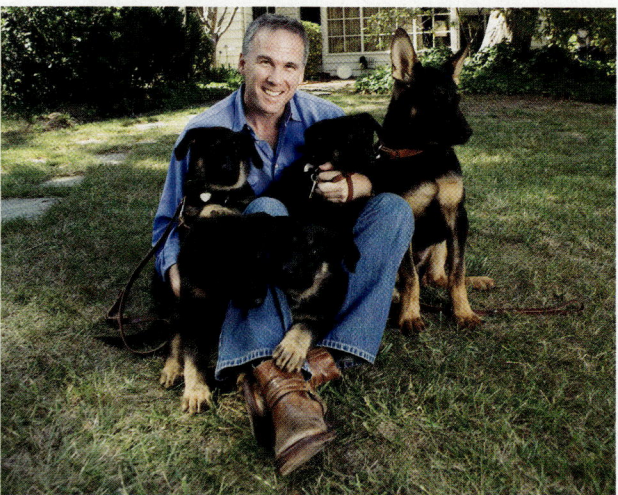

A Liz the championship Holstein cow (*right*) with her clone.

FIGURE 6.10 Examples of animal clones.

Credits: (a) Courtesy of Cyagra, Inc., www.cyagra.com; (b) Ben Glass, courtesy of © BioArts International.

B Trakr's clones with James Symington. Today, the clones are search and rescue dogs for Team Trakr Foundation, an international humanitarian organization that Symington established.

embryos created using cells from people with genetic heart defects are allowing researchers to study how the defect causes developing heart cells to malfunction. Such research may ultimately lead to treatments for people who suffer from fatal diseases. (We return to the topic of stem cells and their potential medical benefits in Chapter 19.) Reproductive cloning of humans is not the intent of such research, but if it were, SCNT would indeed be the first step toward that end.

Take-Home Message

What is cloning?

- Reproductive cloning technologies produce genetically identical individuals.
- The DNA inside a living cell contains all the information necessary to build a new individual.
- In somatic cell nuclear transfer (SCNT), the nuclear DNA of an adult donor is transferred to an egg with no nucleus. The hybrid cell may develop into an embryo that is genetically identical to the donor.

6.6 A Hero Dog's Golden Clones (revisited)

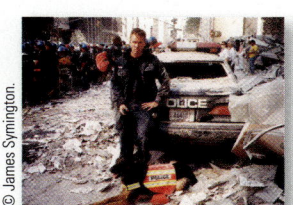

SCNT is not new. The technique first made headlines in 1997, when Scottish geneticist Ian Wilmut and his team produced a clone from the udder cell of an adult sheep. The cloned lamb, named Dolly, looked and acted like a normal sheep at first. However, she died early, suffering from health problems that were most likely an outcome of being a clone. SCNT has also been used to clone mice, rats, rabbits, pigs, cattle, goats, sheep, horses, mules, deer, cats, a camel, a ferret, a monkey, and a wolf. Many of the clones are unusually overweight or have enlarged organs. Cloned mice develop lung and liver problems, and almost all die prematurely. Cloned pigs tend to limp and have heart problems. Some develop without a tail or, even worse, an anus.

WHERE YOU ARE GOING . . .

You will revisit concepts of DNA structure and function many times, particularly when you learn about how genetic information is converted into parts of a cell (Chapter 7). Chromosome structure and cancer will turn up again in context of cell division (Chapter 8). Chapter 12 will show you how mutations are the raw material of evolution. Viruses such as bacteriophages will be explained in more detail in Chapter 13; stem cells, in Chapter 19.

Summary

Section 6.1 Making **clones** (exact genetic copies) of adult animals is now a common practice. The techniques, while improving, are still far from perfect; many attempts are required to produce a live clone, and clones that survive often have health problems. The practice continues to raise ethical questions.

Section 6.2 The DNA of eukaryotes is divided among a characteristic number of **chromosomes** that differ in length and shape. **Histone** proteins that associate with eukaryotic DNA help it pack into a nucleus. When duplicated, a eukaryotic chromosome consists of two **sister chromatids** attached at a **centromere**. **Diploid** cells have two of each type of chromosome. **Chromosome number** is the sum of all chromosomes in a cell of a given species. Members of a pair of **sex chromosomes** differ among males and females. All others are **autosomes**. Autosomes of a pair have the same length, shape, and centromere location. A **karyotype** reveals an individual's complement of chromosomes.

Section 6.3 A DNA molecule consists of two chains of nucleotides coiled into a double helix. A DNA nucleotide has a five-carbon sugar, three phosphate groups, and one of four bases after which the nucleotide is named: adenine, thymine, guanine, or cytosine. Bases of the two DNA strands in a double helix pair in a consistent way: adenine with thymine (A–T), and guanine with cytosine (G–C). The order of the bases (the **DNA sequence**) varies among species and among individuals.

Section 6.4 The DNA sequence of an organism's chromosome(s) is genetic information. A cell passes that information to its offspring by copying its DNA before it divides. In the process of **DNA replication**, **DNA polymerase** copies a molecule of DNA, and two DNA molecules that are identical to the parent are the result. One strand of each DNA molecule is new, and the other is parental; thus the name **semiconservative replication**.

During replication, the double helix unwinds and **primers** form on the exposed single strands of DNA. Starting at the primers, the enzyme DNA polymerase uses each strand as a template to assemble new, complementary strands of DNA from nucleotides. DNA ligase seals any gaps.

Proofreading by DNA polymerases corrects most base-pairing errors as they occur. Uncorrected replication errors become **mutations** (permanent changes in the nucleotide sequence of a cell's DNA). Cancer arises by mutations. Environmental agents such as UV light cause DNA damage that can lead to replication errors, so damaged DNA is normally repaired before replication begins.

Section 6.5 **Reproductive cloning** produces genetically identical individuals (clones). In **somatic cell nuclear transfer** (**SCNT**), a cell from an adult animal is fused with an egg that has had its nucleus removed. The hybrid cell begins dividing and forms an embryo. During development, cells of an embryo become specialized as they begin to use different subsets of their DNA (a process called **differentiation**).

Self-Quiz   Answers in Appendix I

1. Chromosome number _____ .
   a. refers to a particular chromosome pair in a cell
   b. is an identifiable feature of a species
   c. is the number of autosomes in cells of a given type

2. Human body cells are diploid, which means _____ .
   a. they are complete   c. they have two sets of chromosomes
   b. they have autosomes   d. their DNA is in a double helix

3. The _____ is genetic information.
   a. karyotype   c. double helix
   b. sequence of DNA   d. chromosome number

4. One species' DNA differs from others in its _____ .
   a. sugar–phosphate backbone c. DNA sequence
   b. nucleotides   d. all of the above

5. When DNA replication begins, _____ .
   a. the two DNA strands unwind from each other
   b. the two DNA strands condense for base transfers
   c. old strands move to find new strands

6. DNA replication requires _____ .
   a. template DNA   c. primers
   b. nucleotides   d. all of the above

7. All mutations _____ .
   a. cause cancer   c. are caused by radiation
   b. lead to evolution   d. change the DNA sequence

8. Exposure to _____ can lead to mutations.
   a. UV light   c. x-rays
   b. cigarette smoke   d. all of the above

9. _____ is an example of reproductive cloning.
   a. Somatic cell nuclear transfer (SCNT)
   b. Multiple offspring from the same pregnancy
   c. Artificial embryo splitting
   d. a and c
   e. all of the above

10. Match the terms appropriately.
    _____ nucleotide   a. replication enzyme
    _____ clone   b. one old, one new
    _____ autosome   c. copy of an organism
    _____ DNA polymerase   d. chromosome that does not
    _____ mutation   determine sex in humans
    _____ semiconservative   e. monomer of DNA
          replication   f. can cause cancer

Critical Thinking

1. Mutations are the original source of genetic variation. How can mutations accumulate in DNA, given that cells have repair systems that fix mispaired nucleotides or breaks in DNA strands?

2. Woolly mammoths were huge elephant-like mammals. They have been extinct for about 10,000 years, but a few research groups are planning to resurrect one of them by cloning DNA isolated from frozen remains. What are some of the pros and cons, both technical and ethical, of cloning an extinct animal?

3. Show the complementary strand of DNA that forms on this template DNA fragment during replication: GGTTTCTTCAAGAGA.

Digging Into Data

The Hershey–Chase Experiments

By 1950, researchers had discovered bacteriophages, a type of virus that infects bacteria. Like all viruses, these infectious particles carry hereditary information about how to make new viruses. After a virus infects a cell, the cell starts making new virus particles. Bacteriophages inject genetic material into bacteria, but was that material DNA, protein, or both?

Alfred Hershey and Martha Chase found the answer to that question by exploiting the long-known properties of protein (high sulfur content) and DNA (high phosphorus content). They cultured bacteria in a growth medium containing a radioisotope of sulfur (^{35}S). In this medium, the protein (but not the DNA) of bacteriophages that infected the bacteria became labeled with the ^{35}S tracer (Section 2.2). Hershey and Chase allowed the labeled viruses to infect a fresh culture of unlabeled bacteria. They knew from electron micrographs that phages attach to bacteria by their slender tails. They reasoned it would be easy to break this precarious attachment, so they whirled the virus–bacteria mixture in a kitchen blender. The researchers then separated the bacteria from the virus-containing fluid, and measured the ^{35}S content of each separately. The fluid contained most of the ^{35}S. Thus, the viruses had not injected protein into the bacteria (Figure 6.11**A**).

Hershey and Chase repeated the experiment using an isotope of phosphorus, ^{32}P, which labeled the DNA (but not the proteins) of the bacteriophage. This time, they found that the bacteria contained most of the ^{32}P. The viruses had injected DNA into the bacteria (Figure 6.11**B**).

A In one experiment, bacteria were infected with virus particles that had been labeled with a radioisotope of sulfur (^{35}S). The sulfur had labeled only viral proteins. The viruses were dislodged from the bacteria by whirling the mixture in a kitchen blender. Most of the radioactive sulfur was detected in the viruses, not in the bacterial cells. Thus, the viruses had not injected protein into the bacteria.

B In another experiment, bacteria were infected with virus particles that had been labeled with a radioisotope of phosphorus (^{32}P). The phosphorus had labeled only viral DNA. When the viruses were dislodged from the bacteria, the radioactive phosphorus was detected mainly inside the bacterial cells. Thus, the viruses had injected DNA into the cells—evidence that DNA is the genetic material of this virus.

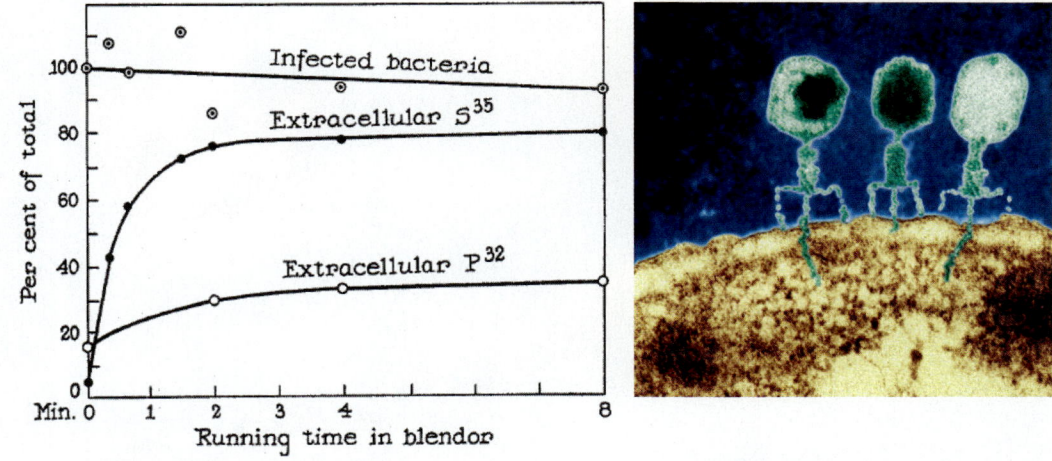

C Detail of Alfred Hershey and Martha Chase's publication describing their experiments with bacteriophage. "Infected bacteria" refers to the percentage of bacteria that survived the blender. The micrograph on the *right* shows three bacteriophage particles injecting DNA into an *E. coli* bacterium.

FIGURE 6.11 Animated! The Hershey–Chase experiments.

Credits: (a,b) © Cengage Learning; (c) left, *Journal of General Physiology*, 36(1), Sept. 20, 1952; right, Eye of Science/ Photo Researchers, Inc..

The graph shown in Figure 6.11**C** is reproduced from Hershey and Chase's original 1952 publication.

1. Why did the amount of radioactivity in each set of samples change over time?

2. After 4 minutes in the blender, what percentage of ^{35}S was outside the bacteria? What percentage of ^{32}P was outside the bacteria?

3. How did the researchers know that the radioisotopes in the fluid came from outside the bacterial cells (extracellular) and not from bacteria that had broken apart?

4. The extracellular concentration of which isotope, ^{35}S or ^{32}P, increased the most with blending? Why do these results imply that bacteriophage viruses inject DNA into bacteria?

Gene Expression and Control 7

7.1 Ricin and Your Ribosomes

The castor-oil plant (*Ricinus communis*) grows wild in tropical regions worldwide, and it is widely cultivated as an ornamental and for its seeds (Figure 7.1). Castor-oil seeds produce and store a protein called ricin. When the seeds germinate, the ricin is broken down and its component amino acids are used to build new proteins. Ricin has an added benefit of being highly toxic to seed-eating beetles, birds, and mammals—including humans. A dose of ricin as small as a few grains of salt can kill an adult human, and there is no antidote.

Humans use castor-oil seeds as a source of castor oil, an ingredient in plastics, cosmetics, paints, soaps, polishes, and many other items. After the oil is extracted from the seeds, the ricin is typically discarded along with the leftover seed pulp. Humans also use ricin for criminal purposes. For example, at the height of the Cold War, the Bulgarian writer Georgi Markov had defected to England and was working as a journalist for the BBC. As he made his way to a bus stop on a London street, an assassin used the tip of a modified umbrella to jam a small, ricin-laced ball into Markov's leg. Markov died in agony three days later.

Ricin's lethal effects were known as long ago as 1888, but using it as a weapon is now banned by most countries under the Geneva Protocol. However, controlling ricin production is impossible, because it takes no special skills or equipment to manufacture the toxin from easily obtained raw materials. Thus, ricin appears periodically in the news. For example, police found pipe bombs and a baby food jar full of ricin in a Tennessee man's shed in 2006. Castor-oil seeds, firearms, and several vials of ricin were found in a Las Vegas motel room after its occupant was hospitalized for ricin exposure in 2008. In 2011, four members of a fringe militia group were arrested after planning to manufacture ricin and release it from a car traveling on an Atlanta interstate.

Ricin is toxic because it inactivates ribosomes, the organelles that assemble amino acids into proteins. Proteins are critical to all life processes, so cells that cannot make them die very quickly. Someone who inhales ricin can die from low blood pressure and respiratory failure within a few days of exposure.

This chapter details how cells convert information encoded in their DNA to an RNA or a protein product. Even though it is extremely unlikely that your ribosomes will ever encounter ricin, protein synthesis is nevertheless worth appreciating for how it keeps you and all other organisms alive.

FIGURE 7.1 Beautiful and deadly: seeds of the castor-oil plant, source of ribosome-busting ricin. Eating eight of these seeds can kill an adult human.

Vaughan Fleming/ SPL/ Photo Researchers, Inc.

7.2 DNA, RNA, and Gene Expression

You learned in Chapter 6 that an individual's chromosomes are like a set of books that provide instructions for building a new individual. You already know the alphabet used to write that book: the four letters A, T, G, and C, for the four nucleotides (adenine, thymine, guanine, and cytosine) in DNA. In this chapter, we investigate the nature of information encoded by the sequence of nucleotides in a DNA strand, and how a cell uses that information.

Information encoded by a chromosome's DNA sequence occurs in hundreds or thousands of units called genes. A **gene** is a DNA sequence that encodes an RNA or protein product. Converting the information encoded by a gene into its product starts with RNA synthesis, or **transcription**, during which enzymes use the gene's DNA sequence as a template to assemble a strand of RNA.

RNA usually occurs in a single-stranded form that is similar to a single strand of DNA (Figure 7.2), but the sugar in an RNA nucleotide (ribose) is slightly different from the sugar in a DNA nucleotide (deoxyribose). Also, three of the bases (adenine, cytosine, and guanine) are the same in RNA and DNA nucleotides, but the fourth base in RNA is uracil (U), not thymine as it is in DNA.

Despite the small differences in structure, DNA and RNA have very different functions. DNA's important but only role is to store a cell's heritable information. By contrast, a cell transcribes several kinds of RNAs, and each kind has a different function. Three types have roles in protein synthesis. **Ribosomal RNA (rRNA)** is the main component of ribosomes (Section 3.4), which assemble amino acids into polypeptide chains (Section 2.9). **Transfer RNA (tRNA)** delivers the amino acids to ribosomes, one by one, in the order specified by a **messenger RNA (mRNA)**. Messenger RNA was named for its function as the "messenger" between DNA and protein. An mRNA's protein-building message is encoded by sets of three nucleotides, "genetic words" that follow one another along its length. Like the words of a sentence, a series of these genetic words can form a meaningful parcel of information—in this case, the sequence of amino acids of a protein. By the process of **translation**, the protein-building information in an mRNA is decoded (translated) into a sequence of amino acids. The result is a polypeptide chain that twists and folds into a protein.

The processes of transcription and translation are part of **gene expression**, a multistep process by which information encoded in a gene guides the assembly of an RNA or protein product. During gene expression, this information flows from DNA to RNA to protein:

$$\text{DNA} \xrightarrow{\text{transcription}} \text{mRNA} \xrightarrow{\text{translation}} \text{protein}$$

A cell's DNA sequence contains the information it needs to make the molecules of life. Each gene encodes an RNA, and RNAs interact to assemble proteins. Proteins (enzymes, in particular) assemble lipids and carbohydrates, replicate DNA, make RNA, and perform many other functions that keep the cell alive.

Take-Home Message

What is the nature of the information carried by DNA?

- A DNA sequence holds information that occurs in units called genes.
- A cell uses the information encoded by a gene to make an RNA or protein product, a process called gene expression.
- A cell transcribes the nucleotide sequence of a gene into RNA.
- Information carried by a messenger RNA (mRNA) can be translated into a protein.

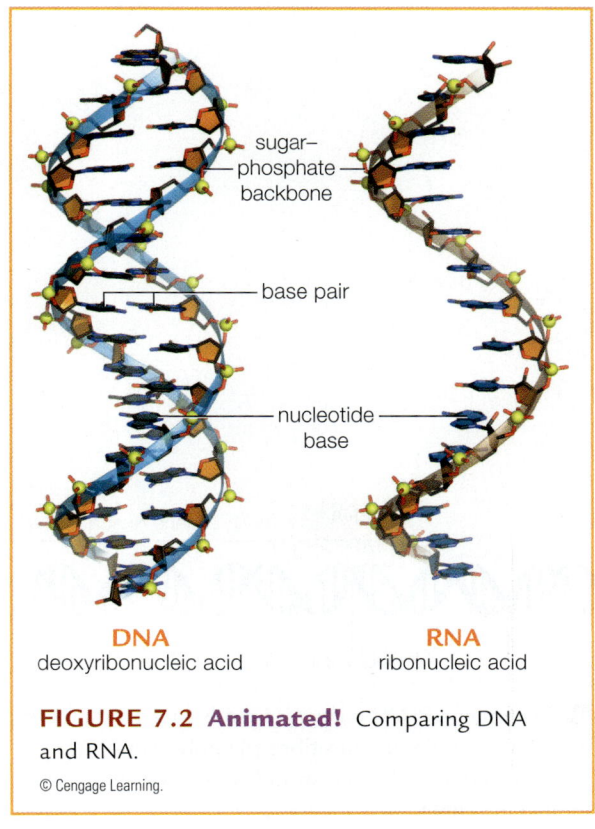

sugar–phosphate backbone

base pair

nucleotide base

DNA
deoxyribonucleic acid

RNA
ribonucleic acid

FIGURE 7.2 Animated! Comparing DNA and RNA.

© Cengage Learning.

gene DNA sequence that encodes an RNA or protein product.

gene expression Process by which the information in a gene guides assembly of an RNA or protein product.

messenger RNA (mRNA) Type of RNA that has a protein-building message.

ribosomal RNA (rRNA) Type of RNA that becomes part of ribosomes.

transcription Process by which enzymes assemble an RNA using the nucleotide sequence of a gene as a template.

transfer RNA (tRNA) Type of RNA that delivers amino acids to a ribosome during translation.

translation Process by which a polypeptide chain is assembled from amino acids in the order specified by an mRNA.

7.3 Transcription: DNA to RNA

> **DNA as a Template** Remember that DNA replication begins with one DNA double helix and ends with two DNA double helices (Section 6.4). The two double helices are identical to the parent molecule because base-pairing rules are followed during DNA replication. A nucleotide can be added to a growing strand of DNA only if it base-pairs with the corresponding nucleotide of the parent strand: G pairs with C, and A pairs with T (Section 6.3):

The same base-pairing rules also govern RNA synthesis in transcription. An RNA strand is structurally so similar to a DNA strand that the two can base-pair if their nucleotide sequences are complementary. In such hybrid molecules, G pairs with C, and A pairs with U (uracil):

During transcription, a strand of DNA acts as a template upon which a strand of RNA is assembled from nucleotides. A nucleotide can be added to a growing strand of RNA only if it is complementary to the corresponding nucleotide of the parent strand of DNA. Thus, each new RNA is complementary in sequence to the DNA strand that served as its template. As in DNA replication, each nucleotide provides the energy for its own attachment to the end of a growing strand.

Transcription is similar to DNA replication in that one strand of a nucleic acid serves as a template for synthesis of another. However, in contrast with DNA replication, only part of one DNA strand (not the whole molecule) serves as the template. The enzyme **RNA polymerase**, not DNA polymerase, adds nucleotides to the end of a growing RNA strand. Also, transcription produces a single strand of RNA, and DNA replication produces two DNA double helices.

1 The enzyme RNA polymerase binds to a promoter in the DNA. The binding positions the polymerase near a gene. Only one of the two strands of DNA will be transcribed into RNA.

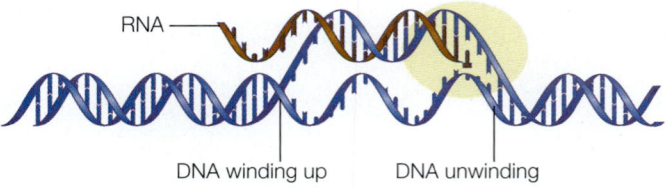

2 The polymerase begins to move along the gene and unwind the DNA. As it does, it links RNA nucleotides in the order specified by the nucleotide sequence of the template DNA strand. The DNA winds up again after the polymerase passes. The structure of the "opened" DNA at the transcription site is called a transcription bubble, after its appearance.

direction of transcription

3 Zooming in on the transcription bubble, we can see that RNA polymerase covalently bonds successive nucleotides into an RNA strand. The new strand is an RNA copy of the gene.

FIGURE 7.3 Transcription. By this process, a strand of RNA is assembled from nucleotides according to a template: a gene region in DNA. **Figure It Out:** After the guanine, what is the next nucleotide that will be added to this growing strand of RNA?

Answer: Another guanine

RNA molecules   DNA molecule

› The Process of Transcription

In eukaryotic cells, transcription occurs in the nucleus; in prokaryotes, it occurs in cytoplasm. The process begins when an RNA polymerase and regulatory proteins attach to the DNA at a site called a **promoter** (Figure 7.3 ❶). Binding positions the polymerase close to the gene that will be transcribed. The polymerase starts moving along the DNA, over the gene region ❷. As it moves, the polymerase unwinds the double helix just a bit so it can "read" the nucleotide sequence of the template DNA strand. The polymerase joins free RNA nucleotides in the order dictated by that DNA sequence.

When the polymerase reaches the end of the gene region, it releases the DNA and the new RNA. RNA polymerase follows base-pairing rules, so the new RNA strand is complementary to the DNA strand from which it was transcribed ❸. It is an RNA copy of a gene, the same way that a paper transcript of a conversation carries the same information in a different format. Typically, many polymerases transcribe a particular gene region at the same time, so many new RNA strands can be produced very quickly (Figure 7.4).

› RNA Modifications

Just as a dressmaker may snip off loose threads or add bows to a dress before it leaves the shop, so do eukaryotic cells tailor their RNA. For example, most eukaryotic genes contain intervening sequences called **introns**. Introns are removed in chunks from a newly transcribed RNA before it leaves the nucleus. Sequences that stay in the RNA are **exons** (Figure 7.5). Exons can be rearranged and spliced together in different combinations, so one gene may encode different proteins. A newly transcribed RNA that will become an mRNA is further tailored after splicing. A tail of 50 to 300 adenines is also added to the end of a new mRNA; hence the name, poly-A tail. Among other functions, the tail is a signal that allows an mRNA to be exported from the nucleus.

Take-Home Message

How is RNA assembled?

- In transcription, RNA polymerase uses the nucleotide sequence of a gene region in a chromosome as a template to assemble a strand of RNA. The new strand is an RNA copy of the gene from which it was transcribed.

- Post-transcriptional modification of RNA occurs in eukaryotes.

7.4 RNA Players in Translation

› mRNA and the Genetic Code

An mRNA is essentially a disposable copy of a gene. Its job is to carry DNA's protein-building information to the other two types of RNA for translation. That protein-building information consists of a linear sequence of genetic "words" spelled with an alphabet of the four nucleotides A, C, G, and U. Each of the genetic "words" carried by an mRNA is

FIGURE 7.4 Typically, many RNA polymerases simultaneously transcribe the same gene, producing a structure called a "Christmas tree" after its shape. Here, three genes next to one another on the same chromosome are being transcribed.

Figure It Out: Are the polymerases transcribing this DNA molecule moving from left to right or from right to left?

Answer: Left to right (the RNAs get longer as the polymerases move along the DNA)

O. L. Miller.

FIGURE 7.5 Animated! Post-transcriptional modification of RNA. Introns are removed and exons spliced together. Messenger RNAs also get a poly-A tail.

© Cengage Learning.

exon Nucleotide sequence that remains in an RNA after post-transcriptional modification.

intron Nucleotide sequence that intervenes between exons and is excised during post-transcriptional modification.

promoter In DNA, a sequence to which RNA polymerase binds.

RNA polymerase Enzyme that carries out transcription.

	second base				
first base	U	C	A	G	third base
U	UUU } phe UUC UUA } leu UUG	UCU UCC } ser UCA UCG	UAU } tyr UAC UAA stop UAG stop	UGU } cys UGC UGA stop UGG trp	U C A G
C	CUU CUC } leu CUA CUG	CCU CCC } pro CCA CCG	CAU } his CAC CAA } gln CAG	CGU CGC } arg CGA CGG	U C A G
A	AUU AUC } ile AUA AUG met	ACU ACC } thr ACA ACG	AAU } asn AAC AAA } lys AAG	AGU } ser AGC AGA } arg AGG	U C A G
G	GUU GUC } val GUA GUG	GCU GCC } ala GCA GCG	GAU } asp GAC GAA } glu GAG	GGU GGC } gly GGA GGG	U C A G

ala	alanine (A)	gly	glycine (G)	pro	proline (P)
arg	arginine (R)	his	histidine (H)	ser	serine (S)
asn	asparagine (N)	ile	isoleucine (I)	thr	threonine (T)
asp	aspartic acid (D)	leu	leucine (L)	trp	tryptophan (W)
cys	cysteine (C)	lys	lysine (K)	tyr	tyrosine (Y)
glu	glutamic acid (E)	met	methionine (M)	val	valine (V)
gln	glutamine (Q)	phe	phenylalanine (F)		

FIGURE 7.6 Animated! The genetic code. Each codon in mRNA is a set of three nucleotides. In the large chart, the *left* column lists a codon's first nucleotide, the *top* row lists the second, and the *right* column lists the third. Sixty-one of the triplets encode amino acids; the remaining three are signals that stop translation. The amino acid names that correspond to abbreviations in the chart are listed *above*.

Figure It Out: Which codons specify lysine (lys)? *Answer: AAA and AAG*

© Cengage Learning.

three nucleotides long, and each is a code—a **codon**—for a particular amino acid. There are four possible nucleotides in each of the three positions of a codon, so there are a total of sixty-four (or 4^3) mRNA codons. Collectively, the sixty-four codons constitute the **genetic code** (Figure 7.6). The sequence of the three nucleotides in a triplet determines which amino acid the codon specifies. For instance, the codon AUG codes for the amino acid methionine (met), and UGG codes for tryptophan (trp).

One codon follows the next along the length of an mRNA, so the order of codons in an mRNA determines the order of amino acids in the polypeptide that will be translated from it. Thus, the nucleotide sequence of a gene is transcribed into the nucleotide sequence of an mRNA, which is in turn translated into an amino acid sequence (Figure 7.7).

Twenty naturally occurring amino acids are encoded by the sixty-four codons in the genetic code. Some amino acids are specified by more than one codon. For instance, GAA and GAG both code for glutamic acid. Other codons signal the beginning and end of a protein-coding sequence. For example, the first AUG in an mRNA is the signal to start translation in most species. AUG also happens to be the codon for methionine, so methionine is always the first amino acid in new polypeptides of such organisms. UAA, UAG, and UGA do not specify an amino acid. They are signals that stop translation, so they are called stop codons. A stop codon marks the end of the protein-coding sequence in an mRNA.

The genetic code is highly conserved, which means that most organisms use the same code and probably always have. Bacteria, archaea, and some protists have a few codons that differ from the typical code, as do mitochondria and chloroplasts—a clue that led to a theory of how these two organelles evolved (we return to this topic in Section 13.3).

a gene region in DNA

transcription

mRNA

translation

amino acid sequence

methionine — tyrosine —— serine ——··
(met) (tyr) (ser)

FIGURE 7.7 Example of the correspondence between DNA, RNA, and proteins. A DNA strand is transcribed into mRNA, and the codons of the mRNA specify a chain of amino acids.

© Cengage Learning.

large subunit   small subunit   intact ribosome

FIGURE 7.8 Animated! Ribosome structure. An intact ribosome consists of a large and a small subunit. Protein components of both subunits are shown in the ribbon models in *green*; rRNA components, in *brown*.

© Cengage Learning.

A Icon and model of the tRNA that carries the amino acid tryptophan. Each tRNA's anticodon is complementary to an mRNA codon. Each also carries the amino acid specified by that codon.

B During translation, tRNAs dock at an intact ribosome (for clarity, only the small subunit is shown, in *tan*). Here, the anticodons of two tRNAs have base-paired with complementary codons on an mRNA (*red*).

FIGURE 7.9 tRNA structure.

© Cengage Learning.

❭ **rRNA and tRNA—The Translators** Ribosomes and transfer RNAs (tRNA) interact to translate an mRNA into a polypeptide. A ribosome has one large and one small subunit (Figure 7.8). Each subunit consists mainly of rRNA, with some associated structural proteins. During translation, a large and a small ribosomal subunit converge as an intact ribosome on an mRNA.

A tRNA has two attachment sites. The first is an **anticodon**, which is a triplet of nucleotides that base-pairs with an mRNA codon (Figure 7.9**A**). The other attachment site binds to an amino acid—the one specified by the codon. Transfer RNAs with different anticodons carry different amino acids. Those tRNAs deliver amino acids, one after the next, to a ribosome during translation of an mRNA (Figure 7.9**B**).

Ribosomal RNA is one example of RNA with enzymatic activity: The rRNA of a ribosome, not the protein, forms a peptide bond between amino acids. As the amino acids are delivered, the ribosome joins them via peptide bonds into a new polypeptide (Section 2.9). Thus, the order of codons in an mRNA—DNA's protein-building message—becomes translated into a new protein.

Take-Home Message

What roles do mRNA, tRNA, and rRNA play during translation?

- mRNA carries protein-building information. The nucleotides in mRNA are "read" in sets of three during protein synthesis. Most of these nucleotide triplets (codons) code for amino acids. The genetic code consists of all sixty-four codons.

- Ribosomes, which consist of two subunits of rRNA and proteins, assemble amino acids into polypeptide chains.

- A tRNA has an anticodon complementary to an mRNA codon, and it has a binding site for the amino acid specified by that codon. tRNAs deliver amino acids to ribosomes during translation.

7.5 Translating the Code: RNA to Protein

Translation, the second part of protein synthesis, occurs in the cytoplasm of all cells. Cytoplasm has many free amino acids, tRNAs, and ribosomal subunits available to participate in the process.

anticodon Set of three nucleotides in a tRNA; base-pairs with mRNA codon.

codon In mRNA, a nucleotide triplet that codes for an amino acid or stop signal during translation.

genetic code Complete set of sixty-four mRNA codons.

FIGURE 7.10 Animated! Translation in eukaryotes.

1 In eukaryotic cells, RNA is transcribed in the nucleus.

2 Finished RNA moves into the cytoplasm through nuclear pores.

3 Ribosomal subunits and tRNA converge on an mRNA.

4 A polypeptide chain forms as the ribosome moves along the mRNA, linking amino acids together in the order dictated by the mRNA codons.

© Cengage Learning 2010.

Figure 7.10 shows an overview of translation as it occurs in eukaryotes. An mRNA is transcribed in the nucleus **1**, and then transported through nuclear pores into the cytoplasm **2**. Translation begins when a small ribosomal subunit binds to the mRNA. Next, the anticodon of a special tRNA called an initiator base-pairs with the first AUG codon of the mRNA. Then, a large ribosomal subunit joins the small subunit **3**.

The ribosome then begins to assemble a polypeptide chain as it moves along the mRNA. Initiator tRNAs carry the amino acid methionine, so the first amino acid of the new polypeptide chain is methionine. Another tRNA brings the second amino acid to the ribosome as its anticodon base-pairs with the second codon in the mRNA. The ribosome joins the first two amino acids by way of a peptide bond:

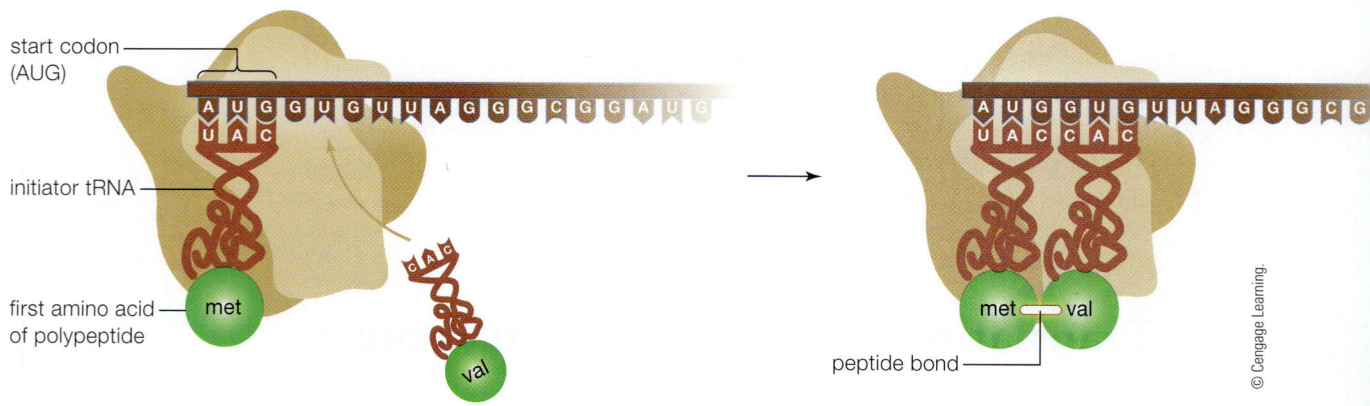

The first tRNA is released and the ribosome moves to the next codon. Another tRNA brings the third amino acid to the ribosome as its anticodon base-pairs with the third codon of the mRNA. The ribosome joins the second and third amino acids by way of a peptide bond:

The second tRNA is released and the ribosome moves to the next codon. Another tRNA brings the fourth amino acid to the ribosome as its anticodon base-pairs with the fourth codon of the mRNA. The ribosome joins the third and fourth amino acids by way of a peptide bond:

The new polypeptide chain continues to elongate ❹ as the ribosome joins amino acids delivered by successive tRNAs.

Translation ends when the ribosome encounters a stop codon in the mRNA. The mRNA and the polypeptide detach from the ribosome, and the ribosomal subunits separate from each other. The new polypeptide will either join the pool of proteins in the cytoplasm, or it will be modified inside rough ER (Section 3.5).

In cells that are making a lot of protein, many ribosomes may simultaneously translate the same mRNA. In bacteria and archaea, transcription and translation both occur in the cytoplasm, and these processes are closely linked in time and space. Translation begins before transcription ends, so a transcription "Christmas tree" is often decorated with ribosome "balls."

Take-Home Message

How is mRNA translated into protein?

- Translation converts the protein-building information carried by an mRNA into a polypeptide.
- During translation, amino acids are delivered to a ribosome by tRNAs in the order dictated by successive mRNA codons. As they arrive, the ribosome joins each to the end of the growing polypeptide chain.

A Hemoglobin, an oxygen-binding protein in red blood cells.

This protein consists of four polypeptides: two alpha globins (*blue*) and two beta globins (*green*). Each globin forms a pocket that cradles a type of cofactor called a heme (*red*). Oxygen gas binds to the iron atom at the center of each heme.

16	17	18	19	20	21	22	23	24	25	26	27	28	29	30
G	G	A	C	T	C	C	T	C	T	T	C	A	G	A
C	C	U	G	A	G	G	A	G	A	A	G	U	C	U

········ pro ——— glu ——— glu ——— lys ——— ser ········

B Part of the DNA (*blue*), mRNA (*brown*), and amino acid sequence (*green*) of human beta globin. Numbers indicate the position of the nucleotide in the coding sequence of the mRNA.

G	G	A	C	A	C	C	T	C	T	T	C	A	G	A
C	C	U	G	U	G	G	A	G	A	A	G	U	C	U

········ pro ——— val ——— glu ——— lys ——— ser ········

C A base-pair substitution replaces a thymine with an adenine. When the altered mRNA is translated, valine replaces glutamic acid as the sixth amino acid of the polypeptide. Hemoglobin with this form of beta globin is called HbS, or sickle hemoglobin.

G	G	A	C	C	C	T	C	T	T	C	A	G	A	C
C	C	U	G	G	G	A	G	A	A	G	U	C	U	G

········ pro ——— gly ——— arg ——— ser ——— leu ········

D A deletion of one nucleotide causes the reading frame for the rest of the mRNA to shift. The protein translated from this mRNA is too short and does not assemble correctly into hemoglobin molecules. The result is beta thalassemia, in which a person has an abnormally low amount of hemoglobin.

G	G	A	C	T	C	C	T	C	T	T	C	C	A	G
C	C	U	G	A	G	G	A	G	A	A	G	G	U	C

········ pro ——— glu ——— glu ——— lys ——— val ········

E An insertion of one nucleotide causes the reading frame for the rest of the mRNA to shift. The protein translated from this mRNA is too short and does not assemble correctly into hemoglobin molecules. As in **D**, the outcome is beta thalassemia.

FIGURE 7.11 Animated! Examples of mutations.

© Cengage Learning.

7.6 Mutated Genes and Their Protein Products

Mutations, remember, are permanent changes in a DNA sequence (Section 6.4). A mutation in which one nucleotide and its partner are replaced by a different base pair is a **base-pair substitution**. Other mutations involve the loss of one or more nucleotides (a **deletion**) or the addition of extra nucleotides (an **insertion**).

Mutations are relatively uncommon events in a normal cell. For example, the chromosomes in a diploid human cell collectively consist of about 6.5 billion nucleotides, any of which may become mutated each time that cell divides. However, only about 175 nucleotides actually do change during DNA replication. On top of that, only about 3 percent of the cell's DNA encodes protein products, so there is a low probability that any of those mutations will be in a protein-coding region.

When a mutation does occur in a protein-coding region, the redundancy of the genetic code offers the cell a margin of safety. For example, a mutation that changes a UCU codon to UCC in an mRNA may not have further effects, because both codons specify serine. Mutations in a gene that have no effect on the gene's product are said to be silent. Other mutations are not silent: They may change an amino acid in a protein, or result in a premature stop codon that shortens it. Both outcomes can have severe consequences for the cell—and the organism.

For example, a mutation can impair health by altering the oxygen-binding properties of hemoglobin, a blood protein. As red blood cells circulate through the lungs, the hemoglobin proteins inside of them bind to oxygen molecules. The cells then travel to other regions of the body, and the hemoglobin releases its oxygen cargo wherever the oxygen level is low. When the red blood cells return to the lungs, the hemoglobin binds to more oxygen.

Hemoglobin's structure allows it to bind and release oxygen. The protein consists of four polypeptides called globins (Figure 7.11**A**). Each globin folds around a heme, which is a cofactor (Section 4.4) with an iron atom at its center. Oxygen molecules bind to hemoglobin at those iron atoms.

In adult humans, two alpha globin chains and two beta globin chains make up each hemoglobin molecule. Defects in either of the polypeptide chains can cause a condition called anemia, in which a person's blood is deficient in red blood cells or in hemoglobin. Both outcomes limit the blood's ability to carry oxygen, and the resulting symptoms range from mild to life-threatening.

Sickle-cell anemia, a type of anemia that is most common in people of African ancestry, arises because of a base-pair substitution in the beta globin gene. The substitution causes the body to produce a version of beta globin in which the sixth amino acid is valine instead of glutamic acid (Figure 7.11**B,C**). Hemoglobin assembled with this altered beta globin chain is called sickle hemoglobin, or HbS.

Unlike glutamic acid, which carries a negative charge, valine carries no charge. As a result of that one base-pair substitution, a tiny patch of the beta globin polypeptide changes from hydrophilic to hydrophobic, which in turn causes the hemoglobin's behavior to change slightly. HbS molecules stick together and form large, rodlike clumps under certain conditions. Red blood cells that contain the clumps become distorted into a crescent (sickle) shape (Figure 7.12). Sickled cells clog tiny blood vessels, thus disrupting blood circulation throughout the body. Over time, repeated episodes of sickling can damage organs and cause death.

A different type of anemia, beta thalassemia, is caused by the deletion of the twentieth nucleotide in the coding region of the beta globin gene (Figure 7.11**D**). Like many other deletions, this one causes the reading frame of the mRNA codons to shift. A frameshift usually has drastic consequences because it garbles the genetic message, just as incorrectly grouping a series of letters garbles the meaning of a sentence:

The fat cat ate the sad rat

T hef atc ata tet hes adr at

The frameshift caused by the beta globin deletion results in a polypeptide that differs drastically from normal beta globin. Hemoglobin molecules do not assemble correctly with the altered polypeptides, which are only 18 amino acids long (the normal beta globin chain has 147 amino acids). This outcome is the source of the anemia.

Beta thalassemia can also be caused by insertion mutations, which, like deletions, often cause frameshifts (Figure 7.11**E**). Insertion mutations are often caused by the activity of transposable elements, which are segments of DNA that can move spontaneously within or between chromosomes. Transposable elements can be hundreds or thousands of nucleotides long, so when one interrupts a gene it becomes a major insertion that changes the gene's product. Transposable elements are common in the DNA of all species; about 45 percent of human DNA consists of transposable elements or their remnants.

Take-Home Message

What happens after a gene becomes mutated?

- Mutations that result in an altered protein can have drastic consequences.

- A base-pair substitution may change an amino acid in a protein, or it may introduce a premature stop codon.

- Frameshifts that occur after an insertion or deletion can change an mRNA's codon reading frame, thus garbling its protein-building instructions.

glutamic acid   valine

A A base-pair substitution results in the abnormal beta globin chain of sickle hemoglobin (HbS). The sixth amino acid in such chains is valine, not glutamic acid. The difference causes HbS molecules to form rod-shaped clumps that distort normally round blood cells into sickle shapes.

sickled cell

normal cell

B *Left*, the sickled cells clog small blood vessels, causing circulatory problems that result in damage to many organs. Destruction of the cells by the body's immune system results in anemia. *Right*, Tionne "T-Boz" Watkins of the music group TLC is a celebrity spokesperson for the Sickle Cell Disease Association of America. She was diagnosed with sickle-cell anemia as a child.

FIGURE 7.12 Animated! How a single base-pair substitution causes sickle-cell anemia.

Credits: (a) © Cengage Learning; (b) left, Dr. Gopal Murti/ SPL/ Photo Researchers, Inc.; right, © Ben Rose/ WireImage/ Getty Images.

7.7 Eukaryotic Gene Controls

All of the cells in your body contain the same DNA with the same genes. However, each cell rarely uses more than 10 percent of its genes at one time. Some of those genes affect structural features and metabolic pathways common to all cells. Others are expressed only by certain subsets of your cells. For example, most of your body cells express genes that encode the enzymes of glycolysis, but only your immature red blood cells express genes that encode globin.

Differentiation, the process by which cells become specialized, occurs as different cell lineages begin to express different subsets of their genes during development. Which genes a cell uses determines the molecules it will produce, which in turn determines what kind of cell it will be.

base-pair substitution Mutation in which a single base pair changes.

deletion Mutation in which one or more nucleotides are lost.

insertion Mutation in which one or more nucleotides become inserted into DNA.

A A transcription factor—the protein product (*gold*) of an insect gene called *antennapedia* attaches to a promoter sequence in a fragment of DNA. In cells of a fly embryo, the binding starts a cascade of cellular events that results in the formation of a leg.

B *Antennapedia* is a homeotic gene whose expression in embryonic tissues of the insect thorax causes legs to form. A mutation that causes *antennapedia* to be expressed in the embryonic tissues of the head causes legs to form there too (*left*). Compare the head of the normal fly on the right.

FIGURE 7.13 An example of gene control.

Control over which genes are expressed at a particular time is crucial for proper development of complex, multicelled bodies. It also allows individual cells to respond appropriately to changes in their environment. The "switches" that turn gene expression on or off are called gene controls—molecules or structures that start, enhance, slow, or stop the individual steps of gene expression. Whether or not a gene is expressed at a given time depends on many factors, such as the type of cell, and conditions in the cytoplasm and extracellular fluid. The factors affect controls governing all steps of gene expression, starting with transcription and ending with delivery of an RNA or protein product to its final destination. Such controls consist of processes that start, enhance, slow, or stop gene expression. For example, proteins called **transcription factors** bind directly to DNA and affect whether and how fast certain genes are transcribed. Some transcription factors prevent RNA polymerase from attaching to a promoter, which in turn prevents transcription of nearby genes. Others help RNA polymerase bind to a promoter, which speeds up transcription (Figure 7.13**A**).

The remainder of this section introduces some specific examples and effects of gene controls in eukaryotes.

❭ Master Genes As an animal embryo develops, its differentiating cells form tissues, organs, and body parts. Some cells that alternately migrate and stick to other cells develop into nerves, blood vessels, and other structures that weave through the tissues. Events like these fill in the body's details, and all are driven by cascades of master gene expression. The products of **master genes** affect the expression of many other genes. Expression of a master gene causes other genes to be expressed, which in turn cause other genes to be expressed, and so on. The final outcome is the completion of an intricate task such as the formation of an eye.

Maternal mRNAs are delivered to opposite ends of an unfertilized egg as it forms. These mRNAs are translated only after the egg is fertilized, and then their protein products diffuse away in gradients that span the entire developing embryo. Depending on where a nucleus falls within the gradients, one or another set of master genes will be transcribed inside of it. The products of those master genes also form in gradients that span the embryo. Still other master

> The "switches" that turn gene expression on or off are called gene controls.

A A fruit fly with a mutation in its *eyeless* gene develops with no eyes.

B Compare the large, round eyes of a normal fruit fly.

C Eyes form wherever the *eyeless* gene is expressed in fly embryos. Abnormal expression of the *eyeless* gene in this fly caused extra eyes to develop on its head and also on its wings.

FIGURE 7.14 *Eyeless:* the eyes have it. Credits: (a,b) David Scharf/ Photo Researchers, Inc.; (c) Eye of Science/ Photo Researchers, Inc.; (d) Left, Courtesy of the Aniridia Foundation International, www.aniridia.net; right, M. Bloch.

D Humans, mice, squids, and other animals have a gene called *PAX6*. In humans, *PAX6* mutations result in missing irises, a condition called aniridia (*left*). Compare a normal iris (*right*). *PAX6* is so similar to *eyeless* that it triggers eye development when expressed in fly embryos.

genes are transcribed depending on where a nucleus falls within these gradients, and so on. Eventually, the products of master genes cause undifferentiated cells to differentiate, and specialized tissues and structures form in specific regions of the embryo.

A homeotic gene is a type of master gene that controls the formation of a body part such as an eye, leg, or wing. These genes are typically named for what happens when a mutation alters their function. For example, fruit flies with a mutation that affects the *antennapedia* gene (*ped* means foot) have legs in place of antennae (Figure 7.13**B**). Flies with a mutated *eyeless* gene develop with no eyes (Figure 7.14**A**). *Dunce* is required for learning and memory. *Wingless, wrinkled,* and *minibrain* are self-explanatory. *Tinman* is necessary for development of a heart. Flies with a mutated *groucho* gene have extra bristles above their eyes (*inset*). One gene was named *toll,* after what its German discoverer exclaimed upon seeing the disastrous effects of the mutation (*toll* is German slang that means "cool!").

groucho

The function of many homeotic genes has been discovered by deliberately manipulating their expression. Researchers can inactivate a gene by introducing a mutation or deleting it entirely, an experiment called a **knockout**. A knockout organism (one that has had a gene knocked out) may differ from normal individuals, and the differences are clues to the function of the missing gene product.

Homeotic genes control development by the same mechanisms in all multicelled eukaryotes, and many are interchangeable among different species. Thus, we can infer that they evolved in the most ancient eukaryotic cells. Consider the *eyeless* gene. Eyes form in embryonic fruit flies wherever this gene is expressed, which, in normal flies, is only in tissues of the head. If the *eyeless* gene is expressed in another part of the developing embryo, eyes form there too (Figure 7.14**C**). Humans, squids, mice, fish, and many other animals have a gene called *PAX6,* which is very similar in DNA sequence to the *eyeless* gene in flies. In humans, mutations in *PAX6* cause eye disorders such as aniridia, in which a person's irises are underdeveloped or missing (Figure 7.14**D**). If the *PAX6* gene from a human or mouse is inserted into an *eyeless* mutant fly, it has the same effect as the *eyeless* gene: An eye forms wherever it is expressed.

knockout An experiment in which a gene is deliberately inactivated in a living organism.

master gene Gene encoding a product that affects the expression of many other genes.

transcription factor Protein that influences transcription by binding to DNA.

A Barr bodies. The photo on the *left* shows the nucleus of five XX cells. Inactivated X chromosomes—Barr bodies—appear as *red* spots. Compare the nucleus of two XY cells in the photo on the *right*.

SRY gene expressed   no *SRY* present

penis

vaginal opening

B An early human embryo appears neither male nor female. *SRY* gene expression determines whether male reproductive organs develop.

FIGURE 7.15 Examples of gene controls associated with sex chromosomes.

Credits: (a) © Dr. William Strauss; (b) After Patten, Carlson, & others.

CH_3

NH_2

cytosine

guanine

FIGURE 7.16 A methyl group (*red*) attached to a nucleotide in DNA. © Cengage Learning.

Barr body Inactivated X chromosome in a cell of a female mammal. The other X chromosome is active.

epigenetic Refers to DNA modifications such as methylation that do not involve a change in DNA sequence.

> **Sex Chromosome Genes** In humans and other mammals, a female's cells contain two X chromosomes, one inherited from her mother, the other one from her father. In each cell, one X chromosome is always tightly condensed (Figure 7.15**A**). We call the condensed X chromosome a **Barr body**, after Murray Barr, who discovered them. Condensation prevents transcription, so most of the genes on a Barr body are not expressed. The inactivation ensures that only one of the two X chromosomes in a female's cells is active. According to a theory called dosage compensation, X chromosome inactivation equalizes expression of X chromosome genes between the sexes. The body cells of male mammals (XY) have one set of X chromosome genes. Body cells of female mammals have two sets, but female embryos do not develop properly when both sets are expressed.

The human X chromosome carries 1,336 genes. Some of those genes are associated with sexual traits, such as the distribution of body fat and hair. However, most of the genes on the X chromosome govern nonsexual traits such as blood clotting and color perception. Such genes are expressed in both males and females. Males, remember, also inherit one X chromosome.

The human Y chromosome carries only 307 genes, but one of them is *SRY*—the master gene for male sex determination in mammals. Its expression in XY embryos triggers the formation of testes, which are male gonads (Figure 7.15**B**). Some of the cells in these primary male reproductive organs make testosterone, a sex hormone that controls the emergence of male secondary sexual traits such as facial hair, increased musculature, and a deep voice. We know that *SRY* is the master gene that controls emergence of male sexual traits because mutations in this gene cause XY individuals to develop external genitalia that appear female. An XX embryo has no Y chromosome, no *SRY* gene, and much less testosterone, so primary female reproductive organs (ovaries) form instead of testes. Ovaries make estrogens and other sex hormones that will govern the development of female secondary sexual traits, such as enlarged, functional breasts, and fat deposits around the hips and thighs.

> **Epigenetics** Transcription is affected by chromosome structure, as when RNA polymerase cannot bind to DNA that is tightly wound around histones (Section 6.2). Certain modifications to histone proteins change the way they interact with the DNA that wraps around them. Some modifications make them release their grip on the DNA; others make them tighten it. For example, adding a methyl group ($—CH_3$) to a histone protein can make the DNA wind more tightly around it, so enzymes that methylate histones discourage transcription.

Methyl groups added to certain atoms of nucleotide bases (Figure 7.16) suppress gene expression in a more permanent way than histone methylation. Once a particular nucleotide has become methylated in a cell's DNA, it will usually stay methylated in all of the cell's descendants. This type of methylation is a necessary part of differentiation, so it begins early in embryonic development: Genes actively expressed in a zygote become silenced as their promoters get methylated. Methylations and other modifications to DNA that affect its function but do not involve changes to the underlying DNA sequence are described as **epigenetic**.

Nucleotide methylation occurs on an ongoing basis, and it is influenced by environmental factors. For instance, individuals conceived during a famine end up with an unusually low number of methyl groups attached to certain genes. One of those genes encodes a hormone that fosters prenatal growth and development. The resulting increase in expression of this gene may offer a survival advantage in a poor nutritional environment.

Methyl groups are added to nucleotides by chance during DNA replication, so cells that divide a lot tend to have more methyl groups in their DNA than inactive cells. Free radicals and toxic chemicals add more methyl groups (Section

6.4). These and other environmental factors that influence nucleotide methylation can have long-term effects. When an organism reproduces, it passes its chromosomes to offspring. The methylation of parental DNA is normally "reset" in the first cell of the new individual, with new methyl groups being added and old ones being removed. This reprogramming does not remove all of the parental methyl groups, however, so methylations acquired during an individual's lifetime can be passed to future offspring.

Epigenetic changes to DNA may persist for generations after an environmental stressor has faded. For example, even when corrected for social factors, grandsons of boys who endured a winter of famine tend to live, on average, about 32 years longer than the grandsons of boys who overate at the same age (Figure 7.17). Nine-year-old boys whose fathers smoked cigarettes before age 11 are very overweight compared with boys whose fathers did not smoke in childhood. In these and other studies, the effect was sex-limited: Boys were affected by lifestyle of individuals in the paternal line; girls, by individuals in the maternal line.

FIGURE 7.17
Epigenetic changes can be heritable.

In 1944, a supply blockade followed by an unusually harsh winter caused a severe famine in the Nazi-occupied Netherlands. Grandsons of boys who endured the famine (such as the one pictured in this photo) can expect to live about 32 years longer than grandsons of boys who ate well during the same winter.

© Marius Meijboom/ Nederlands Fotomuseum, all rights reserved.

Take-Home Message

What is gene control?

- Gene controls regulate all steps of gene expression, from transcription to delivery of the final gene product.
- Many gene controls influence development of complex, multicelled bodies. Others allow cells to respond to changes in their environment.

7.8 Ricin and Your Ribosomes (revisited)

Vaughan Fleming/ SPL/ Photo Researchers, Inc.

Ricin is a ribosome-inactivating protein (RIP). Other RIPs occur in some bacteria, mushrooms, algae, and many plants (including food crops such as tomatoes, barley, and spinach). Many of these proteins are not particularly toxic because they do not cross intact cell membranes very well. Those that do, including ricin, have a polypeptide chain that binds tightly to carbohydrates on plasma membranes (Figure 7.18). Binding causes the cell to take up the RIP by endocytosis (Section 4.6). Once inside the cell, the second polypeptide of the RIP goes to work. This polypeptide is an enzyme that removes a specific adenine base from one of the rRNA chains in the large ribosomal subunit. Once that happens, the ribosome stops working. Protein synthesis grinds to a halt as the ricin inactivates the cell's ribosomes, and the cell quickly dies. One molecule of ricin can inactivate more than 1,000 ribosomes per minute. If enough ribosomes are affected, protein synthesis grinds to a halt. Proteins are critical to all life processes, so cells that cannot make them die very quickly.

Very few people actually encounter ricin. Other RIPs are more prevalent threats to our health. Shiga toxin, made by *Shigella dysenteriae* bacteria, causes dysentery. *E. coli* bacteria, including the strain O157:H7 (Section 3.1), make enterotoxin, an RIP that is the source of symptoms associated with food poisoning.

WHere you aRe GoinG . . .

Chapter 8 discusses the cell cycle and how controls over it go awry in cancer. Chapter 9 explores inheritance; Chapter 10, DNA research. Evolutionary processes involving mutations will be discussed more fully in Chapter 12. Section 13.2 discusses the hypothesis that RNA, not DNA, was genetic material in the distant past. Plant gene controls are discussed in Chapter 28.

ricin   Shiga toxin   *E. coli* enterotoxin

FIGURE 7.18 A few examples of RIPs. These proteins have strikingly similar structures. One polypeptide chain (*red*) helps the molecule cross a cell's plasma membrane. The other chain (*orange*) destroys the cell's capacity for protein synthesis.

© Cengage Learning.

Summary

Section 7.1 The ability to make proteins is critical to all life processes. Molecules such as ricin that inactivate ribosomes can be extremely toxic if they enter cells.

Section 7.2 Information encoded within the nucleotide sequence of DNA occurs in subsets called **genes**. Converting the information in a gene to an RNA or protein product is called **gene expression**. RNA is produced during **transcription**. Ribosomal RNA (**rRNA**) and **transfer RNA** (**tRNA**) interact during **translation** of a **messenger RNA** (**mRNA**) into a protein product:

$$\text{DNA} \xrightarrow{\text{transcription}} \text{mRNA} \xrightarrow{\text{translation}} \text{protein}$$

Section 7.3 During transcription, **RNA polymerase** binds to a **promoter** near a gene region of a chromosome. The polymerase moves over the gene region, linking RNA nucleotides in the order dictated by the nucleotide sequence of the DNA. The new RNA strand is an RNA copy of the gene.

The RNA of eukaryotes is modified before it leaves the nucleus. **Introns** are removed, and the remaining **exons** may be rearranged and spliced in different combinations.

Section 7.4 mRNA carries DNA's protein-building information. The information consists of a series of **codons**, sets of three nucleotides. All sixty-four codons, most of which specify amino acids, constitute the **genetic code**. Each tRNA has an **anticodon** that can base-pair with a codon, and it binds to the amino acid specified by that codon. Enzymatic rRNA and proteins make up the two subunits of ribosomes.

Section 7.5 During translation, genetic information that is carried by an mRNA directs the synthesis of a polypeptide. First, an mRNA, an initiator tRNA, and two ribosomal subunits converge. The intact ribosome then forms a peptide bond between successive amino acids, which are delivered by tRNAs in the order specified by the codons in the mRNA. Translation ends when the ribosome encounters a stop codon.

Section 7.6 **Insertions**, **deletions**, and **base-pair substitutions** are mutations. The activity of transposable elements causes some mutations.

A mutation that changes a gene's product may have harmful effects. For example, a base-pair substitution in the gene for the beta globin chain of hemoglobin causes sickle-cell anemia. Beta thalassemia can be an outcome of a frameshift mutation in the beta globin gene.

Section 7.7 Which genes a cell uses depends on the type of organism, the type of cell, conditions inside and outside the cell, and, in complex multicelled species, the organism's stage of development.

Transcription factors influence transcription by binding to DNA. These and other gene controls guide development in multicelled eukaryotes. All cells of an embryo share the same genes. As different cell lineages use different subsets of those genes during development, they become specialized, a process called differentiation.

The products of **master genes** diffuse through a developing embryo and affect expression of other master genes, which affect the expression of others, and so on. Cells differentiate according to their location on the map. **Knockout** experiments in fruit flies revealed the function of many homeotic genes, which govern development of particular body parts.

In cells of female mammals, one of the two X chromosomes is condensed as a **Barr body**, rendering most of its genes permanently inaccessible. The *SRY* gene determines male sex in humans.

Nucleotide methylations and other **epigenetic** modifications can affect the function of DNA without changing its underlying sequence. Epigenetic changes acquired during an individual's lifetime can be passed to offspring, and may persist for generations.

Self-Quiz   Answers in Appendix I

1. A chromosome contains many genes that are transcribed into different _____ .
   a. proteins   c. RNAs
   b. polypeptides   d. a and b

2. RNAs form by _____ ; proteins form by _____ .
   a. replication; translation
   b. translation; transcription
   c. transcription; translation
   d. replication; transcription

3. The main function of a DNA molecule is to _____ .
   a. store heritable information
   b. carry DNA's genetic message for translation
   c. form peptide bonds between amino acids
   d. carry amino acids to ribosomes

4. The main function of an mRNA molecule is to _____ .
   a. store heritable information
   b. carry DNA's genetic message for translation
   c. form peptide bonds between amino acids
   d. carry amino acids to ribosomes

5. Where does transcription take place in a eukaryotic cell?
   a. nucleus   c. cytoplasm
   b. ribosome   d. b and c are correct

6. Translation takes place in the _____ of all cells.
   a. nucleus   c. cytoplasm
   b. plasma membrane   d. a and c are correct

7. Up to how many amino acids can be encoded by a gene that consists of 45 nucleotides plus a stop codon?

Digging Into Data

Effect of Paternal Grandmother's Food Supply on Infant Mortality

Researchers are investigating long-reaching effects of starvation, in part because historical data on periods of famine are widely available. Before the industrial revolution, a failed harvest in one autumn typically led to severe food shortages the following winter.

A retrospective study has correlated female infant mortality at certain ages with the abundance of food during the paternal grandmother's childhood. Figure 7.19 shows some of the results of this study.

1. Compare the mortality of girls whose paternal grandmothers ate well at age 2 with that of girls whose grandmothers were starving at the same age. Which girl was more likely to die early? About how much more likely was she to die?

2. Children have a period of slow growth around age 9. What trend in this data can you see around that age?

3. There was no correlation between early death of a male child and eating habits of his paternal grandmother, but there was a strong correlation with the eating habits of his paternal grandfather. Does this tell you anything about the location of epigenetic changes that gave rise to this data?

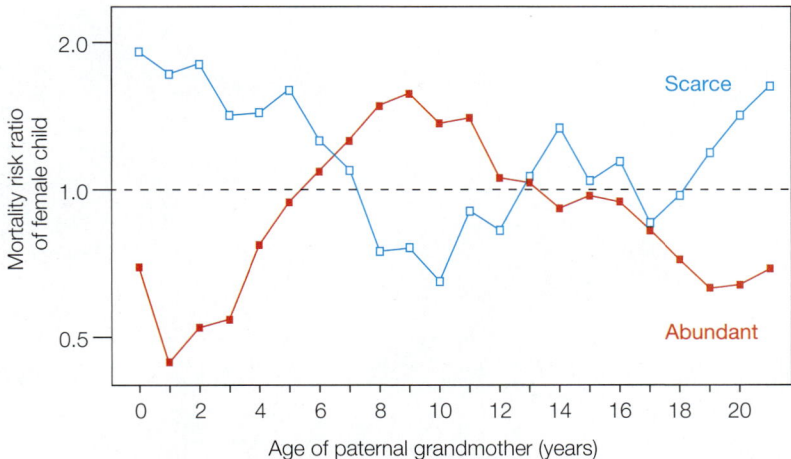

FIGURE 7.19 Graph showing the relative risk of early death of a female child, correlated with the age at which her paternal grandmother experienced a winter with a food supply that was scarce (*blue*) or abundant (*red*) during childhood. The dotted line represents no difference in risk of mortality. A value above the line means an increased risk; one below the line indicates a reduced risk.

Source: Pembrey et al., *European Journal of Human Genetics* (2006) 14, 159–166.

---

8. Most codons specify a(n) _____ .
   a. protein
   b. polypeptide
   c. amino acid
   d. mRNA

9. _____ are mutations.
   a. Transposable elements
   b. Base-pair substitutions
   c. Insertions
   d. Deletions
   e. b, c, and d are correct
   f. all of the above

10. Homeotic gene products _____ .
    a. map out the overall body plan in embryos
    b. control the formation of specific body parts

11. A gene that is knocked out is _____ .
    a. deleted
    b. inactivated
    c. expressed
    d. either a or b

12. Which of the following includes all of the others?
    a. homeotic genes
    b. master genes
    c. *SRY* gene
    d. *PAX6*

13. A cell with a Barr body is _____ .
    a. prokaryotic
    b. a sex cell
    c. from a female mammal
    d. infected by Barr virus

14. True or false? Some gene expression patterns are heritable.

15. Match each term with the most suitable description.
    ____ methylation
    ____ DNA sequence
    ____ ribosome
    ____ mRNA
    ____ genetic code
    ____ differentiation
    ____ transposable element

    a. cells become specialized
    b. gets around
    c. may cause epigenetic effects
    d. linear order of nucleotides
    e. assembles amino acids
    f. set of 64 codons
    g. translated to protein

16. Put the following processes in order of their occurrence during expression of a eukaryotic gene:
    a. mRNA processing
    b. translation
    c. transcription
    d. RNA leaves nucleus

Critical Thinking

1. Use Figure 7.6 to translate this DNA sequence into an amino acid sequence, starting at the first nucleotide:

   GGUUUCUUGAAGAGA

2. Translate the DNA sequence in the previous question, starting at the second nucleotide.

3. An anticodon has the sequence GCG. What amino acid does this tRNA carry? What would be the effect of a mutation that changed the C of the anticodon to a G?

4. Each position of a codon can be occupied by one of four (4) nucleotides. What is the minimum number of nucleotides per codon necessary to specify all 20 of the amino acids that are found in proteins?

5. The termination of prokaryotic transcription often depends on the structure of a newly forming RNA. Transcription stops where the mRNA folds back on itself, forming a hairpin-looped structure like the one shown at *right*. How do you think that this structure stops transcription?

8 How Cells Reproduce

8.1 Henrietta's Immortal Cells

Each human starts out as a fertilized egg. By the time of birth, that cell has given rise to about a trillion cells, all organized as a human body. Even in an adult, billions of cells divide every day as new cells replace worn-out ones. However, despite the ability of human cells to continue dividing as part of a body, they tend to divide a limited number of times and die within weeks when grown in the laboratory.

Researchers started trying to coax human cells to keep dividing outside of the body in the mid-1800s. Immortal cell lineages—cell lines—would allow the researchers to study human diseases (and potential cures for them) without experimenting on people. The quest to create a human cell line continued unsuccessfully until 1951. By this time, researchers George and Margaret Gey had been trying to culture human cells for thirty years. Then their assistant, Mary Kubicek, prepared a new sample of human cancer cells. Mary named the cells HeLa, after the first and last names of the patient, Henrietta Lacks. The HeLa cells began to divide, again and again, every twenty-four hours. They were astonishingly vigorous, quickly coating the inside of their test tube and consuming their nutrient broth. Four days later, there were so many cells that the researchers had to transfer them to more tubes. The cell populations continued to increase at a phenomenal rate, coating the inside of their new tubes within days.

Sadly, cancer cells in the patient were dividing just as fast. Only six months after she had been diagnosed with cervical cancer, malignant cells had invaded tissues throughout her body. Two months after that, Henrietta Lacks, a young African American woman from Baltimore, was dead.

Although Henrietta passed away, her cells lived on in the Geys' laboratory (Figure 8.1). The Geys were able to grow poliovirus in HeLa cells, a practice that enabled them to find out which strains of the virus cause polio. That work was a critical step in the development of polio vaccines, which have since saved millions of lives.

Henrietta Lacks's immortal cells, frozen away in tiny tubes and packed in Styrofoam boxes, continue to be shipped among laboratories all over the world. They are still widely used to investigate cancer, viral growth, protein synthesis, the effects of radiation, and countless other processes important in medicine and research. HeLa cells helped several researchers win Nobel Prizes, and some even traveled into space for experiments on the Discoverer XVII satellite.

Henrietta Lacks was just thirty-one, a wife and mother of five, when runaway cell divisions killed her. Her legacy continues to help people, through her cells that are still dividing, again and again, more than fifty years after she died. Understanding why cancer cells are immortal—and why we are not—begins with understanding the structures and mechanisms that cells use to divide.

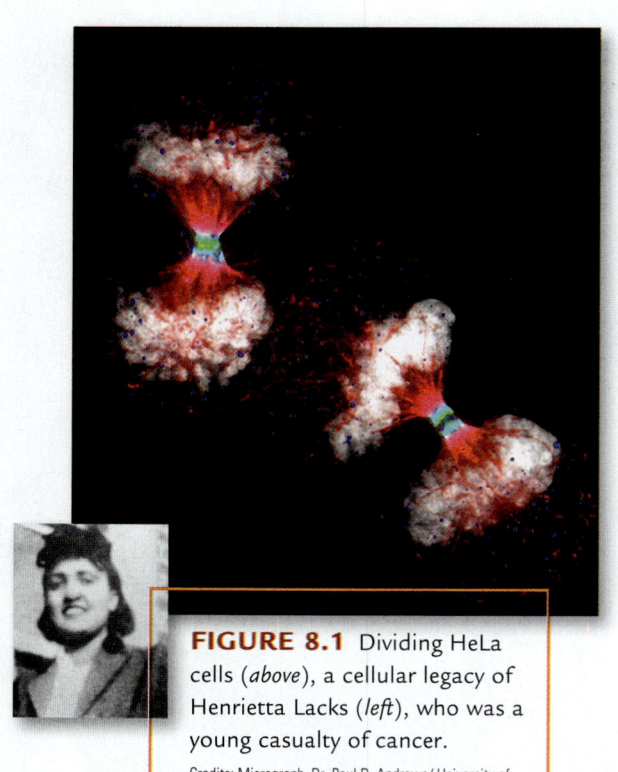

FIGURE 8.1 Dividing HeLa cells (*above*), a cellular legacy of Henrietta Lacks (*left*), who was a young casualty of cancer.

Credits: Micrograph, Dr. Paul D. Andrews/ University of Dundee; inset, Courtesy of the family of Henrietta Lacks.

Table 8.1	Outcomes of Cellular Division Mechanisms		
	Mechanism		
Outcome	Mitosis	Meiosis	Binary Fission
Increases in body size during growth	X		
Replacement of dead or worn-out cells	X		
Repair of damaged tissues	X		
Asexual reproduction	X		X
Sexual reproduction		X	

FIGURE 8.2 A multicelled animal develops from a fertilized egg by repeated cell divisions. These early frog embryos have undergone three divisions.

Figure It Out: Each of these embryos consists of how many cells? *Answer: Eight*

Carolina Biological Supply Company/ Phototake.

8.2 Multiplication by Division

A cell reproduces by dividing, thus producing two cellular offspring. Each of the two descendant cells inherits DNA and a blob of the parent's cytoplasm, which contains enzymes, organelles, and other metabolic machinery that will keep the cell running until it can make its own. The process is a bit more involved than it sounds, however, because each new cell must also end up with a complete set of chromosomes, or it will not grow or function properly.

If a eukaryotic cell simply split in two, only one of its offspring would get the nucleus (and the DNA). Thus, before the cytoplasm divides, the cell duplicates its chromosomes (by DNA replication, Section 6.4). The duplicated chromosomes are then separated and packaged into new nuclei by one of two mechanisms: mitosis or meiosis. These two processes have much in common, but their outcomes differ (Table 8.1).

Mitosis is a nuclear division mechanism that maintains the chromosome number. In multicelled organisms, mitosis and cytoplasmic division are the basis of increases in body size during development (Figure 8.2), and ongoing replacements of damaged or dead cells. Mitosis and cytoplasmic division are also part of **asexual reproduction**, a reproductive mode used by some multicelled eukaryotes and many single-celled ones. (Bacteria and archaea also reproduce asexually, but they do it by a separate mechanism, binary fission, that we consider in Section 13.5.)

Meiosis, a nuclear division mechanism that halves the chromosome number, is central to sexual reproduction. During **sexual reproduction**, two parents contribute genes to offspring. In humans and all other mammals, meiosis in reproductive cells results in sperm and eggs. Cells inside spores, which protect and disperse new generations, form by meiosis during the life cycle of fungi, plants, and many kinds of protists.

Take-Home Message

What is cell division and why does it happen?

- When a cell divides, each descendant cell receives a set of chromosomes and some cytoplasm. In eukaryotic cells, the nucleus divides first, then cytoplasm.
- Mitosis is a nuclear division mechanism that maintains the chromosome number. It is the basis of body size increases, cell replacements, and tissue repair in multicelled eukaryotes; and asexual reproduction in single-celled and some multicelled eukaryotes.
- Meiosis, a nuclear division mechanism that halves the chromosome number, is the basis of sexual reproduction in eukaryotes.

asexual reproduction Reproductive mode by which offspring arise from a single parent only.

meiosis Nuclear division process that halves the chromosome number. Basis of sexual reproduction.

mitosis Nuclear division mechanism that maintains the chromosome number. Basis of body growth, tissue repair and replacement in multicelled eukaryotes; also asexual reproduction in some plants, animals, fungi, and protists.

sexual reproduction Reproductive mode by which offspring arise from two parents and inherit genes from both.

8.3 Mitosis and the Cell Cycle

A life cycle is the collective series of events that an organism passes through during its lifetime. Multicelled organisms and free-living cells have life cycles, but what about cells that make up a multicelled body? Biologists consider such cells to be individually alive, each with its own life that passes through a series of recognizable stages. The events that occur from the time a cell forms until the time it divides are collectively called the **cell cycle** (Figure 8.3).

A typical cell spends most of its life in **interphase** ❶, a stage during which the cell increases its mass, roughly doubles the number of its cytoplasmic components, and replicates its DNA. Interphase is typically the longest stage, and it consists of three stages:

❷ G1, the first interval (or gap) of cell growth, before DNA replication

❸ S, the time of synthesis (DNA replication)

❹ G2, the second interval (or gap), when the cell prepares to divide

G1 and G2 were named gap intervals because outwardly they seem to be periods of inactivity, but they are not. Most cells going about their metabolic business are in G1. Cells preparing to divide enter S, when they copy their DNA. During G2, they make the proteins that will drive mitosis. Once S begins, DNA replication usually proceeds at a predictable rate and ends before mitosis begins ❺.

> **Controls Over the Cell Cycle** The vast majority of differentiated cells in a human body are in G1, and some types never progress past that stage. For example, most of your nerve cells, skeletal muscle cells, heart muscle cells, and fat-storing cells have been in G1 since you were born. When a cell divides—and when it does not—is determined by a host of gene expression controls (Section 7.7). Like the accelerator of a car, some of these controls cause the cell cycle to advance. Like brakes, others stop the cycle from proceeding. Such controls are part of built-in checkpoints that allow problems to be corrected before the cycle proceeds ❻. Protein products of "checkpoint genes" monitor whether a cell's DNA has been copied completely, whether it is damaged, and even whether

Cells are individually alive, even as part of a multicelled body.

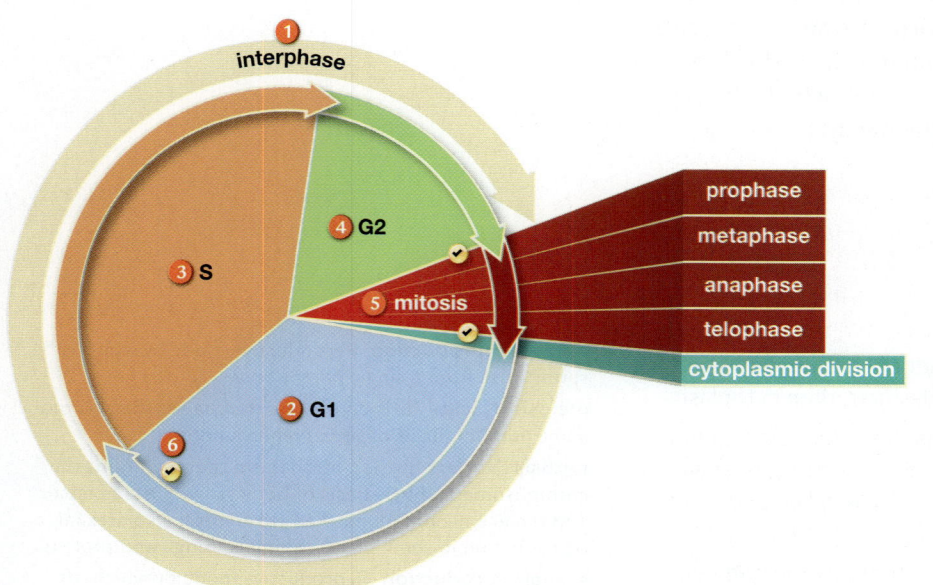

FIGURE 8.3 Animated! The eukaryotic cell cycle. The length of the intervals differs among cells.

❶ A cell spends most of its life in interphase, which includes three stages: G1, S, and G2.

❷ G1 is the interval of growth before DNA replication. The cell's chromosomes are unduplicated.

❸ S is the time of synthesis, during which the cell copies its DNA.

❹ G2 is the interval after DNA replication and before mitosis. The cell prepares to divide during this stage.

❺ The nucleus divides during mitosis, the four stages of which are detailed in the next section. In most cases, the cytoplasm divides after mitosis. The cycle begins anew, in interphase, for each descendant cell.

❻ Built-in checkpoints ✔ stop the cycle from proceeding until certain conditions are met.

© Cengage Learning.

A An unduplicated pair of chromosomes in a cell in G1.

B By G2, each chromosome has been duplicated.

C Mitosis and cytoplasmic division package one copy of each chromosome into each of two new cells.

FIGURE 8.4 How mitosis maintains the chromosome number. For clarity, only one chromosome pair is shown. © Cengage Learning.

enough nutrients to support division are available. These checkpoint proteins interact with cell cycle controls to delay or stop the cell cycle while simultaneously enhancing transcription of genes involved in fixing a problem. After the problem has been corrected, the brakes are lifted and the cell cycle proceeds. If the problem remains uncorrected, other checkpoint proteins may initiate a series of events that eventually cause the cell to self-destruct.

> How Mitosis Maintains the Chromosome Number Remember from Section 6.2 that diploid cells have two sets of chromosomes. For example, human body cells have 46 chromosomes—two of each type. Except for a pairing of sex chromosomes (XY) in males, the chromosomes of each pair are homologous. **Homologous chromosomes** have the same length, shape, and genes (*hom*– means alike). In sexual reproducers like humans, one member of a homologous pair of chromosomes was inherited from the female parent, and the other member was inherited from the male parent.

With mitosis followed by cytoplasmic division, a diploid cell produces two diploid offspring. Both offspring have the same number and kind of chromosomes as the parent. Thus, mitosis maintains the chromosome number. However, if only the total number of chromosomes mattered, then one of the cell's offspring might get, say, two pairs of chromosome 22 and no chromosome 9. A cell cannot function properly without a full complement of DNA, which means it needs to have *a copy of each* chromosome.

Mitosis distributes a complete set of chromosomes into two new nuclei. When a cell is in G1, each of its chromosomes consists of one double-stranded DNA molecule (Figure 8.4**A**). The cell replicates its DNA in S, so by G2, each of its chromosomes consists of two double-stranded DNA molecules (Figure 8.4**B**). These two DNA molecules stay attached at the centromere as sister chromatids (Section 6.2) until mitosis is almost over, and then they are pulled apart and packaged into two separate nuclei.

When sister chromatids are pulled apart, each becomes an individual chromosome that consists of one double-stranded DNA molecule. Thus, each of the two new nuclei that form in mitosis contains a full complement of (unduplicated) chromosomes. When the cytoplasm divides, the two nuclei are packaged into two separate cells (Figure 8.4**C**). Each new cell starts the cell cycle over again in G1 of interphase.

> The Process of Mitosis When a cell is in interphase, its chromosomes are loosened to allow transcription and DNA replication. Loosened chromosomes are spread out, so they are not easily visible under a light microscope

cell cycle A series of events from the time a cell forms until its cytoplasm divides.

homologous chromosomes Paired chromosomes with the same length, shape, and set of genes.

interphase In a eukaryotic cell cycle, the interval between mitotic divisions when a cell grows, roughly doubles the number of its cytoplasmic components, and replicates its DNA.

1 Interphase

Interphase cells are shown for comparison, but interphase is not part of mitosis. The *red* spots in the plant cell nucleus are areas where ribosome subunits are being transcribed and assembled.

pair of centrioles

2 Early Prophase

Mitosis begins. Transcription stops, and the DNA begins to appear grainy as it starts to condense. The centriole pair gets duplicated.

3 Prophase

The duplicated chromosomes become visible as they condense. One of the two centriole pairs moves to the opposite side of the cell. The nuclear envelope breaks up. Spindle microtubules assemble and bind to chromosomes at the centromere. Sister chromatids become attached to opposite centriole pairs.

microtubule of spindle

4 Metaphase

All of the chromosomes are aligned in the middle of the cell.

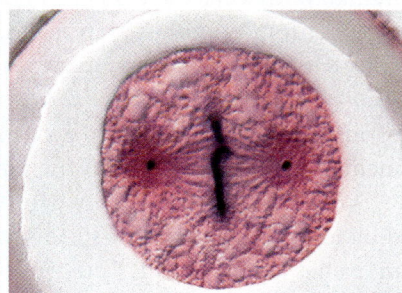

5 Anaphase

Spindle microtubules separate the sister chromatids and move them toward opposite sides of the cell. Each sister chromatid has now become an individual, unduplicated chromosome.

6 Telophase

The chromosomes reach opposite sides of the cell and decondense. Mitosis ends when a new nuclear envelope forms around each cluster of chromosomes.

FIGURE 8.5 Animated! Mitosis. Micrographs show mitosis in the nucleus of a plant cell (onion root, *left*), and an animal cell (fertilized egg of a roundworm, *right*). A diploid (2*n*) animal cell with two chromosome pairs is illustrated.

(Figure 8.5 ❶). In preparation for nuclear division, the chromosomes begin to pack tightly. Transcription and DNA replication stop as the chromosomes condense into their most compact "X" forms ❷. Tight condensation keeps the chromosomes from getting tangled and breaking during nuclear division. A cell reaches **prophase**, the first stage of mitosis, when its chromosomes have condensed so much that they are visible under a light microscope ❸. "Mitosis" is from *mitos*, the Greek word for thread, after the threadlike appearance of the chromosomes during nuclear division.

The cytoplasm of most animal cells contains one closely associated pair of centrioles (Section 3.5). Just before prophase, the cell duplicates this centriole pair, and one of the two resulting pairs moves to the opposite side of the cell. During prophase, microtubules that begin to assemble and extend from the centrioles form a **spindle**, which is an apparatus that moves chromosomes during nuclear division. Plant cells have spindles and structures that organize them, but no centrioles.

The spindle penetrates the nuclear region as the nuclear envelope breaks up. Some microtubules of the spindle stop lengthening when they reach the middle of the cell. Others lengthen until they reach a chromosome and attach to it at the centromere. By the end of prophase, one sister chromatid of each chromosome has become attached to microtubules extending from one centriole pair, and the other sister has become attached to microtubules extending from the other centriole pair.

The opposing sets of microtubules then begin a tug-of-war by adding and losing tubulin subunits. As the microtubules lengthen and shorten, they push and pull the chromosomes. When all the microtubules are the same length, the chromosomes are aligned in the middle of the cell ❹. The alignment marks **metaphase**.

During **anaphase**, the sister chromatids of each duplicated chromosome separate, so each becomes an individual, unduplicated chromosome. The spindle moves the chromosomes toward opposite sides of the cell ❺.

Telophase begins when the two clusters of chromosomes reach opposite ends of the cell ❻. Each cluster has the same number and kinds of chromosomes as the parent cell nucleus had: two of each chromosome, if the parent cell was diploid. A new nucleus forms around each cluster as the chromosomes loosen up again, after which telophase (and mitosis) is over.

Take-Home Message

What happens during the cell cycle?

- A cell's life passes through a series of stages that include interphase, mitosis, and cytoplasmic division. Controls over gene expression can advance, delay, or block this cell cycle in response to internal and external conditions.

- During interphase, the cell increases in mass and replicates its DNA in preparation for nuclear division.

- Mitosis begins as duplicated chromosomes condense and the nuclear envelope breaks up (in prophase). A spindle also forms and attaches to the chromosomes. At metaphase, the spindle has aligned all of the chromosomes in the middle of the cell.

- The sister chromatids of each chromosome separate and are pulled to opposite sides of the cell during anaphase.

- New nuclear envelopes form around the two clusters of chromosomes in telophase. The two new nuclei have the same chromosome number as the parent cell. Mitosis is now over.

anaphase Stage of mitosis during which sister chromatids separate and move toward opposite sides of the cell.

metaphase Stage of mitosis at which all chromosomes are aligned in the middle of the cell.

prophase Stage of mitosis during which chromosomes condense and become attached to a newly forming spindle.

spindle Apparatus that moves chromosomes during nuclear division; consists of dynamically assembled and disassembled microtubules.

telophase Stage of mitosis during which chromosomes arrive at opposite ends of the cell and decondense, and two new nuclei form.

1 After mitosis is completed, the spindle begins to disassemble.

2 At the midpoint of the former spindle, a ring of microfilaments attached to the plasma membrane contracts.

3 This contractile ring pulls the cell surface inward as it shrinks.

4 The ring contracts until it pinches the cell in two.

FIGURE 8.6 Animated! Cytoplasmic division of an animal cell. © Cengage Learning.

5 The future plane of division was established before mitosis began. Vesicles cluster here before mitosis ends.

6 The vesicles fuse with each other, forming a cell plate along the plane of division.

7 The cell plate expands outward along the plane of division. When it reaches the plasma membrane, it attaches to the membrane and partitions the cytoplasm.

8 The cell plate matures as two new cell walls. These walls join with the parent cell wall, so each descendant cell becomes enclosed by its own wall.

FIGURE 8.7 Animated! Cytoplasmic division of a plant cell. © Cengage Learning.

8.4 Cytoplasmic Division Mechanisms

In most eukaryotes, the cell cytoplasm divides between late anaphase and the end of telophase, so two cells form, each with their own nucleus. The mechanism of cytoplasmic division differs between plants and animals.

Typical animal cells pinch themselves in two after nuclear division ends (Figure 8.6 **1**). The cell cortex, which is the mesh of cytoskeletal elements just under the plasma membrane (Section 3.5), includes a band of microfilaments that wraps around the cell's midsection. The band is called a contractile ring because it contracts when its component proteins are energized by ATP. When the ring contracts, it drags the attached plasma membrane inward **2**. The sinking plasma membrane is visible on the outside of the cell as an indentation **3**. The indentation, which is called a **cleavage furrow**, advances around the cell and deepens until the cytoplasm (and the cell) is pinched in two **4**. Each of the two cells formed by this division has its own nucleus and some of the parent cell's cytoplasm, and each is enclosed by a plasma membrane.

Dividing plant cells face a particular challenge because they have stiff cell walls on the outside of their plasma membrane (Section 3.6). Accordingly, plant cells have their own mechanism of cytoplasmic division (Figure 8.7). By the end of anaphase, a set of short microtubules has formed on either side of the future plane of division. These microtubules now guide vesicles from Golgi bodies and the cell surface to the division plane **5**. There, the vesicles and their wall-building contents start to fuse into a disk-shaped cell plate **6**. The plate expands at its edges until it reaches the plasma membrane and attaches to it, thus partitioning the cytoplasm **7**. In time, the cell plate will develop into two new cell walls, so each of the descendant cells will be enclosed by its own plasma membrane and wall **8**.

Take-Home Message

How do eukaryotic cells divide?

- After mitosis, the cytoplasm of the parent cell typically is partitioned into two descendant cells, each with its own nucleus.

- In animal cells, a contractile ring pinches the cytoplasm in two. In plant cells, a cell plate that forms in the middle of the cell partitions the cytoplasm when it reaches and connects to the parent cell wall.

FIGURE 8.8 An oncogene causing a neoplasm. In this section of human breast tissue, activated growth factor receptor is stained *brown*. Normal cells are the ones with lighter staining. The heavily stained cells have formed a neoplasm; the abnormal overabundance of the activated receptor means that mitosis is being continually stimulated in these cells. Cells of most neoplasms have mutations that cause this receptor to be overproduced or overactive.

© From *Expression of the epidermal growth factor receptor (EGFR) and the phosphorylated EGFR in invasive breast carcinomas.* http://breast-cancer research.com/content/10/3/R49.

FIGURE 8.9 Checkpoint genes putting the brakes on cell division. Both of these images show the same nucleus, which has been exposed to radiation that damaged the DNA inside of it (compare Figure 6.8**A**). **A** *Green* dots pinpoint the location of the product of a gene called *53BP1*; **B** *red* dots show the location of the *BRCA1* gene product. Both proteins have clustered around the same chromosome breaks in the same nucleus; both function to recruit DNA repair enzymes. The integrated action of these and other checkpoint gene products blocks mitosis until the DNA breaks are fixed.

© Phillip B. Carpenter, Department of Biochemistry and Molecular Biology, University of Texas–Houston Medical School.

8.5 When Mitosis Becomes Pathological

❭ Cell Division Gone Wrong Sometimes a checkpoint gene mutates so that its protein product no longer works properly. In other cases, a mutation sabotages some gene control, and the cell makes too much or too little of a checkpoint gene's product. When enough checkpoint mechanisms fail, a cell loses control over its cell cycle. The cell may skip interphase, so division occurs over and over with no resting period. Signaling mechanisms that make an abnormal cell die may stop working. The problem is compounded because such mutations are passed along to the cell's descendants, which form a **neoplasm**, an accumulation of abnormally dividing cells.

A neoplasm that forms a lump in the body is called a **tumor**, but the two terms are sometimes used interchangeably. Once a tumor-causing mutation has occurred, the gene it affects is called an oncogene. An **oncogene** is any gene that helps transform a normal cell into a tumor cell (from Greek *onkos*, or bulging mass). Some oncogene mutations can be passed to offspring, which is a reason that some types of tumors tend to run in families.

Genes encoding proteins that promote mitosis are called proto-oncogenes because mutations can turn them into oncogenes (*proto-* means first). Genes involved in producing receptors for growth factors are examples. Growth factors are molecules that stimulate cell division and differentiation, and the plasma membranes of most cells have receptors for them. When a growth factor binds to its receptor, the receptor triggers a series of events that advances the cell cycle from interphase into mitosis. Mutations can result in a receptor that stimulates mitosis even when the growth factor is not present. Most neoplasms carry mutations resulting in an overactivity or overabundance of such receptors (Figure 8.8).

Checkpoint gene products that inhibit mitosis are called tumor suppressors because tumors form when they are missing. The products of the *BRCA1* and *BRCA2* genes are also tumor suppressors; mutations in these genes give rise to neoplasms in the breast, prostate, ovary, and other tissues. *BRCA* gene products are multifunctional proteins that help repair broken or otherwise damaged DNA (Figure 8.9). Cells infected with human papillomavirus (HPV) form skin growths called warts because this virus interferes with tumor suppressors. Some HPV strains are associated with neoplasms that form on the cervix.

cleavage furrow In a dividing animal cell, the indentation where cytoplasmic division will occur.
neoplasm An accumulation of abnormally dividing cells.
oncogene Gene that helps transform a normal cell into a tumor cell.
tumor A neoplasm that forms a lump.

1 Benign neoplasms grow slowly and stay in their home tissue.

2 Cells of a malignant neoplasm can break away from their home tissue.

3 Malignant cells can become attached to the wall of a blood vessel or lymph vessel. They release enzymes that create an opening in the wall, then enter the vessel.

4 The cells creep or tumble along in blood vessels, then leave the bloodstream the same way they got in. They may start growing in other tissues, a process called metastasis.

FIGURE 8.10 Animated! Metastasis. © Cengage Learning.

A Basal cell carcinoma is the most common type of skin cancer. This slow-growing, raised lump may be uncolored, reddish-brown, or black.

B The second most common form of skin cancer is a squamous cell carcinoma. This pink growth, firm to the touch, grows under the surface of skin.

C Melanoma, a malignant neoplasm of skin cells, spreads fastest. Cells form dark, encrusted lumps that may itch or bleed easily.

FIGURE 8.11 Skin cancer can be detected and treated early with periodic screening.

❯ Cancer Benign neoplasms such as ordinary skin moles are not dangerous (Figure 8.10). They grow very slowly, and their cells retain the adhesion proteins (Section 3.3) that keep them properly anchored in their home tissue **1**. A malignant neoplasm gets progressively worse, and is dangerous to health. The disease called **cancer** occurs when the abnormally dividing cells of a malignant neoplasm disrupt body tissues, both physically and metabolically. Malignant cells typically display the following three characteristics:

First, like cells of all neoplasms, malignant cells grow and divide abnormally. Controls that usually keep cells from getting overcrowded in tissues are lost, so their populations may reach extremely high densities with cell division occurring very rapidly. The number of small blood vessels that transport blood to the growing cell mass also increases abnormally.

Second, the cytoplasm and plasma membrane of malignant cells are altered. The cytoskeleton may be shrunken, disorganized, or both. Malignant cells typically have an abnormal chromosome number, with some chromosomes present in multiple copies, and others missing or damaged. The balance of metabolism is often shifted, as in an amplified reliance on ATP formation by fermentation rather than aerobic respiration.

Altered or missing proteins impair the function of the plasma membrane of malignant cells. For example, adhesion proteins in the plasma membrane of malignant cells are often defective or absent, so the cells do not stay anchored properly in tissues **2**. Malignant cells can slip easily into and out of vessels of the circulatory and lymphatic systems **3**. By migrating through these vessels, the cells can establish neoplasms elsewhere in the body **4**. The process in which malignant cells break loose from their home tissue and invade other parts of the body is called **metastasis**. Metastasis is the third hallmark of malignant cells.

Malignant cells can put an individual on a painful road to death. Each year, cancer causes 15 to 20 percent of all human deaths in developed countries. Some cancer-causing mutations are inherited, but lifestyle choices such as not smoking and avoiding exposure of unprotected skin to sunlight can reduce one's risk of acquiring new mutations. Some neoplasms can be detected early with periodic screening such as Pap tests or dermatology exams (Figure 8.11). If detected early enough, many neoplasms can be treated before metastasis occurs.

Take-Home Message

What is cancer?

- Mutations in checkpoint genes can cause neoplasms.
- Cancer is a disease that occurs when the abnormally dividing cells of a malignant neoplasm physically and metabolically disrupt body tissues.
- Although some mutations are inherited, lifestyle choices and early intervention can reduce one's risk of cancer.

8.6 Sex and Alleles

❯ On the Advantages of Sex If the function of reproduction is the perpetuation of one's genes, then an organism that reproduces by mitosis only would seem to win the evolutionary race. In asexual reproduction, a single individual gives rise to offspring. All of the individual's genes are passed to all of its offspring. Sexual reproduction mixes up the genes of two individuals, so only about half of a parent's genetic information is passed to each offspring.

So why sex? Consider that all offspring of asexual reproducers are clones of their parent, so all are adapted the same way to an environment—and all are equally vulnerable to challenges in it. By contrast, the offspring of sexual reproducers differ from parents, and often from one another, in details of their shared, inherited traits. As a group, their diversity offers them a better chance of surviving an environmental challenge than clones. The variation among offspring of sexual reproducers is part of the reason that beneficial mutations spread through sexually reproducing populations more quickly than asexual ones.

Think of a community of predator and prey, say, foxes and rabbits. If one rabbit is better than others at outrunning the foxes, it has a better chance of surviving and producing offspring that have better fox-evading traits. Thus, over generations, rabbits get faster. If one fox is better than others at catching faster rabbits, it has a better chance of surviving and producing offspring that have better rabbit-catching traits. Thus, over generations, foxes get faster. As one species changes, so does the other—an idea called the Red Queen hypothesis, after Lewis Carroll's book *Through the Looking Glass*. In the book, the Queen of Hearts tells Alice, "It takes all the running you can do, to keep in the same place."

Perhaps the most important advantage of sexual reproduction involves the inevitable occurrence of harmful mutations. A population of sexual reproducers has a better chance of weathering the effects of such mutations. With asexual reproduction, individuals bearing a harmful mutation necessarily pass it to all of their offspring. This outcome would be rare in sexual reproduction, because each offspring of a sexual union has a 50 percent chance of inheriting a parent's mutation. Thus, all else being equal, harmful mutations accumulate in an asexually reproducing population more quickly than in a sexually reproducing one.

❯ **Introducing Alleles** In somatic (body) cells of humans and many other sexually reproducing organisms, one chromosome of each homologous pair is maternal, and the other is paternal (Section 6.2). Homologous chromosomes carry the same genes (Figure 8.12**A**). However, the corresponding genes on maternal and paternal chromosomes often vary—just a bit—in DNA sequence. Over evolutionary time, unique mutations accumulate in separate lines of descent, and some of those mutations occur in genes. Thus, the DNA sequence of any gene may differ from the corresponding gene on the homologous chromosome (Figure 8.12**B**). Different forms of the same gene are called **alleles**.

Alleles may encode slightly different forms of the gene's product. Such differences influence thousands of traits shared by a species. For example, the beta globin gene (Section 7.6) has more than 700 alleles: one that causes sickle-cell anemia, several that cause beta thalassemia, and so on. This gene is one of about 20,000 genes in humans, and most of those genes have multiple alleles. Allele differences among individuals are one reason that the members of a sexually reproducing species are not identical. Offspring of sexual reproducers inherit new combinations of alleles, which is the basis of new combinations of traits.

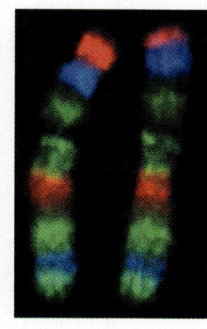

A Corresponding colored patches in this fluorescence micrograph indicate corresponding DNA sequences in a homologous chromosome pair. These chromosomes carry the same set of genes.

Genes occur in pairs on homologous chromosomes.

The members of each pair of genes may be identical, or they may differ slightly, as alleles.

B Homologous chromosomes carry the same series of genes, but the DNA sequence of any one of those genes might differ just a bit from that of its partner on the homologous chromosome.

FIGURE 8.12 Animated! Genes on chromosomes. Different forms of a gene are called alleles.

Credits: (a) Courtesy of Carl Zeiss MicroImaging, Thornwood, NY; (b) © Cengage Learning.

Take-Home Message

Why do populations that reproduce sexually tend to have the most variation in heritable traits?

- Paired genes on homologous chromosomes may vary in DNA sequence as alleles. Alleles arise by mutation.

- Alleles are the basis of differences in shared traits. Offspring of sexual reproducers inherit new combinations of parental alleles—thus new combinations of such traits.

alleles Forms of a gene with slightly different DNA sequences; may encode slightly different versions of the gene's product.

cancer Disease that occurs when a malignant neoplasm physically and functionally disrupts body tissues.

metastasis The process in which cancer cells spread from one part of the body to another.

8.7 Meiosis and the Life Cycle

Meiosis is the process inherent to sexual reproduction that gives rise to new combinations of parental alleles. Meiosis occurs in two stages that parcel chromosomes into new nuclei two times:

Interphase	Meiosis I	Meiosis II
DNA is replicated prior to meiosis I	Prophase I Metaphase I Anaphase I Telophase I	Prophase II Metaphase II Anaphase II Telophase II

The nucleus of a diploid (2*n*) cell contains two sets of chromosomes, one from each parent. DNA replication occurs before meiosis I begins, so each one of the chromosomes consists of two sister chromatids. The first stage of meiosis I is prophase I (Figure 8.13). During this phase, the chromosomes condense, and homologous chromosomes align tightly and swap segments (more about segment-swapping shortly). The pair of centrioles gets duplicated, and one pair moves to the opposite side of the cell as the nuclear envelope breaks up **❶**. A spindle forms, and by the end of prophase I the two chromosomes of each homologous pair are attached to spindle microtubules on opposite sides of the

FIGURE 8.13 Animated! Meiosis.

The micrographs show meiosis in a lily cell. The illustrations show two pairs of chromosomes in a diploid (2*n*) animal cell; homologous chromosomes are indicated in *blue* and *pink*.

Figure It Out: During which phase of meiosis does the number become reduced?

Answer: Anaphase I

Meiosis I   One diploid nucleus to two haploid nuclei

❶ Prophase I Homologous chromosomes condense, pair up, and swap segments. Spindle microtubules attach to them as the nuclear envelope breaks up.

❷ Metaphase I Homologous chromosome pairs are aligned in the middle of the cell. The two chromosomes of each pair are attached to spindle microtubules on opposite sides of the cell.

❸ Anaphase I All of the homologous chromosomes separate and begin heading toward the spindle poles.

❹ Telophase I A complete set of chromosomes clusters at both ends of the cell. A nuclear envelope forms around each set, so two haploid (*n*) nuclei form.

plasma membrane   spindle

nuclear envelope breaking up   centriole pair

one pair of homologous chromosomes

cell. These microtubules grow and shrink, pushing and pulling the chromosomes as they do. At metaphase I, all of the microtubules are the same length, and the chromosomes are aligned in the middle of the cell ❷.

In anaphase I, homologous chromosomes of each pair are pulled apart and toward opposite sides of the cell ❸. During telophase I, a new nuclear envelope forms around each cluster of chromosomes as the DNA loosens up ❹. The two new nuclei have a single set of chromosomes, so they are **haploid** (*n*). The cytoplasm often divides at this point. Each chromosome is still duplicated (it consists of two DNA molecules).

DNA replication does not occur before meiosis II, which proceeds simultaneously in both nuclei that formed in meiosis I. In prophase II, the chromosomes condense and the nuclear envelope breaks up. A new spindle forms. By the end of prophase II, each chromatid is attached to spindle microtubules on one side of the cell, and its sister chromatid is attached to spindle microtubules on the other side ❺. These microtubules push and pull the chromosomes, aligning them in the middle of the cell at metaphase II ❻. In anaphase II, the sister chromatids of each chromosome are pulled apart and toward opposite sides of the cell ❼. Each chromosome now consists of one molecule of DNA. During telophase II, new nuclear envelopes form around the chromosome clusters as the DNA loosens ❽. The cytoplasm often divides at this point to form four haploid (*n*) cells whose nuclei contain one set of (unduplicated) chromosomes.

haploid Having one of each type of chromosome characteristic of the species.

Meiosis II Two haploid nuclei to four haploid nuclei

❺ **Prophase II** The chromosomes condense. Spindle microtubules attach to each sister chromatid as the nuclear envelope breaks up.

❻ **Metaphase II** The (still duplicated) chromosomes are aligned in the middle of the cell.

❼ **Anaphase II** Sister chromatids separate. The now unduplicated chromosomes move toward opposite sides of the cell.

❽ **Telophase II** A complete set of chromosomes clusters at both ends of the cell. A new nuclear envelope forms around each set, so four haploid (*n*) nuclei form.

No DNA replication

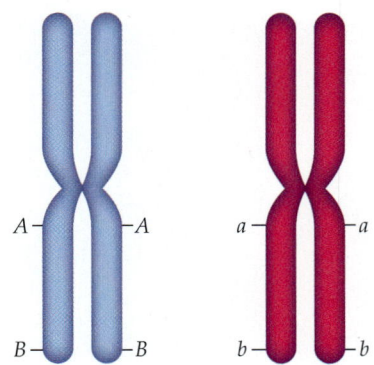

A Here, we focus on only two of the many genes on a chromosome. In this example, one gene has alleles *A* and *a*; the other has alleles *B* and *b*.

B Close contact between homologous chromosomes promotes crossing over between nonsister chromatids, which exchange corresponding pieces.

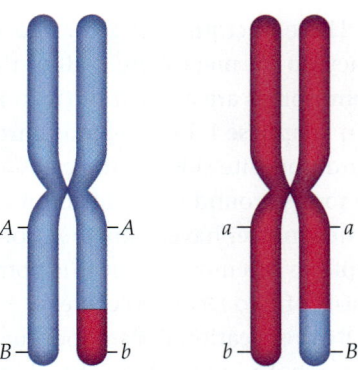

C Crossing over mixes up paternal and maternal alleles on homologous chromosomes.

FIGURE 8.14 Crossing over.

Blue signifies a paternal chromosome, and *pink*, its maternal homologue. For clarity, we show only one pair of homologous chromosomes and one crossover, but more than one crossover may occur in each chromosome pair.

© Cengage Learning.

❯ Meiosis Mixes Alleles Offspring produced by sexual reproduction typically vary in the details of shared traits because they inherit different alleles. Crossing over also contributes to this variation. **Crossing over** is a process by which a chromosome and its homologous partner exchange corresponding pieces of DNA during meiosis (Figure 8.14). When chromosomes condense in prophase I, each is drawn close to its homologous partner, so that the chromatids align along their length. This tight, parallel orientation favors crossing over. Homologous chromosomes may swap any segment of DNA along their length, although crossovers tend to occur more frequently in certain regions.

Crossing over is a normal and frequent process in meiosis. It contributes to variation among sexually reproducing individuals by shuffling maternal and paternal genes, so offspring inherit non-parental combinations of alleles.

❯ From Gametes to Offspring Sexual reproduction involves the fusion of specialized reproductive cells—**gametes**—from two parents. All gametes are haploid, and they arise by division of immature reproductive cells called germ cells. The germ cells of plants are haploid; they form by meiosis. Gametes arise when these cells divide by mitosis. Animal germ cells are diploid; they form during embryonic development and are set aside in special reproductive organs until maturity. We leave details of sexual reproduction for later chapters, but you will need to know a few concepts before you get there.

At fertilization, two haploid gametes fuse and produce a diploid **zygote**, which is the first cell of a new individual. Thus, meiosis halves the chromosome number, and fertilization restores it. If meiosis did not precede fertilization, the chromosome number would double with every generation. If the chromosome number changes, so does the individual's set of genetic instructions. An individual's set of chromosomes is like a fine-tuned blueprint that must be followed exactly, page by page, in order to build a body that functions normally. As you will see in Chapter 9, chromosome number changes can have drastic consequences, particularly in animals.

In a typical animal life cycle, a zygote develops into a multicelled individual (Figure 8.15**A**). Meiosis of germ cells in adult females gives rise to **eggs** (female gametes); meiosis of germ cells in adult males gives rise to **sperm** (male gametes). Two kinds of multicelled bodies form during the life cycle of a plant: sporophytes and gametophytes (Figure 8.15**B**). A sporophyte is diploid, and it produces spores by meiosis in its specialized parts. Spores consist of one or a few haploid cells. These cells undergo mitosis and give rise to haploid gametophytes inside which gametes form. For example, in flowering plants, gametophytes

---

crossing over Process in which homologous chromosomes exchange corresponding segments during meiosis.
egg Female gamete.
gamete Mature, haploid reproductive cell.
sperm Male gamete.
zygote First cell of a new individual; forms by the fusion of two haploid gametes.

form in flowers, which form on specialized reproductive shoots of the sporophyte body.

Fertilization also contributes to the variation that we see among offspring of sexual reproducers. Think about it in terms of human reproduction. Cells that give rise to human gametes have twenty-three pairs of homologous chromosomes. Each time a human germ cell undergoes meiosis, the four gametes that form end up with one of 8,388,608 (or 2^{23}) possible combinations of homologous chromosomes. On top of that, any number of genes may occur as different alleles on the maternal and paternal chromosomes, and crossing over makes mosaics of that genetic information. Then, out of all the male and female gametes that form, which two actually get together at fertilization is a matter of chance. Are you getting an idea of why such fascinating combinations of traits show up among the generations of your own family tree?

Take-Home Message

How does meiosis fit into the life cycle of plants and animals?

- During meiosis, the nucleus of a cell divides twice, so the diploid (2n) chromosome number is reduced to the haploid number (n) for forthcoming gametes.
- The union of two haploid gametes at fertilization results in a diploid zygote.
- Crossing over is recombination between nonsister chromatids of homologous chromosomes during prophase I. It makes new combinations of parental alleles among offspring of sexual reproducers.

8.8 Henrietta's Immortal Cells (revisited)

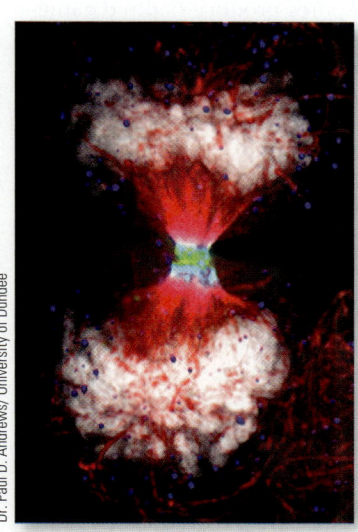

HeLa cells were used in early tests of taxol, a drug that keeps microtubules from disassembling and so interferes with mitosis. Frequent divisions make cancer cells more vulnerable to this poison than normal cells. A more recent example of cancer research is shown in Figure 8.1 (and at *left*). In this micrograph of mitotic HeLa cells, the *blue* dots pinpoint a protein that helps sister chromatids stay attached to one another at the centromere. *Green* identifies an enzyme that helps attach spindle microtubules (*red*) to centromeres; defects in this enzyme or its expression result in descendant cells with too many or too few chromosomes, an outcome that is a hallmark of cancer. At this stage of telophase, both proteins should be closely associated midway between chromosome clusters (*white*). The abnormal distribution is a sign that the chromosomes are not properly attached to the spindle.

These days, physicians and researchers are required to obtain a signed consent form before they take tissue samples from a patient. No such requirement existed in the 1950s, when it was common for doctors to experiment on patients without their knowledge or consent. Thus, the young resident who was treating Henrietta Lacks's cancerous cervix probably never even thought about asking permission to use it for research before he took a sample of it. That sample was the one that the Geys used to establish the HeLa cell line.

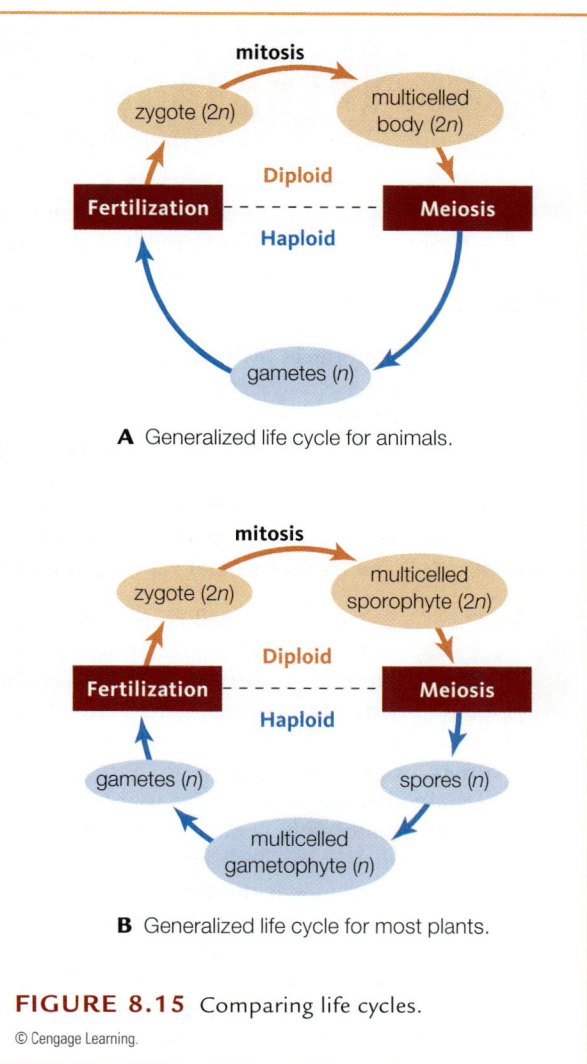

A Generalized life cycle for animals.

B Generalized life cycle for most plants.

FIGURE 8.15 Comparing life cycles.
© Cengage Learning.

WHERE YOU ARE GOING . . .

Mitosis is the basis of reproduction of single-celled eukaryotes (Section 13.6). Sexual reproduction is detailed in Chapters 26 (animals) and 28 (plants). We revisit the variation in traits among individuals of sexually reproducing species in the context of inheritance (Chapter 9) and natural selection (Chapter 12). The HPV virus will come up again as a cause of cervical cancer (Section 22.1), and in context of sexually transmitted diseases (Section 26.5).

Summary

Section 8.1 An immortal line of human cells (HeLa) is a legacy of cancer victim Henrietta Lacks. Researchers all over the world who are trying to unravel the mechanisms of cancer continue to work with these cells.

Section 8.2 A cell reproduces by dividing: nucleus first, then cytoplasm. Each descendant cell receives a complete set of chromosomes and some cytoplasm. Nuclear division mechanisms of **mitosis** and **meiosis** partition chromosomes of a parent cell into new nuclei. Mitosis is the basis of growth, cell replacements, and tissue repair in multicelled species, and **asexual reproduction** in many species. Meiosis is the basis of **sexual reproduction** in eukaryotes.

Section 8.3 A **cell cycle** includes all the stages through which a eukaryotic cell passes during its lifetime; it starts when a new cell forms, and ends when the cell reproduces. Most of a cell's activities, including DNA replication of its **homologous chromosomes**, occur in **interphase**.

Mitosis maintains the chromosome number. During **prophase**, the duplicated chromosomes condense. Microtubules form a **spindle**, and the nuclear envelope breaks up. Spindle microtubules attach to the chromosomes at the centromere. At **metaphase**, all of the chromosomes are aligned in the middle of the cell. During **anaphase**, the sister chromatids of each chromosome detach and move toward opposite sides of the cell. During **telophase**, a nuclear envelope forms around the two clusters of chromosomes. Two new nuclei, each with the parental chromosome number, are the result.

Section 8.4 In most cases, nuclear division is followed by cytoplasmic division. A contractile ring pulls the plasma membrane of an animal cell inward, forming a **cleavage furrow** that eventually pinches the cytoplasm in two. In a plant cell, vesicles merge as a cell plate that expands until it fuses with the parent cell wall and partitions the cytoplasm. The cell plate develops into two new cell walls, so each new cell is surrounded by its own wall.

Section 8.5 The products of checkpoint genes are part of the controls governing the cell cycle. Mutations can disrupt checkpoint gene products or their expression. When all checkpoint mechanisms fail, a cell loses control over its cell cycle, and the cell's descendants form a **neoplasm**. Neoplasms may form lumps called **tumors**. Malignant cells of a neoplasm can break loose from their home tissue and colonize other parts of the body, a process called **metastasis**. **Cancer** arises when malignant cells disrupt body tissues, physically and functionally. Some mutations that turn normal genes into tumor-causing **oncogenes** are inherited, but periodic screening and lifestyle choices can reduce an individual's risk of getting cancer.

Section 8.6 The offspring of sexual reproducers typically vary in shared, inherited traits. This variation in traits can offer an evolutionary advantage over genetically identical offspring produced by asexual reproduction.

Sexual reproduction produces offspring with pairs of chromosomes, one of each homologous pair from the mother and the other from the father. The two chromosomes of a pair carry the same genes. The DNA sequence of paired genes often varies slightly, in which case they are called **alleles**. Alleles are the basis of differences in shared, heritable traits. They arise by mutation.

Section 8.7 DNA replication occurs before meiosis begins, so each chromosome consists of two molecules of DNA (sister chromatids). Two nuclear divisions occur during meiosis. In the first nuclear division (meiosis I), all of the homologous chromosomes line up and then move apart. Two new nuclear envelopes form around the two clusters of chromosomes. This stage reduces the chromosome number from diploid (*2n*) to **haploid** (*n*). The second nuclear division (meiosis II) occurs in both nuclei that formed in meiosis I. Sister chromatids separate in this stage, so at the end of meiosis each chromosome consists of one molecule of DNA. Four haploid nuclei typically form, each with a complete set of (unduplicated) chromosomes.

 During meiosis I, nonsister chromatids of homologous chromosomes exchange segments at the same place along their length. Such **crossing over** mixes up the alleles on maternal and paternal chromosomes, thus giving rise to combinations of alleles not present in either parental chromosome.

Germ cells in the reproductive organs of most animals give rise to haploid **gametes** (**sperm** or **eggs**). Two types of multicelled bodies are typical in life cycles of plants. Meiosis in the diploid sporophyte produces haploid spores. Spores produce haploid gametophytes, which in turn produce gametes.

The fusion of two haploid gametes at fertilization results in a diploid **zygote**. Thus, meiosis halves the chromosome number, and fertilization restores it.

Self-Quiz   Answers in Appendix I

1. Mitosis and cytoplasmic division function in _____ .
   a. asexual reproduction of single-celled eukaryotes
   b. growth and tissue repair in multicelled species
   c. gamete formation in bacteria and archaea
   d. sexual reproduction in plants and animals
   e. both a and b

2. A cell with two of each type of chromosome has a(n) _____ chromosome number.
   a. diploid   c. tetraploid
   b. haploid   d. abnormal

3. How many chromatids does a duplicated chromosome have?
   a. one   b. two   c. three   d. four

4. _____ maintains the chromosome number; _____ halves it.
   a. mitosis; meiosis   b. meiosis; mitosis

5. Except for a pairing of sex chromosomes, homologous chromosomes _____ .
   a. carry the same genes   c. are the same length
   b. are the same shape   d. all of the above

Digging Into Data

HeLa Cells Are a Genetic Mess

HeLa cells continue to be an extremely useful tool in cancer research. One early finding was that HeLa cells can vary widely in chromosome number. The panel of chromosomes in Figure 8.16, originally published in 1989 by Nicholas Popescu and Joseph DiPaolo, shows all of the chromosomes in a single metaphase HeLa cell.

1. What is the chromosome number of this HeLa cell?

2. How many extra chromosomes does this cell have, compared to a normal human body cell?

3. Can you tell that this cell came from a female? How?

FIGURE 8.16 Chromosomes in a HeLa cell.
© Dr. Thomas Ried, NIH and the American Association for Cancer Research.

6. Interphase is the part of the cell cycle when _____ .
   a. a cell ceases to function
   b. a cell forms its spindle apparatus
   c. a cell grows and duplicates its DNA
   d. mitosis proceeds

7. After mitosis, the chromosome number of the two new cells is _____ the parent cell's.
   a. the same as   c. rearranged compared to
   b. one-half of   d. doubled compared to

8. Only _____ is not a stage of mitosis.
   a. prophase   b. interphase   c. metaphase   d. anaphase

9. The main evolutionary advantage of sexual over asexual reproduction is that it produces _____ .
   a. more offspring per individual
   b. more variation among offspring
   c. healthier offspring

10. Alternative forms of the same gene are _____ .
    a. gametes   c. alleles
    b. homologous   d. oncogenes

11. Meiosis functions in _____ .
    a. asexual reproduction of single-celled eukaryotes
    b. growth and tissue repair in multicelled species
    c. gamete formation in bacteria and archaea
    d. sexual reproduction in plants and animals
    e. both a and b

12. Crossing over mixes up _____ .
    a. chromosomes   c. zygotes
    b. alleles   d. gametes

13. The cell illustrated on the *right* is in anaphase I, not anaphase II. I know this because _____ .
    a. crossing over has already occurred
    b. the chromosomes are still duplicated
    c. there are two centriole pairs
    d. sister chromatids have separated

© Cengage Learning.

14. Sexual reproduction in animals requires _____ .
    a. meiosis   c. spore formation
    b. fertilization   d. a and b

15. Match each stage of mitosis with the events listed.
    _____ prophase   a. sister chromatids move apart
    _____ metaphase   b. chromosomes start to condense
    _____ anaphase   c. new nuclei form
    _____ telophase   d. all chromosomes are aligned
                           in the middle of the cell

16. Match each term with the best description.
    _____ cell plate   a. lump of abnormal cells
    _____ spindle   b. made of microfilaments
    _____ tumor   c. divides plant cells
    _____ cleavage furrow   d. produces the spindle
    _____ contractile ring e. migrating, metastatic cells
    _____ cancer   f. made of microtubules
    _____ centriole pair   g. indentation
    _____ gamete   h. haploid
    _____ zygote   i. first cell of new individual

Critical Thinking

1. When a cell reproduces by mitosis and cytoplasmic division, does its life end?

2. The eukaryotic cell in the photo on the *left* is in the process of cytoplasmic division. Is this cell from a plant or an animal? How do you know?

3. Make a simple sketch of meiosis in a cell with a diploid chromosome number of 4. Now try it when the chromosome number is 3.

4. The diploid chromosome number for the body cells of a frog is 26. What would that number be after three generations if meiosis did not occur before gamete formation?

5. Which nuclear division, meiosis I or meiosis II, is most like mitosis? Why?

6. Exposure to radioisotopes or other sources of radiation can damage DNA. Humans exposed to high levels of radiation face a condition called radiation poisoning. Why do you think that hair loss and damage to the lining of the gut are early symptoms of radiation poisoning? Speculate about why exposure to radiation is used as a therapy to treat some kinds of cancers.

Patterns of Inheritance

9

Be sure you understand traits (Section 1.4), chromosome structure and function (6.2), genes (7.2), sexual reproduction and alleles (8.6), and meiosis (8.7) before beginning this chapter. You will revisit sampling error (1.7), the nucleus and other components of eukaryotic cells (3.5), pigments (5.2), mutations (7.6), and gene control and epigenetics (7.7).

9.1 Menacing Mucus

In 1988, researchers discovered a gene that, when mutated, causes cystic fibrosis (CF). Cystic fibrosis is the most common fatal genetic disorder in the United States. The gene, *CFTR*, encodes a protein that actively transports chloride ions out of epithelial cells. Sheets of these cells line the passageways and ducts of the lungs, liver, pancreas, intestines, reproductive system, and skin. When the CFTR protein pumps chloride ions out of these cells, water follows the ions by osmosis. The two-step process maintains a thin film of water on the surface of the epithelial sheets. Mucus slides easily over the wet sheets of cells.

The mutation most commonly associated with CF is a deletion that prevents the CFTR protein from reaching the cell surface. Normally, a newly synthesized CF protein moves from the endoplasmic reticulum (ER) to a Golgi body that routes it to the plasma membrane. The CF deletion leaves the protein stranded in the ER. Although the altered protein functions properly, it never reaches the cell surface to do its job.

Epithelial cells that have no CFTR in their plasma membranes cannot transport chloride ions properly. Too few chloride ions leave these cells. Not enough water leaves them either, so the surfaces of epithelial cell sheets are drier than they should be. Mucus that normally slips and slides through the body's tubes sticks to the walls of the tubes instead. Thick globs of mucus accumulate and clog passageways and ducts throughout the body. Breathing becomes difficult as the mucus obstructs the smaller airways of the lungs.

The CFTR protein also functions as a receptor: It binds to bacteria, and this binding triggers endocytosis in epithelial cells lining the body's ducts and passages. Bacteria that are engulfed by these cells set in motion the immune system's defensive responses. When epithelial cells lack the CFTR protein, this early alert system fails, and disease-causing bacteria have time to multiply before they are detected. Thus, chronic bacterial infections of the intestine and lungs are hallmarks of cystic fibrosis. Daily routines of posture changes and thumps on the chest and back help clear the lungs of some of the thick mucus, and antibiotics help control infections, but there is no cure. Even with a lung transplant, most cystic fibrosis patients live no longer than thirty years, at which time their tormented lungs usually fail (Figure 9.1).

About 1 in 25 people carry the CF mutation in one of their two copies of the *CFTR* gene, but most of them do not realize it because they do not have the symptoms of cystic fibrosis. The disease occurs only when a person inherits two mutated genes, one from each parent. This unlucky event occurs in about 1 of 3,300 births worldwide.

Cody, 23 Jeff, 21 Lindsay, 22

Ben, 23 Savannah, 19 Brandon, 18

FIGURE 9.1 A few of the many victims of cystic fibrosis. At least one young person dies every day in the United States from complications of this disease.

Credits: clockwise form top left, Courtesy of © The Cody Dieruf Benefit Foundation, www.breathinisbelievin.org; Courtesy of © Bobby Brooks and The Family of Jeff Baird; Courtesy of © Steve & Ellison Widener and Breathe Hope, http://breathehope.tamu.edu; Courtesy of The Family of Benjamin Hill, reprinted with permission of © Chappell/Marathonfoto; Courtesy of © The Family of Savannah Brooke Snider; Courtesy of © The Family of Brandon Herriott.

9.2 Tracking Traits

In the nineteenth century, people thought that hereditary material must be some type of fluid, with fluids from both parents blending at fertilization like milk into coffee. However, the idea of "blending inheritance" failed to explain what people could see with their own eyes. Children sometimes have traits such as freckles that do not appear in either parent. A cross between a black horse and a white one does not produce gray offspring.

The naturalist Charles Darwin did not accept the idea of blending inheritance, but he could not come up with an alternative hypothesis even though

The Moravian Museum, Brno

inheritance was central to his theory of natural selection. (We return to Darwin and his theory of natural selection in Chapter 11.) Darwin did not know that hereditary information (DNA) is divided into discrete units (genes), an insight that is critical to understanding how traits are inherited. However, even before Darwin presented his theory, someone had been gathering evidence that would support it. Gregor Mendel (*left*), an Austrian monk, had been carefully breeding thousands of pea plants. By meticulously documenting the passage of traits from one generation to the next, Mendel had been collecting evidence of how inheritance works.

> **Mendel's Experimental Approach** Mendel studied variation in the traits of the garden pea plant. This species is naturally self-fertilizing, which means its flowers produce male and female gametes (Section 8.7) that form viable embryos when they meet up. In order to study inheritance, Mendel had to carry out controlled matings between individuals with specific traits, then observe and document the traits of their offspring. Control over the reproduction of an individual pea plant begins with preventing it from self-fertilizing. Mendel did this by removing a flower's pollen-bearing anthers, then brushing its egg-bearing carpel with pollen from another plant (Figure 9.2**A,B**). He collected the seeds from the cross-fertilized individual, and recorded the traits of the new pea plants that grew from them (Figure 9.2**C,D**).

Many of Mendel's experiments started with plants that "breed true" for particular traits such as white flowers or purple flowers. Breeding true for a trait means that, new mutations aside, all offspring have the same form of the trait as the parent(s), generation after generation. For example, all offspring of pea plants that breed true for white flowers also have white flowers. As you will see in the next section, Mendel discovered that the traits of the offspring of cross-fertilized pea plants often appear in predictable patterns. Mendel's meticulous work tracking pea plant traits led him to conclude (correctly) that hereditary information passes from one generation to the next in discrete units.

> **Inheritance in Modern Terms** DNA was not proven to be hereditary material until the 1950s, but Mendel discovered its units, which we now call genes, almost a century before then. Today, we know that individuals of a species share certain traits because their chromosomes carry the same genes.

Remember from Section 8.6 that body cells of humans and other animals have pairs of genes on pairs of homologous chromosomes; the two genes of a pair may be identical, or they may vary as alleles. Organisms breed true for a trait because they carry identical alleles of genes governing that trait. We say

A The flowers of garden pea plants have reproductive parts called anthers and carpels. Pollen grains that form in anthers produce male gametes; female gametes form in carpels.

B Experimenters can control the transfer of hereditary material from one pea plant to another by snipping off a flower's anthers (to prevent the flower from self-fertilizing), and then brushing pollen from another flower onto its carpel. In this example, pollen from a plant that has purple flowers is brushed onto the carpel of a white-flowered plant.

C Later, seeds develop inside pods of the cross-fertilized plant. An embryo in each seed develops into a mature pea plant.

D Every plant that arises from the cross has purple flowers. Predictable patterns such as this are evidence of how inheritance works.

FIGURE 9.2 Animated!
A breeding experiment with garden pea plants.

Credits: (a) left, Jean M. Labat/ ardea.com; right, Jo Whitworth/Gap Photo/Visuals Unlimited, Inc.; (b,c,d) © Cengage Learning.

genotype ──────────→ phenotype

PP (homozygous
for dominant allele *P*) ──────→

pp (homozygous
for recessive allele *p*) ──────→

Pp (heterozygous
for alleles *P* and *p*) ──────→

FIGURE 9.3 Genotype gives rise to phenotype. In this example, the dominant allele *P* specifies purple flowers; the recessive allele *p*, white flowers.

© Tamara Kulikova/ Shutterstock.

that an individual with two identical alleles of a gene is **homozygous** for the allele. The particular set of alleles that an individual carries is called the individual's **genotype**.

Alleles are the major source of variation in a trait shared by the individuals of a species. A mutation may cause a trait to change, as when a gene that causes flowers to be purple mutates so the resulting flowers are white. "White-flowered" is an example of a **phenotype**, which refers to an individual's observable traits. Any mutated gene is an allele, whether or not it affects phenotype.

We say that an individual with two different alleles of a gene is **heterozygous** (*hetero*– means mixed). The heterozygous phenotype depends on the alleles: In many cases, the effect of one allele influences the effect of the other, and the outcome of this interaction is reflected in the individual's phenotype. An allele is **dominant** when its effect masks that of a **recessive** allele paired with it. Usually, a dominant allele is represented by an italic capital letter such as *A*; a recessive allele, with a lowercase italic letter such as *a*. Consider the purple- and white-flowered pea plants that Mendel studied. In these plants, the allele that specifies purple flowers (let's call it *P*) is dominant over the allele that specifies white flowers (*p*). Thus, a pea plant homozygous for the dominant allele (*PP*) has purple flowers; one homozygous for the recessive allele (*pp*) has white flowers. A heterozygous plant (*Pp*) has purple flowers (Figure 9.3).

Take-Home Message

How do alleles contribute to traits?

- Genotype refers to the particular set of alleles that an individual carries. Genotype is the basis of phenotype, which refers to the individual's observable traits.

- A homozygous individual has two identical alleles of a gene. A heterozygous individual has two nonidentical alleles of the gene.

- A dominant allele masks the effect of a recessive allele paired with it in a heterozygous individual.

9.3 Mendelian Inheritance Patterns

❯ Monohybrid Crosses When homologous chromosomes separate during meiosis (Section 8.7), the gene pairs on those chromosomes separate too. Each gamete that forms carries only one of the two genes of a pair. Let's use our alleles for purple and white flowers in an example (Figure 9.4). Plants homozygous for the dominant allele (*PP*) can only make gametes that carry allele *P* **①**. Plants homozygous for the recessive allele (*pp*) can only make gametes that carry allele *p* **②**. If these homozygous plants are crossed (*PP* × *pp*), only one outcome is possible: A gamete carrying allele *P* meets up with a gamete carrying allele *p*. All of the offspring of this cross will have both alleles—they will be heterozygous (*Pp*) **③**. Because all of the offspring will carry the dominant allele *P*, all will have purple flowers. A grid called a **Punnett square** is helpful for predicting the genetic and phenotypic outcomes of such crosses (Figure 9.5).

Our example illustrated a pattern so predictable that it can be used as evidence of a dominance relationship between alleles. Breeding experiments use such patterns to test genotype. In a testcross, an individual that has a dominant trait (but an unknown genotype) is crossed with an individual known to be homozygous for the recessive allele. The pattern of traits among the offspring of the cross can reveal whether the tested individual is heterozygous or homozy-

dominant Refers to an allele that masks the effect of a recessive allele in heterozygous individuals.

genotype The particular set of alleles that is carried by an individual's chromosomes.

heterozygous Having two different alleles of a gene.

homozygous Having identical alleles of a gene.

monohybrid cross Cross between two individuals identically heterozygous for one gene; for example *Aa* × *Aa*.

phenotype An individual's observable traits.

Punnett square Diagram used to predict the genetic and phenotypic outcome of a cross.

recessive Refers to an allele with an effect that is masked by a dominant allele in heterozygous individuals.

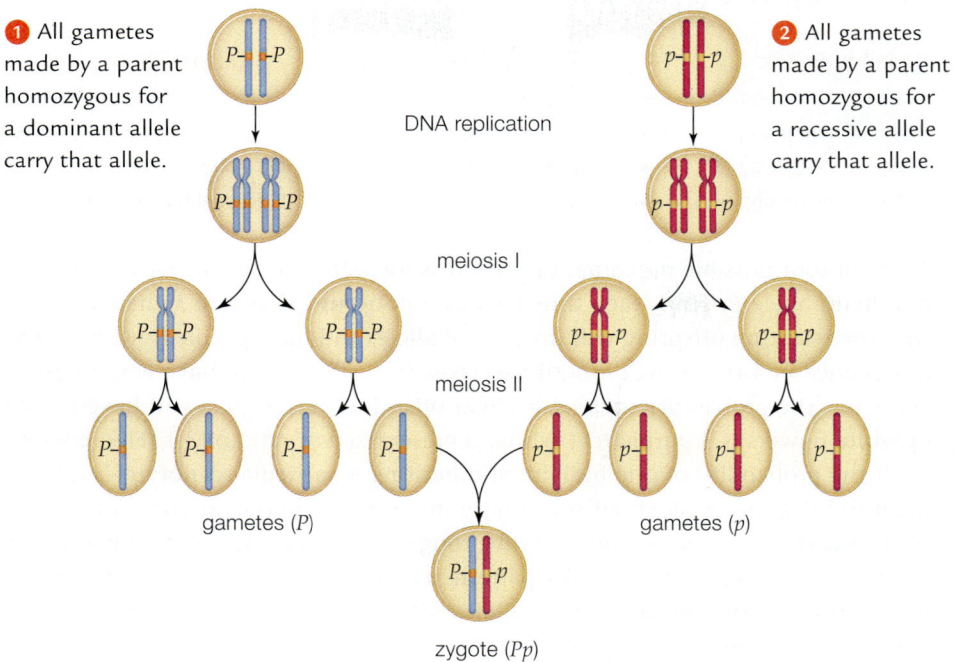

1 All gametes made by a parent homozygous for a dominant allele carry that allele.

DNA replication

meiosis I

meiosis II

gametes (*P*)

2 All gametes made by a parent homozygous for a recessive allele carry that allele.

gametes (*p*)

zygote (*Pp*)

3 If these two parents are crossed, the union of any of their gametes at fertilization produces a zygote with both alleles. All offspring of this cross will be heterozygous.

FIGURE 9.4 Segregation of genes on homologous chromosomes into gametes. Homologous chromosomes separate during meiosis, so the pairs of genes they carry separate too. Each of the resulting gametes carries one of the two members of each gene pair. For clarity, only one set of chromosomes is illustrated. © Cengage Learning.

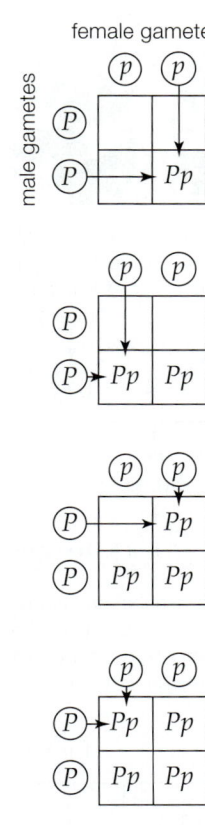

FIGURE 9.5 Making a Punnett square. Parental gametes are listed in circles on the top and left sides of a grid. Each square is filled with the combination of alleles that would result if the gametes in the corresponding row and column met up.

© Cengage Learning.

gous. For example, we may do a testcross between a purple-flowered pea plant (which could have a genotype of either *PP* or *Pp*) and a white-flowered pea plant (*pp*). If all of the offspring of this cross had purple flowers, we could be reasonably certain that the genotype of the purple-flowered parent was *PP*.

Dominance relationships between alleles determine the phenotypic outcome of a **monohybrid cross**, in which individuals identically heterozygous for alleles of one gene—*Pp*, for example—are bred together or self-fertilized. The frequency at which traits associated with the alleles appear among the offspring depends on whether one of the alleles is dominant over the other.

To perform a monohybrid cross, we would start with two individuals that breed true for two different forms of a trait. In garden pea plants, flower color (purple and white) is one example of a trait with two distinct forms, but there are many others. A cross between the two true-breeding individuals yields offspring that are identically heterozygous for the alleles that govern the trait. When these F_1 (first generation) offspring are crossed, the frequency at which the two traits appear in the F_2 (second generation) offspring offers information about a dominance relationship between the two alleles. (F is an abbreviation for filial, which means offspring.)

A cross between purple-flowered heterozygous plants (*Pp*) offers an example of a monohybrid cross. Each plant can make two types of gametes: ones that

A All of the F₁ (first generation) offspring of a cross between two plants that breed true for different forms of a trait are identically heterozygous (*Pp*). These offspring make two types of gametes: *P* and *p*.

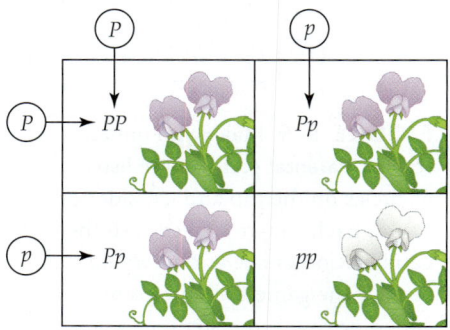

B A cross between two of the identically heterozygous F₁ offspring is a monohybrid cross. In this example, the phenotype ratio among the F₂ (second generation) offspring is 3:1 (three purple to one white).

FIGURE 9.6 Animated! A monohybrid cross.

Figure It Out: In this example, how many possible genotypes are there in the F₂ generation?

Answer: Three (PP, Pp, and pp)

© Cengage Learning.

carry a *P* allele, and ones that carry a *p* allele (Figure 9.6**A**). So, in a monohybrid cross between *Pp* plants (*Pp* × *Pp*), the two types of gametes can meet up in four possible ways at fertilization:

Possible Event	Offspring Genotype	Resulting Phenotype
Sperm *P* meets egg *P* ⟶	zygote genotype is *PP* ⟶	individual has purple flowers
Sperm *P* meets egg *p* ⟶	zygote genotype is *Pp* ⟶	individual has purple flowers
Sperm *p* meets egg *P* ⟶	zygote genotype is *Pp* ⟶	individual has purple flowers
Sperm *p* meets egg *p* ⟶	zygote genotype is *pp* ⟶	individual has white flowers

Three of four possible outcomes of this cross include at least one copy of the dominant allele *P*. Thus, each time fertilization occurs, there are 3 chances in 4 that the resulting offspring will inherit a *P* allele, and have purple flowers. There is 1 chance in 4 that it will inherit two recessive *p* alleles, and have white flowers. Thus, the probability that a particular offspring of this cross will have purple or white flowers is 3 purple to 1 white, represented as a ratio of 3:1 (Figure 9.6**B**).

If the probability of one individual inheriting a particular genotype is difficult to imagine, think about probability in terms of the phenotypes of many offspring. In this example, there will be roughly three purple-flowered plants for every white-flowered one. The 3:1 pattern is an indication that the purple and white flower colors are specified by alleles with a clear dominance relationship: Purple is dominant, and white is recessive.

> **Dihybrid Crosses** A monohybrid cross allows us to track alleles of one gene pair. What about alleles of two gene pairs? In a **dihybrid cross**, individuals identically heterozygous for alleles of two genes (dihybrids) are crossed. As with a monohybrid cross, the pattern of traits seen in the offspring of the cross depends on the dominance relationships between alleles of the genes.

To make a dihybrid cross, we would start with individuals that breed true for two different traits. Let's use a gene for flower color (*P*, purple; *p*, white) and one for plant height (*T*, tall; *t*, short) in an example. Figure 9.7 shows a dihybrid cross starting with a parent plant that breeds true for purple flowers and tall stems (*PPTT*), and one that breeds true for white flowers and short stems (*pptt*). The *PPTT* plant only makes gametes with the dominant alleles (*PT*); the *pptt* plant only makes gametes with the recessive alleles (*pt*) **❶**. So, all offspring from a cross between these two plants (*PPTT* × *pptt*) will be dihybrids (*PpTt*) with purple flowers and tall stems **❷**.

Four combinations of *P* and *T* alleles are possible in the gametes of *PpTt* dihybrids **❸**. If two of these plants are crossed (a dihybrid cross, *PpTt* × *PpTt*), the four types of gametes can combine in sixteen possible ways at fertilization **❹**. Nine of the sixteen genotypes would give rise to tall plants with purple flowers; three, to short plants with purple flowers; three, to tall plants with white flowers; and one, to short plants with white flowers. Thus, the ratio of phenotypes among the offspring of this dihybrid cross would be 9:3:3:1.

Mendel discovered the 9:3:3:1 ratio of phenotypes among offspring of dihybrid crosses, but he had no idea what it meant. He published his results in 1866, but apparently his work was read by few and understood by no one at the time. He died in 1884, never to know that his work would be the starting point for modern genetics.

> **The Contribution of Crossovers** How two gene pairs get sorted into gametes depends partly on whether the two genes are on the same chromosome. When homologous chromosomes separate during meiosis, either member of the pair can end up in either of the two new nuclei that form. Thus, gene

parent plant homozygous for purple flowers and long stems

parent plant homozygous for white flowers and short stems

PPTT

pptt

PT × pt

PpTt dihybrid

PT Pt pT pt

four types of gametes

1 Meiosis in a homozygous individual results in one type of gamete.

2 A cross between the two homozygous individuals yields offspring heterozygous for two genes (dihybrids).

3 Meiosis in dihybrid individuals results in four kinds of gametes.

4 If two of the dihybrid individuals are crossed, the four types of gametes can meet up in 16 possible ways. Out of 16 possible genotypes of the offspring, 9 will result in plants that are purple-flowered and tall; 3, purple-flowered and short; 3, white-flowered and tall; and 1, white-flowered and short. The ratio of phenotypes in this dihybrid cross is 9:3:3:1.

FIGURE 9.7 Animated! A dihybrid cross between plants that differ in flower color and plant height. In this example, *P* and *p* are dominant and recessive alleles for flower color; *T* and *t* are dominant and recessive alleles for height. © Cengage Learning.

pairs on one chromosome tend to sort into gametes independently of gene pairs on other chromosomes.

What about genes on the same chromosome? Mendel studied seven genes in pea plants, which have seven pairs of chromosomes. Was he lucky enough to choose one gene on each of those seven chromosome pairs? As it turns out, some of the genes Mendel studied *are* on the same chromosome. The genes are far enough apart that crossing over occurs between them very frequently—so frequently that they tend to assort into gametes independently, just as if they were on different chromosomes. By contrast, genes that are very close together on a chromosome do not tend to assort independently into gametes, because crossing over does not happen very often between them. Thus, gametes usually end up with parental combinations of alleles of these genes. Such gene linkages were identified by tracking inheritance in human families over several generations. One thing became clear: Crossovers are not at all rare, and are often a required step in order for meiosis to run to completion.

〉 A Human Example: Skin Color The color of human skin begins with skin cell organelles called melanosomes. Melanosomes make two types of melanin: one brownish-black; the other, reddish. Most people have about the same number of melanosomes in their skin cells. Variations in skin color occur because the kinds and amounts of melanins vary among people, as does the formation, transport, and distribution of the melanosomes in the skin.

These variations have a genetic basis. The products of more than 100 genes are involved in the synthesis of melanin, and the formation and deposition of melanosomes. Different alleles of those genes contribute to variations of human skin color. For example, one gene encodes a transport protein in melanosome

dihybrid cross Cross between two individuals identically heterozygous for two genes; for example *AaBb* × *AaBb*.

FIGURE 9.8 Variation in human skin color (*left*) begins with differences in alleles.

Above, fraternal twin girls with their parents. Both of the children's grandmothers are of European descent, and have pale skin. Both of their grandfathers are of African descent, and have dark skin. The twins inherited different alleles of some of the genes that affect skin color from their mixed-race parents, who, given the twins' appearance, must be heterozygous for those alleles.

Credits: left, Richard A. Sturm, "Molecular genetics of human pigmentation diversity," *Human Molecular Genetics*, 2009 Apr 15;18(R1):R9-17, by permission of Oxford University Press; right, © Gary Roberts/worldwidefeatures.com.

membranes. Nearly all people of African, Native American, or east Asian descent carry the same allele of this gene. Between 6,000 and 10,000 years ago, a mutation gave rise to a different allele. The mutation, a single base-pair substitution, changed one amino acid of the transport protein from alanine to threonine. The change results in less melanin—and lighter skin color—than the original African allele does. Today, nearly all people of European descent carry this mutated allele.

A person of mixed ethnicity may make gametes that contain different combinations of alleles for dark and light skin. It is fairly rare that one of those gametes contains all of the alleles for dark skin, or all of the alleles for light skin, but it does happen (Figure 9.8).

Skin color is only one of many human traits that vary as a result of single nucleotide mutations. The small scale of such changes offers a reminder that all of us share the genetic legacy of a common ancestry.

Take-Home Message

How do alleles contribute to traits?

- Diploid cells have pairs of genes, on pairs of homologous chromosomes. The two genes of a pair (which may be identical or not) are separated from each other during meiosis, so they end up in different gametes.

- In most cases, the two genes of a pair are distributed into gametes independently of other gene pairs during meiosis.

All humans share the genetic legacy of a common ancestry.

9.4 Beyond Simple Dominance

The inheritance patterns discussed in the last two sections offer examples of simple dominance, in which the effect of a recessive allele is fully masked by that of a dominant one. Other types of dominance relationships are more complex. In some cases, two alleles affect a trait equally; in others, one is incompletely dominant over the other. Many traits are influenced by multiple genes, and many single genes influence multiple traits.

❯ Codominance With **codominance**, both alleles are fully expressed in heterozygotes, and neither is dominant or recessive. Codominance may occur in multiple allele systems, in which three or more alleles persist in a population. The three alleles of the *ABO* gene offer an example. An enzyme encoded by this gene modifies a carbohydrate on the surface of human red blood cell membranes. The *A* and *B* alleles encode slightly different versions of the enzyme, which in turn modify the carbohydrate differently. The *O* allele has a mutation that prevents its enzyme product from becoming active at all.

Genotypes:
AA
or
AO
AB
BB
or
BO
OO

Phenotypes
(blood type): A   AB   B   O

FIGURE 9.9 Combinations of alleles that are the basis of blood type.

Credits: top, © David Scharf/ Peter Arnold, Inc.; bottom, © Cengage Learning.

homozygous
parent (*RR*) × homozygous
parent (*rr*) → heterozygous
offspring (*Rr*)

A Cross a red-flowered with a white-flowered snapdragon, and all of the offspring will have pink flowers.

B If two of the pink-flowered snapdragons are crossed, the phenotypes of their offspring will occur in a 1:2:1 ratio.

FIGURE 9.10 Animated! Incomplete dominance in heterozygous (pink) snapdragons. One allele (*R*) results in the production of a red pigment; the other (*r*) results in no pigment.

Figure It Out: Is the experiment in **B** a monohybrid or dihybrid cross?

Credits: (a) © JupiterImages Corporation; (b) © Cengage Learning.

Answer: A monohybrid cross

The two alleles you carry for the *ABO* gene determine the form of the carbohydrate on your blood cells, so they are the basis of your blood type (Figure 9.9). The *A* and the *B* alleles are codominant when paired. If your genotype is *AB*, then you have both versions of the enzyme, and your blood is type AB. The *O* allele is recessive when paired with either the *A* or *B* allele. If your genotype is *AA* or *AO*, your blood is type A. If your genotype is *BB* or *BO*, it is type B. If you are *OO*, it is type O.

Receiving incompatible blood cells in a transfusion is very dangerous because the immune system attacks red blood cells bearing molecules that do not occur in one's own body. The attack can cause the blood cells to clump or burst, with potentially fatal consequences. People with type O blood can donate blood to anyone, so they are called universal donors. However, because their body is unfamiliar with the carbohydrates made by people with A or B blood, they can receive type O blood only. People with type AB blood can receive a transfusion of any blood type, so they are called universal recipients.

❭ Incomplete Dominance With **incomplete dominance**, one allele is not fully dominant over the other, so the heterozygous phenotype is somewhere between the two homozygous phenotypes. A gene that influences flower color in snapdragon plants is an example. There is one copy of this gene on each homologous chromosome; both copies are expressed. One allele (*R*) encodes an enzyme that makes a red pigment. The enzyme encoded by a mutated allele (*r*) cannot make any pigment. Plants homozygous for the *R* allele (*RR*) make a lot of red pigment, so they have red flowers. Plants homozygous for the *r* allele (*rr*) do not make any pigment at all, so their flowers are white. Heterozygous plants (*Rr*) make only enough red pigment to color their flowers pink (Figure 9.10**A**). A cross between two pink-flowered heterozygous plants yields red-, pink-, and white-flowered offspring in a 1:2:1 ratio (Figure 9.10**B**).

❭ Epistasis and Pleiotropy Some traits are affected by multiple gene products, an effect called polygenic inheritance or **epistasis**. For example, several

codominance Effect in which two alleles are both fully expressed in heterozygous individuals and neither is dominant over the other.

epistasis Polygenic inheritance, in which a trait is influenced by the products of multiple genes.

incomplete dominance Effect in which one allele is not fully dominant over another, so the heterozygous phenotype is between the two homozygous phenotypes.

FIGURE 9.11 Animated! Epistasis in dogs. Interactions among products of two gene pairs affect coat color in Labrador retrievers. All dogs with an *E* and a *B* allele have black fur. Those with an *E* and two recessive *b* alleles have brown fur. All dogs homozygous for the recessive *e* allele have yellow fur.

Credits: top, © John Daniels/ ardea.com; bottom, © Cengage Learning.

FIGURE 9.12

Marfan syndrome. Basketball star Haris Charalambous died suddenly in 2006 when his aorta burst during warm-up exercises. He was 21.

Charalambous was very tall and lanky, with long arms and legs—traits that are valued in professional athletes such as basketball players. These traits are also associated with Marfan syndrome.

About 1 in 5,000 people are affected by Marfan syndrome worldwide. Like many of them, Charalambous did not realize he had the syndrome.

Courtesy of The Family of Haris Charalambous and the University of Toledo.

gene products affect the coat color of a Labrador retriever, which can be black, yellow, or brown (Figure 9.11). One gene is involved in the synthesis of the pigment melanin. A dominant allele of the gene specifies black fur, and its recessive partner specifies brown fur. A dominant allele of a different gene causes melanin to be deposited in fur, and its recessive partner reduces melanin deposition.

A **pleiotropic** gene is one that influences multiple traits. Mutations in such genes are associated with complex genetic disorders such as sickle-cell anemia (Section 7.6), cystic fibrosis, and Marfan syndrome. With cystic fibrosis, thickened mucus affects the entire body, not just the respiratory tract. The mucus clogs ducts that lead to the gut, which results in digestive problems. Male CF patients are typically infertile because their sperm flow is hampered by the thickened secretions. Marfan syndrome is caused by mutations that affect fibrillin. Long fibers of this protein impart elasticity to tissues of the heart, skin, blood vessels, tendons, and other body parts. In Marfan syndrome, tissues form with defective fibrillin or none at all. The largest blood vessel leading from the heart, the aorta, is particularly affected. The aorta's thick wall is not as elastic as it should be, and it eventually stretches and becomes leaky. Thinned and weakened, the aorta can rupture abruptly during exercise.

Marfan syndrome is particularly difficult to diagnose. Affected people are often tall, thin, and loose-jointed, but there are plenty of tall, thin, loose-jointed people that do not have the syndrome. Symptoms may not be apparent, so many people die suddenly and early without ever knowing they had the disorder (Figure 9.12).

Take-Home Message

Are all alleles dominant or recessive?

- With incomplete dominance, the heterozygous phenotype is a blend of the two homozygous phenotypes. With codominance, the heterozygous phenotype is a combination of both homozygous phenotypes.

- Some traits are influenced by the products of two or more genes. Some genes affect multiple traits.

9.5 Complex Variation in Traits

The pea plant phenotypes that Mendel studied appeared in two or three forms, which made them easy to track through generations. All are single-gene traits (one gene determines the trait). However, many other traits do not appear in distinct forms with predictable inheritance patterns. Such traits are often the result of complex interactions between multiple genes, with environmental influences on top of those interactions.

> **Environmental Effects on Phenotype** The phrase "nature vs. nurture" refers to a centuries-old debate about whether human behavioral traits arise from one's genetics (nature) or from environmental factors (nurture). It turns out that both play a role. The environment affects the expression of many genes, which in turn affects phenotype—including behavioral traits. We can summarize this thinking with the following equation:

$$\text{genotype} + \text{environment} \longrightarrow \text{phenotype}$$

A The color of the snowshoe hare's fur varies by season. In summer, the fur is brown (*left*); in winter, it is white (*right*). The variation offers seasonally appropriate camouflage from predators.

B The body form of the water flea on the *top* develops in environments with few predators. A longer tail spine and a pointy head (*bottom*) develop in response to chemicals emitted by predatory insects.

C The height of a mature yarrow plant depends on the elevation at which it grows.

Height (centimeters) / 60 / 0

Elevation (meters above sea level) / 3060 / 1400 / 30

200 μm

FIGURE 9.13 Examples of environmental effects on phenotype.

Credits: (a) left, JupiterImages Corporation; right, © age fotostock/ SuperStock; (b) © Dr. Christian Laforsch; (c) top, © Pamela Harper/ Harper Horticultural Slide Library; bottom, © Cengage Learning.

For example, environmental cues trigger internal pathways that methylate or demethylate particular regions of DNA, thus suppressing or enhancing gene expression in those regions (Section 7.7). Changes in DNA methylation patterns can be permanent and heritable.

Mechanisms that adjust phenotype in response to external cues are part of an individual's normal ability to adapt to environmental change. For example, seasonal changes in temperature and the length of day affect the production of melanin and other pigments that color the skin and fur of many animals. These species have different color phases in different seasons (Figure 9.13**A**). Hormonal signals triggered by changes in daylength cause the fur to be shed, and different types and amounts of pigments to be deposited in fur that grows back. The resulting change in phenotype offers these animals seasonally appropriate camouflage from predators.

Water fleas have different phenotypes depending on whether the aquatic insects that prey on them are present (Figure 9.13**B**). Individuals also switch between asexual and sexual reproduction depending on environmental conditions. During the early spring, competition is scarce in their freshwater pond habitats. At that time, the fleas reproduce rapidly by asexual means, giving birth to large numbers of female offspring that quickly fill the ponds. Later in the season, competition for resources intensifies as the pond water becomes warmer, saltier, and more crowded. Under these conditions, some of the water fleas start giving birth to males, and then reproducing sexually. The increased genetic diversity of sexually produced offspring may offer the population an advantage in an environment that presents a greater challenge to survival (Section 8.6).

In plants, plasticity of phenotype gives immobile individuals an ability to thrive in diverse habitats. For example, genetically identical yarrow plants grow to different heights at different altitudes (Figure 9.13**C**). More challenging temperature, soil, and water conditions are typically encountered at higher altitudes. Differences in altitude are also correlated with changes in the reproductive mode of yarrow: Plants at higher altitude tend to reproduce asexually, and those at lower altitude tend to reproduce sexually.

Does the environment affect human genes? Consider that certain mutations are associated with schizophrenia, bipolar disorder, depression, and other mood

pleiotropic Refers to a gene whose product influences multiple traits.

FIGURE 9.14 Human iris color, a trait that varies continuously.

Credits: Top row from left, © szefel; © Aaron Amat; © evantravels; © Andrey Armyagov; © Villedieu Christophe; © J. Helgason; Bottom row from left, © Tatiana Makotra; © Vaaka; © rawcaptured; © Tischenko Irina; © lightpoet; © Tatiana Makotra; all from Shutterstock.

A Male biology students at the University of Florida were divided into categories of one-inch increments in height and counted.

B Graphing the resulting data produces a bell-shaped curve, an indication that height varies continuously.

FIGURE 9.15 Animated! Example of continuous variation.

Credits: (a) Courtesy of Ray Carson, University of Florida News and Public Affairs; (b) © Cengage Learning.

disorders. However, not all people with the mutations end up with a mood disorder, so there must be an environmental component too.

Recent discoveries in animal models are beginning to unravel some of the mechanisms by which environment can influence mental state. For example, learning and memory are associated with dynamic and rapid DNA modifications in brain cells. Mood is, too. Stress-induced depression causes methylation-based silencing of a particular nerve growth factor gene. Some antidepressants work by reversing this methylation. As another example, rats whose mothers are not very nurturing end up anxious and having a reduced resilience for stress as adults. The difference between these rats and ones who had nurturing maternal care is traceable to epigenetic DNA modifications that result in a lower than normal level of another nerve growth factor. Drugs can reverse these modifications—and their effects. We do not yet know all of the genes that influence human mental state, but the implication of such research is that future treatments for many disorders will involve deliberate modification of methylation patterns in an individual's DNA.

) Continuous Variation Some traits occur in a range of small differences that is called **continuous variation**. Continuous variation can be an outcome of epistasis, in which multiple genes affect a single trait. The more genes and environmental factors that influence a trait, the more continuous is its variation.

Human skin color varies continuously, as does human eye color (Figure 9.14). The colored part of the eye is a doughnut-shaped structure called the iris. Iris color, like skin color, is the result of interactions among gene products that make and distribute melanins (an example of epistasis). The more melanin deposited in the iris, the less light is reflected from it. Dark irises have dense melanin deposits that absorb almost all light, and reflect almost none. Green and blue eyes have the least amount of melanin, so they reflect the most light.

How do we determine whether a trait varies continuously? Let's use another human trait, height, as an example. First, we divide the total range of phenotypes into measurable categories—inches, in this case (Figure 9.15**A**). Next, we count how many individuals of a group fall into each category; this count gives the relative frequencies of phenotypes across our range of measurable values. Finally, we plot the data as a bar chart (Figure 9.15**B**). A graph line around the top of the bars shows the distribution of values for the trait. If the line is a bell-shaped curve, or **bell curve**, the trait varies continuously.

Take-Home Message

Is genotype the only factor that gives rise to phenotype?

- Phenotype can change in response to environmental cues. Such cues influence gene expression patterns, which in turn influence traits.

- The more genes and other factors that influence a trait, the more continuous is its range of variation.

9.6 Human Genetic Analysis

Some organisms, including pea plants and fruit flies, are ideal for genetic analysis. They have relatively few chromosomes, they reproduce quickly under controlled conditions, and breeding them poses few ethical problems. It does not take long to track a trait through many generations. Humans, however, are a different story. Unlike flies grown in laboratories, we humans live under variable conditions, in different places, and we live as long as the geneticists who study us. Most of us select our own mates and reproduce if and when we want to. Our families tend to be on the small side, so sampling error (Section 1.7) is a major factor in studying them. Because of these and other challenges, geneticists often use historical records to track traits through many generations of a family. These researchers make standardized charts of genetic connections called **pedigrees** (Figure 9.16). Pedigree analyses can reveal whether a trait is associated with a dominant or recessive allele, and whether the allele is on an autosome or a sex chromosome. Pedigree analysis also allows geneticists to determine the probability that a trait will recur in future generations of a family or a population.

> Types of Genetic Variation Some easily observed human traits follow Mendelian inheritance patterns. Like the flower color of pea plants, these traits are controlled by a single gene with alleles that have clear dominance relationships. For example, some people have earlobes that attach at their base, and others have earlobes that dangle free. The allele for unattached earlobes is dominant; the allele for attached earlobes is recessive. Similarly, an allele that specifies a cleft chin is dominant over the allele for a smooth chin, and the allele for dimples is dominant over that for no dimples. Someone who is homozygous for two recessive alleles of the *MC1R* gene makes the reddish kind of melanin but not the brownish-black kind, so this person has red hair.

Single genes on autosomes or sex chromosomes also govern more than 6,000 genetic abnormalities and disorders. A genetic abnormality is a rare or uncommon version of a trait, such as having six fingers on a hand. By contrast, a genetic disorder sooner or later causes medical problems that may be severe. A genetic disorder is often characterized by a specific set of symptoms (a syndrome). Most research in the field of human genetics focuses on disorders, because what we learn may help us develop treatments for affected people.

The next section focuses on inheritance patterns of human single-gene disorders, which affect about 1 in 200 people. Keep in mind that these inheritance patterns are the least common kind. Most human traits are influenced by multiple genes and some have epigenetic contributions or causes. Many genetic disorders are like this, including diabetes, asthma, obesity, cancers, heart disease, and multiple sclerosis. The inheritance patterns of these disorders are complex, and despite intense research our understanding of the genetics behind them remains incomplete. (Appendix IV shows a map of human chromosomes with the locations of some known alleles.)

A Standard symbols used in pedigrees

B A pedigree for polydactyly, which is characterized by extra fingers, toes, or both. The *black* numbers signify the number of fingers on each hand; the *red* numbers signify the number of toes on each foot. Though it occurs on its own, polydactyly is also one of several symptoms of Ellis–van Creveld syndrome.

FIGURE 9.16 Animated! An example of a pedigree.

Credits: photo, Courtesy of Irving Buchbinder, DPM, DABPS, Community Health Services, Hartford CT; art, © Cengage Learning.

bell curve Bell-shaped curve; typically results from graphing frequency versus distribution for a trait that varies continuously in a population.

continuous variation A range of small differences in a shared trait.

pedigree Chart showing the pattern of inheritance of a trait in a family.

Most alleles that give rise to severe genetic disorders are rare in populations, because they compromise the health and reproductive ability of their bearers. Why do they persist? Mutations periodically reintroduce them. In some cases, a codominant allele offers a survival advantage in a particular environment.

Take-Home Message

How do we study inheritance patterns in humans?

- Most human traits are not governed by single genes, so their inheritance does not follow a Mendelian pattern.

- Inheritance patterns in humans are often studied by tracking genetic abnormalities or disorders through family trees.

- A genetic abnormality is a rare version of an inherited trait. A genetic disorder is an inherited condition that causes medical problems.

Inheritance patterns in humans are often studied by tracking genetic disorders through family trees.

9.7 Human Genetic Disorders

> **Autosomal Dominant Disorders** A dominant allele on an autosome is expressed in people who are heterozygous for it as well as those who are homozygous. Traits governed by such alleles (Table 9.1) tend to appear in every generation, and they affect both sexes equally. When one parent is heterozygous, and the other is homozygous for the recessive allele, each of their children has a 50 percent chance of inheriting the dominant allele and having the trait associated with it (Figure 9.17**A**).

A form of hereditary dwarfism called achondroplasia offers an example of an autosomal dominant disorder (one caused by a dominant allele on an autosome). Mutations associated with achondroplasia occur in a gene for a growth hormone receptor. The mutations cause the receptor, a regulatory molecule that slows bone development, to be overly active. About 1 out of 10,000 people are heterozygous for one of these mutations. As adults, these people are, on average, about four feet, four inches (1.3 meters) tall, with abnormally short arms and legs relative to other body parts (Figure 9.17**B**). An allele that causes achondroplasia can be passed to children because its expression does not interfere with reproduction, at least in heterozygous people. The homozygous condition results in severe skeletal malformations that cause early death.

Another autosomal dominant allele causes Huntington's disease, in which involuntary muscle movements increase as the nervous system slowly deteriorates. Mutations associated with this disorder alter a gene for a cytoplasmic protein whose function is still unknown. The mutations are insertions in which the same three nucleotides become repeated many, many times in the gene's sequence. The allele encodes an oversized protein product that is chopped into pieces inside nerve cells of the brain. The pieces clump together, and large aggregates that accumulate in the cytoplasm eventually prevent the cells from functioning properly. Brain cells involved in movement, thinking, and emotion are particularly affected. In the most common form of Huntington's, symptoms do not start until after the age of thirty, and affected people die during their forties or fifties. With this and other late-onset disorders, people tend to reproduce before symptoms appear, so the allele is often passed unknowingly to children.

Hutchinson–Gilford progeria is an autosomal dominant disorder characterized by drastically accelerated aging. It is usually caused by a mutation in the gene for a protein subunit of intermediate filaments that support the nuclear envelope. This protein also has roles in mitosis, DNA synthesis and repair, and transcrip-

Table 9.1	Some Autosomal Dominant Traits in Humans
Disorder or Abnormality	**Main Symptoms**
Achondroplasia	One form of dwarfism
Aniridia	Defects of the eyes
Camptodactyly	Rigid, bent fingers
Huntington's disease	Degeneration of the nervous system
Marfan syndrome	Abnormal or missing connective tissue
Polydactyly	Extra fingers, toes, or both
Progeria	Drastic premature aging

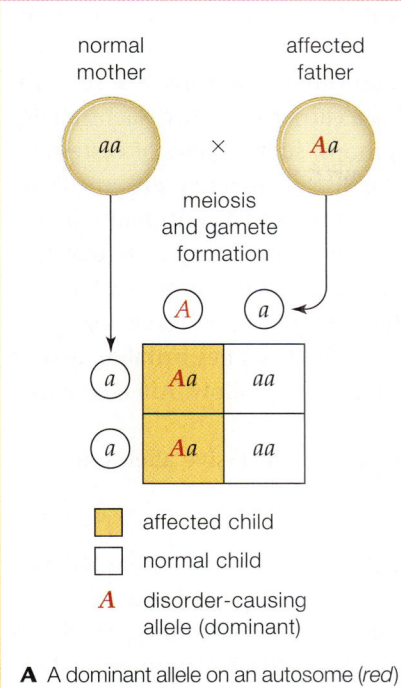

normal mother — *aa* × affected father — *Aa*

meiosis and gamete formation

A *a*

	a	*a*
a	*Aa*	*aa*
a	*Aa*	*aa*

☐ affected child

☐ normal child

A disorder-causing allele (dominant)

A A dominant allele on an autosome (*red*) is fully expressed in heterozygous people.

B Achondroplasia affects Ivy Broadhead (*left*), her brother, father, and grandfather.

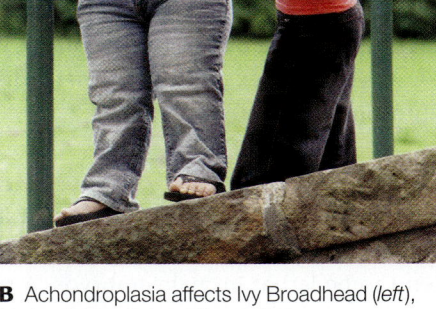

C Symptoms of Hutchinson–Gilford progeria are already evident in Megan Nighbor at age 5.

FIGURE 9.17 Animated! Autosomal dominant inheritance.

tional regulation. The mutation results in a protein that is not processed correctly after translation. Pleiotropic effects include a grossly abnormal nucleus, with nuclear pore complexes that do not assemble properly and membrane proteins localized to the wrong side of the nuclear envelope. The function of the nucleus as protector of chromosomes and gateway of transcription is severely impaired, and DNA damage accumulates quickly. Outward symptoms of the disorder begin to appear before age two. Skin that should be plump and resilient starts to thin, muscles weaken, and bones that should lengthen and grow stronger soften. Premature baldness is inevitable (Figure 9.17**C**). Most people with the disorder die in their early teens as a result of a stroke or heart attack brought on by hardened arteries, a condition typical of advanced age. Progeria does not run in families because affected people do not usually live long enough to reproduce.

> **Autosomal Recessive Disorders** A recessive allele on an autosome is expressed only in homozygous people, so traits associated with the allele tend to skip generations. Both males and females are equally affected. People heterozygous for the allele are carriers, which means that they have the allele but not the trait. Any child of two carriers has a 25 percent chance of inheriting the allele from both parents (Figure 9.18**A**). Being homozygous for the allele, such children would have the trait.

Albinism, a genetic abnormality characterized by an abnormally low level of melanin, is inherited in an autosomal recessive pattern. Mutations associated with the albino phenotype occur in genes involved in the production of melanin. Skin, hair, or eye pigmentation may be reduced or missing. The most dramatic form of the phenotype is caused by mutations that disable an enzyme in the melanin synthesis pathway. The skin is typically very white and does not tan, and the hair is white. The irises lack pigment, so they appear red due to the visibility of the underlying blood vessels (Figure 9.18**B**). Melanin in the retina

carrier mother — *Aa* × carrier father — *Aa*

meiosis and gamete formation

A *a*

A Only people homozygous for a recessive allele on an autosome have the trait associated with the allele.

In this example, both parents are carriers (*red*). Each of their children has a 25 percent chance of inheriting two alleles, and being affected by the trait.

	A	*a*
A	*AA*	*Aa*
a	*Aa*	*aa*

☐ affected child

☐ carrier child

☐ normal child

a disorder-causing allele (recessive)

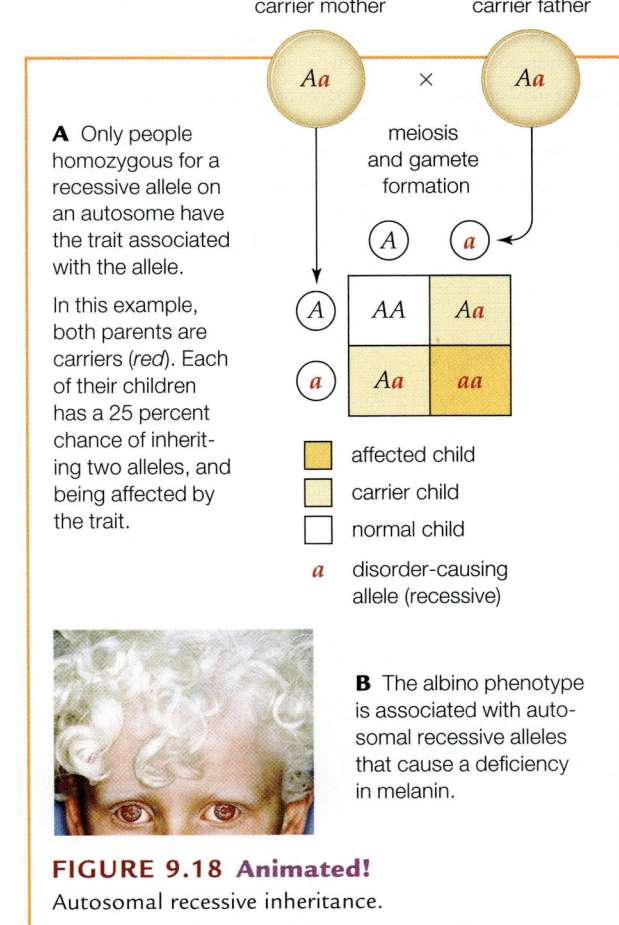

B The albino phenotype is associated with autosomal recessive alleles that cause a deficiency in melanin.

FIGURE 9.18 Animated! Autosomal recessive inheritance.

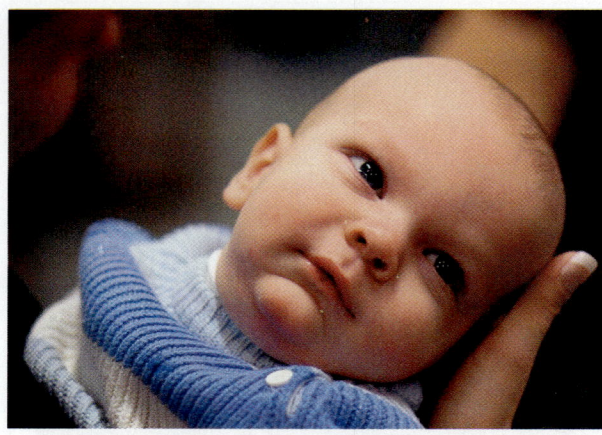

FIGURE 9.19 Conner Hopf, who was diagnosed with Tay–Sachs disease at age 7-1/2 months. He died at 22 months.

Courtesy of © Conner's Way Foundation, www.connersway.com.

Table 9.2	Some Autosomal Recessive Traits in Humans
Trait	Description
Albinism	Absence of pigmentation
Cystic fibrosis	Abnormally thickened secretions cause tissue and organ damage
Ellis–van Creveld syndrome	Dwarfism, heart defects, polydactyly
Friedreich's ataxia	Progressive loss of motor and sensory function
Phenylketonuria (PKU)	Mental impairment
Sickle-cell anemia	Anemia, adverse pleiotropic effects
Tay–Sachs disease	Deterioration of mental and physical abilities; early death

plays a role in vision, so people with this phenotype tend to have reduced visual acuity and other problems with vision.

Tay–Sachs disease is another example of an autosomal recessive disorder. In the general population, about 1 in 300 people is a carrier for a Tay–Sachs allele, but the incidence is ten times higher in some groups, such as Jews of eastern European descent. Mutations associated with the disease affect an enzyme that breaks down a particular type of lipid inside lysosomes. Cells continually make and break down this lipid, and turnover is especially brisk during early development. Affected infants usually seem normal for the first few months, but symptoms appear as the lipid accumulates to higher and higher levels inside their nerve cells. Within three to six months the child becomes irritable, listless, and may have seizures. Blindness, deafness, and paralysis follow. Affected children usually die by age five (Figure 9.19).

A few other examples of genetic disorders caused by recessive alleles on autosomes are listed in Table 9.2.

> **X-Linked Recessive Disorders** Many genetic disorders are associated with alleles on the X chromosome (Table 9.3). Most are inherited in a recessive pattern, probably because those caused by dominant X chromosome alleles tend to be lethal in male embryos. A recessive allele on the X chromosome (an X-linked recessive allele) leaves two clues when it causes a genetic disorder. First, an affected father never passes an X-linked recessive allele to a son, because all children who inherit their father's X chromosome are female (Figure 9.20**A**). Thus, a heterozygous female is always the bridge between an affected male and his affected grandson. Second, the disorder appears in males more often than in females. This is because all males who carry the allele have the disorder, but not all heterozygous females do. Remember that one of the two X chromosomes in each cell of a female is inactivated as a Barr body (Section 7.7). As a result, only about half of a heterozygous female's cells express the recessive allele. The other half of her cells express the dominant, normal allele that she carries on her other X chromosome, and this expression can mask the phenotypic effects of the recessive allele.

Duchenne muscular dystrophy (DMD) is an X-linked recessive disorder characterized by muscle degeneration. It is caused by mutations in the X chromosome gene for dystrophin, a cystoskeletal protein that links actin microfilaments in cytoplasm to a complex of proteins in the plasma membrane. This complex structurally and functionally links the cell to extracellular matrix. When dystrophin is absent, the entire complex is unstable. Muscle cells, which are subject to stretching, are particularly affected. Their plasma membrane is easily damaged, and they become flooded with calcium ions. Eventually, the muscle cells die and become replaced by fat cells and connective tissue.

DMD affects about 1 in 3,500 people, almost all of them boys. Symptoms begin between ages three and seven. Anti-inflammatory drugs can slow the progression of DMD, but there is no cure. When an affected boy is about twelve, he will begin to use a wheelchair and his heart muscle will start to fail. Even with the best care, he will probably die before age thirty, from a heart disorder or respiratory failure (suffocation).

Hemophilia A is an X-linked recessive disorder that interferes with blood clotting. Most of us have a blood clotting mechanism that quickly stops bleeding from minor injuries. That mechanism involves the protein product of a gene on the X chromosome. Bleeding can be prolonged in males who carry a mutation in this gene. Females who have two mutated alleles are also affected (heterozygous females make enough of the protein to have a clotting time that is close to normal). Affected people tend to bruise very easily, but internal bleed-

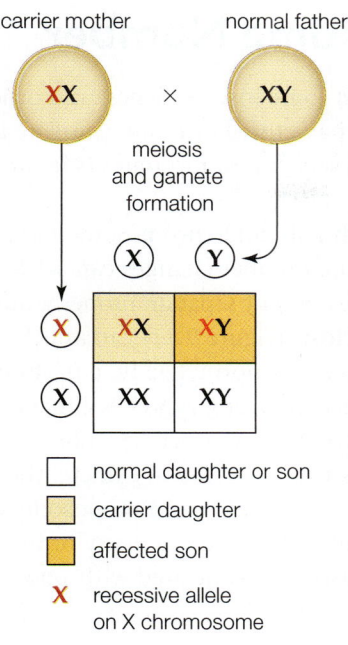

carrier mother · normal father

meiosis and gamete formation

	X	Y
X	XX	XY
X	XX	XY

☐ normal daughter or son

▢ carrier daughter

▣ affected son

X recessive allele on X chromosome

A In this example of X-linked inheritance, the mother carries a recessive allele on one of her two X chromosomes (*red*).

B A view of color blindness. The image on the *left* shows how a person with red–green color blindness sees the image on the *right*. The perception of blues and yellows is normal; red and green appear similar.

You may have one form of red–green color blindness if you see a 7 instead of a 29 in this circle.

You may have another form of red–green color blindness if you see a 3 instead of an 8 in this circle.

C Part of a standardized test for color blindness. A set of 38 of these circles is commonly used to diagnose deficiencies in color perception.

FIGURE 9.20 Animated! X-linked recessive inheritance.

Credits: art, © Cengage Learning; (a,b) Gary L. Friedman, www.FriedmanArchives.com.

ing is their most serious problem. Repeated bleeding inside the joints disfigures them and causes chronic arthritis. About 1 in 7,500 people is affected by the disorder, but that number may be rising because hemophilia is now treatable. More affected people are living long enough to transmit a mutated allele to offspring.

The pattern of X-linked recessive inheritance shows up among individuals who have some degree of color blindness (Figure 9.20**B,C**). The term refers to a range of conditions in which an individual cannot distinguish among some or all colors in the spectrum of visible light. Color vision depends on the proper function of pigment-containing receptors in the eyes. Most of the genes involved in color vision are on the X chromosome, and mutations in those genes often result in altered or missing receptors. Normally, humans can sense the differences among 150 colors. People who have red–green color blindness see fewer than 25 colors because receptors that respond to red and green wavelengths are weakened or absent. Some confuse red and green; others see green as gray.

Table 9.3	Some X-Linked Traits in Humans
Disorder or Abnormality	**Main Symptoms**
Androgen insensitivity syndrome	XY individual has the traits of a female; sterility
Red–green color blindness	Inability to distinguish red from green
Hemophilia	Impaired blood clotting
Muscular dystrophies	Progressive loss of muscle function

Take-Home Message

How do we know when a trait is associated with an allele on an autosome or a sex chromosome?

- Persons heterozygous for an allele inherited in an autosomal dominant pattern have the associated trait. The trait tends to appear in every generation.

- With an autosomal recessive inheritance pattern, only persons homozygous for an allele have the associated trait. The trait tends to skip generations.

- Men who have an X-linked allele have the associated trait, but not all heterozygous women do. Thus, the trait appears more often in men. Men transmit an X-linked allele to their daughters, but not to their sons.

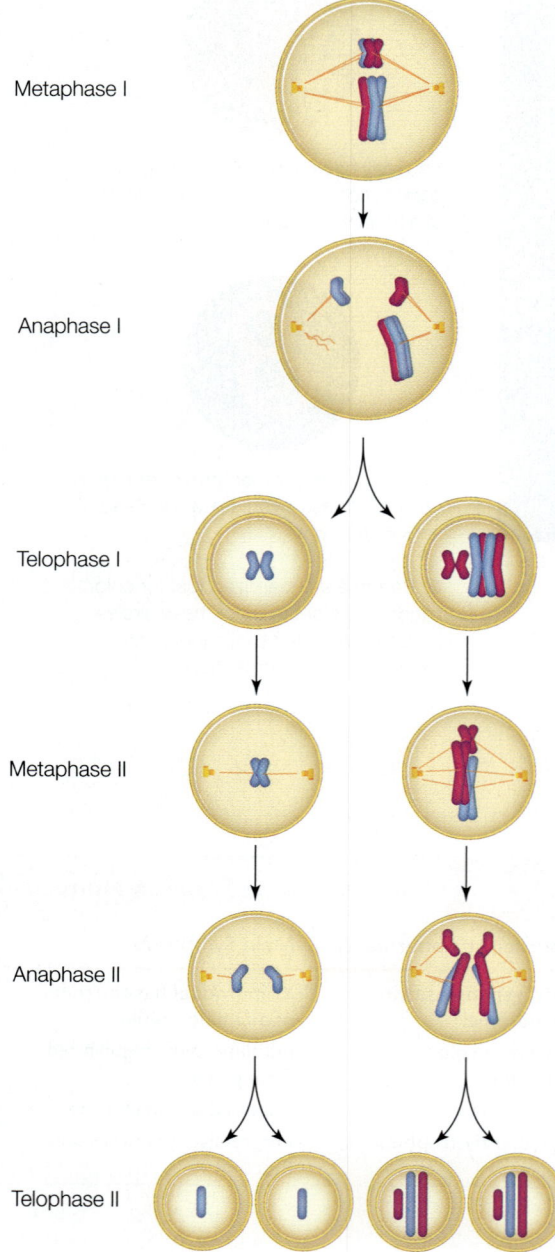

Metaphase I

Anaphase I

Telophase I

Metaphase II

Anaphase II

Telophase II

FIGURE 9.21 Animated! An example of non-disjunction during meiosis.

Of the two pairs of homologous chromosomes shown here, one fails to separate during anaphase I. A zygote that forms from any of the resulting gametes will have an abnormal chromosome number.

Figure It Out: During which stage of meiosis does nonjunction occur in this example?

Answer: Anaphase I

© Cengage Learning.

9.8 Changes in Chromosome Number

About 70 percent of flowering plant species, and some insects, fishes, and other animals, are **polyploid**, which means that they have three or more complete sets of chromosomes. Inheriting more than two full sets of chromosomes is invariably fatal in humans.

Less than 1 percent of children are born with a diploid chromosome number that differs from the normal 46. Chromosome number changes can arise through **nondisjunction**, in which chromosomes do not separate properly during nuclear division. Nondisjunction during meiosis (Figure 9.21) can affect chromosome number at fertilization. For example, if a normal gamete (n) fuses with one that has an extra chromosome ($n+1$), the resulting zygote will have three copies of one type of chromosome and two of every other type ($2n+1$), a condition called trisomy. If an $n-1$ gamete fuses with a normal n gamete, the zygote will be $2n-1$, a condition called monosomy. Trisomy and monosomy are examples of **aneuploidy**, in which an individual's cells have too many or too few copies of a particular chromosome. A few disorders associated with polyploidy and aneuploidy are listed in Table 9.4.

❯ Autosomal Change and Down Syndrome Autosomal aneuploidy is usually fatal in humans, but there are exceptions. A few humans with trisomy are born alive, but only those with trisomy 21 have a high probability of surviving infancy. A newborn with three chromosomes 21 will develop Down syndrome (Figure 9.22). This disorder occurs once in 800 to 1,000 births, and it affects more than 350,000 people in the United States alone. The risk increases with maternal age. Individuals with Down syndrome may have a somewhat flattened facial profile, a fold of skin that starts at the inner corner of each eyelid, white spots on the iris, one deep crease (instead of two shallow creases) across each palm, and other symptoms. Not all of the outward symptoms develop in every individual. That said, people with the disorder tend to have mild to moderate mental impairment and health problems such as heart disease. Their skeleton grows and develops abnormally, so older children have short body parts, loose joints, and misaligned bones of the fingers, toes, and hips. The muscles and reflexes are weak, and motor skills such as speech develop slowly. With medical care, affected individuals live about fifty-five years. Early training can help them learn to care for themselves and to take part in normal activities.

❯ Change in the Sex Chromosome Number Nondisjunction also causes alterations in the number of X and Y chromosomes, with a frequency of about 1 in 400 live births. Most often, such alterations lead to mild difficulties in learning and impaired motor skills such as a speech delay. These problems may be so subtle that the condition is never diagnosed.

Individuals with Turner syndrome have an X chromosome and no corresponding X or Y chromosome (XO). The syndrome is thought to arise most frequently as an outcome of inheriting an unstable Y chromosome from the father. The zygote starts out being genetically male, with an X and a Y chromosome. The Y chromosome breaks up and is lost during early development, so the embryo continues to develop as a female.

There are fewer people affected by Turner syndrome than other chromosome abnormalities: Only about 1 in 2,500 newborn girls has it. The risk does not increase with maternal age. XO individuals grow up well proportioned but short, with an average height of four feet, eight inches (1.4 meters). Their ovaries do not develop properly, so they do not make enough sex hormones to become sexually mature. Development of secondary sexual traits such as breasts is inhibited.

FIGURE 9.22 Down syndrome, genotype and phenotype.
Figure It Out: Is the karyotype from an individual who is male or female? *Answer: Male (XY)*

Credits: left, L. Willatt, East Anglian Regional Genetics Service/ Photo Researchers, Inc.; right, Ciarra, photo by © Michelle Harmon.

A female may inherit multiple X chromosomes, a condition called XXX syndrome. XXX syndrome occurs in about 1 of 1,000 births; as with Down syndrome, risk increases with maternal age. Only one X chromosome is typically active in female cells, so having extra X chromosomes usually does not cause physical or medical problems, but mild mental impairment may occur.

About 1 out of every 500 males has an extra X chromosome (XXY). The resulting disorder, Klinefelter syndrome, develops at puberty. XXY males tend to be overweight, tall, and have mild mental impairment. They make more estrogen and less testosterone than normal males. This hormone imbalance causes affected men to have small testes and prostate glands, low sperm counts, sparse facial and body hair, high-pitched voices, and enlarged breasts. Testosterone injections during puberty can minimize some of these traits.

About 1 in 1,000 males is born with an extra Y chromosome (XYY), a result of nondisjunction of the Y chromosome during sperm formation. Adults tend to be taller than average and have mild mental impairment, but most are otherwise normal. XYY men were once thought to be predisposed to a life of crime. This misguided view was based on sampling error (too few cases in narrowly chosen groups such as prison inmates) and bias (the researchers who gathered the karyotypes also took the personal histories of the participants). That view was disproven in 1976, when a geneticist reported results from his study of 4,139 tall males, all twenty-six years old, who had registered for military service. The military had given the men physical examinations and intelligence tests, and had also recorded social and economic status, education, and any criminal convictions. Only twelve of the males studied were XYY, which meant that the "control group" had more than 4,000 males. The only findings? Mentally impaired, tall males who engage in criminal deeds are more likely to get caught, irrespective of karyotype.

Table 9.4	Some Disorders Associated With Chromosome Number Changes
Disorder or Abnormality	**Main Symptoms**
Down syndrome	Mental impairment; heart defects
Turner syndrome (XO)	Sterility; abnormal ovaries, abnormal sexual traits
Klinefelter syndrome (XXY)	Sterility, mental impairment
XXX syndrome	Minimal abnormalities
XYY syndrome	Mild mental impairment

Take-Home Message

What are the effects of chromosome number changes in humans?

- Polyploidy is fatal in humans, but not in some other organisms.
- Aneuploidy can arise from nondisjunction during meiosis. In humans, most cases of aneuploidy are associated with some degree of mental impairment.

aneuploidy A chromosome abnormality in which there are too many or too few copies of a particular chromosome.
nondisjunction Failure of chromosomes to separate properly during nuclear division.
polyploid Having three or more of each type of chromosome characteristic of the species.

9.9 Genetic Screening

Studying human inheritance patterns has given us many insights into how genetic disorders arise and progress, and how to treat them. Surgery, prescription drugs, hormone replacement therapy, and dietary controls can minimize and in some cases eliminate the symptoms of a genetic disorder. Some disorders can be detected early enough to start countermeasures before symptoms develop. For example, most hospitals in the United States now screen newborns for mutations that cause phenylketonuria, or PKU. The mutations affect an enzyme that converts the amino acid phenylalanine to tyrosine. Without this enzyme, the body becomes deficient in tyrosine, and phenylalanine accumulates to high levels. The imbalance inhibits protein synthesis in the brain, which in turn results in severe neurological symptoms. Restricting all intake of phenylalanine can slow the progression of PKU, so routine early screening has resulted in fewer individuals suffering from the symptoms of the disorder.

Prospective parents can also benefit from human genetics studies. The probability that a child will inherit a genetic disorder can be estimated by checking parental karyotypes and pedigrees, and testing the parents for alleles known to be associated with genetic disorders. Genetic screening is typically done before pregnancy, to help prospective parents make decisions about family planning.

> **Prenatal Diagnosis** Genetic screening can also be done post-conception, in which case it is called prenatal diagnosis. Prenatal means before birth. The term "embryo" applies until eight weeks after fertilization, after which "fetus" is appropriate. Prenatal diagnosis checks for physical and genetic abnormalities in an embryo or fetus. More than 30 conditions are detectable prenatally, including aneuploidy, hemophilia, Tay–Sachs disease, sickle-cell anemia, muscular dystrophy, and cystic fibrosis. If the disorder is treatable, early detection allows the newborn to receive prompt and appropriate treatment. A few defects are even surgically correctable before birth. Prenatal diagnosis also gives parents time to prepare for the birth of an affected child, and an opportunity to decide whether to continue with the pregnancy or terminate it.

As an example of how prenatal diagnosis works, consider a woman who becomes pregnant at age thirty-five. Her doctor will probably perform a procedure called obstetric sonography, in which ultrasound waves directed across the woman's abdomen form images of the fetus's limbs and internal organs (Figure 9.23**A,B**). The images may reveal defects associated with a genetic disorder, in which case a more invasive technique would be recommended for further diagnosis. Fetoscopy yields images of the fetus that are much higher in resolution than ultrasound images (Figure 9.23**C**). With this procedure, sound waves are pulsed from inside the mother's uterus. Samples of tissue or blood are often taken at the same time, and some corrective surgeries can be performed.

Human genetics studies show that our thirty-five-year-old woman has about a 1 in 80 chance that her baby will be born with a chromosomal abnormality, a risk more than six times greater than when she was twenty years old. Thus, even if no abnormalities are detected by ultrasound, she probably will be offered a more thorough diagnostic procedure, amniocentesis, in which a small sample of fluid is drawn from the amniotic sac enclosing the fetus (Figure 9.24). The fluid contains cells shed by the fetus, and those cells can be tested for genetic disorders. Chorionic villus sampling (CVS) can be performed earlier than amniocentesis. With this technique, a few cells from the chorion are removed and tested for genetic disorders. The chorion is a membrane that surrounds the amniotic sac and helps form the placenta, an organ that allows substances to be exchanged between mother and embryo.

A Conventional ultrasound

B 4D ultrasound

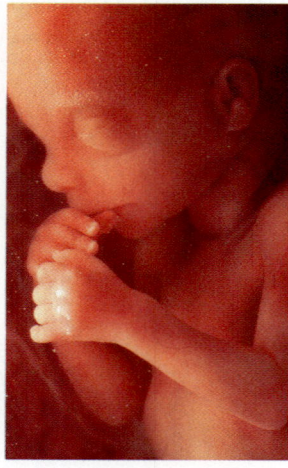

C Fetoscopy

FIGURE 9.23

Three ways of imaging a human fetus.

Credits: (a) en.wikipedia.org/wiki/
File:Embryo_at_14_weeks_profile.jpg;
(b) © LookatSciences/ Phototake, Inc.;
(c) © Howard Sochurek/ The Medical
File/ Peter Arnold, Inc.

An invasive procedure often carries a risk to the fetus. The risks vary by the procedure. Amniocentesis has improved so much that, in the hands of a skilled physician, the procedure no longer increases the risk of miscarriage. CVS occasionally disrupts the placenta's development and thus causes underdeveloped or missing fingers and toes in 0.3 percent of newborns. Fetoscopy raises the miscarriage risk by a whopping 2 to 10 percent, so it is rarely performed unless surgery or another medical procedure is required before the baby is born.

> **Preimplantation Diagnosis** Couples who discover they are at high risk of having a child with a genetic disorder may opt for reproductive interventions such as *in vitro* fertilization. With this procedure, sperm and eggs taken from prospective parents are mixed in a test tube. If an egg becomes fertilized, the resulting zygote will begin to divide. In about forty-eight hours, it will have become an embryo that consists of a ball of eight cells (*inset*). All of the cells in this ball have the same genes, but none has yet committed to being specialized one way or another. Doctors can remove one of these undifferentiated cells and analyze its genes, a procedure called preimplantation diagnosis. The withdrawn cell will not be missed. If the embryo has no detectable genetic defects, it is inserted into the woman's uterus to develop. Most of the resulting "test-tube babies" are born in good health.

FIGURE 9.24 An 8-week-old fetus.

With amniocentesis, fetal cells shed into the fluid inside the amniotic sac are tested for genetic disorders. Chorionic villus sampling tests cells of the chorion, which helps form the placenta.

© Lennart Nilsson/ Bonnierforlagen AB.

Take-Home Message

How do we use what we know about human inheritance?

- Studying inheritance patterns for genetic disorders has helped researchers develop treatments for some of them.

- Genetic testing can provide prospective parents with information about the health of their future children.

9.10 **Menacing Mucus** (revisited)

The allele most commonly associated with cystic fibrosis is eventually lethal in homozygous individuals, but not in those who are heterozygous. This allele is codominant with the normal one, so both copies of the gene are expressed in heterozygous individuals. Such individuals make enough of the normal CFTR protein to have normal chloride ion transport.

The CF allele is at least 50,000 years old and very common: 1 in 25 people carry it in some populations. Why has it persisted for so long and at such high frequency if it is dangerous? The allele may be the lesser of two evils because it offers heterozygous individuals an advantage in surviving certain deadly infectious diseases. The unmutated CFTR protein triggers endocytosis when it binds to bacteria. This process is an essential part of the body's immune response to bacteria in the respiratory tract. However, the same function of CFTR allows bacteria to enter cells of the gastrointestinal tract, where they can be deadly. For example, endocytosis of *Salmonella typhi* (shown at *left*) into epithelial cells lining the gut results in a dangerous infection called typhoid fever. The CF mutation alters the CFTR protein so that bacteria can no longer be taken up by intestinal cells. People that carry it may have a decreased susceptibility to typhoid fever and other bacterial diseases that begin in the intestinal tract.

WHERE YOU ARE GOING . . .

Individual genetic disorders such as muscular dystrophy are discussed in later chapters, in the context of the systems that they affect. Chapter 10 returns to human chromosomes in the context of genomics and genetic engineering. Chapter 12 explores evolutionary adaptations and factors that influence the frequency of alleles in a population. Chapter 26 returns to human reproduction and development; Chapter 28, to plant reproduction.

Summary

Section 9.1 The allele associated with most cases of cystic fibrosis persists at high frequency despite its devastating effects. Only those homozygous for the allele have the disorder.

Section 9.2 Individuals with identical alleles are **homozygous** for the allele. **Heterozygous** individuals have two nonidentical alleles. A **dominant** allele masks the effect of a **recessive** allele on the homologous chromosome. **Genotype** (an individual's particular set of alleles) gives rise to **phenotype**.

Section 9.3 Crossing individuals that breed true for two forms of a trait yields identically heterozygous offspring. A cross between such offspring is a **monohybrid cross**. The frequency at which the traits appear in the offspring of such crosses can reveal dominance relationships among the alleles associated with those traits. **Punnett squares** are useful in determining the probability of the genotype and phenotype of the offspring of crosses. Diploid cells have pairs of genes on homologous chromosomes. The two genes of a pair separate from each other during meiosis, so they end up in different gametes.

Crossing individuals that breed true for two forms of two traits yields offspring that are identically heterozygous for alleles governing those traits. A cross between such offspring is a **dihybrid cross**. Gene pairs on homologous chromosomes tend to sort into gametes independently of other gene pairs during meiosis.

Sections 9.4–9.5 Environmental factors can influence phenotype by altering gene expression.

With **incomplete dominance**, an allele is not fully dominant over its partner on a homologous chromosome, so an intermediate phenotype results. **Codominant** alleles are both fully expressed in heterozygotes; neither is dominant. With **epistasis**, two or more genes affect the same trait. A **pleiotropic** gene affects two or more traits.

A trait that is influenced by the products of multiple genes often occurs in a range of small increments of phenotype called **continuous variation**. Continuous variation is indicated by a **bell curve** in the range of values.

Section 9.6 Geneticists use **pedigrees** to track traits through generations of a family. Such studies can reveal inheritance patterns for alleles that can be predictably associated with specific phenotypes, especially genetic abnormalities or disorders. A genetic abnormality is an uncommon version of a heritable trait that does not result in medical problems. A genetic disorder is a heritable condition that sooner or later results in mild or severe medical problems.

Section 9.7 A trait associated with a dominant allele on an autosome occurs in heterozygous individuals, male and female. Such traits appear in every generation of families that carry the allele. A trait associated with a recessive allele on an autosome appears only in individuals homozy-

gous for the allele. Such traits occur in both males and females, but they can skip generations.

An allele is inherited in an X-linked pattern when it occurs on the X chromosome. Most X-linked inheritance disorders are recessive, and these tend to appear in men more often than in women. Heterozygous women have a dominant, normal allele that can mask the effects of the recessive one; men do not. Men can transmit an X-linked allele to their daughters, but not to their sons. Only a woman can pass an X-linked allele to a son.

Section 9.8 Changes in chromosome number are usually an outcome of **nondisjunction**, in which chromosomes fail to separate properly during meiosis. In **aneuploidy**, an individual's cells have too many or too few copies of a chromosome. Except for trisomy 21, which causes Down syndrome, most cases of autosomal aneuploidy are lethal. Some organisms are **polyploid** (they have three or more sets of chromosomes).

Section 9.9 Prospective parents can estimate their risk of transmitting a harmful allele to offspring with genetic screening, in which their pedigrees and genotype are analyzed by a genetic counselor. Prenatal genetic testing can reveal a genetic disorder before birth.

Self-Quiz   Answers in Appendix I

1. A heterozygous individual has a _____ for a trait being studied.
   a. pair of identical alleles
   b. pair of nonidentical alleles
   c. haploid condition, in genetic terms

2. An organism's observable traits constitute its _____ .
   a. phenotype   c. genotype
   b. variation   d. pedigree

3. The offspring of the cross $AA \times aa$ are _____ .
   a. all AA c. all Aa
   b. all aa d. 1/2 AA and 1/2 aa

4. Assuming all alleles have a clear dominant/recessive relationship, a dihybrid cross leads to a phenotypic ratio in offspring that is typically close to _____ .
   a. 3:1   b. 1:2:1   c. 1:1:1:1   d. 9:3:3:1

5. The probability of a crossover occurring between two genes on the same chromosome _____ .
   a. is unrelated to the distance between them
   b. decreases with the distance between them
   c. increases with the distance between them

6. _____ alleles are both fully and equally expressed.
   a. Dominant   c. Pleiotropic
   b. Codominant   d. a and b

7. A bell curve indicates _____ in a trait.
   a. pleiotropy   c. continuous variation
   b. crossing over   d. aneuploidy

8. Constructing a pedigree is particularly useful when studying inheritance patterns in organisms that _____ .
   a. produce many offspring per generation
   b. produce few offspring per generation
   c. have a very large chromosome number
   d. have a fast life cycle

Digging Into Data

The Cystic Fibrosis Mutation and Typhoid Fever

The CF mutation disables the receptor function of the CFTR protein, so it inhibits the endocytosis of bacteria into epithelial cells. Endocytosis is an important part of the respiratory tract's immune defenses against common *Pseudomonas* bacteria, which is why *Pseudomonas* infections are a chronic problem in cystic fibrosis patients.

The CF mutation also inhibits endocytosis of *Salmonella typhi* into cells of the gastrointestinal tract, where internalization of this bacteria can cause typhoid fever. Typhoid fever is a common worldwide disease. Its symptoms include extreme fever and diarrhea, and the resulting dehydration causes delirium that may last several weeks. If untreated, it kills up to 30 percent of those infected. Most of the 600,000 people who die annually from typhoid fever are children.

In 1998, Gerald Pier and his colleagues compared the uptake of *S. typhi* by different types of epithelial cells: those homozygous for the normal allele, and those heterozygous for the CF mutation (Figure 9.25). (Cells homozygous for the mutation do not take up *S. typhi* bacteria.)

1. Regarding the Ty2 strain of *S. typhi*, about how many more bacteria were able to enter normal cells (those expressing unmutated *CFTR*) than cells expressing the gene with the CF mutation?

2. Which strain of bacteria entered normal epithelial cells most easily?

3. The CF mutation inhibited the entry of all three *S. typhi* strains into epithelial cells. Can you tell which strain was most inhibited?

FIGURE 9.25 Effect of the CF mutation on uptake of three different strains of *Salmonella typhi* bacteria by epithelial cells. © Cengage Learning.

---

9. A female child inherits one X chromosome from her mother and one from her father. What sex chromosome does a male child inherit from each of his parents?

10. Nondisjunction at meiosis can result in _____ .
    a. duplications   c. crossing over
    b. aneuploidy   d. pleiotropy

11. True or false? All traits are inherited in a predictable way.

12. If one parent is heterozygous for a dominant autosomal allele and the other parent does not carry the allele, a child of theirs has a _____ chance of being heterozygous.
    a. 25 percent   c. 75 percent
    b. 50 percent   d. no chance; it will die

13. Klinefelter syndrome (XXY) is most easily diagnosed by _____ .
    a. pedigree analysis   c. karyotyping
    b. aneuploidy   d. phenotypic treatment

14. Match each example with the most suitable description.
    _____ dihybrid cross   a. *bb*
    _____ monohybrid cross   b. *AaBb* × *AaBb*
    _____ homozygous condition   c. *Aa*
    _____ heterozygous condition   d. *Aa* × *Aa*

15. Match the terms appropriately.
    _____ polyploid   a. symptoms of a genetic disorder
    _____ syndrome   b. extra sets of chromosomes
    _____ aneuploidy   c. gametes end up with the wrong
    _____ nondisjunction   chromosome number
          during meiosis   d. two genes affect the same trait
    _____ epistasis   e. one extra chromosome
    _____ genotype   f. varies continuously
    _____ human eye color   g. an individual's alleles

Critical Thinking

1. Mendel crossed a true-breeding pea plant with green pods and a true-breeding pea plant with yellow pods. All offspring had green pods. Which color is recessive?

2. Assuming that independent assortment occurs during meiosis, what type(s) of gametes will form in individuals with the following genotypes?
    a. *AABB*   b. *AaBB*   c. *Aabb*   d. *AaBb*

3. Refer to problem 2. Determine the frequencies of each genotype among offspring from an *AABB* × *aaBB* mating.

4. Explain the difference between codominance and incomplete dominance.

5. Suppose you identify a new gene in mice. One of its alleles specifies white fur, another specifies brown. How would you discover the dominance relationship between the two alleles?

6. Certain genes are vital for development. When mutated, they are lethal in homozygous recessives. Even so, heterozygotes can perpetuate recessive, lethal alleles. The allele *Manx* (M^L) in cats is an example. Homozygous cats (M^LM^L) die before birth. In heterozygotes (M^LM), the spine develops abnormally, and the cats end up with no tail. Two M^LM cats mate. What is the probability that any one of their surviving kittens will be heterozygous?

7. As you read earlier, Duchenne muscular dystrophy arises through the expression of a recessive X-linked allele. Usually, symptoms start to appear in childhood. Gradual, progressive loss of muscle function leads to death, usually by age twenty or so. Unlike color blindness, the disorder is nearly always restricted to males. Suggest why.

10 Biotechnology

This chapter builds on your understanding of DNA structure (Sections 6.2, 6.3) and replication (6.4). Clones (6.1), gene expression (7.2), and knockouts (7.7) are part of genetic engineering. You will revisit tracers (2.2), triglycerides (2.8), lipoproteins (2.9), β-carotene (5.2), mutations (7.6), cancer (8.5), alleles (8.6), environmental influences on human traits (9.5), genetic disorders (9.6, 9.7), and genetic screening (9.9).

FIGURE 10.1 A SNP-chip, used by personal DNA testing companies to analyze their customers' chromosomes for SNPs. This chip, shown actual size, reveals which versions of 1,140,419 SNPs occur in the DNA of four individuals at a time. Only about 1 percent of the 3 billion nucleotides in a person's DNA are unique to the individual.

Courtesy of © Illumina, Inc., www.illumina.com.

10.1 Personal DNA Testing

About 99 percent of your DNA is exactly the same as everyone else's. If you compared your DNA with your neighbor's, about 2.97 billion nucleotides of the two sequences would be identical; the remaining 30 million or so nonidentical nucleotides are sprinkled throughout your chromosomes. The sprinkling is not entirely random because some regions of DNA vary less than others. Such conserved regions are of particular interest to researchers because they are the ones most likely to have an essential function. When a conserved sequence does vary among people, the variation tends to be in particular nucleotides.

A nucleotide difference carried by a measurable percentage of a population, usually above 1 percent, is called a **single-nucleotide polymorphism**, or **SNP** (pronounced "snip"). Alleles of most genes differ by single nucleotides, and differences in alleles are the basis of the variation in human traits that makes each individual unique (Section 8.6). Thus, SNPs account for many of the differences in the way humans look, and they also have a lot to do with differences in the way our bodies work—how we age, respond to drugs, weather assaults by pathogens and toxins, and so on.

Consider the lipoprotein particles that carry fats and cholesterol through our bloodstreams. These particles consist of variable amounts and types of lipids and proteins, one of which is specified by the gene *APOE*. About one in four people carries an allele of this gene that has a cytosine for thymine substitution at nucleotide 4,874. The allele is called *ε4*, and the SNP it carries alters the gene's protein product by one amino acid. How the change affects lipoprotein particles remains unclear, but we do know that carrying the *ε4* allele increases one's risk of developing Alzheimer's disease later in life, particularly people who are homozygous for it.

About 4.5 million SNPs in human DNA have been identified, and that number is growing every day. A few companies are now offering to determine some of the SNPs you carry. The companies extract your DNA from the cells in a few drops of spit, then analyze it for SNPs (Figure 10.1).

Personal genetic testing may soon revolutionize medicine by allowing physicians to customize treatments on the basis of an individual's genetic makeup. For example, an allele associated with a heightened risk of a particular medical condition could be identified long before symptoms actually appear. People with that allele could then be encouraged to make lifestyle changes known to delay the onset of the condition. For some conditions, treatment that begins early enough may prevent symptoms from developing at all. Physicians could design treatments to fit the way a condition is likely to progress in the individual, prescribing only those drugs that will work in the person's body.

① A restriction enzyme recognizes a specific nucleotide sequence (*orange* boxes) in DNA from two sources.

② The enzyme cuts the DNA into fragments. This enzyme leaves single-stranded tails ("sticky ends") on cut DNA.

③ When the DNA fragments from the two sources are mixed together, matching sticky ends base-pair with each other.

④ DNA ligase joins the base-paired DNA fragments. Molecules of recombinant DNA are the result.

FIGURE 10.2 Animated! Making recombinant DNA.

© Cengage Learning.

10.2 Finding Needles in Haystacks

> **Cutting and Pasting DNA** In the 1950s, excitement over the discovery of DNA's structure gave way to frustration: No one could determine the order of nucleotides in a molecule of DNA. Identifying a single nucleotide among thousands or millions of others turned out to be a huge technical challenge. Research in a seemingly unrelated field yielded a solution when Werner Arber, Hamilton Smith, and their coworkers discovered why some bacteria are resistant to viral infection. Special enzymes inside these bacteria chop up any injected viral DNA before it has a chance to integrate into the bacterial chromosome. The enzymes restrict viral growth; hence their name, **restriction enzymes**. A restriction enzyme cuts DNA wherever a specific nucleotide sequence occurs (Figure 10.2). For example, the enzyme *Eco*RI (named after *E. coli*, the bacteria from which it was isolated) cuts DNA at the nucleotide sequence GAATTC ①. Other restriction enzymes cut at different sequences.

The discovery of restriction enzymes allowed researchers to cut chromosomal DNA into manageable chunks. It also allowed them to combine DNA fragments from different organisms. How? Many restriction enzymes, including *Eco*RI, leave single-stranded tails on DNA fragments ②. Researchers realized that complementary tails will base-pair, regardless of the source of the DNA ③. The tails are called "sticky ends," because two DNA fragments stick together when their matching tails base-pair. The enzyme DNA ligase (Section 6.4) can be used to seal the gaps between base-paired sticky ends, so continuous DNA strands form ④.

Thus, using appropriate restriction enzymes and DNA ligase, researchers can cut and paste DNA from different sources. The result, a hybrid molecule that consists of genetic material from two or more organisms, is called **recombinant DNA**. Making recombinant DNA is the first step in **DNA cloning**, a set of laboratory methods that uses living cells to mass-produce specific DNA fragments.

Researchers often clone DNA fragments by inserting them into plasmids. These small circles of nonchromosomal DNA are common in bacteria. A bacterium copies all of its DNA before it divides, so its offspring inherit plasmids along with chromosomes. If a plasmid carries a fragment of foreign DNA, that fragment gets copied and distributed to descendant cells along with the plasmid. Thus, plasmids can be used as **cloning vectors**, which are DNA molecules that can be used to carry foreign DNA into host cells (Figure 10.3). A host cell

FIGURE 10.3 A commercial cloning vector. Restriction enzyme recognition sequences on this plasmid are indicated on the *right* by the name of the enzyme that cuts them. Foreign DNA can be inserted into the vector at these sites. Bacterial genes (*gold*) help researchers identify host cells that take up a vector with inserted DNA.

© Cengage Learning.

cloning vector A DNA molecule that can accept foreign DNA and be replicated inside a host cell.

DNA cloning Set of procedures that uses living cells to make many identical copies of a DNA fragment.

recombinant DNA A DNA molecule that contains genetic material from more than one organism.

restriction enzyme Type of enzyme that cuts DNA at a specific nucleotide sequence.

single-nucleotide polymorphism (SNP) One-nucleotide DNA sequence variation carried by a measurable percentage of a population.

A A restriction enzyme cuts a specific nucleotide sequence in chromosomal DNA and in a plasmid cloning vector.

B A fragment of chromosomal DNA and the plasmid base-pair at their sticky ends. DNA ligase joins the two pieces of DNA.

C The recombinant plasmid is inserted into a host cell. When the cell reproduces, it copies the plasmid along with its chromosome. Each descendant cell receives a plasmid.

FIGURE 10.4 Animated! An example of cloning. Here, a fragment of chromosomal DNA is inserted into a plasmid.
© Cengage Learning.

Thirty PCR cycles may amplify the number of template DNA molecules a billionfold.

into which a cloning vector has been inserted can be grown in the laboratory (cultured) to yield a huge population of genetically identical cells, or clones (Section 6.1). Each clone contains a copy of the vector (Figure 10.4). A DNA fragment carried by the vector can be collected in large quantities by harvesting it from the clones.

> DNA Libraries The entire set of genetic material—the **genome**—of most organisms consists of thousands of genes. To study or manipulate a single gene, researchers must first separate the gene from all of the others. They often begin by cutting an organism's DNA into pieces, and then cloning all the pieces. The result is a set of clones that collectively contain all of the DNA in a genome.

A set of cells that hosts various cloned DNA fragments is called a **DNA library**. In such libraries, a cell that contains a particular DNA fragment of interest is mixed up with thousands or millions of others that do not—a needle in a haystack. One way to find that one clone among the others involves **probes**, which are fragments of DNA or RNA labeled with a tracer (Section 2.2). For example, researchers may synthesize a short chain of nucleotides based on the known DNA sequence of a gene, then attach a radioactive phosphate group to it. The nucleotide sequences of the probe and the gene are complementary, so the two can base-pair. Base pairing between DNA (or DNA and RNA) from more than one source is called **nucleic acid hybridization**. When the probe is mixed with DNA from a library, it will base-pair with (hybridize to) the gene, but not to other DNA. Researchers can pinpoint a clone that hosts the gene by detecting the label on the probe. That clone is isolated and cultured, and its DNA can be extracted in bulk for research or other purposes.

> PCR The polymerase chain reaction (**PCR**) is a technique used to mass-produce copies of a particular section of DNA without having to clone it in living cells (Figure 10.5). The reaction can transform a needle in a haystack—that one-in-a-million fragment of DNA—into a huge stack of needles with a little bit of hay in it.

DNA library Collection of cells that host different fragments of foreign DNA, often representing an organism's entire genome.

genome An organism's complete set of genetic material.

nucleic acid hybridization Base pairing between DNA or RNA from different sources.

PCR Polymerase chain reaction. Method that rapidly generates many copies of a specific DNA fragment.

probe Short fragment of DNA designed to hybridize with a nucleotide sequence of interest and labeled with a tracer.

targeted section

FIGURE 10.5 Animated! Two rounds of PCR. Each cycle of this reaction can double the number of copies of a targeted sequence of DNA. Thirty cycles can make a billion copies. © Cengage Learning.

1 DNA (*blue*) with a targeted sequence is mixed with primers (*pink*), nucleotides, and heat-tolerant *Taq* DNA polymerase.

The starting material for PCR is any sample of DNA with at least one molecule of a targeted sequence. It might be DNA from, say, a mixture of 10 million different clones, a sperm, a hair left at a crime scene, or a mummy—essentially any sample that has DNA in it.

The PCR reaction is based on DNA replication (Section 6.4). First, the starting (template) DNA is mixed with DNA polymerase, nucleotides, and primers. In PCR, two primers are made; each base-pairs with one end of the section of DNA to be amplified, or mass-produced **1**. Next, the reaction mixture is exposed to repeated cycles of high and low temperatures. High temperature disrupts the hydrogen bonds that hold the two strands of a DNA double helix together (Section 6.3), so every molecule of DNA unwinds and becomes single-stranded. As the temperature of the reaction mixture is lowered, the single DNA strands hybridize with the primers **2**.

2 When the mixture is heated, the double-stranded DNA separates into single strands. When the mixture is cooled, some of the primers base-pair with the DNA at opposite ends of the targeted sequence.

The DNA polymerases of most organisms denature at the high temperature required to separate DNA strands. The kind that is used in PCR reactions, *Taq* polymerase, is from *Thermus aquaticus*. This bacterial species lives in hot springs and other high-temperature environments, so its DNA polymerase necessarily tolerates heat. *Taq* polymerase recognizes hybridized primers as places to start DNA synthesis **3**. Synthesis proceeds along the template strand until the temperature rises and the DNA separates into single strands **4**. The newly synthesized DNA is a copy of the targeted section.

3 *Taq* polymerase begins DNA synthesis at the primers, so it produces complementary strands of the targeted DNA sequence.

When the mixture cools, more primers hybridize, and DNA synthesis begins again. The number of copies of the targeted section of DNA can double with each cycle of heating and cooling **5**. Thirty PCR cycles may amplify that number a billionfold.

4 The mixture is heated again, so all double-stranded DNA separates into single strands. When it is cooled, primers base-pair with the targeted sequence in the original template DNA and in the new DNA strands.

Take-Home Message

What techniques allow researchers to study DNA?

- DNA cloning uses living cells to mass-produce particular DNA fragments. Restriction enzymes cut DNA into fragments, then DNA ligase seals the fragments into cloning vectors. Recombinant DNA molecules result.

- A cloning vector that holds foreign DNA can be introduced into a living cell. When the host cell divides, it gives rise to huge populations of genetically identical cells (clones), each of which contains a copy of the foreign DNA.

- Researchers can isolate one gene from the many others in a genome by making DNA libraries. A probe can be used to identify one clone that hosts a DNA fragment of interest among many other clones in a DNA library.

- PCR quickly mass-produces copies of a targeted section of DNA.

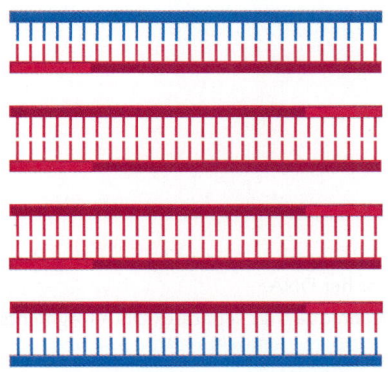

5 Each round of PCR reactions can double the number of copies of the targeted DNA section.

10.3 Studying DNA

> Sequencing the Human Genome Researchers use a technique called **sequencing** to determine the order of nucleotides in a fragment of DNA that has typically been isolated by cloning or PCR. The most common method is similar to DNA replication (Section 6.4). The DNA to be sequenced (the template) is mixed with nucleotides, a primer, and DNA polymerase. Starting at the primer, the polymerase joins the nucleotides into a new strand of DNA, in the order dictated by the sequence of the template.

The sequencing reaction produces millions of DNA fragments of different lengths, all incomplete copies of the starting DNA. These fragments are then separated by length, using a technique called **electrophoresis** in which an electric field pulls DNA fragments through a semisolid gel. DNA fragments of different sizes move through the gel at different rates. The shorter the fragment, the faster it moves, because shorter fragments slip through the tangled molecules of the gel faster than longer fragments do. All fragments of the same length move through the gel at the same speed, so they gather into bands. The order of the bands in the gel reflects the sequence of the template DNA (Figure 10.6).

The sequencing method we have just described was invented in 1975. Ten years later, it had become so routine that people were debating about sequencing the entire human genome. Knowing the sequence would have huge potential payoffs for medicine and research, but sequencing 3 billion nucleotides was a daunting proposition at the time. It would require at least 6 million sequencing reactions, a task that would have taken 50 years to complete given available techniques. Opponents said the effort would divert too much attention and funding from more urgent research. However, sequencing techniques kept getting better, and with each improvement more nucleotides could be sequenced in less time. Automated (robotic) DNA sequencing and PCR had just been invented. Both were still too cumbersome and expensive to be useful in routine applications, but they would not be so for long. Waiting for faster technologies seemed the most efficient way to sequence the genome, but just how fast did they need to be before the project should begin?

A few privately owned companies decided not to wait, and started sequencing. One of them intended to patent the sequence after it was determined, a development that provoked widespread outrage but also spurred commitments in the public sector. In 1988, the National Institutes of Health (NIH) effectively annexed the project by hiring James Watson (of DNA structure fame) to head an official Human Genome Project, and providing $200 million per year to fund it. A consortium formed between the NIH and international institutions that were sequencing different parts of the genome. Watson set aside 3 percent of the funding for studies of ethical and social issues arising from the research. He later resigned over a patent disagreement, and geneticist Francis Collins took his place.

Amid ongoing squabbles over patent issues, Celera Genomics formed in 1998. With biologist Craig Venter at its helm, the company intended to commercialize genetic information. Celera invented faster techniques for sequencing genomic DNA, because the first to have the complete sequence had a legal basis for patenting it. The competition motivated the public consortium to move its efforts into high gear.

FIGURE 10.6 Sequencing, a method of determining the nucleotide sequence of a DNA molecule.

On this computer screen, the four nucleotides (adenine, thymine, guanine, and cytosine) are color-coded green, red, yellow, and blue, respectively. The order of colored bands in each vertical "lane" represents part of the sequence of the template DNA.

Patrick Landmann/ Photo Researchers, Inc.

DNA profiling Identifying an individual by analyzing the unique parts of his or her DNA.

electrophoresis Technique that separates DNA fragments by size.

genomics The study of genomes.

sequencing Method of determining the order of nucleotides in a DNA molecule.

```
758  GATAATCCTGTTTTGAACAAAAGGTCAAATTGCTGAATAGAAA-GTCTTGATTAACTAAAAGATGTACAAAGTGGAATTA 836  Human
752  GATAATCCTGTTTTGAACAAAAGGTCAAATTGCTGAATAGAAA-GTCTTGATTAACTAAAAGATGTACAAAGTGGAATTA 830  Mouse
751  GATAATCCTGTTTTGAACAAAAGGTCAAATTGCTGAATAGAAA-GTCTTGATTAACTAAAAGATGTACAAAGTGGAATTA 829  Rat
754  GATAATCCTGTTTTGAACAAAAGGTCAAATTGCTGAATAGAAA-GTCTTGATTAACTAAAAGATGTACAAAGTGGAATTA 832  Dog
782  GATAATCCTGTTTTGAACAAAAGGTCAAATTGCTGAATAGAAA-GTCTTGATTAACTAAAAGATGTACAAAGTGGAATTA 860  Chicken
758  GATAATCCTGTTTTGAACAAAAGGTCAAATTGCTGAATAGAAA-GTCTTGATTAAGTAAAAGATGTACAAAGTGGAATTA 836  Frog
823  GATAATCCTGTTTTGAACAAAAGGTCAGATTGCTGAATAGAAAAGGCTTGATTAAAGCAGAGATGTACAAAGTGGACGCA 902  Zebrafish
763  GATAATCCTGTTTTGAACAAAAGGTCAAATTGTTGAATAGAGACGCTTTGATAAAGCGGAGGAGGTACAAAGTGGGACC- 841  Pufferfish
```

Then, in 2000, U.S. President Bill Clinton and British Prime Minister Tony Blair jointly declared that the sequence of the human genome could not be patented. Celera kept sequencing anyway. Celera and the public consortium separately published about 90 percent of the sequence in 2001. By 2003, fifty years after the discovery of the structure of DNA, the sequence of the human genome was officially completed.

> **Genomics** It took 15 years to sequence the human genome for the first time, but the techniques have improved so much that sequencing an entire genome now takes about a day. Anyone can now pay to have their genome sequenced. However, despite our ability to determine the sequence of an individual's genome, it will be a long time before we understand all the information coded within that sequence.

The human genome contains a massive amount of seemingly cryptic data. One way to decipher it is by comparing it to the genomes of other species, the premise being that all organisms are descended from shared ancestors, so all genomes are related to some extent. We see evidence of such genetic relationships simply by comparing the raw sequence data, which, in some regions, is extremely similar across many species (Figure 10.7).

The study of genomes is called **genomics**, a broad field that encompasses whole-genome comparisons, structural analysis of gene products, and surveys of small-scale variations in sequence. Genomics is providing powerful insights into evolution, and it has many medical benefits. We have learned the function of many human genes by studying their counterpart genes in other species. For instance, researchers comparing human and mouse genomes discovered a human version of a mouse gene, *APOA5*, that encodes a lipoprotein. Mice with an *APOA5* knockout have four times the normal level of triglycerides (Section 2.8) in their blood. The researchers then looked for—and found—a correlation between *APOA5* mutations and high triglyceride levels in humans. High triglycerides are a risk factor for coronary artery disease.

> **DNA Profiling** As you learned in Section 10.1, only about 1 percent of your DNA is unique. The shared part is what makes you human; the differences make you a unique member of the species. In fact, those differences are so unique that they can be used to identify you. Identifying an individual by his or her DNA is called **DNA profiling**.

One type of DNA profiling involves SNP-chips (an example is shown in Figure 10.1). A SNP-chip has a tiny glass plate with microscopic spots of DNA stamped on it. The DNA sample in each spot is a short, synthetic single strand with a unique SNP sequence. When an individual's genomic DNA is washed over a SNP-chip, it hybridizes only with DNA spots that have a matching SNP sequence. Probes reveal where the genomic DNA has hybridized—and which SNPs are carried by the individual (Figure 10.8).

FIGURE 10.7 Alignment of genomic DNA from various species. This is a region of the gene for a DNA polymerase. Differences are highlighted. The chance that any two of these sequences would randomly match is about 1 in 10^{46}.

© Cengage Learning.

FIGURE 10.8 A SNP-chip analysis.

A Each spot is a region where the individual's genomic DNA has hybridized with one SNP. A *red* or *green* dot means that the individual is homozygous for a SNP; a combined signal (*yellow* dot) indicates heterozygosity.

B The entire chip tests for 550,000 SNPs. The small *white* box indicates the magnified portion shown in **A**.

Credits: (a) The Sanger Institute. Wellcome Images; (b) Wellcome Trust Sanger Institute.

Another method of DNA profiling involves analysis of **short tandem repeats**, sections of DNA in which a series of 4 or 5 nucleotides is repeated several times in a row. Short tandem repeats tend to occur in predictable spots, but the number of repeats in each spot differs among individuals. For example, one person's DNA may have fifteen repeats of the nucleotides TTTTC at a certain spot on one chromosome. Another person's DNA may have this sequence repeated only twice in the same location. Such repeats slip spontaneously into DNA during replication, and their numbers grow or shrink over generations. Unless two people are identical twins, the chance that they have identical short tandem repeats in even three regions of DNA is 1 in a quintillion (10^{18}), which is far more than the number of people who have ever lived. Thus, an individual's array of short tandem repeats is, for all practical purposes, unique.

Analyzing a person's short tandem repeats begins with PCR, which is used to copy ten to thirteen particular regions of chromosomal DNA known to have repeats. The lengths of the copied DNA fragments differ among most individuals, because the number of tandem repeats in those regions also differs. Thus, electrophoresis can be used to reveal an individual's unique array of short tandem repeats (Figure 10.9).

Short tandem repeat analysis will soon be replaced by full genome sequencing, but for now it continues to be a common DNA profiling method. Geneticists compare short tandem repeats on Y chromosomes to determine relationships among male relatives, and to trace an individual's ethnic heritage. They also track mutations that accumulate in populations over time by comparing DNA profiles of living humans with those of ancient ones. Such studies are allowing us to reconstruct population dispersals that happened in the ancient past.

Short tandem repeat profiles are routinely used to resolve kinship disputes, and as evidence in criminal cases. Within the context of a criminal or forensic investigation, DNA profiling is called DNA fingerprinting. As of January 2012, the database of DNA fingerprints maintained by the Federal Bureau of Investigation (the FBI) contained the short tandem repeat profiles of 10.3 million offenders, and had been used in over 160,000 criminal investigations. DNA fingerprints have also been used to identify the remains of over 400,000 people, including the individuals who died in the World Trade Center on September 11, 2001.

A *Gray* boxes indicate which regions of the individual's DNA were tested.

B The number of repeats is shown in a box below each peak. A peak's location on the x-axis corresponds to the length of the DNA fragment amplified (a measure of the number of repeats). Peak size reflects the amount of DNA.

FIGURE 10.9 Animated! An individual's (partial) short tandem repeat profile. Remember, human body cells are diploid. Double peaks appear on a profile when the two members of a chromosome pair carry a different number of repeats.

Figure It Out: How many repeats does this individual have at the Penta D region?

Answer: 12 on one chromosome, and 14 on the other

© Cengage Learning.

Take-Home Message

How do we use what researchers discover about DNA?

- Improvements in DNA sequencing techniques and worldwide efforts allowed the human genome sequence to be determined.
- Analysis of the human genome sequence is yielding new information about our genes and how they work.
- DNA profiling identifies individuals by the unique parts of their DNA.

---

genetic engineering Process by which deliberate changes are introduced into an individual's genome.

genetically modified organism (**GMO**) Organism whose genome has been modified by genetic engineering.

short tandem repeats In chromosomal DNA, sequences of 4 or 5 nucleotides repeated multiple times in a row.

transgenic Refers to a genetically modified organism that carries a gene from a different species.

10.4 Genetic Engineering

Traditional cross-breeding methods can alter genomes, but only if individuals with the desired traits will interbreed. Genetic engineering takes gene-swapping to an entirely different level. **Genetic engineering** is a laboratory process by which an individual's genome is deliberately modified. A gene from one species may be transferred to another to produce an organism that is **transgenic**, or a gene may be altered and reinserted into an individual of the same species. Both methods result in a **genetically modified organism**, or **GMO**.

❯ **Genetically Modified Microorganisms** The most common GMOs are bacteria and yeast. These cells have the metabolic machinery to make complex organic molecules, and they are easily modified (Figure 10.10). Bacteria and yeast have been modified to produce medically important proteins. People with diabetes were among the first beneficiaries of such organisms. Insulin for their injections was once extracted from animals, but it provoked an allergic reaction in some people. Human insulin, which does not provoke allergic reactions, has been produced by transgenic *E. coli* since 1982. Slight modifications of the gene have also yielded fast-acting and slow-release forms of human insulin.

Engineered microorganisms also produce proteins used in food manufacturing. For example, cheese is traditionally made with an extract of calf stomachs, which contain the enzyme chymotrypsin. Most cheese manufacturers now use chymotrypsin produced by genetically engineered bacteria. Other enzymes produced by GMOs improve the taste and clarity of beer and fruit juice, slow bread staling, or modify fats.

❯ **Designer Plants** As crop production expands to keep pace with human population growth, it places unavoidable pressure on ecosystems everywhere. Irrigation leaves mineral and salt residues in soils. Tilled soil erodes, taking topsoil with it. Runoff clogs rivers, and fertilizer in it causes algae to grow so fast that fish suffocate. Pesticides can be harmful to humans and beneficial insects such as bees.

Pressured to produce more food at lower cost and with less damage to the environment, many farmers have begun to rely on genetically modified crop plants. Genes can be introduced into plant cells by way of electric or chemical shocks, by blasting them with microscopic DNA-coated pellets, or by using *Agrobacterium tumefaciens* bacteria. *A. tumefaciens* carries a plasmid with genes that cause tumors to form on infected plants; hence the name Ti plasmid (for Tumor-inducing). Researchers replace the tumor-inducing genes with foreign or modified genes, then use the plasmid as a vector to deliver the desired genes into plant cells. Whole plants can be grown from plant cells that integrate a recombinant plasmid into their chromosomes (Figure 10.11).

FIGURE 10.10 Genetically modified bacteria. A green fluorescent protein from jellyfish has been inserted into these *E. coli* cells. The cells are genetically identical, so the visible variation in fluorescence among them reveals differences in gene expression. Such differences may help us discover why some bacteria of a population become dangerously resistant to antibiotics, and others do not.

Courtesy of Systems Biodynamics Lab, P. I. Jeff Hasty, UCSD Department of Bioengineering, and Scott Cookson.

A A Ti plasmid carrying a foreign gene is inserted into an *Agrobacterium tumefaciens* bacterium.

B The bacterium infects a plant cell and transfers the Ti plasmid into it. The plasmid DNA becomes integrated into one of the cell's chromosomes.

C The plant cell divides, and its descendants form an embryo. Several embryos are sprouting from this mass of cells.

D Each embryo develops into a transgenic plant that expresses the foreign gene. The glowing tobacco plant in the photo on the *right* is expressing a gene from fireflies.

FIGURE 10.11 Animated! Using the Ti plasmid to make a transgenic plant.

Credits: (a–c) © Cengage Learning; (d) left, © Lowell Georgis/ Corbis; right, Keith V. Wood.

FIGURE 10.12 Genetically modified crops can help farmers use less pesticide.

Top, the *Bt* gene conferred insect resistance to the genetically modified plants that produced this corn. *Bottom*, corn produced by unmodified plants is more vulnerable to insect pests.

The Bt and Non-Bt corn photos were taken as part of field trial conducted on the main campus of Tennessee State University at the Institute of Agriculture and Environmental Research. The work was supported by a competitive grant from the CSREES, USDA titled "Southern Agricultural Biotechnology Consortium for Underserved Communities," (2000–2005). Dr. Fisseha Tegegne and D. Ahmad Aziz served as Principal and Co-principal Investigators respectively to conduct the portion of the study in the State of Tennessee.

gene therapy Treating a genetic defect or disorder by transferring a normal or modified gene into the affected individual.

xenotransplantation Transplantation of an organ from one species into another.

Many genetically modified crops carry genes that impart resistance to devastating plant diseases and pests. GMO crops such as Bt corn and soy help farmers use smaller amounts of toxic pesticides. Organic farmers often spray their crops with spores of Bt (*Bacillus thuringiensis*), a bacterial species that makes a protein toxic only to some insect larvae. Researchers transferred the gene encoding the Bt protein into plants. The engineered plants produce the Bt protein, but otherwise they are essentially identical to unmodified plants. Larvae die shortly after eating their first and only GMO meal. Farmers can use much less pesticide on crops that make their own (Figure 10.12).

Genetic modifications can also make food plants more nutritious. For example, rice plants have been engineered to make β-carotene, an orange photosynthetic pigment (Section 5.2) that cells of the small intestine remodel into vitamin A. These rice plants carry two genes in the β-carotene synthesis pathway: one from corn, the other from bacteria. One cup of the engineered rice seeds—grains of Golden Rice—has enough β-carotene to satisfy a child's daily need for vitamin A.

The USDA Animal and Plant Health Inspection Service (APHIS) regulates the introduction of GMOs into the environment. At this writing, APHIS has deregulated eighty-six crop plants, which means the plants are approved for unregulated use in the United States. Worldwide, more than 330 million acres are currently planted in GMO crops, the majority of which are corn, sorghum, cotton, soy, canola, and alfalfa engineered for resistance to the herbicide glyphosate. Rather than tilling the soil to control weeds, farmers can spray their fields with glyphosate, which kills the weeds but not the engineered crops.

After long-term, widespread use of glyphosate, weeds resistant to the herbicide are becoming more common. The engineered gene is also appearing in wild plants and in nonengineered crops, which means that recombinant genes can (and do) escape into the environment. The genes are probably being transferred to nontransgenic plants via pollen carried by wind or insects.

> Biotech Barnyards Traditional cross-breeding has produced animals so unusual that transgenic animals may seem a bit mundane by comparison (Figure 10.13**A**). Cross-breeding is also a form of genetic manipulation, but many transgenic animals would never have occurred without laboratory intervention.

The first genetically modified animals were mice. Today, engineered mice are commonplace, and they are invaluable in research (Figure 10.13**B**). For example, we have discovered the function of human genes (including the *APOA5* gene discussed in Section 10.3) by inactivating their counterparts in mice. Genetically modified mice are also used as models of human diseases. For example, researchers inactivated the molecules involved in the control of glucose metabolism, one by one. Studying the effects of the knockouts in mice has resulted in much of our current understanding of how diabetes works in humans.

Genetically modified animals also make proteins that have medical and industrial applications. Various transgenic goats produce proteins used to treat cystic fibrosis, heart attacks, blood clotting disorders (Figure 10.13**C**), and even nerve gas exposure. Milk from goats transgenic for lysozyme, an antibacterial protein in human milk, may protect infants and children in developing countries from acute diarrheal disease. Goats transgenic for a spider silk gene produce the silk protein in their milk; researchers can spin this protein into nanofibers that are useful in medical and electronics applications. Rabbits make human interleukin-2, a protein that triggers divisions of immune cells. Genetic engineering has also given us extra-large sheep, dairy goats with heart-healthy milk, pigs with heart-healthy fat and environmentally friendly low-phosphate feces, and cows that are resistant to mad cow disease.

› Knockout Cells and Organ Factories Millions of people suffer with organs or tissues that are damaged beyond repair. In any given year, more than 80,000 of them are on waiting lists for an organ transplant in the United States alone. Human donors are in such short supply that illegal organ trafficking is now a common problem. Pigs are a potential source of organs for transplantation, because pig and human organs are about the same in both size and function. However, the human immune system battles anything it recognizes as nonself. It rejects a pig organ at once, because it recognizes proteins and carbohydrates on the plasma membrane of pig cells. Within a few hours, blood coagulates inside the organ's vessels and dooms the transplant. Drugs can suppress the immune response, but they also render organ recipients particularly vulnerable to infection. Researchers have produced genetically modified pigs that lack the offending molecules on their cells. The human immune system may not reject tissues or organs transplanted from these pigs.

Pig-to-human transplants are an example of **xenotransplantation**, the transfer of an organ from one species into another. Critics of xenotransplantation are concerned that, among other things, such transplants would invite pig viruses to infect humans, perhaps with catastrophic results. Their concerns are not unfounded: All human pandemics have been caused by animal viruses that adapted to replicate in people.

Tinkering with the genes of animals raises a host of ethical dilemmas. For example, mice, monkeys, and other animals that have been genetically modified to carry mutations associated with certain human diseases often suffer the same terrible symptoms of these conditions as humans do. However, these animals are allowing researchers to study—and test treatments for—conditions such as multiple sclerosis, cystic fibrosis, diabetes, cancer, and Huntington's disease without experimenting on humans.

Take-Home Message

What is genetic engineering?

- Genetic engineering is the directed alteration of an individual's genome, and it results in a genetically modified organism (GMO).

- A transgenic organism carries a gene from a different species. Transgenic organisms, including bacteria and yeast, are used in research, medicine, and industry.

- Genetically modified crop plants can help farmers be more productive while reducing their costs. However, widespread use of these crops is having unintended environmental effects.

- Most genetically engineered animals are used for medical applications and medical research.

10.5 Genetically Modified Humans

We know of more than 15,000 serious genetic disorders. Collectively, they cause 20 to 30 percent of infant deaths each year, and account for half of all mentally impaired patients and a fourth of all hospital admissions. They also contribute to many age-related disorders, including cancer, Parkinson's disease, and diabetes. Drugs and other treatments can minimize the symptoms of some genetic disorders, but gene therapy is the only cure. **Gene therapy** is the transfer of recombinant DNA into an individual's body cells, with the intent to correct a genetic defect or treat a disease. The transfer, which occurs by way of lipid

A The genes of this chicken were modified using a traditional method: cross-breeding. Featherless chickens can survive in deserts where cooling systems are not an option.

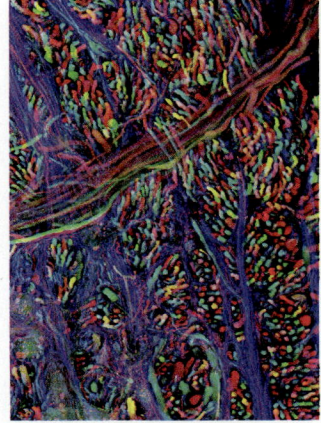

B Mice transgenic for multiple pigments ("brainbow mice") are allowing researchers to map the complex neural circuitry of the brain. Individual nerve cells in the brain stem of a brainbow mouse are visible in this fluorescence micrograph.

C This transgenic goat produces milk that contains human antithrombin III, a protein that inhibits blood clotting. Antithrombin III purified from transgenic goats' milk is used as a medication to treat people who have a hereditary deficiency of this protein.

FIGURE 10.13 Examples of genetically modified animals.

Credits: (a) © Adi Nes, Dvir Gallery Ltd.; (b) Courtesy of © Dr. Jean Levit. The Brainbow technique was developed in the laboratories of Jeff W. Lichtman and Joshua R. Sanes at Harvard University. This image has received the Bioscape imaging competition 2007 prize; (c) Transgenic goat produced using nuclear transfer at GTC Biotherapeutics. Photo used with permission.

FIGURE 10.14 Rhys Evans, who was born with SCID-X1. His immune system has been permanently repaired by gene therapy.

© Jeans for Gene Appeal.

clusters or genetically engineered viruses, inserts an unmutated gene into an individual's chromosomes.

Human gene therapy is a compelling reason to embrace genetic engineering research. It is now being tested as a treatment for AIDS, muscular dystrophy, heart attack, sickle-cell anemia, cystic fibrosis, hemophilia A, Parkinson's disease, Alzheimer's disease, several types of cancer, and inherited diseases of the eye, the ear, and the immune system. The results are encouraging. For example, little Rhys Evans (Figure 10.14) was born with SCID-X1, a severe X-linked genetic disorder that stems from a mutated allele of the *IL2RG* gene. The gene encodes a receptor for an immune signaling molecule. Children affected by this disorder can survive only in germ-free isolation tents, because they cannot fight infections. In the late 1990s, researchers used a genetically engineered virus to insert unmutated copies of *IL2RG* into cells taken from the bone marrow of twenty boys with SCID-X1. Each child's modified cells were infused back into his bone marrow. Within months of their treatment, eighteen of the boys left their isolation tents for good. Rhys was one of them. Gene therapy had permanently repaired their immune systems.

Manipulating a gene within the context of a living individual is unpredictable even when we know its sequence and location on a chromosome. No one, for example, can predict where a virus-injected gene will become integrated into a chromosome. Its insertion might disrupt other genes. If it interrupts a gene that is part of the controls over cell division, then cancer might be the outcome. Five of the twenty boys treated with gene therapy for SCID-X1 have since developed a type of bone marrow cancer called leukemia, and one of them has died. The researchers had wrongly predicted that cancer related to the gene therapy would be rare. Research now implicates the very gene targeted for repair, especially when combined with the virus that delivered it. Apparently, integration of the modified viral DNA activated nearby proto-oncogenes (Section 8.5) in the children's chromosomes.

❭ Eugenics The idea of selecting the most desirable human traits, **eugenics**, is an old one. It has been used as a justification for some of the most horrific episodes in human history, including the genocide of 6 million Jews during World War II. Thus, it continues to be a hotly debated social issue. For example, using gene therapy to cure human genetic disorders seems like a socially acceptable goal to most people. However, imagine taking this idea a bit further. Would it also be acceptable to engineer the genome of an individual who is within a normal range of phenotype in order to modify a particular trait? Researchers have already produced mice that have improved memory, enhanced learning ability, bigger muscles, and longer lives. Why not people?

Given the pace of genetics research, the eugenics debate is no longer about how we would engineer desirable traits, but how we would choose the traits that are desirable. Realistically, cures for many severe but rare genetic disorders will not be found, because the financial return will not even cover the cost of the research. Eugenics, however, might just turn a profit. How much would potential parents pay to be sure that their child will be tall or blue-eyed? Would it be okay to engineer "superhumans" with breathtaking strength or intelligence? How about a treatment that can help you lose that extra weight, and keep it off permanently? The gray area between interesting and abhorrent can be very different depending on who is asked. In a survey conducted in the United States, more than 40 percent of those interviewed said it would be fine to use gene therapy to make smarter and cuter babies. In one poll of British parents, 18 percent would be willing to use it to keep a child from being aggressive, and 10 percent would use it to keep a child from growing up to be homosexual.

eugenics Idea of deliberately improving the genetic qualities of the human race.

Some people are concerned that gene therapy puts us on a slippery slope that may result in irreversible damage to ourselves and to the biosphere. We as a society may not have the wisdom to know how to stop once we set foot on that slope; one is reminded of our peculiar human tendency to leap before we look. And yet, something about the human experience allows us to dream of such things as wings of our own making, a capacity that carried us into space. In this brave new world, the questions before you are these: What do we stand to lose if serious risks are not taken? And, do we have the right to impose the consequences of taking such risks on those who would choose not to take them?

Take-Home Message

Do we genetically modify people?

- Genes can be transferred into a person's cells to correct a genetic defect or treat a disease. However, the outcome of altering a person's genome remains unpredictable given our current understanding of how the genome works.

- We as a society continue to work our way through the ethical implications of applying DNA technologies.

10.6 Personal DNA Testing (revisited)

The results of SNP analysis by a personal DNA testing company also include estimated risks of developing conditions associated with your particular set of SNPs. For example, the test will probably determine whether you are homozygous for one allele of the *MC1R* gene. If you are, the company's report will tell you that you have red hair. Very few SNPs have such a clear cause-and-effect relationship as the *MC1R* allele for red hair, however. Most human traits are polygenic, and many are also influenced by environmental factors such as lifestyle (Section 9.5). Thus, although a DNA test can reliably determine the SNPs in an individual's genome, it cannot reliably predict the effect of those SNPs on the individual.

For example, if you are heterozygous for the ε4 allele of the *APOE* gene, a DNA testing company cannot tell you whether you will develop Alzheimer's disease later in life. Instead, the company will report your lifetime risk of developing the disease, which is about 29 percent, as compared with about 9 percent for someone who has no ε4 allele.

What does a 29 percent lifetime risk of developing Alzheimer's disease mean? The number is a probability statistic; it means that, on average, 29 of every 100 people who have the ε4 allele eventually get the disease. Having an increased risk of developing Alzheimer's does not mean you are certain to end up with the disease, however. Not everyone who develops the disease has the ε4 allele, and not everyone with the ε4 allele develops Alzheimer's disease. Other factors, including epigenetic modifications of DNA, contribute to the disease.

The plunging cost of genetic testing has spurred an explosion of companies offering SNP profiling and personal genome sequencing. In some cases, such testing is already in clinical use, for example in diagnosis of early-onset genetic disorders, or in predicting whether chemotherapy will extend the life of an individual who has breast cancer. However, we still have only a limited understanding of how genes contribute to most conditions, particularly age-related disorders such as Alzheimer's disease. Geneticists believe that it will be at least five to ten more years before we can use genotype to accurately predict an individual's risk of these conditions.

WHERE YOU ARE GOING . . .

DNA sequence comparisons are helping researchers study evolutionary relationships (Section 11.6). The escape of transgenic genes into the environment is an example of gene flow, an evolutionary process described in Section 12.4. In nature, plasmids facilitate gene transfers among bacteria (13.5). Many vaccines (22.8) are genetically engineered, and genetically engineered plants are useful in phytoremediation (Section 27.1). Research using specific genetically engineered animals is explained in relevant chapters.

Courtesy of © Illumina, Inc., www.illumina.com.

Corbis/SuperStock

Summary

Section 10.1 Personal DNA testing companies identify a person's unique array of **single-nucleotide polymorphisms**, or **SNP**s. Personal genetic testing may soon revolutionize the way medicine is practiced.

Section 10.2 In **DNA cloning**, researchers use **restriction enzymes** to cut a sample of DNA into pieces, and then DNA ligase to splice the pieces into plasmids or other **cloning vectors**. The resulting molecules of **recombinant DNA** are inserted into host cells such as bacteria. When a host cell divides, it forms huge populations of genetically identical descendant cells (clones), each with a copy of the cloned DNA fragment.

pDrive Cloning Vector
3.85 kb

Kanamycin

P_lac lacZ

Ampicillin

- Kpn I
- Sph I
- Pst I
- Bam HI
- Eco RI
- Sal I
- Acc I
- Xho I
- Xba I
- Bst XI
- Sac I
- Not I

A **DNA library** is a collection of cells that contain different fragments of DNA, often representing an organism's entire **genome**. Researchers can use **probes** to identify cells that carry a specific fragment of DNA. Base pairing between nucleic acids from different sources is called **nucleic acid hybridization**. The polymerase chain reaction (**PCR**) uses primers and a heat-resistant DNA polymerase to rapidly increase the number of copies of a targeted section of DNA.

Section 10.3 A technique called **sequencing** reveals the order of nucleotides in DNA. DNA polymerase is used to partially replicate a DNA template. The reaction produces a mixture of DNA fragments of all different lengths. **Electrophoresis** separates the fragments by length into bands.

Genomics is providing insights into the function of the human genome. Similarities between genomes of different organisms are evidence of evolutionary relationships, and can be used as a predictive tool in research. **DNA profiling** identifies a person by the unique parts of his or her DNA. An example is the determination of an individual's array of **short tandem repeats**. Within the context of a criminal investigation, a DNA profile is called a DNA fingerprint.

Section 10.4 Recombinant DNA technology is the basis of **genetic engineering**, the directed modification of an organism's genetic makeup with the intent to modify its phenotype. A gene from one species is inserted into an individual of a different species to make a **transgenic** organism, or a gene is modified and reinserted into an individual of the same species. The result of either process is a **genetically modified organism** (GMO).

Transgenic bacteria and yeast produce medically valuable proteins. Transgenic crop plants are helping farmers produce food more efficiently, but they are also having unintended environmental consequences. Genetically modified animals produce human proteins, and may one day provide a source of organs and tissues for **xenotransplantation** into humans.

Section 10.5 With **gene therapy**, a gene is transferred into body cells to correct a genetic defect or treat a disease. Potential benefits of genetically modifying humans must be weighed against potential risks. The practice raises ethical issues such as whether **eugenics** is desirable in some circumstances.

Self-Quiz
Answers in Appendix I

1. _____ cut(s) DNA molecules at specific sites.
   a. DNA polymerase   c. Restriction enzymes
   b. DNA probes   d. DNA ligase

2. A _____ is a small circle of bacterial DNA that is separate from the bacterial chromosome.
   a. plasmid   c. nucleus
   b. chromosome   d. double helix

3. For each species, all _____ in the complete set of chromosomes is/are the _____ .
   a. genomes; genotype   c. SNPs; DNA profile
   b. DNA; genome   d. DNA; library

4. A set of cells that host various DNA fragments collectively representing an organism's entire set of genetic information is a _____ .
   a. genome   c. DNA library
   b. clone   d. GMO

5. PCR can be used to _____ .
   a. increase the number of specific DNA fragments
   b. check DNA fingerprints
   c. modify a human genome
   d. a and b are correct

6. Fragments of DNA can be separated by electrophoresis according to _____ .
   a. sequence   b. length   c. species

7. *Taq* polymerase is used for PCR because it _____ .
   a. tolerates the high temperature needed to separate DNA strands
   b. is an enzyme from a bacterium
   c. does not require primers

8. _____ is a technique to determine the order of nucleotides in a fragment of DNA.
   a. PCR   c. Electrophoresis
   b. Sequencing   d. Nucleic acid hybridization

9. Which of the following can be used to carry foreign DNA into host cells? Choose all correct answers.
   a. RNA   e. lipid clusters
   b. viruses   f. blasts of microscopic pellets
   c. PCR   g. xenotransplantation
   d. plasmids   h. sequencing

10. A transgenic organism _____ .
   a. carries a gene from another species
   b. has been genetically modified
   c. can pass a foreign gene to offspring
   d. all of the above

11. _____ can correct a genetic defect in an individual.
   a. Cloning vectors   c. Xenotransplantation
   b. Gene therapy   d. a and b

Digging Into Data

Enhanced Spatial Learning Ability in Mice with an Autism Mutation

Autism is a neurobiological disorder with a range of symptoms that include impaired social interactions and stereotyped patterns of behavior such as hand-flapping or rocking. A relatively high proportion of autistic people—around 10 percent—have an extraordinary skill or talent such as greatly enhanced memory.

Mutations in neuroligin 3, an adhesion protein (Section 3.3) that connects brain cells to one another, have been associated with autism. One mutation changes amino acid 451 from arginine to cysteine. Mouse and human neuroligin 3 are very similar. In 2007, Katsuhiko Tabuchi and his colleagues genetically modified mice to carry the same arginine-to-cysteine substitution in their neuroligin 3. The mutation caused an increase in transmission of some types of signals between brain cells. Mice with the mutation had impaired social behavior, and, unexpectedly, enhanced spatial learning ability (Figure 10.15).

1. In the first test, how many days did unmodified mice need to learn to find the location of a hidden platform within 10 seconds?

2. Did the modified or the unmodified mice learn the location of the platform faster in the first test?

3. Which mice learned faster the second time around?

4. Which mice had the greatest improvement in memory?

A Mice were tested in a water maze, in which a platform is submerged slightly below the surface of a deep pool of warm water. The platform is not visible to swimming mice.

Mice do not particularly enjoy swimming, so they locate the hidden platform as fast as they can. When tested again in the same pool, they use visual cues around the pool's edge to remember the platform's location.

B How quickly mice remember the location of a hidden platform in a water maze is a measure of spatial learning ability.

FIGURE 10.15 Enhanced spatial learning ability in mice engineered to carry a particular mutation in neuroligin 3 (*R451C*), compared with unmodified (wild-type) mice.

Credits: (a) Courtesy of Dr. S. Thuret, Kings College London; (b) © Cengage Learning.

---

12. Put the following tasks in the order they would occur during a cloning experiment.
    a. using DNA ligase to seal DNA fragments into vectors
    b. using a probe to identify a clone in the library
    c. using DNA polymerase to sequence the DNA of the clone
    d. making a DNA library of clones
    e. cutting genomic DNA with restriction enzymes

13. Match the terms with the best description.
    _____ DNA profile
    _____ Ti plasmid
    _____ nucleic acid hybridization
    _____ eugenics
    _____ SNP
    _____ transgenic
    _____ GMO

    a. GMO with a foreign gene
    b. alleles commonly contain them
    c. a person's unique collection of short tandem repeats
    d. base pairing of DNA or DNA and RNA from different sources
    e. selecting "desirable" traits
    f. genetically modified
    g. used in some gene transfers

Critical Thinking

1. Restriction enzymes in bacterial cytoplasm cut injected bacteriophage DNA wherever certain sequences occur. Why do you think these enzymes do not chop up the bacterial chromosome, which is exposed to the enzymes in cytoplasm?

2. The *FOXP2* gene encodes a transcription factor associated with vocal learning in mice, bats, birds, and humans. In humans, loss-of-function mutations in *FOXP2* result in severe speech and language disorders. In mice, they hamper brain function and vocalizations.

The chimpanzee, gorilla, and rhesus *FOXP2* proteins are identical; the human version differs in only 2 amino acids, an evolutionary change that may have contributed to the development of spoken language. Mice genetically engineered to carry the human version of *FOXP2* show changes in vocal patterns, and more growth and greater adaptability of neurons involved in memory and learning. Biologists do not anticipate that a similar experiment with *FOXP2* in chimpanzees would confer the ability to speak, because spoken language is a polygenic trait. However, if future genetic engineering research produces a talking chimp, how do you think the debate about using animals for research would change?

3. In 1918, an influenza pandemic that originated with avian flu killed 50 million people. Researchers isolated samples of that virus from bodies of infected people preserved in Alaskan permafrost since 1918. From the samples, they sequenced the viral genome, then reconstructed the virus. The reconstructed virus is 39,000 times more infectious than modern influenza strains, and 100 percent lethal in mice.

Understanding how this virus works can help us defend ourselves against other deadly influenza strains that arise. For example, discovering what makes it so infectious would help researchers design more effective vaccines. Critics of the research are concerned: If the virus escapes the containment facilities (even though it has not done so yet), it might cause another pandemic. Worse, terrorists could use the published DNA sequence and methods to make the virus for horrific purposes. Do you think this research makes us more or less safe?

11 Evidence of Evolution

...Where you Have Been

You may wish to review critical thinking (Section 1.6) before reading this chapter, which explores a clash between belief and science (1.8). What you know about alleles (8.6) will help you understand natural selection. We revisit taxonomy (1.5), radioisotopes (2.2), DNA sequences (6.3) and sequencing (10.3), the genetic code (7.4), and master genes (7.7).

11.1 Reflections of a Distant Past

A The Barringer Crater in Arizona.

B The K–T boundary layer, a unique layer of rock that formed 65.5 million years ago, worldwide. The red pocketknife gives an idea of scale.

FIGURE 11.1 Evidence to inference.

A What made the Barringer Crater? Rocky evidence points to a 300,000-ton asteroid that collided with Earth 50,000 years ago.

B The K–T boundary layer marks an abrupt transition in the fossil record that implies a mass extinction.

Credits: (a) © Brad Snowder; (b) © David A. Kring, NASA/ Univ. Arizona Space Imagery Center.

How do you think about time? Perhaps you can conceive of a few hundred years of human events, maybe a few thousand, but how about a few million? Envisioning the very distant past requires an intellectual leap from the familiar to the unknown. One way to make that leap involves, surprisingly, asteroids. Asteroids are small planets hurtling through space. They range in size from 1 to 1,500 kilometers (roughly 0.5 to 1,000 miles) across. Millions of them orbit the sun between Mars and Jupiter—cold, stony leftovers from the formation of our solar system. Asteroids are difficult to see even with the best telescopes, because they do not emit light. Many cross Earth's orbit, but most of those pass us by before we know about them. Some have not passed us at all.

The mile-wide Barringer Crater in Arizona is difficult to miss (Figure 11.1**A**). An asteroid 45 meters (150 feet) wide made this impressive pockmark in the desert sandstone when it slammed into Earth 50,000 years ago. The impact was 150 times more powerful than the bomb that leveled Hiroshima. No humans were in North America at the time of the impact. If there were no witnesses, how can we know anything about what happened? We often reconstruct history by studying physical evidence of events that took place long ago. Geologists were able to infer the most probable cause of the Barringer Crater by analyzing tons of meteorites, melted sand, and other rocky clues at the site.

Similar evidence points to even larger asteroid impacts in the more distant past. For example, fossil hunters have long known about a **mass extinction**, or permanent loss of major groups of organisms, that occurred 65.5 million years ago. The event is marked by an unusual, worldwide layer of rock called the K–T boundary layer (Figure 11.1**B**). There are plenty of dinosaur fossils below this layer. Above it, in rock layers that were deposited more recently, there are no dinosaur fossils, anywhere. A gigantic impact crater—274 kilometers (170 miles) across and 1 kilometer (3,000 feet) deep—off the coast of what is now the Yucatán Peninsula dates to about 65.5 million years ago. Coincidence? Many scientists say no. The asteroid that made the Yucatán crater had to be at least 20 kilometers (12 miles) wide when it hit. The impact of an asteroid that size would have been *40 million times* more powerful than the one at Barringer. The scientists infer that the impact caused a global catastrophe of sufficient scale to wipe out the dinosaurs.

You are about to make an intellectual leap through time, to places that were not even known a few centuries ago. We invite you to launch yourself from this premise: Natural phenomena that occurred in the past can be explained by the same physical, chemical, and biological processes that operate today. That premise is the foundation for scientific research into the history of life. The research represents a shift from experience to inference—from the known to what can only be surmised—and it gives us astonishing glimpses into the distant past.

A Emu, native to Australia   **B** Rhea, native to South America   **C** Ostrich, native to Africa

FIGURE 11.2 Similar-looking, related species native to distant geographic realms. These birds are unlike most others in several unusual features, including long, muscular legs and an inability to fly. All are native to open grassland regions about the same distance from the equator.

Credits: (a) © Earl & Nazima Kowall/ Corbis; (b,c) © Wolfgang Kaehler/ Corbis.

11.2 Early Beliefs, Confusing Discoveries

The seeds of biological inquiry were taking hold in the Western world more than 2,000 years ago, as the Greek philosopher Aristotle made connections between observations in an attempt to explain the order of the natural world. Like few others of his time, Aristotle viewed nature as a continuum of organization, from lifeless matter through complex plants and animals. By the fourteenth century, Aristotle's ideas about nature had been transformed into a rigid view of life, in which a "great chain of being" extended from the lowest form (snakes), through humans, to spiritual beings. Each link in the chain was a species, and each was said to have been forged at the same time, in one place and in a perfect state. The chain itself was complete and continuous. Because everything that needed to exist already did, there was no room for change.

European naturalists that embarked on globe-spanning survey expeditions brought back tens of thousands of plants and animals from Asia, Africa, North and South America, and the Pacific Islands. Each newly discovered species was carefully catalogued as another link in the chain of being. By the late 1800s, naturalists were seeing patterns in where species live and how they might be related, and had started to think about the natural forces that shape life. These naturalists were pioneers in **biogeography**, the study of patterns in the geographic distribution of species and communities. Some of the patterns raised questions that could not be answered within the framework of prevailing belief systems. For example, globe-trotting explorers had discovered plants and animals living in extremely isolated places. The isolated species looked suspiciously similar to species living across vast expanses of open ocean, or on the other side of impassable mountain ranges (Figure 11.2). Could different species be related? If so, how could the related species end up geographically isolated? Naturalists of the time also had trouble classifying organisms that are very similar in some features, but different in others (Figure 11.3).

A African milk barrel (*Euphorbia horrida*), native to the Great Karoo desert of South Africa   **B** Saguaro cactus (*Carnegiea gigantea*), native to the Sonoran Desert of Arizona

FIGURE 11.3 Similar-looking, unrelated species.

Credits: (a) © Richard J. Hodgkiss, www.succulent-plant.com; (b) © Marka/ SuperStock.

biogeography Study of patterns in the geographic distribution of species and communities.

mass extinction Simultaneous loss of many lineages from Earth.

A Pythons and boa constrictors have tiny leg bones, but snakes do not walk.

B We humans use our legs, but not our coccyx (tail bones).

FIGURE 11.4 Animated! Vestigial body parts.

Credits: (a) © Dr. John Cunningham/ Visuals Unlimited; (b) Gary Head.

Such observations are part of **comparative morphology**, the study of anatomical patterns—similarities and differences among the body plans of organisms. Comparative morphology is an important part of taxonomy (Section 1.5).

Organisms that are outwardly similar may be quite different internally; think, for example, of fishes and porpoises. Organisms that differ greatly in outward appearance may be similar in underlying structure. For example, a human arm, a porpoise flipper, and a bat wing have comparable internal bones, as Section 11.6 explains. Comparative morphology in the nineteenth century revealed body parts that have no apparent function, an idea that added to the naturalists' confusion. If every species had been created in a perfect state, then why were there useless parts such as leg bones in snakes (which do not walk), or the vestiges of a tail in humans (Figure 11.4)?

Fossils were puzzling too. A **fossil** is physical evidence—remains or traces—of an organism that lived in the ancient past. Geologists mapping rock formations exposed by erosion or quarrying had discovered identical sequences of rock layers in different parts of the world. Deeper layers held fossils of simple marine life. Layers above those held similar but more complex fossils. In higher layers, fossils that were similar but even more complex resembled modern species. The photos on the *right* show one such series, ten fossils of shelled protists, each from a successive layer of stacked rocks. What did these sequences mean?

Fossils of many animals unlike any living ones were also being unearthed. If these animals had been perfect at the time of creation, then why had they become extinct?

Taken as a whole, the accumulating findings from biogeography, comparative morphology, and geology did not fit with prevailing beliefs of the nineteenth century. If species had not been created in a perfect state (and extinct species, fossil sequences, and "useless" body parts implied that they had not), then perhaps species had indeed changed over time.

Courtesy of Daniel C. Kelley, Anthony J. Arnold, and William C. Parker, Florida State University Department of Geological Science.

Take-Home Message

How did observations of the natural world change our thinking in the nineteenth century?

■ Increasingly extensive observations of nature in the nineteenth century did not fit with prevailing belief systems.

■ Cumulative findings from biogeography, comparative morphology, and geology led naturalists to question traditional ways of interpreting the natural world.

11.3 A Flurry of New Theories

〉 Squeezing New Evidence Into Old Beliefs In the nineteenth century, naturalists were faced with increasing evidence that life on Earth, and even Earth itself, had changed over time. Around 1800, Georges Cuvier, an expert in zoology and paleontology, was trying to make sense of the new information. He had observed abrupt changes in the fossil record, and knew that many fossil spe-

comparative morphology The scientific study of anatomical patterns in body plans.

evolution Change in a line of descent.

fossil Physical evidence of an organism that lived in the ancient past.

lineage Line of descent.

cies seemed to have no living counterparts. Given this evidence, he proposed an idea startling for the time: Many species that had once existed were now extinct. Cuvier also knew about evidence that Earth's surface had changed. For example, he had seen fossilized seashells on mountaintops far from modern seas. Like most others of his time, he assumed Earth's age to be in the thousands, not billions, of years. He reasoned that geological forces unlike those operating in the present would have been necessary to raise seafloors to mountains in this short time span. Catastrophic geological events would have caused extinctions, after which surviving species repopulated the planet. Cuvier's idea came to be known as catastrophism. We now know it is incorrect; geological processes have not changed over time.

Another scholar, Jean-Baptiste Lamarck, was thinking about processes that drive **evolution**, or change in a line of descent. A line of descent is also called a **lineage**. Lamarck thought that a species gradually improved over generations because of an inherent drive toward perfection, up the chain of being. The drive directed an unknown "fluida" into body parts needing change. By Lamarck's hypothesis, environmental pressures cause an internal need for change in an individual's body, and the resulting change is inherited by offspring.

Try using Lamarck's hypothesis to explain why a giraffe's neck is very long. You might predict that some short-necked ancestor of the modern giraffe stretched its neck to browse on leaves beyond the reach of other animals. The stretches may have even made its neck a bit longer. By Lamarck's hypothesis, that animal's offspring would inherit a longer neck. The modern giraffe would have been the result of many generations that strained to reach ever loftier leaves. Lamarck was correct in thinking that environmental factors affect traits, but his understanding of how traits are passed to offspring was incomplete.

❯ Darwin and the HMS Beagle

After an attempt to study medicine in college, Charles Darwin earned a degree in theology from Cambridge. All through school, however, he had spent most of his time with faculty members and other students who embraced natural history. In 1831, botanist John Henslow arranged for the 22-year-old Darwin to become a naturalist aboard the *Beagle*, a ship about to embark on a survey expedition to South America (Figure 11.5). The young man who had no formal training in science quickly became an enthusiastic naturalist. During the *Beagle*'s five-year voyage, Darwin found many unusual fossils, and saw diverse species living in environments that ranged from the sandy shores of remote islands to plains high in the Andes. Along the way, he read the first volume of a new and popular book, Charles Lyell's *Principles of Geology*. Lyell was a proponent of what became known as the theory of uniformity, the idea that gradual, repetitive change had shaped Earth. For many years, geologists had been chipping away at the sandstones, limestones, and other types of rocks that form from accumulated sediments in lakebeds, river bottoms, and ocean floors. These rocks held evidence that gradual processes of geological change operating in the present were the same ones that operated in the distant past.

The theory of uniformity held that strange catastrophes were not necessary to explain Earth's surface. Gradual, everyday geological processes such as erosion could have sculpted Earth's current landscape over great spans of time. The theory challenged the prevailing belief that Earth was 6,000 years old. According to traditional scholars, people had recorded everything that happened in those 6,000 years—and in all that time, no one had mentioned seeing a species evolve. However, by Lyell's calculations, it must have taken millions of years to sculpt Earth's surface. Darwin's exposure to Lyell's ideas gave him insights into the history of the regions he would encounter on his journey. Was millions of years enough time for species to evolve? Darwin thought that it was.

A Depiction of the HMS *Beagle*. With Darwin aboard as ship's naturalist, the vessel set sail in 1831 to map the coast of South America, but ended up circumnavigating the globe over a period of five years.

B Charles Darwin. The geology, fossils, plants, and animals he encountered on the *Beagle*'s expedition changed the way he thought about evolution.

C A page from Darwin's 1836 journal, "Transmutation of Species."

Text reads as follows:

"Let a pair be introduced and increase slowly, from many enemies, so as often to intermarry who will dare say what result. According to this view animals on separate islands ought to become different if kept long enough apart with slightly differing circumstances. . . ."

FIGURE 11.5 Charles Darwin and the voyage of the HMS *Beagle*.

Credits: (a) © Gordon Chancellor; (b) Painting by George Richmond; (c) Cambridge University Library.

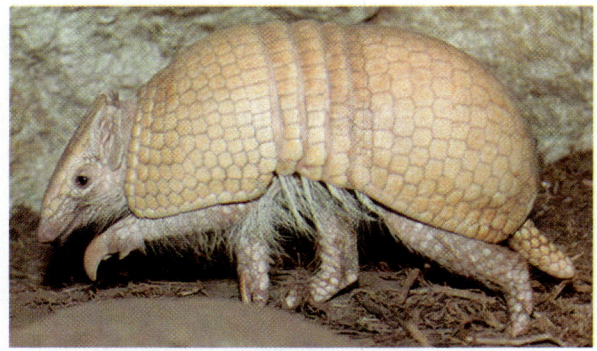

A A modern armadillo, about a foot long.

B Fossil of a glyptodon, an automobile-sized mammal that existed from 2 million to 15,000 years ago.

FIGURE 11.6 Ancient relatives: the armadillo and the glyptodon. Though widely separated in time, these animals share a restricted distribution and unusual traits, including a shell and helmet of keratin-covered bony plates—a material similar to crocodile and lizard skin. (The fossil in **B** is missing its helmet.) Their unique shared traits were a clue that helped Darwin develop a theory of evolution by natural selection.

Credits: (a) © John White; (b) 2004 Arent.

> **A Key Insight—Variation in Traits** Darwin sent to England thousands of specimens he collected on his voyage. Among them were fossil glyptodons. These armored mammals are extinct, but they have many traits in common with modern armadillos (Figure 11.6). For example, both glyptodons and armadillos have helmets and protective shells that consist of unusual bony plates. Armadillos also live only in places where glyptodons once lived. Could the odd shared traits and restricted distribution mean that glyptodons were ancient relatives of armadillos? If so, perhaps traits of their common ancestor had changed in the line of descent that led to armadillos. But why would such changes occur?

Back in England, Darwin pondered his notes and fossils. He read an essay by one of his contemporaries, economist Thomas Malthus. Malthus had correlated increases in the size of human populations with episodes of famine, disease, and war. He proposed the idea that humans run out of food, living space, and other resources because they tend to reproduce beyond the capacity of their environment to sustain them. When that happens, the individuals of a population must either compete with one another for the limited resources, or develop technology to increase their productivity. Darwin realized that Malthus's ideas had wider application: All populations, not just human ones, must have the capacity to produce more individuals than their environment can support.

Reflecting on his journey, Darwin started thinking about how individuals of a species are not always identical; they often vary a bit in the details of shared traits such as size, coloration, and so on. He saw such variation among many of the finch species that live on isolated islands of the Galápagos archipelago. This island chain is separated from South America by 900 kilometers (550 miles) of open ocean, so most species living on the islands did not have the opportunity for interbreeding with mainland populations. The Galápagos island finches resembled finch species in South America, but many of them had unique traits that suited them to their particular island habitat.

Darwin was familiar with dramatic variations in traits that breeders of dogs and horses had produced through hundreds of years of selective breeding, or **artificial selection**. He recognized that an environment could similarly select traits that make individuals of a population suited to it. It dawned on Darwin that having a particular form of a shared trait might give an individual a survival or reproductive advantage over competing members of its species. Darwin realized that in any population, some individuals have forms of shared traits that make them better suited to their environment than others. In other words, individuals of a natural population vary in fitness. We define **fitness** as the degree of adaptation to a specific environment, and measure it as relative genetic contribution to future generations. A trait that enhances an individual's fitness is called an evolutionary **adaptation**, or **adaptive trait**.

Over many generations, individuals with the most adaptive traits tend to survive longer and reproduce more than their less fit rivals. Darwin understood that this process, which he called **natural selection**, could be a mechanism by which evolution occurs. If an individual has a form of a trait that makes it better suited to an environment, then it is better able to survive. If an individual is better able to survive, then it has a better chance of living long enough to produce offspring. If individuals that bear an adaptive, heritable trait produce more offspring than those that do not, then the frequency of that trait will tend to increase in the population over successive generations. Table 11.1 summarizes this reasoning.

> **Great Minds Think Alike** Darwin wrote out his ideas about natural selection, but let ten years pass without publishing them. In the meantime, Alfred Wallace, a naturalist who had been studying wildlife in the Amazon basin and the Malay Archipelago, wrote an essay and sent it to Darwin for advice.

Table 11.1 Principles of Natural Selection, in Modern Terms

Observations about populations

- Natural populations have an inherent reproductive capacity to increase in size over time.

- As a population expands, resources that are used by its individuals (such as food and living space) eventually become limited.

- When resources are limited, the individuals of a population compete for them.

Observations about genetics

- Individuals of a species share certain traits.

- Individuals of a natural population vary in the details of those shared traits.

- Shared traits have a heritable basis, in genes. Slightly different forms of those genes (alleles) give rise to variation in shared traits.

Inferences

- A certain form of a shared trait may make its bearer better able to survive.

- Individuals of a population that are better able to survive tend to leave more offspring.

- Thus, an allele associated with an adaptive trait tends to become more common in a population over time.

FIGURE 11.7 Alfred Wallace, codiscoverer of natural selection.

Down House and The Royal College of Surgeons of England.

Wallace's essay outlined evolution by natural selection—the very same hypothesis as Darwin's. Wallace had written earlier letters to Darwin and Lyell about patterns in the geographic distribution of species; he too had connected the dots. Wallace is now called the father of biogeography (Figure 11.7).

In 1858, just weeks after Darwin received Wallace's essay, the hypothesis of evolution by natural selection was presented at a scientific meeting. Both Darwin and Wallace were credited as authors. Wallace was still in the field and knew nothing about the meeting, which Darwin did not attend. The next year, Darwin published *On the Origin of Species*, which laid out detailed evidence in support of natural selection. Many people had already accepted the idea of descent with modification, or evolution. However, there was a fierce debate over the idea that evolution occurs by natural selection. Decades would pass before experimental evidence from the field of genetics led to its widespread acceptance as a theory by the scientific community.

As you will see in the remainder of this chapter, the theory of evolution by natural selection is supported by and helps explain the fossil record as well as similarities in the form, function, and biochemistry of living things.

Take-Home Message

How did the way that naturalists thought about the history of life change in the nineteenth century?

- Evidence found in the 1800s led to the idea that Earth and the species on it had changed over very long spans of time. This idea set the stage for Darwin's theory of evolution by natural selection.

- Natural selection is a process in which individuals of a population survive and reproduce with differing success depending on the details of their shared, heritable traits.

- Traits favored in a particular environment are adaptive. An adaptive trait increases the chances that an individual bearing it will survive and reproduce.

adaptation (**adaptive trait**) A heritable trait that enhances an individual's fitness in a particular environment.

artificial selection Process whereby humans alter traits of a domestic species by selective breeding over generations.

fitness Degree of adaptation to an environment, as measured by an individual's relative genetic contribution to future generations.

natural selection Differential survival and reproduction of individuals of a population based on differences in shared, heritable traits. Driven by environmental pressures.

11.4 Fossil Evidence

Even before Darwin's time, fossils were recognized as stone-hard evidence of earlier forms of life (Figure 11.8). Most fossils consist of mineralized bones, teeth, shells, seeds, spores, or other hard body parts. Trace fossils such as footprints and other impressions, nests, burrows, trails, eggshells, or feces are evidence of an organism's activities.

The process of fossilization typically begins when an organism or its traces become covered by sediments or volcanic ash. Mineral-rich groundwater seeps into the remains, filling spaces around and inside of them. Metal ions and inorganic compounds dissolved in the water gradually replace minerals in bones and other hard tissues. Mineral particles that crystallize and settle out of the groundwater inside cavities and impressions form detailed imprints of internal and external structures. Sediments that slowly accumulate on top of the site exert increasing pressure. After a very long time, extreme pressure transforms the mineralized remains into rock.

Most fossils are found in layers of sedimentary rock such as sandstone and shale. These types of rocks form as rivers wash silt, sand, volcanic ash, and other materials from land to sea. Mineral particles in the materials settle on seafloors in horizontal layers that vary in thickness and composition. After many millions of years, the layers of sediments become compacted into layered sedimentary rock.

Biologists study layers of sedimentary rock in order to understand the historical context of fossils in them (Figure 11.9**A**). Usually, the deepest layers in a particular formation of rock were the first to form, and those closest to the surface formed most recently. Thus, in general, the deeper the layer, the older the fossils it contains. A layer's composition relative to other layers holds information about local and global events that were occurring as it formed; the K–T boundary layer discussed in Section 11.1 is one example. As you will see shortly, minerals in the layers allow us to estimate the age of fossils in them. The relative thicknesses of the different layers can provide other clues. For instance, layers of sedimentary rock deposited during ice ages tend to be thinner than other layers. Why? During the ice ages, tremendous volumes of water froze and became locked in glaciers. Sedimentation slowed as rivers and seas dried up. When the climate warmed, the glaciers melted, and rivers began to flow again. As sedimentation resumed, the layers became thicker.

A Fossil skeleton of an ichthyosaur that lived about 200 million years ago. These marine reptiles were about the same size as modern porpoises, breathed air like them, and probably swam as fast, but the two groups are not closely related.

B Extinct wasp (*Leptofoenus pittfieldae*) encased in amber. This 9-mm-long insect lived about 20 million years ago.

C Fossilized leaf of a 260-million-year-old seed fern (*Glossopteris*).

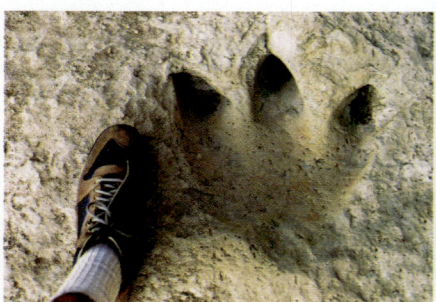

D Fossilized footprint of a theropod, a name that means "beast foot." This group of carnivorous dinosaurs, which includes the familiar *Tyrannosaurus rex*, arose about 250 million years ago. Modern birds are descended from them.

E Coprolite (fossilized feces). Fossilized food remains and parasitic worms inside coprolites offer clues about the diet and health of extinct species. A fox-like animal excreted this one.

FIGURE 11.8 Examples of fossils.

Credits: (a) Jonathan Blair; (b) © Dr. Michael Engel, University of Kansas; (c) Martin Land/ Photo Researchers, Inc.; (d) © Louie Psihoyos/ Getty Images; (e) Courtesy of Stan Celestian/ Glendale Community College Earth Science Image Archive.

> The Fossil Record We have fossils for more than 250,000 known species. Considering the current range of biodiversity, there must have been many millions more, but we will never know all of them. Why not? The odds are against finding evidence of an extinct species, because fossils are relatively rare.

A Cindy Looy and Mark Sephton climb the walls of Butterloch Canyon, Italy, to look for fossilized fungal spores in a 251-million-year-old layer of rock.

B High in the Andes, a scientist measures the stride of a carnivorous dinosaur that left footprints in an ancient shallow seabed.

> **FIGURE 11.9** Fossils are most often found in layered sedimentary rock. This type of rock forms over hundreds of millions of years, often at the bottom of a sea. Geological processes can tilt the rock and lift it far above sea level, where the layers become exposed by the erosive forces of water and wind.
>
> Credits: (a) © Jonathan Blair/ Corbis; (b) © Louie Psihoyos/ Getty Images.

When an organism dies, its remains are often quickly obliterated by scavengers. Organic materials decompose in the presence of moisture and oxygen, so remains that escape scavenging can endure only if they dry out, freeze, or become encased in an air-excluding material such as sap, tar, or mud. Remains that do become fossilized are often deformed, crushed, or scattered by erosion and other geological assaults.

In order for us to know about an extinct species that existed long ago, we have to find a fossil of it. At least one specimen had to be buried before it decomposed or something ate it. The burial site had to escape destructive geological events, and it had to be a place accessible enough for us to investigate (Figure 11.9**B**).

Most ancient species had no hard parts to fossilize, so we do not find much evidence of them. For example, there are many fossils of bony fishes and mollusks with hard shells, but few fossils of the jellyfishes and soft worms that were probably much more common. Also think about relative numbers of organisms. Fungal spores and pollen grains are typically released by the billions. By contrast, the earliest humans lived in small bands and few of their offspring survived. The odds of finding even one fossilized human bone are much smaller than the odds of finding a fossilized fungal spore. Finally, imagine two species, one that existed only briefly and the other for billions of years. Which is more likely to be represented in the fossil record?

Despite these challenges, the fossil record is substantial enough to help us reconstruct large-scale patterns in the history of life.

❯ Radiometric Dating Remember from Section 2.2 that a radioisotope is a form of an element with an unstable nucleus. Atoms of a radioisotope become atoms of other elements—daughter elements—as their nucleus disintegrates. This radioactive decay is not influenced by temperature, pressure, chemical bonding state, or moisture; it is influenced only by time. Thus, like the ticking of a perfect clock, each type of radioisotope decays at a constant rate. The time

B Long ago, trace amounts of ^{14}C and a lot more ^{12}C were incorporated into the tissues of a nautilus. The carbon atoms were part of organic molecules in the nautilus's food. ^{12}C is stable and ^{14}C decays, but the proportion of the two isotopes in the nautilus's tissues remained the same. Why? As long as it was alive, the nautilus continued to gain both types of carbon atoms in the same proportions from its food.

C When the nautilus died, it stopped eating, so its body stopped gaining carbon. The ^{12}C atoms already in its tissues were stable, but the ^{14}C atoms (represented as *red* dots) were decaying into nitrogen atoms. Thus, over time, the amount of ^{14}C decreased relative to the amount of ^{12}C. After 5,370 years, half of the ^{14}C had decayed; after another 5,370 years, half of what was left had decayed, and so on.

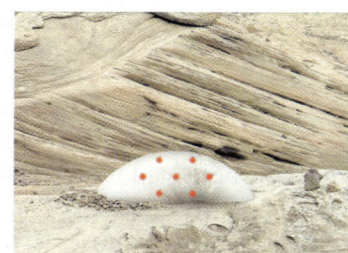

D Fossil hunters discover the fossil and measure its ^{14}C and ^{12}C content—the number of atoms of each isotope. The ratio of those numbers can be used to calculate how many half-lives passed since the organism died. For example, if the ^{14}C to ^{12}C ratio is one-eighth of the ratio in living organisms, then three half-lives must have passed since the nautilus died. Three half-lives of ^{14}C is 16,110 years.

FIGURE 11.10 Animated! Radiometric dating.

A *Above*, half-life is the time it takes for half of the atoms in a sample of radioisotope to decay.

B–D *Right*, an example of carbon 14 (^{14}C) dating. ^{14}C that forms continually in the atmosphere offsets its radioactive decay into ^{14}N, so the ratio of ^{14}C to ^{12}C atoms in the atmosphere remains constant. Both carbon isotopes combine with oxygen to become CO_2, which enters food chains by way of photosynthesis.

Figure It Out: What percentage of a radioisotope remains after two of its half-lives have passed? *Answer: 25 percent.*

Credits: (a,c,d) © Cengage Learning; (b) © PhotoDisc/ Getty Images.

---

it takes for half of the atoms in a sample of radioisotope to decay is called a **half-life** (Figure 11.10**A**). Half-life is a characteristic of each radioisotope. For example, radioactive uranium 238 decays into thorium 234, which decays into something else, and so on until it becomes lead 206. The half-life of the decay of uranium 238 to lead 206 is 4.5 billion years..

The predictability of radioactive decay can be used to find the age of a volcanic rock (the date it solidified). Rock deep inside Earth is hot and molten, so atoms swirl and mix in it. Rock that reaches the surface cools and hardens. As the rock cools, minerals crystallize in it. Each kind of mineral has a characteristic structure and composition. For example, the mineral zircon (shown at *left*) consists primarily of ordered arrays of zircon silicate molecules ($ZrSiO_4$). Some of the molecules in a zircon crystal have uranium atoms substituted for zirconium atoms, but never lead atoms. Thus, new zircon crystals that form as molten rock cools contain no lead. However, uranium decays into lead at a predictable rate. Thus, over time, uranium atoms disappear from a zircon crystal, and lead atoms accumulate in it. The ratio of uranium atoms to lead atoms in a zircon crystal can be measured precisely. That ratio can be used to calculate how long ago the crystal formed (its age).

We have just described **radiometric dating**, a method that can reveal the age of a material by measuring its content of a radioisotope and daughter elements. The oldest known terrestrial rock, a tiny zircon crystal found in Australia, formed 4.404 billion years ago.

Recent fossils that still contain organic molecules can be dated by measuring their carbon 14 content (Figure 11.10**B–D**). Most of the ^{14}C in a fossil will

zircon

Courtesy of Stan Celestian/ Glendale Community College Earth Science Image Archive.

half-life Characteristic time it takes for half of a quantity of a radioisotope to decay.
radiometric dating Method of estimating the age of a rock or fossil by measuring the content and proportions of a radioisotope and its daughter elements.

have decayed after about 60,000 years. The age of fossils older than that can be estimated by dating volcanic rock formations above and below the fossil-containing layer.

› Finding a Missing Link

The discovery of intermediate forms of cetaceans (an order of animals that includes whales, dolphins, and porpoises) provides an example of how scientists use fossil finds to piece together evolutionary history. For some time, evolutionary biologists predicted that the ancestors of modern cetaceans walked on land, then took up life in the water. Evidence in support of this line of thinking includes a set of distinctive features of the skull and lower jaw that cetaceans share with some kinds of ancient carnivorous land animals. DNA sequence comparisons indicate that the ancient land animals were probably artiodactyls, hooved mammals with an even number of toes (two or four) on each foot (Figure 11.11**A**). Modern representatives of the artiodactyl lineage include hippopotamuses, camels, pigs, deer, sheep, and cows.

Until recently, we had no fossils demonstrating gradual changes in skeletal features that accompanied a transition of whale lineages from terrestrial to aquatic life. Researchers knew there were intermediate forms because they had found a representative fossil of an ancient whalelike skull, but without a complete skeleton the rest of the story remained speculative.

Then, in 2000, Philip Gingerich and his colleagues found two of the missing links when they unearthed complete skeletons of two ancient whales: a fossil *Rodhocetus kasrani* excavated from a 47-million-year-old rock formation in Pakistan (Figure 11.11**B**); and a fossil *Dorudon atrox*, from 37-million-year-old sandstone in Egypt (Figure 11.11**C**). Whalelike skull bones and intact ankle bones were present in the same skeletons. The ankle bones of both fossils have distinctive features in common with those of extinct and modern artiodactyls. Modern cetaceans do not have even a remnant of an ankle bone (Figure 11.11**D**). With their in-between ankle bones, *Rodhocetus* and *Dorudon* were probably offshoots of the ancient artiodactyl-to-modern-whale lineage as it transitioned from life on land to life in water. The proportions of limbs, skull, neck, and thorax indicate *Rodhocetus* swam with its feet, not its tail. Like modern whales, the 5-meter (16-foot) *Dorudon* was clearly a fully aquatic tail-swimmer: The entire hindlimb was only about 12 centimeters (5 inches) long, much too small to have supported the animal out of water.

Take-Home Message

What do rocks and fossils have to do with biology?

- Fossils are evidence of organisms that lived in the remote past, a stone-hard historical record of life.

- The fossil record will never be complete, but even so it is substantial enough to help us reconstruct patterns and trends in the history of life.

- The predictability of radioisotope decay can be used to estimate the age of rock layers and fossils in them. Radiometric dating helps evolutionary biologists retrace changes in ancient lineages.

A 30-million-year-old *Elomeryx*. This small terrestrial mammal was a member of the same artiodactyl group that gave rise to hippopotamuses, pigs, deer, sheep, cows, and whales.

B *Rodhocetus kasrani*, an ancient whale, lived about 47 million years ago. Its distinctive ankle bones point to a close evolutionary connection to artiodactyls.

Compare the ankle bones of a modern artiodactyl, a pronghorn antelope (*top*), with those of a *Rodhocetus* (*bottom*). Artiodactyls are defined by the unique "double-pulley" shape of the bone that forms the lower part of their ankle joint.

C *Dorudon atrox*, an ancient whale that lived about 37 million years ago. Its tiny, artiodactyl-like ankle bones were much too small to have supported the weight of its body on land, so this mammal had to be fully aquatic.

D Modern cetaceans such as the sperm whale have remnants of a pelvis and leg, but no ankle bones.

FIGURE 11.11 Links in the ancient lineage of whales. The ancestors of whales and other cetaceans probably walked on land. Comparable hindlimb bones are highlighted in *blue*.

Credits: (a) W. B. Scott (1894); (b) top, Doug Boyer in P. D. Gingerich et al. (2001) © American Association for Advancement of Science; below, John Klausmeyer, University of Michigan Exhibit of Natural History; bottom, © P. D. Gingerich, University of Michigan. Museum of Paleontology; (c) top, © P. D. Gingerich and M. D. Uhen (1996), ©University of Michigan. Museum of Paleontology; bottom, © Cengage Learning 2010; (d) © Cengage Learning.

11.5 Drifting Continents, Changing Seas

Wind, water, and other erosive forces continuously sculpt Earth's surface, but they are only part of a much bigger picture of geological change. Earth itself also changes dramatically. For example, all continents that exist today were once part of a bigger supercontinent—**Pangea**—that split into fragments and drifted apart. This idea, originally called continental drift, was proposed in the early 1900s. The theory was consistent with many observations that had no better explanation, such as why the Atlantic coasts of South America and Africa seem to "fit" like jigsaw puzzle pieces, and why the same types of fossils occur in identical rock formations on both sides of the Atlantic Ocean. It also explained why the magnetic poles of gigantic rock formations point in different directions on different continents. Rock forms when molten lava solidifies on Earth's surface. Some types of iron-rich minerals become magnetic as they solidify, and their magnetic poles align with Earth's poles when they do. If the continents never moved, then all of these ancient rocky magnets should be aligned north-to-south, like compass needles. Indeed, the magnetic poles of the rock formations are aligned—but not north-to-south. Either Earth's magnetic poles veer dramatically from their north–south axis, or the continents must wander.

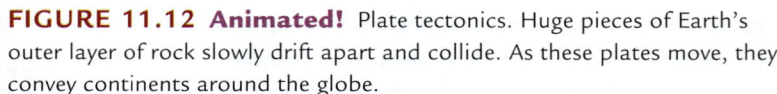

FIGURE 11.12 Animated! Plate tectonics. Huge pieces of Earth's outer layer of rock slowly drift apart and collide. As these plates move, they convey continents around the globe.

❶ At oceanic ridges, plumes of molten rock welling up from Earth's interior drive the movement of tectonic plates. New crust spreads outward as it forms on the surface, forcing adjacent tectonic plates away from the ridge and into trenches elsewhere.

❷ At trenches, the advancing edge of one plate plows under an adjacent plate and buckles it.

❸ Faults are ruptures in Earth's crust where plates meet. The diagram shows a rift fault, in which plates move apart. The aerial photo (*left*) shows about 4.2 kilometers (2.6 miles) of the San Andreas Fault, which extends 1,300 km (800 miles) through California. The San Andreas Fault is a boundary between two tectonic plates slipping against one another in opposite directions.

❹ Plumes of molten rock rupture a tectonic plate at what are called "hot spots." The Hawaiian Islands have been forming from molten rock that continues to erupt from a hot spot under the Pacific (tectonic) Plate.

Credits: top, USGS; bottom, © Cengage Learning.

Initially, most scientists were skeptical of the continental drift idea because no accepted geological mechanism explained how or why continents would move. That mechanism would emerge in the late 1950s, when deep-sea explorers discovered that ocean floors are not as static and featureless as had once been assumed. Immense ridges stretch thousands of kilometers across the seafloor (Figure 11.12). Molten rock spewing from the ridges pushes old seafloor outward in both directions ❶, then hardens into new seafloor as it cools. Elsewhere, old seafloor plunges into deep trenches ❷. These discoveries supported a plausible mechanism for continental drift, which by then had been renamed **plate tectonics**. Earth's relatively thin outer layer of rock is cracked into immense plates, like a huge cracked eggshell. Molten rock streaming out from an undersea ridge or continental rift at one edge of a plate pushes old rock at the opposite edge into a trench. The movement is like that of a colossal conveyor belt that transports continents on top of it to new locations. The plates move no more than 10 centimeters (4 inches) a year—about half as fast as your toenails grow—but it is enough to carry a continent all the way around the world after 40 million years or so (Figure 11.13).

Evidence of continental movement is all around us, in faults ❸ and various other geological features of our landscapes. For example, volcanic island chains (archipelagos) form as a plate moves across an undersea hot spot. These hot spots are places where a narrow plume of molten rock wells up from deep inside Earth and ruptures a tectonic plate ❹.

The fossil record also provides evidence in support of plate tectonics. Consider an unusual geological formation that occurs in a belt across Africa. The sequence of rock layers in this formation is so complex that it is quite unlikely to have formed more than once, but identical sequences also occur in huge belts that span India, South America, Africa, Madagascar, Australia, and Antarctica. The most probable explanation for the wide distribution is that the formations were deposited together on a single continent that later broke up. This explanation is supported by fossils in the rock layers: the remains of the seed fern *Glossopteris* (pictured in Figure 11.8**C**), whose seeds were too heavy to float or to be windblown over an ocean; and an early reptile (*Lystrosaurus*), whose body was not built for swimming between continents.

Glossopteris disappeared in a mass extinction event 251 million years ago, and *Lystrosaurus* disappeared 6 million years after that. Both organisms were extinct millions of years before Pangea formed. This and other evidence suggests that both organisms evolved together on a different supercontinent, one that predated Pangea. The older supercontinent, which we now call **Gondwana**, included most of the landmasses that currently exist in the Southern Hemisphere as well as India and Arabia. Many modern species, including the ratite birds pictured in Figure 11.2, live only in places that were once part of Gondwana. After Gondwana formed about 500 million years ago, it wandered across the South Pole, then drifted north until it merged with other continents to form Pangea about 270 million years ago.

We now know that at least five times since Earth's outer layer of rock solidified 4.55 billion years ago, a single supercontinent formed and then split up again. As you will see in later chapters, the resulting changes have had a profound impact on the course of life's evolution. A continent's climate changes—often dramatically so—along with its position on Earth. Colliding continents physically separate organisms living in oceans, and bring together those that had been living apart on land. As continents break up, they separate organisms living on land, and bring together ones that had been living in separate oceans. Such changes are a major driving force of evolution, a topic that we return to in the next chapter.

600 mya

430 mya — Gondwana

340 mya

240 mya — Pangea

200 mya

150 mya

65 mya

present

FIGURE 11.13
A series of reconstructions of the drifting continents. © Ron Blakey and Colorado Plateau Geosystems, Inc.

Gondwana Supercontinent that existed before Pangea, more than 500 million years ago.

Pangea Supercontinent that formed about 270 million years ago.

plate tectonics Theory that Earth's outer layer of rock is cracked into plates, the slow movement of which rafts continents to new locations over geological time.

Eon	Era	Period	Epoch	mya	Major Geologic and Biological Events
Phanerozoic	Cenozoic	Quaternary	Recent	0.01	Modern humans evolve. Major extinction event is now under way.
			Pleistocene	1.8	
		Tertiary	Pliocene	5.3	Tropics, subtropics extend poleward. Climate cools; dry woodlands and grasslands emerge. Adaptive radiations of mammals, insects, birds.
			Miocene	23.0	
			Oligocene	33.9	
			Eocene	55.8	
			Paleocene	65.5 ◄	Major extinction event, perhaps precipitated by asteroid impact. Mass extinction of all dinosaurs and many marine organisms.
	Mesozoic	Cretaceous	Late		
				99.6	
			Early		Climate very warm. Dinosaurs continue to dominate. Important modern insect groups appear (bees, butterflies, termites, ants, and herbivorous insects including aphids and grasshoppers). Flowering plants originate and become dominant land plants.
				145.5	
		Jurassic			Age of dinosaurs. Lush vegetation; abundant gymnosperms and ferns. Birds appear. Pangea breaks up.
				199.6 ◄	Major extinction event
		Triassic			Recovery from the major extinction at end of Permian. Many new groups appear, including turtles, dinosaurs, pterosaurs, and mammals.
				251 ◄	Major extinction event
	Paleozoic	Permian			Supercontinent Pangea and world ocean form. Adaptive radiation of conifers. Cycads and ginkgos appear. Relatively dry climate leads to drought-adapted gymnosperms and insects such as beetles and flies.
				299	
		Carboniferous			High atmospheric oxygen level fosters giant arthropods. Spore-releasing plants dominate. Age of great lycophyte trees; vast coal forests form. Ears evolve in amphibians; penises evolve in early reptiles (vaginas evolve later, in mammals only).
				359 ◄	Major extinction event
		Devonian			Land tetrapods appear. Explosion of plant diversity leads to tree forms, forests, and many new plant groups including lycophytes, ferns with complex leaves, seed plants.
				416	
		Silurian			Radiations of marine invertebrates. First appearances of land fungi, vascular plants, bony fishes, and perhaps terrestrial animals (millipedes, spiders).
				443 ◄	Major extinction event
		Ordovician			Major period for first appearances. The first land plants, fishes, and reef-forming corals appear. Gondwana moves toward the South Pole and becomes frigid.
				488	
		Cambrian			Earth thaws. Explosion of animal diversity. Most major groups of animals appear (in the oceans). Trilobites and shelled organisms evolve.
				542	
Proterozoic					Oxygen accumulates in atmosphere. Origin of aerobic metabolism. Origin of eukaryotic cells, then protists, fungi, plants, animals. Evidence that Earth mostly freezes over in a series of global ice ages between 750 and 600 mya.
				2,500	
Archaean and earlier					3,800–2,500 mya. Origin of bacteria and archaea.
					4,600–3,800 mya. Origin of Earth's crust, first atmosphere, first seas. Chemical, molecular evolution leads to origin of life (from protocells to anaerobic single cells).

FIGURE 11.14 Animated! The geologic time scale, correlated with sedimentary rock formations exposed by erosion in the Grand Canyon. Transitions between the formations mark the boundaries of great time spans in Earth's history.

mya: millions of years ago. Dates are from the International Commission on Stratigraphy, 2007.

Credits: (a) © Cengage Learning; (b) © Michael Pancier.

A *Above*, we can reconstruct some of the events in the history of life by studying rocky clues in the layers. *Blue* triangles mark times of great mass extinctions. "First appearance" refers to appearance in the fossil record, not necessarily the first appearance on Earth; we often discover fossils that are significantly older than previously discovered specimens.

〉 Putting Time Into Perspective Radiometric dating and fossils allow us to recognize similar sequences of sedimentary rock layers around the world (Figure 11.14). Transitions between these rock formations mark boundaries of great intervals of time in the **geologic time scale**, which is a chronology of Earth's history. Each layer's composition offers clues about conditions on Earth during the time the layer was deposited. Fossils in the layers are a record of life during that period of time.

geologic time scale Chronology of Earth history.

Kaibab Limestone

Toroweap Formation

Permian

Coconino Sandstone

Hermit Shale

Esplanade Sandstone

Wescogame Formation

Manakacha Formation

Carboniferous

Watahomigi Formation

Redwall Limestone

Temple Butte Formation

Muav Limestone

Cambrian

Bright Angel Shale

Tapeats Sandstone

Chuar Group*

Nankoweap Formation*

Unkar Group*

Proterozoic

Vishnu Basement Rocks

* Layers not visible in this view of the Grand Canyon

B Each rock layer has a composition and set of fossils that reflect events during its deposition. For example, Coconino Sandstone, which stretches from California to Montana, is mainly weathered sand. Ripple marks and reptile tracks are the only fossils in it. Many think it is the remains of a vast sand desert, like the Sahara is today.

Take-Home Message

How has Earth changed over geological time?

- Over great spans of time, movements of Earth's crust have caused dramatic changes in the global distribution of continents and oceans. These changes profoundly influenced the course of life's evolution.

- The geologic time scale chronicles Earth's history. It correlates geological and evolutionary events of the ancient past.

11.6 Evidence in Form and Function

To biologists, remember, evolution means change in a line of descent. How do they reconstruct evolutionary events that occurred in the ancient past? Evolutionary biologists are a bit like detectives, using clues to piece together a history that they did not witness in person. Fossils provide some clues. Others are encoded in the body form and function of organisms that are alive today.

> **Morphological Divergence** Comparative morphology can be used to unravel evolutionary relationships in many cases. Body parts that appear similar in separate lineages because they evolved in a common ancestor are called **homologous structures** (*hom*– means "the same"). Homologous structures may be used for different purposes in different groups, but the very same genes direct their development.

A body part that outwardly appears very different in separate lineages may be homologous in underlying form. For example, even though vertebrate forelimbs vary in size, shape, and function from one group to the next, they clearly are alike in the structure and positioning of bony elements, and in their internal patterns of nerves, blood vessels, and muscles.

Populations that are not interbreeding tend to diverge genetically, and in time these genetic divergences give rise to changes in body form. Change from the body form of a common ancestor is an evolutionary pattern called **morphological divergence**. Consider the limb bones of modern vertebrate animals. Fossil evidence suggests that many vertebrates are descended from a family of ancient "stem reptiles" that crouched low to the ground on five-toed limbs. Descendants of this ancestral group diversified over millions of years, and eventually gave rise to modern reptiles, birds, and mammals. A few lineages that had become adapted to walking on land even returned to life in the seas. During this time, five-toed limbs became adapted for many different purposes (Figure 11.15). They became modified for flight in extinct reptiles called pterosaurs and in bats and most birds. In penguins and porpoises, the limbs are now flippers useful for swimming. In humans, five-toed forelimbs became arms and hands with four fingers and an opposable thumb. Among elephants, the limbs are now strong and pillarlike, capable of supporting a great deal of weight. Limbs degenerated to nubs in pythons and boa constrictors, and to nothing at all in other snakes.

> **Morphological Convergence** Body parts that appear similar in different species are not always homologous; they sometimes evolve independently in lineages subject to the same environmental pressures. The independent evolution of similar body parts in different lineages is **morphological convergence**. Structures that are similar as a result of morphological convergence are called **analogous structures**. Analogous structures look alike but did not evolve in a shared ancestor; they evolved independently after the lineages diverged.

For example, bird, bat, and insect wings all perform the same function, which is flight. However, several clues tell us

pterosaur

chicken

penguin

porpoise

bat

human

elephant

stem reptile

FIGURE 11.15 Animated! Morphological divergence among vertebrate forelimbs, starting with the bones of an ancient stem reptile.

The number and position of many skeletal elements were preserved when these diverse forms evolved; notice the bones of the forearms. Certain bones were lost over time in some of the lineages (compare the digits numbered 1 through 5). Drawings are not to scale.

© Cengage Learning.

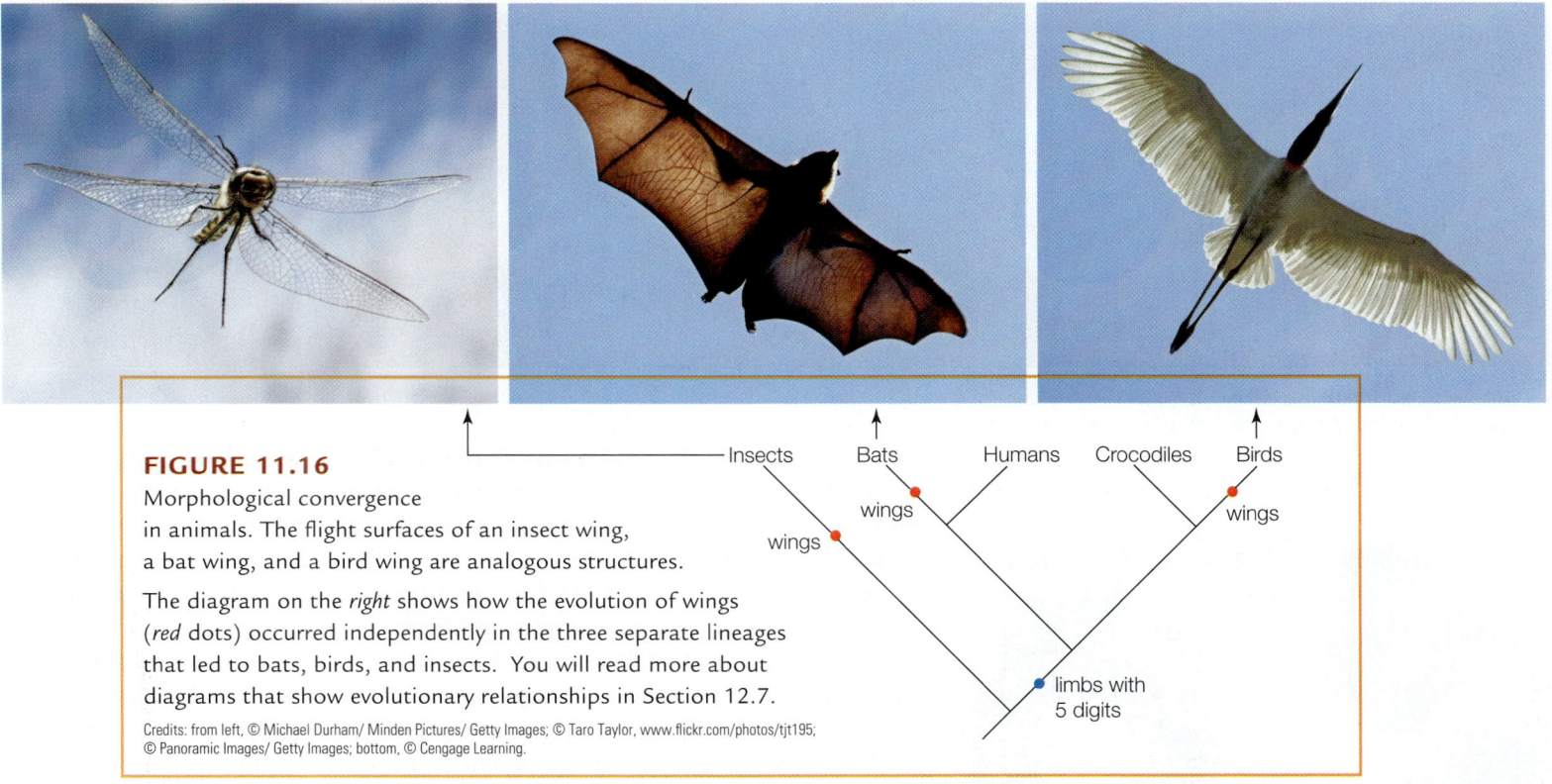

FIGURE 11.16

Morphological convergence in animals. The flight surfaces of an insect wing, a bat wing, and a bird wing are analogous structures.

The diagram on the *right* shows how the evolution of wings (*red* dots) occurred independently in the three separate lineages that led to bats, birds, and insects. You will read more about diagrams that show evolutionary relationships in Section 12.7.

Credits: from left, © Michael Durham/ Minden Pictures/ Getty Images; © Taro Taylor, www.flickr.com/photos/tjt195; © Panoramic Images/ Getty Images; bottom, © Cengage Learning.

that the wing surfaces are not homologous. The wings are adapted to the same physical constraints that govern flight, but each is adapted in a different way. In the case of birds and bats, the limbs themselves are homologous, but the adaptations that make those limbs useful for flight differ. The surface of a bat wing is a thin, membranous extension of the animal's skin. By contrast, the surface of a bird wing is a sweep of feathers, which are specialized structures derived from skin. Insect wings differ even more. An insect wing forms as a saclike extension of the body wall. Except at forked veins, the sac flattens and fuses into a thin membrane. The sturdy, chitin-reinforced veins structurally support the wing. Unique adaptations for flight are evidence that wing surfaces of birds, bats, and insects are analogous structures that evolved after the ancestors of these modern groups diverged (Figure 11.16).

As another example of morphological convergence, the similar external structures of American cacti and African euphorbias (see Figure 11.3) are adaptations to similarly harsh desert environments where rain is scarce. Distinctive accordion-like pleats allow the plant body to swell with water when rain does come. Water stored in the plants' tissues allows them to survive long dry periods. As the stored water is used, the plant body shrinks, and the folded pleats

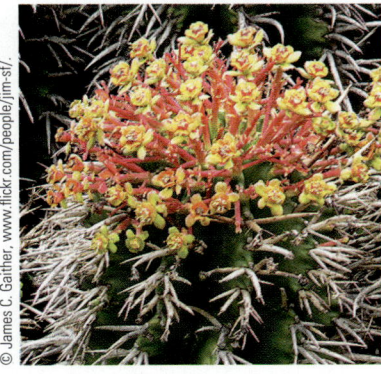

provide it with some shade in an environment that typically has none. Despite these similarities, however, a closer look reveals many differences that indicate the two types of plants are not closely related. For example, cactus spines have a simple fibrous structure; they are modified leaves that arise from dimples on the plant's surface. Euphorbia spines project smoothly from the plant surface, and they are not modified leaves: In many species the spines are actually dried flower stalks (*left*).

analogous structures Similar body structures that evolved separately in different lineages.

homologous structures Body structures that are similar in different lineages because they evolved in a common ancestor.

morphological convergence Evolutionary pattern in which similar body parts evolve separately in different lineages.

morphological divergence Evolutionary pattern in which a body part of an ancestor changes in its descendants.

A Visual comparison of vertebrate embryos. All vertebrates go through an embryonic stage in which they have four limb buds, a tail, and divisions called somites along their back. From *left* to *right*: human, mouse, bat, chicken, alligator.

B How differences in body form arise from differences in master gene expression. Expression of the *Hoxc6* gene is indicated by *purple* stain in two vertebrate embryos, chicken (*left*) and garter snake (*right*). Expression of this gene causes a vertebra to develop ribs as part of the back. Chickens have 7 vertebrae in their back and 14 to 17 vertebrae in their neck; snakes have upwards of 450 back vertebrae and essentially no neck.

FIGURE 11.17 Comparative embryology.

Credits: (a) from left, © Lennart Nilsson/ Bonnierforlagen AB; Courtesy of Anna Bigas, IDIBELL-Institut de Recerca Oncologica, Spain; From "Embryonic staging system for the short-tailed fruit bat, *Carollia perspicillata*, a model organism for the mammalian order Chiroptera, based upon times pregnancies in captive-bred animals," C. J. Cretekos et al., *Developmental Dynamics* Volume 233, Issue 3, July 2005, Pages: 721–738. Reprinted with permission of Wiley-Liss, Inc. a subsidiary of John Wiley & Sons, Inc.; Courtesy of Prof. D. G. Elisabeth Pollerberg, Institut für Zoologie, Universität Heidelberg, Germany; USGS; (b) Courtesy of Ann C. Burke, Wesleyan University.

> **Similarities in Patterns of Development** The embryos of many vertebrate species develop in similar ways. For example, all vertebrates go through a stage in which they have four limb buds, a tail, and a series of somites—divisions of the body that give rise to the backbone and associated skin and muscle (Figure 11.17**A**). Such similarities arise because the same master genes orchestrate development across different lineages. Because a mutation in a master gene can unravel development, such genes tend to be highly conserved. Highly conserved genes have changed very little or not at all over evolutionary time, even among lineages that diverged a very long time ago.

If the same genes direct development in all vertebrate lineages, how do the adult forms end up so different? Part of the answer is that there are differences in the onset, rate, or completion of early steps in development. These differences are brought about by variations in master gene expression patterns. Consider how homeotic gene expression determines the identity of particular zones of the body. Homeotic genes called *Hox* occur in clusters of ten. You have already read about one *Hox* gene, *antennapedia*, in Section 7.7. *Antennapedia* determines the identity of the thorax (the body part with legs) in insects and other arthropods. The vertebrate version of *antennapedia* is called *Hoxc6*, and it determines the identity of the back (as opposed to the neck or tail). Expression of the *Hoxc6* gene causes ribs to develop on a vertebra (Figure 11.17**B**). Vertebrae of the neck and tail normally develop with no *Hoxc6* expression, and no ribs.

> **Similarities in Biochemistry** Similarities in the nucleotide sequence of a gene or the amino acid sequence of a protein can be evidence of an evolutionary relationship. Such comparisons are often used in conjunction with morphological comparisons.

Over time, inevitable mutations change a genome's DNA sequence. Most of these mutations are neutral. Neutral mutations have no effect on an individual's survival or reproduction, so we can assume they accumulate at a constant rate. For example, a nucleotide substitution that changes one codon from AAA to AAG in a protein-coding region would probably not affect the protein product, because both codons specify lysine (Section 7.4). Other neutral mutations can change the amino acid sequence but not the function of a protein product.

Mutations alter the DNA of a lineage independently of all other lineages. The more recently two lineages diverged, the less time there has been for unique neutral mutations to accumulate in the DNA of each one. That is why the genomes of closely related species tend to be more similar than those of distantly related ones—a general rule that can be used to estimate relative times of divergence.

The DNA from nuclei, mitochondria, or chloroplasts can be used in nucleotide comparisons. Mitochondrial DNA can also be used to compare different

```
honeycreepers (10) . . .CRDVQFGWLIRNLHANGASFFFICIYLHIGRGIYYGSYLNK--ETWNIGVILLLTLMATAFVGYVLPWGQMSFWG...
       song sparrow . . .CRDVQFGWLIRNLHANGASFFFICIYLHIGRGIYYGSYLNK--ETWNVGIILLLALMATAFVGYVLPWGQMSFWG...
   Gough Island finch . . .CRDVQFGWLIRNIHANGASFFFICIYLHIGRGLYYGSYLYK--ETWNVGVILLLTLMATAFVGYVLPWGQMSFWG...
          deer mouse . . .CRDVNYGWLIRYMHANGASMFFICLFLHVGRGMYYGSYTFT--ETWNIGIVLLFAVMATAFMGYVLPWGQMSFWG...
    Asiatic black bear . . .CRDVHYGWIIRYMHANGASMFFICLFMHVGRGLYYGSYLLS--ETWNIGIILLFTVMATAFMGYVLPWGQMSFWG...
       bogue (a fish) . . .CRDVNYGWLIRNLHANGASFFFICIYLHIGRGLYYGSYLYK--ETWNIGVVLLLVMGTAFVGYVLPWGQMSFWG...
              human . . .TRDVNYGWIIRYLHANGASMFFICLFLHIGRGLYYGSFLYS--ETWNIGIILLLATMATAFMGYVLPWGQMSFWG...
    thale cress (a plant) . . .MRDVEGGWLLRYMHANGASMFLIVVYLHIFRGLYHASYSSPREFVWCLGVVIFLLMIVTAFIGYVLPWGQMSFWG...
        baboon louse . . .ETDVMNGWMVRSIHANGASWFFIMLYSHIFRGLWVSSFTQP--LVWLSGVIILFLSMATAFLGYVLPWGQMSFWG...
        baker's yeast . . .MRDVHNGYILRYLHANGASFFFMVMFMHMAKGLYYGSYRSPRVTLWNVGVIIFTLTIATAFLGYCCVYGQMSHWG...
```

individuals of the same sexually reproducing animal species. Mitochondria are inherited intact from a single parent, usually the mother, and they contain their own DNA. Thus, any differences in mitochondrial DNA sequences between maternally related individuals are due to mutations, not genetic recombination during fertilization.

It is useful to remember that coincidental homologies are statistically more likely to occur between DNA sequences than amino acid sequences, because there are only four nucleotides in DNA (versus twenty amino acids in proteins). Thus, getting useful information from comparing DNA requires a lot more data than comparing proteins (Figure 11.18). However, DNA sequencing has become so fast that there is a lot of data available to compare. Genomics studies with such data have shown us (for example) that about 30 percent of the genes in yeast have counterparts in the human genome. So do 40 percent of roundworm genes, and 50 percent of fruit fly genes.

FIGURE 11.18 Alignment of part of the amino acid sequence of mitochondrial cytochrome *b* from twenty species. This protein is a crucial component of mitochondrial electron transfer chains.

The honeycreeper sequence is identical in ten species of honeycreeper; amino acids that differ in the other species are shown in *red*. Dashes indicate gaps in the alignment.

Figure It Out: Based on this comparison, which species is the most closely related to the honeycreepers?   *Answer: The song sparrow*

© Cengage Learning.

Take-Home Message

How are comparisons in form and function used to study evolution?

- Body parts are often modified differently in different lines of descent. Some body parts that appear alike evolved independently in different lineages.

- Similarities in patterns of embryonic development are the result of master genes that have been conserved over evolutionary time.

- Lineages that diverged long ago generally have more differences between their DNA and proteins than do lineages that diverged more recently.

11.7 Reflections of a Distant Past
(revisited)

The K–T boundary was named after the periods it separates: the Cretaceous (K) and Tertiary (T). K stands for *Kreide*, a German word for chalk (the Cretaceous layer is especially abundant in this sedimentary rock). The K–T boundary layer is unusually rich in iridium, an element rare on Earth's surface but common in asteroids. After finding the iridium, researchers looked for evidence of an asteroid big enough to cover the entire Earth with its debris. They looked for—and found—the Yucatán Peninsula impact crater. It is so big that no one had realized it was a crater before. The K–T boundary layer also contains shocked quartz (*left*) and small glass spheres called tektites—rocks that form when quartz or sand (respectively) undergoes a sudden, violent application of extreme pressure. As far as we know, the only processes on Earth that produce shocked quartz and tektites are atomic bomb explosions and meteorite impacts.

© U.S. Geological Survey.

WHERE YOU ARE GOING . . .

The next chapter continues the theme of evolutionary processes, including how natural selection works (12.1–12.4) and how tectonic plate movements can affect evolution (12.5). A continent's climate is affected by its position on the globe, as you will see in Chapter 18. Chapter 26 returns to animal development.

Summary

Section 11.1 Events of the ancient past can be explained by the same physical, chemical, and biological processes that operate today. An asteroid impact may have caused a **mass extinction** 65.5 million years ago.

Section 11.2 Expeditions by nineteenth-century explorers yielded increasingly detailed observations of nature. Geology, **biogeography**, and **comparative morphology** of organisms and their **fossils** led to new ways of thinking about the natural world.

Section 11.3 Prevailing belief systems may influence interpretation of the underlying cause of a natural event. Nineteenth-century naturalists attempted to reconcile traditional belief systems with physical evidence of **evolution**, or change in a **lineage** over time.

Humans select desirable traits in animals by selective breeding, or **artificial selection**. Charles Darwin and Alfred Wallace independently came up with a theory of how environments also select traits, stated here in modern terms: A population tends to grow until it exhausts environmental resources. As that happens, competition for those resources intensifies among the population's individuals. Individuals with forms of shared, heritable traits that make them more competitive for the resources tend to produce more offspring. Thus, **adaptive traits** (**adaptations**) that impart greater **fitness** to an individual become more common in a population over generations. The process in which environmental pressures result in the differential survival and reproduction of individuals of a population is called **natural selection**. It is one of the processes that drives evolution.

Section 11.4 Fossils are typically found in stacked layers of sedimentary rock, with younger fossils in the layers deposited more recently. The fossil record will always be incomplete. With **radiometric dating**, the **half-life** of a radioisotope is used to determine the age of rocks and fossils. A sedimentary rock layer's age is estimated by dating volcanic rock above and below it.

Section 11.5 According to the **plate tectonics** theory, Earth's crust is cracked into giant plates that carry landmasses to new positions as they move. Earth's landmasses have periodically converged as supercontinents such as **Gondwana** and **Pangea**. Transitions in the fossil record are the boundaries of great intervals of the **geologic time scale**, a chronology of Earth's history that correlates geological and evolutionary events.

Section 11.6 Comparative morphology can reveal evidence of evolutionary connections among lineages. **Homologous structures** are similar body parts that, by **morphological divergence**, became modified differently in different lineages. Such parts are evidence of a common ancestor. **Analogous structures** are body parts that look alike in different lineages but did not evolve in

a common ancestor. By the process of **morphological convergence**, they evolved separately after the lineages diverged.

We can discover and clarify evolutionary relationships through comparisons of nucleic acid and protein sequences, because lineages that diverged recently tend to share more sequences than ones that diverged long ago. Master genes that affect development tend to be highly conserved, so similarities in patterns of embryonic development reflect shared ancestry that can be evolutionarily ancient.

```
ASFFFICIYLH
ASMFFICLFLH
ASMFFICLFMH
ASFFFICIYLH
ASMFFICLFLH
ASMFLIVVYLH
ASWFFIMLYSH
```

Self-Quiz   Answers in Appendix I

1. The number of species on an island usually depends on the size of the island and its distance from a mainland. This statement would most likely be made by _____ .
    a. an explorer   c. a geologist
    b. a biogeographer   d. a philosopher

2. The bones of a bird's wing are similar to the bones in a bat's wing. This observation is an example of _____ .
    a. uniformity   c. comparative morphology
    b. evolution   d. a lineage

3. Evolution _____ .
    a. is natural selection
    b. is heritable change in a line of descent
    c. can occur by natural selection
    d. b and c are correct

4. A trait is adaptive if it _____ .
    a. arises by mutation   c. is passed to offspring
    b. increases fitness   d. occurs in fossils

5. In which type of rock are you more likely to find a fossil?
    a. basalt, a dark, fine-grained volcanic rock
    b. limestone, composed of sedimented calcium carbonate
    c. slate, a volcanically melted and cooled shale
    d. granite, which forms by crystallization of molten rock below Earth's surface
    e. all are equally likely to hold fossils

6. Which of the following is a fossil?
    a. an insect encased in 10-million-year-old tree sap
    b. a woolly mammoth frozen in Arctic permafrost for the last 50,000 years
    c. mineral-hardened remains of a whalelike animal found in an Egyptian desert
    d. an impression of a plant leaf in a rock
    e. all of the above can be considered fossils

7. If the half-life of a radioisotope is 20,000 years, then a sample in which three-quarters of that radioisotope has decayed is _____ years old.
    a. 15,000   b. 26,667   c. 30,000   d. 40,000

8. Did Pangea or Gondwana form first?

9. The Cretaceous ended _____ million years ago.

10. Through _____ , a body part of an ancestor is modified differently in different lines of descent.

11. Homologous structures among major groups of organisms may differ in _____ .
    a. size   b. shape   c. function   d. all of the above

12. By altering steps in the program by which embryos develop, a mutation in a _____ may lead to major differences in body form between related lineages.
    a. derived trait
    b. homeotic gene
    c. homologous structure
    d. all of the above

13. All of the following data types can be used as evidence of shared ancestry except similarities in _____ .
    a. amino acid sequences
    b. DNA sequences
    c. fossil morphologies
    d. embryonic development
    e. form due to convergence
    f. all are appropriate

14. Match the terms with the most suitable description.
    ____ fitness
    ____ fossils
    ____ natural selection
    ____ half-life
    ____ catastrophism
    ____ uniformity
    ____ analogous structures
    ____ sedimentary rock
    ____ homologous structures
    ____ neutral mutation

    a. does not affect fitness
    b. geological change occurs continuously
    c. geological change occurs in sudden major events
    d. good for finding fossils
    e. survival of the fittest
    f. characteristic of a radioisotope
    g. insect wing and bird wing
    h. human arm and bird wing
    i. evidence of life in distant past
    j. measured by relative genetic contribution to future generations

Critical Thinking

1. Radiometric dating does not measure the age of an individual atom. It is a measure of the age of a quantity of atoms—a statistic. As with any statistical measure, its values may deviate around an average (sampling error, Section 1.8). Imagine that one sample of rock is dated ten different ways. Nine of the tests yield an age close to 225,000 years. One test yields an age of 3.2 million years. Do the nine consistent results imply that the one that deviates is incorrect, or does the one odd result invalidate the nine that are consistent?

11:21:10 a.m. mammals, dinosaurs
11:37:18 a.m. flowering plants
11:59:59 a.m. first humans
10:40:57 a.m. early fishes
12:00:00 a.m. Earth's crust solidifies
Phanerozoic
Archaean and earlier
Proterozoic
2:05:13 a.m. archaea, bacteria
5:28:41 a.m. eukaryotes
© Cengage Learning.

2. If you think of geological time spans as minutes, life's history might be plotted on a clock such as the one shown on the *right*. According to this clock, the most recent epoch started in the last 0.1 second before noon. Where does that put you?

Digging Into Data

Abundance of Iridium in the K–T Boundary Layer

In the late 1970s, geologist Walter Alvarez was investigating the composition of the 1-centimeter-thick layer of clay that marks the Cretaceous–Tertiary (K–T) boundary all over the world. He asked his father, Nobel Prize–winning physicist Luis Alvarez, to help him analyze the elemental composition of the layer. The photo (*right*) shows Luis and Walter Alvarez with a section of the K–T boundary layer.

The Alvarezes and their colleagues tested samples of the layer taken from formations in Italy and Denmark. The researchers discovered that the K–T boundary layer had a much higher iridium content than the surrounding rock layers. Some of their results are shown in the table in Figure 11.19.

Iridium belongs to a group of elements that are much more abundant in asteroids and other solar system materials than they are in Earth's crust. The Alvarez group concluded that the K–T boundary layer must have originated with extraterrestrial material. They calculated that an asteroid 14 kilometers (8.7 miles) in diameter would contain enough iridium to account for the extra iridium in the K–T boundary layer.

1. What was the iridium content of the K–T boundary layer?

2. How much higher was the iridium content of the boundary layer than the sample taken 0.7 meters above the layer?

Sample Depth	Average Abundance of Iridium (ppb)
+ 2.7 m	< 0.3
+ 1.2 m	< 0.3
+ 0.7 m	0.36
boundary layer	41.6
− 0.5 m	0.25
− 5.4 m	0.30

FIGURE 11.19 Abundance of iridium in and near the K–T boundary layer in Stevns Klint, Denmark. Many rock samples taken from above, below, and at the boundary layer were tested for iridium content. Depths are given as meters above or below the boundary layer.

The iridium content of an average Earth rock is 0.4 parts per billion (ppb) of iridium. An average meteorite contains about 550 ppb of iridium.

Processes of Evolution 12

This chapter discusses taxonomy (1.5) in the context of evolutionary concepts (Section 11.3). A review of gene mutations (7.6), alleles (8.6, 8.7), genetics (9.2–9.5), and comparative morphology and biochemistry (11.6) will be helpful. We revisit populations (1.2), sampling error (1.7), polyploidy (9.8), DNA profiling (10.3), transgenic plants (10.4), evidence vs. inference (11.1), and plate tectonics and the geologic time scale (11.5).

12.1 Rise of the Super Rats

Slipping in and out of the pages of human history are rats—*Rattus*—the most notorious of mammalian pests. Rats thrive in urban centers, where garbage is plentiful and natural predators are not. The average city in the United States sustains about one rat for every ten people. Part of their success stems from an ability to reproduce very quickly: Rat populations can expand within weeks to match the amount of garbage available for them to eat.

Unfortunately for humans, rats can carry pathogens and parasites associated with infectious diseases such as bubonic plague and typhus. They chew through walls and wires, and eat or foul 20 to 30 percent of our total food production. In total, rats cost us about $19 billion annually.

© Rollin Verlinde/ Vilda.

For decades, people have been fighting back with dogs, traps, ratproof storage facilities, and poisons that include arsenic and cyanide. Baits laced with warfarin, an organic compound that interferes with blood clotting, became popular in the 1950s. Rats that ate the poisoned baits died within days after bleeding internally or losing blood through cuts or scrapes. Warfarin was extremely effective, and its impact on harmless species was much lower than that of other rat poisons. It quickly became the rat poison of choice. By 1980, however, about 10 percent of rats in urban areas were resistant to warfarin. What happened?

To find out, researchers compared rats that were resistant to warfarin with those who were not. They traced the difference to a gene: A particular mutation in that gene was common among warfarin-resistant rat populations, but rare among vulnerable ones. Warfarin inhibits the gene's product, an enzyme that recycles vitamin K after it has been used to activate blood clotting factors. The mutations made the enzyme less active, but also insensitive to warfarin. "What happened" was evolution by natural selection. As warfarin exerted pressure on rat populations, the warfarin-resistance allele became adaptive, and the rat populations changed. Rats that had an unmutated allele died after eating warfarin. The lucky ones that had a warfarin-resistance allele survived and passed it to their offspring. The rat populations recovered quickly, and a higher proportion of rats in the next generation carried the mutated allele. With each onslaught of warfarin, the frequency of this allele in the rat populations increased.

Selection pressures can and often do change. When warfarin resistance increased in rat populations, people stopped using warfarin. The frequency of the warfarin-resistance allele in rat populations declined, probably because rats that carry the allele are not as healthy as ones that do not. Now, savvy exterminators in urban areas know that the best way to control a rat infestation is to exert another kind of selection pressure: Remove their source of food, which is usually garbage. Then the rats will eat each other.

FIGURE 12.1 Some phenotypic variation among humans. Variation in shared traits is an outcome of differences in alleles that influence those traits.

Credits: third from left, © Roderick Hulsbergen/ www.photography.euweb.nl; all others, © JupiterImages Corporation.

12.2 Making Waves in the Gene Pool

Section 1.2 introduced a **population** as a group of interbreeding individuals of the same species in some specified area. The individuals of a population (and a species) share morphological, physiological, and behavioral traits because they have the same genes. Almost every shared trait varies a bit among members of sexually-reproducing species (Figure 12.1). This variation arises because different individuals inherit different combinations of alleles (Sections 8.6 and 8.7).

Some traits occur in distinct forms, or morphs. A trait with only two forms is dimorphic (*di–* means two). Purple and white flower color in the pea plants that Gregor Mendel studied is an example of a dimorphic trait (Section 9.2). Dimorphic flower color occurs in this case because the interaction of two alleles with a clear dominance relationship gives rise to the trait. Traits with more than two distinct forms are polymorphic (*poly–*, many). Human blood type, which is determined by the codominant *ABO* alleles, is an example (Section 9.4). The genetic basis of traits that vary continuously (Section 9.5) is often quite complex. Any or all of the genes that influence such traits may have multiple alleles.

In earlier chapters, you learned about genetic events that contribute to variation in shared traits (Table 12.1). Mutation is the original source of new alleles. Other events shuffle alleles into different combinations, and what a shuffle that is! There are $10^{116,446,000}$ possible combinations of human alleles. Not even 10^{10} people are living today. Unless you have an identical twin, it is extremely unlikely that another person with your precise genetic makeup has ever lived, or ever will.

> **An Evolutionary View of Mutations** Being the original source of new alleles, mutations are worth another look, this time in context of their impact on populations. We cannot predict when or in which individual a particular gene will mutate. We can, however, predict the average mutation rate of a species, which is the probability that a mutation will occur in a given interval. In the human species, that rate is about 2.2×10^{-9} mutations per base pair per year. In other words, about 70 nucleotides in the human genome sequence change every decade.

Many mutations give rise to structural, functional, or behavioral alterations that reduce an individual's chances of surviving and reproducing. Even one biochemical change may be devastating. Consider collagen, a fibrous protein that is a primary structural component of tissues that compose skin, bones, tendons, lungs, blood vessels, and other parts of the vertebrate body. If one of the genes for collagen mutates in a way that changes the protein's structure, the entire

Table 12.1	Sources of Variation in Traits Among Individuals of a Species
Genetic Event	**Effect**
Mutation	Source of new alleles
Crossing over at meiosis I	Introduces new combinations of alleles into chromosomes
Independent assortment at meiosis I	Mixes maternal and paternal chromosomes
Fertilization	Combines alleles from two parents
Changes in chromosome number or structure	Often dramatic changes in structure and function

population A group of organisms of the same species that live in a specific location and breed with one another more often than they breed with members of other populations.

body may be affected. A mutation such as this can change phenotype so drastically that it results in death, in which case it is called a **lethal mutation**.

A neutral mutation changes the nucleotide sequence in DNA, but the alteration has no effect on survival or reproduction (Section 11.6). Occasionally, a change in the environment favors a mutation that had previously been neutral or even somewhat harmful. The warfarin-resistance mutation in rats is an example. Even if a beneficial mutation bestows only a slight advantage, its frequency tends to increase in a population over time. This is because natural selection operates on traits that have a genetic basis. With natural selection, remember, environmental pressures result in an increase in the frequency of an adaptive trait in a population over generations (Section 11.3).

Mutations have been altering genomes for billions of years, and they are still at it. Cumulatively, they have given rise to Earth's staggering biodiversity. Think about it: The reason you do not look like an apple or an earthworm or even your next-door neighbor began with mutations that occurred in different lines of descent.

❯ Allele Frequencies Together, all the alleles of all the genes of a population comprise a pool of genetic resources—a **gene pool**. The members of a population breed with one another more often than they breed with members of other populations, so their gene pool is more or less isolated. We refer to the abundance of any particular allele among members of a population as its **allele frequency**. Any change in allele frequency in the gene pool of a population (or a species) is called **microevolution**.

A theoretical reference point called genetic equilibrium occurs when the allele frequencies of a population do not change (in other words, the population is not evolving). Genetic equilibrium can only occur if every one of the following five conditions are met:

1. Mutations never occur.

2. The population is infinitely large.

3. The population is isolated from all other populations of the species.

4. Mating is random.

5. All individuals survive and produce the same number of offspring.

As you can imagine, all five conditions are never met in nature, so natural populations are never in equilibrium.

Microevolution is always occurring in natural populations because, as you will see in the next sections, processes that drive it—mutation, natural selection, and genetic drift—are always operating. Remember, even though we can recognize patterns of evolution, none of them are purposeful. Evolution simply fills the nooks and crannies of opportunity.

Evolution is not purposeful. It simply fills the nooks and crannies of opportunity.

---

allele frequency Abundance of a particular allele among members of a population.

directional selection Mode of natural selection in which a phenotype at one end of a range of variation is favored.

gene pool All the alleles of all the genes in a population; a pool of genetic resources.

lethal mutation Mutation that alters phenotype so drastically that it causes death.

microevolution Change in an allele's frequency in a population or species.

Take-Home Message

What is microevolution?

- Individuals of a natural population share morphological, physiological, and behavioral traits characteristic of the species. Alleles are the basis of differences in the details of those shared traits.

- All alleles of all individuals in a population make up the population's gene pool. An allele's abundance in the gene pool is called its allele frequency.

- Microevolution is change in allele frequency. It is always occurring in natural populations because processes that drive it are always operating.

12.3 Modes of Natural Selection

Natural selection influences the frequency of alleles in a population by operating on phenotypes with a heritable, genetic basis. How phenotype is affected depends on the species and the selection pressures in the environment.

❯ Directional Selection With **directional selection**, phenotypes at one end of a range of variation become more common over time (Figure 12.2). The following examples show how field observations provide evidence of directional selection.

The peppered moth is a species of insect found in temperate zones around the world. These moths feed and mate at night, then rest motionless on trees during the day. Coloration in peppered moths is determined by a single gene. Individuals that carry a dominant allele of this gene are black; those homozygous for a recessive allele are white with black speckles. In preindustrial England, the vast majority of peppered moths were light-colored, and the black form was extremely rare. At this time, the air was clean, and light-gray lichens grew on the trunks and branches of most trees. Light-colored individuals were well camouflaged from moth-eating birds when they rested on the lichens, but dark-colored individuals were not (Figure 12.3**A**).

By the 1850s, the dark moths had become much more common than the light moths. Why? The industrial revolution had begun, and smoke emitted by coal-burning factories was beginning to change the environment. Air pollution was killing the lichens. Dark moths were better camouflaged on lichen-free, soot-darkened trees (Figure 12.3**B**).

In the 1950s, H. B. Kettlewell bred dark and light moths in captivity, marked them for easy identification, then released them in several areas. His team recaptured more of the dark moths in the polluted areas and more light ones in the less polluted ones. They also observed predatory birds eating more light-colored moths in soot-darkened forests, and more dark-colored moths in cleaner, lichen-rich forests. Dark-colored moths were at a selective advantage in industrialized areas.

Pollution controls went into effect in 1952. As a result of improved environmental standards, tree trunks gradually became free of soot, and lichens made a comeback. Kettlewell observed that moth phenotypes shifted too: Wherever pollution decreased, the frequency of dark moths decreased as well.

Recent research has confirmed Kettlewell's results implicating birds and soot as selective agents of peppered moth coloration.

A Light-colored peppered moths on a nonsooty tree trunk (*top*) are hidden from predators; dark ones (*bottom*) stand out.

B In places where soot darkens tree trunks, the dark color (*top*) provides more camouflage than the light color (*bottom*).

FIGURE 12.3 Animated!
Adaptive value of two color forms of the peppered moth.
J. A. Bishop, L. M. Cook.

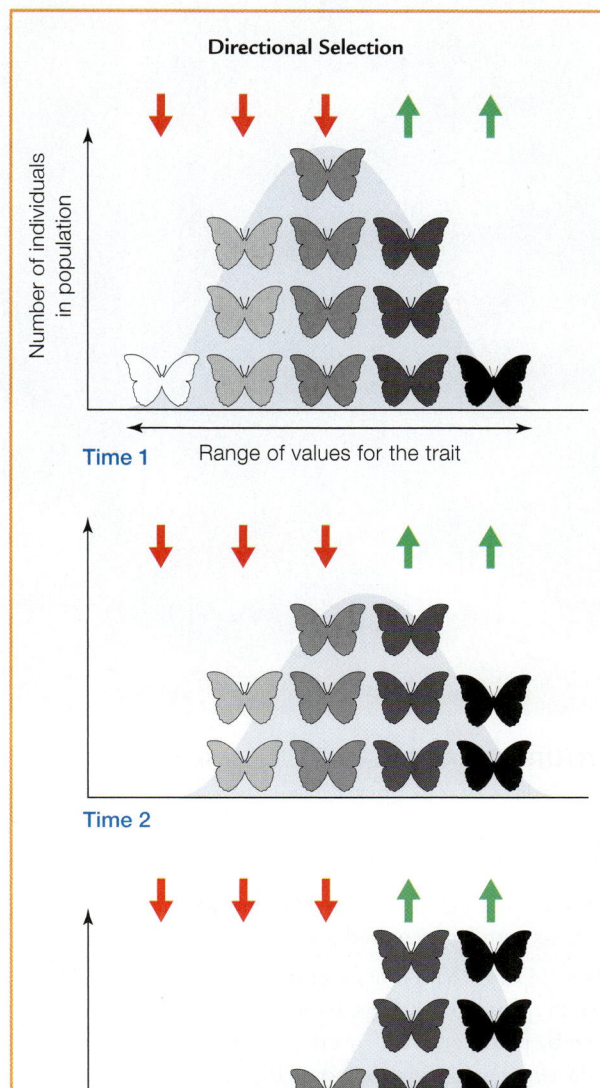

Directional Selection

Time 1   Range of values for the trait

Time 2

Time 3

FIGURE 12.2 Animated! With directional selection, a form of a trait at one end of a range of variation is favored.

The bell-shaped curves indicate continuous variation in a butterfly wing-color trait. *Red* arrows indicate which forms are being selected against; *green*, forms that are being favored.

© Cengage Learning.

FIGURE 12.4 Directional selection in populations of rock pocket mice.

1 Mice with dark fur are more common in areas with dark basalt rock.

2 Mice with light fur are more common in areas with light-colored granite.

3 Mice with coat colors that do not match their surroundings are more easily seen by predators, so they are preferentially eliminated from the populations.

© Cengage Learning 2010.

Directional selection also affects the color of rock pocket mice in Arizona's Sonoran Desert. Rock pocket mice are small mammals that spend the day sleeping in underground burrows, emerging at night to forage for seeds (Figure 12.4). The environment is dominated by light brown granite. There are also patches of dark basalt, the remains of ancient lava flows. Most of the mice in populations that inhabit the dark rock have dark gray coats **1**. Most of the mice in populations that inhabit the light brown rock have light brown coats **2**. The difference arises because mice that match the rock color in each habitat are camouflaged from their natural predators. Night-flying owls more easily see mice that do not match the rocks **3**, and they preferentially eliminate easily seen mice from each population. Thus, in both habitats, selective predation has resulted in a directional shift in the frequency of alleles that affect coat color.

Human attempts to control the environment can result in directional selection, as is the case with the warfarin-resistant rats. The use of antibiotics is another example. Prior to the 1940s, scarlet fever, tuberculosis, and pneumonia caused one-fourth of the annual deaths in the United States. Since the 1940s, we have been relying on antibiotics such as penicillin to fight these and other dangerous bacterial diseases. We also use them in other, less dire circumstances. Antibiotics are used preventively in humans, and they are part of the daily rations of millions of cattle, pigs, chickens, fish, and other animals raised on factory farms.

Bacteria evolve at a much accelerated rate compared with humans, in part because they reproduce very quickly. For example, the common intestinal bacteria *E. coli* can divide every 17 minutes. Each new generation is an opportunity

stabilizing selection Mode of natural selection in which an intermediate form of a trait is favored over extreme forms.

for mutation, so the gene pool of a bacterial population varies greatly. Thus, in any population of bacteria, some cells are likely to carry alleles that allow them to survive an antibiotic treatment. As susceptible individuals die and the survivors reproduce, the frequency of antibiotic-resistance alleles increases in the population. A typical two-week course of antibiotics can potentially exert selection pressure on over a thousand generations of bacteria, and antibiotic-resistant strains may be the outcome.

Antibiotic-resistant bacteria have plagued hospitals for many years, and now they are becoming similarly common in schools. Even as researchers scramble to find new antibiotics, this trend is bad news for the millions of people each year who contract a dangerous bacterial disease such as cholera or tuberculosis.

> **Stabilizing Selection** With **stabilizing selection**, an intermediate form of a trait is favored over extreme forms. This mode of natural selection tends to preserve the midrange phenotypes in a population (Figure 12.5). For example, stabilizing selection maintains an intermediate body mass in populations of sociable weaver birds (Figure 12.6). These birds build large communal nests in areas of the African savanna, and their body mass has a genetic basis. Between 1993 and 2000, Rita Covas and her colleagues investigated selection pressures that operate on sociable weaver body mass by capturing and weighing thousands of birds before and after the breeding seasons. Covas's results indicated that optimal body mass in sociable weavers is a trade-off between the risks of starvation and predation. Birds that carry less fat are more likely to starve than fatter birds. However, birds that carry more fat spend more time eating, which in this species means foraging in open areas where predators can find them. Fatter birds are also more attractive to predators, and not as agile when escaping. Thus, predators select against the fattest individuals. Birds of intermediate weight have the selective advantage, and they make up the bulk of sociable weaver populations.

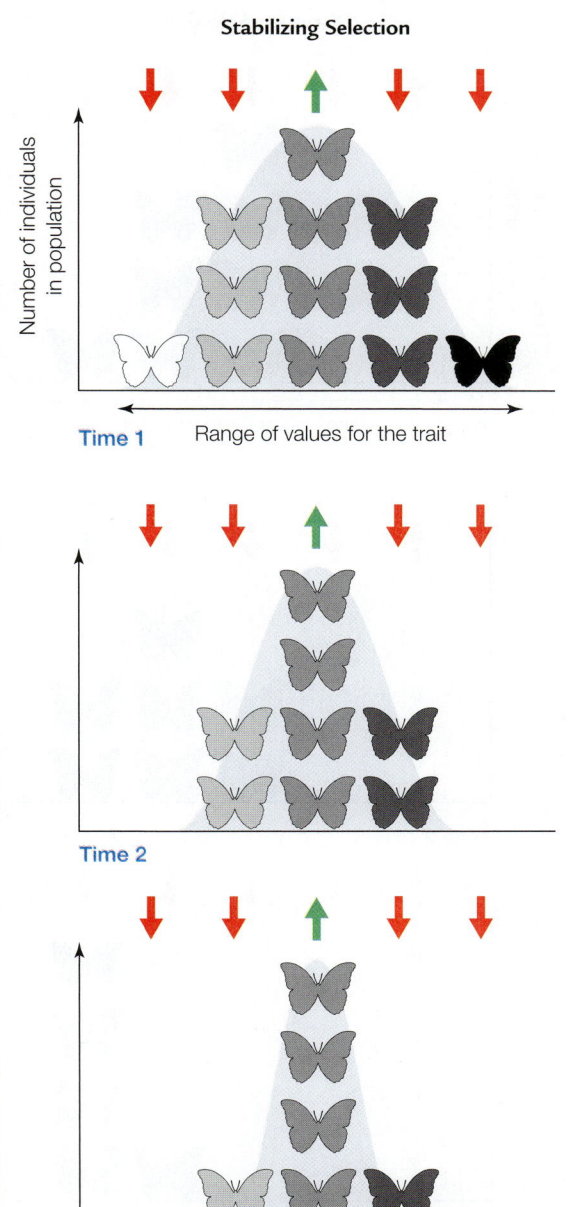

FIGURE 12.5 Animated! Stabilizing selection eliminates extreme forms of a trait, and maintains an intermediate form. *Red* arrows indicate which forms are being selected against; *green*, the form that is being favored. Compare the data set from a field experiment, shown in Figure 12.6.

© Cengage Learning.

FIGURE 12.6 Stabilizing selection in sociable weavers. These birds (*top*) build large communal nests (*middle*) in the African savanna. The graph shows the number of birds (out of 977) that survived a breeding season.

Figure It Out: What is the optimal mass of a sociable weaver?

Answer: About 29 grams

Disruptive Selection

Number of individuals in population

Time 1

Range of values for the trait

Time 2

Time 3

FIGURE 12.7 Animated! Disruptive selection eliminates midrange forms of a trait, and maintains extreme forms.

© Cengage Learning.

› **Disruptive Selection** With **disruptive selection**, forms of a trait at both ends of a range of variation are favored. This mode of natural selection tends to reduce the frequency of intermediate forms of a trait (Figure 12.7).

Consider the black-bellied seedcracker, a colorful finch species native to Cameroon, Africa. In these birds, there is a genetic basis for bill size. The bill of a typical black-bellied seedcracker, male or female, is either 12 millimeters wide, or wider than 15 millimeters (Figure 12.8). Birds with a bill size between 12 and 15 millimeters are uncommon. Seedcrackers with the large and small bill forms inhabit the same geographic range, and they breed randomly with respect to bill size. It is as if every human adult were 4 feet or 6 feet tall, with no one of intermediate height.

Environmental factors that affect feeding performance maintain the dimorphism in seedcracker bill size. The finches feed mainly on the seeds of two types of sedge, a grasslike plant. One sedge produces hard seeds; the other, soft seeds. Small-billed birds are better at opening the soft seeds, but large-billed birds are better at cracking the hard ones. All seeds are abundant during Cameroon's semiannual wet seasons, and all seedcrackers feed on both types of seeds. However, sedge seeds become scarce during the region's dry seasons. When competition for food intensifies, each bird focuses on eating the seeds that it opens most efficiently: Small-billed birds feed mainly on soft seeds, and large-billed birds feed mainly on hard seeds. Birds with intermediate-sized bills cannot open either type of seed as efficiently as the other birds, so they are less likely to survive the dry seasons.

lower bill 12 mm wide

lower bill 15 mm wide

FIGURE 12.8 Animated!
Disruptive selection in African seedcracker populations maintains a distinct dimorphism in bill size.

Competition for scarce food during dry seasons favors birds with bills that are either 12 millimeters wide (*top*) or 15 to 20 millimeters wide (*bottom*). Birds with bills of intermediate size are selected against.

Thomas Bates Smith.

Take-Home Message

How does natural selection drive evolution?

- Natural selection occurs in different patterns depending on the species involved and the pressures in their particular environment.
- With directional selection, a phenotype at one end of a range of variation is favored.
- With stabilizing selection, an intermediate phenotype is favored, and extreme forms are selected against.
- With disruptive selection, an intermediate form of a trait is selected against, and extreme phenotypes are favored.

A Male northern elephant seals fight for sexual access to a cluster of females. This type of competition favors aggressive males with large body size.

B A male bird of paradise engaged in a flashy courtship display has caught the eye (and, perhaps, the sexual interest) of a female. Female birds of paradise are choosy; a male mates with any female that accepts him.

C Stalk-eyed flies cluster on aerial roots to mate. Females prefer males with the longest eyestalks, a trait that provides no obvious survival advantage other than sexual attractiveness. This male's very long eyestalks (*top*) have captured the interest of the three females below him.

> **FIGURE 12.9** Sexual selection in action.
>
> Credits: (a) © Ingo Arndt/ Nature Picture Library; (b) Bruce Beehler; (c) Courtesy of Gerald Wilkinson.

12.4 Factors That Affect Diversity

❯ **Nonrandom Mating** Not all evolution is driven by selection for traits that influence survival. For example, competition for mates is also a selective pressure. Consider how individuals of many sexually reproducing species have a distinct male or female phenotype. Individuals of one sex (often males) tend to be more colorful, larger, or more aggressive than individuals of the other sex. These traits seem puzzling because they take energy and time away from activities that enhance survival, and some actually hinder an individual's ability to survive. Why, then, do they persist?

The answer is **sexual selection**, in which the genetic winners outreproduce others of a population because they are better at securing mates. With this mode of natural selection, the most adaptive forms of a trait are those that help individuals defeat rivals for mates, or are most attractive to the opposite sex.

For example, the females of some species cluster in defensible groups when they are sexually receptive, and males compete for sole access to the groups. Competition for the ready-made harems favors combative males (Figure 12.9**A**). As another example, males or females that are choosy about mates act as selective agents on their own species. The females of some species shop for a mate among males that display species-specific cues such as a specialized appearance or courtship behavior (Figure 12.9**B**). The cues often include flashy body parts or movements, traits that tend to attract predators and in some cases are a physical hindrance. However, to a female member of the species, a flashy male's survival despite his obvious handicap implies health and vigor, two traits that are likely to improve her chances of bearing healthy, vigorous offspring. Selected males pass alleles for their attractive traits to the next generation of males, and females pass alleles that influence mate preference to the next generation of females. Sexual selection can give rise to highly exaggerated traits (Figure 12.9**C**).

disruptive selection Mode of natural selection that favors forms of a trait at the extremes of a range of variation; intermediate forms are selected against.

sexual selection Mode of natural selection in which some individuals outreproduce others because they are better at securing mates.

☐ 0%–2%	
☐ 2%–4%	
☐ 4%–6%	
☐ 6%–8%	
☐ 8%–10%	
☐ 10%–12%	
☐ 12%–14%	
☐ >14%	

A Distribution (by percentage) of people who carry the sickle-cell allele.

B Distribution of malaria cases (*orange*) reported in Africa, Asia, and the Middle East in the 1920s, before programs to control the mosquitoes that transmit this disease began. Notice the correlation with the distribution of the sickle-cell allele shown in **A**. The photo shows a physician searching for mosquito larvae in Southeast Asia.

> **FIGURE 12.10** Malaria and sickle-cell anemia.
>
> Credits: (a,b left) After Ayala and others; (b right) © Michael Freeman/ Corbis.

bottleneck Drastic reduction in population size; a result of severe selection pressure.

fixed Refers to an allele for which all members of a population are homozygous.

founder effect After a small group of individuals found a new population, allele frequencies in the new population differ from those in the original population.

genetic drift Change in allele frequencies in a population due to chance alone.

inbreeding Mating among close relatives.

› Maintaining Multiple Alleles Any mode of natural selection may keep two or more alleles circulating at relatively high frequency in a population's gene pool, a state called balanced polymorphism. For example, sexual selection maintains multiple alleles that govern eye color in populations of *Drosophila* fruit flies. Female flies prefer to mate with rare white-eyed males, until the white-eyed males become more common than red-eyed males, at which point the red-eyed flies are again preferred.

Balanced polymorphism can also arise in an environment that favors heterozygous individuals (Section 9.2). Consider the gene that encodes the beta globin chain of hemoglobin. *HbA* is the normal allele; the codominant *HbS* allele carries a mutation that causes sickle-cell anemia (Section 7.6). Individuals homozygous for the *HbS* allele often die in their teens or early twenties from complications of the disorder.

Despite being so harmful, the *HbS* allele persists at very high frequency among human populations in tropical and subtropical regions of Asia, Africa, and the Middle East. Why? Populations with the highest frequency of the *HbS* allele also have had the highest historical incidence of malaria (Figure 12.10). Mosquitoes transmit the parasitic protist that causes malaria, *Plasmodium*, to human hosts. *Plasmodium* multiplies in the liver and then in red blood cells, which rupture and release new parasites during recurring bouts of severe illness.

It turns out that people who make both normal and sickle hemoglobin are more likely to survive malaria than people who make only normal hemoglobin. In *HbA/HbS* heterozygous individuals, *Plasmodium*-infected red blood cells sometimes sickle. The abnormal shape brings the cells to the attention of the immune system, which destroys them along with the parasites they harbor. By contrast, *Plasmodium*-infected red blood cells of individuals homozygous for the normal *HbA* allele do not sickle, so the parasite may remain hidden from the immune system.

In areas where malaria is common, the persistence of the *HbS* allele is a matter of relative evils. Malaria and sickle-cell anemia are both potentially deadly. Heterozygous individuals are not completely healthy, but they do have a better chance of surviving malaria than those who carry two normal alleles. With or without malaria, people with both alleles are more likely to live long enough to reproduce than individuals who are homozygous for the *HbS* allele. The result is that nearly one-third of people from the most malaria-ridden regions of the world are heterozygous for the *HbS* allele.

> **Genetic Drift** **Genetic drift** is random change in an allele's frequency over time, brought about by chance alone. We explain genetic drift in terms of probability—the chance that some event will occur. Sample size is important in probability. Remember from Section 1.7 that each time you flip a coin, there is a 50 percent chance it will land heads up. With 10 flips, the proportion of times heads actually land up may be very far from 50 percent. With 1,000 flips, that proportion is more likely to be near 50 percent.

The same rule holds for populations: the larger the population, the smaller the impact of random changes in allele frequencies. Imagine two populations, one with 10 individuals, the other with 100. If allele *X* occurs in both populations at a 10 percent frequency, then only one person carries the allele in the small population. If that individual dies without reproducing, allele *X* will be lost from that population. However, ten individuals in the large population carry the allele. All ten would have to die without reproducing for the allele to be lost. Thus, the chance that the small population will lose allele *X* is greater than that for the large population. This effect is a general one: The loss of genetic diversity is possible in all populations, but it is more likely in small ones (Figure 12.11). When all individuals of a population are homozygous for an allele, we say that the allele is **fixed**. The frequency of a fixed allele will not change unless mutation or another process introduces a new allele into the population.

> **Bottlenecks and the Founder Effect** A drastic reduction in population size, which is called a **bottleneck**, can greatly reduce genetic diversity. For example, northern elephant seals (shown in Figure 12.9**A**) underwent a bottleneck during the late 1890s, when hunting reduced their population size to twenty or fewer individuals. Hunting restrictions have since allowed the population to recover, but genetic diversity among its members has been greatly reduced. Today, all northern elephant seals are homozygous for every allele that has been tested. The bottleneck and subsequent genetic drift eliminated many alleles that were previously present in the population.

A loss of genetic diversity can also occur when a small group of individuals establishes a new population. If the founding group is not representative of the original population in terms of allele frequencies, then the new population will not be representative of it either, an outcome called the **founder effect**. Consider that all three *ABO* alleles for blood type (Section 9.4) are common in most human populations. Native Americans are an exception, with the majority of individuals being homozygous for the *O* allele. Native Americans are descendants of early humans that migrated from Asia between 14,000 and 21,000 years ago, across a narrow land bridge that once connected Siberia and Alaska. Analysis of DNA from ancient skeletal remains reveals that most early Americans were also homozygous for the *O* allele. Modern Siberians have all three alleles. Thus, the humans who first populated the Americas were probably members of a small group that had reduced genetic diversity compared with the general population.

Founding populations are often necessarily inbred. **Inbreeding** is nonrandom breeding or mating between close relatives of a population. Close relatives share more alleles than nonrelatives do, so inbred populations tend to have unusually high numbers of individuals homozygous for recessive alleles—many of which are harmful. This is one reason why most human societies discourage or forbid incest (mating between parents and children or between siblings).

A The size of these populations of beetles was maintained at 10 individuals. Allele b^+ was lost in one population (one graph line ends at 0).

B The size of these populations of beetles was maintained at 100 individuals. Compare the genetic drift in **A**.

FIGURE 12.11 Animated! Genetic drift in flour beetles (shown *below* on a flake of cereal). Randomly selected beetles heterozygous for alleles b^+ and b were maintained in populations of **A** 10 or **B** 100 individuals for 20 generations.

Graph lines in **B** are smoother than in **A**, indicating that drift was greatest in the sets of 10 beetles and least in the sets of 100. Allele b^+ was lost in one population (one graph line ends at 0). Notice that the average frequency of allele b^+ rose at the same rate in both groups, an indication that natural selection was at work too: Allele b^+ was weakly favored.

Figure It Out: In how many populations did allele b^+ become fixed?

Answer: Six

Credits: (a,b) Adapted from S. S. Rich, A. E. Bell, and S. P. Wilson, "Genetic drift in small populations of Tribolium," *Evolution* 33:579–584, Fig. 1, p. 580, © 1979 by John Wiley and Sons. Used by permission of the publisher; below, Photo by Peggy Greb/ USDA.

10 mm

The Old Order Amish in Lancaster County, Pennsylvania, offer an example of the effects of inbreeding within human populations. Amish people marry only within their community. Intermarriage with other groups is not permitted, and no "outsiders" are allowed to join the community. As a result, Amish populations are moderately inbred, and many of their individuals are homozygous for harmful recessive alleles. The Lancaster population has an unusually high frequency of an allele that causes Ellis–van Creveld syndrome (Figure 12.12). This allele has been traced to a man and his wife, two of a group of 400 Amish who immigrated to the United States in the mid-1700s. As a result of the founder effect and inbreeding since then, about 1 of 8 people in the Lancaster population is now heterozygous for the allele, and 1 in 200 is homozygous for it.

> **Gene Flow** Individuals tend to mate or breed most frequently with other members of their own population. However, not all populations of a species are completely isolated from one another, and nearby populations may occasionally interbreed. Also, individuals sometimes leave one population and join another. **Gene flow**, the movement of alleles between populations, occurs in both cases. Gene flow tends to stabilize allele frequencies, countering the evolutionary effects of mutation, natural selection, and genetic drift in a population.

FIGURE 12.12 An Amish child with Ellis–van Creveld syndrome. The syndrome is characterized by dwarfism, polydactyly, and heart defects, among other symptoms. The recessive allele that causes it is unusually common in the Old Order Amish of Lancaster County, an outcome of the founder effect and moderate inbreeding.

Dr. Victor A. McKusick.

Gene flow is typical among populations of animals, but it also occurs in less mobile organisms. Consider the acorns that jays disperse when they gather nuts for the winter (*right*). Every fall, the birds visit acorn-bearing oak trees repeatedly, then bury acorns in the soil of home territories that may be as much as a mile away. The jays transfer acorns (and the alleles carried by these seeds) among populations of oak trees that may otherwise be genetically isolated.

© Ashok Khosla, www.seeingbirds.com.

Gene flow also occurs when wind or animals transfer pollen from one plant to another, often over great distances (we return to the topic of pollination in Chapter 28). Many opponents of genetic engineering cite gene flow from transgenic crop plants into wild populations via pollen transfer. For example, herbicide-resistance genes and the *Bt* gene (Section 10.4) are now commonly found in weeds and unmodified crop plants. The long-term effects of this gene flow are currently unknown.

Take-Home Message

What mechanisms maintain or reduce phenotypic variation in natural populations?

- With sexual selection, a trait is adaptive if it gives an individual an advantage in securing mates. Sexual selection that reinforces differences between males and females sometimes results in exaggerated physical or behavioral traits.

- Any form of natural selection can maintain multiple alleles at high frequency in a population's gene pool.

- Genetic drift, or random change in allele frequencies, can reduce a population's genetic diversity. Its effect is greatest in small populations, such as one that endures a bottleneck.

- Gene flow is the physical movement of alleles between populations. It tends to oppose the evolutionary effects of mutation, natural selection, and genetic drift in a population by stabilizing allele frequencies.

gene flow The movement of alleles into and out of a population.

reproductive isolation The end of gene flow between populations.

speciation Evolutionary process in which new species arise.

12.5 Speciation

Mutation, natural selection, and genetic drift operate on all natural populations, and they do so independently in populations that are not interbreeding. When gene flow does not keep two populations alike, different genetic changes accumulate in each one. Over time, the populations may become so different that we call them different species. The evolutionary process in which new species arise is called **speciation**.

Evolution is a dynamic, extravagant, messy, and ongoing process that can be challenging for people who like their categories neat. Speciation offers a perfect example, because it rarely occurs at a precise moment in time: Individuals often continue to interbreed even as populations are diverging, and populations that have already diverged may come together and interbreed again. Every time speciation happens, it happens in a unique way, which means that each species is a product of its own unique evolutionary history. However, there are recurring patterns. For example, reproductive isolation is always part of speciation. **Reproductive isolation**, the end of gene flow between populations, is part of the process by which sexually reproducing species attain and maintain their separate identities. Mechanisms that prevent successful interbreeding reinforce differences between diverging populations (Figure 12.13).

> **Mechanisms of Reproductive Isolation** Some closely related species cannot interbreed because the timing of their reproduction differs. The periodical cicada (*inset*) offers an example. Cicadas feed on roots as they mature underground, then emerge to reproduce. Three species of cicada reproduce every 17 years. Each has a sibling species with nearly identical form and behavior, except that the siblings emerge on a 13-year cycle instead of a 17-year cycle. Sibling species have the potential to interbreed, but they can only get together once every 221 years!

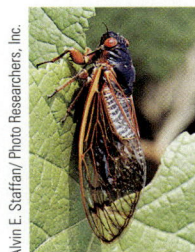

Alvin E. Staffan / Photo Researchers, Inc.

Closely related species that are adapted to different microenvironments in the same region may be ecologically isolated. For example, two species of manzanita (a plant) native to the Sierra Nevada mountain range rarely hybridize. One species that is better adapted for conserving water inhabits dry, rocky hillsides high in the foothills. The other lives on lower slopes where water stress is not as intense. The physical separation makes cross-pollination unlikely.

In animals, behavioral differences can prevent gene flow between species. For instance, individuals of some species engage in courtship displays before sex. A female recognizes vocalizations and movements of a male of her species as an overture to sex, but females of different species usually do not (Figure 12.14).

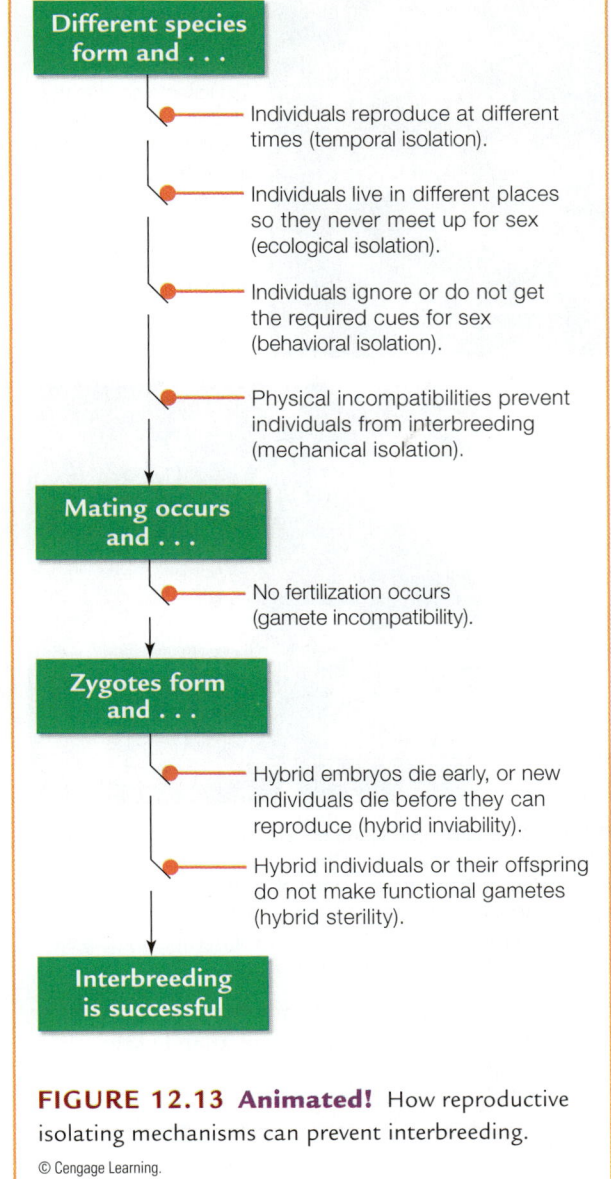

Different species form and . . .
- Individuals reproduce at different times (temporal isolation).
- Individuals live in different places so they never meet up for sex (ecological isolation).
- Individuals ignore or do not get the required cues for sex (behavioral isolation).
- Physical incompatibilities prevent individuals from interbreeding (mechanical isolation).

Mating occurs and . . .
- No fertilization occurs (gamete incompatibility).

Zygotes form and . . .
- Hybrid embryos die early, or new individuals die before they can reproduce (hybrid inviability).
- Hybrid individuals or their offspring do not make functional gametes (hybrid sterility).

Interbreeding is successful

FIGURE 12.13 Animated! How reproductive isolating mechanisms can prevent interbreeding.

© Cengage Learning.

FIGURE 12.14 Behavioral isolation.

A male peacock spider approaches a female, signaling his intent to mate with her with by raising and waving colorful flaps, and gesturing his legs in time with abdominal vibrations. If the female does not recognize his species-specific courtship display (or if it fails to impress her), she will probably kill and eat him.

© Jürgen Otto.

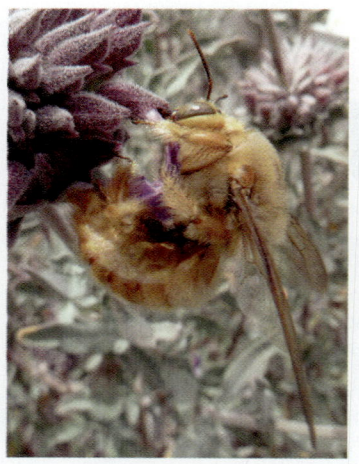

A The flowers of black sage are too delicate to support larger insects. Big insects access the nectar of small sage flowers only by piercing from the outside, as this carpenter bee is doing. When they do so, they avoid touching the flower's reproductive parts.

anther

stigma

B The reproductive parts (anthers and stigma) of white sage flowers are too far away from the petals to be brushed by honeybees, so honeybees are not efficient pollinators of this species. White sage is pollinated mainly by larger bees and hawkmoths, which brush the flower's stigma and anthers as they pry apart the petals to access nectar.

FIGURE 12.15 Mechanical isolation in sage.

Credits: (a) Courtesy of © Ron Brinkmann, www.flickr.com/photos/ ronbrinkmann; (b) © David Goodin.

The size or shape of an individual's reproductive parts may prevent it from mating with members of closely related species. For example, plants called black sage and white sage grow in the same areas, but hybrids rarely form because the flowers of these two related species have become specialized for different pollinators (Figure 12.15).

Even if gametes of different species do meet up, they often have molecular incompatibilities that prevent a zygote from forming. For example, the molecular signals that trigger pollen germination in flowering plants are species-specific (we return to pollen germination and other aspects of flowering plant reproduction in Section 28.2). Gamete incompatibility may be the primary speciation route among animals that release their eggs and free-swimming sperm into water.

Genetic changes are the basis of divergences in form, function, and behavior. Even chromosomes of species that diverged relatively recently may be different enough that a hybrid zygote ends up with extra or missing genes, or genes with incompatible products. Such outcomes typically disrupt embryonic development. Hybrids that do survive embryonic development often have reduced fitness. For example, hybrid offspring of lions and tigers have more health problems and a shorter life expectancy than individuals of either parent species.

Some interspecies crosses produce robust but sterile offspring. For example, mating a female horse (64 chromosomes) with a male donkey (62 chromosomes) produces a mule. Mules are healthy, but their 63 chromosomes cannot pair up evenly during meiosis, so this animal makes few viable gametes.

If hybrids are fertile, their offspring usually have lower and lower fitness with each successive generation. Incompatible nuclear and mitochondrial DNA may be the cause (mitochondrial DNA is inherited from the mother only).

❯ Allopatric Speciation Genetic changes that lead to a new species can begin with physical separation between populations. With **allopatric speciation**, a physical barrier arises and separates two populations, ending gene flow between them (*allo–* means different; *patria*, fatherland). Then, reproductive isolating mechanisms evolve so even if the populations do meet up again later, their individuals could not interbreed.

Gene flow between populations separated by distance is usually intermittent. Whether a geographic barrier can completely block that gene flow depends on how the species travels (such as by swimming, walking, or flying), and how it reproduces (for example, by internal fertilization or by pollen dispersal).

Mexico

Pacific Ocean

Isthmus of Panama

Columbia

0 (km) 1500
0 (mi) 1000

Alpheus nuttingi (Atlantic)

Alpheus millsae (Pacific)

FIGURE 12.16 Animated!

An example of allopatric speciation.

When the Isthmus of Panama formed 4 million years ago, it cut off gene flow among ocean-dwelling populations of snapping shrimp. Today, shrimp species on opposite sides of the isthmus are so similar that they might interbreed, but they are behaviorally isolated: Instead of mating when they are brought together, they snap their claws at one another aggressively. The photos show two of the many closely related species that live on opposite sides of the isthmus.

Credits: left, © Cengage Learning; right, © Arthur Anker.

Triticum urartu (wild einkorn)	Aegilops (wild goatgrass, unknown species)	Triticum (hybrid)	Triticum turgidum (emmer)	Aegilops tauschii (goatgrass)	Triticum aestivum (bread wheat)
14 AA ×	14 BB →	14 AB →	28 AABB ×	14 DD →	42 AABBDD

A About 11,000 years ago, a diploid wheat (einkorn) hybridized with a diploid species of wild goatgrass.

B Tetraploid (4*n*) emmer arose when the chromosome number of the resulting hybrid doubled.

C Common bread wheat is the result of a hybridization between emmer and a diploid goatgrass.

FIGURE 12.17 Animated!
Sympatric speciation in wheat.

The wheat genome, which consists of seven chromosomes, occurs in slightly different forms called *A*, *B*, *C*, *D*, and so on.

Many wheat species are polyploid, carrying more than two copies of the genome. Chromosome number is indicated for each species. For example, modern bread wheat (*Triticum aestivum*) is hexaploid, with six copies of the wheat genome: two each of genomes *A*, *B*, and *D* (or 42 *AABBDD*).

Photo by © J. Honegger, courtesy of S. Stamp, E. Merz, www.sortengarten/ethz.ch.

A geographic barrier can arise in an instant, or over an eon. The Great Wall of China is an example of a barrier that arose abruptly. As it was being built, the wall interrupted gene flow among nearby populations of insect-pollinated plants. DNA sequence comparisons show that trees, shrubs, and herbs on either side of the wall are diverging genetically. Geographic isolation usually occurs much more slowly. For example, it took millions of years of tectonic plate movements (Section 11.5) to bring the two continents of North and South America close enough to collide. The land bridge where the two continents now connect is called the Isthmus of Panama. When this isthmus formed about 4 million years ago, it cut off the flow of water—and gene flow among populations of aquatic organisms—as it separated one large ocean into what are now the Pacific and Atlantic Oceans (Figure 12.16).

〉 Sympatric Speciation With **sympatric speciation**, populations inhabiting the same geographic region speciate in the absence of a physical barrier between them (*sym–* means together). Sympatric speciation can occur in a single generation if the chromosome number multiplies spontaneously, a mechanism that is not uncommon in plants. Polyploidy (Section 9.8) typically arises when an abnormal nuclear division during meiosis or mitosis doubles the chromosome number. For example, if the nucleus of a somatic cell in a flowering plant fails to divide during mitosis, the resulting polyploid cell may proliferate and give rise to shoots and flowers. If the flowers can self-fertilize, a new polyploid species may result. Common bread wheat originated after related species hybridized, and then the chromosome number of the hybrid offspring doubled (Figure 12.17). Today, almost all ferns and the majority of flowering plant species are polyploid, as well as a few conifers, insects and other arthropods, mollusks, fishes, amphibians, and reptiles.

Sympatric speciation can also occur with no change in chromosome number. The mechanically isolated sage plants you just learned about speciated with no physical barrier to gene flow. As another example, more than 500 species of cichlid, a freshwater fish, arose in the shallow waters of Lake Victoria. This large freshwater lake sits isolated from river inflow on an elevated plain in Africa's Great Rift Valley. Since Lake Victoria formed about 400,000 years ago, it has dried up three times. DNA sequence comparisons indicate that almost all of the

allopatric speciation Speciation pattern in which a physical barrier that separates members of a population ends gene flow between them.

sympatric speciation Pattern in which speciation occurs in the absence of a physical barrier.

FIGURE 12.18 Red fish, blue fish: Males of four closely related species of cichlid native to Lake Victoria, Africa.

Hundreds of cichlid species arose by sympatric speciation in this lake. Mutations in genes that affect females' perception of the color of ambient light in deeper or shallower regions of the lake also affect their choice of mates. Female cichlids prefer to mate with brightly colored males of their own species.

Figure It Out: Which form of natural selection is driving sympatric speciation in these cichlids?

Answer: Sexual selection

Kevin Bauman, www.african-cichlid.com.

cichlid species in this lake arose since the last dry spell, which was 12,400 years ago. How could hundreds of species arise so quickly? In this case, the answer begins with differences in the color of ambient light and water clarity in different parts of the lake. The light in the lake's shallower, clear water is mainly blue; the light that penetrates the deeper, muddier water is mainly red. The cichlids vary in color and in patterning (Figure 12.18). Outside of captivity, female cichlids rarely mate with males of other species. Given a choice, they prefer to mate with brightly colored males of their own species. Their preference has a basis in genes that encode light-sensitive pigments of the retina (part of the eye). Retinal pigments made by species that live mainly in shallow areas of the lake are more sensitive to blue light. The males of these species are also the bluest. Retinal pigments made by species that live mainly in deeper areas of the lake are more sensitive to red light. Males of these species are redder. In other words, the colors that a female cichlid sees best are the same colors displayed by males of her species. Thus, mutations in genes that affect color perception are likely to affect a female's choice of mates. Such mutations are probably the way sympatric speciation occurs in these fish.

Take-Home Message

How do species attain and maintain separate identities?

- Speciation is an evolutionary process by which new species form. It varies in its details and duration, but reproductive isolation is always part of the process.
- With allopatric speciation, a physical barrier that arises prevents gene flow between populations. Genetic divergences then give rise to new species.
- With sympatric speciation, new species arise in the absence of a physical barrier to gene flow.

12.6 Macroevolution

Microevolution is change in allele frequencies within a single species or population. **Macroevolution** is our name for evolutionary patterns on a larger scale: grand evolutionary trends such as land plants evolving from green algae, the dinosaurs disappearing in a mass extinction, a burst of divergences from a single species, and so on.

> Stasis With the simplest macroevolutionary pattern, **stasis**, lineages persist for millions of years with little or no change. Consider coelacanths, an order of ancient lobe-finned fish that had been assumed extinct for at least 70 million years until a fisherman caught one in 1938. Modern coelacanths are very similar to fossil specimens hundreds of millions of years old (Figure 12.19).

adaptive radiation A burst of genetic divergences from a lineage gives rise to many new species.

exaptation Adaptation of an existing structure for a completely new purpose.

extinct Refers to a species that no longer has living members.

key innovation An evolutionary adaptation that gives its bearer the opportunity to exploit a particular environment more efficiently or in a new way.

macroevolution Large-scale evolutionary patterns and trends.

stasis Evolutionary pattern in which a lineage persists with little or no change over evolutionary time.

FIGURE 12.19

An example of stasis.

The coelacanth lineage has changed very little over evolutionary time.

Top, a fossil of a coelacanth that lived 320 million years ago.

Bottom, a live coelacanth found near Sulawesi in 1998.

Credits: top, Courtesy of The Virtual Fossil Museum, www.fossilmuseum .net; bottom, © Arnaz Mehta.

> **Exaptation** Major evolutionary novelties often stem from the adaptation of an existing structure for a completely new purpose. This macroevolutionary pattern is called **exaptation**. For example, the feathers that allow modern birds to fly are derived from feathers that first evolved in some dinosaurs. Those dinosaurs could not have used their feathers for flight, but they probably did use them for insulation. Thus, we say that flight feathers in birds are an exaptation of insulating feathers in dinosaurs.

> **Mass Extinctions** By current estimates, more than 99 percent of all species that ever lived are now **extinct**, which means they no longer have living members. In addition to continuing small-scale extinctions, the fossil record indicates that there have been more than twenty mass extinctions, which are simultaneous losses of many lineages. These include five catastrophic events in which the majority of species on Earth disappeared (Section 11.5).

> **Adaptive Radiation** With **adaptive radiation**, one lineage rapidly diversifies into several new species. Adaptive radiation can occur after individuals colonize a new environment that has a variety of different habitats with few or no competitors. The adaptation of populations to different regions of the new environment produces many new species. This is how Hawaiian honeycreepers arose (Figure 12.20). More than 4 million years ago, a small population of finches migrated to the Hawaiian Islands across thousands of miles of open ocean (a substantial barrier to later gene flow). No predators had preceded the birds, but tasty insects and plants that bore tender leaves, nectar, seeds, and fruits were already there. The finches thrived in their isolation, and their descendants spread into habitats along the coasts, through dry lowland forests, and into highland rain forests. Over many generations, populations living in the different habitats became hundreds of separate species—the honeycreepers—as unique forms and behaviors evolved in them. These unique traits helped the birds exploit special opportunities presented by their particular island habitats.

Adaptive radiation may occur after a key innovation evolves. A **key innovation** is a new trait that allows its bearer to exploit a habitat more efficiently or in a novel way. The evolution of lungs offers an example, because lungs were a key innovation that opened the way for an adaptive radiation of vertebrates on land.

Adaptive radiation can also occur after geologic or climatic events eliminate some species from a habitat. The surviving species can then exploit resources from which they had previously been excluded. This is the way mammals were able to undergo an adaptive radiation after the dinosaurs disappeared.

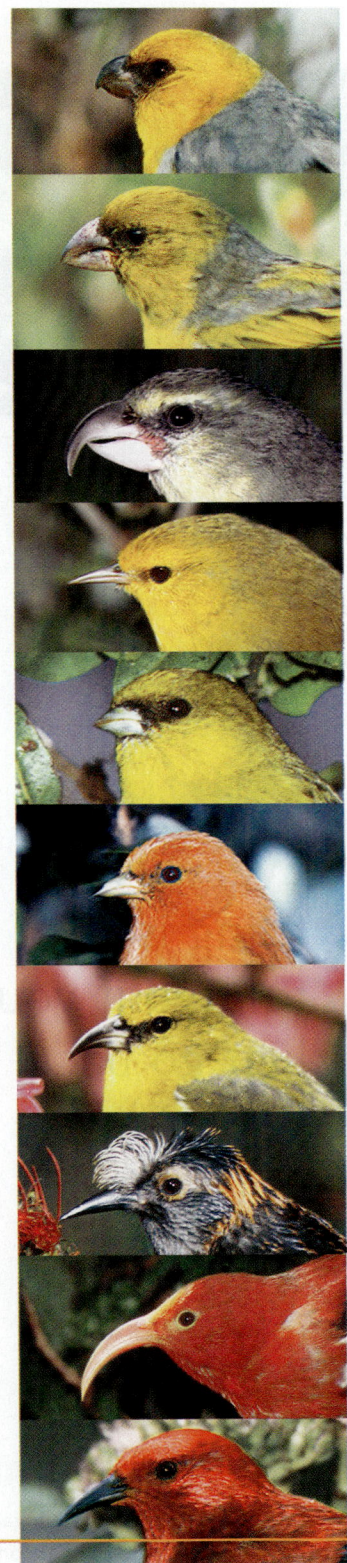

FIGURE 12.20 Adaptive radiation gave rise to the Hawaiian honeycreepers, whose diversity is represented here by a sampling of the species that have not yet become extinct. The bills of these birds are adapted to feed on insects, seeds, fruits, nectar in floral cups, and other foods. All honeycreeper species are descended from a common ancestor that probably resembled the housefinch shown at *left*.

© Jack Jeffrey Photography.

A To a *Myrmica sabuleti* ant, a honey-exuding, hunched-up *Maculinea arion* caterpillar appears to be an ant larva. This deceived ant is preparing to carry the caterpillar back to its nest, where the caterpillar will eat ant larvae for the next 10 months until it pupates.

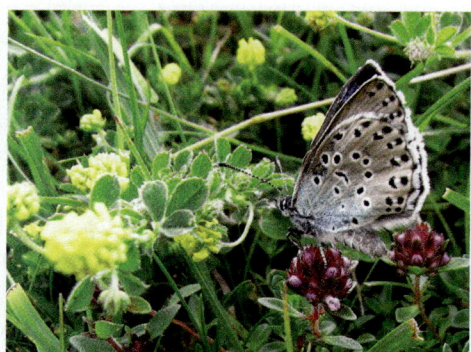

B A *Maculinea arion* butterfly emerges from the pupa and lays eggs on wild thyme flowers. Larvae that emerge from the eggs will survive only if a colony of *Myrmica sabuleti* ants adopts them.

FIGURE 12.21 An example of coevolved species: *Myrmica sabuleti* and *Maculinea arion*.

Credits: (a) © Jeremy Thomas/ Natural Visions; (b) © Brian Raine, www.flickr.com/people/25801055@N00.

> Every living thing is related if you just go back far enough in time.

character Quantifiable, heritable characteristic or trait.

clade A group whose members share one or more defining derived traits.

cladistics Making hypotheses about evolutionary relationships among clades.

coevolution The joint evolution of two closely interacting species; each species is a selective agent for traits of the other.

derived trait A character present in a clade but not in the clade's ancestors.

monophyletic group An ancestor in which a derived trait evolved, together with all of its descendants.

phylogeny Evolutionary history of a species or group of species.

> **Coevolution** The process by which close ecological interactions between two species cause them to evolve jointly is called **coevolution**. One species acts as an agent of selection on the other; each adapts to changes in the other. Over evolutionary time, the two species may become so interdependent that they can no longer survive without one another.

Relationships between coevolved species can be quite intricate. Consider the large blue butterfly (*Maculinea arion*), a parasite of ants (Figure 12.21). After hatching, the larvae (caterpillars) feed on wild thyme flowers and then drop to the ground. An ant that finds a caterpillar strokes it, which makes the caterpillar exude honey. The ant eats the honey and continues to stroke the caterpillar, which secretes more honey. This interaction continues for hours, until the caterpillar suddenly hunches itself up into a shape that appears (to an ant) very much like an ant larva. The deceived ant then picks up the caterpillar and carries it back to the ant nest, where, in most cases, other ants kill it—except, however, if the ants are of the species *Myrmica sabuleti*. Secretions of the caterpillar fool these ants into treating it just like a larva of their own. For the next 10 months, the caterpillar lives in the nest and grows to gigantic proportions by feeding on ant larvae. After it metamorphoses into a butterfly, the insect emerges from the ground to mate. Eggs are deposited on wild thyme near another *M. sabuleti* nest, and the cycle starts anew. This relationship between ant and butterfly is typical of coevolved relationships in that it is extremely specific. Any increase in the ants' ability to identify a caterpillar in their nest selects for caterpillars that better deceive the ants, which in turn select for ants that can better identify the caterpillars. Each species exerts directional selection on the other.

> **Evolutionary Theory** Biologists do not doubt that macroevolution occurs, but many disagree about how it occurs. However we choose to categorize evolutionary processes, the very same genetic change may be at the root of all evolution—fast or slow, large-scale or small-scale. Dramatic jumps in morphology, if they are not artifacts of gaps in the fossil record, may be the result of mutations in homeotic or other regulatory genes. Macroevolution may include more processes than microevolution, or it may not. It may be an accumulation of many microevolutionary events, or it may be an entirely different process. Evolutionary biologists may disagree about these and other hypotheses, but all of them are trying to explain the same thing: how all species are related by descent from common ancestors.

Take-Home Message

What is macroevolution?

- Macroevolution comprises large-scale patterns of evolutionary change such as adaptive radiation, the origin of major groups, and mass extinctions.

12.7 Phylogeny

Classifying life's tremendous diversity into a series of taxonomic ranks (Section 1.5) is a useful endeavor, in the same way that it is useful to organize a telephone book or contact list in alphabetical order. Today, however, reconstructing evolutionary relationships among organisms has become at least as important as classifying them. Thus, biologists often focus on unraveling **phylogeny**, the evolutionary history of a species or a group of them. Phylogeny is a kind of genealogy that follows a lineage's evolutionary relationships through time.

Humans were not around to witness the evolution of most species, but we do use evidence to understand events in the past (Section 11.1). Each species bears traces of its own unique evolutionary history in its characters. A **character** is a quantifiable, heritable trait, such as the nucleotide sequence of ribosomal RNA or the presence of wings (Table 12.2).

Traditional classification schemes group organisms based on shared characters: Birds have feathers, cacti have spines, and so on. By contrast, evolutionary biology tries to fit each species into a bigger picture of evolution, because every living thing is related if you just go back far enough in time. Evolutionary biologists pinpoint what makes the organisms share the characters in the first place: a common ancestor. They determine common ancestry by identifying derived traits. A **derived trait** is a character present in a group under consideration, but not in the group's ancestors. A group whose members share one or more defining derived traits is called a **clade**. By definition, a clade is a **monophyletic group**—one that consists of an ancestor (in which a derived trait evolved) together with any and all of its descendants.

Each species is a clade. Many higher taxonomic rankings are also equivalent to clades—flowering plants, for example, are both a phylum and a clade—but some are not. For example, the traditional Linnaean class Reptilia ("reptiles") includes crocodiles, alligators, tuataras, snakes, lizards, turtles, and tortoises. While it is convenient to classify these animals together, they would not constitute a clade unless birds are also included, as you will see in Chapter 15.

It is the recent nature of a derived trait that defines a clade. Consider how alligators look a lot more like lizards than birds. In this case, the similarity in appearance does indicate shared ancestry, but it is a more distant relationship than alligators have with birds. Evolutionary biologists discovered that alligators and birds share a more recent common ancestor than alligators and lizards do. Derived traits—a gizzard and a four-chambered heart—evolved in the lineage that gave rise to alligators and birds, but not in the one that gave rise to lizards.

A species' ancestry remains the same no matter how it evolves. However, as with traditional taxonomy, we can make mistakes grouping organisms into clades if the information we have is incomplete. A clade is necessarily a hypothesis, and which organisms comprise it may change when new discoveries are made. As with all hypotheses, the more data that support a cladistic grouping, the less likely it is to require revision.

⟩ Cladistics In the big picture of evolution, all clades are interconnected; an evolutionary biologist's job is to figure out where the connections are. Making hypotheses about evolutionary relationships among clades is called **cladistics**. One way of doing this involves the logical rule of simplicity: When there are several possible ways that a group of clades can be connected, the simplest evolutionary pathway is probably the correct one. By comparing all of the possible connections among the clades, we can identify the simplest: the one in which the defining derived traits evolved the fewest number of times. The process of finding the simplest pathway is called parsimony analysis (Figure 12.22).

Table 12.2	Examples of Characters		
	Bird	Bat	Dolphin
Warm-blooded	Y	Y	Y
Hair	N	Y	Y
Milk	N	Y	Y
Teeth	N	Y	Y
Wings	Y	Y	N
Feathers	Y	N	N

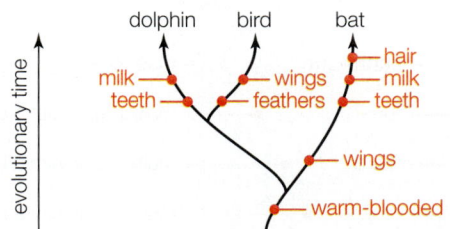

A If the bird and dolphin are most closely related, the derived traits would have evolved nine times in total.

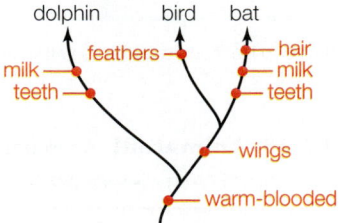

B If the bird and bat are most closely related, the derived traits would have evolved eight times in total.

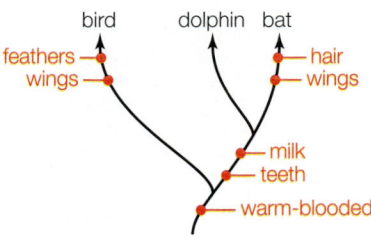

C If the dolphin and bat are most closely related, the derived traits would have evolved seven times in total.

FIGURE 12.22 An example of cladistics, using parsimony analysis with characters in Table 12.2.

A, **B**, and **C** show the three possible evolutionary pathways that could connect birds, bats, and dolphins; *red* indicates the evolution of a derived trait.

The pathway most likely to be correct (**C**) is the simplest—the one in which the derived traits would have had to evolve the fewest number of times.

© Cengage Learning.

A Evolutionary connections among clades are represented as lines on a cladogram. Sister groups emerge from a node, which represents a common ancestor.

B A cladogram can be viewed as "sets within sets" of derived traits.

FIGURE 12.23 Animated! An example of a cladogram.

(a left, b) © Cengage Learning; (a right) © National Geographic/ SuperStock.

The result of a cladistic analysis is a **cladogram**, a type of evolutionary tree that diagrams a summary of our best data-supported hypotheses about how a group of clades evolved (Figure 12.23). We use cladograms to visualize evolutionary trends and patterns. Data from an outgroup (a species not closely related to any member of the group under study) may be included in order to "root" the tree. Each line in a cladogram represents a lineage, which may branch into two lineages at a node. The node represents a common ancestor of two lineages. Every branch of a cladogram is a clade; the two lineages that emerge from a node on a cladogram are called **sister groups**.

❯ Applications Studies of phylogeny reveal how species relate to one another and to species that are now extinct. In doing so, they inform our understanding of how shared ancestry interconnects all species—including our own. Such studies also have a variety of practical applications.

The story of the Hawaiian honeycreepers offers an example of how finding ancestral connections can help species that are still living. The first Polynesians arrived on the Hawaiian Islands sometime before 1000 A.D., and Europeans followed in 1778. Hawaii's rich ecosystem was hospitable to all newcomers, including the settlers' domestic animals and crops. Escaped livestock began to eat and trample rain forest plants that had provided Hawaiian honeycreepers with food and shelter. Entire forests were cleared to grow imported crops, and plants that escaped cultivation began to crowd out native plants. Mosquitoes accidentally introduced in 1826 spread diseases from imported chickens to native bird species. Stowaway rats and snakes ate their way through populations of native birds and their eggs. Mongooses deliberately imported to eat the rats and snakes preferred to eat birds and bird eggs.

The very isolation that had spurred adaptive radiations also made the honeycreepers vulnerable to extinction. The birds had no built-in defenses against predators or diseases of the mainland. Specializations such as extravagantly elongated beaks became hindrances when the birds' habitats suddenly changed or disappeared. Thus, at least 43 honeycreeper species that had thrived on the islands before humans arrived were extinct by 1778. Conservation efforts began in the 1960s, but 26 more species have since disappeared. Today, 35 of the remaining 68 species are endangered (Figure 12.24). They are still being pressured by established populations of invasive, nonnative species of plants and animals. Rising global temperatures are also allowing mosquitoes to invade high-altitude habitats that had previously been too cold for the insects, so honeycreeper species remaining in these habitats are now succumbing to avian malaria and other mosquito-borne diseases.

As more and more honeycreeper species become extinct, the group's reservoir of genetic diversity dwindles. The lowered diversity means the group as a whole is less resilient to change, and more likely to suffer catastrophic species losses. Deciphering their phylogeny can tell us which honeycreeper species are most different from the others—and those are the ones most valuable in terms of preserving genetic diversity. Such research allows us to concentrate our resources and conservation efforts on those species that hold the best hope for the survival of the entire group. For example, we now know the poouli (Figure

cladogram Evolutionary tree diagram that shows a network of evolutionary relationships among clades.
sister groups The two lineages that emerge from a node on a cladogram.

12.24**C**) to be the most distantly related member of the genus. Unfortunately, the knowledge came too late; the poouli is probably extinct. Its extinction means the loss of a large part of evolutionary history of the group: One of the longest branches of the honeycreeper family tree is gone forever.

Cladistics is also used to correlate past evolutionary divergences with behavior and dispersal patterns of existing populations. Such studies are useful in conservation efforts. For example, a decline in antelope populations in African savannas is at least partly due to competition with domestic cattle. A cladistic analysis of mitochondrial DNA sequences suggested that current populations of blue wildebeest are genetically less similar than they should be, based on other antelope groups of similar age. Combined with behavioral and geographic data, the analysis helped conservation biologists realize that a patchy distribution of preferred food plants is preventing gene flow among blue wildebeest populations. The absence of gene flow can lead to a catastrophic loss of genetic diversity in populations under pressure. Restoring appropriate grasses in intervening, unoccupied areas of savanna would allow isolated wildebeest populations to reconnect.

Researchers also study the evolution of viruses and other infectious agents by grouping them into clades based on biochemical characters. Viruses can mutate every time they infect a host, so their genetic material changes over time. Consider the H5N1 strain of influenza virus, which infects birds and other animals. Humans infected with H5N1 have a very high mortality rate, but to date, human-to-human transmission has been rare. However, the virus replicates in pigs without causing symptoms. Pigs transmit the virus to other pigs—and apparently to humans too. A phylogenetic analysis of H5N1 isolated from pigs showed that the virus "jumped" from birds to pigs at least three times since 2005, and that one of the isolates had acquired the potential to be transmitted among humans. An increased understanding of how this virus adapts to new hosts is helping researchers design more effective vaccines for it.

Take-Home Message

How is studying phylogeny useful?

- Evolutionary biologists study phylogeny in order to understand how all species are connected by shared ancestry.

- Among other applications, phylogeny research can help us focus our conservation efforts, and understand the spread of infectious diseases.

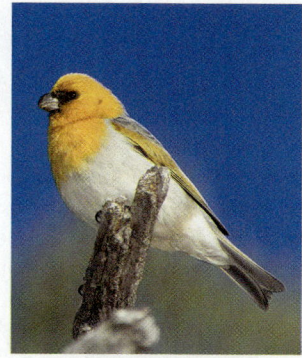

A The palila has an adaptation that allows it to feed on the seeds of a native Hawaiian plant, which are toxic to most other birds. The one remaining palila population is declining because these plants are being trampled by cows and gnawed to death by goats and sheep. Only about 1,200 palila remained in 2010.

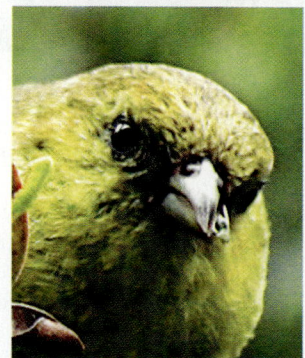

B The unusual lower bill of the akekee points to one side, allowing this bird to pry open buds that harbor insects. Avian malaria carried by mosquitoes to higher altitudes is decimating the last population of this species. Between 2000 and 2007, the number of akekee plummeted from 7,839 birds to 3,536.

C This poouli—rare, old, and missing an eye—died in 2004 from avian malaria. There were two other poouli alive at the time, but neither has been seen since then.

FIGURE 12.24 Three honeycreeper species: going, going, and gone. The genetic diversity of Hawaiian honeycreepers is dwindling along with their continued extinctions. Deciphering their evolutionary connections may help us preserve the remaining species.

Credits: (a) © Jack Jeffrey Photography; (b) Courtesy of © Lucas Behnke; (c) Bill Sparklin/Ashley Dayer.

12.8 **Rise of the Super Rats** (revisited)

The warfarin-resistance allele is clearly adaptive in populations of rats exposed to warfarin. Heterozygous rats require extra vitamin K, but being mildly deficient in vitamin K is not so bad compared with being dead from rat poison. However, rats homozygous for the allele are at a serious disadvantage: They tend to die early because they cannot easily obtain enough vitamin K from their diet to sustain normal blood clotting and bone formation. Thus, in the absence of warfarin, the allele's frequency in a rat population declines quickly. Periodic exposure to the poison keeps the allele circulating—an example of how a selection pressure can maintain a balanced polymorphism.

© Rollin Verlinde/Vilda.

WHERE YOU ARE GOING . . .

You will revisit cladograms throughout Unit Three as you learn about life's diversity. Chapter 16 returns to population ecology; Chapter 17, to ecosystems. Both concepts are important for understanding human impacts on the biosphere (Chapter 18). Mitigating that impact includes conservation efforts aimed at maintaining biodiversity. You will see more examples of natural selection in populations (16.4) and coevolved species (17.3). Plant reproduction returns in Section 28.2; polyploid plants, in 28.5.

Summary

Section 12.1 Populations tend to change along with the selection pressures that operate on them. Our efforts to control pests such as rats have selected for resistant populations.

Section 12.2 The individuals of a **population** share physical, behavioral, and physiological

traits. Alleles, which arise by mutation, are the basis of differences in the forms of shared traits. All alleles of all genes in a population form a pool of genetic resources called a **gene pool**. Mutations may be **lethal**, neutral, or adaptive.

Microevolution, or changes in **allele frequency** within a population's gene pool, occurs constantly in natural populations by processes of mutation, natural selection, and genetic drift. We use deviations from a theoretical genetic equilibrium to study how populations evolve.

Section 12.3 Natural selection, the process in which environmental pressures result in the differential survival and reproduction of individuals of a population, occurs in patterns. With **directional selection**, a phenotype at one end of a range of variation is favored. **Stabilizing selection** acts on extreme forms of a trait, so intermediate forms are favored. With **disruptive selection**, intermediate forms of a trait are selected against, and forms at the extremes are favored.

Section 12.4 **Sexual selection** is a mode of natural selection in which the adaptive traits are those that make their bearers better at securing mates. Any mode of natural selection can give rise to a balanced polymorphism, in which two or more alleles are maintained at relatively high frequency in a population.

Random change in allele frequencies—**genetic drift**—can lead to the loss of genetic diversity in a population by causing alleles to become **fixed**. Genetic drift is pronounced in small or **inbreeding** populations. The **founder effect** may occur after an evolutionary **bottleneck**. **Gene flow** can counter the effects of mutation, natural selection, and genetic drift.

Section 12.5 The details of **speciation** differ every time it occurs, but **reproductive isolation**, the end of gene flow between populations, is always a part of the process. With **allopatric speciation**, a geographic barrier arises and interrupts gene flow between populations. After gene flow ends, genetic divergences that occur independently in the populations result in separate species. Speciation can also occur with no barrier to gene flow, a pattern called **sympatric speciation**. Polyploid species of many plants (and a few animals) have originated by sympatric speciation.

Section 12.6 **Macroevolution** refers to large-scale patterns of evolution such as stasis, adaptive radiation, coevolution, and mass extinction. Major evolutionary novelty often arises by **exaptation**, in which a lineage uses a structure for a different purpose than its ancestor did. With **stasis**, a lineage changes very little over evolutionary time. A **key innovation** can result in an **adaptive radiation**, or rapid diversification of a lineage into several new species. **Coevolution** occurs when two species act as agents of selection upon one another. A lineage that has no more living members is said to be **extinct**.

Section 12.7 Evolutionary biologists reconstruct evolutionary history (**phylogeny**) by comparing physical, behavioral, and biochemical traits, or **characters**, among species. They define a **clade** as a **monophyletic group** that consists of an ancestor in which one or more **derived traits** evolved, together with all of its descendants.

Making hypotheses about the evolutionary history of a group of clades is called **cladistics**. Evolutionary tree diagrams are based on the premise that all organisms are connected by shared ancestry. In a **cladogram**, each line represents a lineage. A lineage branches into two **sister groups** at a node, which represents a shared ancestor.

Besides offering knowledge about how all species are interrelated, deciphering phylogeny helps us preserve endangered species. Phylogeny research is also used for studying the spread of viruses and other agents of infectious diseases.

Self-Quiz   Answers in Appendix I

1. _____ is the original source of new alleles.
   a. Mutation   d. Gene flow
   b. Natural selection e. All are original sources of
   c. Genetic drift   new alleles

2. Individuals don't evolve; _____ do.

3. Evolution can only occur when _____ .
   a. mating is random
   b. there is selection pressure
   c. neither is necessary

4. Match the modes of natural selection with their best descriptions.
   _____ stabilizing   a. eliminates extreme forms of a trait
   _____ directional   b. eliminates midrange forms of a trait
   _____ disruptive   c. shifts allele frequencies in one
                            direction

5. Sexual selection, such as competition between males for access to fertile females, frequently influences aspects of body form and can lead to _____ .
   a. male/female differences   c. exaggerated traits
   b. male aggression   d. all of the above

6. The persistence of the sickle allele at high frequency in a population is a case of _____ .
   a. bottlenecking   c. natural selection
   b. balanced   d. inbreeding
      polymorphism   e. both b and d

7. _____ tends to keep different populations of a species similar to one another.
   a. Genetic drift
   b. Gene flow
   c. Mutation
   d. Natural selection

8. The theory of natural selection does not explain _____ .
   a. genetic drift
   b. the founder effect
   c. gene flow
   d. how mutations arise
   e. inheritance
   f. any of the above

9. A fire devastates all trees in a wide swath of forest. Populations of a species of tree-dwelling frog on either side of the burned area diverge to become separate species. This is an example of _____ .
   a. allopatric speciation
   b. adaptive radiation
   c. sympatric speciation
   d. an evolutionary bottleneck

10. Sex in many birds is typically preceded by an elaborate courtship dance. If a male's movements are unrecognized by the female, she will not mate with him. This is an example of _____ .
   a. reproductive isolation
   b. behavioral isolation
   c. sexual selection
   d. all of the above

11. Cladistics _____ .
   a. is a way of reconstructing evolutionary history
   b. may involve parsimony analysis
   c. is based on derived traits
   d. all of the above are correct

12. In cladistics, the only taxon that is always correct as a clade is the _____ .
   a. genus   b. family   c. species   d. kingdom

13. In evolutionary trees, each node represents a(n) _____ .
   a. single lineage
   b. extinction
   c. point of divergence
   d. adaptive radiation

14. In cladograms, sister groups are _____ .
   a. inbred
   b. the same age
   c. represented by nodes
   d. members of the same family

15. Match the evolution concepts.
   _____ gene flow
   _____ sexual selection
   _____ extinct
   _____ genetic drift
   _____ cladogram
   _____ adaptive radiation
   _____ coevolution
   _____ phylogeny

   a. can lead to interdependent species
   b. changes in a population's allele frequencies due to chance alone
   c. alleles enter and leave a population
   d. evolutionary history
   e. adaptive traits make their bearers better at securing mates
   f. burst of divergences from one lineage into many
   g. no more living members
   h. diagram of sets within sets

Critical Thinking

1. Species have been traditionally characterized as "primitive" and "advanced." For example, mosses were considered to be primitive, and flowering plants advanced; crocodiles were primitive and mammals were advanced. Why do most biologists of today think it is incorrect to refer to any modern species as primitive?

2. Rama the cama, a llama–camel hybrid, was born in 1997. The idea was to breed an animal that has the camel's strength and endurance, and the llama's gentle disposition. However, instead of being large, strong, and sweet, Rama is smaller than expected

Digging Into Data

Resistance to Rodenticides in Wild Rat Populations

Beginning in 1990, rat infestations in northwestern Germany started to intensify despite continuing use of rat poisons. In 2000, Michael H. Kohn and his colleagues tested wild rat populations around Munich. For part of their research, they trapped wild rats in five towns, and tested those rats for resistance to warfarin and the more recently developed poison bromadiolone. The results are shown in Figure 12.25.

1. In which of the five towns were most of the rats susceptible to warfarin?

2. Which town had the highest percentage of poison-resistant wild rats?

3. What percentage of rats in Olfen were warfarin resistant?

4. In which town do you think the application of bromadiolone was most intensive?

FIGURE 12.25 Resistance to rat poisons in wild populations of rats in Germany, 2000.

© Cengage Learning.

and has a camel's short temper. The breeders plan to mate him with Kamilah, a female cama. What potential problems with this mating should the breeders anticipate?

3. Two species of antelope, one from Africa, the other from Asia, are put into the same enclosure in a zoo. To the zookeeper's surprise, individuals of the different species begin to mate and produce healthy, hybrid baby antelopes. Explain why a biologist might not view these offspring as evidence that the two species of antelope are in fact one.

4. Some people think that many of our uniquely human traits arose by sexual selection. Over thousands of years, women attracted to charming, witty men perhaps prompted the development of human intellect beyond what was necessary for mere survival. Men attracted to women with juvenile features may have shifted the species as a whole to be less hairy and softer featured than any of our simian relatives. Can you think of a way to test these hypotheses?

13 Early Life Forms and the Viruses

. . . WHERE YOU HAVE BEEN

This chapter explains how organic compounds (Section 2.6) could have originated and assembled to form cells. We take another look at bacteria and archaea (3.4), and how eukaryotic traits (3.5) could have evolved. The chapter draws on your understanding of photosynthesis (5.2–5.5), aerobic respiration (5.6), and the geologic time scale (11.5).

FIGURE 13.1 Animated!

Analyzing wild chimpanzee feces for the presence of SIV. Rebecca Rudicell was part of a team that combined laboratory and field studies to investigate the effects of this virus on wild chimpanzees. SIV is the ancestor of HIV (*right*), the virus that causes AIDS in humans.

Credits: © Steve Wood–UAB Creative; inset, NIBSC/ Photo Researchers, Inc.

13.1 Evolution and Disease

The term "microorganisms" refers to a variety of single-celled organisms (bacteria, archaea, and some protists) as well as to the viruses, which are not cells. Microorganisms existed for billions of years before more complex multicelled organisms arose, and most of them are harmless or beneficial to humans. However, a small minority are human **pathogens**, meaning they cause disease.

All microorganisms, including viruses, evolve. Like larger, more complex life forms, they have genes that can mutate. Therefore, natural selection adapts them to their environment. In the case of human pathogens, that environment is our body. Thus, studying the evolutionary history of pathogens can help us find ways to ward off infections or minimize their effects.

In recent years, scientists have learned quite a bit about the origin and evolution of HIV (human immunodeficiency virus). HIV causes AIDS (acquired immune deficiency syndrome), a disease that kills more than 2 million people each year. Section 22.7 discusses in detail the fatal effects of AIDS on the human immune system. In this chapter we focus on the virus itself.

HIV was first isolated in the early 1980s. Since then, gene sequence comparisons have shown that the most common strain (HIV-1) evolved from simian immunodeficiency virus (SIV), which infects chimpanzees in Africa. A recent study that combined field observations of a wild chimpanzee population with laboratory analysis (Figure 13.1) showed the effects of SIV on chimpanzee health. In the wild, SIV-infected chimpanzees leave fewer offspring and die earlier than unaffected animals.

Knowing about HIV's ancestry may help us develop new weapons against the virus. For example, although SIV does harm chimpanzees, its effects are nowhere near as devastating as those of untreated HIV in humans. Determining how the chimpanzee immune system fights SIV may provide insights that we can put to use in our own fight against AIDS.

Understanding of evolutionary history may also help us defeat malaria, a disease that causes almost a million deaths each year. Although the protist that causes malaria lives, feeds, and reproduces inside human cells, it retains a chloroplast-like organelle called an apicoplast. The apicoplast is an evolutionary legacy passed down from a photosynthetic ancestor that swam in ancient seas. The apicoplast no longer functions in photosynthesis, but it retains some metabolic activities essential to the protist. Knowing this, researchers hope to develop drugs that kill the malaria-causing protist by interfering with apicoplast function. Humans have no photosynthetic ancestors, so we do not have apicoplasts or similar organelle. Thus, drugs that target the apicoplast should have few or no negative side effects on human health.

pathogen Disease-causing agent, such as a bacterium or virus.

13.2 Before There Were Cells

> **Conditions on the Early Earth** Our solar system began assembling when asteroids (large rocks) orbiting the sun collided, forming bigger rocky objects. The heavier these pre-planets became, the more gravitational pull they had, and the more material they gathered. By about 4.6 billion years ago, this process had formed Earth and the other planets that orbit our sun.

Planet formation did not clear out all the rocky debris from space, so the early Earth received a constant hail of meteorites upon its still molten surface. More molten rock and gases spewed from volcanoes. Gases released by meteorites and volcanoes were the main source of the early atmosphere.

Studies of volcanic eruptions, meteorites, ancient rocks, and other planets suggest that Earth's early atmosphere probably held water vapor, carbon dioxide, and gaseous hydrogen and nitrogen—but very little oxygen gas. Iron in rocks rusts when exposed to oxygen gas, and the oldest rocks with rust date to about 2.3 billion years ago. The near absence of oxygen gas in Earth's early atmosphere probably made the origin of life possible. If significant amounts of the gas had been present, the molecules of life would not have been able to form on their own or persist for very long.

At first, any water falling on Earth's molten surface evaporated immediately. As the surface cooled, rocks formed. Later, rains washed mineral salts out of these rocks and the salty runoff pooled in early seas (Figure 13.2). It was in these seas that life began.

FIGURE 13.2 Artist's depiction of conditions on the early Earth.
© Cengage Learning.

> **Origin of the Building Blocks of Life** Until the early 1800s, chemists thought that organic molecules possessed a special "vital force" and could only be made inside living organisms. Then, in 1825, a German chemist synthesized urea, a molecule abundant in urine. Later, another chemist made alanine, an amino acid. These synthetic reactions demonstrated that nonliving mechanisms could yield organic molecules.

There are three main hypotheses about the source of the organic building blocks of Earth's first life.

1. *Lightning fueled atmospheric reactions.* In 1953, Stanley Miller and Harold Urey tested the hypothesis that lightning could have powered synthesis reactions in Earth's early atmosphere. To simulate this process, they filled a reaction chamber with methane, ammonia, and hydrogen gas, and zapped it with sparks from electrodes (Figure 13.3). Within a week, a variety of organic molecules had formed, including some amino acids that occur in living organisms. This result was hailed as a breakthrough, the discovery of the first steps down the road to life. However, scientists continually revisit ideas in light of new findings. We now know that the mixture of gases used in the Miller–Urey simulation probably did not represent Earth's early atmosphere. More accurate simulations with different mixes of gases produce a far lower yield of amino acids.

2. *Delivery from space via meteorites.* The presence of amino acids, sugars, and nucleotide bases in meteorites that fell to Earth suggests an alternative origin for life's building blocks. Perhaps these molecules formed in interstellar clouds of ice, dust, and gases and were delivered to Earth by meteorites. Keep in mind that during Earth's early years, meteorites fell to Earth thousands of times more frequently than they do today.

FIGURE 13.3 Animated! Diagram of the apparatus that Stanley Miller used to test whether organic building blocks of life could have formed under conditions that prevailed on early Earth. Miller circulated water vapor, hydrogen gas (H_2), methane (CH_4), and ammonia (NH_3) in a glass chamber to simulate the first atmosphere. Sparks provided by an electrode simulated lightning.

Figure It Out: Which compound in Miller's mixture provided the nitrogen for the amino group in the amino acids?   *Answer: Ammonia*

© Cengage Learning.

Simulations and experiments cannot prove how life or cells began, but they can show us what is plausible.

FIGURE 13.4 A hydrothermal vent on the seafloor. At such vents, mineral-rich water heated by geothermal energy streams out into cold ocean water. As the water cools, dissolved minerals such as iron sulfides come out of solution, forming a chimney-like structure around the vent. Rocks at these vents are covered with cell-sized chambers.

Courtesy of the University of Washington.

hydrothermal vent Underwater opening from which mineral-rich water heated by geothermal energy streams out.

iron–sulfur world hypothesis Hypothesis that life began in rocks rich in iron sulfide near deep-sea hydrothermal vents.

protocells Membranous sac that contains interacting organic molecules; hypothesized to have formed prior to the earliest life forms.

RNA world hypothesis Hypothesis that RNA served as the first material of inheritance.

3. *Reactions at deep-sea hydrothermal vents.* By another hypothesis, synthesis of life's building blocks occurred on the seafloor near hydrothermal vents. A **hydrothermal vent** is like an underwater geyser, a place where mineral-rich water heated by geothermal energy streams out through a rocky opening in the seafloor (Figure 13.4). Researchers Günter Wächtershäuser and Claudia Huber showed that amino acids form spontaneously in a simulated vent environment.

❯ Origin of Metabolism Modern cells take up small organic molecules, concentrate them, and assemble them into larger organic polymers. Before there were cells, a nonbiological process that concentrated organic subunits would have increased the chance of polymer formation.

By one hypothesis, organic polymers began to form on clay-rich tidal flats. Clay particles have a slight negative charge, so positively charged molecules in seawater stick to them. At low tide, evaporation would have concentrated the subunits even more, and sunlight might have provided energy that speeded the formation of polymers. Amino acids do form short chains under simulated tidal flat conditions.

Günter Wächtershäuser's **iron–sulfur world hypothesis** proposes that early metabolic reactions took place on the surface of rocks around hydrothermal vents. These rocks are porous, with many tiny chambers about the size of cells. Metabolism may have begun when iron sulfide in the rocks donated electrons to dissolved carbon monoxide (CO), setting in motion reactions that formed larger organic compounds. In simulations of vent conditions, organic compounds such as pyruvate do form and accumulate. In addition, iron–sulfur clusters serve as cofactors in all modern organisms. The clusters function as electron donors in essential metabolic reactions. The universal requirement for iron–sulfur cofactors may be a legacy of life's rocky beginnings.

❯ Origin of Genetic Material DNA is the genetic material in all modern cells. Cells pass copies of their DNA to descendant cells, which use instructions encoded in DNA to build proteins. Some of these proteins aid synthesis of new DNA, which is passed along to descendant cells, and so on. Protein synthesis depends on DNA, which is built by proteins. How did this cycle begin?

In the 1960s, Francis Crick and Leslie Orgel addressed this dilemma by proposing the **RNA world hypothesis**: Early on, RNA served a dual role, functioning both as a genome and as a catalyst. Evidence that RNA can both store genetic information and function like an enzyme in protein synthesis supports this hypothesis. RNAs that function as enzymes (called ribozymes) are common in living cells. For example, some ribozymes cut noncoding bits (introns) out of newly formed RNAs (Section 7.3), and the rRNA in ribosomes speeds formation of peptide bonds during protein synthesis (Section 7.5).

If the earliest self-replicating genetic systems were RNA-based, then why do all organisms now have a genome of DNA? The structure of DNA may hold the answer. Compared to a double-stranded DNA molecule, a single-stranded RNA breaks more easily and is more prone to replication errors. Thus, a switch from RNA to DNA would have made larger, more stable genomes possible.

❯ Origin of Cell Membranes Self-replicating molecules and products of other early synthetic reactions would have floated away from one another unless something enclosed them. In modern cells, a plasma membrane serves this function. If the first reactions took place in tiny rock chambers, the rock would have acted as a boundary. Over time, lipids produced by reactions inside such a chamber could have accumulated and lined the chamber wall, forming a proto-cell. A **protocell** is a membrane-enclosed collection of interacting molecules that

A Illustration of a laboratory-produced protocell with a bilayer membrane of fatty acids and strands of RNA inside.

© Janet Iwasa.

B Micrograph of a laboratory-formed protocell with RNA-coated clay (red) surrounded by fatty acids and alcohols.

From Hanczyc, Fujikawa, and Szostak, "Experimental Models of Primitive Cellular Compartments: Encapsulation, Growth, and Division"; www.sciencemag.org, *Science* 24 October 2003; 302;529, Fig. 2, p. 619.

C Field-testing a hypothesis about protocell formation. David Deamer pours a mix of small organic molecules and phosphates into a pool heated by volcanic activity.

Photo by Tony Hoffman, courtesy of David Deamer.

FIGURE 13.5 Protocells. Scientists test hypotheses about protocell formation through laboratory simulations and field experiments.

can take up material and replicate. Scientists hypothesize that protocells were the ancestors of cellular life.

Researchers have combined organic molecules in the laboratory to produce synthetic protocells that have some of the properties we associate with life. For example, vesicle-like spheres of fatty acid that hold RNA (Figure 13.5**A**, **B**) "grow" by incorporating fatty acids and nucleotides from their surroundings. Mechanical force causes the spheres to divide.

Researchers also carry out field experiments to determine what type of conditions favor or interfere with protocell formation. For example, David Deamer has tested what happens when small organic compounds such as those found on meteorites are added to pools heated by volcanic activity (Figure 13.5**C**). His results show that such a hot, acidic environment does not favor formation of membrane-like structures.

Simulations and experiments cannot prove how life or cells began, but they can show us what is plausible. A variety of investigations by many researchers tell us this: Chemical and physical processes that operate today can produce simple organic compounds, concentrate them, and assemble them into protocells (Figure 13.6). Billions of years ago, the same processes operating in the same way may have opened the way to the first life.

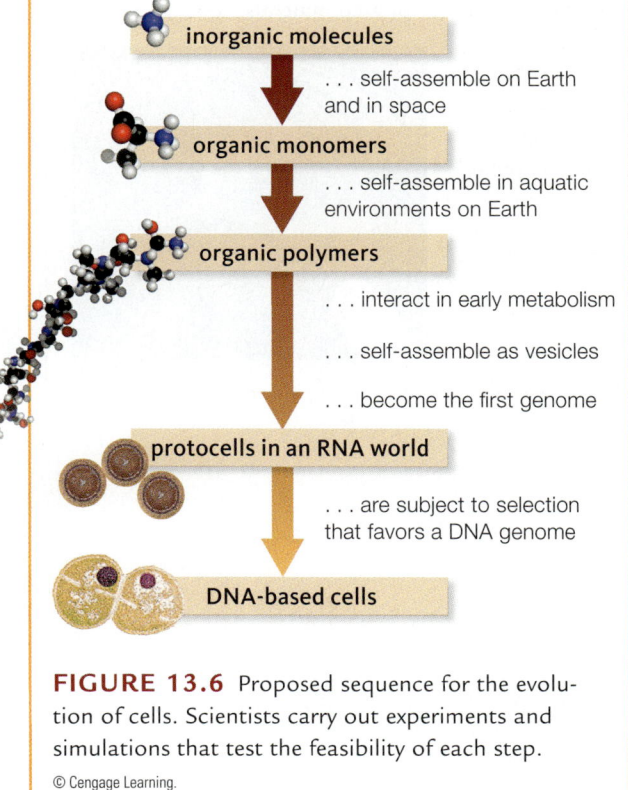

inorganic molecules

. . . self-assemble on Earth and in space

organic monomers

. . . self-assemble in aquatic environments on Earth

organic polymers

. . . interact in early metabolism

. . . self-assemble as vesicles

. . . become the first genome

protocells in an RNA world

. . . are subject to selection that favors a DNA genome

DNA-based cells

FIGURE 13.6 Proposed sequence for the evolution of cells. Scientists carry out experiments and simulations that test the feasibility of each step.

© Cengage Learning.

Take-Home Message

What do scientific studies reveal about the origin of life?

- Small organic subunits could have formed on the early Earth, or formed in space and fallen to Earth on meteorites.

- Complex organic molecules could have self-assembled from simpler ones.

- The first genetic material may have been RNA rather than DNA.

- Protocells—chemical-filled membranous sacs that grow and divide—may have been the ancestors of the first cells.

13.3 Origin of the Three Domains

› Evidence of Early Life When did life on Earth begin? Different studies provide different answers, although all indicate that life has existed for billions of years. By 4.3 billion years ago, the liquid water required for life had begun to pool on Earth's surface. Some scientists think that organic compounds found in 3.85-billion-year-old sedimentary rocks could have been produced by early cells. However, the oldest fossils widely accepted to be cells are from rocks that date to 3.4 billion years ago. These spherical, walled fossil cells do not have a nucleus and most are 5 to 25 μm in diameter, making them similar in size to modern bacteria and archaea. Given the low level of oxygen gas at the time they lived, the cells were probably anaerobic. The fact that these fossil cells occur in association with pyrite (iron sulfide) suggests that they may have produced energy by stripping electrons from this mineral.

› Origin of Bacteria and Archaea Two domains of single-celled organisms, Bacteria and Archaea, diverged very early in the history of life, perhaps as early as 3.5 billion years ago. Shortly after this divergence, some bacteria began to capture and use light energy in photosynthetic pathways that did not produce oxygen. Many modern bacteria still rely on these pathways. Then, by 2.7 billion years ago, one lineage of bacteria (cyanobacteria) began to carry out photosynthesis by the oxygen-releasing pathway (Section 5.3). These aquatic bacteria grew as dense mats that trapped sediments. Over many years, cell growth and sediment deposition formed dome-shaped, layered structures called **stromatolites**, some of which were preserved as fossils (Figure 13.7).

As a result of oxygen-producing photosynthesis, oxygen began to accumulate in the air and water. The rising oxygen levels had two important consequences:

1. Oxygen created a new selective pressure, putting organisms that thrived in higher-oxygen conditions at an advantage. The pathway of aerobic respiration (Section 5.6) evolved and became widespread. This pathway requires oxygen, and it is far more efficient at releasing energy from organic molecules than other pathways. Aerobic respiration would later meet the high-energy needs of multi-celled eukaryotes.

2. Ozone gas (O_3) formed and accumulated as an **ozone layer** in the upper atmosphere. The ozone layer prevents much of the sun's ultraviolet (UV) radiation from reaching Earth's surface. Such radiation can damage DNA and other biological molecules (Section 6.4). Water screens out some UV radiation, but without the ozone layer to protect it, life could not have moved onto land.

› Origin of Eukaryotes Eukaryotes, cells with a nucleus and other organelles (Section 3.4), first appear in the fossil record about 1.8 million years ago. Gene sequence comparisons indicate that eukaryotes have both archaeal and bacterial ancestors. Some nuclear genes in eukaryotes are similar to archaeal genes and others to bacterial genes. Eukaryotic genes that govern genetic processes (DNA replication, transcription, and translation) tend to be similar to some archaeal genes. Eukaryotic genes that govern metabolism and membrane formation tend to be similar to some bacterial genes.

In all eukaryotes, the DNA resides in a nucleus. The outer layer of the nucleus, the nuclear envelope, is a double layer of membrane with protein-lined pores that control the flow of material into and out of the nucleus. By contrast, the DNA of archaea and bacteria typically lies unenclosed in the cytoplasm.

The nucleus and endomembrane system probably evolved from infoldings of the plasma membrane (Figure 13.8). Such infoldings can be selectively advanta-

FIGURE 13.7 Stromatolites. *Above*, artist's depiction of stromatolites in an ancient sea. The surface of each stromatolite consists of a mat of living photosynthetic bacteria. Beneath it are layers of earlier generations of bacteria that trapped sediment and precipitated minerals. Chase Studios/ Photo Researchers, Inc.

Right, fossil stromatolite showing the many-layered structure.

© Dr. J Bret Bennington/ Hofstra University.

geous because they increase the surface area available for membrane-associated reactions. Membrane infoldings serve this purpose in a few modern bacteria. For example, the marine bacterium *Nitrosococcus oceani* obtains energy by using enzymes embedded in its highly folded internal membranes to break down nitrogen-containing compounds. Membrane infoldings can also divide the cytoplasm into compartments inside which conditions favor specific reactions. Membrane folds that extend around the DNA to form a nuclear envelope help protect the cell's genetic material.

Some eukaryotic organelles are thought to be descended from bacteria. Mitochondria and chloroplasts resemble bacteria in their structure and genome, so these organelles probably evolved through **endosymbiosis**. By this process, one species enters another, then lives and replicates inside it. (*Endo*– means within; *symbiosis* means living together.) Endosymbionts inside a cell can be passed along to the cell's descendants when the cell divides.

Mitochondria resemble some modern bacteria that are aerobic heterotrophs, so biologists infer that mitochondria share an ancestor with these bacteria. Presumably, the bacteria invaded or were engulfed by an early eukaryote and began living inside it. Over generations, the bacteria came to rely on their eukaryotic host for raw materials, while the host used ATP produced by its bacterial guests.

Similarly, genes of chloroplasts resemble those of modern oxygen-producing photosynthetic bacteria. As a result, biologists infer that chloroplasts evolved by endosymbiosis from relatives of these bacteria. The bacteria would have provided their host with sugars and received shelter and carbon dioxide in return.

The endosymbiont hypothesis provides a possible explanation for the observed composite genome of eukaryotes. Suppose an early eukaryote with a genome derived from an archaeal ancestor engulfed or otherwise partnered with a bacterial cell. Over time, bacterial genes related to metabolism and membrane formation migrated into the nucleus and the corresponding archaeal genes were lost. The result would be a composite genome of the type we observe today.

> **Eukaryotic Diversification** The first eukaryotes were also the first protists. The earliest protist fossil that we can assign to any modern group is *Bangiomorpha pubescens*, a red alga that lived about 1.2 billion years ago (Figure 13.9). It was multicelled and had a cellular division of labor. Some of the cells held the alga in place, and others produced two types of sexual spores. Apparently, *B. pubescens* was one of the earliest sexually reproducing organisms.

Fossils of the earliest animals known date back to 570 million years ago. These early animals shared the oceans with bacteria, archaea, fungi, and protists, including the lineage of green algae that would later give rise to land plants. Animal diversity increased tremendously during a great adaptive radiation in the Cambrian, 543 million years ago. When that period finally ended, all of the major animal lineages, including vertebrates (animals with backbones), were represented in the seas. On the next two pages, Figure 13.10 summarizes the events discussed in this section. We then devote the remainder of this chapter to discussions of modern viruses, bacteria, archaea, and protists.

Take-Home Message

What was early life like and how did it change Earth?

- Life arose by 3–4 billion years ago; it was initially anaerobic and prokaryotic.
- An early divergence separated ancestors of modern bacteria from the lineage that would lead to archaea and eukaryotic cells.
- Evolution of oxygen-producing photosynthesis in bacteria led to formation of the ozone layer, and favored organisms that carried out aerobic respiration.

FIGURE 13.8 Animated! Proposed origin of a eukaryotic nuclear envelope and ER membrane by infolding of the plasma membrane of a prokaryotic ancestor. © Cengage Learning.

infolding of plasma membrane in prokaryotic ancestor

ER

nuclear envelope of early eukaryote

20 µm

FIGURE 13.9 Fossil eukaryotes. This multicelled red alga lived 1.2 billion years ago.

© N.J. Butterfield, University of Cambridge.

endosymbiosis One species lives inside another.

ozone layer Atmospheric layer with a high concentration of ozone that prevents much UV radiation from reaching Earth's surface.

stromatolites Dome-shaped structures composed of layers of prokaryotic cells and sediments; form in shallow seas.

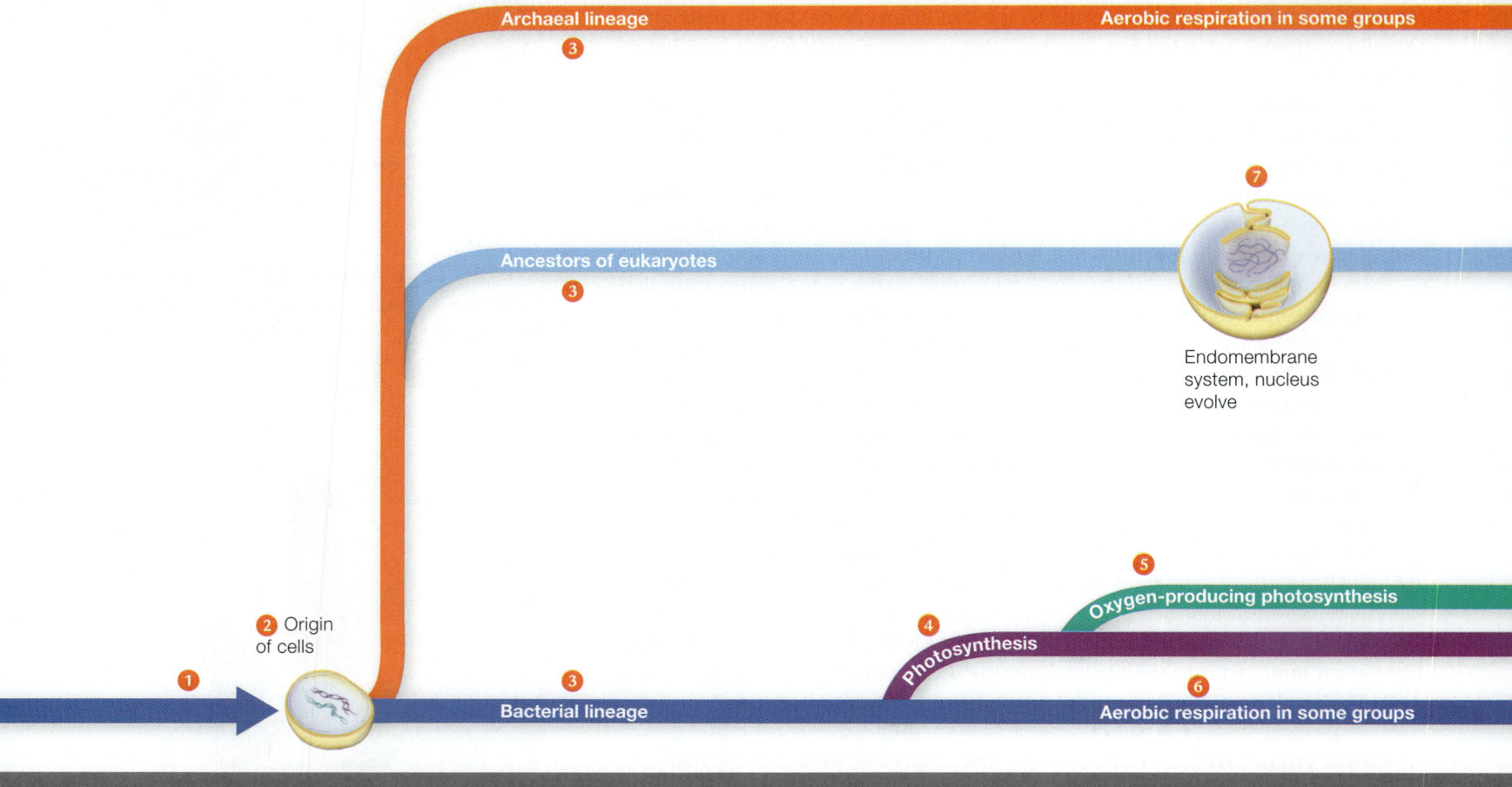

Hydrogen-rich, oxygen-poor atmosphere

Atmospheric oxygen level begins to increase

Archaeal lineage — ③ — Aerobic respiration in some groups

⑦

Ancestors of eukaryotes — ③

Endomembrane system, nucleus evolve

⑤

Oxygen-producing photosynthesis

② Origin of cells

④

Photosynthesis

①

③ Bacterial lineage — Aerobic respiration in some groups ⑥

| 3.8 billion years ago | 3.2 billion years ago | 2.7 billion years ago |

Steps Preceding Cells

❶ Lipids, proteins, nucleic acids, and complex carbohydrates formed from the simple organic compounds present on early Earth.

Origin of Cells

❷ The first cells did not have a nucleus or other organelles. Oxygen was scarce, so the first cells made ATP by anaerobic pathways.

Three Domains of Life

❸ An early divergence, separated the bacteria from the common ancestor of archaeal and eukaryotic cells. Not long after that, the archaea and eukaryotic cells diverged.

Photosynthesis and Aerobic Respiration Evolve

❹ Photosynthetic pathways that did not produce oxygen evolved in some bacterial lineages.

❺ Oxygen-producing photosynthesis evolved in a branch from this lineage, and oxygen began to accumulate.

❻ Aerobic respiration became the predominant metabolic pathway in some bacteria and archaea.

Endomembrane System and Nucleus Evolve

❼ Cell sizes and the amount of genetic information continued to expand in ancestors of what would become the eukaryotic cells. The endomembrane system, including the nuclear envelope, arose through the modification of cell membranes between 3 and 2 billion years ago.

FIGURE 13.10 Animated! Milestones in the history of life, based on the most widely accepted hypotheses. This figure also shows the evolutionary connections among all groups of organisms. The time line is not to scale.

Figure It Out: Which organelle evolved first, mitochondria or chloroplasts?

Answer: Mitochondria

© Cengage Learning.

Atmospheric oxygen reaches current levels; ozone layer gradually forms

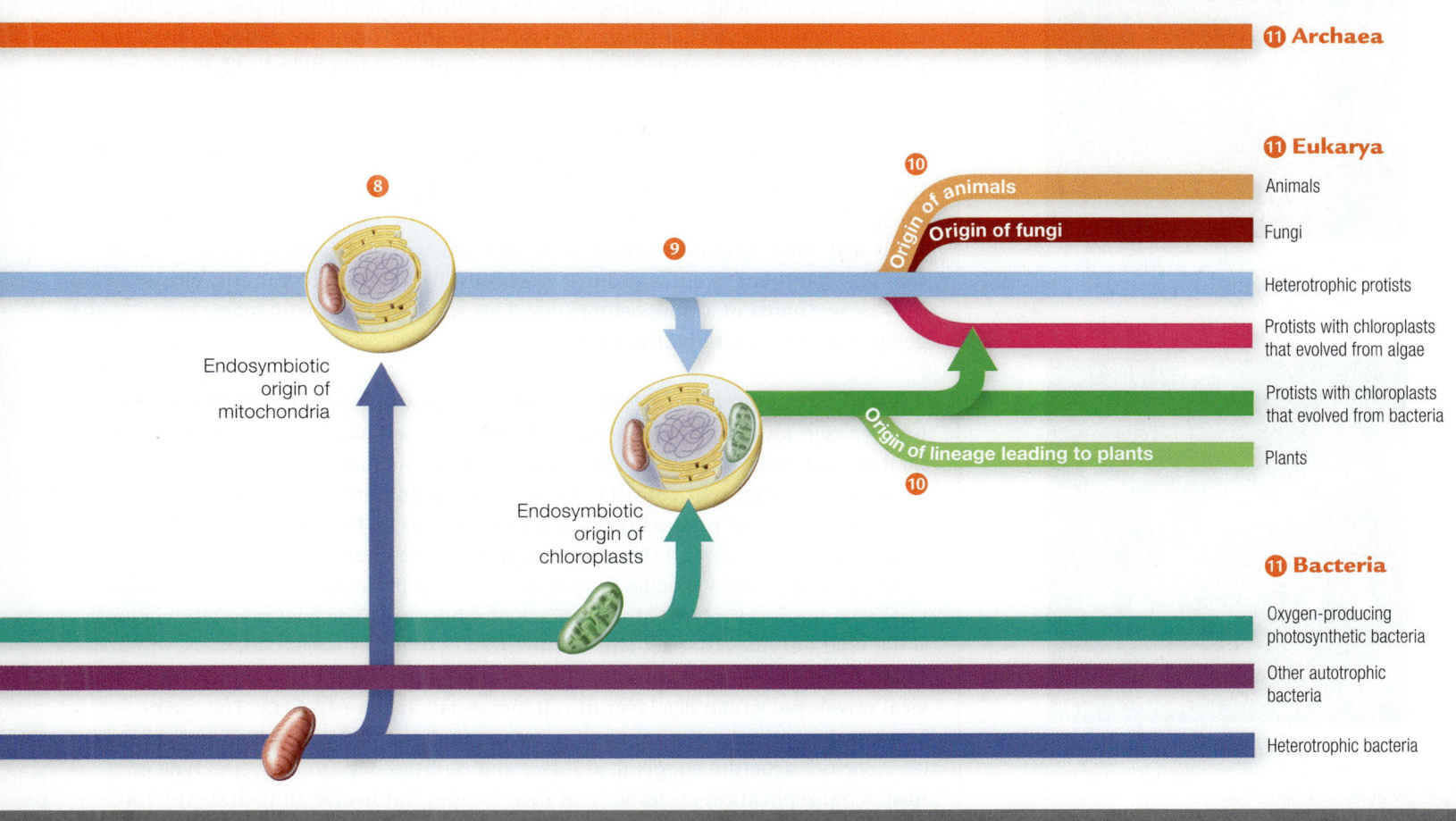

11 **Archaea**

11 **Eukarya**

8

10 Origin of animals

Animals

Origin of fungi

Fungi

Heterotrophic protists

9

Protists with chloroplasts that evolved from algae

Endosymbiotic origin of mitochondria

Protists with chloroplasts that evolved from bacteria

Origin of lineage leading to plants

Plants

10

Endosymbiotic origin of chloroplasts

11 **Bacteria**

Oxygen-producing photosynthetic bacteria

Other autotrophic bacteria

Heterotrophic bacteria

1.2 billion years ago	900 million years ago	435 million years ago

Endosymbiotic Origin of Mitochondria

8 Before about 1.2 billion years ago, aerobic bacteria entered and lived inside a eukaryotic cell. Over generations, descendants of these bacteria evolved into mitochondria.

Endosymbiotic Origin of Chloroplasts

9 An oxygen-producing, photosynthetic bacterial cell entered a eukaryotic cell. Over generations, bacterial descendants evolved into chloroplasts.

Plants, Fungi, and Animals, Evolve

10 By 900 million years ago, representatives of all major lineages—including fungi, animals, and the algae that would give rise to plants—had evolved in the seas.

Lineages That Have Endured to the Present

11 Modern organisms are related by descent and share certain traits. However, each lineage also has characteristic traits that evolved in response to the unique selective pressures it experienced.

A Tobacco mosaic virus, a helical virus that infects tobacco and related plants.

After Stephen L. Wolfe.

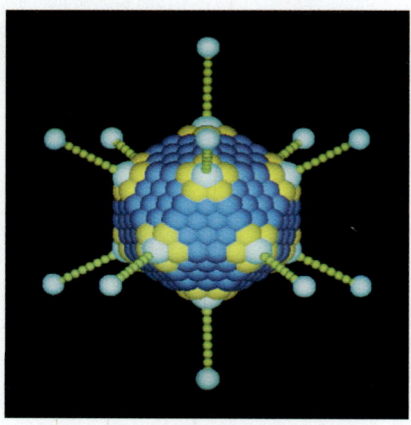

B An adenovirus, a polyhedral virus that infects animals. The 20-sided coat encloses double-stranded DNA.

© Dr. Richard Feldmann/ National Cancer Institute.

C A herpesvirus, an enveloped virus that infects animals. The envelope is derived from a host cell.

© Russell Knightly/ Photo Researchers, Inc.

FIGURE 13.11 Animated!

Examples of virus structure.

13.4 The Viruses

In the late 1800s, biologists studying diseased tobacco plants discovered a previously unknown pathogen. It was smaller than even the smallest cells, and could not be seen with a light microscope. The scientists called this unseen infectious entity a virus, a term that means "poison" in Latin. Today, we define a **virus** as a noncellular infectious particle that replicates only inside a living cell, which we call the virus's host.

The fact that viruses can only replicate inside living cells suggests that they may be derived from cells. Viruses could, for example, have evolved from transposable elements. Recall from Section 7.6 that transposable elements are nucleotide sequences that can move into and out of a genome within a cell. Virus may have originated when such elements acquired genes for coat proteins that allowed them to enter and exit host cells. By another hypothesis, viruses may be remnants of a time before cells. This would explain why many viral genes have no counterpart in modern cells.

❭ Viral Structure and Replication A free viral particle (one that is not inside a cell) always includes a viral genome enclosed within a protein coat. The viral genome may be RNA or DNA, and it may be single-stranded or double-stranded. The viral coat consists of many protein subunits that bond together in a repeating pattern, producing a helical rod (Figure 13.11**A**) or many-sided (polyhedral) shape (Figure 13.11**B**). The coat protects the viral genetic material and assists its entry into a specific type of host cell. In all viruses, components of the viral coat bind to proteins at the surface of a host cell. The coat may also enclose some viral enzymes that act within the host.

In many viruses that infect animals, the protein coat is enclosed within a **viral envelope**, which is a layer of cell membrane (Figure 13.11**C**). The membrane is derived from the host cell in which the viral particle formed.

A virus's structure adapts it to infect and replicate within a specific type of host. Viral replication cycles vary in their details, but nearly all include the following steps. The virus first attaches to an appropriate host cell by binding to a specific protein or proteins in the host's plasma membrane. Once a virus comes into contact with and attaches to an appropriate host cell, the viral genome, and in some cases other viral components, are taken up by that cell.

A viral infection is like a cellular hijacking. Viral genes take over a host's cellular machinery. They direct the cell to replicate viral DNA or RNA and to build viral proteins. These viral components self-assemble to form new viral particles. The particles may be released when the infected host cell bursts (lyses) or they may bud from the host cell, taking some of its plasma membrane with them.

❭ Bacteriophages **Bacteriophages**, sometimes called phages, are the oldest viral lineage. These nonenveloped viruses infect bacteria. One type of bacteriophage, called lambda, has a complex structure. A headlike protein coat encloses the viral DNA. Other protein components of the virus allow it to bind to a bacterium, pierce its cell wall, and inject DNA into the cell. You learned earlier how Hershey and Chase used this type of bacteriophage to identify DNA as the genetic material of all organisms (Chapter 6 Digging Into Data).

Bacteriophages replicate in bacteria by two pathways. Both pathways begin when a bacteriophage attaches to a bacterial cell and injects its DNA (Figure 13.12). In the lytic pathway, viral genes are expressed immediately ❶. The infected host first produces viral components that self-assemble as virus particles. Then, a viral-encoded enzyme breaks down the host's cell wall. Breakdown of the cell wall kills the cell and releases viral particles into the environment.

E Lysis of host cell lets new virus particles escape.

A Virus particle binds, injects genetic material.

A1 Viral DNA is inserted into host chromosome by viral enzyme action.

A2 Chromosome and integrated viral DNA are replicated.

1 Lytic Pathway

2 Lysogenic Pathway

D Accessory parts are attached to viral coat.

B Host replicates viral genetic material, builds viral proteins.

A3 Cell divides; recombinant DNA is in each descendant cell.

C Viral proteins self-assemble into a coat around viral DNA.

A4 Viral enzyme excises viral DNA from chromosome.

FIGURE 13.12 Animated! The two bacteriophage replication pathways.

Credits: left, Photo Researchers, Inc.; right, © Cengage Learning.

In the lysogenic pathway, viral DNA becomes integrated into the host cell's genome and viral genes are not immediately expressed, so the cell remains healthy **2**. When the cell reproduces, viral DNA is copied and passed to the cell's descendants along with the host's genome. Like miniature time bombs, the viral DNA inside the new cells awaits a signal to enter the lytic pathway.

Some bacteriophages can only replicate by the lytic pathway. They always kill their host cell quickly and are not passed from one bacterial generation to the next. Others embark upon either the lytic or lysogenic pathway, depending on conditions in the host cell.

❯ Plant Viruses Plant viruses are typically nonenveloped, with a helical structure and a genome of single-stranded RNA. The tobacco mosaic virus, illustrated in Figure 13.11**A**, is an example.

Plant cells have a thick wall, so plants usually become infected only after insects, pruning, or some other mechanical injury creates a wound that allows the virus into a cell. Sucking insects such as aphids and whiteflies are the vector for many viral plant diseases. A **disease vector** is an organism that transmits a pathogen from one host to the next.

Symptoms of a viral infection include spotted or streaked leaves and flowers, curled leaves, yellowed leaves, and stunted growth. The virus often alters the tiny channels (plasmodesmata) that connect the cytoplasm of one plant cell to another, allowing viral particles to move more easily between cells.

Once a plant has become infected by a virus, little can be done to treat it. Thus, protecting crop plants from viruses depends mainly upon controlling insect vectors of viral diseases and breeding plants that are virus-resistant. Genetic engineering can also be used to transfer genes that confer viral resistance into crop plants. For example, introducing the gene that encodes the coat protein of the tobacco mosaic virus into a plant enhances that plant's ability to fight off tobacco mosaic virus. The presence of the viral protein triggers the plant's own antiviral defenses before infection occurs.

> A viral infection is like a cellular hijacking. Viral genes take over a host's cellular machinery.

bacteriophage Virus that infects bacteria.

disease vector Organism that carries a pathogen from one host to the next.

viral envelope A layer of cell membrane derived from the host cell in which an enveloped virus was produced.

virus A noncellular infectious particle with a protein coat and a genome of RNA or DNA; replicates only in living cells.

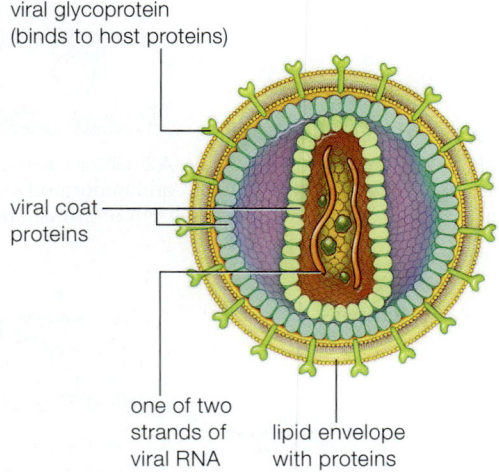

viral glycoprotein
(binds to host proteins)

viral coat
proteins

one of two
strands of
viral RNA

lipid envelope
with proteins

HIV DNA

2

3 reverse
transcription

4

5 transcription

HIV RNA

6 translation 7

8

9

1

HIV

FIGURE 13.13 Animated! Replication of HIV, an enveloped RNA virus (*above*).

❶ Viral protein binds to proteins at the surface of a white blood cell.

❷ Viral RNA and enzymes enter the cell.

❸ Viral reverse transcriptase uses viral RNA to make double-stranded viral DNA.

❹ Viral DNA enters the nucleus and becomes integrated into the host genome.

❺ Transcription produces viral RNA.

❻ Some viral RNA is translated to produce viral proteins.

❼ Other viral RNA forms the new viral genome.

❽ Viral proteins and viral RNA self-assemble at the host plasma membrane.

❾ New virus buds from the host cell, with an envelope of host plasma membrane.

Figure It Out: What is the product of reverse transcription of HIV RNA? *Answer: Double-stranded DNA*

© Cengage Learning.

> **Viruses and Human Disease** Most viral diseases of humans produce mild symptoms and trouble us only briefly. For example, some rhinoviruses infect membranes of our upper respiratory system and cause common colds. Such a cold ends when the immune system eliminates all virus-infected cells.

A minority of viral diseases are more persistent. Herpesviruses cause cold sores, genital herpes, mononucleosis, or chicken pox. Typically the initial infection causes symptoms for only a short time. However, the virus remains in the body in a latent state, and can reawaken later on. The herpes simplex virus 1 (HSV-1) can remain latent in nerve cells for years. When activated, the virus replicates and causes painful "cold sores" on the edge of the lips.

Measles, mumps, rubella (German measles), and chicken pox are viral childhood diseases that, until recently, were common worldwide. Today, most children in developed countries are protected against these illnesses because they have been vaccinated. There is also a new vaccine against human papillomavirus (HPV), a sexually transmitted virus that can cause cancers of the cervix, penis, anus, or mouth. Administering a vaccine primes the body to fight off a specific pathogen, a process explained in detail in Section 22.8.

> **HIV—The AIDS Virus** HIV (**human immunodeficiency virus**) is an enveloped RNA virus that replicates inside human white blood cells (Figure 13.13). It attaches to a cell via a glycoprotein that extends out beyond the viral envelope ❶. After attachment, the viral envelope fuses with the blood cell's plasma membrane, releasing viral enzymes and RNA into the cell ❷.

A viral enzyme called reverse transcriptase uses viral RNA as a template to synthesize a double-stranded DNA ❸. This DNA enters the nucleus together with another viral enzyme that inserts the DNA into one of the host's chromosomes ❹. Once integrated, the viral DNA is replicated and transcribed along with the host genome ❺. Some of the resulting viral RNA is translated into viral proteins ❻ and some becomes the genetic material of new HIV particles ❼. The particles self-assemble at the plasma membrane ❽. As the virus buds from the host cell, some of the host's plasma membrane becomes the viral envelope ❾. Each new virus can then infect another white blood cell. New HIV-infected cells are also produced when an infected cell replicates.

Drugs designed to fight HIV take aim at steps in viral replication. Some interfere with the way HIV binds to a host cell. Others impair reverse transcription or assembly of new virus particles. These antiviral drugs lower the number of HIV particles, so a person stays healthier. Lowering the concentration of HIV in body fluids also reduces the risk of passing the virus to others.

〉 New Flus Like HIV, the influenza viruses that cause flus are enveloped RNA viruses. Mutation alters all viral genomes, but RNA viruses mutate especially quickly. These viruses have a reverse transcriptase that makes frequent replication errors. Once made, the errors remain uncorrected because repair enzymes and proofreading activity do not operate during reverse transcription.

To keep up with ongoing mutations in influenza viruses, scientists produce a new flu shot every year. Unfortunately, determining which flu strains will be circulating is not an exact science. Even after a flu shot, a person is susceptible to a virus that is not targeted by the vaccine.

Influenza subtypes, such as H1N1, are named based on the structure of two viral surface proteins: hemagglutinin (H) and neuraminidase (N). In April of 2009, a new version of the H1N1 subtype of influenza appeared unexpectedly. The media referred to this virus as "swine flu," but it had genes from a human flu virus, a bird flu virus, and two different swine flu viruses. Such composite genomes arise as a result of **viral reassortment**, the swapping of genes between related viruses that infect a host at the same time (Figure 13.14). The 2009 H1N1 virus was discovered when it caused an epidemic in Mexico. An **epidemic** is an outbreak of disease in a limited region. Within months, there was a **pandemic**, an outbreak of disease that simultaneously affects people throughout the world. Fortunately, governments released reserves of antiviral drugs to treat those infected, and a vaccine was created to prevent new infections. The World Health Organization declared the H1N1 pandemic over in August 2010, although cases still occur among the unvaccinated.

Another strain of influenza, the H5N1 subtype, is a bird flu that occasionally infects people who have direct contact with birds. When the virus does infect people, the death rate is disturbingly high, about 60 percent. Fortunately, person-to-person transmission of the H5N1 virus is exceedingly rare.

Health officials continue to carefully monitor H5N1 and H1N1 influenza. Either virus could mutate, and their coexistence raises the possibility of a potentially disastrous gene exchange. If H1N1 picked up genes from avian H5N1, the result could be a flu virus that is both easily transmissible and deadly.

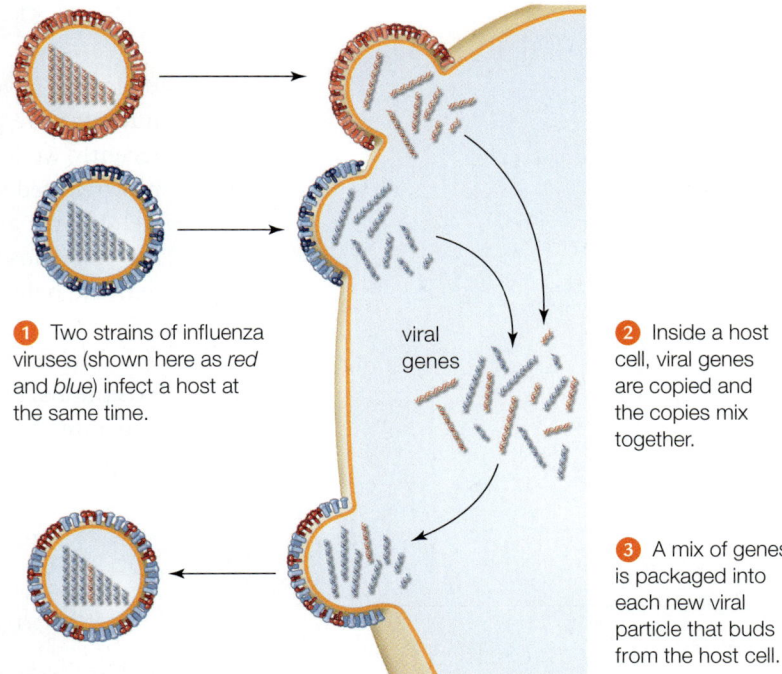

❶ Two strains of influenza viruses (shown here as *red* and *blue*) infect a host at the same time.

viral genes

❷ Inside a host cell, viral genes are copied and the copies mix together.

❸ A mix of genes is packaged into each new viral particle that buds from the host cell.

FIGURE 13.14 Viral reassortment. When a host cell is infected by two viruses of the same type, such as two influenza viruses, viral genes reassort to form viruses with new gene combinations. © Cengage Learning.

Take-Home Message

What are viruses and how do they affect us?

- Viruses are noncellular particles that consist of genetic material wrapped in a protein coat. They replicate only inside living cells.
- Viral structure varies. Each type of virus infects and replicates inside a specific type of host.
- A virus harms and eventually kills a host cell. Viral genes direct the host cell's metabolic machinery to produce new viral particles.
- Viral genomes can be altered by mutation. Viruses with new combinations of genes also arise as a result of viral reassortment.

epidemic Disease outbreak that occurs in a limited region.
HIV (**human immunodeficiency virus**) Enveloped RNA virus that causes AIDS.
pandemic Disease outbreak with cases worldwide.
viral reassortment Two viruses of the same type infect an individual at the same time and swap genes.

13.5 Bacteria and Archaea

Biologists have historically divided all life into two groups. Cells without a nucleus were prokaryotes and those with a nucleus were eukaryotes. More recently we learned that "prokaryotes" actually constitute two distinct lineages, now referred to as the domains Bacteria and Archaea. **Bacteria** are the more well-known and widespread group of cells that do not have a nucleus. **Archaea** are more closely related to eukaryotes than to bacteria and many types live in extreme habitats.

〉 Structure and Function Bacteria and archaea are small and, with rare exceptions, cannot be seen without a light microscope. Section 3.4 described the structure of prokaryotic cells, which we review briefly here. Figure 13.15 shows a typical bacterial cell. There is no nucleus or membrane-enclosed organelles like those of eukaryotes. The prokaryotic chromosome (a ring of DNA) lies in the cytoplasm, as do the ribosomes.

Nearly all bacteria and archaea have a porous cell wall around their plasma membrane. The wall gives the cell its shape, which may be spherical, spiral, or rod-shaped. We describe a spherical cell as a coccus, a spiral-shaped one as a spirillum, and a rod-shaped one as a bacillus. The cell depicted in Figure 13.15 is a bacillus. Like many bacteria, it has a capsule of secreted material around its cell wall.

Most bacteria can move from place to place. Some have one or more bacterial flagella that rotate like a propeller. Other bacteria glide along surfaces by using thin protein filaments called pili (singular, pilus) as grappling hooks. The pilus is extended out to a surface, sticks to it, then shortens, drawing the cell forward. Pili can also be used to lock a cell in place, or to draw cells together prior to the exchange of genetic material, as described below.

〉 Reproduction and Gene Transfers Bacteria and archaea have staggering reproductive potential. Some divide every twenty minutes. Division most commonly occurs by **binary fission**, a mechanism of asexual reproduction that yields two equal-sized, genetically identical descendant cells (Figure 13.16).

Although bacteria and archaea do not reproduce sexually, they do transfer genetic material among existing individuals. By a process called **conjugation**, one cell transfers a small circle of DNA, called a **plasmid**, to another. A plasmid is separate from the larger bacterial chromosome and usually has only a few genes. Cells get together for conjugation when one cell extends a sex pilus out to a prospective partner and reels it in (Figure 13.17). Once the cells are close together, the cell that made the sex pilus passes a copy of its plasmid to its partner. The cells then separate. Afterward, each cell has a copy of the plasmid. Each will duplicate the plasmid and pass it along to its descendants, and each can also transfer a copy of the plasmid to other cells by conjugation.

Two other mechanisms also alter bacterial or archaeal genomes. First, a cell can take up DNA from its environment. Second, viruses that infect bacteria and archaea sometimes move genes from one cell to another.

The ability of bacteria to acquire new genetic information by conjugation or other methods has important health implications. Suppose a gene for antibiotic resistance arises in one bacterial cell. Not only can this gene be passed on to that cell's descendants, but it can also be transferred to other existing cells. Such transfers speed the rate at which a gene spreads through a population.

pilus

DNA
plasma membrane
cell wall
secreted capsule

bacterial flagellum

FIGURE 13.15 Animated! Generalized prokaryotic body plan. © Cengage Learning.

archaea Prokaryotes most closely related to eukaryotes; many live in extreme environments.
autotroph Organism that uses carbon dioxide as its carbon source; obtains energy from light or breakdown of minerals.
bacteria Most diverse and well-known lineage of single-celled organisms that lack a nucleus.
binary fission Method of asexual reproduction in which a prokaryote divides into two identical descendant cells.
conjugation One bacterium transfers a plasmid to another.
decomposer Organism that breaks down organic material into its inorganic subunits.
heterotroph Organism that obtains both carbon and energy by breaking down organic compounds.
plasmid Of many prokaryotes, a small ring of nonchromosomal DNA replicated independently of the chromosome.

› Identifying Species, Investigating Diversity In eukaryotes, a species is defined largely on the basis of the ability of its members to mate and produce fertile offspring (Section 12.5). This definition does not apply to organisms such as bacteria and archaea that reproduce only asexually. In these groups, a species is defined as a collection of individuals that share an ancestor and have a high degree of similarity in many independently inherited traits.

The relatively new field of metagenomics is devoted to investigating microbial diversity by analyzing DNA in samples collected from an environment. Metagenomic studies often reveal a remarkable degree of species diversity. For example, air samples collected in Texas cities contained about 1,800 different kinds of bacteria. Another study of bacteria that live on the skin of the human inner elbow turned up nearly 200 species.

Metabolic diversity gives bacteria and archaea the collective ability to live just about anywhere there is a source of energy and carbon. Organisms harvest energy and nutrients from the environment in four different ways (Table 13.1). All four nutritional modes occur among bacteria or archaea.

Autotrophs are producers (Section 1.3) that build their own food using carbon dioxide (CO_2) as their carbon source. Autotrophs obtain energy in two ways. Photoautotrophs use light energy; they are photosynthetic. Many bacteria are photoautotrophs, as are plants and photosynthetic protists such as algae. By contrast, chemoautotrophs obtain energy by removing electrons from inorganic molecules such as hydrogen sulfide or methane. They use energy released by this process to build food from CO_2. Chemoautotrophic bacteria and archaea are the main producers in dark environments such as the seafloor.

Heterotrophs cannot use inorganic sources of carbon. Instead, they obtain carbon by taking up organic molecules from their environment. As with autotrophs, there are two types. Photoheterotrophs harvest energy from light, and carbon from alcohols, fatty acids, or other small organic molecules. Heliobacteria that live in the soils of rice paddies are an example. Chemoheterotrophs obtain both energy and carbon by breaking down carbohydrates, lipids, and proteins. Most bacteria and some archaea are chemoheterotrophs, as are animals, fungi, and nonphotosynthetic protists. All pathogenic bacteria are chemoheterotrophs that extract the organic compounds they need to live from their host. Other bacterial chemoheterotrophs serve as **decomposers**, meaning they convert organic molecules into inorganic ones. By their actions, decomposers make nutrients that were tied up in wastes and remains accessible to producers.

Most eukaryotic organisms are aerobic, meaning they rely on aerobic respiration (Section 5.6) and thus require oxygen. By contrast, many bacteria and most archaea are anaerobes, which means they can tolerate an oxygen-free environment. Some are obligate anaerobes, meaning oxygen either slows their growth or kills them outright. Anaerobes are harmed by oxygen because oxidation reactions damage their biological molecules and, unlike aerobic cells, they do not have enzymes that can repair that damage. We find obligate anaerobes in aquatic sediments and the animal gut. They also can infect deep wounds.

 1 A bacterium has one circular chromosome that attaches to the inside of the plasma membrane.

 2 The cell duplicates its chromosome, attaches the copy beside the original, and adds membrane and wall material between them.

 3 When the cell has just about doubled in size, a new membrane and wall are deposited across its midsection.

 4 Two genetically identical cells result.

FIGURE 13.16 Animated! Binary fission, a mechanism of asexual reproduction in bacteria and archaea. © Cengage Learning.

FIGURE 13.17 A sex pilus from one bacterium contacts another prior to conjugation, a mechanism of gene exchange. © Dr. Dennis Kunkel/ Visuals Unlimited.

Table 13.1	How Organisms Obtain Energy and Carbon	
Nutritional Type	Energy Source	Carbon Source
Photoautotroph	Light	Carbon dioxide
Chemoheterotroph	Organic compounds	Organic compounds
*Photoheterotroph	Light	Organic compounds
*Chemoautotroph	Inorganic compounds	Carbon dioxide

* Prokaryotes only

A Thermally heated waters. Pigmented archaea color rocks in waters of this Nevada hot spring.

B Highly salty waters. Pigmented extreme archaea color brine in this California lake.

C The gut of many animals. Cows belch to expel methane produced by archaea in their digestive system.

FIGURE 13.18 Examples of archaeal habitats.

Credits: (a) © Savannah River Ecology Laboratory; (b) Courtesy of Benjamin Brunner; (c) Dr. John Brackenbury/ Photo Researchers, Inc.

> **Archaeal Diversity and Ecology** Archaea were first discovered in the 1970s. Before that, all prokaryotes were thought to be bacteria. However, archaea differ from bacteria in the composition of their cell wall and plasma membrane. Like eukaryotes, archaea organize their DNA around histone proteins, which bacteria do not have. The first sequencing of an archaeal genome provided definitive evidence that archaea and bacteria are distinct lineages. You are more closely related to archaea than to bacteria.

Many archaea thrive in seemingly hostile habitats. Some are **extreme thermophiles**, organisms that live in very hot places. For example, archaea have been found in scalding hot water near deep-sea hydrothermal vents and in thermal springs (Figure 13.18**A**).

Other archaea are **extreme halophiles**, organisms that live in highly salty water (Figure 13.18**B**). Some salt-loving archaea have purple pigments that they use to capture light energy for ATP synthesis. These photoheterotrophs use the ATP to break down organic compounds, rather than making their own food from carbon dioxide and water as photosynthetic cells do.

Many archaea, including some extreme halophiles and thermophiles, are **methanogens**, which means "methane makers." These chemoautotrophs release methane gas as a by-product of their metabolic reactions. Methanogenic archaea abound in sewage, marsh sediments, and the animal gut (Figure 13.18**C**). All are strictly anaerobic, meaning oxygen kills them.

As biologists continue to explore archaeal diversity, they are finding that archaea are not restricted to extreme environments. Archaea live alongside bacteria nearly everywhere. They are particularly abundant in the ocean's deep waters. So far, scientists have not found any archaea that pose a major threat to human health. However, some that live in the mouth may encourage gum disease, and some that live in the gut may encourage weight gain.

> **Beneficial Bacteria** Many bacteria have important roles in nutrient cycles. Photosynthesis evolved in many bacterial lineages, but only the cyanobacteria (Figure 13.19**A**) use a pathway that produces oxygen as a by-product. Biologists infer that ancient cyanobacteria were the ancestors of modern chloroplasts. Thus, we have cyanobacteria and their chloroplast relatives to thank for nearly all of the oxygen that we breathe.

Cyanobacteria are also among the bacteria that make nitrogen available to plants and photosynthetic protists. There is a lot of nitrogen gas in the air, but eukaryotes cannot use it. Of all organisms, only a small minority of bacteria can carry out **nitrogen fixation**. In this process, bacteria make ammonia by combining hydrogen with nitrogen atoms they get from nitrogen gas. Plants and other photosynthetic protists can then take up the ammonia and use it to build essential molecules such as amino acids.

Some nitrogen-fixing bacteria live as free cells in the soil. Others enter and live inside the roots of legumes, a group of plants that includes peas, alfalfa, and clover. The plants benefit from the presence of the bacteria, which provide them with ammonia. The bacteria benefit by living in the shelter of the roots and receiving sugar from the plant.

Bacteria also help cycle nutrients by acting as decomposers. Together with fungal decomposers, bacteria ensure that nutrients in wastes and remains of organisms return to the soil in a form that plants can use.

Lactate-fermenting bacteria are among the decomposers. Sometimes these bacteria get into our food and spoil it, as when they cause milk to go sour. On the other hand, we use some lactate fermenters to make sauerkraut, pickles, cheese, and yogurt (Figure 13.19**B**). Lactate fermenters are also among the estimated 100 trillion bacterial cells in a healthy human

FIGURE 13.19 Helpful bacteria.

A Aquatic cyanobacteria. These cells release oxygen as a by-product of photosynthesis. Some of them also fix nitrogen.

B Lactate-fermenting bacteria used to produce yogurt. Other lactate-fermenting bacteria live in the human gut or serve as decomposers in the soil.

Credits: (a) © R. Calentine/ Visuals Unlimited; (b) SciMAT / Photo Researchers, Inc.

intestine. The acidity of the lactate these bacteria produce helps keep disease-causing organisms from taking hold in our gut. Other intestinal bacteria benefit us by producing essential vitamins or by breaking down materials we could not otherwise digest. For example, most of the vitamin K you need is produced by *Escherichia coli* bacteria in your large intestine.

The bacteria and other microorganisms that normally live in or on our body are referred to as our **normal flora**. Their importance can become apparent when we take antibiotics. These drugs kill beneficial bacteria as well as harmful ones, so they can disrupt the balance among microorganisms. When this happens in the intestine, it can cause diarrhea. When it happens in the vagina, the result is an overgrowth of yeast (fungal) cells that cause vaginitis.

❯ Harmful Bacteria Bacteria cause many common diseases (Table 13.2). Usually, bacterial infections can be halted with antibiotics; however, some types of bacteria have evolved resistance to one or more antibiotic drugs.

Bacterial diseases such as whooping cough (also called pertussis) and tuberculosis spread when a person with an active infection coughs or sneezes, distributing bacteria-laden droplets into the environment. Whooping cough once caused many infant and early childhood deaths, but a vaccine against the disease has made it rare in the United States. Tuberculosis remains a major cause of death, killing about 1.7 million people worldwide each year.

Impetigo, a skin disease, is spread by direct contact with sores, or with objects such as towels that came into contact with the sores. It occurs when *Streptococcus* and *Staphylococcus* bacteria infect outer skin layers. These bacteria also cause infections inside the body. *Streptococcus* causes strep throat, and can damage heart valves if an infection goes untreated. *Staphylococcus aureus* can cause invasive "staph" infections. Unfortunately, staph infections involving antibiotic-resistant *S. aureus* are on the rise.

Gonorrhea, syphilis, and chlamydia are sexually transmitted bacterial diseases. All can threaten fertility if untreated. Syphilis can also spread throughout the body and cause brain damage and blindness.

Bacteria sometimes enter our body in tainted food or water. In the United States, the rod-shaped bacterium *Salmonella* commonly causes food poisoning. Infection by *Salmonella* results in nausea, abdominal cramps, and diarrhea, but is rarely life-threatening. In contrast, the worldwide death toll from cholera is about 100,000 people per year. *Vibrio cholerae* bacteria infect the small intestine and produce a toxin that causes diarrhea and vomiting. The disease spreads when bacteria-tainted feces contaminate drinking water.

Table 13.2	Examples of Bacterial Diseases
Disease	**Description**
Whooping cough	Childhood respiratory disease
Tuberculosis	Respiratory disease
Impetigo, boils	Blisters, sores on skin
Strep throat	Sore throat, can damage heart
Cholera	Diarrheal illness
Syphilis	Sexually transmitted disease
Gonorrhea	Sexually transmitted disease
Chlamydia	Sexually transmitted disease
Lyme disease	Rash, flulike symptoms, spread by ticks
Botulism, tetanus	Muscle paralysis by bacterial toxin

extreme halophile Organism that lives where the salt concentration is high.

extreme thermophile Organism that lives where the temperature is very high.

methanogen Organism that produces methane gas as a metabolic by-product.

nitrogen fixation Process of combining nitrogen gas with hydrogen to form ammonia.

normal flora Collection of microorganisms that normally live in or on a healthy animal or person.

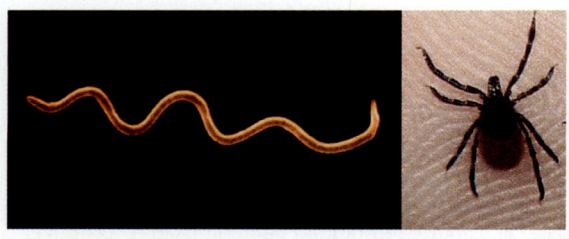

FIGURE 13.20 A pathogenic spirochete bacterium *Borrelia burgdorferi* (*left*) and its vector, a tick (*right*). *B. burgdorferi* causes Lyme disease.

Credits: left, Stem Jems/ Photo Researchers, Inc.; right, California Department of Health Services.

Some bacteria form resting structures called endospores that can survive boiling, irradiation, and drying out. When conditions improve, an endospore germinates and a bacterium emerges. Endospores that germinate in the human body can be lethal. In 2001, *Bacillus anthracis* endospores sent in the mail caused fatal cases of anthrax in five people who inhaled them. *Clostridium tetani* endospores from the soil sometimes get into wounds, resulting in a disease called tetanus. A bacterial toxin interferes with nervous control of muscle, so muscles remain locked in contraction, an effect referred to as "lockjaw."

Lyme disease is one of the vector-borne bacterial diseases. Ticks carry the unwalled spirochete bacteria that causes the disease between vertebrate hosts (Figure 13.20). Lyme disease may initially cause a bull's eye–shaped rash at the site of the tick bite. Later, flulike symptoms occur. The disease typhus is transmitted by lice that carry rickettsias, tiny bacteria that live inside cells of their host.

Take-Home Message

What are prokaryotes?

- Prokaryotes are cells that do not have a nucleus. They reproduce mainly by binary fission and they swap genes by conjugation and other processes.

- Collectively, prokaryotes are the most metabolically diverse organisms. They live almost anywhere there is carbon and energy.

- Archaea are the most recently discovered domain of life. Many archaea live in extremely hot or salty habitats.

- Bacteria include species that play important roles in nutrient cycles. Some live in or on the human body, either as normal flora or as pathogens.

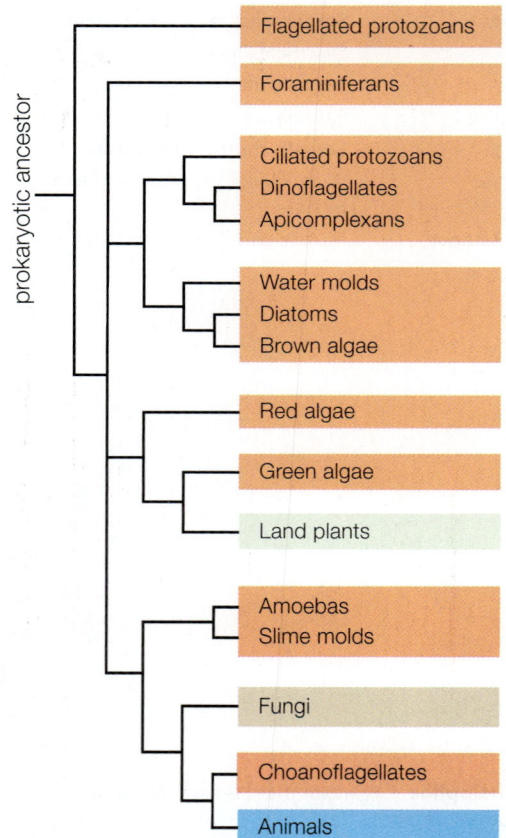

FIGURE 13.21 Evolutionary tree for the eukaryotes. *Orange* boxes indicate the lineages that were traditionally considered protists.

© Cengage Learning.

13.6 The Protists

Of all existing species, **protists** are the most like the first eukaryotic cells. Unlike prokaryotes, protist cells have a nucleus and a cytoskeleton that includes microtubules. Most also have mitochondria, endoplasmic reticulum, and Golgi bodies. All protists have multiple chromosomes, each consisting of DNA with proteins attached. Protists reproduce asexually by mitosis, sexually by meiosis, or both.

Most protists are single-celled species. But there are also colonial forms and multicelled species. Many protists are photoautotrophs that have chloroplasts. Others are predators, parasites, or decomposers.

Until recently, protists were defined largely by what they are *not*—not prokaryotes, not plants, not fungi, not animals. They were lumped together in a kingdom between prokaryotes and the "higher" forms of life. Thanks to improved comparative methods, protists are now being assigned to groups that reflect evolutionary relationships. Figure 13.21 shows where the protists that we cover in this book fit in the eukaryote family tree. There are many other protist lineages, but this sampling will demonstrate their diversity and importance. Notice that protists are not a single lineage. Some of the protists are actually more closely related to plants, animals, or fungi than they are to other protists.

❯ Flagellated Protozoans "Protozoans" is the general term for heterotrophic protists that live as single cells. **Flagellated protozoans** are single, unwalled cells with one or more flagella (structures described in Section 3.5). All groups are entirely or mostly heterotrophic. A **pellicle**, a layer of elastic proteins just beneath the plasma membrane, helps the cells retain their shape.

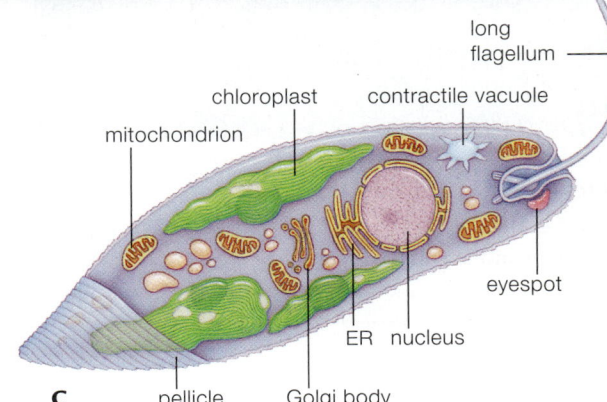

FIGURE 13.22 Animated! Flagellated protozoans.
A *Trichomonas*, a sexually transmitted parasite of humans.
B The trypanosome that causes African sleeping sickness.
C Structure of a photosynthetic euglenoid.

Credits: (a) © Dr. Dennis Kunkel/ Visuals Unlimited; (b) Oliver Meckes/ Photo Researchers, Inc.; (c) © Cengage Learning.

Some protozoans have multiple flagella that they use to swim through the body fluids of animals. For example, *Trichomonas* (Figure 13.22**A**) swims through the human reproductive tract. It causes trichomoniasis, one of the most common sexually transmitted diseases among college students. Another multiflagellated protozoan, *Giardia*, lives as an intestinal parasite inside humans, cattle, and some wild animals. *Giardia* cells can survive outside the host body in a resting stage called a cyst. If you drink water from a stream contaminated with *Giardia* cysts, the resulting infection causes the disease giardiasis. Symptoms range from mild cramps to severe diarrhea that can persist for weeks.

Trypanosomes are another group of parasitic flagellates. These long, tapered cells have a single flagellum that runs the length of the body inside a membrane. Biting insects act as vectors for trypanosomes that parasitize humans. In sub-Saharan Africa, tsetse flies spread the trypanosome that causes African sleeping sickness (Figure 13.22**B**). This fatal disease affects the nervous system. Infected people become drowsy during the day and often cannot sleep at night.

Euglenoids have a single long flagella and most live in ponds and lakes. As in other freshwater protists, one or more **contractile vacuoles** counter the tendency of water to diffuse into the cell. Excess water collects in the contractile vacuoles, which contract and expel it to the outside (Figure 13.22**C**). Euglenoids are related to trypanosomes, and the early ones were heterotrophs. Many still are, but others such as the one depicted in Figure 13.22**C** have chloroplasts that evolved when a heterotrophic cell partnered with a green alga.

〉 Foraminifera Foraminifera, or forams, are single-celled predators that secrete a shell containing calcium carbonate. Threadlike cytoplasmic extensions protrude through openings in the cell. Most forams live on the seafloor, where they probe the water and sediments for prey. Others are part of marine **plankton**, the mostly microscopic organisms that drift or swim in the open sea. Planktonic forams often have photosynthetic protists that live in their cytoplasm (Figure 13.23). Long ago, the calcium-rich remains of forams and other protists with calcium carbonate shells began to accumulate on the seafloor. Over great spans of time, these deposits became limestone and chalk.

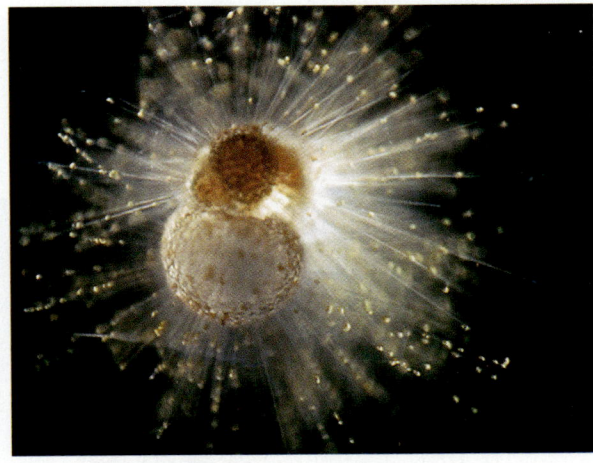

FIGURE 13.23 A planktonic foraminiferan. *Gold* dots on its cytoplasmic extensions are algal cells.

Courtesy of Allen W. H. Bé and David A. Caron.

contractile vacuole In freshwater protists, an organelle that collects and expels excess water.

flagellated protozoan Member of a heterotrophic lineage of protists that have one or more flagella.

foraminiferan Heterotrophic protist that secretes a calcium carbonate shell.

pellicle Layer of proteins that gives shape to many unwalled, single-celled protists.

plankton Community of mostly microscopic drifting or swimming organisms.

protist A eukaryote that is not a fungus, plant, or animal.

FIGURE 13.24 Cilates. These micrographs show a barrel-shaped ciliate with tufts of cilia (*Didinium*), catching and engulfing a *Paramecium*, another ciliate.

Gary W. Grimes and Steven L'Hernault.

> **The Ciliates** Ciliated protozoans, or ciliates, are close relatives of dinoflagellates and apicomplexans. **Ciliates** are unwalled cells that have many cilia (structures described in Section 3.5). Most ciliates are predators in seawater or fresh water. They feed on bacteria, algae, and one another (Figure 13.24). Other ciliates live in the gut of mammalian grazers such as cattle and sheep. Like some bacteria, these ciliates help their host digest plant material. Only one species of ciliate (*Balantidium coli*) is a known human pathogen. It also infects pigs, and people become infected when pig feces containing a resting form of the ciliate taints drinking water. Infection causes nausea and diarrhea.

> **Dinoflagellates** The name **dinoflagellate** means "whirling flagellate." These single-celled protists typically have two flagella, one at the cell's tip and the other running in a groove around its middle like a belt (Figure 13.25**A**). Combined action of the two flagella causes the cell to rotate as it moves forward. Most dinoflagellates deposit cellulose just beneath their plasma membrane, and these deposits form thick protective plates.

Some dinoflagellates live in fresh water and others in the oceans. Some prey on bacteria, others are parasites of animals, and still others have chloroplasts that evolved from algal cells. A few photosynthetic species live inside cells of reef-building corals. They supply their coral host, which is an invertebrate animal, with essential sugars.

Some marine dinoflagellates are **bioluminescent**, meaning they convert ATP energy into light (Figure 13.25**B**). Emitting light may protect a cell by startling a predator that was about to eat it. By another hypothesis, the flash of light acts like a car alarm. It attracts the attention of other organisms, including predators that pursue would-be eaters of dinoflagellate.

In nutrient-enriched water, free-living photosynthetic dinoflagellates or other aquatic protists sometimes undergo great increases in population size,

FIGURE 13.25 Dinoflagellates.

Credits (a) © Dr. David Phillips/ Visuals Unlimited;
(b) © Frank Borges Llosa/ www.frankly.com.

A Two dinoflagellate cells, each with a flagellum at its tip and another wrapped around its middle. Action of the flagella causes the cells to rotate as they move along.

B Tropical waters shimmer with light when countless bioluminescent dinoflagellates emit light in response to a disturbance.

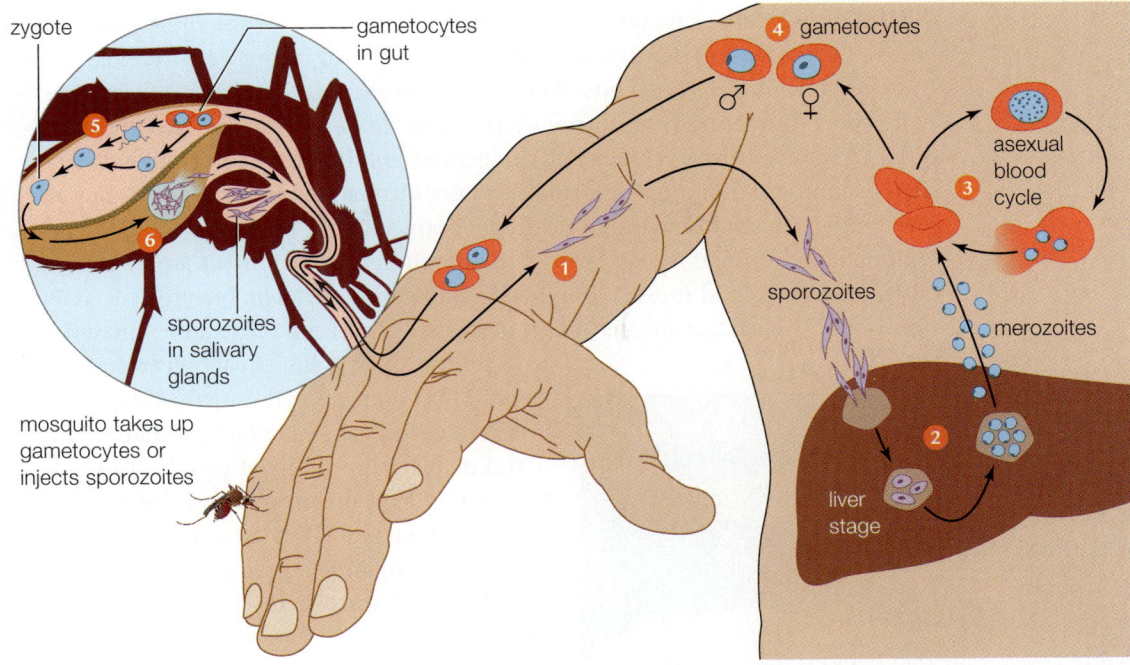

zygote

gametocytes in gut

sporozoites in salivary glands

mosquito takes up gametocytes or injects sporozoites

gametocytes

asexual blood cycle

sporozoites

merozoites

liver stage

FIGURE 13.26 Animated! Life cycle of a malaria-causing apicomplexan.

1 Infected mosquito bites a human. Sporozoites enter the blood, which carries them to the liver.

2 Sporozoites reproduce asexually in liver cells, then mature into merozoites. Merozoites leave the liver and enter the bloodstream, where they infect red blood cells.

3 Inside some red blood cells, merozoites reproduce asexually. These cells burst and release more merozoites into the bloodstream.

4 Inside other red blood cells, merozoites develop into male and female gametocytes.

5 A female mosquito bites and sucks blood from the infected person. Gametocytes in red blood cells enter her gut and mature into gametes, which fuse to form zygotes.

6 Zygotes develop into sporozoites that migrate to the mosquito's salivary glands.

Based on Fig. 1 from "Genetic linkage and association analyses for trait mapping in Plasmodium falciparum," by Xinzhuan Su, Karen Hayton & Thomas E. Wellems, *Nature Reviews Genetics 8*, 497–506 (July 2007).

a phenomenon known as an **algal bloom**. Algal blooms can harm other organisms. When the cells die, aerobic bacteria that feed on their remains use up all the oxygen in the water, so aquatic animals suffocate. Also, some dinoflagellates produce toxins that can kill aquatic organisms directly and sicken people.

❯ Apicomplexans **Apicomplexans** are parasitic protists that spend part of their life inside cells of their hosts. Their name refers to a complex of microtubules at their apical (top) end that allows them to enter a host cell. They are also sometimes called sporozoans. As noted in Section 13.1, although modern apicomplexans are all parasites, they evolved from free-living photosynthetic cells and retain remnants of chloroplasts. These chloroplast-derived organelles may be a useful target for drugs that fight apicomplexan disease.

Apicomplexans infect a variety of animals, from worms and insects to humans. Their life cycle often involves more than one host species. For example, *Plasmodium*, the apicomplexan that causes malaria, requires two hosts to complete the three stages of its life cycle (Figure 13.26). A female *Anopheles* mosquito transmits an infective *Plasmodium* cell (called the sporozoite) to a human when she bites **1**. The sporozoite travels in blood vessels to liver cells, where it reproduces asexually **2**. Some of the resulting offspring, called merozoites, enter red blood cells, where they reproduce asexually to produce more merozoites **3**. Other merozoites enter into red blood cells and develop into immature gametes, or gametocytes **4**.

When a mosquito bites an infected person, it takes up gametocytes along with blood. The gametocytes mature in the mosquito's gut, then fuse to form zygotes **5**. Zygotes develop into new sporozoites that migrate to the insect's salivary glands, where they await transfer to a new vertebrate host **6**.

Malaria symptoms usually start a week or two after a bite, when the infected liver cells rupture and release merozoites and cellular debris into blood. After the initial episode, symptoms may subside, but continued infection damages the body and is eventually fatal. Malaria kills more than a million people each year.

algal bloom Population explosion of single-celled aquatic organisms such as dinoflagellates.

apicomplexan Parasitic protist that enters and lives inside the cells of its host.

bioluminescent Able to use ATP to produce light.

ciliate Single-celled, heterotrophic protist with many cilia.

dinoflagellates Single-celled, aquatic protist typically with cellulose plates and two flagella; may be heterotrophic or photosynthetic.

Chapter 13 Early Life Forms and the Viruses   **255**

> **Water Molds, Diatoms, and Brown Algae** **Water molds** are heterotrophs known to scientists as oomycotes. The term means "egg fungus," and these organisms were once mistakenly grouped with the fungi. Like fungi, the oomycotes form a mesh of nutrient-absorbing filaments, but the two groups differ in many structural traits and are genetically distinct. Most oomycotes help decompose organic debris and dead organisms in aquatic habitats, but a few are parasites that have significant economic effects. Some grow as fuzzy white patches on fish in fish farms and aquariums. Others infect land plants, destroying crops and forests. Members of the genus *Phytophthora* are especially notorious. Their name means "plant destroyer" and they cause an estimated $5 billion in crop losses each year. In the mid-1800s, one species destroyed Irish potato crops, causing a famine that killed and displaced millions of people. Today, another *Phytophthora* species is causing an epidemic of sudden oak death in Oregon, Washington, and California. Millions of oaks have already died.

The closest relatives of the water molds are two photosynthetic groups: diatoms and brown algae (Figure 13.27). Both have chloroplasts that include a brownish accessory pigment (fucoxanthin) that tints them olive green, golden, or dark brown.

Diatoms have a two-part silica shell, with upper and lower parts that fit together like a shoe box. Some cells live individually, and others form chains ❶. Most diatoms float near the surface of seas or lakes, but some live in moist soil or in water droplets that cling to mosses in damp environments.

Marine diatoms have lived and died in the oceans for many millions of years, and their remains form vast deposits on the sea floor. In places where these deposits have been uplifted onto land, silica-rich diatomaceous earth is quarried for use in filters, abrasive cleaners, and as an insecticide that is not harmful to vertebrates.

FIGURE 13.27 Artist's depiction of two related protist groups common in California's coastal waters, ❶ microscopic diatoms with a silica shell and ❷ a large multicelled kelp (a brown alga).

© Cengage Learning 2010.

Brown algae are multicelled inhabitants of temperate or cool seas. In size, they range from microscopic filaments to giant kelps that stand 30 meters (100 feet) tall. Giant kelps form forestlike stands in coastal waters of the Pacific Northwest ❷. Like trees in a forest, kelps shelter a wide variety of other organisms. The Sargasso Sea in the North Atlantic Ocean is named for its abundance of *Sargassum*. This kelp forms vast, floating mats up to 9 meters (30 feet) thick. The mats provide food and shelter to fish, sea turtles, and invertebrates.

Sargassum and other brown algae have commercial uses. Alginic acid from the cell walls of brown algae is used to produce algins, which serve as thickeners, emulsifiers, and suspension agents. Algins are used to manufacture ice cream, pudding, jelly beans, toothpaste, cosmetics, and other products.

Although some large brown algae have a plantlike form, this similarity is an example of morphological convergence, rather than evidence of shared ancestry. Brown algae evolved from a different single-celled ancestor than the lineage that includes red algae, green algae, and land plants.

> **Red Algae** Some **red algae** are single cells, but most are multicelled forms that live in tropical seas. Most commonly they have a branching structure, but some form thin sheets (Figure 13.28). Coralline algae (red algae with cell walls hardened by calcium carbonate) are a component of tropical coral reefs. Red algae are tinted red to black by accessory pigments called phycobilins. These pigments absorb the blue-green light that penetrates deep into water. Phycobilins allow red algae to carry out photosynthesis at greater depths than other algae.

Red algae have many commercial uses. Nori, the sheets of seaweed used to wrap some sushi, is a red alga that is grown commercially. Agar and carrageenan are valuable products extracted from the cell walls of other red algae. Agar keeps baked goods and cosmetics moist, helps jellies set, and is used to make capsules that hold medicines. Carrageenan is added to soy milk, dairy foods, and the fluid that is sprayed on airplanes to prevent ice formation.

> **Green Algae** **Green algae** include single-celled, colonial, and multi-celled species (Figure 13.29). Most live in fresh water, but some are marine, and some grow on soil, trees, or other damp surfaces. A few single-celled species partner with a fungus to form a lichen.

The single-celled alga *Chlorella* is cultivated in ponds, dried, and sold in powdered or pill form as a nutritional supplement. *Chlorella* is also a promising candidate for biofuel production because it has a high oil content. Many ciliates, including some *Paramecium*, already get an energy boost from *Chlorella*. They feed on sugars produced by *Chlorella* cells that live inside them.

Red algae, green algae, and land plants share a variety of unique traits, including chloroplasts containing a particular type of chlorophyll and a cell wall of cellulose. These similarities are taken as evidence that these three groups share a common ancestor. Both red algae and green algae evolved from an ancestral protist that had chloroplasts descended from cyanobacteria. After the two algal lineages diverged, land plants evolved from one lineage of green algae.

FIGURE 13.28 Branching and sheetlike red algae growing 75 meters (225 ft) beneath the sea surface in the Gulf of Mexico.

Image courtesy of FGB-NMS/UNCW-NURC.

brown alga Multicelled, photosynthetic protist with brown accessory pigments.

diatom Single-celled photosynthetic protist with brown accessory pigments and a two-part silica shell.

green alga Single-celled, colonial, or multicelled photosynthetic protist belonging to the group most closely related to land plants.

red alga Single-celled or multicelled photosynthetic protist with red accessory pigment.

water mold Heterotrophic protist that forms a mesh of nutrient-absorbing filaments.

A Two desmids, a type of single-celled green alga that lives mainly in fresh water.

B *Volvox*, a colonial, freshwater green alga. Each sphere is a colony made up of flagellated cells linked by thin cytoplasmic strands.

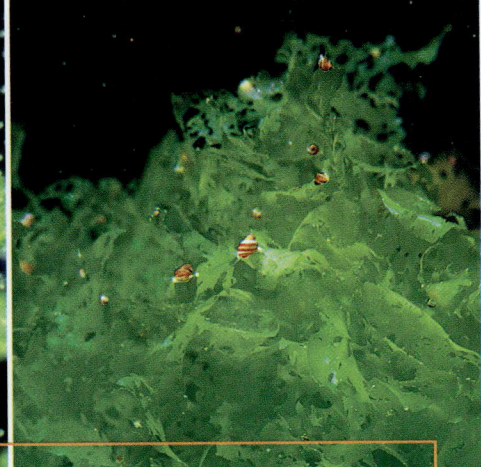

C Sheets of sea lettuce (*Ulva*), a multi-celled marine green alga. Although the sheets can be longer than your arm, they are thinner than a human hair.

FIGURE 13.29 Green algae.

Credits: (a) © Wim van Egmond/ Visuals Unlimited; (b) © Kim Taylor/ Bruce Coleman, Inc. Photoshot; (c) © Lawson Wood/ Corbis.

FIGURE 13.30 An amoeba, an unwalled, shape-shifting cell that moves by extending pseudopods (lobes of cytoplasm).

M I Walker/ Photo Researchers, Inc.

›Amoebas and Slime Molds The free-living amoebas and the slime molds are grouped together as amoebozoans. Members of this group do not have a cell wall, shell, or pellicle, so they continually change shape. A compact blob of a cell can extend lobes of cytoplasm called pseudopods (Section 3.5) to move about and to capture food.

Amoebas such as *Amoeba proteus* (Figure 13.30) always live and feed as solitary cells. Most amoebas are predators in freshwater habitats. Others live inside animals, and some cause human disease. Each year, about 50 million people suffer from amebic dysentery after drinking water contaminated by *Entamoeba histolytica* cysts. Inadequate sterilization of contact lenses or swimming with lenses can result in an eye infection by *Acanthamoeba*, an amoeba common in soil, standing water, and even tap water.

Slime molds are sometimes described as "social amoebas." There are two types, cellular slime molds and plasmodial slime molds. **Cellular slime molds** spend the bulk of their existence as individual haploid amoeboid (amoeba-like) cells. *Dictyostelium discoideum* is an example (Figure 13.31). Each cell eats bacteria and reproduces by mitosis ❶. When food runs out, thousands of cells aggregate to form a multicelled mass ❷. Environmental gradients in light and moisture induce the mass to crawl along as a cohesive unit often referred to as a "slug" ❸. When the slug reaches a suitable spot, its component cells differentiate to form a fruiting body. Some cells become a stalk, and others become spores atop it ❹. When a spore germinates, it releases a cell that starts the life cycle anew ❺.

Plasmodial slime molds spend most of their life cycle as a multinucleated mass called a plasmodium. The plasmodium forms when a diploid cell divides its nucleus repeatedly by mitosis, but does not undergo cytoplasmic division. The resulting mass can be as big as a dinner plate. It streams out along the forest floor feeding on microbes and organic matter (Figure 13.32). When food supplies dwindle, a plasmodium develops into many spore-bearing fruiting bodies.

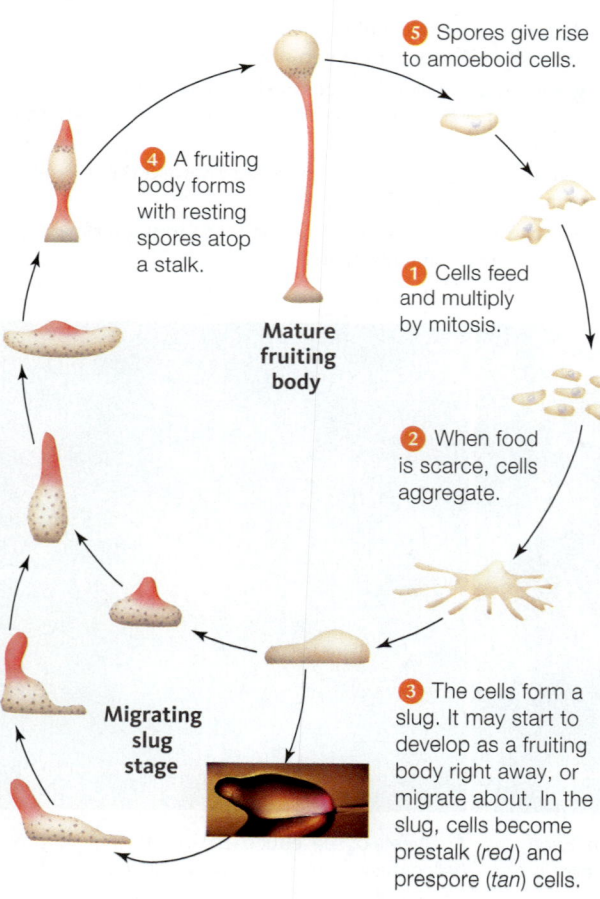

❺ Spores give rise to amoeboid cells.

❹ A fruiting body forms with resting spores atop a stalk.

Mature fruiting body

❶ Cells feed and multiply by mitosis.

❷ When food is scarce, cells aggregate.

❸ The cells form a slug. It may start to develop as a fruiting body right away, or migrate about. In the slug, cells become prestalk (*red*) and prespore (*tan*) cells.

Migrating slug stage

FIGURE 13.31 Animated! Life cycle of the cellular slime mold *Dictyostelium discoideum*.

Credits: Art, © Cengage Learning; Photo, Carolina Biological Supply Company.

FIGURE 13.32 Plasmodial slime mold streaming across a log.

Edward S. Ross.

›Choanoflagellates The **choanoflagellates** are the protist group with genes most similar to those of animals. Choanoflagellates also look a lot like the feeding cells of sponges, which are among the simplest animals. As a result, choanoflagellates are thought to be the closest living relatives of animals. Note that they are not considered ancestors of animals, but rather a group that shared a common single-celled ancestor with animals long ago.

Choanoflagellate means "collared flagellate" and it aptly describes this group. Each cell has a long flagellum surrounded by a ring (or collar) of tiny filaments reinforced with the protein actin (Figure 13.33). Movement of the flagellum creates a current that draws water through the filaments. After tiny bits of food become entrapped, the cell extends pseudopods to capture them. The food is then digested within the cell body.

Most choanoflagellates live as single cells, but some form colonies. The colonies arise when cells divide, then descendant cells stick together with the help of adhesion proteins. Choanoflagellate adhesion proteins are similar to those found in animals, and researchers have discovered that even solitary choanoflagellates have them. By one hypothesis, the common ancestor of animals and choanoflagellates was a single-celled protist with adhesion proteins that helped it capture prey. Later, these proteins were put to use in a new context, helping choanoflagellates stick together to form multicelled colonies. Later still, the proteins allowed animal cells to adhere to one another in multicelled bodies. This modification in the use of adhesion proteins ia an example of exaptation, an evolutionary process by which a trait that evolves with one function later takes on a different function (Section 12.6).

FIGURE 13.33 Drawing of a solitary choanoflagellate, the modern protist group most closely related to animals. © Cengage Learning.

Take-Home Message

What are the protists?

- The protists are a diverse collection of eukaryotic lineages, some of which are only distantly related to one another.

- Most protists live as single cells, but there are colonial and multicelled species.

- We find protists that are photosynthesizers, predators, and decomposers in lakes, seas, and damp places on land. Protists also live inside the bodies of other eukaryotes, including humans. Some of these protists are helpful, but others are parasites and pathogens.

- Green algae are the closest protist relatives of land plants, and choanoflagellates are the closest protist relatives of animals.

amoeba Solitary heterotrophic protist that feeds and moves by extending pseudopods.

cellular slime mold Heterotrophic protist that usually lives as a single-celled, amoeba-like predator. When conditions are unfavorable, cells aggregate into a cohesive group that can form a fruiting body.

choanoflagellates Heterotrophic protists with a collared flagellum; protist group most closely related to animals.

plasmodial slime mold Heterotrophic protist that moves and feeds as a multinucleated mass; forms a fruiting body when conditions are unfavorable.

13.7 Evolution and Disease (revisited)

A virus cannot move itself, so it cannot seek out its host. However, some viruses alter a host's behavior in a way that increases the chance that viral particles will encounter a new host. For example, a viral infection may cause sneezing, which distributes viral particles into the air, from which they can be inhaled.

Other pathogens increase the odds of successful transfer by manipulating their host. For example, a dog infected by the rabies virus accumulates viral particles in its saliva. The virus also affects areas of the dog's brain, causing the dog to become less fearful and more aggressive. The changes in behavior increase the odds that the dog will bite another animal, thus passing along the rabies-causing virus.

Plasmodium, the protist that causes malaria, alters the feeding behavior of the mosquitoes that carry it in a way that helps spread the disease. A mosquito that is carrying sporozoites (the infectious form of the protist) bites a greater number of people per night than a mosquito without sporozoites.

WHERE YOU ARE GOING . . .

The next chapter discusses the evolution of plants and fungi from protist ancestors, and Section 15.2 returns to the relationship between choanoflagellates and animals. The role of bacteria in the nitrogen cycle and in plant nutrition is covered in Sections 17.6 and 27.5. We return to the subject of the human normal flora in Section 22.3, and to human gut bacteria in particular in Section 23.3. Immune effects of HIV infection are discussed in Section 22.7, and Section 26.5 considers the sexual transmission of this and other pathogens.

Summary

Section 13.1 We share Earth with countless microorganisms, some of which are **pathogens**. Like their larger counterparts, microorganisms evolve in response to natural selection and their traits adapt them to their environments.

Section 13.2 Laboratory simulations provide indirect evidence that organic subunits can self-assemble under certain conditions. They also show how complex organic compounds and **protocells** may have formed on the early Earth. The **iron–sulfur world hypothesis** holds that these events took place near **hydrothermal vents**. The **RNA world hypothesis** holds that the first genome was RNA-based.

Section 13.3 Fossil **stromatolites** are evidence of early bacterial life. An early branching separated the bacteria from the archaea. Production of oxygen by some photosynthetic bacteria altered Earth's atmosphere and allowed formation of a protective **ozone layer**. Eukaryotes have a composite ancestry with both bacterial and archaeal genes. Mitochondria and chloroplasts are thought to have evolved from bacteria by **endosymbiosis**.

Section 13.4 **Viruses** consist of RNA or DNA inside a protein coat. Some also have a **viral envelope**. Viruses replicate only in living cells, as when **bacteriophages** multiply in bacteria. Mutation and **viral reassortment** produce new types of viruses. Insects serve as **disease vectors** that spread some viral diseases of plants. **HIV** infects human cells. It began as an **epidemic** and is now a **pandemic**.

Section 13.5 **Archaea** and **bacteria** were historically lumped together as "prokaryotes." They lack a nucleus, reproduce asexually by **binary fission**, and swap **plasmids** by **conjugation**.

Bacteria and archaea may be aerobic or anaerobic, and they show great metabolic diversity. **Autotrophs** use carbon dioxide as their carbon source. They include photoautotrophs such as cyanobacteria and chemoautotrophs such as the archaea that are **methanogens**. By contrast, **heterotrophs** obtain carbon from organic compounds. Chemoheterotrophs that serve decomposers and pathogens obtain energy by breaking down organic compounds.

Bacteria serve as **decomposers**, release oxygen into the air, and carry out **nitrogen fixation**. Many are part of our **normal flora**, but others are human pathogens.

Archaea include heat-loving **extreme thermophiles** or salt-loving **extreme halophiles**. Others live in less extreme environments.

Section 13.6 The "protist kingdom" is being reorganized to reflect new understanding of evolutionary relationships. Most **protists** are single-celled, but some lineages include multicelled species. Nearly all live in water or moist habitats, including host tissues.

Flagellated protozoans are single cells with a translucent **pellicle** that constrains their shape. Most are heterotrophs, although some euglenoids have chloroplasts that evolved from green algae. Freshwater species use a **contractile vacuole** to expel excess water.

Foraminifera are single-celled heterotrophs with calcium carbonate shells. Most live on the seafloor, but some drift as marine **plankton**. Shells of foraminifera contribute to limestone and chalk.

Ciliates, dinoflagellates, and apicomplexans are close relatives. **Ciliates** are single-celled heterotrophs that use cilia to move and feed. **Dinoflagellates** are single cells that move with a whirling motion and can be heterotrophs or photosynthetic. Some are **bioluminescent** and others cause **algal blooms**. **Apicomplexans**, such as the species that cause malaria, are parasites that spend part of their life in cells of their host.

Water molds are heterotrophs that grow as filaments. Some are pathogens of fish or plants. **Diatoms** are single-celled, silica-shelled, aquatic producers. **Brown algae** include the giant kelps, which are the largest protists.

Red algae have accessory pigments called phycobilins that allow them to live at greater depths than other algae. Red algae share a common ancestor with the green algae. The land plants evolved from one lineage of **green algae**.

Amoebas are shape-shifting cells that extend pseudopods to feed and move. They live in aquatic habitats and animal bodies. The related **cellular slime molds** spend part of their life as a single cell and part in a cohesive group that can migrate and differentiate to form a spore-bearing body. **Plasmodial slime molds** feed as a giant multinucleated mass, then form spores when food runs out. The **choanoflagellates** are the protists most closely related to animals.

Self-Quiz   Answers in Appendix I

1. An increase in _____ in the atmosphere allowed the formation of an ozone layer that protects against UV radiation.

2. Stanley Miller's experiment demonstrated _____ .
   a. the great age of Earth
   b. that amino acids can assemble under some conditions
   c. that oxygen is necessary for life
   d. all of the above

3. The evolution of a new type of _____ resulted in an increase in the amount of oxygen in the atmosphere.
   a. binary fission   c. aerobic respiration
   b. sexual reproduction   d. photosynthesis

4. Mitochondria are most likely descendants of _____ .
   a. archaea   c. cyanobacteria
   b. aerobic bacteria   d. anaerobic bacteria

5. Bacteria transfer genes among themselves by _____ .
   a. binary fission   c. conjugation
   b. the lytic pathway   d. mitosis

6. The first eukaryotes were _____ .
   a. bacteria   b. protists   c. fungi   d. animals

7. True or false? Some protists are more related to plants than to other protists.

8. What protist group is most closely related to animals?

Digging Into Data

How *Plasmodium* Summons Mosquitoes

Dr. Jacob Koella and his associates hypothesized that *Plasmodium* might benefit by making its human host more attractive to hungry mosquitoes when gametocytes were available in the host's blood. Gametocytes can be taken up by the mosquito and mature into gametes inside its gut (Figure 13.26).

To test their hypothesis, the researchers recorded the response of mosquitoes to the odor of *Plasmodium*-infected children and uninfected children over the course of 12 trials on 12 separate days. Figure 13.34 shows their results.

1. On average, which group of children was the most attractive to mosquitoes?

2. Which group of children attracted the fewest mosquitoes on average?

3. What percentage of the total number of mosquitoes were attracted to the most attractive group?

4. Did the data support Dr. Koella's hypothesis?

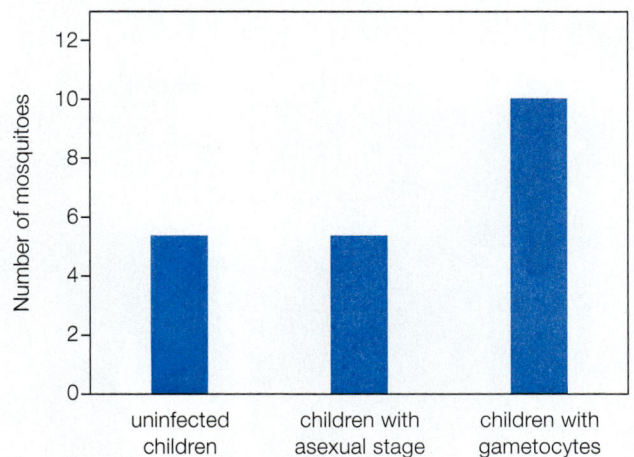

FIGURE 13.34 Number of mosquitoes (out of 100) attracted to uninfected children, children with the asexual stage of *Plasmodium*, and children with gametocytes in their blood. The bars show the average number of mosquitoes attracted to that category of child over the course of 12 separate trials.

© Cengage Learning.

9. The _____ are parasitic eukaryotes that live in other cells.
   a. viruses
   c. euglenoids
   e. both a and b
   b. apicomplexans
   d. slime molds
   f. all are correct

10. Remains of _____ form chalk and limestone deposits.
   a. ciliates
   c. foraminifera
   b. diatoms
   d. dinoflagellates

11. Some of the _____ are human pathogens.
   a. slime molds
   c. flagellated protozoans
   b. archaea
   d. both a and c

12. Plants descended from a lineage of _____ algae.
   a. red
   b. brown
   c. green

13. Silica-rich remains of _____ are used as an insecticide.
   a. dinoflagellates
   c. foraminiferans
   b. diatoms
   d. choanoflagellates

14. The genetic material of a _____ may be DNA or RNA.
   a. bacteria
   c. ciliate
   b. dinoflagellate
   d. virus

15. Match these terms with the appropriate definition.
   _____ green algae
   a. protist population explosion
   _____ virus
   b. social amoeba
   _____ bacteria
   c. most diverse prokaryotes
   _____ brown algae
   d. noncellular infectious agent
   _____ endospore
   e. include the largest protists
   _____ euglenoid
   f. flagellate with chloroplasts
   _____ algal bloom
   g. closest relative of plants
   _____ dinoflagellate
   h. layered prokaryotes and sediment
   _____ slime mold
   i. resistant resting stage
   _____ stromatolite
   j. whirling cell

Critical Thinking

1. Researchers looking for fossils of the earliest life forms face many hurdles. For example, few sedimentary rocks date back more than 3 billion years. Review what you learned about plate tectonics (Section 11.5). Explain why so few remaining samples of these early rocks remain.

2. Craig Venter has produced what he calls a "synthetic cell." He inserted a laboratory-made copy of the genetic material from one bacterial species (*Mycoplasma mycoides*) into a cell of a related species (*Mycoplasma capricolum*). The introduced genome took over and the recipient cell acted and replicated like natural *M. mycoides*. By altering the gene sequence inserted into a cell, this technique could be used to create organisms suited to specific tasks. Do you consider an organism created in this way to be a new life form? What properties would you look at to make this decision?

3. Viruses that do not have a lipid envelope tend to remain infectious outside the body longer than enveloped viruses. "Naked" viruses are also less likely to be rendered harmless by soap and water. Can you explain why?

4. Diatoms are the main producers in polar seas. Like other producers, they require dissolved nitrogen and phosphorus to grow. Unlike most producers, their reproduction can be limited by the lack of dissolved silica. Explain why a lack of silica would interfere with diatom reproduction.

5. Many bacteria and most archaea are anaerobic, whereas most eukaryotes require oxygen. What are the advantages of relying solely on aerobic respiration? What are the disadvantages?

Plants and Fungi

14

. . . WHERE YOU HAVE BEEN

This chapter continues the story of how plants evolved from green algae (Section 13.6). We return to the topic of plant life cycles, first introduced in Section 8.7. We also reconsider the yeasts, fungi whose role in making bread and alcoholic drinks was discussed in Section 5.7.

14.1 Fungal Threats to Crops

Plants are the main producers on land and thus serve as an important food source for both animals and fungi. As a result, we sometimes find ourselves in competition with fungi for plant foods. For example, wheat is both an important agricultural crop and the host for wheat stem rust fungus (*Puccinia graminis*). Wheat stem rust fungus is an obligate plant parasite, meaning it can grow and reproduce only in living plant tissue. Fungi disperse by releasing microscopic spores that travel on the wind. A wheat stem rust infection begins when a fungal spore lands on the leaf of a wheat plant. The spore germinates (becomes active), and a fungal filament grows into the plant. As fungal filaments extend through the plant's tissues, they take up photosynthetic sugars that the plant would normally use to meet its own needs. As a result, an affected plant is stunted and produces little or no wheat. About a week after infection, ten of thousands of rust-colored spore sacs appear on the infected plant's stem (Figure 14.1**A**,**B**). Each spore can disperse and infect a new plant.

Outbreaks of wheat stem rust disease routinely destroyed crops worldwide until the 1960s, when a plant breeding program headed by Norman Borlaug produced resistant strains of wheat. In 1970, Borlaug received the Nobel Peace Prize for his role in preventing food shortages that contribute to global instability. Planting of rust-resistant wheats provided a respite from outbreaks of wheat stem rust for decades. Then, in 1999, scientists discovered a new strain of wheat stem rust in Uganda. This strain, Ug99, has mutations that allow it to infect most wheat varieties that were previously resistant to wheat stem rust.

Ug99 is now spreading. Microscopic fungal spores can lodge in crevices on dust particles. When winds lift these particles aloft, spores go along for the ride. The dustborne fungal spores can disperse long distances riding winds that swirl high above Earth's surface. By 2009, windblown spores of Ug99 had reached Kenya, Ethiopia, and Sudan, crossed the Red Sea to Yemen, and from there crossed the Persian Gulf to Iran. India, the world's second largest wheat producer, is expected to be affected soon. Most likely, winds will eventually distribute Ug99 throughout the world. Fungicides can minimize the damage, but are too expensive for farmers in developing nations.

With one of the world's most important food crops at risk, international teams of plant scientists quickly set to work (Figure 14.1**C**). The scientists' goal is to develop new varieties of wheat with genes that provide resistance against Ug99. Scientists have already bred some resistant wheats. Now the challenge is multiplying enough seed and distributing it to farmers who currently grow susceptible varieties. This task is complicated by the fact that wheat is not of commercial interest to private seed companies, because farmers save their own seed and resow it. Thus, there is no central source to distribute resistant strains.

A Wheat stem covered with fungal spore chambers.

B Micrograph of spores escaping into the air.

C Plant breeders Peter Njau, Godwin Macharia, and Davinder Singh (*left* to *right*) are part of an international team of scientists racing to develop and field test Ug99-resistant strains of wheat.

FIGURE 14.1 Wheat stem rust fungus, a current threat to world food supplies.

14.2 Plant Traits and Evolution

Plants are land-dwelling, multicelled photosynthetic eukaryotes. They evolved from a lineage of freshwater green algae (the charophyte algae), and share many traits with this group. Figure 14.2 shows one of the modern charophyte algae thought to be a close relative of land plants. The defining trait that sets plants apart from algae is a multicelled embryo that forms, develops within, and is nourished by tissues of the parental plant.

The story of plant evolution is a tale of adaptation to increasingly drier environments. A multicelled green alga lives surrounded by water, so it can absorb water and dissolved nutrients across its entire body surface. Water also buoys its parts, thus helping it stand upright. Land plants, however, face the constant threat of drying out and must hold themselves upright.

FIGURE 14.2 *Coleochaete orbicularis*, a freshwater charophyte alga that is closely related to land plants. It grows as multicellular disk attached to a surface. © Dr. Ralf Wagner, www.dr-ralf-wagner.de.

FIGURE 14.3 Animated! Alternation of generations in the plant life cycle. © Cengage Learning.

> **Life Cycles** In the animal life cycle, a multicelled diploid individual produces haploid cells (eggs or sperm) that combine to produce a new diploid individual. In contrast, plants have two multicelled stages in their life cycle, which is described as an alternation of generations (Figure 14.3). The diploid generation, the **sporophyte**, produces spores by meiosis ❶. A plant spore is a nonmotile cell that undergoes mitosis and develops into the multicelled, haploid generation—a **gametophyte** ❷ that produces gametes by mitosis ❸. Gametes unite at fertilization to form a zygote ❹ that develops into a new sporophyte ❺.

The relative size, complexity, and longevity of the sporophyte and gametophyte vary among the land plants. In the oldest surviving plant lineages, collectively referred to as nonvascular plants or **bryophytes**, the haploid gametophyte is larger and longer-lived than the diploid sporophyte. In all other plants, sporophytes dominate the life cycle and gametophytes are tiny.

> **Structural Adaptations to Life on Land** Most bryophytes and all other plants have a waxy **cuticle**, a secreted layer that lessens evaporative water loss from the body surface. Adjustable pores called **stomata** (singular, stoma) extend across the cuticle. Depending on environmental conditions, these pores either open to allow gas exchange or close to conserve water.

Bryophytes have threadlike structures that hold them in place, but they do not have true roots. Such roots not only anchor plants, they also take up water with dissolved mineral ions from the soil. Plants with true roots have a vascular system with two tissues that serve as internal pipelines. **Xylem** is the vascular tissue that carries water and mineral ions. **Phloem** is the vascular tissue that distributes the sugars produced by photosynthetic cells. More than 90 percent of modern plants have xylem and phloem and are known as **vascular plants**.

Vascular tissues also provide structural support. An organic compound called **lignin** stiffens the walls of xylem. Evolution of lignin-stiffened tissue allowed

bryophyte Member of an early evolving plant lineage that does not have vascular tissue; for example, a moss.

cuticle Secreted covering at a body surface.

gametophyte Haploid gamete-forming body that forms in a plant life cycle.

lignin Compound that stiffens walls of some cells (including xylem) in vascular plants.

phloem Vascular tissue that distributes dissolved sugars.

plant Multicelled, photosynthetic organism; develops from an embryo that forms on the parent and is nourished by it.

sporophyte Diploid spore-forming body that forms in a plant life cycle.

stomata Adjustable pores in a plant cuticle.

vascular plant A plant that has xylem and phloem.

xylem Vascular tissue that distributes water and dissolved mineral ions.

FIGURE 14.4 Animated! Cross-section of a leaf, showing some traits that adapt plants to land. © Cengage Learning.

vascular tissue (a leaf vein)

layer of waxy cuticle

xylem

phloem

cuticle

stoma

vascular plants to stand taller than nonvascular ones and to branch. Leaves, organs that enhance photosynthesis and gas exchange, also give many vascular plants a competitive edge. Figure 14.4 shows a leaf of a vascular plant.

> **Reproduction and Dispersal** All bryophytes, and some vascular plants such as ferns, produce flagellated sperm that swim to eggs. The more recently evolved seed-bearing vascular plants, or seed plants, make pollen grains. A **pollen grain** is a walled, immature male gametophyte. Winds or animals typically carry pollen grains to a female gametophyte.

Bryophytes and seedless vascular plants release spores, but seed-bearing plants protect spores within their tissues and release seeds. A plant spore is a single haploid cell with a thick wall. By contrast, a **seed** consists of an embryo sporophyte and food to support it, enclosed within a protective coat.

There are two kinds of seed plants, gymnosperms and angiosperms. Angiosperms are the only plants that make flowers and disperse seeds inside of fruits. Figure 14.5 summarizes relationships among plant groups.

Take-Home Message

What adaptive traits allow plants to live on land and in dry places?

- Plants are multicelled, typically photosynthetic organisms that protect and nourish their multicelled embryos. They evolved from a lineage of green algae.

- Adaptations to life on land include a waxy cuticle with stomata, true roots, and vascular tissues that distribute materials and provide structural support.

- The plant life cycle alternates between two multicelled generations: a haploid gametophyte generation and a diploid sporophyte generation.

- The most recently evolved plant lineages produce pollen grains and seeds. These adaptations allowed the seed plants to spread into diverse habitats.

pollen grain Male gametophyte of a seed plant.
seed Embryo sporophyte of a seed-bearing plant packaged with nutritive tissue inside a protective coat.

FIGURE 14.5 Animated! Proposed relationships among major groups of plants. Relationships among the nonvascular plants remain under investigation. © Cengage Learning.

Nonvascular Plants (Bryophytes): liverworts, mosses, hornworts

Seedless Vascular Plants: club mosses, ferns, horsetails

Gymnosperms

Angiosperms (Flowering plants)

flowers, fruits

pollen, seeds

sporophyte dominance, vascular tissue

protected embryo

ancestral green alga

14.3 Nonvascular Plants

Modern bryophytes (nonvascular plants) include 24,000 species belonging to three lineages: mosses, hornworts, and liverworts. Some mosses have tubes that conduct water and sugar, but no bryophyte has the lignin-stiffened vascular pipelines that define vascular plants. As a result, few bryophytes stand more than 20 centimeters (8 inches) tall.

> **Mosses** **Mosses** are the most diverse and familiar nonvascular plants. We will use the life cycle of the moss *Polytrichum* to illustrate a bryophyte life cycle (Figure 14.6). Like all bryophytes, this moss has a gametophyte-dominated life cycle. A moss gametophyte has leaflike green parts that grow from a central stalk **1**. Threadlike structures (rhizoids) hold the gametophyte in place.

The moss sporophyte consists of a stalk with a spore-producing chamber (a sporangium) at its tip **2**. The sporophyte is not photosynthetic and so depends on the gametophyte to which it is attached for nourishment. Meiosis of cells inside a spore chamber yields haploid spores **3**. After dispersal by the wind, a spore germinates and grows into a gametophyte. Multicellular gamete-producing structures (gametangia) develop in or on the gametophyte. The moss we are using as our example has separate sexes, with each gametophyte producing eggs or sperm **4**. Other bryophytes are bisexual. In either case, rain triggers the release of flagellated sperm that swim through a film of water to eggs **5**. Some moss gametangia facilitate sperm movement by attracting mites and crawling insects that unknowingly pick up sperm and move them to adjacent plants. Fertilization inside the egg chamber produces a zygote **6** that develops into a new sporophyte **7**. Mosses also reproduce asexually by fragmentation when a bit of gametophyte breaks off and develops into a new plant.

Mosses include 14,000 or so species. Among these, 350 or so species of peat mosses (*Sphagnum*) are of great ecological and commercial importance. Peat mosses are the main plants in peat bogs that cover more than 350 million acres in high-latitude regions of Europe, Asia, and North America. Freshly harvested peat moss is an important commercial product. It is dried and added to planting mixes to help soil retain moisture.

Many peat bogs have existed for thousands of years, and layer upon layer of partially decayed plant remains have become compressed to form deposits of a carbon-rich material called peat. Blocks of peat are cut, dried, and used as a clean-burning fuel, especially in Ireland. Occasionally, people cutting peat for fuel come across well-preserved human remains. The anaerobic and highly acidic conditions in peat bogs slow decay, so the bodies sometimes retain intact clothing, skin, and hair even after hundreds of years.

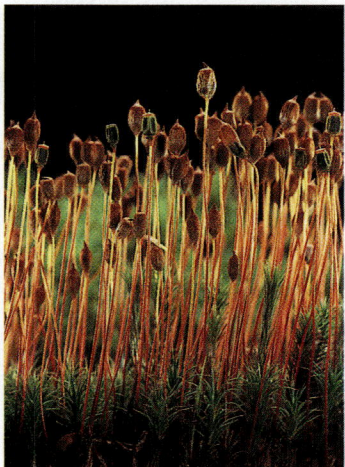

FIGURE 14.6 Animated! Life cycle of the moss *Polytrichum*, shown at *left*.

1 The leafy green part of a moss is the haploid gametophyte.

2 The diploid sporophyte has a stalk and a capsule (sporangium). It is not photosynthetic.

3 Haploid spores form by meiosis in the capsule, are released, and drift with the winds.

4 Spores germinate and develop into male or female gametophytes with gametangia that produce eggs or sperm by mitosis.

5 Sperm swim to eggs.

6 Fertilization occurs in the egg chamber on the female gametophyte and produces a zygote.

7 The zygote grows and develops into a sporophyte while remaining attached to and nourished by its female parent.

Credits: top, © Cengage Learning; bottom, Jane Burton/ Bruce Coleman Ltd.

mosses Most diverse group of bryophytes (nonvascular plants). Low-growing plants with flagellated sperm disperse by producing spores.

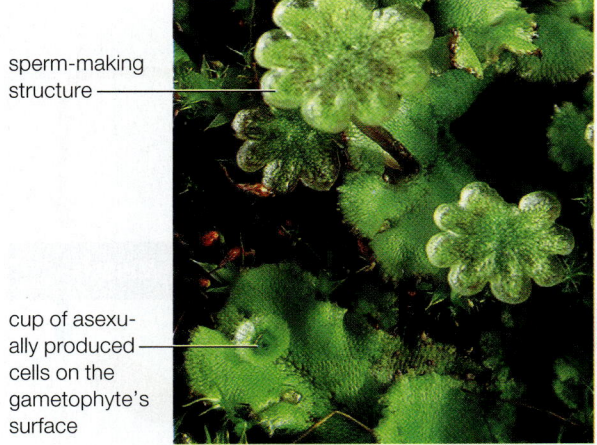

sperm-making structure

cup of asexually produced cells on the gametophyte's surface

A Male gametophyte with structures for both sexual and asexual reproduction.

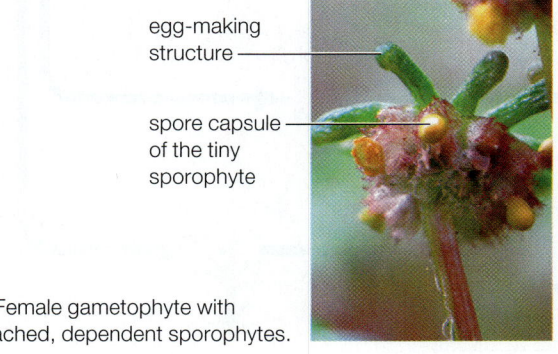

egg-making structure

spore capsule of the tiny sporophyte

B Female gametophyte with attached, dependent sporophytes.

FIGURE 14.7 Liverwort reproduction. The umbrella liverwort (*Marchantia polymorpha*) reproduces asexually by dispersing cell clumps that form in cups on the gametophyte surface. It reproduces sexually by producing gametes on umbrella-like gametangia (gamete-producing structures).

Figure It Out: Do spores of liverworts develop into sporophytes or gametophytes?

Answer: All plant spores develop into gametophytes.

Credits: (a) © Todd Boland/ Shutterstock; (b) © Dr. Annkatrin Rose, Appalachian State University.

› Liverworts and Hornworts Liverworts and hornworts are less familiar nonvascular lineages that often live beside mosses in damp places.

Spores that date to 470 million years ago and resemble those of modern liverworts provide the earliest evidence of land plants. These fossils, together with evidence from genetic comparisons, suggest that liverworts are the oldest surviving land plant lineage. The umbrella liverwort (*Marchantia polymorpha*) is a modern species common worldwide. Its flat, ground-hugging gametophytes can reproduce asexually by producing clusters of cells in cups that form on its surface (Figure 14.7**A**). The splashdown of raindrops disperses the cell clusters, which develop into new gametophytes. When conditions favor sexual reproduction, umbrella-shaped structures form on the gametophyte surface and produce eggs or sperm. Sexes are separate, so sperm must swim from the male gametophyte that produced them to a female gametophyte. After an egg is fertilized, it develops into a tiny nonphotosynthetic sporophyte that remains attached to the umbrella-like portion of a female gametophyte (Figure 14.7**B**).

Hornworts have flattened, ribbonlike gametophytes (Figure 14.8). Fertilization of eggs that form on the gametophyte body produces a zygote that develops into a tall, horn-shaped sporophyte. Hornwort sporophytes, unlike those of mosses and liverworts, contain chloroplasts and grow continually. They can be several centimeters tall. These sporophyte traits, together with evidence from gene comparisons, indicate that hornworts are probably the closest living nonvascular relatives of the vascular plants.

sporophyte

gametophyte

FIGURE 14.8 A hornwort.

© University of Wisconsin–Madison, Department of Biology, Anthoceros CD.

Take-Home Message

What are bryophytes?

- Bryophytes include three lineages of low-growing plants (mosses, hornworts, and liverworts). All have flagellated sperm and disperse by releasing spores.

- Bryophytes are the only modern plants in which the gametophyte dominates the life cycle and the sporophyte is dependent upon it.

- Liverworts are probably the oldest plant lineage. Hornworts are most likely the closest living relatives of vascular plants.

14.4 Seedless Vascular Plants

Seedless vascular plants include ferns, club mosses, and horsetails. Like bryophytes, these plants have flagellated sperm that require a film of water to swim to eggs. Also like bryophytes, they disperse by releasing spores.

Seedless vascular plants differ from bryophytes in other aspects of their life cycle and structure. The gametophyte is reduced in size and is relatively short-lived. Although the sporophyte develops on the gametophyte body, it can live independently after the gametophyte dies. Lignin stiffens a sporophyte's body, and a system of vascular tissue distributes water, sugars, and minerals through it. These innovations in support and plumbing allow seedless vascular sporophytes to be large and structurally complex, with roots, stems, and leaves.

epiphyte Plant that grows on the trunk or branches of another plant but does not harm it.

ferns Most diverse lineage of seedless vascular plants.

rhizome Stem that grows horizontally along or just below the ground.

sorus Cluster of spore-forming chambers on a fern frond.

FIGURE 14.9 Animated!

Life cycle of a common North American fern (*Woodwardia*).

1. The familiar leafy form is the diploid sporophyte.

2. Meiosis in cells on the underside of fronds produces haploid spores.

3. After their release, the spores germinate and grow into tiny gametophytes that produce eggs and sperm.

4. Sperm swim to eggs and fertilize them, forming a zygote.

5. The sporophyte begins its development attached to the gametophyte, but it continues to grow and live independently after the gametophyte dies.

Credits: art, © Cengage Learning; photos, A. & E. Bomford/ Ardea, London.

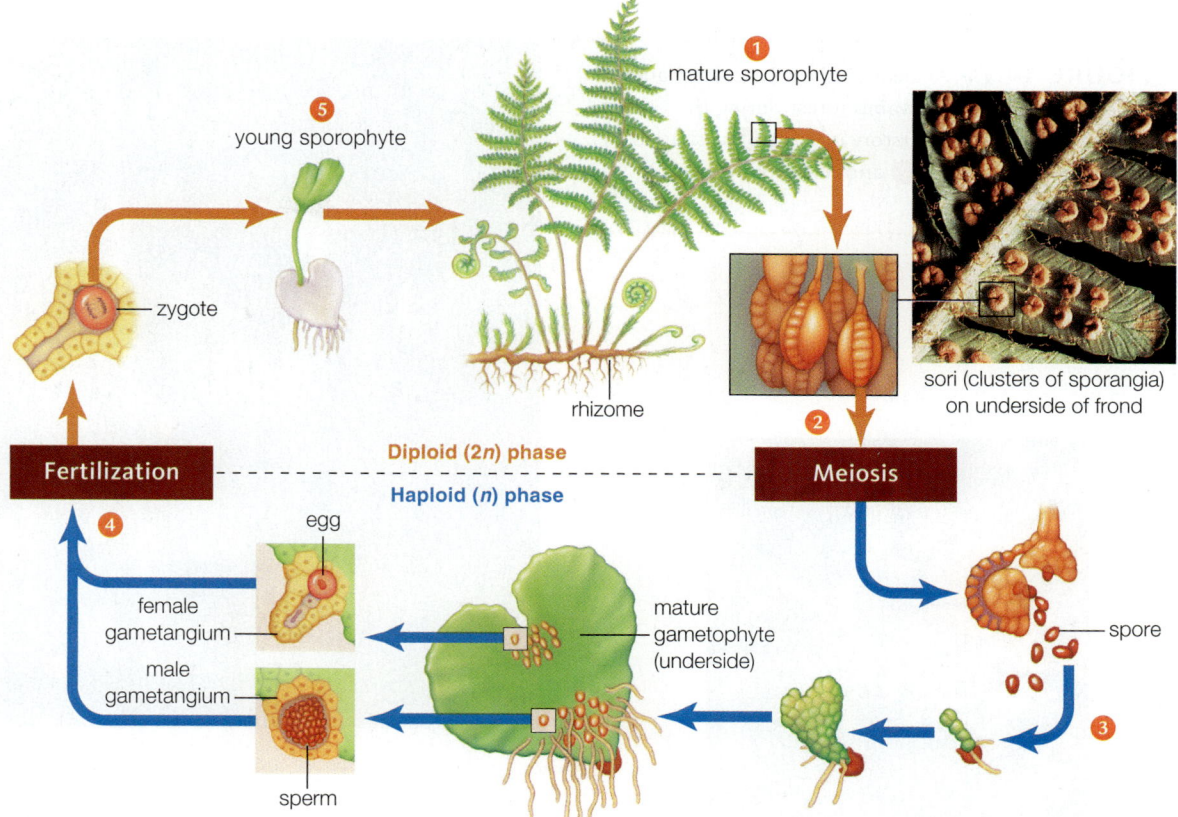

mature sporophyte

young sporophyte

zygote

rhizome

sori (clusters of sporangia) on underside of frond

Fertilization

Diploid (2n) phase

Haploid (n) phase

Meiosis

egg

female gametangium

male gametangium

mature gametophyte (underside)

spore

sperm

> **Ferns** We begin our survey of the seedless vascular plants with the most diverse and familiar lineage, the **ferns**. Figure 14.9 shows the life cycle of a typical fern. The leafy plant we envision when we think of a fern is a sporophyte 1. In most ferns, roots and fronds (leaves) sprout from **rhizomes**, stems that grow along or just below the ground. Spores form by meiosis in capsules that cluster as **sori** (singular, sorus) on the underside of leaves 2. When the capsule pops open, wind disperses the spores. A spore develops into a gametophyte a few centimeters across that forms eggs and sperm in chambers on its underside 3. Sperm must swim to eggs to fertilize them 4. The resulting zygote develops into a new sporophyte, and its parental gametophyte dies 5.

In many ferns, asexual reproduction occurs more frequently than sexual reproduction. New shoots and roots develop from a rhizome as it grows through the soil. Then the connection to the parent plant breaks, and that segment of rhizome becomes an independent plant.

Fern sporophytes vary enormously in their size and form. Some float on freshwater ponds and have fronds less than 1 centimeter wide (Figure 14.10**A**). Many tropical ferns are **epiphytes**, plants that live attached to the trunk or branches of another plant but do not withdraw nutrients from it (Figure 14.10**B**). The largest ferns are tree ferns that can be 25 meters (80 feet) tall.

A The floating fern *Azolla pinnata* is not as wide as a finger. Chambers in its fronds shelter nitrogen-fixing cyanobacteria. Southeast Asian farmers grow this species in rice fields as a natural alternative to synthetic nitrogen fertilizers.

© S. Navie.

B Bird's nest fern (*Asplenium nidus*), one of the epiphytes. It has undivided fronds that can be more than a meter in length.

David C. Clegg/ Photo Researchers, Inc.

FIGURE 14.10 Variations in fern form.

FIGURE 14.11 Ancient seedless vascular plants. An artist's depiction of a swamp forest during the Carboniferous period. An understory of ferns ❶ is shaded by tree-sized club mosses ❷ and horsetails ❸.
© Cengage Learning 2010.

FIGURE 14.12 Modern seedless vascular plants.

A Club moss (*Lycopodium*), about 20 centimeters (8 inches) tall. Spores form in the cone-shaped structures at branch tips.

B Horsetail (*Equisetum*) stems. Most modern species stand less than one meter high.

❯ Club Mosses, Horsetails, and the Coal Forests

Ferns are currently the most diverse seedless vascular plants, but relatives of the modern club mosses and horsetails were the dominant plants in swamp forests during the Carboniferous period from 360 to 300 million years ago (Figure 14.11). The remains of these lush forests became compacted and were transformed over time to coal. When we burn coal, we are releasing the energy of sunlight captured by these plants hundreds of millions of years ago.

Modern club mosses and horsetails are far smaller than their Carboniferous counterparts. Club mosses are common on the floor of temperate forests (Figure 14.12**A**). Their branching form makes them look like tiny pine trees and they are often collected for use in holiday wreaths. Spores form in a conelike reproductive structure or at the tips of branches. The spores have a waxy coating that makes them ignite easily. They were used to create flashes for early photography and to produce special effects in theaters.

Horsetails thrive along streams and roadsides, and in disturbed areas (Figure 14.12**B**). Their stems contain silica, a gritty mineral that helps these plants resist insect and snail predators. Before the invention of modern abrasive cleansers, the silica-rich stems of some *Equisetum* species were used to scrub pots and polish metals. These species are commonly referred to as scouring rushes.

Take-Home Message

What are seedless vascular plants?

- Seedless vascular plants include ferns, club mosses, and horsetails.

- These plants disperse by releasing spores, and the life cycle is dominated by a sporophyte that has vascular tissue and lignin. The gametophyte is small and relatively short-lived.

- Like bryophytes, seedless vascular plants have flagellated sperm that must swim through a film of water to reach eggs.

14.5 Rise of the Seed Plants

Seed plants evolved from a lineage of seedless vascular plants about 400 million years ago. They survived alongside bryophytes and seedless nonvascular plants until the late Carboniferous, then rose to dominance as the climate became drier. Unique traits of seed plants give them a competitive advantage over seedless plants in places where water is scarce.

Gametophytes of seedless vascular plants are free-living; they develop from spores released into the environment. By contrast, gametophytes of seed plants develop within the protection of spore-forming chambers (pollen sacs and ovules) that form on a sporophyte body (Figure 14.13).

Meiosis of a cell in a **pollen sac** produces four microspores ❶. **Microspores** develop into sperm-producing male gametophytes (pollen grains) ❷.

In **ovules**, meiosis and unequal cytoplasmic division produce three small cells that die and one large **megaspore** ❸. The megaspore develops into an egg-producing female gametophyte ❹.

A seed plant releases pollen grains, but holds onto its eggs. Wind or animals can deliver pollen from one seed plant to the ovule of another, a process known as **pollination** ❺. Because the sperm of seed plants do not need to swim through a film of water to reach eggs, these plants can reproduce even in very dry environments.

After pollination, a pollen tube grows into the ovule and delivers a nonmotile sperm to the egg ❻. Fertilization produces a zygote within the ovule, which matures into a seed ❼. Releasing seeds puts seed plants at an advantage over plants that release spores. A seed contains a multicelled embryo sporophyte and stored food that the embryo can draw on during early development. By contrast, a plant spore is a single cell that does not have any food reserves.

Structural traits also gave seed plants an advantage. Some seed plants undergo **secondary growth** (growth in diameter) and produce wood. **Wood** is lignin-stiffened tissue that strengthens and protects older stems and roots. The giant nonvascular plants that lived in Carboniferous forests did undergo secondary growth. However, because their trunks were softer and more flexible than those of woody seed plants, they could not grow as tall as modern trees do.

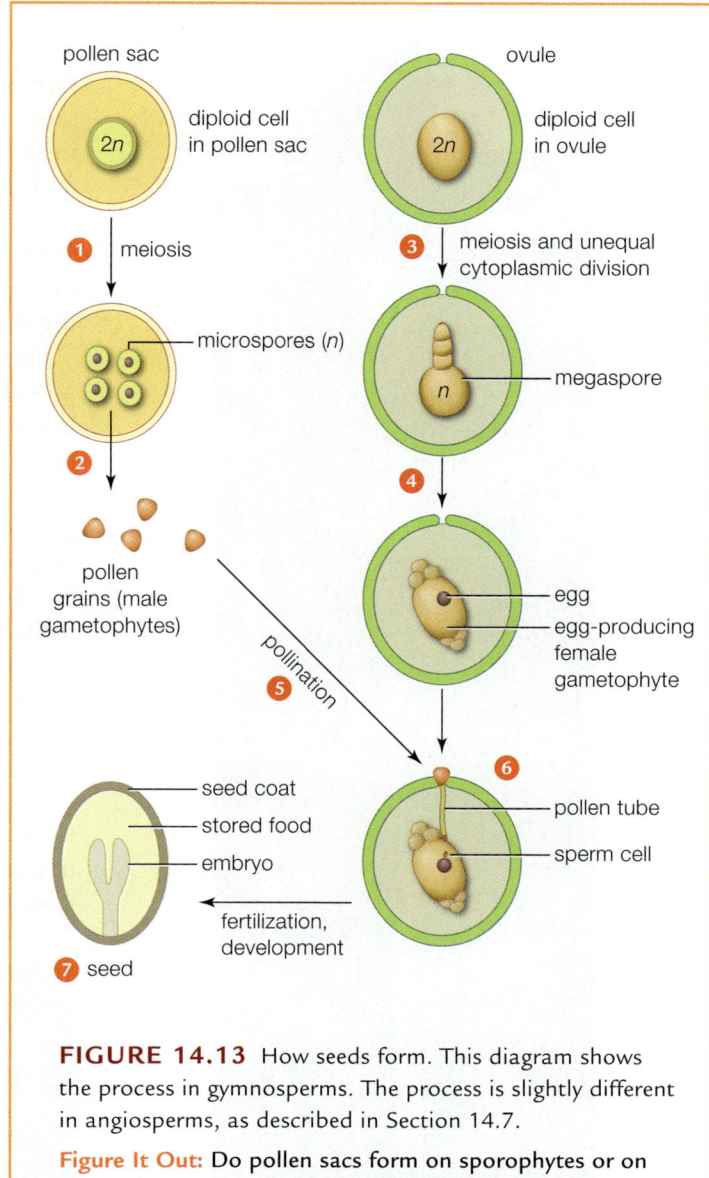

FIGURE 14.13 How seeds form. This diagram shows the process in gymnosperms. The process is slightly different in angiosperms, as described in Section 14.7.

Figure It Out: Do pollen sacs form on sporophytes or on gametophytes?

Answer: Pollen sacs form on sporophytes. They produce spores that develop into male gametophytes.

© Cengage Learning.

Take-Home Message

Which characteristics gave seed plants an adaptive advantage over spore-bearing plants?

- Seed plants do not release spores. Instead, their spores develop into gametophytes within the protection of the sporophyte body.
- Seed plants package their male gametophytes as pollen grains that wind or animals can disperse even under dry conditions.
- The female gametophyte of a seed plant forms inside an ovule that becomes a seed after fertilization. A seed contains a plant embryo and a food supply that it can draw upon during its early development.
- Some seed plants undergo secondary growth (they thicken) and become woody. Woody plants can grow taller than nonwoody ones.

megaspore In seed plants, a haploid cell that gives rise to a female gametophyte.

microspore In seed plants, a haploid cell that gives rise to a male gametophyte (pollen grain).

ovule Of seed plants, chamber inside which megaspores form and develop into female gametophytes; after fertilization this chamber becomes a seed.

pollen sac Of seed plants, chamber in which microspores form and develop into male gametophytes (pollen grains).

pollination Delivery of pollen to female part of a plant.

secondary growth Increase in diameter of a plant part.

wood Lignin-stiffened secondary growth of some seed plants.

14.6 Gymnosperms

Gymnosperms are seed plants that produce seeds on the surface of ovules. Their seeds are said to be "naked," because unlike those of angiosperms they are not inside a fruit. (*Gymnos* means naked and *sperma* is taken to mean seed.) However, many gymnosperms enclose their seeds in a fleshy or papery covering.

〉 Conifers **Conifers** are the most diverse gymnosperms. Most are woody trees or shrubs with needlelike or scalelike leaves. Most conifers are evergreen, meaning they do not shed their leaves all at once.

Evergreen conifers are the main plants in cool Northern Hemisphere forests. A conical shape helps many conifers shed snow easily. Needlelike leaves with a heavy wax coating help them minimize water loss during a long winter when soil is frozen, or during a dry hot season.

Conifers include two record-setting species. Bristlecone pines (*Pinus longaeva*) are among the longest-lived plants known (Figure 14.14). Coast redwoods (*Sequoia sempervirens*) include some of the world's tallest trees (Figure 14.15).

Many conifers are of great economic importance. For example, pines are the main source of lumber for building homes, and we use fir bark as mulch in our gardens. Some pines make a sticky resin that deters insects from boring into them. We use this resin to make turpentine, which is used as a paint thinner and solvent. We use oils from cedar in cleaning products, and we eat the seeds, or "pine nuts," of pinyon pines.

A pine tree's life cycle is typical of conifers. The tree is a sporophyte and it forms specialized spore-bearing structures that we call cones. There are two types of cones: small, soft pollen cones and large, woody, ovulate cones (Figure 14.16). Both have scales (modified leaves) arranged around a central axis. Pollen made

FIGURE 14.14 Bristlecone pine (*Pinus longaeva*) in the Sierra Nevada Mountains. The oldest living members of this species are more than 4,600 years old.

© Dave Cavagnaro/ Peter Arnold, Inc.

FIGURE 14.15 Composite photo of one of the world's tallest trees, a coast redwood (*Sequoia sempervirens*) that stands 116 meters (379 feet) high in a Northern California forest. Can you spot the three blue-shirted people climbing among its branches? © James Balog/ Aurora Photos.

ovule, which may become seed

FIGURE 14.16 Animated! Pine cones. **A** Pollen cone. **B** Cross section of an ovulate cone. Credits: (a) R. J. Erwin/Photo Researchers, Inc.; (b) © Stan Elems/ Visuals Unlimited.

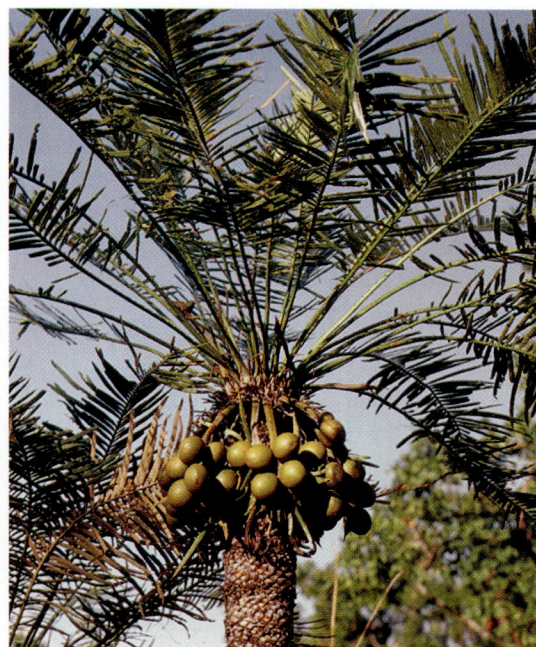

A Palmlike Australian cycad with its fleshy seeds.

B Fan-shaped leaves and fleshy seeds of a *Ginkgo biloba* tree.

FIGURE 14.17 Gymnosperm diversity.

Credits: (a) © M. Fagg, Australian National Botanic Gardens; (b) © Georgia Silvera Seamans, localecology.org.

by a pollen cone is released and drifts on the wind. Ovulate cones produce a sticky substance that traps pollen grains. After pollen is trapped, a tube grows into the ovulate cone scale and delivers a nonflagellated sperm cell to an egg in an ovule inside it. After fertilization, the ovule develops into a seed.

> Living Fossils Cycads and ginkgos are sometimes described as living fossils. Both were diverse in dinosaur times. They are the only modern gymnosperms that have flagellated sperm. In both groups, a male plant produces pollen that is carried by the wind to a female plant. As in conifers, the ovule of the female plant secretes fluid that traps pollen grains. Sperm emerge from pollen grains, then swim through the secreted fluid to eggs in the plant's ovule.

About 130 species of cycads survived to the present, mainly in tropical and subtropical regions. Cycads look like palms or ferns but are not close relatives of either (Figure 14.17**A**). "Sago palms" commonly used in landscaping and as houseplants are actually cycads.

The only living ginkgo is *Ginkgo biloba*, the maidenhair tree. Ginkgos are native to China, but their pretty fan-shaped leaves (Figure 14.17**B**) and resistance to air pollution make them popular in American cities. Ginkgos are deciduous, meaning they drop their leaves seasonally. Typically, male trees are planted because female trees make seeds with a fleshy covering that has a strong, unpleasant odor. Gingko extracts help treat symptoms of Alzheimer's disease.

Take-Home Message

What are gymnosperms?

- Gymnosperms are one of the two lineages of seed-bearing vascular plants. Seeds form on the surface of cones or other spore-producing structures.

- Conifers, the most diverse gymnosperm group, are evergreen trees with needlelike leaves. They dominate cool, high-latitude forests.

- Cycads and ginkgos are more ancient lineages. Their pollen grains produce flagellated sperm.

conifer Cone-bearing gymnosperm such as a pine.
gymnosperm Seed plant that produces "naked" seeds, which are not encased by a fruit.

Figure 14.18 — Floral structure

stamen
- filament
- anther

carpel
- stigma
- style
- ovary

petal

sepal

receptacle

ovule (forms within ovary)

FIGURE 14.18 Animated! Floral structure.
© Cengage Learning.

14.7 Angiosperms—Flowering Plants

› Floral Structure and Function **Angiosperms** are vascular seed plants, and the only plants that make flowers and fruits. A **flower** is a specialized reproductive shoot (Figure 14.18). Sepals, which usually have a green leaflike appearance, ring the base of a flower and enclose it until it opens. They surround a ring of petals, which are often brightly colored.

Stamens, a flower's pollen-producing parts, surround a **carpel** that captures pollen and produces eggs. Typically a stamen consists of a tall stalk, called the filament, topped by an **anther** that holds two pollen sacs.

A carpel has a sticky **stigma**, a region specialized for receiving pollen, at its tip. The stigma is located atop a stalk, called the **style**. At the base of the style is an **ovary**, a chamber containing one or more egg-producing ovules. After fertilization, an ovule matures into a seed and the ovary becomes the **fruit**. The name angiosperm refers to the fact that seeds form within the protection of the ovary. (*Angio*– means enclosed chamber, and *sperma*, seed.)

› A Flowering Plant Life Cycle Figure 14.19 shows a generalized life cycle for a flowering plant. A flower forms on the sporophyte body. Pollen sacs in the anthers hold diploid cells ❶ that produce microspores by meiosis ❷. The microspores develop into pollen grains (immature male gametophytes) ❸.

FIGURE 14.19 Animated! Angiosperm life cycle. © Cengage Learning.

❶ An anther has two pollen sacs with diploid cells that give rise to microspores.

pollen sac of anther (cutaway view)

2n

❹ An ovule forms on the ovary wall. It contains a diploid cell that will undergo meiosis.

ovule

Meiosis

❷ Meiosis produces microspores.

microspores (n)

❸ Microspores develop into pollen grains (male gametophytes).

pollen grain (n)

❾ The ovule develops into a seed.

seed coat (2n)
embryo (2n)   } seed
endosperm (3n)

Double fertilization

Meiosis

❺ Meiosis produces megaspores.

megaspores (n)

❻ A megaspore develops into a female gametophyte that includes an egg and six other cells, including a cell with two nuclei.

mature female gametophyte in ovule

cell with 2 nuclei

egg

pollen tube delivers 2 sperm to ovule

❼ Pollination occurs, and the pollen grain germinates. A pollen tube grows to and through ovary tissue to the ovule, where it releases two sperm.

pollination

mature male gametophyte

pollen tube
sperm

❽ During double fertilization, one sperm fertilizes the egg to form a zygote; the other fertilizes the cell with two nuclei to form a triploid (3n) cell.

Ovules form on the wall of an ovary at the base of a carpel ❹. Meiosis of cells in ovules yields haploid megaspores ❺. A megaspore develops into a female gametophyte consisting of a haploid egg, a cell with two nuclei, and a few other cells ❻.

Pollination occurs when a pollen grain arrives on a receptive stigma, the uppermost part of the carpel ❼. The pollen grain germinates, and a pollen tube grows through the style (the structure that elevates the stigma) to the ovary at the base of the carpel. Two nonflagellated sperm form inside the pollen tube as it grows.

Double fertilization occurs when a pollen tube delivers the two sperm into the ovule ❽. One sperm fertilizes the egg to create a zygote. The other sperm fuses with the cell that has two nuclei, forming a triploid (3*n*) cell. After double fertilization, the ovule matures into a seed ❾. The zygote develops into an embryo sporophyte and the triploid cell develops into **endosperm**, a nutritious tissue that will serve as a source of food for the developing embryo.

FIGURE 14.20 A pollinator. A bee unknowingly transfers pollen from flower to flower as it gathers pollen and sips nectar.

Courtesy of Christine Evers.

❭ Keys to Angiosperm Diversity

Flowering plants constitute 90 percent of all modern plant species. They survive in nearly every land habitat. What accounts for angiosperm success? For one thing, they tend to grow faster than gymnosperms. Think of how a plant like a dandelion or a grass can grow from a seed and produce seeds of its own within a few months. In contrast, most gymnosperms take years to mature and produce seeds.

After pollen-producing plants evolved, some insects began feeding on the plants' highly nutritious pollen. Plants gave up some pollen but gained a reproductive edge when insects unknowingly moved pollen between plants or between plant parts, thus facilitating pollination.

Many traits of flowering plants are adaptations that attract **pollinators**, animals that move pollen of one plant species onto female reproductive structures of the same species. Insects are the most common pollinators (Figure 14.20), but birds, bats, and other animals also serve in this role. Brightly colored petals, sugary nectar, or a strong fragrance can attract pollinators. The wind-pollinated plants tend to have small, pale, unscented flowers that lack nectar. For example, you have probably never noticed the small, pale flowers of grasses, which are wind pollinated.

Over time, plants coevolved with their animal pollinators. Again, coevolution refers to two or more species jointly evolving as a result of a close ecological interaction. Inherited changes in one species exert selective pressure on the other species, which also evolves.

A variety of fruit structures helped angiosperms disperse their seeds. Some fruits float in water, ride the winds, stick to animal fur, or survive a trip through an animal's gut. Gymnosperm seeds have less diverse dispersal mechanisms.

❭ Major Groups

The vast majority of flowering plants belong to one of two lineages. The 80,000 or so **monocots** include orchids, palms, lilies, and grasses, such as rye, wheat, corn, rice, sugarcane, and other valued plants. The **eudicots** include most herbaceous (nonwoody) plants such as tomatoes, cabbages, roses, daisies, most flowering shrubs and trees, and cacti. Monocots and eudicots evolved different structural details such as the arrangement of their vascular tissues and the number of flower petals. Also, some eudicots put on secondary growth and become woody. No monocots produce true wood. Section 27.2 describes the differences between eudicots and monocots in detail.

angiosperm Seed plant that produces flowers and fruits.

anther Part of the stamen that contains pollen sacs.

carpel Ovule-containing part of a flower.

double fertilization In flowering plants, one sperm fertilizes the egg, forming the zygote, and another fertilizes a diploid cell, forming what will become endosperm.

endosperm Nutritive tissue in an angiosperm seed.

eudicots Largest lineage of angiosperms; includes herbaceous plants, woody trees, and cacti.

flower Specialized reproductive shoot of a flowering plant.

fruit Mature ovary tissue that encloses a seed or seeds.

monocots Lineage of angiosperms that includes grasses, orchids, and palms.

ovary Of flowering plants, a floral chamber that holds one or more ovules.

pollinator Animal that moves pollen from one plant to another, thus facilitating pollination.

stamen Pollen-producing part of a flower. Consists of an anther that contains pollen sacs, atop a filament.

stigma Pollen-receiving part of a carpel.

style Elongated portion of a carpel that holds the stigma above the ovary.

A Mechanized harvesting of wheat. **B** A field of cotton ready for harvest. **C** Marijuana found growing illegally in Oregon.

FIGURE 14.21 Growing angiosperms for food, fabric, or drugs. Credits: (a) Photo USDA; (b) Photo by Scott Bauer, USDA/ARS; (c) Courtesy of Linn County, Oregon Sheriff's Office.

› Ecology and Human Uses of Angiosperms It would be nearly impossible to overestimate the importance of the angiosperms. As the dominant plants in most land habitats, they provide food and shelter for land animals. They also supply many products that meet human needs (Figure 14.21).

Angiosperms provide nearly all of our food, directly or as feed for livestock. Cereal crops are the most widely planted. The United States devotes more acreage to corn than to any other plant. Worldwide, rice feeds more people than any other crop. Wheat, barley, and sorghum are other widely grown grains. All are grasses. Legumes are the second most important source of human food. They can be paired with grains to provide all the amino acids the human body needs to build proteins. Soybeans, lentils, peas, and peanuts are examples of legumes.

Name a part of a plant, and humans probably eat it. In addition to the seeds of grains and legumes, we dine on leaves of lettuce and spinach, stems of asparagus, developing flowers of broccoli, modified roots of potatoes, carrots, and beets, and fruits of tomatoes, apples, and blueberries. Stamens of crocus flowers provide the spice saffron, and the bark of a tropical tree provides cinnamon.

Fibers used to make clothing come from two main sources, petroleum and plants. Plant fibers include cotton, flax, ramie, and hemp. We also use fibers from flowering plants to weave rugs, and in many places to thatch roofs. Oak and other hardwoods derived from angiosperms provide flooring and furniture.

We extract medicines and psychoactive drugs from plants. Aspirin is derived from a compound discovered in willows. Digitalis from foxglove strengthens a weak heartbeat. Coffee, tea, and tobacco are widely used stimulants. Marijuana, illegal in the United States, is one of the country's most valuable cash crops. Worldwide, cultivation of opium poppies (the source of heroin) and coca (the source of cocaine) have wide-reaching health, economic, and political effects.

Take-Home Message

What are angiosperms?

- Angiosperms are plants in which seeds develop in ovaries that become fruit.
- Angiosperms are the most diverse plants. Adaptations that contributed to their success include short life cycles, coevolution with insect pollinators, and a variety of fruit structures that aid in dispersal of seeds.

14.8 Fungal Traits and Diversity

> **Yeasts, Molds, and Mushrooms** With this section, we begin our survey of another major lineage of eukaryotes, the fungi. Fungi are more closely related to animals than to plants. Fungi and animals are most closely related to amoebozoan protists (Section 13.6), whereas plants are descended from green algae. **Fungi** are spore-producing heterotrophs that have cell walls made of chitin, a polysaccharide also found in the external skeletons of insects and crabs.

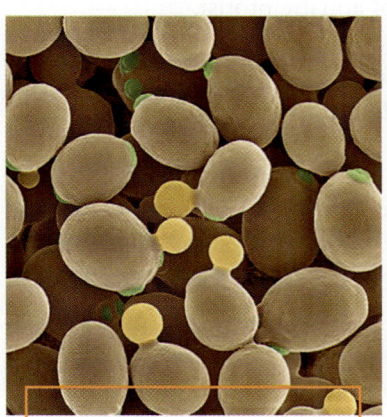

FIGURE 14.22 A yeast. Yeasts live as single cells and reproduce by budding.

© Dr. Dennis Kunkel/ Visuals Unlimited.

All fungi obtain nutrients by extracellular digestion and absorption. As a fungus grows in or over organic matter, it secretes digestive enzymes, then absorbs the resulting breakdown products.

Fungi that live as single cells are commonly called yeasts (Figure 14.22). Yeasts sold for baking bread are fungi, as are the organisms that cause yeast infections of the mouth or vagina. In many yeasts, new cells bud from existing ones.

Mushrooms and molds are multicelled fungi. You may have seen mushrooms growing in the wild (Figure 14.23**A**), and you probably have encountered them in salads and on pizzas. Molds are familiar to anyone who has left produce or other food sitting around too long (Figure 14.23**B**).

A multicelled fungus grows as a mesh of threadlike branching filaments collectively called a **mycelium** (plural, mycelia). Each filament is a **hypha** (plural, hyphae) consisting of cells arranged end to end (Figure 14.23**C**). Fungi do not have vascular tissue, but walls between cells in a hypha are porous, so materials flow among cells. As a result, nutrients or water taken up in one part of the mycelium are shared with cells in other regions.

Some fungal bodies are enormous. The largest organism we know about is a soil fungus in Oregon. Its hyphae extend through the soil over an area of about 2,200 acres, and it is still growing. Given its growth rate and size, researchers estimate that this individual has been alive for 8,000 years.

Most fungi are free-living decomposers that help keep nutrients cycling in ecosystems. Other fungi live in or on other organisms. Some of these, such as the wheat stem rust discussed in Section 14.1, are parasites and cause disease. Other fungi benefit their host or have no effect. Many fungi can partner with cyanobacteria or green algae to form a composite organism that we call a lichen. We return to relationships between fungi and other species in Section 14.9.

FIGURE 14.23 Multicelled fungi.

Credits: (a) Robert C. Simpson/ Nature Stock; (b) Photo by Scott Bauer/ USDA; (c) Micrograph Garry T. Cole, University of Texas, Austin/ BPS.

Fungi are more closely related to animals than to plants.

fungus Spore-producing heterotroph with cell walls of chitin that feeds by extracellular digestion and absorption.
hypha A single filament in a fungal mycelium.
mycelium Mass of threadlike filaments (hyphae) that make up the body of a multicelled fungus.

A Scarlet hood mushroom.

B Green mold on grapefruit.

— one cell (part of a hypha in a mycelium)

C Micrograph of a mycelium. A multicelled fungus is composed of a mycelium, a mesh of many filaments called hyphae.

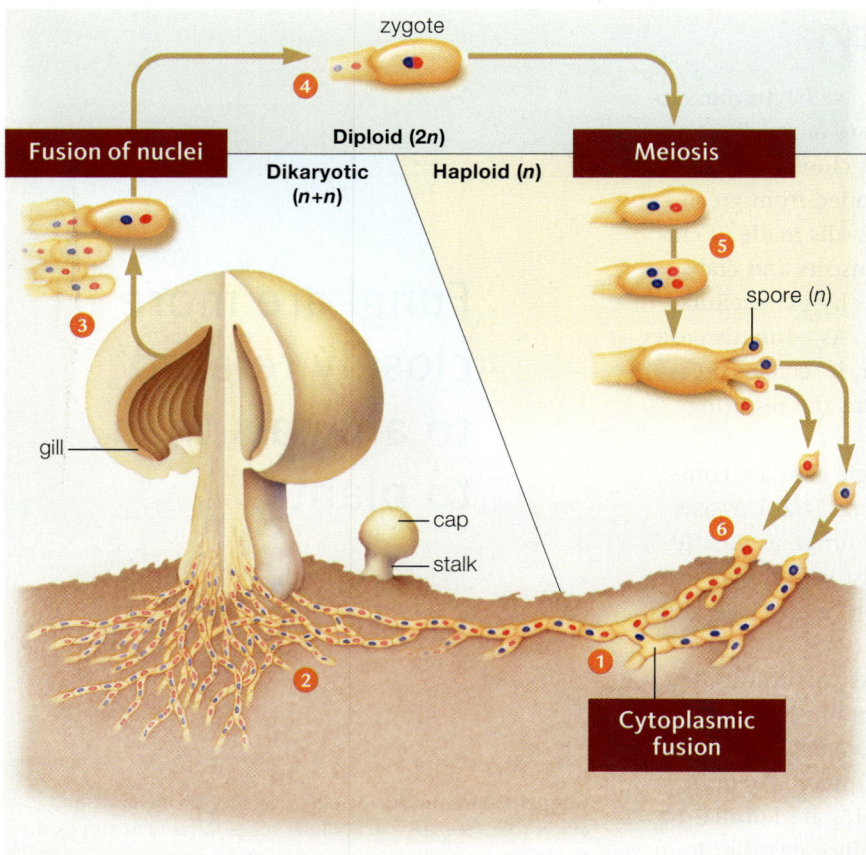

zygote

Fusion of nuclei

Diploid (2n)

Dikaryotic (n+n) **Haploid (n)**

Meiosis

spore (n)

gill

cap

stalk

Cytoplasmic fusion

FIGURE 14.24 Animated!
Generalized life cycle for a club fungus. Hyphae of different mating strains often grow through the same patch of soil.

❶ Two haploid hyphal cells meet and their cytoplasm fuses, forming a dikaryotic (n+n) cell.

❷ Mitotic cell divisions form a mycelium that produces a mushroom.

❸ Spore-making cells form at the edges of the mushroom's gills.

❹ Inside these dikaryotic cells, nuclei fuse, making the cells diploid (2n).

❺ The diploid cells undergo meiosis, forming haploid (n) spores.

❻ Spores are released and give rise to a new haploid mycelium.

Figure It Out: Are cells that make up the stalk of a mushroom haploid, diploid, or dikaryotic?

After T. Rost, et al., Botany, Wiley, 1979.   *Answer: Dikaryotic*

> **Fungal Diversity** About 56,000 species of fungi have been named, and there may be a million more. Fungi have been diversifying for a long time. Some fossil fungi date to 500 million years ago. The fossils resemble chytrids, the only modern fungi that are mainly aquatic and have a flagellated stage in their life cycle. The fossils and the existence of the aquatic, flagellated chytrids are taken as evidence that fungi, like plants and animals, evolved from some type of aquatic protist.

Three major lineages of fungi became established by 300 million years ago. They are the club fungi, zygote fungi, and sac fungi. We focus on these groups for the remainder of this chapter.

> **Fungal Life Cycles** In all multicelled fungi, sexual reproduction begins with the fusion of haploid hyphae from two individuals of different mating strains. What happens next varies.

Figure 14.24 shows the life cycle of a mushroom-producing club fungus. Haploid hyphae of two different mating strains of club fungus (represented in the figure by cells with different colored nuclei) sometimes meet in the soil. If they do, their cytoplasm fuses and a dikaryotic cell forms ❶. Dikaryotic means "having two nuclei." (*Di–* means two and *karyon* means nucleus, as in eukaryotic.) We represent the dikaryotic state as *n+n*. Mitotic divisions produce a dikaryotic mycelium that grows through the soil ❷. Multicelled club fungi spend most of their life cycle growing as a dikaryotic mycelium. When conditions favor reproduction, a sudden burst of hyphal growth produces a mushroom that emerges above the ground. Mushrooms are the reproductive parts, or fruiting bodies, of club fungi and they exist only briefly. A typical mushroom has a stalk and a cap, with thin sheets of tissue called gills on the cap's underside ❸. Fusion of haploid nuclei inside cells at the edges of the gills forms diploid zygotes ❹. Each zygote undergoes meiosis, producing haploid spores ❺. After a spore germinates, repeated mitotic divisions give rise to a new haploid mycelium ❻.

Like club fungi, multicelled sac fungi form fruiting bodies. Some of these fruiting bodies resemble mushrooms, but they do not have gills with club-shaped cells. Instead, spores form in sac-like cells on the fruiting body. Hence the name "sac" fungus.

Zygote fungi, such as the bread mold *Rhizopus*, most often produce spores asexually. Specialized hyphae develop and spores form by mitosis in cells at their tips (Figure 14.25). When sexual reproduction does occur, hyphae fuse, gametes form, and fertilization produces a diploid spore called a zygospore. The zygospore is a resting stage and can survive unfavorable conditions. When conditions favor fungal growth, the zygospore undergoes meiosis and releases haploid spores that grow into a new mycelium.

> **Human Uses of Fungi** Fungi are important as food crops. Mushroom farms produce many varieties of club fungi by inoculating sterilized compost with spores of the desired crop species. Morels and truffles (Figure 14.26) are tasty fruiting bodies of sac fungi. Morels form above ground like the mushrooms of club fungi. Truffles are fruiting bodies of fungi that associate with tree roots and thus are difficult to cultivate. The truffle forms underground. When truffle

FIGURE 14.25 Black bread mold, a zygote fungus. The tiny black dots are spore sacs at the tips of special hyphae, as shown in the inset micrograph. Credits: © Micrograph J. D. Cunningham/ Visuals Unlimited; inset, Micrograph Ed Reschke.

FIGURE 14.26 Sac fungus fruiting bodies prized by gourmets. **A** Morels produce spores in pits on their surface. **B** Truffles grow underground. Internally formed spores are released by an animal that digs up and eats the truffle. Both truffles and morels grow in association with tree roots. Credits: (a) © ostromec/ Shutterstock.com; (b) © age fotostock/ SuperStock.

spores mature, the fungus gives off a scent like that of a male pig seeking a mate. Female pigs that catch a whiff disperse truffle spores as they root through the soil in search of this seemingly subterranean suitor. Dogs can also be trained to snuffle out truffles. In 2010, two Italian truffles with a combined weight of 1.5 kilograms (about 3 pounds) were sold at auction for $330,000.

We use fungal fermentation reactions to produce food and drinks. A package of baker's yeast contains spores of a sac fungus (*Saccharomyces cerevisiae*). Set bread dough out to rise, and yeast cells carry out fermentation reactions that produce carbon dioxide, causing the dough to expand (rise). *Saccharomyces* also helps produce beer and wine. Fermentation by another fungus (*Aspergillus*) produces soy sauce and the citric acid used to preserve and flavor soft drinks. Still other fungi are responsible for the tangy blue veins in cheeses such as Roquefort.

Some of our most important medicines are compounds first isolated from fungi. Most famously, the initial source of the antibiotic penicillin was the sac fungal mold *Penicillium*. Another antibiotic, cephalosporin, was first isolated from *Cephalosporium*. Fungi have also yielded drugs used to lower blood pressure, reduce cholesterol levels, or to prevent rejection of transplanted organs.

Some mushroom toxins alter mood and cause hallucinations. The drug LSD was first isolated from ergot, a club fungus that infects grains. The so-called "magic mushrooms" contain psilocybins that produce LSD-like effects. Use of LSD and psilocybin-containing mushrooms is illegal in the United States.

Mushroom toxins evolved as a defense against mushroom-eating animals, and some can have deadly effects if eaten. Mushrooms gathered from the wild should always be identified by an experienced mushroom forager before they are eaten because many edible species have poisonous look-alikes.

Take-Home Message

What are fungi?

- Fungi are heterotrophs that absorb nutrients from their environment. They live as single cells or as a multicelled mycelium and disperse by producing spores.
- Mushrooms are spore-producing fruiting bodies of club fungi. Some sac fungi also produce large fruiting bodies that we eat.
- Zygote fungi grow as molds that most often produce spores asexually.

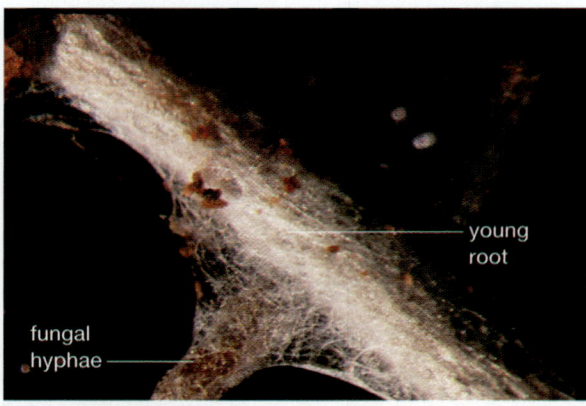

FIGURE 14.27 Mycorrhiza formed by a fungus and the roots of a hemlock tree. © Gary Braasch.

young root

fungal hyphae

lichen Composite organism consisting of a fungus and a single-celled photosynthetic alga or bacterium.
mutualism Species interaction that benefits both species.
mycorrhiza Fungus–plant root partnership.

14.9 Fungal Ecology

❯ Fungi as Partners Nearly all plants form mutually beneficial relationships, or **mutualisms**, with fungi. Many soil fungi, including truffles, live in or on tree roots in a partnership known as a **mycorrhiza** (plural, mycorrhizae). In some cases, fungal hyphae form a dense net around roots but do not penetrate them (Figure 14.27). In other cases, hyphae grow inside root cell walls.

Hyphae of both kinds of mycorrhizae grow through soil and serve to increase the absorptive surface area of their plant partner. The fungus shares minerals it takes up with the plant, and the plant gives up sugars to the fungus. It is a beneficial trade; many plants do poorly without a fungal partner.

Lichens are composite organisms consisting of a fungus and a single-celled photosynthetic species, either a green alga or a cyanobacterium. The fungus makes up most of the lichen's mass and shelters the photosynthetic species, which shares nutrients with the fungus.

Lichens grow on many exposed surfaces (Figure 14.28). They are ecologically important as colonizers in places that are too hostile for other organisms. By releasing acids and retaining water that freezes and thaws, lichens help break down rocks and form soil. Formation of soil allows plants to move in and take root. Millions of years ago, lichens may have preceded plants onto the land.

Lichens take up water and nutrients across their surface, but they also absorb toxic air pollutants. Where air pollution levels are high, lichens can become poisoned and die. For this reason, researchers sometimes monitor the health of lichen populations as an indicator of air quality.

❯ Fungi as Decomposers and Pathogens Fungi are important decomposers. They secrete digestive enzymes onto wastes and remains, then absorb some of the breakdown products. Because not all nutrients are absorbed, some remain in the soil to nourish other organisms. Club fungi play an especially important role as decomposers in forests. They are the only fungi that can break down lignin, which is abundant in wood.

Some fungi do not wait until an organism is dead to feed on it. Such parasitic fungi can be important plant pathogens. You have already learned about wheat stem rust disease (Section 14.1). Powdery mildew, a whitish powder that appears on leaves, is a sac fungus that feeds on leaf tissues. Another fungus caused the demise of the American chestnut. Chestnut trees were abundant in North America's eastern forests until the early 1900s. Then, a pathogenic fungus introduced from Asia caused mature trees to die back to the ground. Today,

FIGURE 14.28 Animated! Lichens.
A Leaflike lichen on a birch tree.
B Encrusting lichens on granite. Acids secreted by lichens help break down rocks to create soil.
Credits: (a) Gary Head; (b) © Mark E. Gibson/ Visuals Unlimited.

a few chestnut trees still sprout from old root systems, but they cannot reach mature size.

Human fungal infections most frequently involve body surfaces. Typically a fungus feeds on the outer layers of skin, secreting enzymes that dissolve keratin, the main skin protein. Infected areas become raised, red, and itchy. For example, several species of fungus infect skin between the toes and on the sole of the foot, causing "athlete's foot" (Figure 14.29**A**). Fungi also cause skin infections misleadingly known as "ringworm." No worm is involved. The ring-shaped lesion (Figure 14.29**B**) is caused by the growth of hyphae outward from the initial infection. Fungal vaginitis, a vaginal yeast infection, occurs when single-celled fungi that normally live in the vagina in low numbers undergo a population explosion. Symptoms include itching or burning sensations, a thick, odorless, whitish vaginal discharge, and pain during intercourse.

Fungi seldom cause systemwide disease in otherwise healthy people, but infections can be life-threatening in people whose immune system is impaired by AIDS, chemotherapy, or drugs that must be taken after an organ transplant.

B Ringworm. The lesion expands as the fungus grows outward from the initial site of infection.

A Athlete's foot.

FIGURE 14.29 Fungal skin infections.
Credits: (a) Dr. P. Marazzi/ Photo Researchers, Inc. (b) © ISM / Phototake, Inc.

Take-Home Message

How do fungi interact with other species?

- Mycorrhizal fungi live in or on plant roots in a mutually beneficial relationship. Fungi also live with single-celled photosynthetic cells as lichens.

- Fungi benefit other organisms when they feed on wastes and remains, releasing nutrients into the soil. They harm other organisms, including humans, by infecting and feeding on their tissues.

14.10 **Fungal Threats to Crops** (revisited)

Current agricultural practices increase the ease with which fungal infections of crop plants spread. Farmers often plant a single variety of a crop in dense stands that cover an extensive area. The proximity of many genetically identical host plants allows a fungus to spread quickly through a field. In addition, farmers in different parts of the world often plant the same few varieties of a crop.

During the 1960s, a fungal disease nearly wiped out the global banana crop, which consisted largely of a single susceptible variety, called Gros Michel. Today, nearly all bananas sold in the United States are a variety called Cavendish, which is immune to the fungal strain that caused the 1960s crop failure. Unfortunately, a new fungal strain that does infect Cavendish arose in the 1990s and has already decimated banana plantations in much of Asia. Given the extent of global trade and the ability of fungal spores to travel long distances on the winds, this new fungus, like wheat stem rust fungus, is likely to spread worldwide.

© Africa Studio/ Shutterstock.

Protecting our food supply requires maintaining the genetic diversity of our crop plants. Having many varieties of a crop increases the chance that at least one variety will be immune to a new disease. That variety can be planted or used to create new disease-resistant varieties, either through traditional plant breeding or by using genetic engineering to transfer disease-resistance genes.

WHERE YOU ARE GOING . . .

We delve more deeply into the structure and function of vascular plants, especially the angiosperms, in Chapters 27 and 28. Fungal pathogens of humans come up again when we discuss symptoms of AIDS (Section 22.7). We return to the role of fungi as decomposers when we investigate food webs and nutrient cycling (Sections 17.5 and 17.6).

Summary

Section 14.1 Plants are food for both people and fungi. Winds spread spores of fungal pathogens, so outbreaks of fungal disease can often spread to crops over a wide area.

Section 14.2 **Plants** evolved from freshwater green algae. Their life cycle includes two multicelled forms, a haploid **gametophyte** and a diploid **sporophyte**. The gametophyte dominates the life cycle of **bryophytes**, but the sporophyte dominates in **vascular plants**.
Key adaptations to dry habitats include a waterproof **cuticle** with **stomata**, and internal pipelines of **xylem** and **phloem**. **Lignin**-reinforced xylem helps vascular plants stand upright. **Seeds** and male gametophytes that can be dispersed without water (**pollen grains**) evolved in seed plants.

sporophyte

gametophyte

Section 14.3 Bryophytes include three lineages of low-growing plants: **mosses**, liverworts, and hornworts. Their flagellated sperm reach eggs by swimming through films or droplets of water that cling to the plant. The sporophyte begins development inside gametophyte tissues. It remains attached to and often dependent upon the gametophyte even when mature.

Section 14.4 **Ferns** are seedless vascular plants. Sporophytes dominate their life cycle and produce spores in **sori**. Gametophytes produce flagellated sperm. Ferns grow from **rhizomes** (horizontal stems). Some live on trees as **epiphytes**. Other seedless vascular plants include club mosses and horsetails. Coal formed from the remains of ancient nonvascular seed plants.

Section 14.5 Seed-bearing vascular plants make two types of spores. **Microspores** give rise to pollen grains in a **pollen sac**. **Megaspores** form in **ovules**, and give rise to egg-producing female gametophytes. Even in the absence of water, winds or pollinators can move pollen, thus facilitating **pollination**. The seed is a mature ovule, with an embryo sporophyte and some nutritive tissue inside it. Some seed plants undergo **secondary growth** and produce **wood**.

Section 14.6 **Conifers**, cycads, and ginkgos are among the **gymnosperms**. They are adapted to dry climates and bear seeds on exposed surfaces of spore-bearing structures. In conifers, these spore-bearing structures are distinctive cones.

Section 14.7 **Angiosperms** are the dominant land plants. They alone have **flowers**. Pollen forms in **anthers**, the part of a **stamen** that holds pollen sacs. Many flowering plants coevolved with **pollinators** that deliver pollen to a receptive **stigma**. After pollination, a pollen tube grows through a **style** of the flower's **carpel** to the **ovary** at its

base and **double fertilization** occurs. The ovary becomes a **fruit** containing one or more seeds. A flowering plant seed includes an embryo sporophyte and **endosperm**, a nutritious tissue. Most crops are angiosperms. There are two main lineages of flowering plants. **Monocots** include grasses and palm trees. **Eudicots** include most flowering trees and shrubs, as well as most herbaceous plants.

Section 14.8 **Fungi** include single-celled and multicelled heterotrophs. They secrete enzymes onto organic material and absorb the resulting breakdown products. Multicelled fungi grow as a **mycelium** composed of many filaments called **hyphae**. Fungi produce spores both sexually and asexually. A mushroom is a spore-producing body of a club fungus. Fungi of some sac fungi are also used as food. Bread mold is an example of a zygote fungus.

Section 14.9 A **mycorrhiza** is a **mutualism** in which plants rely on a fungus that lives in or on their roots and help them take up food. Fungi also partner with photosynthetic cells and form **lichens**. Fungi are important decomposers, especially of wood. Some fungi infect plants or animals, causing disease.

Self-Quiz   Answers in Appendix I

1. Which of the following statements is not correct?
   a. Gymnosperms were the earliest flowering plants.
   b. Mosses are nonvascular plants.
   c. Ferns and angiosperms are vascular plants.
   d. Only angiosperms produce fruits.

2. Which does *not* apply to gymnosperms or angiosperms?
   a. vascular tissues   c. single spore type
   b. diploid dominance   d. all of the above

3. Bryophytes have independent _____ and dependent _____ .
   a. sporophytes; gametophytes
   b. gametophytes; sporophytes

4. Ferns are classified as _____ plants.
   a. multicelled aquatic   c. seedless vascular
   b. nonvascular seed   d. seed-bearing vascular

5. The _____ produce flagellated sperm.
   a. ferns   c. monocots
   b. conifers   d. a and c

6. The _____ produced in the male cones of a conifer develop into pollen grains.
   a. ovules   c. megaspores
   b. ovaries   d. microspores

7. A seed is _____ .
   a. a female gametophyte   c. a mature pollen tube
   b. a mature ovule   d. an immature spore

8. Match the terms appropriately.
   _____ gymnosperm   a. gamete-producing body
   _____ sporophyte   b. help control water loss
   _____ horsetail   c. "naked" seeds
   _____ bryophyte   d. spore-producing body
   _____ gametophyte   e. nonvascular land plant
   _____ stomata   f. seedless vascular plant
   _____ angiosperm   g. flowering plant

9. All fungi _____ .
   a. are multicelled
   b. form flagellated spores
   c. are heterotrophs
   d. all of the above

10. Fungal decomposers derive nutrients from _____ .
    a. organic wastes and remains
    b. living plants
    c. living animals
    d. photosynthesis

11. A mushroom is _____ .
    a. the food-absorbing part of a fungus
    b. the only part of the fungal body not made of hyphae
    c. a reproductive structure that releases sexual spores
    d. the longest-lived part of the fungal life cycle

12. Human fungal infections most commonly involve
    a. the brain
    b. the heart
    c. the digestive system
    d. body surfaces

13. A _____ is a composite organism composed of a fungus and a single-celled photosynthetic species.
    a. mycorrhiza
    b. lichen
    c. decomposer
    d. ringworm

14. Cell walls of fungi are composed of _____ .
    a. cellulose
    b. keratin
    c. lignin
    d. chitin

15. Match the terms appropriately.
    _____ decomposer
    _____ yeast
    _____ mushroom
    _____ chytrid
    _____ hypha
    _____ mycelium
    _____ dikaryotic

    a. filament made of walled cells
    b. club fungus fruiting body
    c. fungus with flagellated spores
    d. mesh of fungal filaments
    e. having two nuclei in a cell
    f. single-celled fungus
    g. breaks down organic matter

Critical Thinking

1. Early botanists admired ferns but found their life cycle perplexing. In the 1700s, they learned to propagate them by sowing what appeared to be tiny dustlike "seeds" from the undersides of fronds. Despite many attempts, the scientists could not find the pollen source, which they assumed must stimulate the "seeds" to develop. Imagine you could write to one of these botanists. Compose a note that would clear up their confusion.

2. Today, the tallest bryophytes reach a maximum height of 20 centimeters (8 inches) or so. So far as we know from fossils, there were no giants among their ancestors. Lignin and vascular tissue first evolved in relatives of club moss, and some extinct species stood 40 meters (130 feet) high. Among modern seed plants, *Sequoia* (a gymnosperm) and *Eucalyptus* (an angiosperm) can be more than 100 meters (330 feet) high. Explain why evolution of vascular tissues and lignin would have allowed such a dramatic increase in plant height. How might being tall give one plant species a competitive advantage over another?

3. Some poisonous mushrooms such as those shown at the *right* have bright, distinctive colors that mushroom-eating animals learn to recognize. Once sickened, an animal avoids these species. Other toxic mushrooms are as dull-looking as edible ones, but they have an unusually strong odor. Some scientists hypothesize that the strong odors aid in defense against mushroom-eating animals that are active at night. Devise an experiment to test this hypothesis.

Digging Into Data

Removing Stumps to Save Trees

The club fungus *Armillaria ostoyae* infects living trees and acts as a parasite, withdrawing nutrients from them. When the tree dies, the fungus continues to dine on its remains. Fungal hyphae grow out from the roots of infected trees and roots of dead stumps. If these hyphae contact roots of a healthy tree, they can invade and cause a new infection.

Canadian forest pathologists hypothesized that removing stumps after logging could help prevent tree deaths. To test this hypothesis, they carried out an experiment. In half of a forest, they removed stumps after logging. In a control area, they left stumps behind. For more than 20 years, they recorded tree deaths and whether *A. ostoyae* caused them. Figure 14.30 shows the results.

1. Which tree species was most often killed by *A. ostoyae* in control forests? Which was least affected by the fungus?

2. For the species most affected, what percentage of deaths did *A. ostoyae* cause in control and in experimental forests?

3. Do the overall data support the hypothesis that stump removal helps protect living trees from infection by *A. ostoyae*?

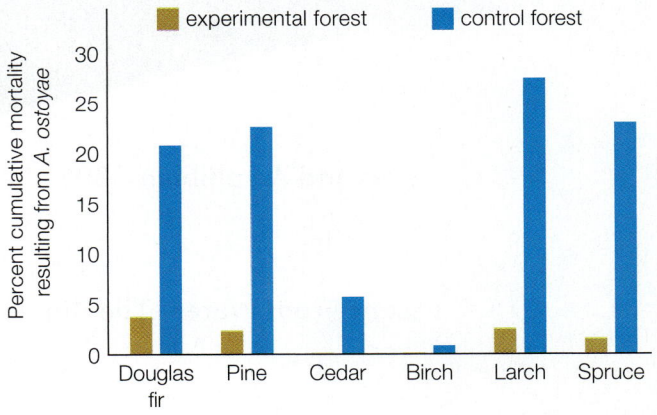

FIGURE 14.30 Results of a long-term study of how logging practices affect tree deaths caused by the fungus *A. ostoyae*. In the experimental forest, whole trees—including stumps—were removed (*brown* bars). The control half of the forest was logged conventionally, with stumps left behind (*blue* bars).

After graph from www.pfc.forestry.ca.

15 Animal Evolution

Section 13.6 introduced the closest protist relative of animals. Section 3.5 described animal cells, and Section 8.7 discussed animal life cycles. This chapter also draws on your knowledge of homeotic genes (7.7) and of comparative embryology and genomics (11.6).

FIGURE 15.1 Feathered fossil species. The *Archaeopteryx* fossil was discovered in Germany. Drawings of *Sinosauropteryx* and *Confuciusornis* are based on fossils unearthed in China.

A *Archaeopteryx*

B *Confuciusornis sanctus*

C *Sinosauropteryx prima*

15.1 Early Birds

In Darwin's time, the presumed absence of transitional fossils kept many people from accepting his new theory of evolution by natural selection. Skeptics wondered, if new species evolve from existing ones, then where are fossils that document these transitions? Where are fossils that bridge major groups? In fact, one such "missing link" fossil was unearthed by workers at a limestone quarry in Germany just one year after Darwin's *On the Origin of Species* was published.

The fossil from the German quarry was about the size of a large crow, and it looked like a small meat-eating dinosaur. Like other dinosaurs, it had a long, bony tail, clawed digits, and a heavy jaw with short, spiky teeth. However, the fossil also had feathers (Figure 15.1**A**). The new fossil species was named *Archaeopteryx* (ancient winged one). To date, seven *Archaeopteryx* fossils have been unearthed. Radiometric dating (Section 11.4) indicates that the species lived about 150 million years ago.

Archaeopteryx is the most widely known transitional fossil in the bird lineage, but there are many others. *Confuciusornis sanctus* (sacred Confucius bird) fossils 120 million years old were discovered in China. Like modern birds, and unlike *Archaeopteryx*, the Confucius bird had a toothless beak and short birdlike tail with long feathers (Figure 15.1**B**). However, its dinosaur ancestry remains apparent. Unlike modern bird wings, the wings of *C. sanctus* had grasping digits with claws at their tips.

Another fossil that takes us back even farther in time provides a clue as to which group of dinosaurs gave rise to the birds. In 1994, a farmer in China discovered a fossil of a small dinosaur-like animal with short forelimbs and a long tail (Figure 15.1**C**). Unlike most dinosaurs, this one had areas covered with tiny filaments that resemble downy feathers of modern birds. Researchers named the farmer's fuzzy find *Sinosauropteryx prima*, meaning first Chinese feathered dragon. Given the animal's shape and the lack of long feathers, *S. prima* was certainly flightless. If the fuzzy filaments are indeed feathers, they probably served as insulation, in the same way that downy feathers do in modern birds.

Scientists interested in the evolutionary transitions that led to modern animal diversity look to fossils for physical evidence of these transitions, and use radiometric dating to determine when transitional species lived. The structure, biochemistry, and genetic traits of living organisms also provide information about the branchings that gave rise to modern animal groups.

Interpretations of fossil data and gene comparisons can vary, and biologists sometimes disagree about relationships among lineages, both living and extinct. However, these disagreements do not call into question Darwin's theory that all animal lineages arose by descent with modification from a common ancestor. Rather, arguments concern the details of where, when, and how the various branchings took place. By using new methods to compare genes, analyzing new fossil finds, or reanalyzing old ones, scientists seek the evidence necessary to support currently held hypotheses, revise them, or discard them for new ones.

A Choanoflagellate colony. Such colonies consist of genetically identical cells. Animals are thought to have evolved from a similar colonial protist.

15.2 Animal Origins, Traits, and Trends

Animals are multicelled heterotrophs that take food into their body, where they digest it and absorb the released nutrients. An animal has a few to hundreds of types of unwalled cells. These cells become specialized as the individual develops from an embryo (an early developmental stage) to an adult. Most animals reproduce sexually, some reproduce asexually, and some do both. Nearly all animals are motile (can move from place to place), during part or all of their life cycle.

⟩ Animal Origins The **colonial theory of animal origins** proposes that animals evolved from a heterotrophic protist that formed colonies. At first, all cells in the colony were similar. Each could survive and reproduce on its own. Later, mutations produced cells that specialized in some tasks and did not carry out others. Perhaps some cells captured food more efficiently, but did not make gametes, whereas other made gametes but did not catch food. The division of labor among interdependent cells made them more efficient, allowing the colonies with mutations to obtain more food and produce more offspring. Over time, additional specialized cell types evolved, producing the first animal.

Studies of modern protists support the colonial theory. Choanoflagellates are the modern protists most closely related to animals. Some species can live either as individual cells or as a colony (Figure 15.2**A**). Colonies form when a cell undergoes mitosis, but descendant cells stay together. Thus, cells of a choanoflagellate colony are, like the cells of an animal body, genetically identical.

Modern members of the oldest animal lineages can give us an idea of what the earliest animals may have been like. For example, placozoans represent an early branch on the animal family tree. They have the fewest genes and simplest body plan of any living animal. A placozoan's tiny flattened body consists of four cell types (Figure 15.2**B**). By contrast, your body has more than 200 different kinds of cells. Ciliated cells on a placozoan's lower surface allow it to move in search of bacteria and single-celled algae. Gland cells, also on its lower surface, secrete enzymes onto this food, then take up breakdown products.

The first animals may have evolved in the seas as early as 1 billion years ago. However, fossils show that most animal lineages arose during an explosion of diversity in the Cambrian (542–488 million years ago). During this time, the oxygen concentration in seawater increased dramatically. All animals carry out aerobic respiration, so abundant oxygen would have allowed larger, more active animals to evolve. At the same time, early supercontinents were breaking up. Movement of land masses cut off gene flow among populations, encouraging speciation (Section 12.5). Species interactions may have also encouraged evolutionary innovations. For example, once the first predators arose, mutations that produced protective hard parts were favored in prey animals. Changes in genes that regulate body plans (homeotic genes, Section 7.7) may have sped things along. Beneficial mutations in these genes would have allowed adaptive changes to body form in response to predation or other selective forces.

B A placozoan. Placozoans are the simplest living animals and one of the oldest animal lineages. This individual is the size of a head of a pin. Its red color comes from red algal cells it has eaten.

animal A eukaryotic heterotroph that is made up of unwalled cells and develops through a series of stages. Most ingest food, reproduce sexually, and move.

colonial theory of animal origins Well-accepted hypothesis that animals evolved from a colonial protist.

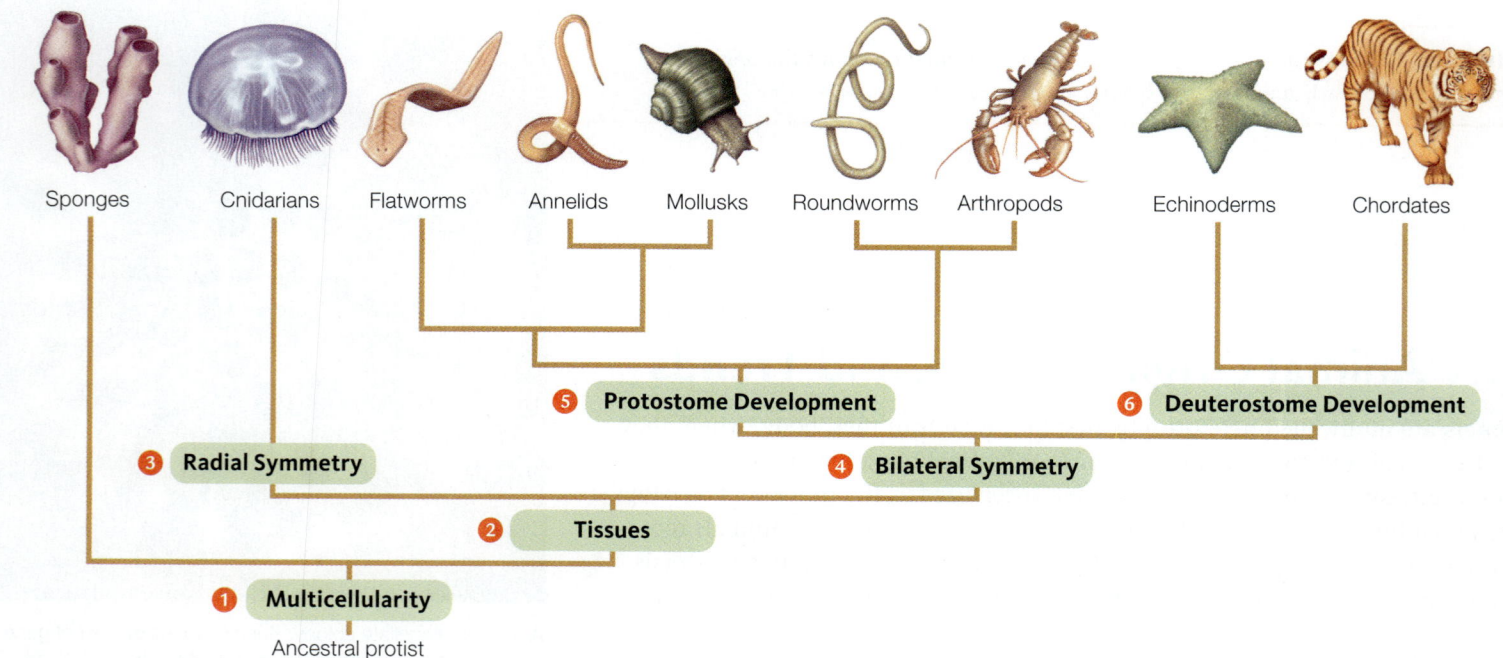

Sponges   Cnidarians   Flatworms   Annelids   Mollusks   Roundworms   Arthropods   Echinoderms   Chordates

5 Protostome Development

6 Deuterostome Development

3 Radial Symmetry

4 Bilateral Symmetry

2 Tissues

1 Multicellularity

Ancestral protist

FIGURE 15.3 Family tree for major animal phlya based on body form and genetic comparisons. Vertebrates (animals with a backbone) are a subgroup of the chordates.

Figure It Out: What type of body symmetry is most common in animals?

© Cengage Learning.

Answer: Bilateral symmetry

Of about 2 million named animals, only about 50,000 are vertebrates.

> **Major Animal Groups and Evolutionary Trends** Figure 15.3 shows relationships among the major animal groups covered in this book. All animals are descended from a common multicelled ancestor **1**. The earliest animals were aggregations of cells, and sponges still show this level of organization. However, most animals have tissues **2**. A **tissue** consists of one or more types of cells that are organized in a specific pattern and that carry out a particular task. In the early animal lineages, embryos had two tissue layers: an outer ectoderm and an inner endoderm. In later lineages, cell movements produced a middle embryonic layer called mesoderm. The evolution of a three-layer embryo allowed an increase in structural complexity.

The structurally simplest animals such as sponges are asymmetrical—you cannot divide their body into halves that are mirror images. Sea anemones and other cnidarians have **radial symmetry**: Body parts are repeated around a central axis, like the spokes of a wheel **3**. Radial animals have no front or back end. They attach to an underwater surface or drift along, so their food can arrive from any direction. Most animals have **bilateral symmetry**. They have a right and left half, with body parts repeated on either side of the body **4**. Bilateral animals have a distinctive "head end" that has a concentration of nerve cells.

Animals with a three-layer embryo are classified into two groups that differ in the way they develop as embryos. In protostomes, the first opening that forms on an embryo becomes the mouth **5**. *Proto–* means first and *stoma* means opening. In deuterostomes, the second opening becomes the mouth **6**.

All animals take in and digest food, then expel digestive wastes, but details of the process vary. In sponges, digestion is intracellular. Cnidarians and flatworms digest food inside a saclike gut called a **gastrovascular cavity**. Food enters this cavity through the same opening that expels wastes. Most bilateral animals have a tubular gut, or **complete digestive tract**, with an opening at either end. A tubular gut has advantages. Parts of the tube can be specialized for taking in food, digesting food, absorbing nutrients, or compacting waste. Unlike a saclike cavity, a tubular gut carries out all these tasks simultaneously.

A mass of tissues and organs surrounds a flatworm's gut (Figure 15.4**A**). However, most animals have a "tube within a tube" body plan, in which a fluid-filled body cavity surrounds the gut. Typically, this cavity is lined with tissue derived

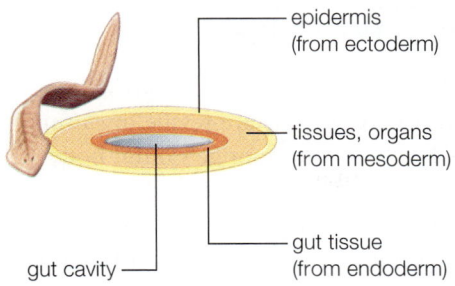

A Acoelomate body plan of a flatworm.

epidermis (from ectoderm)

tissues, organs (from mesoderm)

gut cavity

gut tissue (from endoderm)

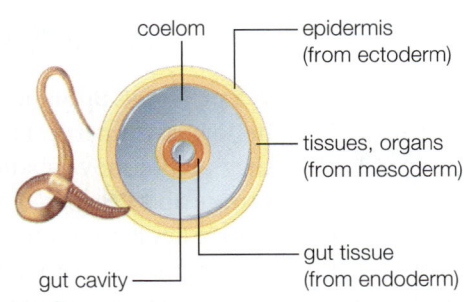

B Coelomate body plan of an earthworm.

coelom

epidermis (from ectoderm)

tissues, organs (from mesoderm)

gut cavity

gut tissue (from endoderm)

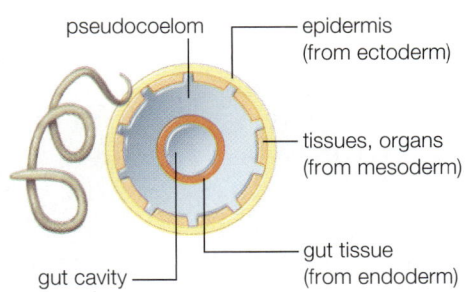

C Pseudocoelomate body plan of a roundworm.

pseudocoelom

epidermis (from ectoderm)

tissues, organs (from mesoderm)

gut cavity

gut tissue (from endoderm)

from mesoderm, and is called a **coelom**. Earthworms have this type of body plan (Figure 15.4**B**). A few invertebrates, such as roundworms, have a pseudocoelom, with only a partial mesodermal lining (Figure 15.4**C**).

Evolution of a fluid-filled body cavity provided a number of benefits. First, materials can diffuse through the fluid to body cells. Second, muscles can redistribute the fluid to alter the shape of body parts in ways that allow movement. Finally, internal organs were not hemmed in by a mass of tissue, so they could become larger and move more freely.

Gases and nutrients diffuse quickly through the body of a small animal. However, diffusion alone cannot move substances through a large body fast enough to keep its cells alive. In most animals, a circulatory system speeds the distribution of substances through the body. In an open circulatory system, blood is pumped out of vessels into internal spaces, from which it is taken up again by a heart. A closed system allows for faster blood flow than an open one. In a closed circulatory system, a heart or hearts propel blood through a continuous system of vessels. Materials carried by the blood diffuse out of vessels and into cells, and vice versa.

Segmentation is common in bilateral animals, meaning similar units are repeated along the length of the body. We clearly see body segments in annelids such as earthworms. We also find clues to our own origins in the segmented body of an early human embryo. Segmentation opened the way to evolutionary innovations in body form. When many segments have organs that carry out the same function, some segments can become modified without endangering the animal's survival.

Of about 2 million named animals, only about 50,000 are **vertebrates**: animals that have a backbone. These include fishes, amphibians, reptiles, birds, and mammals. The vast majority of animals are **invertebrates**: animals that do not have a backbone. Invertebrates are the starting point for our survey of animal diversity beginning in the next section.

FIGURE 15.4 Animated! Variations in body plans among bilateral animals. The diagrams show a cross-section through the body. Relative width of tissue layers is not shown to scale.

Figure It Out: Which of the animals shown above develops from an embryo with three tissue layers?

© Cengage Learning.

Answer: *All three do.*

Take-Home Message

What are animals?

- Animals are multicelled heterotrophs that typically ingest food. Their cells are unwalled.

- Animals reproduce sexually and, in many cases, asexually. They go through a period of embryonic development, and most move about during at least part of the life cycle.

- Animal bodies differ in their symmetry, type of digestive cavity, type of body cavity (if any), and degree of segmentation.

- Most animals are invertebrates (they do not have a backbone).

bilateral symmetry Having right and left halves with similar parts, and a front and back that differ.

coelom A body cavity completely lined by tissue derived from mesoderm.

complete digestive tract Tubular gut.

gastrovascular cavity Saclike gut.

invertebrate Animal without a backbone.

radial symmetry Having parts arranged around a central axis, like spokes around a wheel.

tissue One or more types of cells that are organized in a specific pattern and that carry out a particular task.

vertebrate Animal with a backbone.

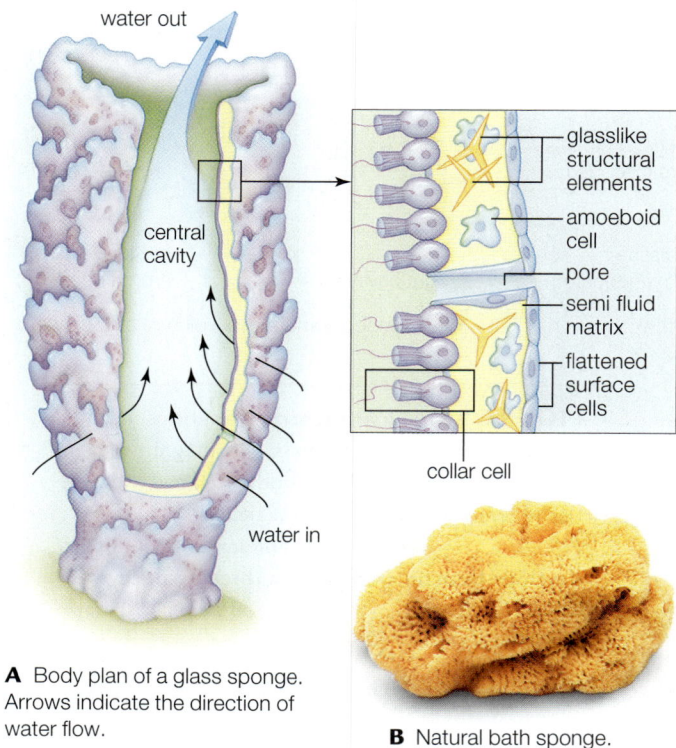

A Body plan of a glass sponge. Arrows indicate the direction of water flow.

water out

central cavity

water in

glasslike structural elements

amoeboid cell

pore

semi fluid matrix

flattened surface cells

collar cell

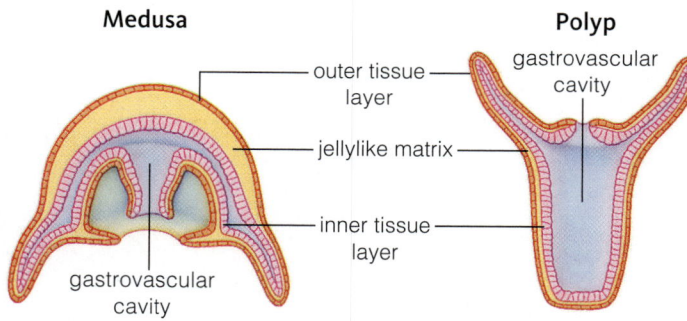

B Natural bath sponge.

FIGURE 15.5 Animated! Sponges, animals that have no body symmetry or tissues.

Credits: (a) © Cengage Learning; (b) Image © ultimathule/ Shutterstock.

Medusa

Polyp

outer tissue layer

gastrovascular cavity

jellylike matrix

inner tissue layer

gastrovascular cavity

A Structure of the two body plans.

B Jelly, a medusa.

C Sea anemone, a polyp.

FIGURE 15.6 Animated! Cnidarian body plans.

Credits: (a) © Cengage Learning; (b) © Boris Pamikov/ Shutterstock; (c) © Brandon D. Cole/ Corbis.

15.3 Invertebrate Diversity

❯ Sponges Sponges (phylum Porifera) are aquatic animals with a hollow, asymmetrical body (Figure 15.5**A**). Adults are sessile, meaning they do not move about. Flattened cells cover a sponge's outer body surface and flagellated collar cells line its internal cavities. A jellylike matrix lies between these cell layers. Movement of the collar cells' flagella causes water to flow in through many pores, then out through one or more larger openings.

In sponges, unlike most animals, digestion is intracellular. Collar cells filter bits of food from the water, engulf them by endocytosis, then digest them. Amoeba-like cells in the matrix receive digested food from collar cells and distribute it to other cells. In many species, cells in the matrix also secrete fibrous proteins or glassy spikes (called spicules) that structurally support the body and discourage predators. Some protein-rich sponges are harvested from the sea, dried, cleaned, and bleached, then sold for bathing and cleaning (Figure 15.5**B**).

A typical sponge is a **hermaphrodite**: an individual that produces both eggs and sperm. Usually sperm are released into the water, and eggs are retained by the parent. After fertilization, the zygote develops into a ciliated larva. A **larva** (plural, larvae) is a free-living, sexually immature form in an animal life cycle. Sponge larvae exit the parental body, swim briefly, then settle and develop into adults.

❯ Cnidarians Cnidarians (phylum Cnidaria) are aquatic, radially symmetrical animals with stinging tentacles. The two common body plans, medusa and polyp, both consist of two tissue layers with a secreted jellylike matrix between them (Figure 15.6**A**). A bell-shaped **medusa** (plural, medusae) swims or drifts about. Jellies (jellyfish) are medusae (Figure 15.6**B**). A **polyp**, such as a sea anemone, is tubular, and one end usually attaches to a surface (Figure 15.6**C**). In both polyps and medusae, tentacles ring the entrance to a gastrovascular cavity that takes in and digests food, expels wastes, and functions in gas exchange.

The phylum name Cnidaria is derived from *cnidos*, the Greek word for nettle, a kind of stinging plant. The name refers to unique stinging

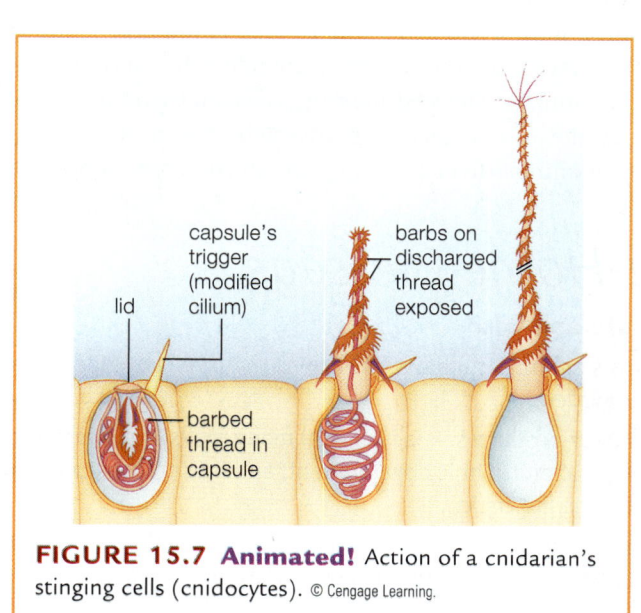

lid

capsule's trigger (modified cilium)

barbs on discharged thread exposed

barbed thread in capsule

FIGURE 15.7 Animated! Action of a cnidarian's stinging cells (cnidocytes). © Cengage Learning.

cells (cnidocytes) at the surface of the tentacles. The cells have a capsule-like organelle (a nematocyst) with coiled thread inside it (Figure 15.7). Touch causes the capsule to pop open, forcing the thread outward. The thread entangles prey or pierces it and delivers venom. Tentacles push prey through the mouth into the gastrovascular cavity, where gland cells secrete enzymes that digest it.

Reef-building corals are polyps that capture prey with their tentacles, but also rely on sugars made by photosynthetic protists (dinoflagellates) that live in their tissues. Each polyp secretes a hard, calcium-rich skeleton around its base. Over time, skeletal remains of many generations of polyps accumulate as a reef, with a layer of living polyps at its surface. Coral reefs are of great ecological importance because they provide food and shelter for many types of animals.

⟩ Flatworms Flatworms (phylum Platyhelminthes) are flat-bodied worms that do not have a body cavity. They are the simplest animals that develop from a three-layered embryo. Many flatworms are aquatic, but a few live in damp places on land. Still others live as parasites inside animals.

Planarians are free-living flatworms common in ponds. Cilia on their surface allow them to glide along. A muscular tube (a pharynx) sucks food into a highly branched gastrovascular cavity (Figure 15.8). Nutrients and oxygen diffuse from the fine branches to body cells. Clusters of nerve cells in the head serve as a simple brain. The head also has chemical receptors and light-detecting eyespots.

Tapeworms are hermaphroditic flatworms that infect the vertebrate gut. A tapeworm has a segmented body, and it adds new segments as it grows. Figure 15.9 shows the life cycle of the beef tapeworm, which can infect people who eat undercooked beef. Eating undercooked pork or fish can also cause a human tapeworm infection.

Flukes, also parasites, have an unsegmented body with suckers. Some tropical freshwater snails carry blood flukes that infect human intestinal veins and cause schistosomiasis, a potentially fatal disease. People become infected when they wade in water containing larval flukes, which enter the body through a cut.

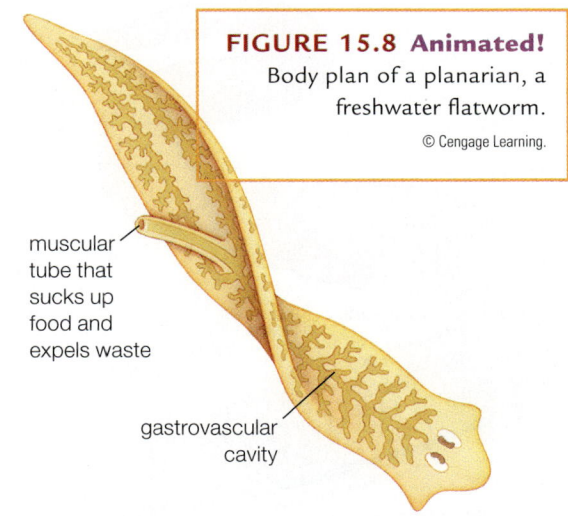

FIGURE 15.8 Animated!
Body plan of a planarian, a freshwater flatworm.
© Cengage Learning.

muscular tube that sucks up food and expels waste

gastrovascular cavity

cnidarian Radially symmetrical invertebrate with two tissue layers; uses tentacles with stinging cells to capture food.
flatworm Bilaterally symmetrical invertebrate with organs but no body cavity; for example, a planarian or tapeworm.
hermaphrodite Animal that makes both eggs and sperm.
larva Preadult stage in some animal life cycles.
medusa Bell-shaped, free-swimming cnidarian body form.
polyp Tubular, typically sessile, cnidarian body form.
sponge Aquatic invertebrate that has no tissues or organs and filters food from the water.

FIGURE 15.9 Animated!
Life cycle of the beef tapeworm.

❶ A person becomes infected by eating undercooked beef that contains the resting stage of the tapeworm.

❷ In the human intestine, a barbed scolex at the worm's front end attaches it to the intestinal wall. The worm grows by adding new body units called proglottids. Over time, it can become many meters long.

❸ Each proglottid produces both eggs and sperm, which combine. Proglottids with fertilized eggs leave the body in feces.

❹ Cattle eat grass contaminated with proglottids or early larvae.

❺ The larval tapeworm forms a cyst in the cattle muscle.

Credits: art, © Cengage Learning; right, Andrew Syred/ Photo Researchers, Inc.

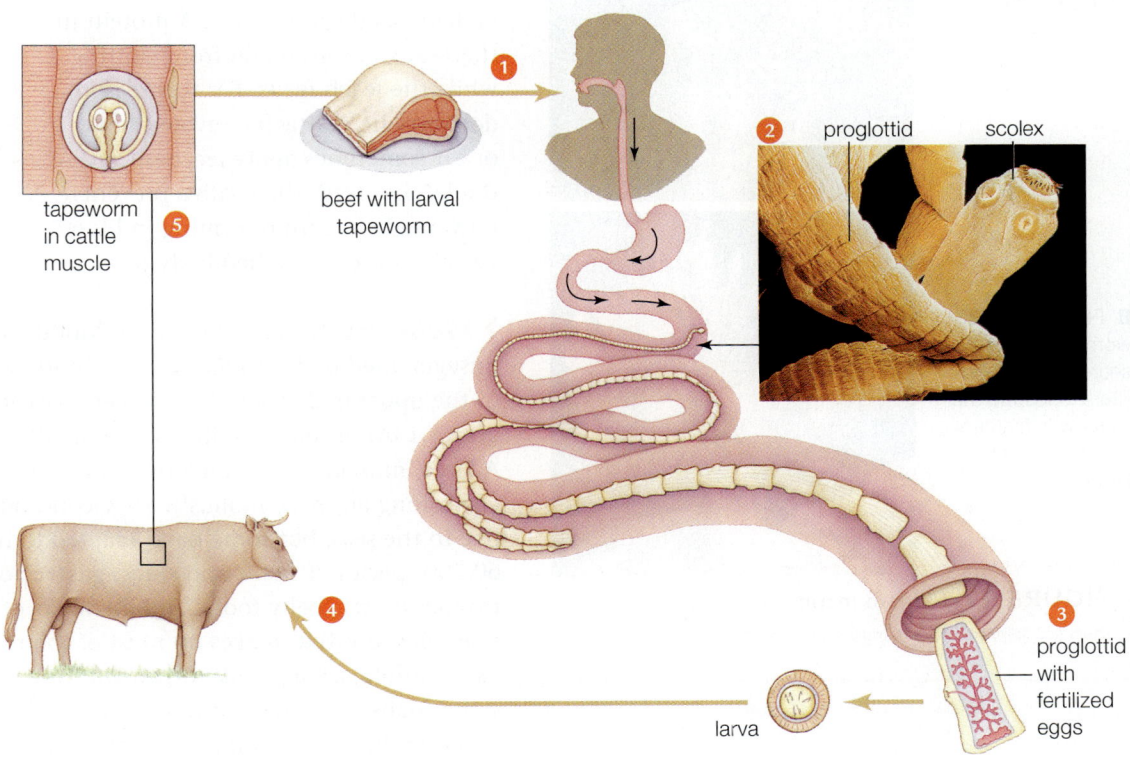

tapeworm in cattle muscle

beef with larval tapeworm

proglottid   scolex

larva

proglottid with fertilized eggs

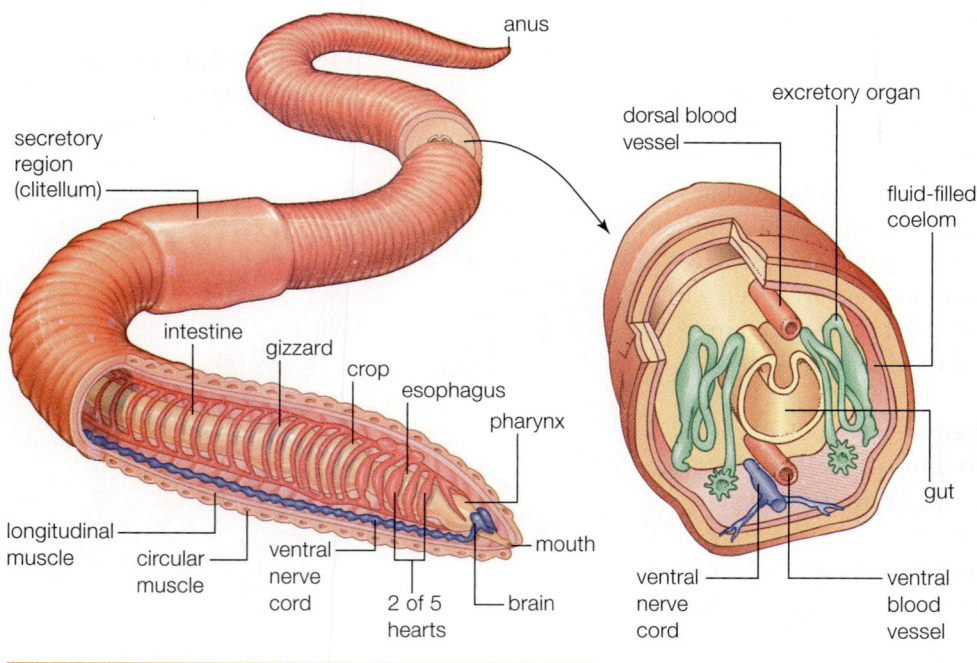

anus

secretory region (clitellum)

intestine
gizzard
crop
esophagus
pharynx

longitudinal muscle
circular muscle
ventral nerve cord
2 of 5 hearts
brain
mouth

dorsal blood vessel
excretory organ
fluid-filled coelom
gut
ventral nerve cord
ventral blood vessel

FIGURE 15.10 Animated! Earthworm body plan.

After Solomon, 8th edition, p. 624, figure 29-4, © Brooks/Cole Cengage Learning.

> **Annelids** Annelids (phylum Annelida) are segmented worms with a coelom, a complete digestive system, and a closed circulatory system. The phylum name comes from the Greek word *annulus*, which means ringed.

Earthworms, the most familiar annelids, live on land and most have more than one hundred segments (Figure 15.10). An earthworm eats its way through the soil, digesting any organic material that it takes in.

Each segment has a fluid-filled coelomic chamber with paired excretory organs that remove waste from the fluid. A digestive tract with specialized regions runs through the coelom. A nerve cord extends the length of the body and connects to a simple brain. Multiple hearts pump blood through vessels.

Earthworms are hermaphrodites. Mucus produced by a secretory region, the clitellum, glues worms together while they exchange sperm. Later, the clitellum secretes a silky case for the fertilized eggs the worm deposits in the soil.

Polychaetes are a lineage of mostly marine annelids that typically have many bristles per segment. (*Poly*– means many and *chaete* means bristle.) Some polychaetes, including the sandworms often sold as bait, are active predators (Figure 15.11**A**). Others have appendages specialized for filtering food from currents (Figure 15.11**B**).

Leeches, another annelid lineage, typically live in fresh water, although some are found in damp places on land. Most are scavengers and predators of invertebrates. A few suck blood from vertebrates (Figure 15.12). A protein in their saliva keeps blood from clotting while the leech feeds. For this reason, doctors who reattach a severed finger or ear sometimes apply leeches to it. As the leeches feed, their saliva prevents unwanted clots from forming in blood vessels of the reattached body part.

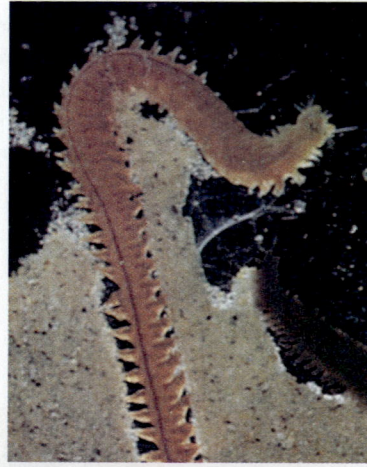

A Sandworm (*Nereis*), an active predator that patrols the intertidal zone of sandy shores.

B Feather duster worm. It lives in a secreted tube and filters food from the water with feathery tentacles on its head.

FIGURE 15.11 Two marine polychaetes.

Credits: (a) J. Solliday/ BPS; (b) Jon Kenfield/ Bruce Coleman Ltd.

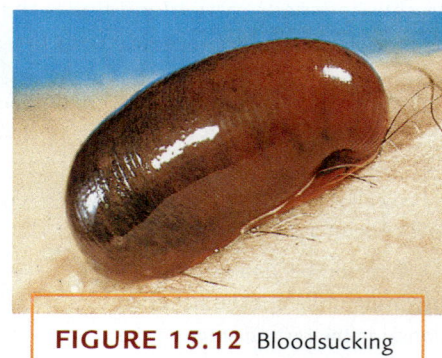

FIGURE 15.12 Bloodsucking leech. J. A. L. Cooke/ Oxford Scientific Films.

> **Mollusks** Mollusks (phylum Mollusca) have a small coelom and a soft, unsegmented body. *Mulluscus* is Latin for soft. The **mantle**, a skirtlike extension of the upper body wall, drapes over a visceral mass that contains the organs (Figure 15.13**A**). In most mollusks, the mantle secretes a hard, calcium-rich shell. A large, muscular foot functions in locomotion.

Among animals, mollusks are second only to arthropods in diversity. Most live in the seas, but some have adapted to life in fresh water or on land. With 60,000 species of snails and slugs, **gastropods** are the largest subgroup. Gastropod means "belly foot," and members of this group glide about on a broad muscular foot that makes up most of their lower body mass. Most gastropods use a **radula**, a tonguelike organ with bits of hard chitin, to scrape up algae, but some snails have adapted to a predatory lifestyle. For example, cone snails use a harpoonlike radula to inject a paralyzing venom into prey such as small fish.

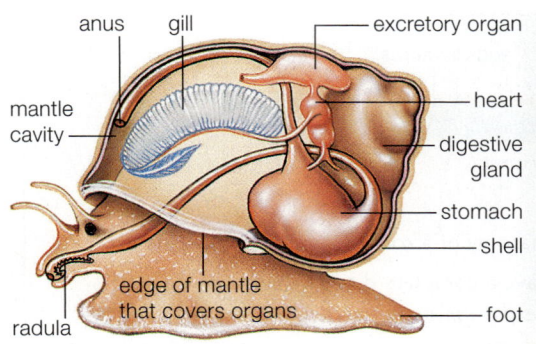

anus — gill — excretory organ

mantle cavity

heart

digestive gland

stomach

shell

edge of mantle that covers organs

radula

foot

A Body plan of an aquatic snail (a gastropod).

B Nudibranchs (sea slugs), gastropods that do not have a shell.

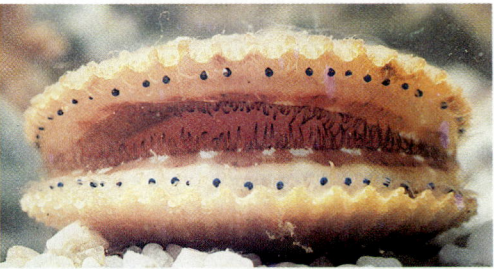

C Scallop, a bivalve, with a two-part shell. Blue dots are eyes at the edge of the mantle.

FIGURE 15.13 Animated! Mollusk structure and diversity.

Credits: (a) Cengage Learning; (b,d) Alex Kirstitch; (c) Frank Park/ ANT Photo Library; (e) NURC/UNCW and NOAA/FGBNMS.

Some snails and slugs have adapted to life on land. Their mantle cavity encloses a lung that allows them to breathe air. Both land slugs and sea slugs have lost their shell. Most defend themselves by secreting substances that predators find distasteful. The sea slugs shown in Figure 15.13**B** have a more interesting defense. They feed on cnidarians and store undischarged nematocysts from their prey in feathery outgrowths on their backs. A predator that attacks the nudibranch gets stung by secondhand nematocysts.

Most mollusks that end up on dinner plates, including mussels, oysters, clams, and scallops, are **bivalves** (Figure 15.13**C**). About 15,000 species of bivalves live in fresh water and the seas. Their hinged, two-part shell encloses a body that does not have a distinct head or a radula. Bivalves typically attach to a surface or burrow into sediment. They feed by drawing water into the mantle cavity and filtering out bits of food. Being filter-feeders, bivalves occasionally take in toxins or pathogens that can sicken people who eat them. Risk of illness is greatest when bivalves are eaten raw or only partially cooked.

Cephalopods include cuttlefish, squids, nautiluses (Figure 15.13**D**), and octopuses (Figure 15.13**E**). Cephalopod means "head-footed," and their foot has been modified into tentacles and/or arms that extend from the head. All cephalopods are predators and most have beaklike, biting mouthparts in addition to a radula. Cephalopods move by jet propulsion, drawing water into their mantle cavity, then forcing it out through a funnel-shaped siphon. Of all mollusks, only cephalopods have a closed circulatory system.

Five hundred million years ago, cephalopods with a long conelike shell were the top predators in the seas. Some had shells 5 meters (15 feet) long. Today, nautiluses have a coiled external shell, but other cephalopods have a highly reduced shell or none at all. The evolution of jawed fishes 400 million years ago was most likely the impetus for this shift in body form. When jawed fishes began to compete with cephalopods for prey, fast-moving, agile cephalopods were at an advantage. Reduction or loss of the shell improved speed and agility.

Competition with fishes also favored improved eyesight. Like vertebrates, cephalopods have eyes with a lens that focuses light. Because mollusks and vertebrates are not closely related, this type of eye is assumed to have evolved independently in these groups.

Cephalopods include the fastest (squids), biggest (giant squid), and smartest (octopuses) invertebrates. Of all invertebrates, octopuses have the largest brain relative to body size, and show the most complex behavior.

D Chambered nautilus, the only modern cephalopod enclosed within a shell.

E Octopus, a cephalopod. Individually controlled suckers on its eight arms allow it to hold onto objects, and to stick to and walk on surfaces.

annelid Segmented worm with a coelom, complete digestive system, and closed circulatory system.

bivalve Mollusk with a hinged two-part shell.

cephalopod Predatory mollusk with a closed circulatory system; moves by jet propulsion.

gastropod Mollusk that moves about on its enlarged foot.

mantle Skirtlike extension of tissue in mollusks; covers the mantle cavity and secretes the shell in species that have a shell.

mollusk Invertebrate with a reduced coelom and a mantle.

radula Tonguelike organ of many mollusks.

A Intestinal parasite, *Ascaris*, passed by a child in Kenya.

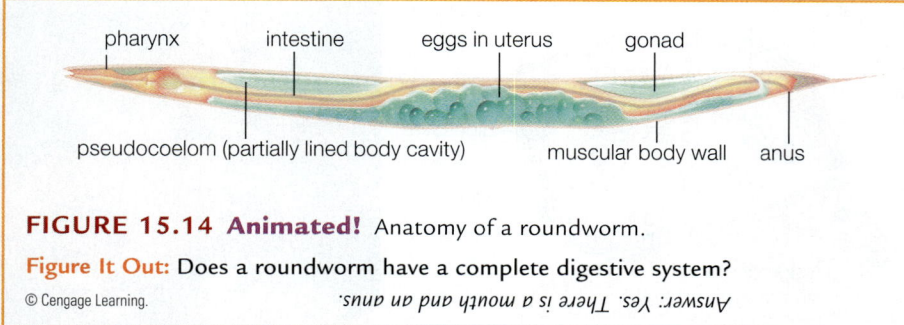

pharynx   intestine   eggs in uterus   gonad

pseudocoelom (partially lined body cavity)   muscular body wall   anus

FIGURE 15.14 Animated! Anatomy of a roundworm.

Figure It Out: Does a roundworm have a complete digestive system?

© Cengage Learning.

Answer: Yes. There is a mouth and an anus.

B Lymphatic filariasis damages lymph vessels so fluid pools in lower body parts, as in this man's left leg.

FIGURE 15.15 Roundworms and disease.

Credits: (a) CDC/ Henry Bishop; (b) Courtesy of © Emily Howard Staub and The Carter Center.

> **Roundworms** **Roundworms**, or nematodes (phylum Nematoda), are unsegmented, cylindrical worms with a pseudocoelom (Figure 15.14). Roundworms have a complete digestive system, excretory organs, and a nervous system, but no circulatory or respiratory organs. Like insects, roundworms secrete a protective body covering, or cuticle, that is periodically molted (shed) and replaced as the worm increases in size.

Roundworms live in the seas, in fresh water, in damp soil, and inside other animals. Most of the nearly 20,000 species are free-living decomposers less than a millimeter long. Parasitic roundworms tend to be larger, and some that live inside whales can extend for meters.

Like the fruit fly, the soil roundworm *Caenorhabditis elegans* is commonly used in scientific studies. It serves as a useful model for developmental processes because it has the same tissue types as more complex organisms, but it is transparent, has less than 1,000 body cells, and reproduces fast. In addition, its genome is about 1/30 the size of the human genome. Such traits make it easy for scientists to monitor each cell's fate during development. In 2002, the Nobel Prize in Medicine was awarded to scientists whose studies of *C. elegans* revealed how genes control the selective death of cells during normal development.

Several kinds of parasitic roundworms live inside humans. The giant intestinal roundworm, *Ascaris lumbricoides* (Figure 15.15A), currently infects more than 1 billion people. Most live in developing tropical nations, but infections also occur in rural parts of the American Southeast. Typically an infected person has no symptoms, but a heavy infection can block the intestine.

Roundworms transmitted by mosquitoes cause another tropical disease called lymphatic filariasis. The worms travel through the body's lymph vessels. Repeated infections can injure valves in these vessels, causing lymph to pool in lower limbs (Figure 15.15B). Elephantiasis, the common name for this disease, refers to fluid-filled, "elephant-like" legs. The swelling is permanent because lymph vessels remain damaged even after the worms have been eliminated.

Pinworms (*Enterobius vermicularis*) are the most common infectious roundworms in the United States. Children are most often affected. The small worms, about the size of a staple, live in the rectum. At night, females crawl out and lay eggs on the skin around the anus. Their movement produces an itching sensation and scratching to relieve the itch puts eggs onto fingertips and under fingernails. If swallowed, they start a new infection.

Roundworms also affect our livestock, pets, and crop plants. A roundworm that infects pigs can also infect humans who eat undercooked pork, causing trichinosis. Dogs are susceptible to heartworms transmitted by mosquitoes. Cats usually become infected by eating an infected rodent that serves as an intermediate host. Many crop plants are susceptible to nematode parasites. In some cases, the worms suck on plant roots and in others they actually enter the plant. Either way, the infection stunts growth and lowers yields.

> **Arthropod Traits** **Arthropods** (phylum Arthropoda) are invertebrates with jointed legs. They have a complete digestive system and an open circulatory system. Researchers have identified more than a million living species and many more remain to be discovered.

Arthropods secrete a cuticle hardened with chitin, the same material in the mollusk radula. The cuticle serves as an external skeleton—an **exoskeleton**. It helps fend off predators and serves as a point of attachment for muscles. Among land arthropods, the exoskeleton also helps conserve water and supports the animal's weight. A hardened exoskeleton does not restrict growth, because, like roundworms, arthropods molt their cuticle after each growth spurt. A new cuticle forms under the old one, which is shed (Figure 15.16).

If an arthropod's cuticle were uniformly hard like a plaster cast, it would prevent movement. However, the cuticle thins at joints, regions where two hard body parts meet. "Arthropod" means jointed leg. Body parts move when the muscles that attach to the exoskeleton on either side of a joint contract.

In early arthropods, body segments were distinct and all appendages were alike. In many groups, segments later fused into structural units such as a head, thorax (midsection), and abdomen (hind section). Specialized appendages such as wings developed on some segments.

A typical arthropod has one or more pairs of eyes. Insects and crustaceans have **compound eyes** that consist of many individual units, each with its own lens. Such eyes are highly sensitive to movement. Most arthropods also have one or two pairs of **antennae**, sensory structures on the head that detect touch, odor, and vibrations.

Most arthropods undergo a change in body form during their life cycle. They undergo **metamorphosis**: The body plan is remodeled as larvae develop into adults. The monarch butterfly is an example. Its larvae are caterpillars that crawl about and feed on milkweed plants (Figure 15.17**A**). The caterpillar grows and molts several times, before attaching its hind end to a branch and molting its final larval cuticle to become a pupa (Figure 15.17**B**). Tissue remodeling then produces the winged adult, a monarch butterfly that sips floral nectar (Figure 15.17**C**). The caterpillar and the butterfly each have a body plan that adapts them to different tasks. Having such different bodies prevents adults and larvae from competing for the same resources.

> **Arthropod Diversity and Ecology** Horseshoe crabs have lived in Earth's oceans for more than 400 million years, with little change in form. One of the four modern species lives in waters along North America's Atlantic and Gulf coasts. The group's common name refers to a horseshoe-shaped covering over the head and thorax (Figure 15.18). A long spine that extends from the last abdominal segment helps the animal right itself after a wave flips it over. Horseshoe crabs are bottom feeders that eat clams and worms.

FIGURE 15.16 Centipede molting its old cuticle and revealing a new red one beneath.

Jane Burton/ Bruce Coleman, Ltd.

A Larva (caterpillar).   **B** Pupa.   **C** Adult.

FIGURE 15.17 Complete metamorphosis in the monarch butterfly. A leaf-eating larvae becomes a pupa. Tissue remodeling during the pupal stage yields the adult, a butterfly.

Credits: (a,b) © Jacob Hamblin/ Shutterstock; (c) © Laurie Barr/ Shutterstock.

FIGURE 15.18
Horseshoe crab. It is not a true crab, but rather a close relative of spiders and scorpions.

© Satin/ Shutterstock.

antenna Of some arthropods, sensory structure on the head that detects touch and odors.

arthropod Invertebrate with jointed legs and a hardened exoskeleton that is periodically molted.

compound eye Eye that consists of many individual units, each with its own lens.

exoskeleton External skeleton.

metamorphosis Dramatic remodeling of body form during the transition from larva to adult.

roundworm Unsegmented worm with a pseudocoelom and a cuticle that is molted as the animal grows.

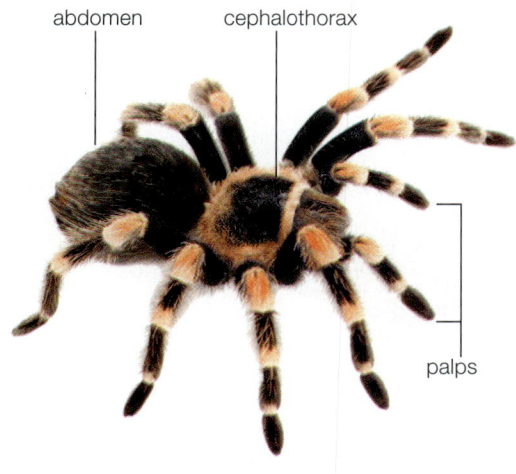

abdomen   cephalothorax

palps

A Spider (tarantula). It kills prey with a venomous bite.

palps

stinger

B Scorpion. It delivers venom through a stinger.

C Tick. It sucks blood from vertebrates.

> **FIGURE 15.19** Arachnids. All have four pairs of legs.
>
> Credits: (a) © Eric Isselée/ Shutterstock.com; (b) © Frans Lemmens/ The Image Bank/ Getty Images; (c) CDC/ Dr. Christopher Paddock.

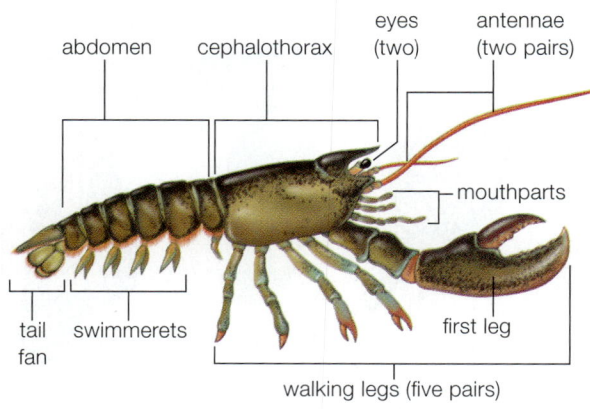

abdomen cephalothorax eyes (two) antennae (two pairs)

mouthparts

tail fan swimmerets first leg

walking legs (five pairs)

A Body plan of an American lobster.

B Krill swimming in Antarctic waters.

C Barnacles filtering food from the water with their feathery legs.

> **FIGURE 15.20**
>
> Marine crustaceans.
>
> Credits: (a) Cengage Learning; (b) © David Tipling/ Photographer's Choice/ Getty Images; (c) © Peter Parks/ Imagequestmarine.com.

Arachnids have four pairs of walking legs, a pair of touch-sensitive palps, and no antennae. They include spiders, scorpions, ticks, and mites. Spiders and scorpions are venomous predators. Spiders deliver venom through fanglike mouthparts. They have a two-part body, with the cephalothorax (fused head and thorax) separated from the abdomen by a narrow "waist" (Figure 15.19**A**). The abdomen contains silk-making glands. Scorpions catch their prey with claw-like palps and subdue it with a venom-producing stinger on the final abdominal segment (Figure 15.19**B**). They eat insects, spiders, and even small lizards. Ticks suck blood from vertebrates (Figure 15.19**C**). Some serve as vectors for bacterial diseases, such as Lyme disease (Section 13.5). Mites, the smallest arachnids, are usually less than a millimeter long. Many are scavengers, but some are parasites. A mite that burrows beneath the skin causes scabies in humans and mange in dogs. Some larval mites, commonly called chiggers, crawl into a hair follicle, secrete an enzyme that breaks down proteins, then suck up liquefied tissues.

Crustaceans are mostly marine arthropods that have two pairs of antennae. Humans harvest many decapod (ten-footed) crustaceans such as shrimps, crabs, and lobsters (Figure 15.20**A**) as food. All are bottom-feeding scavengers that have five pairs of legs. Other animals also depend on crustaceans for food. Shrimplike crustaceans called krill abound in cool ocean waters (Figure 15.20**B**). Each is only a few centimeters long, but they are so abundant and nutritious that a 100-ton blue whale can subsist mainly on the krill that it filters from seawater.

Barnacles are marine crustaceans and the only arthropods that secrete an external shell (Figure 15.20**C**). The calcium-rich shell closes up tight to protect the animal from predators, drying winds, and surf. Barnacle larvae swim, then settle and develop into adults that attach by their heads to rocks, piers, boats, and even whales. They feed by filtering food from seawater with their feathery legs. Given that a barnacle is glued in place at its head end, you might think finding a sexual partner could be challenging. But barnacles tend to cluster in groups and most are hermaphrodites. A barnacle extends its penis, which can be several times the body length, out to its neighbors.

Isopods are a mostly marine group of crustaceans, but you are most likely to be familiar with the species that have adapted to life on land. Commonly called sowbugs and pill bugs, they live in damp places and feed on organic debris or on soft, young plant parts. Some can be pests of agricultural crops. The species commonly known as pill bugs defend themselves from threats by rolling into a ball (*right*).

Centipedes and millipedes belong to another arthropod lineage. These nocturnal ground dwellers have a long body with many similar segments (Figure

© Chris Howey/ Shutterstock.

15.21). The head has paired antennae and two simple eyes. Centipedes are speedy venomous predators with a flat, low-slung body and one pair of legs per segment. The slower-moving millipedes eat decaying vegetation. Their cylindrical body has a cuticle hardened by calcium carbonate and two legs per segment.

Insects, the most diverse arthropod group, have a three-part body: (1) a head with a single pair of antennae, (2) a thorax with three pairs of legs, and (3) an abdomen (Figure 15.22**A**). The four most diverse subgroups all have two pairs of wings on the thorax. There are about 150,000 species of flies, and at least as many beetles, or coleopterans. The 130,000 hymenopterans include bees, wasps, and ants. The moths and butterflies contribute about 120,000 species. As a comparison, consider that there are about 4,500 species of mammals.

Most flowering plants are pollinated by a member of one of the four most diverse insect groups (Figure 15.22**B**). Interactions between these pollinators and flowering plants likely contributed to an increased rate of speciation in both.

Insects have other important ecological roles. For example, they serve as food for many kinds of wildlife. Larval moths and butterflies, commonly called caterpillars or inchworms, feed songbird nestlings. Aquatic larvae of some insects such as dragonflies serve as food for fish. Most amphibians and reptiles feed mainly on insects. In addition, insects dispose of wastes and remains. Flies and beetles are quick to find an animal corpse or a pile of feces (Figure 15.22**C**). They lay eggs in or on this organic material, and the larvae that hatch devour it. By their actions, these insects help distribute nutrients through the ecosystem.

On the other hand, some insects are pests. Insects are our main competitors for crop plants. For example, Mediterranean fruit flies, or medflies, damage citrus crops (Figure 15.22**D**). Insects also transmit dangerous diseases. Mosquitoes spread malaria and carry parasitic roundworms between human hosts. Fleas that bite rats and then bite humans can transmit bubonic plague. Body lice can transmit typhus. Bedbugs (Figure 15.22**E**) do not transmit disease, but their bites itch and infestations can have severe psychological and economic effects.

A Centipede, a speedy predator.

B Millipede, a scavenger of decaying plant material.

FIGURE 15.21 Centipede and millipede.

© Eric Isselée/ Shutterstock.

arachnids Land-dwelling arthropods with four pairs of walking legs and no antennae; for example, a spider, scorpion, or tick.

crustaceans Mostly marine arthropods with two pairs of antennae; for example, a shrimp, crab, lobster, or barnacle.

insects Land-dwelling arthropods with a pair of antennae, three pairs of legs, and—in the most diverse groups—wings.

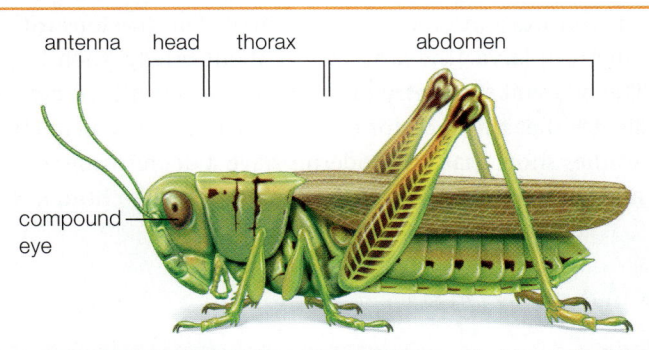

antenna   head   thorax   abdomen

compound eye

FIGURE 15.22 Insect structure and diversity.

Credits: (a) From RUSSELL/WOLFE/HERTZ/STARR, Biology, 1E. © 2008 Brooks/Cole, a part of Cengage Learning, Inc. Reproduced by permission. www.cengage.com/permissions; (b) Photo by Jack Dykinga, USDA, ARS; (c) Gregory G. Dimijian, M.D./ Photo Researchers, Inc.; (d) Photo by Scott Bauer/ USDA; (e) CDC/ Piotr Naskrecki.

A Insect body plan, as illustrated by a grasshopper. Three pairs of legs extend from the thorax, as do two pairs of wings (shown folded). The head has two antennae and two compound eyes.

B Bee serving as a pollinator.   **C** Dung beetle gathering feces.   **D** Medfly, a threat to citrus.   **E** Bedbug, a human parasite.

FIGURE 15.23 Echinoderms.

A Sea star, and a close-up of its arm with little tube feet.

B Sea urchin, which moves about on spines and a few tube feet. Sea urchins typically graze on algae.

C Sea cucumber, with rows of tube feet along its elongated body. The hard parts have been reduced to microscopic plates embedded in a soft body.

Credits: (a) Herve Chaumeton/ Agence Nature; (b) © Derek Holzapfel/ photos.com; (c) Andrew David, NOAA/NMFS/SEFSC Panama City, Lance Horn, UNCW/ NURC-Phantom II ROV operator.

> **Echinoderms** Echinoderms (phylum Echinodermata) include about 6,000 marine invertebrates such as sea stars, sea urchins, and sea cucumbers (Figure 15.23). Their phylum name means spiny-skinned and refers to interlocking spines and plates of calcium carbonate embedded in their skin. The plates form an internal skeleton called an **endoskeleton**. Adults have a radially symmetrical body plan, but larvae are bilateral, suggesting the group had a bilateral ancestor.

Sea stars (sometimes called starfish) are the most familiar echinoderms. They do not have a brain, but they do have a decentralized nervous system. Eyespots at the tips of their arms detect light and movement. A typical sea star moves about on tiny, fluid-filled tube feet (Figure 15.23**A**). Tube feet are part of a **water–vascular system**, a system of fluid-filled tubes unique to echinoderms.

Most sea stars prey on bivalve mollusks. To feed, the sea star slides its stomach out through its mouth and into a bivalve's shell. The stomach secretes acid and enzymes that kill the mollusk and begin to digest it. Partially digested food enters the stomach and digestion is completed with the aid of digestive glands in the arms. A sea star's reproductive organs are also in its arms. Eggs and sperm are released into the water. Fertilization produces an embryo that develops into a ciliated, bilaterally symmetrical larva. The larva swims about briefly, then develops into an adult. The bilateral symmetry of the larva, along with evidence from genetic studies, indicates that the ancestor of echinoderms was a bilateral animal. Developmental studies show that echinoderms have a deuterostome developmental pattern, and thus are members of the same lineage as chordates.

Take-Home Message

What are the traits of the main invertebrate groups?

- Sponges are filter-feeders that have specialized cells but no tissues. Cnidarians have two tissue layers and a radial body plan. Most other invertebrates have a bilateral body plan with three tissue layers that form organ systems.

- Three groups of "worms" differ in their traits and are not closely related. Flatworms do not have a coelom. Annelids have a coelom and a segmented body. Roundworms have a pseudocoelom and are the only worms that molt.

- Mollusks and arthropods are the two most diverse invertebrate phyla. Mollusks are soft-bodied, although many make a shell. Arthropods have a hardened exoskeleton and jointed legs. Insects, an arthropod subgroup, are the most diverse invertebrates and the only winged ones.

- Echinoderms are radial animals with bilateral larvae. They are on the same branch of the animal family tree as the chordates.

chordates Animal phylum characterized by a notochord, dorsal nerve cord, pharyngeal gill slits, and a tail that extends beyond the anus. Includes invertebrate and vertebrate groups.

echinoderms Invertebrates with a water–vascular system and an endoskeleton made of hardened plates and spines.

endoskeleton Internal skeleton.

lancelets Invertebrate chordates that have a fishlike shape and retain their defining chordate traits into adulthood.

notochord Stiff rod of connective tissue that runs the length of the body in chordate larvae or embryos.

tunicates Invertebrate chordates that lose their defining chordate traits during the transition to adulthood.

water–vascular system Of echinoderms, a system of fluid-filled tubes and tube feet that function in locomotion.

15.4 Introducing the Chordates

> ## Chordate Traits
Chordates (phylum Chordata) are defined by four traits seen in their embryos: (1) A **notochord**, a rod of stiff but flexible connective tissue, extends the length of the body and provides support. (2) A dorsal, hollow nerve cord parallels the notochord. (3) Gill slits open across the wall of the pharynx (throat region). (4) A muscular tail extends beyond the anus. Depending on the chordate group, some, all, or none of these traits persist in the adult.

> ## Invertebrate Chordates
There are two groups of invertebrate chordates, tunicates and lancelets. Both are marine and both filter food from currents of water that pass through gill slits in their pharynx.

Lancelets (subphylum Cephalochordata) have a fishlike shape and are 3 to 7 centimeters long (Figure 15.24). Lancelets retain all characteristic chordate traits as adults. The dorsal nerve cord extends into the head, where a single eyespot at its tip detects light. However, the head does not have a brain or any paired sensory organs like those of fishes.

Tunicates (subphylum Urochordata) have larvae with typical chordate traits (Figure 15.25**A**). The larvae swim about briefly, then undergo metamorphosis to the adult form (Figure 15.25**B**). Of the four typical chordate traits, the adult retains only the pharynx with gill slits. A secreted carbohydrate-rich covering or "tunic" encloses the adult body and gives tunicates their common name. Most adult tunicates attach to an undersea surface and filter food from the water. As water flows in an oral opening and past gill slits, bits of food stick to mucus on the gills and are then sent to a gut. Water leaves through another body opening.

Which invertebrate chordate is most closely related to vertebrates? An adult lancelet looks more like a fish than an adult tunicate does, but such superficial similarities are sometimes deceiving. Studies of developmental processes and gene sequences suggest that tunicates are the closest invertebrate relatives of vertebrates. Keep in mind that neither tunicates nor lancelets are ancestors of vertebrates. These groups share a recent common relative, but each has unique traits that put it on a separate branch of the animal family tree.

A Dorsal, hollow nerve cord **B** Notochord **C** Pharynx with gill slits

eyespot

D Tail extends beyond anus

anus

FIGURE 15.24 Animated!
A lancelet. These marine invertebrates retain all chordate traits into adulthood. As the photo *below* shows, lancelets live in sandy sediments, and filter food from the water with their pharynx.

Credits: top, © Cengage Learning; bottom, Runk & Schoenberger/ Grant Heilman, Inc.

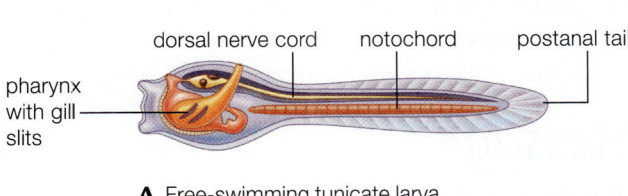

pharynx with gill slits

dorsal nerve cord notochord postanal tail

A Free-swimming tunicate larva.

B Photo and diagram of an adult tunicate.

water flows in

water flows out

pharynx with gill slits

secreted "tunic"

1 cm

FIGURE 15.25 Animated! Tunicates. Larval tunicates are free-swimming and have all the characteristic chordate features. After metamorphosis, the adult retains only the pharynx with gill slits.

Credits: (a, b right) From RUSSELL/WOLFE/HERTZ/STARR. Biology, 1E. © 2008 Brooks/Cole, a part of Cengage Learning, Inc. Reproduced by permission. www.cengage.com/permissions; (b left) © California Academy of Sciences.

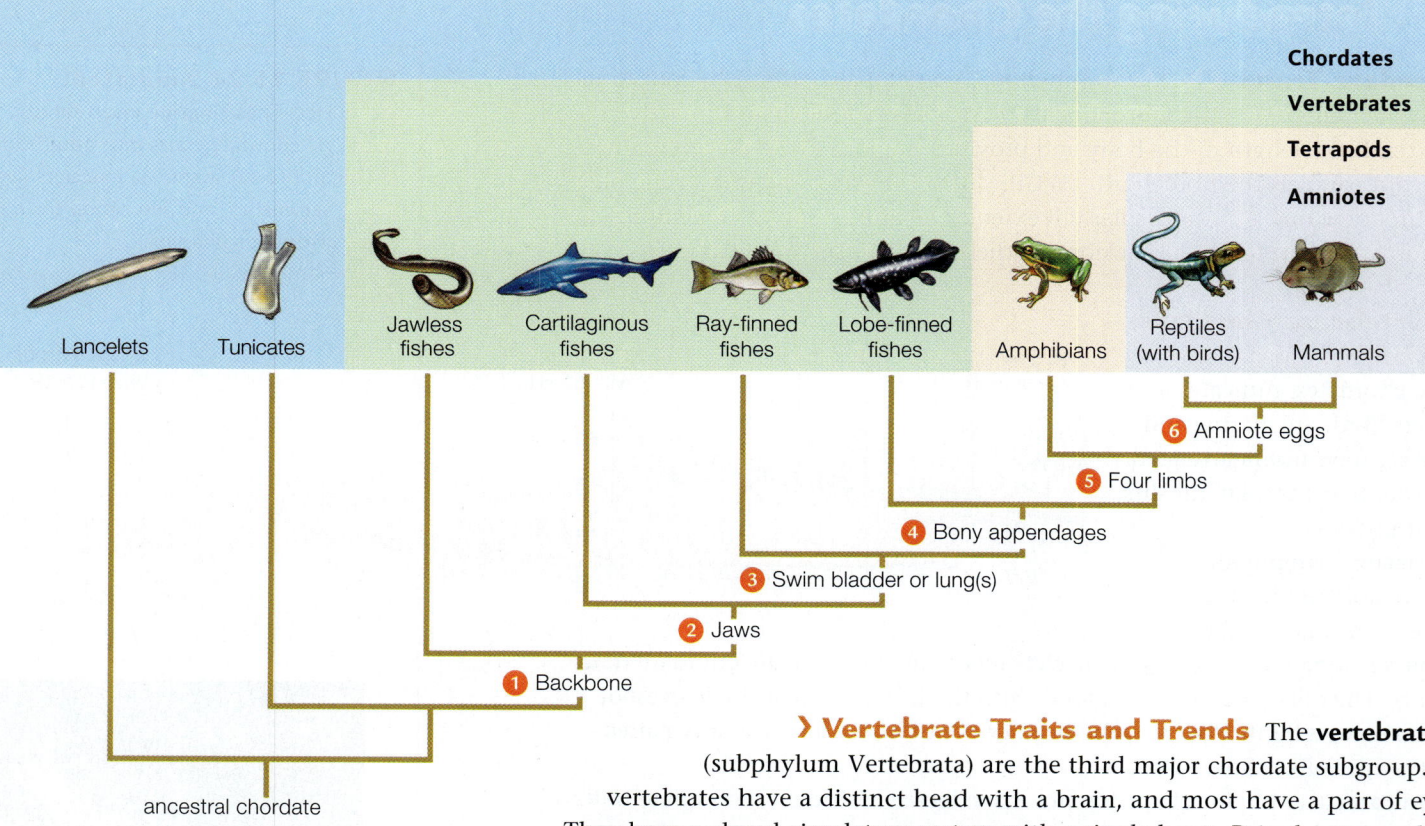

Chordates

Vertebrates

Tetrapods

Amniotes

Lancelets Tunicates Jawless fishes Cartilaginous fishes Ray-finned fishes Lobe-finned fishes Amphibians Reptiles (with birds) Mammals

6 Amniote eggs

5 Four limbs

4 Bony appendages

3 Swim bladder or lung(s)

2 Jaws

1 Backbone

ancestral chordate

FIGURE 15.26 Animated!
The chordate family tree. The vast majority of the chordates are vertebrates.
Figure It Out: Do the cartilaginous fishes have a backbone?
© Cengage Learning.   *Answer: Yes*

> **Vertebrate Traits and Trends** The **vertebrates** (subphylum Vertebrata) are the third major chordate subgroup. All vertebrates have a distinct head with a brain, and most have a pair of eyes. They have a closed circulatory system with a single heart. Paired organs called **kidneys** filter the blood, adjust its volume and solute composition, and eliminate wastes. The digestive system is complete.

Figure 15.26 shows the chordate family tree and notes the innovations that define various lineages. The first major was innovation a backbone **1**, a **vertebral column** that develops from the notochord of a vertebrate embryo. In most vertebrates, this flexible but sturdy structure encloses the spinal cord that develops from the nerve cord. The vertebral column is part of an endoskeleton.

The next major innovation was jaws, hinged skeletal elements used in feeding **2**. Jaws evolved by expansion of bony parts that structurally supported the gill slits of early jawless fishes. They opened up new feeding opportunities allowing an adaptive radiation. The vast majority of modern fishes have jaws.

Some jawed fish evolved internal air sacs **3**. In some species, such a sac functioned as a simple **lung**, a respiratory organ in which blood exchanges gases with the air. One lineage of fish with lungs also had bony paired fins **4**. Their descendants evolved bony paired limbs, becoming the first **tetrapods**, or four-legged walkers **5**. A special type of egg (the amniote egg) allowed one tetrapod group to lay eggs on land **6**. Members of that lineage, the amniotes (reptiles, birds, and mammals), are the most successful tetrapods on land.

Take-Home Message

What trends shaped chordate evolution?

- The earliest chordates were invertebrates, and two groups of invertebrate chordates (tunicates and lancelets) still exist. They are aquatic filter-feeders.

- Vertebrates evolved from an invertebrate chordate ancestor. All have a vertebral column, or backbone.

- Jaws, lungs, limbs, and waterproof eggs were key innovations that led to the adaptive radiation of vertebrates, first in the seas and then on the land.

15.5 Fishes and Amphibians

We begin our survey of vertebrate diversity with **fishes**, the first vertebrate lineage to evolve and the most fully aquatic.

> **The Jawless Fishes** Modern **jawless fishes** have a smooth, elongated body without paired fins. Their skeletal elements are made of cartilage, the same type of connective tissue that supports your nose and ears. Figure 15.27**A** shows the distinctive mouth of one jawless fish, a lamprey. As adults, most lampreys feed on other fish. Lacking jaws, a lamprey cannot bite. Instead, the lamprey attaches itself to a fish using an oral disk ringed by horny teeth made of the protein keratin. Once attached, the lamprey secretes enzymes and scrapes up bits of its host's flesh with a tooth-covered tongue. The host fish often dies from blood loss or a resulting infection.

> **Jawed Fishes** Most jawed fishes have paired fins and **scales**: hard, flattened structures that grow from and often cover the skin. Scales and an internal skeleton make a fish denser than water and prone to sinking. Highly active swimmers have fins with a shape that helps lift them, something like the way that wings help lift up an airplane. Water resists movements through it, so speedy swimmers typically have a streamlined body that reduces friction.

There are two groups of jawed fishes: cartilaginous fishes and bony fishes (Figure 15.27**B,C**). As their name implies, **cartilaginous fishes** are jawed fishes with a skeleton made of cartilage. They include 850 species, with sharks being the most well known. Some sharks are predators that swim in upper ocean waters. Others strain plankton from the water or suck up food from the seafloor. Human surfers and swimmers resemble typical prey of some predatory sharks, and rare attacks by a few species give the group an undeserved bad reputation. Worldwide, shark attacks kill about 25 people a year. For comparison, dogs kill about 30 people each year in the United States alone.

cartilaginous fish Fish that has jaws, paired fins, and a skeleton made of cartilage; for example, a shark.

fish Aquatic vertebrate of the oldest and most diverse vertebrate group.

jawless fish Fish that has a skeleton of cartilage, but no jaws or paired fins; for example, a lamprey.

kidney Organ of the vertebrate urinary system that filters blood and adjusts its composition.

lung Saclike organ inside which blood exchanges gases with the air.

scales Hard, flattened elements that cover the skin of reptiles and some fishes.

tetrapod Vertebrate with four limbs.

vertebral column Backbone.

vertebrate Animal with a backbone; a fish, amphibian, reptile, bird, or mammal.

A Jawless fish. A lamprey with no paired fins. It attaches to other fishes with its oral disk and scrapes at their flesh.

paired pelvic fins

paired pectoral fins

B Cartilaginous fish. A shark with jaws and paired fins. This species is a swift predator.

FIGURE 15.27 Animated! Major groups of fishes.

Credits: (a) Heather Angel/ Natural Visions; (b) © Luiz A. Rocha/ Shutterstock; (c) After E. Solomon, L. Berg, and D.W. Martin, Biology, Seventh Edition, © Brooks/Cole Cengage Learning.

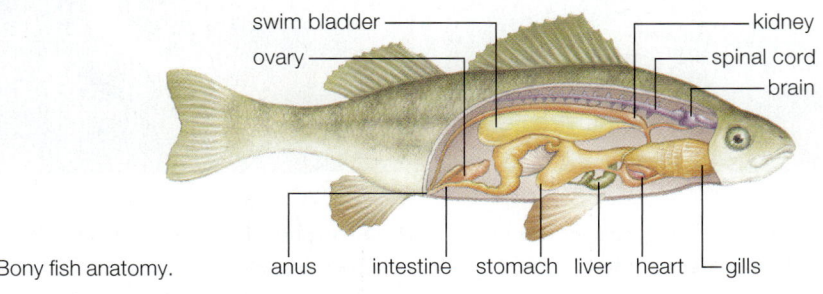

swim bladder — kidney
ovary — spinal cord
— brain

C Bony fish anatomy.

anus   intestine   stomach   liver   heart   gills

A Goldfish (carp), a ray-finned fish. Its fins consist of a web of skin supported by thin spines.

B Lungfish, a lobe-finned fish. Its thick, fleshy fins have sturdy bony supports.

FIGURE 15.28 Ray-finned and lobe-finned fishes.

Credits: (a) © iStockphoto.com/GlobalP; (b) © Wernher Krutein/ photovault.com.

In **bony fishes**, an embryonic skeleton of cartilage becomes transformed to an adult skeleton consisting mainly of bone. Both cartilaginous fishes and bony fishes have paired fins, but only bony fishes can move those fins. Bony fishes also differ from other fishes in having their gill slits hidden beneath a gill cover. In jawless and cartilaginous fish, gill slits are visible at the body surface.

There are two lineages of bony fishes. **Ray-finned fishes** have flexible fins supported by thin rays derived from skin. A gas filled swim bladder helps them adjust their buoyancy. With about 30,000 species, ray-finned fishes constitute nearly half of the vertebrates. They include most familiar freshwater fish (Figure 15.28**A**) as wells as marine species such as tuna, halibut, and cod.

Modern **lobe-finned fishes** include coelacanths (Section 12.6) and lungfishes (Figure 15.28**B**). These fish have thick, fleshy pelvic and pectoral fins supported by the bones inside them. As the name suggests, lungfishes have one or two lungs in addition to their gills. The fish inflates its lungs by gulping air, then air in the lungs exchanges gases with blood. Having lungs in addition to gills allows lungfishes to survive in low-oxygen waters.

> **Early Tetrapods** Tetrapods are descendants of a freshwater lobe-finned fish. Fossils reveal how the skeleton became modified as fishes adapted to swimming evolved into four-legged walkers (Figure 15.29). The bones inside a lobe-finned fish's pelvic and pectoral fins are homologous (Section 11.6) with amphibian limb bones. However, the transition to land was not only a matter of skeletal changes. Division of the heart into three chambers allowed blood to flow in two circuits, one to the body and one to the increasingly important lungs. Changes in the inner ear improved the ability to detect airborne sounds, and eyelids prevented the eyes from drying out.

What was the advantage of living on land? An ability to survive out of water is useful in seasonally dry places. Individuals on land were also safe from aquatic predators and had access to a new food—insects—which had recently evolved.

1 Fish (*Eusthenopteron*) with fins and no ribs.

2 Fish (*Tiktaalik*) with ribs and modified fins.

3 Early amphibian (*Icthyostegna*) with ribs and limbs.

FIGURE 15.29 Fossil species from the late Devonian show how vertebrates made the transition from water to land. Credits: left, © Cengage Learning; (1,3 right) © P. E. Ahlberg; (2 right) Illustration by © Kalliopi Monoyios.

FIGURE 15.30
Amphibians.
All have a
scaleless body.

Credits: (a) Photo by James Bettaso, US Fish and Wildlife Service; (b) Stephen Dalton/ Photo Researchers, Inc.; (c) © iStockphoto.com/ Tommounsey.

A Salamander, with equal-sized front and back limbs.

B Frog, with long, muscular legs and short forelimbs.

〉 Modern Amphibians **Amphibians** are scaleless tetrapods that spend time on land, but require water to breed. Fertilization is typically external. Eggs and sperm are released into water through a **cloaca**, a body opening that also serves as the exit for urinary and digestive wastes. (Sharks, reptiles, birds, and egg-laying mammals also have a cloaca.) Amphibian larvae are aquatic and have gills. Most species lose their gills and develop lungs during the transition to adulthood. The adults are active predators with a three-chambered heart.

Of all amphibians, the 530 or so species of salamanders and newts most closely resemble early tetrapods in body form. Their forelimbs and back limbs are similarly sized, and they have a long tail (Figure 15.30**A**). Frogs and toads belong to the most diverse amphibian lineage, with more than 5,000 species. Long, muscular hindlimbs allow the tailless adults to swim, hop, and make spectacular leaps (Figure 15.30**B**). The much smaller forelimbs help absorb the impact of landings.

Larvae of salamanders look like small adults, except for the presence of gills. In contrast, frog and toad larvae differ markedly from adults. The larvae, commonly called tadpoles, have gills and a tail but no limbs (Figure 15.30**C**).

C Tadpole, a swimming frog larva with gills and a tail.

Biologists are alarmed by ongoing declines in many amphibian populations. Shrinking or deteriorating habitats are part of the problem. People commonly fill low-lying areas where pooling of seasonal rains could provide breeding grounds for amphibians. Climate change causes additional harm by fostering the spread of pathogens and parasites, and increasing the amount of UV radiation that reaches Earth's surface. Amphibians are also highly sensitive to water pollution. Their thin skin, unprotected by scales, allows waste carbon dioxide to diffuse out of their body. Unfortunately, it also allows chemical pollutants to enter.

Take-Home Message

What are the traits of fishes and amphibians?

- Lampreys are modern fishes that do not have jaws or paired fins.
- Cartilaginous fishes such as sharks have jaws and immobile paired fins. Like lampreys, they have a skeleton made of cartilage.
- Bony fishes are the most diverse fishes. They have paired movable fins.
- The lobe-finned fishes, a subgroup of the bony fishes, are the closest living relatives of the tetrapods.
- Amphibians are tetrapods with a three-chambered heart and a scaleless body. Fertilization typically takes place in water. Gilled larve typically develop into adults that have lungs.

amphibian Tetrapod with a three-chambered heart and scaleless skin that typically develops in water, then lives on land as a carnivore with lungs.

bony fish Jawed fish with a skeleton composed mainly of bone.

cloaca Of some vertebrates, body opening that releases urinary and digestive waste, and functions in reproduction.

lobe-finned fish Bony fish with bony supports in its fins.

ray-finned fish Bony fish with fins supported by thin rays derived from skin.

FIGURE 15.31 Hognose snakes hatching from their leathery amniote eggs.

Z. Leszczynski/ Animals Animals.

15.6 Escape From Water—The Amniotes

〉 Amniote Innovations About 135 million years ago, in the early Carboniferous, the **amniotes** branched off from an amphibian ancestor. This new lineage became better adapted to life in dry places. Fertilization is internal, meaning a male puts sperm into a female's body. **Amniote eggs** have internal membranes that allow embryos to develop away from water (Figure 15.31). In addition, amniote skin is rich in the protein keratin, which makes it waterproof. A pair of well-developed, highly efficient kidneys help the body conserve water.

An early branching of the amniote lineage separated ancestors of mammals from the common ancestor of all modern reptiles. You probably do not think of birds as reptiles, but from an evolutionary perspective they are. To a biologist, **reptiles** include modern turtles, lizards, snakes, crocodilians, and birds, as well as the extinct dinosaurs.

Mammals   Turtles   Lizards, snakes   Crocodilians   Birds

Reptiles

The ability to regulate internal body temperature evolved in some early amniotes, including the ancestors of birds and mammals. Fish, amphibians, turtles, lizards, and snakes are **ectotherms**, which means "heated from outside." Ectotherms adjust their internal temperature by altering their behavior. They bask on a warm rock to heat up, or go into a burrow to cool off. In contrast, birds and mammals are **endotherms** that produce their own heat through metabolic processes. Because endotherms use more energy keeping warm, they require more food than ectotherms. A bird or mammal requires far more calories than a lizard or snake of the same weight. However, because endotherms warm themselves, they can remain active at lower temperatures than ectotherms.

〉 Reptile Diversity Lizards are the most diverse reptiles. The smallest can fit on a dime (Figure 15.32**A**). Most are predators, although iguanas eat plants. Lizards typically lay eggs that develop outside the body, but some species give birth to live young. In the live-bearers, eggs develop inside the mother, but they are not nourished by her tissues.

Snakes evolved from ancestral lizards and some retain bony remnants of ancestral hindlimbs. All snakes are carnivores. Many have flexible jaws that help

A From the Dominican Republic, the smallest known species of lizard.

FIGURE 15.32 Reptiles.

Credits: (a) © S. Blair Hedges, Pennsylvania State University; (b) © Kevin Schafer/ Corbis; (c) Kevin Schafer/ Tom Stack & Associates.

B Galápagos tortoise.

C Spectacled caiman, a crocodilian.

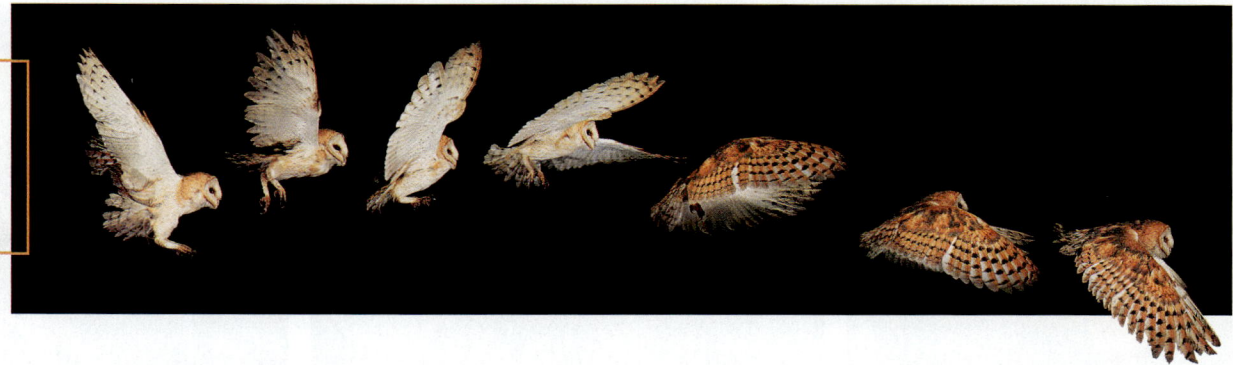

them swallow prey whole. Rattlesnakes and other fanged species bite and subdue prey with venom made in modified salivary glands. Like lizards, most snakes lay eggs, but some hold eggs inside the body and give birth to live young.

Turtles and tortoises have a bony, scale-covered shell that connects to the backbone. They do not have teeth. Instead a "beak" made of keratin covers their jaws. Tortoises are plant eaters adapted to a life spent on land (Figure 15.32**B**). Turtles spend most of their time in or near water. Sea turtles emerge from the ocean only to lay their eggs on beaches. All sea turtles are now endangered.

Crocodilians live in or near water and include crocodiles, alligators, and caimans. They are predators with powerful jaws, a long snout, and sharp teeth (Figure 15.32**C**). Crocodilians also have a highly efficient four-chambered heart, like mammals and birds. They are the closest living relatives of birds and like most birds they lay eggs, then protect and care for their young.

The Jurassic and Cretaceous periods (200–65 million years ago) are sometimes referred to as the "Age of Reptiles." During this time the dinosaurs underwent a great adaptive radiation and became the dominant animals on land. They then became extinct about 65 million years ago, during a mass extinction most likely caused by an asteroid impact (Section 11.1). Birds, which had earlier evolved from feathered dinosaurs, survived this extinction event.

Birds are the only modern amniotes with feathers and their body is adapted to flight. Bird wings are homologous to our arms. Contraction of one set of muscles produces a powerful downstroke that lifts the bird. A less powerful set of muscles contracts to raise the wing (Figure 15.33). Birds have the most efficient respiratory system of any vertebrate. This system provides the oxygen necessary to produce as steady supply of ATP that can power flight. A four-chambered heart pumps oxygen-poor blood to the lungs and oxygen-rich blood to the rest of the body in separate circuits. Flying also requires good eyesight and a great deal of coordination. Compared to a lizard of a similar body mass, a bird has much larger eyes and a larger brain. Most birds are surprisingly lightweight. Air cavities inside a bird's bones keep its body weight low, as does the lack of a bladder (an organ that stores urinary waste in many other vertebrates). Rather than heavy, bony teeth, bird jaws are covered with a lightweight beak made of keratin. The structure of the beak adapts a bird to feed on a particular type of food.

In birds, as in other reptiles, fertilization is internal. Unlike most male reptiles, male birds do not typically have a penis. Thus, to inseminate a female, a male bird must press his cloaca against hers, in a maneuver poetically described as a cloacal kiss (Figure 15.34).

Many birds migrate, flying from one region to another in response to a seasonal change. Such flights can require remarkable endurance. One bar-tailed godwit (a shorebird) flew nonstop from its nesting site in Alaska to New Zealand. It traveled 11,500 kilometers (7,145 miles), without landing to feed or rest.

FIGURE 15.34 Mating house sparrows. The male does not have a penis. To inseminate his partner, he must balance on her back and bend his body so his cloaca meets hers.

amniote Vertebrate that produces amniote eggs; a reptile, bird, or mammal.

amniote egg Egg with four membranes that allows an embryo to develop away from water.

bird Modern amniote with feathers.

ectotherm Animal that gains heat from the environment; commonly called "cold-blooded."

endotherm Animal that produces its own heat; commonly called "warm-blooded."

reptile Amniote subgroup that includes lizards, snakes, turtles, crocodilians, and birds.

FIGURE 15.35 Mammal mothers.

A Egg-laying mammal, the platypus. Her young hatched from eggs laid outside her body. They lick milk that oozes from her skin.

B Marsupial, an opossum. Her four offspring developed to an early stage in her body, then climbed into a pouch on her belly to nurse and complete development.

C Placental mammal, a hamster. Her young developed to a late stage inside her body. After birth, they suck milk from nipples on her lower surface.

Credits: (a) Jean Phillipe Varin/ Jacana/ Photo Researchers, Inc.; (b) Jack Dermid; (c) © Minden Pictures/ SuperStock.

❯ Mammals **Mammals** are the only amniotes in which females nourish their offspring with milk secreted from mammary glands. The group name is derived from the Latin *mamma*, meaning breast. Mammals are also the only animals that have hair or fur. Both are modifications of scales. Like birds, mammals are endotherms. A coat of fur or head of hair helps them maintain their core temperature. Mammals have a single lower jawbone and most have more than one type of tooth. By contrast, reptiles have a hinged two-part jaw, and all teeth are similarly shaped. Having a variety of different kinds of teeth allows mammals to eat more types of foods than most other vertebrates.

Mammals evolved early in the Jurassic, and early mouselike species coexisted with dinosaurs. By 130 million years ago, three lineages had evolved: **monotremes** (egg-laying mammals), **marsupials** (the pouched mammals), and **placental mammals** (mammals in which an organ called the placenta provides nutrients to developing offspring). Figure 15.35 shows examples of each group.

Only three species of monotremes survive, and most pouched mammals live in Australia or New Zealand. In contrast, placental mammals occur worldwide. What gives placental mammals their competitive edge? They have a higher metabolic rate, better body temperature control, and a more efficient way to nourish embryos. Compared to other mammals, placental mammals develop to a far more advanced stage inside their mother's body.

Rats and bats are the most diverse mammals. About half the 4,000 species of placental mammals are rodents and, of those, about half are rats. The next most diverse group is the bats, with about 375 species. Bats are the only flying mammals. Although some may look like flying mice, bats are more closely related to carnivores such as wolves and foxes than to rodents.

Take-Home Message

What are amniotes?

- Amniotes are vertebrates that are adapted to life on land by their unique eggs, waterproof skin, and highly efficient kidneys.

- One amniote lineage includes snakes, lizards, turtles, and crocodiles as well as the birds. Birds are the only amniotes with feathers. Like mammals, birds are endotherms; metabolic heat maintains their internal temperature.

- Mammals are amniotes that nourish their young with milk. Most have hair or fur. The three lineages are egg-laying monotremes, pouched marsupials, and placental mammals. Placental mammals are the most widespread and diverse.

mammal Vertebrate that nourishes its young with milk from mammary glands.

marsupial Mammal in which young complete development in a pouch on the mother's body.

monotreme Egg-laying mammal.

placental mammal Mammal in which young are nourished within the mother's body by way of a placenta.

primate Mammalian group with grasping hands; includes lemurs, tarsiers, monkeys, apes, and humans.

15.7 Human Evolution

❭ Primate Traits **Primates** are an order of placental mammals that includes humans, apes, monkeys, and their close relatives. Primates first evolved in tropical forests, and many of the group's characteristic traits arose as adaptations to life among the branches. Primate shoulders have an extensive range of motion that facilitates climbing (Figure 15.36). Unlike most mammals, a primate can extend its arms out to its sides, reach above its head, and rotate its forearm at the elbow. With the exception of humans, all living primates have both hands and feet capable of grasping. Mammals often have claws, but the tips of primate fingers and toes typically have touch-sensitive pads protected by flattened nails.

Most mammals have widely spaced eyes set toward the side of the skull, but primate eyes tend to be at the front of the head. As a result, both eyes view the same area, each from a slightly different vantage point. The brain integrates the differing signals it receives from the two eyes to produce a three-dimensional image. A primate's excellent depth perception adapts it to a life spent leaping or swinging from limb to limb.

Compared to other mammals, primates have a large brain for their body size. The regions of the brain devoted to vision and to information processing are expanded, and the area devoted to smell is reduced.

Primates have a varied diet, and their teeth reflect this lack of specialization. They have teeth for tearing flesh, as well as for grinding material.

Most primates spend their life in a social group that includes adults of both sexes. Females usually give birth to only one or two young at a time and provide care for an extended period after birth.

❭ Modern Primates The oldest existing primate lineages include the lemurs and tarsiers (Figure 15.37). Like dogs and most other mammals, lemurs

FIGURE 15.36 Adapted to climbing. A female orangutan (an Asian ape) demonstrates her wide range of shoulder motion and the grasping ability of her hands and feet. Eyes situated at the front of her face provide excellent depth perception.

© Thomas Marent/ ardea.com.

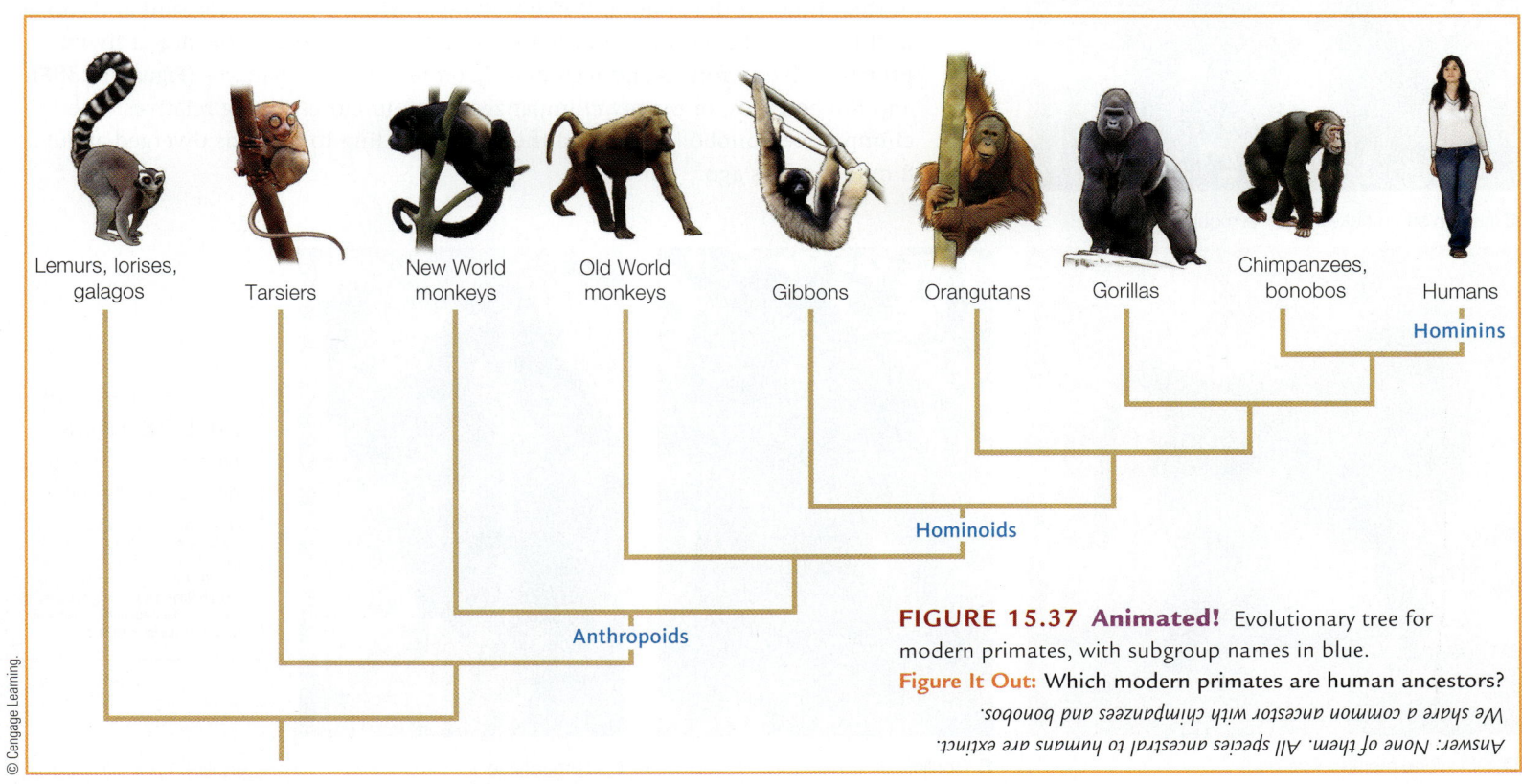

Lemurs, lorises, galagos | Tarsiers | New World monkeys | Old World monkeys | Gibbons | Orangutans | Gorillas | Chimpanzees, bonobos | Humans

Hominins

Hominoids

Anthropoids

FIGURE 15.37 Animated! Evolutionary tree for modern primates, with subgroup names in blue.

Figure It Out: Which modern primates are human ancestors?

Answer: None of them. All species ancestral to humans are extinct. We share a common ancestor with chimpanzees and bonobos.

© Cengage Learning.

A Lemur

B Tarsier

have a moist nose and an upper lip that attaches tightly to the underlying gum (Figure 15.38**A**). A dry nose and a movable, noncleft upper lip evolved in the common ancestor of tarsiers (Figure 15.38**B**) and all other modern groups. This innovation allows a wider range of facial expressions and vocalizations.

Anthropoids include monkeys, apes, and humans; "anthropoid" means human-like. Nearly all anthropoids are diurnal (active during the day) and have good eyesight, including color vision.

New World monkeys (Figure 15.38**C**) climb through forests of Central and South America in search of fruits. They have a flat face and a nose with widely separated nostrils. A long tail helps them maintain balance. In many species, the tail is prehensile, meaning it grasps things.

Old World monkeys live in Africa, the Middle East, and Asia. They tend to be larger than New World monkeys and have a longer nose with closely set nostrils. Some are tree-climbing forest dwellers. Others, such as baboons (Figure 15.38**D**), spend most of their time on the ground in grasslands and deserts. Not all Old World monkeys have a tail, but in those that do the tail is typically short and never prehensile.

Tailless nonhuman primates are commonly called apes. About 15 species of small apes called gibbons inhabit Southeast Asian forests. Gibbons are sometimes referred to as "lesser apes," in comparison with the larger apes, or "great apes." The forest-dwelling orangutan of Sumatra and Borneo is the only surviving Asian great ape (Figure 15.36). All African great apes (gorillas, chimpanzees, and bonobos) are native to central Africa, live in social groups, and spend most of their time on the ground. When walking, African apes lean forward and support their weight on their knuckles (Figure 15.38**E**). Gorillas, the largest living primates, live in forests and feed mainly on leaves. Chimpanzees (Figure 15.38**F**) and the bonobos, or pygmy chimpanzees, are our closest living relatives. The chimpanzee/bonobo lineage and the lineage leading to humans diverged about 8 million years ago.

C New World monkey (squirrel monkey)

D Old World monkey (baboon)

E Gorilla

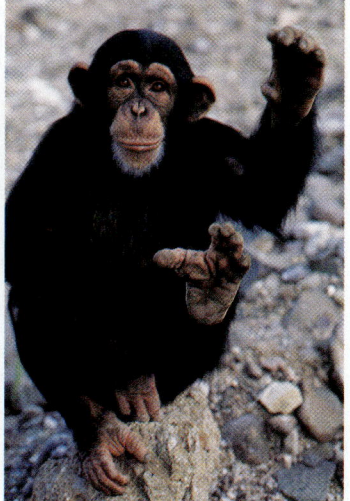

F Chimpanzee

FIGURE 15.38
Diversity of modern, nonhuman primates.

Credits: (a) © iStockphoto.com/ Catharina van den Dikkenberg; (b) © iStockphoto.com/ Robin O'Connel; (c) Photobucket; (d) © iStockphoto.com/ JasonRWarren; (e) © Dallas Zoo, Robert Cabello; (f) Kenneth Garrett/ National Geographic Image Collection.

> **Early Bipedal Species** Taxonomists now call humans and their closest extinct relatives **hominins** (previously called hominids). The defining trait of hominins is **bipedalism**—habitual upright walking—so researchers studying human origins look for fossil evidence that a species walked upright.

Footprints in Tanzania document the passage of a bipedal species 3.6 million years ago (Figure 15.39**A**). We can tell from the prints that the walkers' feet had a pronounced arch and a big toe in line with the other toes, two adaptations to upright walking. The walkers were probably **australopiths**, hominins that lived in Africa from about 4 million to 1.2 million years ago. Australopiths include some likely human ancestors such as *Australopithicus afarensis*, a species best known from a largely complete fossil skeleton nicknamed Lucy (Figure 15.39**B**). Lucy was about the size of a chimpanzee, with a chimpanzee-sized skull and brain, and she had long dangling arms. However, the structure of her pelvis and backbone is more like that of a human. Chimpanzees have a long, narrow pelvis, whereas Lucy and modern humans have a shorter, wider one. Chimpanzees who support themselves with their hands as they walk have a C-shaped backbone, whereas Lucy and modern humans have an S-shaped one. Fossil skulls of other *A. afarensis* individuals also support the hypothesis that this species walked upright. These skulls, like those of modern humans, have an opening at the base that connects to the backbone. In animals that walk on all fours, this opening is located at the rear of the skull.

> **The Early Humans** **Humans** are members of the genus *Homo*, which first appeared in Africa more than 2 million years ago. The most ancient species assigned to the genus is **Homo habilis**. The name means "handy man." The first *H. habilis* fossil discovered was given this name because of simple stone tools found in the same vicinity. A typical *H. habilis* had a somewhat larger brain and smaller face than australopiths. Like australopiths, it had long apelike forearms, but its fingers resemble those of modern humans.

Homo erectus appears in the fossil record beginning about 1.8 million years ago. Its name means "upright man," and like modern humans, *H. erectus* stood on legs that were longer than its arms and it had a relatively big brain (Figure 15.40). *H. erectus* put its brain to use in crafting tools from stone. Early stone tools were simple rock chips. Toolmaking improved dramatically about 1.4 million years ago, when African *H. erectus* began to use a succession of carefully placed strikes to produce tools in a variety of specific shapes. Marks on fossilized bones of prey animals indicate that *H. erectus* used these stone tools to cut up scavenged animal carcasses, scrape meat from bones, and extract marrow.

FIGURE 15.40 *Homo erectus*. A 1.53-million-year-old skeleton of a juvenile male from Kenya.

Science VU/NMK/Visuals Unlimited, Inc.

A Footprints made 3.6 million years ago by bipedal hominins walking across volcanic ash. The walkers had a shorter stride than modern humans.

B *Australopithecus afarensis*. This individual, known as Lucy, lived 3.5 million years ago. Her species is a likely human ancestor.

FIGURE 15.39 Evidence of early hominins.

Credits: (a) top, Louise M. Robbins; bottom, © Kenneth Garrett/ National Geographic Image Collection; (b) Dr. Donald Johanson, Institute of Human Origins.

anthropoids Monkeys, apes, and humans.

australopiths Informal name for a lineage of chimpanzee-sized hominins that lived in Africa between 4 million and 1.2 million years ago.

bipedalism Habitually walking upright.

hominin Humans and extinct humanlike species.

Homo erectus Human species that dispersed out of Africa.

Homo habilis Earliest named human species.

humans Members of the genus *Homo*.

People of non-African ancestry have a Neanderthal in their family tree.

Chimpanzees use gestures and calls to communicate, and early humans probably did the same. However, it is unlikely that either *H. habilis* or *H. erectus* had a spoken language. Evolution of a capacity for speech involved changes in the brain, remodeling of the larynx (voice box), and an improved ability to control airflow through the vocal tract. Thus far, scientists have not found fossil evidence of these anatomical modifications in either early species of *Homo*.

Homo erectus was probably the first human species to extend its range beyond its African birthplace. By 1.7 million years ago, *H. erectus* populations had been established in places as far from Africa as the island of Java and the Republic of Georgia. At the same time, African populations continued to thrive. Over thousands of generations, geographically separated groups adapted to local environmental conditions. Some populations became so different from *H. erectus* that we call them new species.

> **Origin of Our Species** A 195,000-year-old fossil found in Ethiopia is the earliest evidence of modern humans, called **Homo sapiens**. Compared to *H. erectus*, we have a higher, rounder skull, a larger brain, and a flatter face with smaller jawbones and teeth. We also have a chin, a protruding area of thickened bone in the middle of our lower jawbone.

Two competing models describe how *H. sapiens* evolved from *H. erectus*. According to the multiregional model, populations of *H. erectus* in Africa and other regions evolved into populations of *H. sapiens* gradually, over more than a million years. Gene flow among populations maintained the species through the transition to fully modern humans (Figure 15.41**A**). By this model, variations now seen among modern Africans, Asians, and Europeans began to accumulate soon after their ancestors branched from an ancestral *H. erectus* population. The replacement model holds that *H. sapiens* evolved within the past 200,000 years from a population of *H. erectus* living in sub-Saharan Africa. Later, bands of *H. sapiens* entered regions occupied by *H. erectus* populations, and replaced them (Figure 15.41**B**). This model emphasizes the similarities among all living humans.

In support of the replacement model, genetic studies place modern Africans closest to the root of our family tree. Studies of the Y chromosome indicate that all modern men can trace their ancestry to a male individual who lived in Africa approximately 60,000 years ago. However, as we will discuss shortly, there is some evidence of long-standing regional differences in human genomes.

The range of modern humans expanded as successive generations traveled along the coasts of Africa, then Eurasia and Australia. We know from fossilized feces deposited in an Oregon cave that modern humans were living in North America by about 14,000 years ago.

> **Neanderthals—Our Closest Extinct Relatives** Neanderthals, members of **Homo neanderthalensis**, are our closest extinct relatives. Our lineages diverged 500,000 years ago and we both trace our ancestry to *H. erectus*:

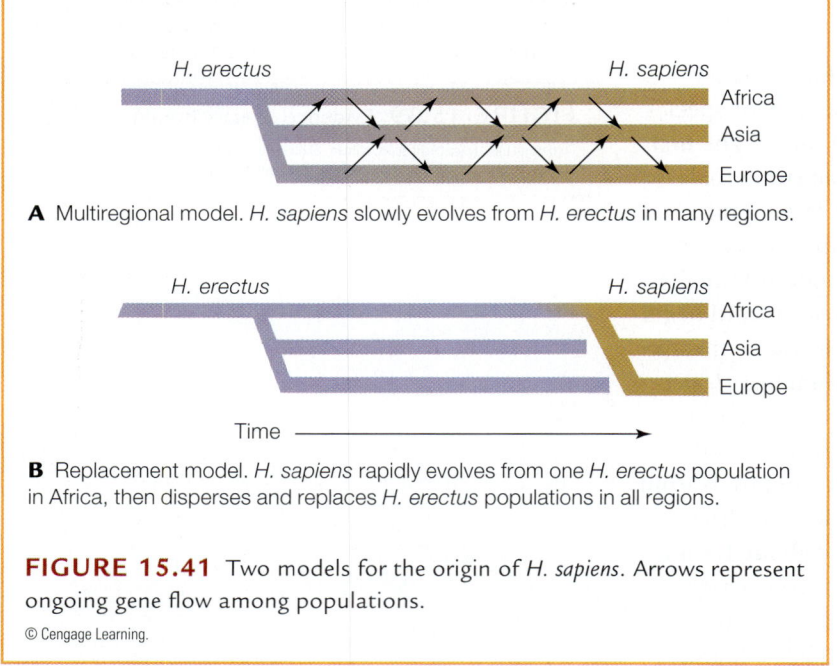

A Multiregional model. *H. sapiens* slowly evolves from *H. erectus* in many regions.

Time ⟶

B Replacement model. *H. sapiens* rapidly evolves from one *H. erectus* population in Africa, then disperses and replaces *H. erectus* populations in all regions.

FIGURE 15.41 Two models for the origin of *H. sapiens*. Arrows represent ongoing gene flow among populations.

© Cengage Learning.

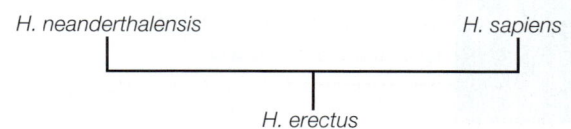

Homo neanderthalensis Neanderthals. Closest extinct relatives of modern humans; had large brain, stocky body.
Homo sapiens Modern humans; evolved in Africa, then expanded their range worldwide.

Compared to *H. sapiens*, Neanderthals had a shorter, stockier build, with thicker bones and bulkier muscles. A reconstruction of a Neanderthal male, based on material from multiple fossils, stands about 164 centimeters (5'4") tall (Figure 15.42). Neanderthals had a braincase that was longer and lower than that of modern humans, but their brain was as big or bigger than ours. Their face had pronounced brow ridges, a large nose with widely spaced nostrils, and no chin.

Neanderthals lived in the Middle East, Europe, and in central Asia as far east as Siberia. Fossils of Neanderthal individuals who survived despite disabilities such as the loss of a limb testify to a compassionate social structure. Some simple burials suggest possible symbolic thought. Several lines of evidence suggest Neanderthals were able to speak.

About 60,000 years ago, humans dispersing out of Africa met and interbred with a Neanderthal population living in the Middle East. How do we know? Modern humans native to regions outside of Africa have distinctive Neanderthal genes not seen in Africans. There is no genetic evidence of later interbreeding when modern humans expanded into Europe 40,000 years ago. The last traces of a Neanderthal population are from caves in Gibraltar and date to 28,000 years ago. Competition with our species may have contributed to their demise, but disease and habitat changes probably played a role as well.

Take-Home Message

How did humans evolve?

- Humans, like other primates, have grasping hands and shoulders adapted to climbing. Unlike most primates, they walk upright.

- Australopiths are a group of upright-walking species that probably include human ancestors. They evolved in Africa by about 4 million years ago.

- The first humans, or members of the genus *Homo*, also arose in Africa. *Homo erectus* migrated out of Africa and into Europe and Asia.

- Modern humans (*Homo sapiens*) and the extinct Neanderthals are descendants of *Homo erectus*. By the currently favored model, modern humans evolved in Africa, then migrated worldwide.

Homo neanderthalensis   *Homo sapiens*

FIGURE 15.42 Skeletal anatomy of Neanderthal and modern human males. The Neanderthal skeleton is a reconstruction based on multiple fossils. Each color denotes a different fossil.

Courtesy of @ Blaine Maley, Washington University, St. Louis.

15.8 Early Birds (revisited)

How do we know what color a feathered dinosaur or bird was? Some fossil feathers such as those of the dinosaur *Sinosauropteryx* contain microscopic granules of the type that contain pigment in modern bird feathers. In birds, different shaped granules contain different shades of the pigment of melanin, which comes in red to brown shades. By studying the distribution and shape of pigment granules in *Sinosauropteryx* fossils, scientists concluded that this animal had a reddish color and a tail with alternating red and white bands.

© Cengage Learning.

If a fossil is recent enough to contain some intact DNA, genes can also provide information about coloring. For example, analysis of DNA extracted from Neanderthal bones revealed that some individuals had a mutation in one of their melanin-encoding genes. This mutation would have affected their hair color, making them redheads.

WHERE YOU ARE GOING . . .

Section 17.3 discusses animals as predators and prey, parasites and hosts. You will learn about animal circulatory systems in Section 21.2, respiratory systems in 21.7, and digestive systems in 23.3. Section 24.3 describes both invertebrate and vertebrate nervous systems, and 24.6 explains how animals sense the world. We return to animal reproduction and development in 26.2. Sections 28.1 and 28.3 address the importance of animals as pollinators and in seed dispersal.

Summary

Section 15.1 Fossils tell us about the relationships between lineages and how the traits that define each lineage evolved.

Section 15.2 **Animals** are multicelled heterotrophs that ingest food and move about. According to the **colonial theory of animal origins**, they evolved from a colonial protist. Most animals are **invertebrates**, rather than **vertebrates** with a backbone. Animals with **tissues** have either **radial symmetry** or **bilateral symmetry**. Most animals have a **coelom**, a lined body cavity, around their gut, which may be a saclike **gastrovascular cavity** or a tubelike **complete digestive tract**.

Section 15.3 **Sponges** are sessile filter-feeders with no body symmetry or tissues. Each individual is a **hermaphrodite**; it makes eggs and sperm. The sponge **larva** is free-swimming.

The radially symmetrical **cnidarians** have two body types: **medusa** (as in jellies) and **polyp** (as in sea anemones). Stinging cells help cnidarians capture prey.

Flatworms have simple organ systems, a gastrovascular cavity, and no coelom. Planarians are free-living flatworms, and tapeworms and flukes are parasites.

Annelids are coelomate, segmented worms with a complete digestive system and a closed circulatory system. Earthworms are the best known. The lineage also includes leeches and polychaetes.

Mollusks have a **mantle**, a skirtlike extension of the body. In many groups, the mantle secretes a shell. **Gastropods** (such as snails) and **cephalopods** (such as octopuses) have a head with a **radula** used in feeding. **Bivalves** such as clams are filter-feeders.

Roundworms are unsegmented worms with a complete gut and an incompletely lined coelom. They may be free-living or parasitic.

Arthropods have a jointed **exoskeleton**. Horseshoe crabs and most **crustaceans** are aquatic, whereas **arachnids**, centipedes and millipedes, and insects live on land. Insects have paired **antennae** and **compound eyes**, and most undergo **metamorphosis**. They are the most diverse animals and the only winged invertebrates.

Echinoderms such as sea stars have an **endoskeleton** of spines, spicules, or plates of calcium carbonate. A **water–vascular system** with tube feet allows adults to move about. Adults are radial, but bilateral ancestry is evident in their larval stages and other features.

Section 15.4 Four embryonic traits define the **chordates**: a **notochord**, a dorsal hollow nerve cord (which becomes a brain and spinal cord), a pharynx with gill slits, and a tail extending past the anus. Depending on the group, some or all of the features persist in adults.

Lancelets and **tunicates** are invertebrate filter-feeding chordates. Most chordates are vertebrates; they have a **vertebral column** (backbone) of cartilage or bone. Jaws, **lungs**, limbs, and waterproof eggs were key innovations that made the adaptive radiation of **vertebrates** possible. **Tetrapods** have four limbs. All vertebrates have a complete digestive system, closed circulatory system, and **kidneys**.

Section 15.5 The earliest **fishes** were **jawless fishes**. **Cartilaginous fishes** and **bony fishes** have jaws, **scales**, and paired fins. The two bony fish lineages are the highly diverse **ray-finned fishes** and the **lobe-finned fishes**. **Amphibians** evolved from a lobe-finned fish. Existing groups include salamanders, frogs, and toads. Fertilization is external, with eggs and sperm exiting the body through a **cloaca** that also expels digestive and urinary wastes.

Section 15.6 **Amniotes** were the first vertebrates that did not need external water for reproduction. Their skin and kidneys conserve water, and they produce **amniote eggs**. Mammals are one amniote lineage. **Reptiles**, including **birds** (the only living animals with feathers), are another. Most reptiles are **ectotherms**, but birds are **endotherms** (produce their own heat), as are mammals. There are three lineages of **mammals**: egg-laying **monotremes**, pouched **marsupials**, and placental mammals. **Placental mammals** are the most diverse group.

Section 15.7 **Primates** are a lineage adapted to climbing and all have hands capable of grasping. Lemurs have a moist, doglike snout, but tarsiers and **anthropoids** (monkeys, apes, humans) have a dry nose and movable upper lip. Our closest living relatives are the chimpanzees and bonobos. **Bipedalism** defines the **hominins**, which include **humans** and their extinct close relatives. **Australopiths** were early hominins. The first named members of our genus, *Homo habilis*, resembled australopiths. *Homo erectus* had a larger brain and migrated out of Africa. Modern humans (*Homo sapiens*) arose in Africa. As they dispersed, some mated with the now extinct Neanderthals (*Homo neanderthalensis*).

Self-Quiz   Answers in Appendix I

1. True or false? Animal cells do not have walls.

2. The colonial theory of animal origins states that _____ .
   a. animals are more closely related to plants than to fungi
   b. animals evolved from a colonial protist
   c. most animals live in colonies
   d. all animals have a backbone

3. A body cavity that is fully lined with tissue derived from mesoderm is a _____ .
   a. pseudocoelom   c. coelom
   b. kidney   d. gastrovascular cavity

4. Most animal bodies have _____ symmetry.
   a. radial   b. bilateral   c. no

5. Earthworms are most closely related to _____ .
   a. insects   c. leeches
   b. tapeworms   d. roundworms

6. The _____ have a cuticle and molt as they grow.
   a. roundworms   c. arthropods
   b. annelids   d. both a and c

7. List the four distinguishing chordate traits. Which of these traits are retained by an adult tunicate?

Digging Into Data

Sustaining Ancient Horseshoe Crabs

Atlantic horseshoe crabs, *Limulus polyphemus*, are sometimes called living fossils. For more than a million years, their eggs have fed migratory shorebirds. More recently, people began to harvest horseshoe crabs for use as bait. More recently still, people started using horseshoe crab blood to test injectable drugs for potentially deadly bacterial toxins. To keep horseshoe crab populations stable, blood is extracted from captured animals, which are then returned to the wild. Concerns about the survival of animals after bleeding led researchers to do an experiment. They compared survival of animals captured and maintained in a tank with that of animals captured, bled, and kept in a similar tank. Figure 15.43 shows the results.

1. In which trial did the most control crabs die? In which did the most bled crabs die?

2. Looking at the overall results, how did the mortality of the two groups differ?

3. Based on these results, would you conclude that bleeding harms horseshoe crabs more than capture alone does?

	Control Animals		Bled Animals	
Trial	Number of crabs	Number that died	Number of crabs	Number that died
1	10	0	10	0
2	10	0	10	3
3	30	0	30	0
4	30	0	30	0
5	30	1	30	6
6	30	0	30	0
7	30	0	30	2
8	30	0	30	5
Total	200	1	200	16

FIGURE 15.43 Mortality of young male horseshoe crabs kept in tanks for the two weeks after their capture. Blood was drawn from half of the animals on the day of their capture. Control animals were handled, but not bled. The experiment was repeated eight times with eight different sets of horseshoe crabs.

Data: Walls, E., Berkson, J., Fish. Bull. 101:457–459 (2003).

Jane Burton/ Bruce Coleman, Ltd.

---

8. True or false? A shark's vertebral column is made of bone.

9. All vertebrates are _____ but only some are _____ .
   a. tetrapods; mammals
   b. chordates; amniotes
   c. amniotes; hominins
   d. bipedal; australopiths

10. Amniote adaptations to land include _____ .
    a. waterproof skin
    b. internal fertilization
    c. highly efficient kidneys
    d. specialized eggs
    e. a and c
    f. all of the above

11. True or false? Australopiths belong to the species *Homo*.

12. Birds and placental mammals _____ .
    a. are ectotherms
    b. lay eggs
    c. have mammary glands
    d. have a four-chambered heart

13. Match the organisms with the appropriate description.
    _____ sponges
    _____ cnidarians
    _____ flatworms
    _____ roundworms
    _____ annelids
    _____ arthropods
    _____ mollusks
    _____ echinoderms
    _____ amphibians
    _____ fishes
    _____ birds
    _____ mammals

    a. most diverse vertebrates
    b. no true tissues, no organs
    c. jointed exoskeleton
    d. mantle over body mass
    e. segmented worms
    f. tube feet, spiny skin
    g. have specialized stinging cells
    h. lay amniote eggs
    i. feed young secreted milk
    j. unsegmented, molting worms
    k. first terrestrial tetrapods
    l. saclike gut, no coelom

14. Arrange the events in order, from most ancient to most recent.
    _____ 1
    _____ 2
    _____ 3
    _____ 4
    _____ 5
    _____ 6

    a. Cambrian explosion of diversity
    b. Origin of animals
    c. Tetrapods move onto the land
    d. Extinction of dinosaurs
    e. *Homo erectus* leaves Africa
    f. First jawed vertebrates evolve

Critical Thinking

1. In the summer of 2000, only 10 percent of the lobster population in Long Island Sound survived after a massive die-off. Many lobstermen in New York and Connecticut lost small businesses that their families had owned for generations. Some believe the die-off followed heavier sprays of pesticides to control mosquitoes that carry the West Nile virus. Explain why a chemical substance that targets mosquitoes might also harm lobsters but not fish.

2. In 1798, a stuffed platypus specimen was delivered to the British Museum. At the time, many biologists were sure it was a fake, assembled by a clever taxidermist. Soft brown fur and beaverlike tail put the animal firmly in the mammalian camp. But a ducklike bill and webbed feet suggested an affinity with birds. Reports that the animal laid eggs only added to the confusion.

   We now know that platypuses burrow in riverbanks and forage for prey under water. Webbing on their feet can be retracted to reveal claws. The highly sensitive bill allows the animal to detect prey even with its eyes and ears tightly shut.

   To modern biologists, a platypus is clearly a mammal. Like other mammals, it has fur and the females produce milk. Young animals have more typical mammalian teeth that are replaced by hardened pads as the animal matures. Why do you think modern biologists can more easily accept the idea that a mammal can have some reptile-like traits, such as laying eggs? What do they know that gives them an advantage over scientists living in 1798?

3. Humans belong to the genus *Homo* and chimpanzees to the genus *Pan*. Yet studies of primate genes show that chimpanzees and humans are more closely related to one another than each is to any other ape. In light of this result, some researchers suggest that chimpanzees should be renamed as members of the genus *Homo*. What practical, scientific, and ethical issues might be raised by such a change in naming?

16 Population Ecology

This chapter considers the structure and growth of populations (Section 1.2). We refer back to sampling error (1.8), directional selection (12.3), and the origin and migration of modern humans (15.7).

16.1 A Honkin' Mess

Visit a grassy park or a golf course in many parts of the United States and you may need to watch where you step. Wide expanses of grass with a nearby body of water often attract large numbers of Canada geese, or *Branta canadensis* (Figure 16.1). These plant-eating birds produce an abundance of slimy, green feces that can soil shoes, stain clothes, and discourage picnics. Goose feces also wash into ponds and waterways. The nutrients they add encourage bacteria and algae to grow, clouding the water and making swimming unappealing and possibly dangerous. Goose feces sometimes contain microbes that can sicken humans.

The number of Canada geese in the United States has increased dramatically. For example, Michigan had about 9,000 in 1970, and has 300,000 today. Controlling their number is challenging because several different Canada goose populations spend time in the United States. A **population** is a group of organisms of the same species who live in a specific location and breed with one another more often than they breed with members of other populations. In the past, nearly all Canada geese seen in the United States were migratory. They nested in northern Canada, flew to the United States to spend the winter, then returned to Canada. The common name of the species reflects this behavior.

Most Canada geese still migrate, but some populations have become permanent residents of the United States. The geese breed where they grew up, and the nonmigratory birds are generally descendants of geese deliberately introduced to a park or hunting preserve. During the winter, migratory birds often mingle with nonmigratory ones. For example, a bird that breeds in Canada and flies to Virginia for the winter finds itself alongside geese that have never left Virginia.

Life is harder for migratory geese than for nonmigratory ones. Flying hundreds of miles to and from a northern breeding area takes lots of energy and is dangerous. A bird that does not migrate can devote more energy to producing young than a migratory one can. If the nonmigrant lives in a suburban or urban area, it also benefits from an unnatural abundance of food (grass) and an equally unnatural lack of predators. Not surprisingly, the greatest increases in Canada geese have been among nonmigratory birds living where humans are plentiful.

Migratory birds are protected under federal law and by international treaties. However, in 2006, increasing complaints about Canada geese led the U.S. Fish and Wildlife Service to exempt this species from some protections. The agency encouraged wildlife managers to look for ways to reduce nonmigratory Canada goose populations, without unduly harming migratory birds. To do so, these biologists need to know about the traits that characterize different goose populations, as well as how these populations interact with one another, with other species, and with their physical environment.

These sorts of questions are the focus of the science of **ecology**, the study of interactions among organisms, and between organisms and their physical environment. Ecology is not the same as environmentalism, which is advocacy for protection of the environment. However, environmentalists often cite the results of ecological studies when drawing attention to environmental concerns.

FIGURE 16.1 Canada geese overrun a California park during the summer.

Courtesy of Joel Peter.

demographics Statistics that describe a population.

ecology The study of interactions among organisms, and between organisms and their environment.

population A group of organisms of the same species who live in a specific location and breed with one another more often than they breed with members of other populations.

population density Number of members of a population in a given area.

population distribution The way in which members of a population are dispersed in their environment.

population size Number of individuals in a population.

16.2 Characteristics of Populations

When studying a population, ecologists collect information about its gene pool, reproductive traits, and behavior of its component individuals. They also look at **demographics**—vital statistics that describe the population.

❯ Demographic Traits **Population size** refers to the number of individuals of a species in a population. **Population density** is the number of individuals in some specified area or volume of a habitat, such as the number of frogs per acre of rain forest or the number of amoebas per liter of pond water.

Population distribution describes the location of individuals relative to one another. Members of most populations have a clumped distribution, meaning they tend to be closer to one another than would be predicted by chance alone. A patchy distribution of resources encourages clumping. For example, Canada geese tend to congregate in places with suitable food such as grass and a nearby body of water. Limited dispersal ability also causes clumping. Many plant seeds fall and sprout beneath a parent plant. Asexual reproduction can also lead to clusters, as when strawberry plants arise from runners or coral polyps reproduce by dividing in two. Finally, many animals benefit by living in a social group. Figure 16.2**A** shows chimpanzees using sticks to capture termites, a behavior one chimp learns from another in its group. Group members can also warn one another of threats and share food.

Competition for resources can produce a near-uniform distribution, with individuals more evenly spaced than would be expected by chance. Creosote bushes in deserts of the American Southwest grow in this pattern. Competition for water among the plants' root systems prevents them from growing in close proximity. Similarly, seabirds in breeding colonies often show a near-uniform distribution. Each bird aggressively repels others that get within reach of its beak as it sits atop its nest (Figure 16.2**B**).

A random population distribution is rare in nature. Random distribution arises when resources are uniformly available, and proximity to others neither benefits nor harms individuals. For example, when wind-dispersed dandelion seeds land on the uniform environment of a suburban lawn, dandelions grow in a random pattern (Figure 16.2**C**). Wolf spider burrows are also randomly distributed relative to one another. When seeking a burrow site, the spiders neither avoid one another nor seek one another out.

The scale of a study area and timing of the study can influence the observed pattern of distribution. For example, although seabirds may be spaced almost uniformly at a nesting site, nesting sites are clustered along a shoreline. Also, these birds crowd together in the breeding season, but disperse at other times.

Age structure refers to the distribution of individuals among various age categories. As you will learn later in this chapter, a population's age structure affects its capacity for future growth. A population with a relatively large number of young individuals who have not yet begun reproducing has a greater potential for growth than one composed mainly of older individuals.

❯ Collecting Demographic Data Scientists often cannot make a direct count of all the members of a population. Instead, they sample the population, then use data from that sample to estimate the size and other characteristics of the population as a whole.

Plot sampling estimates the total number of individuals in an area on the basis of direct counts in a fraction of the area. For example, to determine the number of daisies in a prairie or clams in a mudflat, ecologists might begin by measuring the number of individuals in several 1 meter by 1 meter square plots.

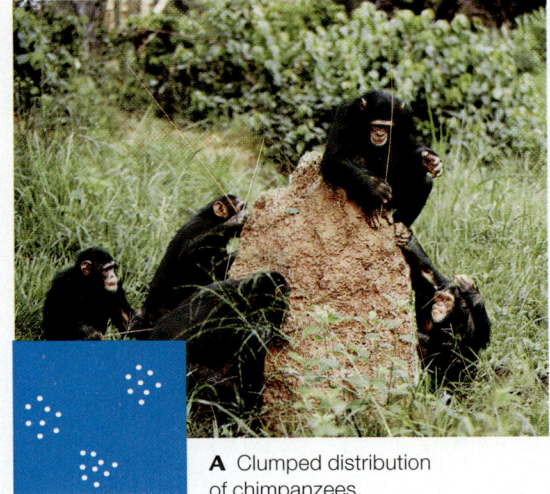

A Clumped distribution of chimpanzees.

B Near-uniform distribution of nesting seabirds.

C Random distribution of dandelions.

FIGURE 16.2 Population distributions.

Credits: art, Cengage Learning; (a) © Steve Bloom/ stevebloom.com; (b) © Eric and David Hosking/ Corbis; (c) Elizabeth A. Sellers/ life.nbii.gov.

A Florida Key deer, a member of an endangered species, marked with a neck ring for a population study.

$$\frac{\text{marked individuals in sampling at time 2}}{\text{total captured in sampling 2}} = \frac{\text{marked individuals in sampling at time 1}}{\text{total population size}}$$

B Equation used to estimate total population size on the basis of mark-recapture sampling.

FIGURE 16.3 Animated! Mark–recapture studies.

Figure It Out: Suppose scientists catch, mark, and release 100 deer. Later, they return and again catch 100 deer. In this latter sample, 50 deer are marked. How many deer are in the total population?

Answer: 50/100 = 100/total size, so total size must be 200.

Photo © Cynthia Bateman, Bateman Photography.

To calculate the total population size, researchers multiply the average number of individuals in the sample plots by the number of plots in the area the population inhabits. Size estimates from plot sampling are most accurate for organisms that do not move about and live in an area where conditions are uniform.

Biologists use another sampling method, called mark–recapture sampling, to study mobile animals. Researchers capture animals, mark them (Figure 16.3**A**), then release them. Some time later, the animals are captured again. The proportion of marked animals in the second sample is taken to be representative of the proportion marked in the whole population (Figure 16.3**B**).

Information about the traits of individuals in a sample plot or capture group can be used to infer properties of the population as a whole. For example, if half of the deer recaptured in a mark–recapture study are of reproductive age, half of the population is assumed to share this trait. This sort of extrapolation is based on the assumption that the individuals in the sample are representative of the general population in terms of the traits under investigation.

Any study that draws conclusions based on only a sample of a population is susceptible to sampling error (Section 1.8). The larger the sample, the more likely that conclusions drawn from that sample will be accurate.

> Using Demographic Data Wildlife managers use demographic information to decide how best to manage populations. For example, when the U.S. Fish and Wildlife Service wanted to set up a plan for management of nonmigratory Canada geese, its wildlife managers began by evaluating the size, density, and distribution of the nonmigratory populations. Based on this information, the Service decided to allow destruction of some eggs and nests, along with increased hunting opportunities during times when migratory Canada geese are least likely to be present. Scientists will continue to monitor the demographics of the populations to determine the effects of these measures.

Take-Home Message

How do scientists describe populations?

- Members of a population live in the same area and breed with one another.

- Populations can be described in terms of their size, density, and how their members are distributed through their environment. Most populations have a clumped distribution.

- Population studies typically utilize sampling methods. A sample of the population is studied, then that information is used to infer the size and traits of the population as a whole. Such methods run the risk of sampling error, which can be minimized by using a large sample size.

16.3 Population Growth

> Exponential Growth A population grows when its birth rate exceeds its death rate. Ecologists typically measure births and deaths per individual, or per capita. For example, if the birth rate in a population of 2,000 mice is 1,000 young per month, then the birth rate is 1,000/2,000, or 0.5 per mouse per month. Subtract a population's per capita death rate from its per capita birth rate and you have its **per capita growth rate**. If the death rate in our population of 2,000 mice is 200 per month (0.1 per mouse per month), then the per capita growth rate is $0.5 - 0.1 = 0.4$ per mouse per month.

FIGURE 16.4 Animated!
Exponential growth in population of mice with a per capita growth rate (*r*) of 0.4 per mouse per month and an initial population size of 2,000.

Credits: left, Jeff Lepore/ Photo Researchers, Inc.;
(a,b) © Cengage Learning.

A

	Starting Population Size		Net Monthly Increase	New Population Size
$G = r \times$	2,000	=	800	2,800
$r \times$	2,800	=	1,120	3,920
$r \times$	3,920	=	1,568	5,488
$r \times$	5,488	=	2,195	7,683
$r \times$	7,683	=	3,073	10,756
$r \times$	10,756	=	4,302	15,058
$r \times$	15,058	=	6,023	21,081
$r \times$	21,081	=	8,432	29,513
$r \times$	29,513	=	11,805	41,318
$r \times$	41,318	=	16,527	57,845
$r \times$	57,845	=	23,138	80,983
$r \times$	80,983	=	32,393	113,376
$r \times$	113,376	=	45,350	158,726
$r \times$	158,726	=	63,490	222,216
$r \times$	222,216	=	88,887	311,103
$r \times$	311,103	=	124,441	435,544
$r \times$	435,544	=	174,218	609,762
$r \times$	609,762	=	243,905	853,667
$r \times$	853,667	=	341,467	1,195,134

B

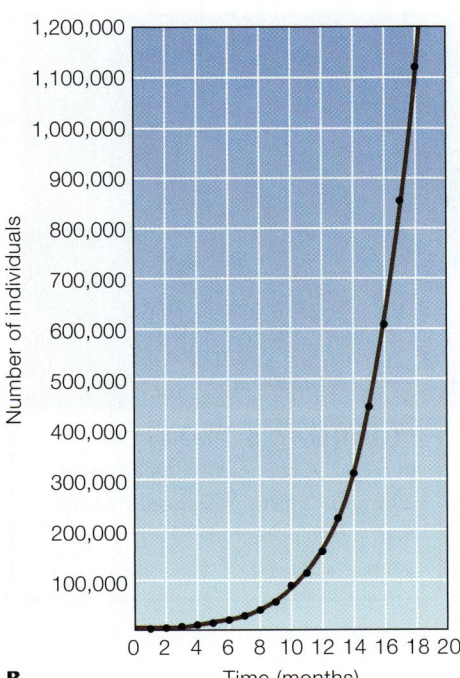

The **exponential model of population growth** describes how a population's size changes over time if its per capita growth rate is constant and its resources are unlimited. Under these theoretical conditions, the population growth in any interval (*G*) can be calculated as follows:

$$G = r \times N$$
$$\text{(per capita growth rate)} \quad \text{(number of individuals)}$$

Suppose we apply this to our population of 2,000 mice with their per capita growth rate of 0.4 per month. In the first month, the population grows by 2,000 mice × 0.4. This brings the size to 2,800. In the next month, 2,800 × 0.4, or 1,120 mice, are added, and so on (Figure 16.4**A**). At this growth rate, the number of mice would rise from 2,000 to more than 1 million in under two years! Plotting the size of this population against time produces a J-shaped, or "hockey stick," curve, characteristic of exponential growth (Figure 16.4**B**).

Exponential population growth is analogous to the compounding of interest on a bank account that pays a fixed rate of return. Although the interest does not change, the amount of interest paid continually increases. Each year, the annual interest paid into the account adds to the size of the balance, and the following year's interest payment is based on that higher balance.

The exponential model of population growth assumes that resources are unlimited, so it cannot accurately predict the long-term growth of most populations. Resources are never unlimited. However, this model does provide insight into the expected short-term growth of a population with plentiful resources. For example, when a few individuals of a species colonize a new habitat, the resulting population often grows exponentially for a period.

❭ Biotic Potential The exponential growth rate for a population under ideal conditions is its **biotic potential**. This is a theoretical value that would hold if shelter, food, and other essential resources were unlimited and there were no predators or pathogens. Microorganisms such as bacteria have the highest

biotic potential Maximum possible population growth under optimal conditions.

exponential model of population growth Model for population growth when resources are unlimited. The per capita growth rate remains constant as population size increases.

per capita growth rate The number of individuals added during some interval divided by the initial population size.

FIGURE 16.5 Dividing *E. coli* bacteria. A capacity to divide every 20 minutes gives this species a very high biotic potential.

CNRI/ Photo Researchers, Inc.

biotic potentials. A population of the gut bacteria *Escherichia coli* can double in size in 20 minutes (Figure 16.5). Large mammals have some of the lowest biotic potentials. Elephants cannot reproduce until they are 15 or so years old, so their populations double much more slowly. Populations seldom reach their biotic potential because of limiting factors, which we discuss below.

> Carrying Capacity and Logistic Growth In the real world, the resources that organisms need to survive and reproduce are always limited. The **logistic model of population growth** addresses this limitation. With this model, the rate of population growth does not remain constant, but rather declines as population density increases. When there are few individuals relative to the amount of resources, the population grows exponentially (Figure 16.6).

As the number of individuals rises, **density-dependent limiting factors** such as competition begin to put the brakes on growth. As individuals become more and more crowded, they must compete with one another for food, hiding places, nesting sites, and other essential resources. Parasitism and disease also increase with population density, and both hinder population growth. As as result of density-dependent limiting factors, population growth begins to slow ❷.

Population growth continues until it eventually levels off at the environment's carrying capacity ❸. **Carrying capacity** is the maximum number of individuals of a species that a particular environment can sustain indefinitely. The carrying capacity for a species is not constant; rather, it depends on physical and biological factors that can change over time. For example, a prolonged drought can lower the carrying capacity for a plant species and for any animals that depend on that plant. The population size of one species can also affect the carrying capacity of another species with similar resource needs. For example, a grassland can support only so many grazers, so the presence of more than one grass-eating species lowers the carrying capacity for all of them.

> Density-Independent Factors The logistic model of population growth describes what happens when density-dependent limiting factors influence population growth. However, other factors can affect population size, including harsh weather such as hurricanes, and natural disasters such as tsuna-

FIGURE 16.6 One example of logistic growth: what happens when a few deer are introduced to a new habitat with finite resources.

❶ When the population is small, individuals have access to all the resources they require and the population grows exponentially.

❷ As population size grows, the growth rate begins to slow as density-dependent limiting factors begin to have an effect.

❸ Eventually, population size levels off. Population size plotted against time produces an S-shaped curve.

© Cengage Learning.

FIGURE 16.7 Changes in the size of a reindeer population. The reindeer were introduced to an island off the coast of Alaska in 1944.

Credits: left, © Cengage Learning; right, © Jacques Langevin/ Corbis Sygma.

mis or landslides. Such **density-independent limiting factors** increase the death rate in crowded and uncrowded populations alike; they do not arise as an effect of crowding.

In nature, density-dependent and density-independent factors often interact to determine the fate of a population. Consider what happened after the 1944 introduction of 29 reindeer to St. Matthew Island, an uninhabited island off the coast of Alaska. When biologist David Klein first visited the island in 1957, he found 1,350 well-fed reindeer munching on lichens (Figure 16.7). When Klein returned in 1963, he counted 6,000 reindeer. The population had soared far above the island's carrying capacity. Although a population can temporarily exceed an environment's carrying capacity, the high density cannot be sustained. Klein observed that density-dependent negative effects were already apparent. For example, the average body size of the reindeer had decreased.

When Klein returned in 1966, only 42 reindeer survived. The single male had abnormal antlers, and was thus unlikely to breed. There were no fawns. Klein figured out that thousands of reindeer had starved to death during the winter of 1963–1964. That winter was unusually harsh, with low temperatures, high winds, and 140 inches of snow. Most reindeer, already in poor condition as a result of increased competition, starved when deep snow covered their food. A population decline had been expected—a population that exceeds its carrying capacity usually shrinks and falls below that capacity—but bad weather magnified the extent of the crash. By the 1980s, there were no reindeer on the island.

> No population can grow forever. Limiting factors put the brakes on population growth.

Take-Home Message

What factors affect population growth?

- The exponential model of population growth describes the growth of a population with unlimited resources. In this idealized circumstance, per capita growth rate remains constant, but the population grows faster and faster.

- The logistic model of population growth describes the growth of a population affected by density-dependent limiting factors. Such a population grows exponentially at first, then its growth rate declines as a result of competition for resources, infectious disease, and other negative effects of crowding.

- A population undergoing logistic growth levels off at the environment's carrying capacity for that species.

- Density-independent factors such as harsh weather are not addressed by models for population growth, but they too affect natural populations.

carrying capacity Maximum number of individuals of a species that a specific environment can sustain.

density-dependent limiting factor Factor whose negative effect on growth is felt most in dense populations; for example, infectious disease or competition for food.

density-independent limiting factor Factor that limits growth in populations regardless of their density; for example a natural disaster or harsh weather.

logistic model of population growth Model for growth of a population limited by density-dependent factors; numbers increase exponentially at first, then the growth rate slows and population size levels off at carrying capacity.

16.4 Life History Patterns

Species, and even populations within a species, vary in their **life history**, which is a set of heritable traits such as rate of development, age at first reproduction, number of breeding events, and life span. In this section, we look at how such life history traits vary and the evolutionary basis for this variation.

Table 16.1	Annual Plant* Life Table		
Age Interval (days)	Survivorship (number surviving at start of interval)	Number Dying During Interval	Death Rate (number dying/ number surviving)
0–63	996	328	0.329
63–124	668	373	0.558
124–184	295	105	0.356
184–215	190	14	0.074
215–264	176	4	0.023
264–278	172	5	0.029
278–292	167	8	0.048
292–306	159	5	0.031
306–320	154	7	0.045
320–334	147	42	0.286
334–348	105	83	0.790
348–362	22	22	1.000
362–	0	0	0
		996	

* *Phlox drummondii;* data from W. J. Leverich and D. A. Levin, 1979.

> Survivorship Each species has a characteristic life span, but only a few individuals survive to the maximum age possible. Death is more likely at some ages than others. To gather information about age-specific risk of death, researchers focus on a **cohort**, a group of individuals that are all born at about the same time. Members of the cohort are tracked from birth until the last one dies. Mortality data from a cohort study can be summarized in a life table. Table 16.1 provides an example of a life table.

A **survivorship curve** plots how many members of a cohort remain alive over time, providing information about age-specific death rates. Ecologists describe three generalized types of survivorship curves. A type I curve is convex (bulges outward), indicating survivorship is high until late in life (Figure 16.8**A**). This pattern is characteristic of humans and other large mammals that produce one or two young and care for them. A diagonal type II curve indicates that the death rate of the population does not vary much with age (Figure 16.8**B**). A type II curve is characteristic of lizards, small mammals, and large birds. In these groups, old individuals are as likely to die of disease or predation as young ones. A type III curve is concave (bulges inward), indicating that the death rate for a population is highest early in life (Figure 16.8**C**). Marine animals that release eggs into water have this type of curve, as do plants that release enormous numbers of tiny seeds.

> Evolution of Life Histories To produce offspring, an individual must invest resources that it could otherwise use to grow and maintain itself. Species differ in the manner in which they distribute their parental investment among offspring and over the course of their lifetime. Some invest little in each of many offspring, others invest a lot in only a few offspring. Some reproduce

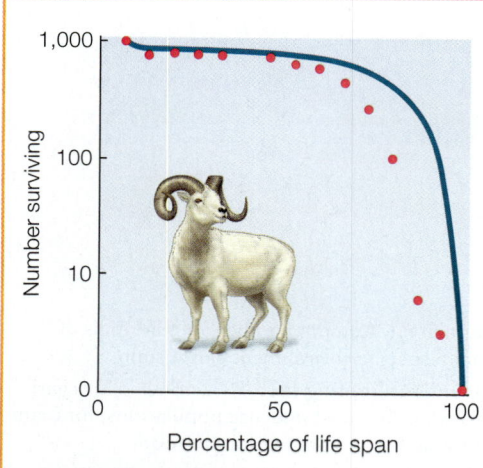

A Type I curve. Mortality is highest very late in life. Data is for Dall sheep (*Ovis dalli*).

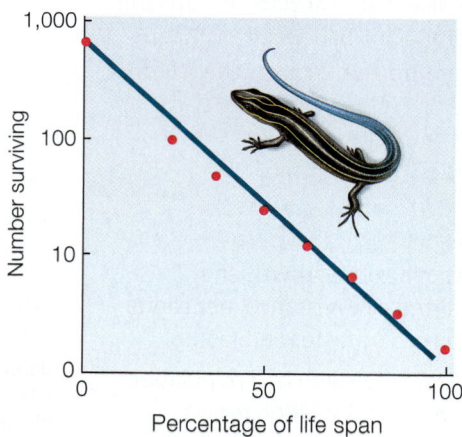

B Type II curve. Mortality does not vary with age. Data is for a small lizard (*Eumeces fasciatus*).

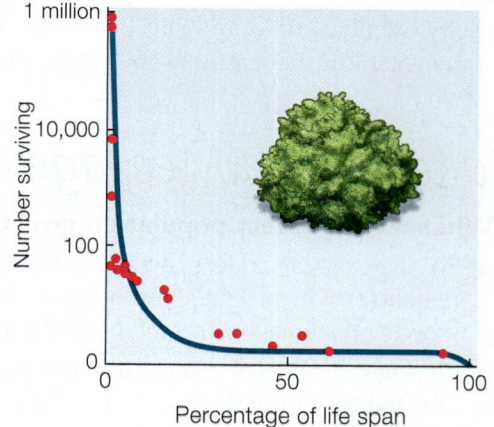

C Type III curve. Mortality is highest early in life. Data is for a desert shrub (*Cleome droserifolia*).

FIGURE 16.8 Animated! Types of survivorship curves.

© Cengage Learning.

Figure It Out: Which type of curve best fits the plant mortality data shown in Table 16.1? *Answer: Type III. Mortality is highest early in life.*

Opportunistic life history	Equilibrial life history
shorter development	longer development
early reproduction	later reproduction
fewer breeding episodes, many young per episode	more breeding episodes, few young per episode
less parental investment per young	more parental investment per young
higher early mortality, shorter life span	low early mortality, longer life span

FIGURE 16.9 Opportunistic and equilibrial life histories.
A,B Dandelions and flies are opportunists. A dandelion releases hundreds of tiny seeds and a fly lays many tiny eggs.
C,D Whales and coconut palms are equilibrial species. They produce few offspring at a time and invest heavily in each one.

Credits: (a) © Carly Rose Hennigan/ Shutterstock; (b) © Richard Baker; (c) © holbox/ Shutterstock; (d) Florida Fish and Wildlife Conservation Commission/NOAA; art, © Cengage Learning.

once, others many times. In studying this variation, ecologists have come to recognize two general life history strategies (Figure 16.9). Both strategies maximize the number of offspring that will be produced and survive to adulthood, but each does so under different environmental conditions.

When a species lives where conditions vary in an unpredictable manner, its populations seldom reach their carrying capacity. As a result, there is usually little competition for resources. Mortality occurs mainly as a result of density-independent factors. Such conditions favor an **opportunistic life history**, in which individuals produce as many offspring as possible, as quickly as possible. Opportunistic species (also called *r*-selected species) tend to have a short generation time and small body size. Because parental investment is spread across many offspring, each offspring receives a relatively small share. Opportunistic species usually have a type III survivorship curve, with mortality heaviest early in life. Annual plants such as dandelions (Figure 16.9**A**) have an opportunistic life history. They mature within weeks, produce many tiny seeds, then die. Flies are opportunistic animals. A female fly can lay hundreds of small eggs (Figure 16.9**B**) in a temporary food source such as a rotting tomato or a pile of feces.

When a species lives in a stable environment, its populations are often near their carrying capacity for that environment. Under these circumstances, competition for resources can be fierce and an **equilibrial life history**, in which parents produce a few, high-quality offspring, is adaptive. Equilibrial species (also called *K*-selected species) tend to have a large body and a long generation time. Large mammals such as whales have this sort of life history. They take years to reach adult body size and begin reproducing. When mature, a female whale produces only one large calf at a time (Figure 16.9**D**), and she continues to invest in the calf by nursing it after its birth. Similarly, a coconut palm grows for years before beginning to produce a few coconuts at a time (Figure 16.9**C**). In both whales and coconut palms, a mature individual produces young for many years.

〉 Evidence of Life History Evolution Often, different populations within a species live in slightly different environments, and their life history traits reflect these differences. Consider one long-term study in which biologists David Reznick and John Endler documented the evolutionary effects of predation on the life history traits of *Poecilia reticulata*, a species of guppy (a small

cohort Group of individuals born during the same interval.
equilibrial life history Life history favored in stable environments; individuals grow large, then invest a lot in each of a few offspring.
life history Set of traits related to growth, survival, and reproduction such as life span, age-specific mortality, age at first reproduction, and number of breeding events.
opportunistic life history Life history favored in unpredictable environments; individuals reproduce while young and invest little in each of many offspring.
survivorship curve Graph showing the decline in numbers of a cohort over time.

FIGURE 16.10 Animated! Experimental evidence that predation affects life history traits in guppies.

Credits: (a) Helen Rodd, inset David Reznick/ University of California–Riverside, computer enhanced by Lisa Starr; (b,c) Hippocampus Bildarchiv; (d) © Cengage Learning based on data from Reznick D.A., Bryga H., and Endler J.A. (1990) *Nature* 346: 357–359.

fish). The study began with fieldwork in the mountains of Trinidad, an island in the southern Caribbean Sea. Here, guppies live in shallow freshwater streams (Figure 16.10**A**). Waterfalls in the streams function as barriers that keep guppies from moving from one part of the stream to another. As a result, guppy populations are genetically isolated. Waterfalls also restrict the movement of predatory fish, so different predators prey upon different guppy populations.

The two kinds of guppy predators that also live in the streams differ in their body size and feeding habits. Killifish (Figure 16.10**B**), which are relatively small, prey on small, immature guppies but ignore full-grown adults. Pike cichlids

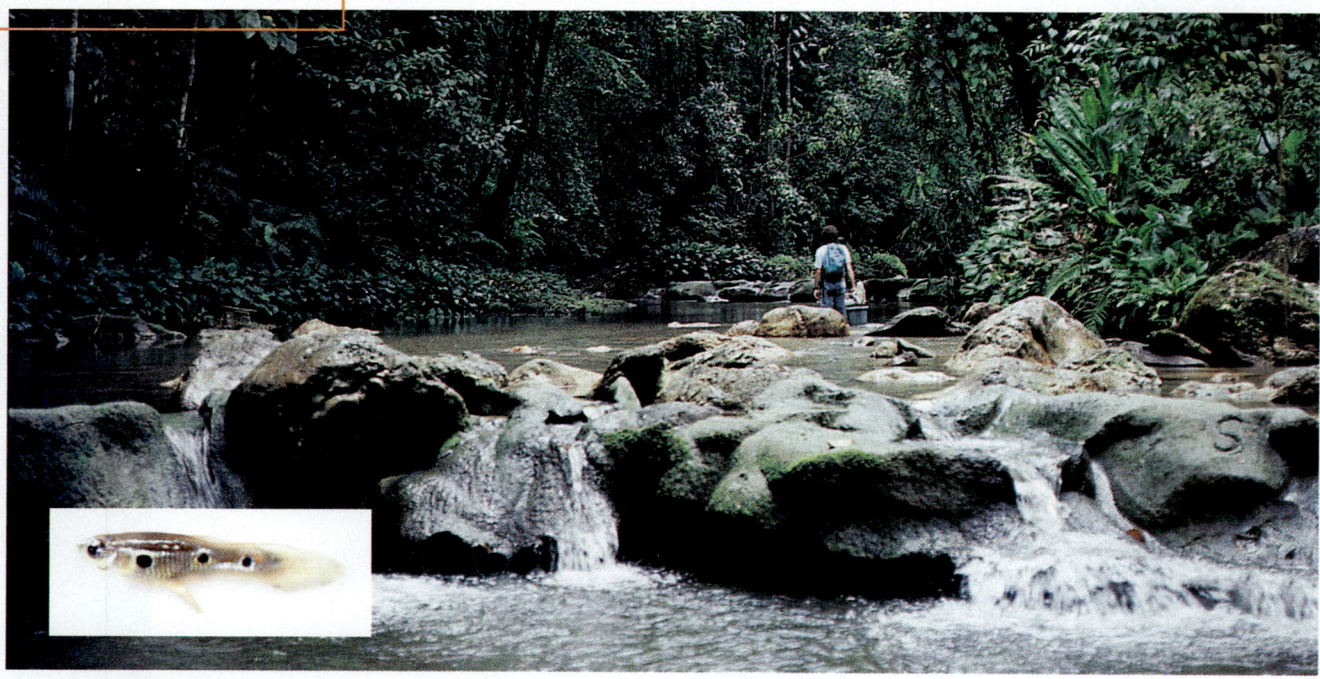

A Biologist David Reznick at the study site, a freshwater stream in Trinidad. The inset photo shows a guppy, the fish Reznick was studying.

B Killifish, preys on small guppies.

C Pike cichlid, preys on large guppies.

D Results of an experiment in which some guppies from a pool with pike cichlids were moved to a pool with killifish, but no other guppies. Other guppies were left in their original pool as a control. After 11 years, the life history traits of both groups were compared.

Life History Trait (All values are averages)	Control Group (fish that remained in a pool with pike cichlids)	Experimental Group (fish that were moved to a pool with killifish)
Male age at maturity	48.5 days	58.2 days
Male weight at maturity	67.5 milligrams	76.1 milligrams
Female age at maturity	85.7 days	93.6 days
Female weight at maturity	161.5 milligrams	185.6 milligrams

(Figure 16.10**C**) are larger and tend to pursue big, mature guppies, while ignoring small, young ones. Many of the streams have one of these predators, but not the other. Reznick and Endler discovered a correlation between the type of predator and the life history traits of the preyed-upon guppy population. Guppies living with pike cichlids alone tended to breed at a smaller body size than guppies in streams containing only killifish. Individuals in these guppy populations reproduce earlier, have more offspring at one time, and breed more often.

Based on their observations, Reznick and Endler suspected that predation shapes guppy life history patterns through natural selection, and they devised an experiment to test this hypothesis. They found a pool that held guppies and pike cichlids, but no killifish. They left some of these guppies in place as a control group, and moved others to a pool that contained only killifish, predicting that exposure to this unfamiliar predator would cause life history traits of the experimental population to evolve.

Eleven years later, researchers who revisited the stream found that the guppy population at the experimental site had indeed evolved. Its members differed from the control population in their average age and size at first reproduction (Figure 16.10**D**), as well as other life history traits. Selective pressure exerted by a new predator that eats small fishes had favored individuals that put their energy into growth rather than reproduction, until they were too big to be eaten.

The evolution of life history traits in response to predation is not merely of theoretical interest. It has economic importance. Just as guppies evolved in response to predators, a population of Atlantic codfish (*Gadus morhua*) evolved in response to human fishing pressure. From the mid-1980s to early 1990s, fishing pressure on the North Atlantic population of codfish increased. As it did, the age of sexual maturity shifted, with fishes that reproduced while young and small becoming an increased proportion of the population. Such individuals were at an advantage because both commercial fisherman and sports fishermen preferentially caught and kept larger fish (Figure 16.11). Fishing pressure continued to rise, until 1992, when declining cod numbers caused the Canadian government to ban cod fishing in some areas. That ban, and later restrictions, came too late to stop the Atlantic cod population from crashing. In some areas, the population declined by 97 percent and has not recovered.

Looking back, it is clear that life history changes were an early sign that the North Atlantic cod population was in trouble. Had biologists recognized what was happening, they might have been able to save the fishery and protect the livelihood of more than 35,000 fishers and associated workers. Ongoing monitoring of the life history data for other economically important fishes may help prevent similar disastrous crashes in the future.

FIGURE 16.11 A fisherman with a large Atlantic codfish. A human preference for big codfish selected for fish that matured while still young and small.

© Bruce Bornstein, www.captbluefin.com.

Take-Home Message

What factors shape life history patterns?

- Life history traits such as the age at which an organism first reproduces and the number of offspring it produces at one time have a heritable basis and are subject to natural selection.

- Survivorship curves and life tables provide information about mortality of members of a cohort over time.

- Rapid production of many offspring is most adaptive when environmental factors keep population density low. Producing offspring that are fewer in number but better able to compete is adaptive if populations are often near carrying capacity.

- In some cases, predators influence the life history traits of their prey.

Estimated population	
10,440 years ago	5 million
By 1804	1 billion
By 1927	2 billion
By 1960	3 billion
By 1974	4 billion
By 1987	5 billion
By 1999	6 billion
By 2011	7 billion

domestication of plants, animals 9000 B.C. (about 11,000 years ago)

agriculturally based urban societies

beginning of industrial, scientific revolutions

FIGURE 16.12 Growth curve (*red*) for the world human population. The *orange* box lists how long it took for the human population to increase from 5 million to 7 billion. NASA.

16.5 Human Populations

> **Population Size and Growth Rate** For most of its history, the human population grew very slowly (Figure 16.12). The growth rate began to pick up about 10,000 years ago, and during the past two centuries, it soared. Three trends promoted the large increases. First, humans migrated into new habitats and expanded into new climate zones. Second, they developed technologies that increased the carrying capacity of existing habitats. Third, they sidestepped some limiting factors that typically restrain population growth.

Modern humans evolved in Africa by about 200,000 years ago, and by 43,000 years ago, their descendants were established in much of the world (Section 15.7). Few species can expand into such a broad range of habitats, but our large brains allowed us to master a variety of skills. Humans learned how to start fires, build shelters, make clothing, manufacture tools, and cooperate in hunts. With the advent of language, knowledge of such skills did not die with the individual.

The invention of agriculture about 11,000 years ago provided a more dependable food supply than traditional hunting and gathering. A pivotal factor was the domestication of wild grasses, including species ancestral to wheat and rice. In the middle of the eighteenth century, people learned to harness energy in fossil fuels to operate machinery. This innovation opened the way to high-yielding mechanized agriculture and to improved food distribution systems. Food production was further enhanced in the early 1900s, when chemists discovered a way to convert gaseous nitrogen to ammonia. Previously, this process had been carried out primarily by nitrogen-fixing bacteria. Use of synthetic nitrogen fertilizers dramatically increased crop yields. Invention of synthetic pesticides in the mid-1900s also contributed to increased food production.

Disease has historically dampened human population growth. For example, during the mid-1300s, one-third of Europe's population was lost to a pandemic known as the Black Death. Beginning in the mid-1800s, an increased understanding of the link between microorganisms and illness led to improvements

age structure Of a population, distribution of individuals among various age groups.
total fertility rate Average number of children born to females of a population over the course of their lifetimes.

in food safety, sanitation, and medicine. People began to pasteurize foods and drinks, heating them to reduce the numbers of harmful bacteria. They also began to protect their drinking water. The first modern sewer system was constructed in London, England, in the late 1800s. By diverting wastewater downstream of the city's water source, the system lowered the incidence of waterborne diseases such as cholera and typhoid fever. Beginning in the early 1900s, chlorination and other methods of sterilizing drinking water contributed to a further decline in these diseases in the most industrialized nations.

Advances in sanitation also lowered the death rate associated with medical treatment. In the mid-1800s, Ignaz Semmelweis, a physician in Vienna, began urging doctors to wash their hands between patients. His advice was largely ignored until after his death, when Louis Pasteur popularized the idea that unseen organisms cause disease. Acceptance of this idea also revolutionized surgery, which had previously been carried out with little regard for cleanliness.

Vaccines and antibiotics also helped lower death rates. Vaccinations became widespread in developed countries during the 1800s. Antibiotics are a more recent development. Large-scale production of penicillin, the first antibiotic to be widely used, did not begin until the 1940s.

A worldwide decline in death rates without an equivalent drop in birth rates is responsible for the ongoing explosion in human population size. It took more than 100,000 years for the human population to reach 1 billion in number. Since then, the rate of increase has risen steadily. The population is now about 7 billion and it is expected to reach 9 billion by 2050.

⟩ Fertility Rates and Future Growth Most governments now recognize that population growth, resource depletion, pollution, and quality of life are interconnected. Many offer family planning programs. The United Nations Population Division estimates that more than 60 percent of married women worldwide use family planning methods.

The **total fertility rate** is the average number of children born to the females of a population during their reproductive years. In 1950, the worldwide total fertility rate averaged 6.5 children per woman. By 2011, it had declined to 2.5 but it remains above the replacement level, which is the number of children a woman must bear to ensure that two children grow to maturity and replace her and her partner. At present, this replacement level is 2.1 for developed countries and as high as 2.5 in some developing countries. (It is higher in developing countries because more female children die before reaching reproductive age.)

At present, China (with 1.3 billion people) and India (with 1.1 billion) dwarf other countries. Together, they hold 38 percent of the world population. Next in line is the United States, with 305 million. Compare the **age structure**, or distribution of individuals among age groups, for these three countries (Figure 16.13). Notice especially the size of the age groups that will be reproducing during the next fifteen years. The broader the base of an age structure diagram, the greater the proportion of young people, and the greater the expected growth. Government policies that favor couples who have only one child have helped China to narrow its pre-reproductive base.

Even if every couple now alive decides to bear no more than two children, world population growth will not slow for many years, because more than one-third of the world population is now in the broad pre-reproductive base. About 1.9 billion people are about to enter the reproductive age bracket.

While some countries face overpopulation, others have a declining birth rate and an increasing average age. In some developed countries, the decreasing total fertility rate and increasing life expectancy have resulted in a high proportion of older adults. In Japan, people over 65 currently make up about 20 percent of the

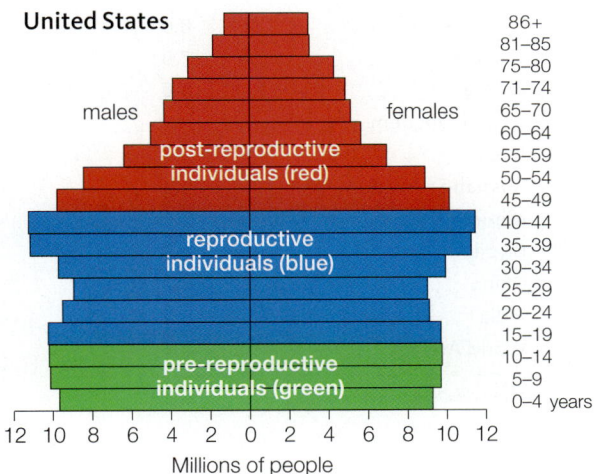

FIGURE 16.13 Animated! Age structure diagrams for the world's three most populous countries. The width of each bar represents the number of individuals in a 5-year age group. *Green* bars represent people in their pre-reproductive years. The *left* side of each chart indicates males; the *right* side, females.

Figure It Out: Which country has the largest number of people in the 45 to 49 age group?   *Answer: China*

© Cengage Learning.

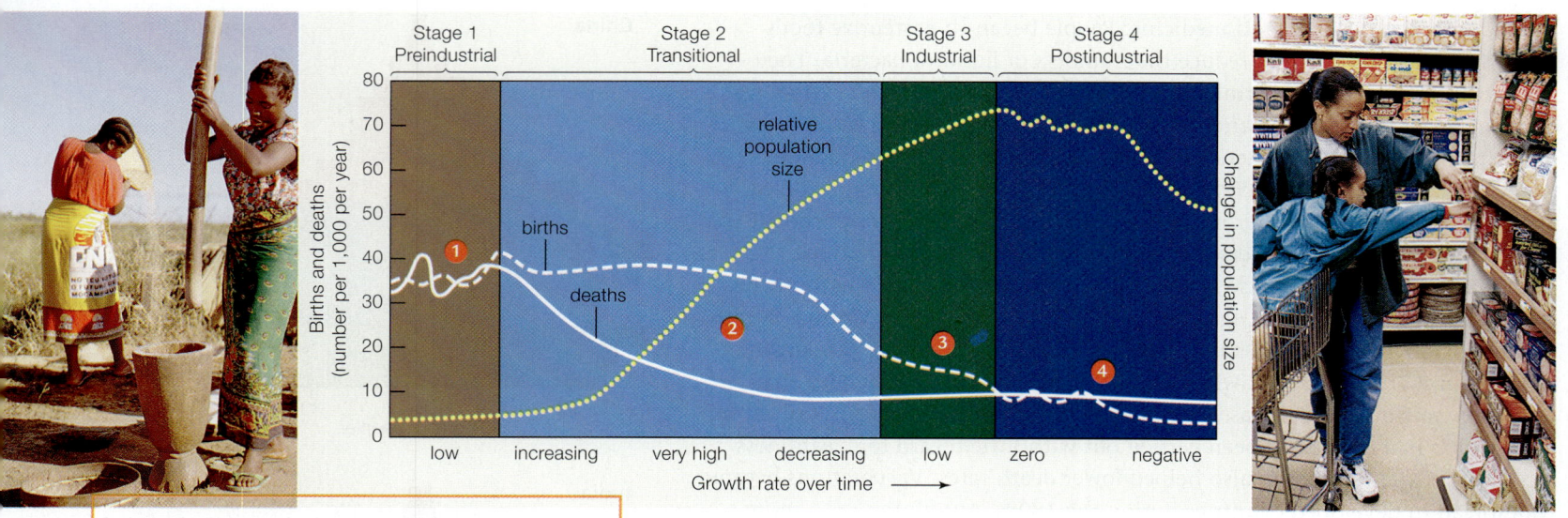

Stage 1 Preindustrial | Stage 2 Transitional | Stage 3 Industrial | Stage 4 Postindustrial

births

deaths

relative population size

Births and deaths (number per 1,000 per year)

Change in population size

low | increasing | very high | decreasing | low | zero | negative

Growth rate over time ⟶

FIGURE 16.14 Animated! Demographic transition model for changes in population growth rates and sizes, correlated with long-term changes in the economy.

Figure It Out: In which stage are death rates highest?

Answer: The preindustrial stage

Credits: left, © Adrian Arbib/ Corbis; middle, © Cengage Learning; right, © Don Mason/ Corbis.

Table 16.2	Ecological Footprints*
Country	Hectares per Capita
United States	8.0
Canada	7.0
France	5.0
United Kingdom	4.9
Japan	4.7
Russian Federation	4.4
Mexico	3.0
Brazil	2.9
China	2.2
India	0.9
World Average	2.7

* 2010 data from www.footprintnetwork.org.

demographic transition model Model describing the changes in human birth and death rates that occur as a region becomes industrialized.

ecological footprint Area of Earth's surface required to sustainably support a particular level of development and consumption.

population. In the United States, the proportion of people over 65 is projected to reach this level by 2030.

❯ **A Demographic Transition** Demographic factors vary among countries, with the most highly developed countries having the lowest fertility rate, the lowest infant mortality, and the highest life expectancy. The **demographic transition model** describes how changes in birth and death rates in human populations change over the course of four stages of economic development (Figure 16.14). Living conditions are harshest during the preindustrial stage, before technological and medical advances become widespread. Birth and death rates are both high, so the growth rate is low ❶. In the transitional stage, industrialization begins, and food production and health care improve. The death rate drops fast, but the birth rate declines more slowly ❷. As a result, the population growth rate increases rapidly. India is in this stage. During the industrial stage, industrialization is in full swing and the birth rate declines. People move from the country to cities, where couples tend to want smaller families. The birth rate moves closer to the death rate, and the population grows less rapidly ❸. Mexico is currently in this stage. In the postindustrial stage, a population's growth rate becomes negative. The birth rate falls below the death rate, and population size slowly decreases ❹.

The demographic transition model was developed to explain the social changes that occurred when western Europe and North America industrialized. Whether the model will accurately predict changes in modern developing countries remains to be seen. Global economic conditions have changed dramatically since Europe and North America began their industrial revolutions in the late 1800s. Today, most developing countries receive economic aid from nations that have already become fully industrialized, but they also must compete with these nations in a global market.

❯ **Resource Consumption** Resource consumption rises with economic and industrial development. Ecological footprint analysis is one method of comparing resource use. An **ecological footprint** is the amount of Earth's surface required to support a particular level of development and consumption in a sustainable fashion. Table 16.2 shows the per capita global footprint data for 2010. The average person in the United States has an ecological footprint nearly three times that of an average world citizen, and about nine times that of an average person living in India.

Moreover, a global shift toward industrialized lifestyles strains Earth's finite resources. The average person in an industrialized nation uses far more nonrenewable resources than one in a less developed country. For example, the United States accounts for about 4.6 percent of the world's population, yet it uses about 25 percent of the world's minerals and energy supply. Billions of people living in India, China, and other less developed nations would like to own the same kinds of goods that people in developed countries enjoy. However, Earth may not have the resources to make that possible. The World Resources Institute estimates that for everyone now alive to have an average American lifestyle would require the resources of four Earths. Finding ways to meet the wants and needs of expanding populations with limited resources will be a challenge.

> For everyone now alive to live like an average American would require the resources of four Earths.

Take-Home Message

What factors affect human population growth?

- Through expansion into new regions, invention of agriculture, and technological innovation, the human population has sidestepped environmental resistance to growth. Its size has been skyrocketing since the industrial revolution.

- Historically, death rates, and later birth rates, have fallen as nations have become more industrialized.

- Earth's resources are limited, so the current exponential growth of the human population is unsustainable.

16.6 A Honkin' Mess (revisited)

In January of 2009, both engines of a US Airways flight failed shortly after the plane took off from a New York City airport. Fortunately, the pilot was able to land the plane in the nearby Hudson River (Figure 16.15), where boats unloaded all 155 people aboard. Afterwards, investigators from the Federal Aviation Agency were asked to find out what caused the engine failure. The pilot had reported a bird strike, and the investigators found bits of feather, bone, and muscle in the plane's wing flaps and engines. Samples of this tissue were sent to the Smithsonian Institution, which analyzed the DNA. Unique sequences in the DNA identified the tissue in both engines as Canada goose. One engine had female goose DNA. The other had male and female DNA, so at least three birds were involved. Researchers were even able to tell which population of geese these unlucky birds belonged to. The mix of hydrogen isotopes varies with latitude, so the isotope mix in a feather provides information about where that feather developed. The geese were migratory; their feathers had developed in Canada, not in New York.

WHERE YOU ARE GOING . . .

Section 17.3 looks at the effect of interspecific interactions on population size. Section 22.8 discusses vaccines, which have lowered human death rates, and Section 26.5 discusses birth control methods, which have lowered human birth rates. The effects of human population growth on other species are considered in depth in Chapter 18.

FIGURE 16.15 A commercial airline floats in New York's Hudson River. Both of the plane's engines failed after sucking in Canada geese.

AP Images/ Steven Day.

Summary

Section 16.1 A **population** is a group of individuals that live in a given area and tend to mate with one another. Human actions have influenced the growth of some Canada goose populations. The study of populations is one aspect of the field of biology known as **ecology**.

Section 16.2 Populations vary in their **demographics**, or characteristics such as **population size**, **population density**, and **population distribution**. Populations most commonly have a clumped distribution because of limited dispersal, a need for resources that are clumped, and/or the benefits of living in a social group.

Section 16.3 Birth and death rates determine how fast a population grows. A constant, positive **per capita growth rate** results in **exponential growth**, in which the population increases by a fixed percentage of the whole with each successive interval. As a result, a plot of numbers over time produces a J-shaped curve. The maximum theoretical rate of population growth is the **biotic potential**.

An essential resource that is in short supply is a **limiting factor** for growth. The **logistic growth** model describes how population growth is affected by **density-dependent factors**, such as disease or competition for resources. The population slowly increases in size, goes through a rapid growth phase, then stabilizes once carrying capacity is reached. **Carrying capacity** is the maximum number of individuals that can be sustained indefinitely by the resources available in their environment. Adverse weather and other **density-independent factors** can affect any population regardless of its size.

Section 16.4 Age at maturity, number of reproductive events, offspring number per event, and life span are aspects of a **life history**. Life histories are often studied by following a **cohort**, a group of individuals that were born in the same time interval.

Three types of **survivorship curves** are common: a high death rate late in life, a constant rate at all ages, or a high rate early in life. Life histories have a genetic basis and are subject to natural selection. Depending on the environment, a population may be more successful if its individuals reproduce once, or many times. At low population density, species with an **opportunistic life history** (its individuals make many offspring fast) have an advantage. At a higher population density, species with an **equilibrial life history** (its individuals invest more in fewer, higher-quality offspring) are at an advantage.

Predation can also alter life history patterns because predators (including humans) act as selective agents that affect the traits of prey populations.

Section 16.5 The human population has now surpassed 7 billion. Expansion into new habitats and the invention of agriculture allowed early increases. Later, improved sanitation and technologi-cal innovations raised the carrying capacity and minimized many limiting factors that negatively affect populations of other species.

A population's **total fertility rate** is the average number of children born to women during their reproductive years. The global total fertility rate is declining and most countries have family planning programs of some sort. A population's **age structure** influences its growth. The pre-reproductive base of the human population is so large that it is expected to grow for many years to come.

The **demographic transition model** predicts how human population growth rates change with industrialization. Historically, both the death rate and birth rates have fallen with increased industrialization, but it remains to be seen if this holds true today.

People in developed nations have a much larger **ecological footprint** than those in developing nations. Earth may not have enough resources to support the current population in the style of the most developed nations.

Self-Quiz
Answers in Appendix I

1. Most commonly, individuals of a population have a _____ distribution.
   - a. clumped
   - b. random
   - c. nearly uniform
   - d. none of the above

2. All members of a population _____ .
   - a. are the same age
   - b. reproduce
   - c. belong to the same species
   - d. all of the above

3. The exponential model of population growth assumes _____ .
   - a. the death rate declines as population density increases
   - b. per capita growth rate does not change
   - c. industrialization causes a fall in birth rates
   - d. resources are limited

4. For a given species, the maximum rate of population increase under ideal conditions is the _____ .
   - a. biotic potential
   - b. carrying capacity
   - c. environmental resistance
   - d. density control

5. Competition for resources and disease are _____ controls on population growth rates.
   - a. density-independent
   - b. density-dependent

6. An increase in the population of a prey species would most likely _____ the carrying capacity for that species' predators.
   - a. increase
   - b. decrease
   - c. not affect
   - d. stabilize

7. Members of a species with an _____ life history have many offspring and invest little in each one.
   - a. equilibrial
   - b. opportunistic

8. The logistic model of population growth takes into account _____ , but not _____ .
   - a. density-dependent factors; density-independent factors
   - b. density-independent factors; density-dependent factors

9. The human population is now about _____ .
   - a. 7 billion   b. 7 million   c. 70 billion   d. 70 million

10. Compared to the less developed countries, the highly developed ones have a higher _____ .
   - a. death rate
   - b. birth rate
   - c. total fertility rate
   - d. resource consumption rate

11. True or false? A population can never exceed its carrying capacity.

12. The total fertility rate of a population is _____ .
    a. always higher than the replacement fertility rate
    b. the average number of offspring a female has in her lifetime
    c. the maximum number of children a woman could have if her resources were unlimited

13. If an exponentially growing population of 1,000 mice has a per capita growth rate (*r*) of 0.3 mice per month, how many mice will there be one month from now?
    a. 3,000   c. 1,300
    b. 3,300   d. 300

14. Match each term with its most suitable description.
    _____ carrying capacity   a. change in birth and death rates with industrialization
    _____ logistic growth   b. group of individuals born during the same period of time
    _____ exponential growth   c. population growth plots out as an S-shaped curve
    _____ demographic transition   d. largest number of individuals sustainable by the resources in a given environment
    _____ limiting factor   e. population growth plots out as a J-shaped curve
    _____ cohort   f. essential resource that restricts population growth when scarce

Critical Thinking

1. The age structure diagrams for two hypothetical human populations are shown at *left*. Describe the growth rate of each population and discuss the current and future social and economic problems that each is likely to face.

2. Each summer, the giant saguaro cacti in deserts of the American Southwest produce tens of thousands of tiny black seeds apiece. Most die, but a few land in a sheltered spot and sprout the following spring. The saguaro is a CAM plant (Section 5.5) and it grows very slowly. After 15 years, a saguaro may be only knee high, and it will not flower until it is about age 30. The cactus may survive to age 200. Saguaros share their desert habitat with annuals such as poppies, which sprout just after the seasonal rains, produce seeds, and die in just a few weeks. How would you describe these two life history patterns? How could such different life histories both be adaptive in this environment?

3. Most species consist of many populations. When determining whether a species need to be protected, the U.S. Fish and Wildlife Service holds hearings to determine whether or not specific populations are endangered. Imagine that you are a member of the panel at such a hearing. A biologist comes before you to discuss a population of whales that spend their summer in a nearby bay. What would you want to know about this whale population to decide whether it is threatened?

Digging Into Data

Monitoring Iguana Populations

In 1989, Martin Wikelski started a long-term study of marine iguana populations in the Galápagos Islands. He marked the iguanas on two of the islands—Genovesa and Santa Fe—and collected data on how their body size, survival, and reproductive rates varied over time. The iguanas eat algae and have no predators, so deaths are usually the result of food shortages, disease, or old age. His studies showed that the iguana populations decline during El Niño events, when water surrounding the islands heats up.

In January 2001, an oil tanker ran aground and leaked a small amount of oil into the waters near Santa Fe. Figure 16.16 shows the number of marked iguanas that Wikelski and his team counted in their census of study populations just before the spill and about a year later.

1. Which island had more marked iguanas at the time of the first census?

2. How much did the population size on each island change between the first and second census?

3. Wikelski concluded that changes on Santa Fe were the result of the oil spill, rather than sea temperature or other climate factors common to both islands. How would the census numbers be different from those he observed if an adverse event had affected both islands?

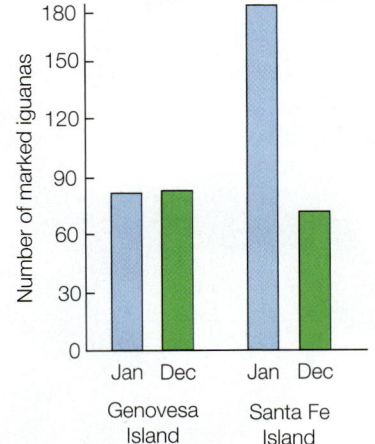

FIGURE 16.16 Shifting numbers of marked marine iguanas on two Galápagos islands. An oil spill occurred near Santa Fe just before the January 2001 census (*blue* bars). A second census was carried out in December 2001 (*green* bars).

Credits: top, © Reinhard Dirscherl/www.bciusa.com; right, © Cengage Learning.

17 Communities and Ecosystems

This chapter shows how species interactions can influence population size (Section 16.2) and cause directional selection (Section 12.3) and coevolution (12.6). Our discussion of energy flow in ecosystems builds on the introduction to this topic in Section 4.2. In covering nutrient cycles, we emphasize once again the essential roles of bacteria (13.5).

17.1 Fighting the Foreign Fire Ants

Red imported fire ants, or RIFAs (Figure 17.1**A**), are native to South America but, with a bit of human assistance, they have dramatically expanded their range. The ants first arrived in the southeastern United States in the 1930s, probably as stowaways on a cargo ship. They flourished in their new environment and eleven states now have well-established populations. In addition, the ants have used the United States as a springboard for dispersal to the Caribbean, Australia, New Zealand, and several Asian countries. Scientists know the RIFAs in these newly colonized regions came from the United States because their recently established populations all share a genetic marker that is very rare in South American RIFAs, but predominates among populations in the United States.

Red imported fire ants are considered pests because of their negative health and economic effects. The ants live in the ground and form moundlike nests. Accidentally step on one of these nests and large numbers of ants will rush out to bite and sting you. The venom injected by a RIFA's stinger causes an extremely painful burning sensation, followed by formation of a raised itchy bump at the site of the sting. As a result, a field or lawn colonized by red imported fire ants is inhospitable to humans, pets, and livestock. The ants' attraction to electricity also causes problems. For unknown reasons, large numbers of RIFAs sometimes congregate inside electrical motors, appliances, or switches, causing these devices to malfunction. Efforts to prevent the spread of red imported fire ants center on quarantines—prohibitions against moving soil that could contain the ants from affected areas to unaffected ones. Thus, the arrival of RIFAs can spell disaster for commercial plant or sod growers.

Fire ants also have a negative impact on natural communities. A biological **community** includes all the species in a region. Competition with red imported fire ants typically causes a region's native ant populations to decline, and the resulting change in species composition can harm ant-eating animals. For example, the Texas horned lizard feeds mainly on native harvester ants, but it does not eat the red imported fire ants that have largely replaced its natural prey. Red imported fire ants also harm native species by feeding on their eggs and feeding on or stinging their young. Ground-nesting animals such as quail are especially vulnerable to fire ant predation (Figure 17.1**B**). The presence of RIFAs can even affect native plants. The ants interfere with pollination by displacing or preying on native pollinators such as ground-nesting bees. They also impede dispersal of native plants whose seeds would normally be spread by native ant species that they have replaced.

Given all the problems RIFAs cause in the United States, you might wonder what things are like in their native South America. The ants are not considered much of a concern there, in part because they are far less common. In South America, parasites, predators, and diseases keep RIFA numbers far lower than those in affected parts of the United States. When some RIFAS left South America, they benefited by leaving their many natural enemies behind.

A One red imported fire ant worker. A colony can contain thousands of such workers, each with a stinger.

B Red imported fire ants attack and kill eggs and hatchlings of ground-nesting birds such as quail.

FIGURE 17.1 Red imported fire ant (*Solenopsis invicta*), an introduced threat to native species.

Credits: (a) Alex Wild/ Visuals Unlimited, Inc.; (b) © James Mueller.

17.2 Factors That Shape Communities

Communities vary in size and often nest one inside another. For example, we find a community of microbial organisms inside the gut of a termite. That termite is part of a larger community of organisms that live on a fallen log. This community is in turn part of a still larger forest community.

Even communities that are similar in scale differ in their species diversity. There are two components to **species diversity**: The first, species richness, refers to the number of species that are present; the second is species evenness, or the relative abundance of each species. A pond in which five fish species occur in nearly equal numbers has a higher species evenness, and thus a higher species diversity, than a pond with one abundant fish species and four rare ones.

Community structure is also dynamic, which means that the array of species and their relative abundances change over time. Communities change over a long time span as they form and then age. Some change suddenly as a result of natural or human-induced disturbances.

> **Nonbiological Factors** Factors related to geography and climate affect community structure. These factors include soil quality, sunlight intensity, rainfall, and temperature, which vary with latitude, elevation, and—for aquatic habitats—depth. Tropical regions receive the most sunlight energy and have the most even temperature. For most plants and animal groups, the number of species is greatest in the tropical regions near the equator, and declines as you move toward the poles. Tropical rain forest communities are highly diverse (Figure 17.2), and forest communities in temperate regions are less so. Similarly, tropical reef communities are more diverse than comparable marine communities farther from the equator.

> **Biological Factors** The evolutionary history and adaptations of the various species in a community can also influence community structure. Each species evolved in and is adapted to a specific **habitat**, the type of place where it typically occurs. All species of a community share the same habitat, the same "address," but each also has a unique ecological role that sets it apart. This role is the species' **niche**, which we describe in terms of the conditions, resources, and interactions necessary for survival and reproduction. Aspects of an animal's niche include temperatures it can tolerate, the kinds of foods it can eat, and the types of places where it can breed or hide. A description of a plant's niche would include details of its requirements for soil, water, light, and pollinators.

Species interactions also affect community structure. Actions of one species often affect other species. In some cases, the effect is indirect. For example, when songbirds eat caterpillars, the birds indirectly benefit the trees that the caterpillars feed on, while directly reducing the abundance of caterpillars. Other species interactions are more direct, and we consider these type of interactions and their consequences in detail in the next section.

FIGURE 17.2
Tropical diversity. Two of the 12 or so fruit-eating pigeon species that live in New Guinea's tropical rain forests (*left*). *Top*, a Victoria crowned pigeon, which is the size of a turkey. *Below*, the smaller superb crowned fruit pigeon.

Credits: left, Donna Hutchins; right, top, © Martin Harvey, Gallo Images/ Corbis; bottom, © Len Robinson, Frank Lane Picture Agency/ Corbis.

Take-Home Message

What factors shape a community?

- Communities vary in their species diversity as a result of nonbiological factors such as differences in incoming sunlight, temperature, and soil quality.

- Biological factors such as species requirements for survival and reproduction, as well as interactions with other species, also influence community structure.

community All populations of all species in some area.
habitat The type of place in which a species lives.
niche The role of a species in its community.
species diversity The number of species and their relative abundance.

17.3 Species Interactions in Communities

We recognize five types of direct interactions among species in a community: commensalism, mutualism, competition, predation, and parasitism (Table 17.1). Three of these—parasitism, commensalism, and mutualism—can be types of **symbiosis**, which means "living together." Symbiotic species, also known as symbionts, spend most or all of their life cycle in close association with each other. An endosymbiont is a species that lives inside its partner.

Regardless of whether one species helps or hurts another, two species that interact closely may coevolve over generations. In **coevolution**, each species is a selective agent that shifts the range of variation in the other (Section 12.6).

Table 17.1	Interspecific Interactions	
Type of Interaction	Direct Effect on Species 1	Direct Effect on Species 2
Commensalism	Benefits	None
Mutualism	Benefits	Benefits
Competition	Harmed	Harmed
Predation	Benefits	Harmed
Parasitism	Benefits	Harmed

> **Commensalism and Mutualism** **Commensalism** benefits one species and does not affect the other. For example, some orchids that live attached to a tree trunk or branch (Figure 17.3) benefit by getting a perch in the sun, while the tree is unaffected. As another example, commensal bacteria that live in the gut of many animals benefit by having a warm, nutrient-rich place to live, and their presence neither helps nor harms their host.

Other gut bacteria assist their host by aiding in digestion or synthesizing vitamins. An interaction that benefits both species is a **mutualism**. Flowering plants take part in mutualistic relationships with animals that pollinate them or disperse their seeds. Plants also have a mutualistic relationship with fungi that live on or in their roots. The fungi absorb mineral ions from the soil and share them with the plant. In return the fungi get sugars from the plant.

Mutualism is not a cozy cooperative venture, but rather a case of mutual exploitation. If there is a cost to participating, individuals who minimize that cost will be at a selective advantage. For example, a plant that lures a pollinator with a small nectar reward has an advantage over one that provides more nectar.

Mutualisms can be essential to one or both partners. A pink anemone fish will be eaten by a predator unless it has a sea anemone to hide in (Figure 17.4). The anemone's tentacles are covered by stinging cells that do not affect the anemone fish, but help keep predators that would eat the fish or its eggs away. The anemone can survive on its own, but benefits by having a partner that chases away the few fish species that are able to feed on anemone tentacles.

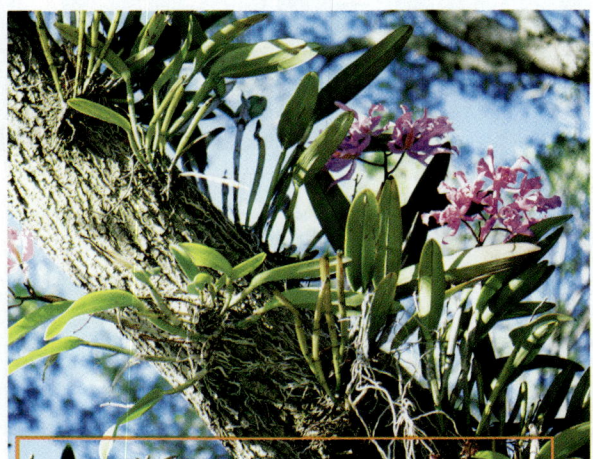

FIGURE 17.3 Commensalism. This tree provides orchids with an elevated perch from which they can capture sunlight. The presence of the orchids has no effect on the tree.
© John Mason/ ardea.com.

FIGURE 17.4 Mutualism. An anemone fish nestles among the tentacles of a sea anemone. In this mutually beneficial partnership, each species protects the other.
© Thomas W. Doeppner.

FIGURE 17.5 Competition among scavengers. **A** Golden eagle and a red fox face off over a moose carcass. **B** The eagle attacks the fox with its talons. After this attack, the fox retreated, leaving the eagle to exploit the carcass.
© Pekka Komi.

As a final example, mitochondria probably evolved from aerobic bacteria that were endosymbionts inside early eukaryotes (Section 13.3). The bacteria entered host cells or were taken in as food. Over generations, the bacteria lost their ability to live on their own and their hosts came to rely on ATP that these symbionts produced. Similarly, chloroplasts probably evolved from photosynthetic bacteria.

❯ **Competitive Interactions** As Malthus and Darwin understood, competition among members of a species can be fierce and such intraspecific competition often drives natural selection (Section 11.3). In this chapter, we focus on **interspecific competition**, which is competition between members of different species. Usually interspecific competition is not as intense as competition among members of the same species. Although the resource requirement of two species may overlap somewhat, their needs are never as similar those of two members of the same species.

Members of one species sometimes actively prevent members of another species from using a resource. For example, scavengers such as eagles and foxes fight over carcasses (Figure 17.5). Plants can also interfere with their competitors. For example, a sagebrush plant secretes chemicals that taint the soil around it, thus preventing potential competitors from taking root.

In other cases, competing species do not directly interfere with one another. Instead, all scramble for a share of a limited resource. For example, blue jays, deer, and squirrels eat acorns in oak forests. Although these animals do not fight over acorns, they do compete. By eating or storing acorns, each species reduces the number of acorns available to the others.

When species compete for a resource, each obtains less of that resource than it would if it lived alone. In this way, competition has a negative impact on all competitors. The more alike two species are, the more intensely they compete. If both species require the same limiting resource, the superior competitor will drive the lesser competitor to extinction in their shared habitat. We refer to this outcome of competition as **competitive exclusion**.

The more alike two species are, the more intensely they will compete.

coevolution Joint evolution of two closely interacting species; each species is a selective agent that shifts the range of variation in the other.

commensalism Species interaction that benefits one species and neither helps nor harms the other.

competitive exclusion When two species compete for a limiting resource, one drives the other to local extinction.

interspecific competition Two species compete for a limited resource and both are harmed by the interaction.

mutualism Species interaction that benefits both species.

symbiosis One species lives on or inside another in a commensal, mutualistic, or parasitic relationship.

Paramecium

A Time (days)   **B** Time (days)   **C** Time (days)

FIGURE 17.6 Animated! Competitive exclusion. Two species of the cilated protozan *Paramecium*, *P. caudatum* and *P. aurelia*, both feed on bacteria. When grown in separate test tubes, each does well **A**, **B**. When grown together **C**, one species drives the other to extinction.

Figure It Out: Which species of *Paramecium* was the superior competitor?   *Answer: P. aurelia*

Credits: left, Michael Abbey/ Photo Researchers, Inc.; (a–c) © Cengage Learning.

The concept of competitive exclusion arose out of experiments carried out by G. Gause in the 1930s. Gause grew two species of *Paramecium* separately and then in the same culture (Figure 17.6). Both of these protists ate bacteria and they competed intensely for food. When the two species were grown together, one species outgrew the other, driving it to extinction.

Competitors whose resource needs are not exactly the same can coexist. However, competition suppresses the population growth of all competing species. In each species, the individuals who differ most from competing species will be at a selective advantage. As a result, competing species often evolve in a way that lowers the intensity of competition between them. The result of this process is **resource partitioning**, a subdividing of an essential resource, which reduces competition among species that use it. For example, the two pigeons shown in Figure 17.2 are among the twelve or so species that feed on fruit in the forests of New Guinea. Although all of these species eat fruit, the birds can coexist because they eat fruits of different sizes and types.

FIGURE 17.7 Predation. Predators such as this lynx catch, kill, and devour their prey, in this case a snowshoe hare.

Ed Cesar/ Photo Researchers, Inc.

❯ Predator–Prey Interactions In **predation**, one free-living species captures, kills, and eats another (Figure 17.7). Predators exerts selective pressure on prey, favoring those with the best anti-predator defenses. Prey defenses in turn favor those predators best able to overcome them. As a result, predators and prey may engage in an evolutionary arms race that continues over many generations.

You have already learned about some defensive adaptations. Many prey species have hard or sharp parts that make them difficult to eat. Think of a snail's shell or a porcupine's quills. Others have chemicals that taste bad or sicken predators. Most defensive toxins in animals come from the plants they eat. For example, a monarch butterfly caterpillar feeds on and takes up chemicals from the milkweed plant. If a bird eats a monarch caterpillar or butterfly, these plant-derived chemicals will sicken it.

Many well-defended prey have **warning coloration**, a conspicuous pattern or color that predators learn to avoid. For example, stinging wasps and bees typically have black and yellow stripes (Figure 17.8**A**). Their similar appearance is a type of **mimicry**, an evolutionary pattern in which one species comes to resemble another. In one type of mimicry, well-defended species benefit by looking alike. In another type of mimicry, prey masquerade as a species that has a defense that they lack. For example, some flies that cannot sting resemble stinging bees or wasps (Figure 17.8**B**). Such a fly benefits when predators avoid it after a run-in with the better-defended look-alike species.

A The coloration of this yellow jacket wasp warns predators that it can sting.

B This fly, which cannot sting, benefits by mimicking the color pattern of wasps.

A The color and form of this praying mantis helps it hide from its prey—insects attracted to the flowers that it resembles.

B Earthy colors and a rounded shape help stone plants (*Lithops*) hide from herbivores.

Some prey have a last-chance trick that can startle an attacking predator. Section 1.7 described how eyespots and a hissing sound protect some butterflies. A lizard's tail may detach from the body and wiggle a bit as a distraction, while the lizard runs off. Skunks and some beetles squirt foul-smelling, irritating repellents at potential predators.

Camouflage is a body shape, color pattern, or behavior that allows an individual to blend into its surroundings and avoid detection. Prey benefit when camouflage hides them from predators, and predators benefit when it hides them from prey (Figure 17.9).

Many predators have evolved sharp teeth and claws that can pierce protective hard parts. Speedy prey select for faster predators. For example, the cheetah, the fastest land animal, can run 114 kilometers per hour (70 mph). Its preferred prey, Thomson's gazelles, run 80 kilometers per hour (50 mph).

> Plants and Herbivores In **herbivory**, an animal feeds on a plant, which may or may not die as a result. Some plants can withstand loss of their parts and quickly grow replacements. For example, grasses are seldom killed by herbivores. They have a fast growth rate and store enough resources in their roots to replace the shoots lost to grazers.

Camouflage helps some plants hide from herbivores. Succulent plants in the genus *Lithops* are commonly called stone plants or living stones because of their resemblance to the desert stones that they hide among (Figure 17.9).

Other plants have traits that fend off herbivores. Physical deterrents include spines, thorns, and fibrous, difficult-to-chew leaves. Plants can also produce compounds that taste bad to herbivores or sicken them. Ricin, a toxin made by castor bean plants (Section 7.1), makes herbivores ill. Caffeine in coffee beans and nicotine in tobacco leaves defend plants against insects. Capsaicin, a compound that makes some peppers "hot," defends seeds against seed-eating mammals. Rodents find capsaicin-rich pepper fruits unpalatable, leaving them to be eaten by birds, which do not taste capsaicin. A pepper benefits by deterring rodent seed eaters because rodents chew up and kill seeds, whereas birds excrete the seeds alive and intact.

> Parasites and Parasitoids **Parasites** obtain nutrients from a living host. Earlier chapters discussed some examples. Some bacteria, protists, and fungi are parasites. Tapeworms and flukes are parasitic annelid worms.

camouflage Body shape, pattern, or behavior that helps a plant or animal blend into its surroundings.

herbivory An animal feeds on a plant, which may or may not die as a result.

mimicry Two or more species come to resemble one another.

parasite A species that withdraws nutrients from another species (its host), usually without killing it.

predation One species (the predator) captures, kills, and feeds on another (its prey).

resource partitioning Use of different portions of a limited resource; allows species with similar needs to coexist.

warning coloration Distinctive color or pattern that makes a well-defended prey species easy to recognize.

A Dodder (*Cuscuta*), also known as strangleweed or devil's hair. The golden leafless stems twine around a host plant (in this case a grass). Its modified roots take up water and nutrients from the host's vascular tissue.

B Cowbird chick with its foster parent. A female cowbird is a brood parasite who minimizes her cost of parental care by laying her eggs in the nests of other bird species.

C Parasitoid as a biological control agent: A commercially raised parasitoid wasp about to deposit a fertilized egg in an aphid. The wasp larva will devour the aphid from the inside.

> **FIGURE 17.10** Parasites and parasitoids.
>
> Credits: (a) © The Samuel Roberts Noble Foundation, Inc.; (b) E. R. Degginger/ Photo Researchers, Inc.; (c) © Peter J. Bryant/ Biological Photo Service.

Some roundworms, insects, and crustaceans are parasites, as are all ticks. Even a few plants are parasitic. Plants known as dodders have little chlorophyll (Figure 17.10**A**). Their modified roots pierce the stem of a host plant, then withdraw sugars from it. Mistletoes are another group of parasitic plants.

Many parasites are pathogens (organisms that cause disease in their hosts). Even when a parasite does not cause obvious symptoms, infection can weaken the host, making it more vulnerable to predation or less attractive to potential mates. Some parasitic infections cause sterility.

Adaptations to a parasitic lifestyle include traits that allow the parasite to locate hosts and to feed undetected. For example, ticks that feed on mammals or birds move toward a source of heat and carbon dioxide, which are likely signs of a potential host. A chemical in tick saliva acts as a local anesthetic, preventing the host from noticing the feeding tick. Parasites that live inside other organisms often have adaptations that help them evade a host's immune defenses.

Hosts' defenses against parasites include immune responses (a topic we consider in detail in Chapter 22) and behavioral responses. For example, many primates take turns removing ticks and other parasites from one another. Birds kill external parasites by preening their feathers, and their bill shape reflects this function, as well as its role in feeding. In pigeons, even a slight experimental modification of bill shape that has no effect on feeding can result in an increase in parasite numbers.

Brood parasites do not feed on their hosts, but rather steal parental care. They lay their eggs in another animal's nest and thus trick it into raising their young. European cuckoos and North American cowbirds (Figure 17.10**B**) are examples. Freed from the constraints imposed by parental care, a female cowbird can lay as many as thirty eggs in a single season. Some other birds, fish, and insects are also brood parasites.

Parasitoids take the concept of forcing another to care for one's young to an even higher level; these insects lay their eggs inside other insects. Their eggs hatch into larvae that devour the host from the inside, eventually killing it.

Biological pest control, the use of a pest's natural enemies to reduce its numbers, often makes use of commercially raised parasites and parasitoids. Biological pest control provides an alternative to chemical pesticides, which typically kill or harm a wide variety of non-target species. Species chosen for use as biological pest control agents target only a specific type of host or prey species. The parasitoid wasp in Figure 17.10**C** is an example of a biological control agent. It lays eggs only in aphids, sucking insects that damage many crop plants.

Take-Home Message

How do species interactions affect a community?

- In commensalism, one species benefits and the other is unaffected.

- In mutualism, two species exploit one another in a way that benefits both.

- Competition for resources has a negative effect on both competitors. If both depend on the same limited resource, the stronger competitor may drive the weaker one to local extinction, a process called competitive exclusion.

- Interspecific competition favors individuals of both species whose resource needs are most unlike those of the competing species. Over time, competition alters traits related to resource use and leads to resource partitioning.

- Predators benefit at the expense of their prey, and parasites benefit at the expense of their hosts. As a result, predators and parasites select for defensive traits in prey and hosts. These defenses in turn select for traits that help the predators and parasites overcome them.

brood parasite An animal that tricks another species into raising its young, for example a cowbird.

parasitoid An insect that lays eggs in another insect, and whose young devour their host from the inside.

17.4 How Communities Change

› Ecological Succession Community structure—the kinds of species and their relative abundance in a habitat—is in constant flux. In a process called **ecological succession**, the array of species gradually shifts over time as organisms alter their own habitat. One array of species is replaced by another, which is in turn replaced by another, and so on.

Primary succession occurs in habitats that lack soil and have few or no existing species. For example, a rocky area exposed by retreat of a glacier undergoes primary succession (Figure 17.11). At first, no multicellular organisms are present ❶. The community begins to change as pioneer species gain a foothold ❷. **Pioneer species** colonize new or vacated habitats. They often include lichens, mosses, and hardy annual plants with wind-dispersed seeds. As generations of pioneers live and die, they help build and improve the soil. In doing so, they set the stage for their own replacement. Seeds of shrubby species take root in the mats of pioneers ❸. Over time, organic wastes and remains build up and, by adding volume and nutrients to soil, this material allows tall trees to take hold ❹.

Secondary succession occurs after a natural or human disturbance removes the natural array of species, but not the soil. We observe this kind of succession after a fire destroys a forest or a plowed field is abandoned and wild species move in and take over.

The 1980 eruption of Mount Saint Helens, a volcano in Washington State, gave scientists an opportunity to observe succession in action (Figure 17.12). The eruption showered the area around the volcano with volcanic rock and ash, wiping out the existing plant life and covering the mature soil. Since then, plant life has colonized the area and succession is under way.

Such field studies of succession have led to a change in the way the process is viewed. The concept of ecological succession was first developed in the late 1800s. At that time, it was viewed as a predictable and directional process that culminated in a "climax community," an array of species that persists over time and is reconstituted in the event of a disturbance. Physical factors such as climate, altitude, and soil type were thought to determine which species appeared

FIGURE 17.11 Animated! Artist's depiction of how primary succession in a previously glaciated area can lead to establishment of a forest community.
© Cengage Learning 2010.

ecological succession A gradual change in a community in which one array of species replaces another.

pioneer species Species that can colonize a new habitat.

primary succession Ecological succession occurs in an area where there was previously no soil.

secondary succession Ecological succession occurs in an area where a community previously existed and soil remains.

A Mount Saint Helens erupted in 1980. Volcanic ash completely buried the community that had previously existed at the base of this volcano.

B In less than a decade, numerous pioneer species had become established.

C Twelve years after eruption, Douglas fir seedlings were taking hold in soils enriched by volcanic ash.

FIGURE 17.12 Ecological succession after a volcanic eruption.
Credits: (a) R. Barrick/ USGS; (b) USGS; (c) P. Frenzen, USDA Forest Service.

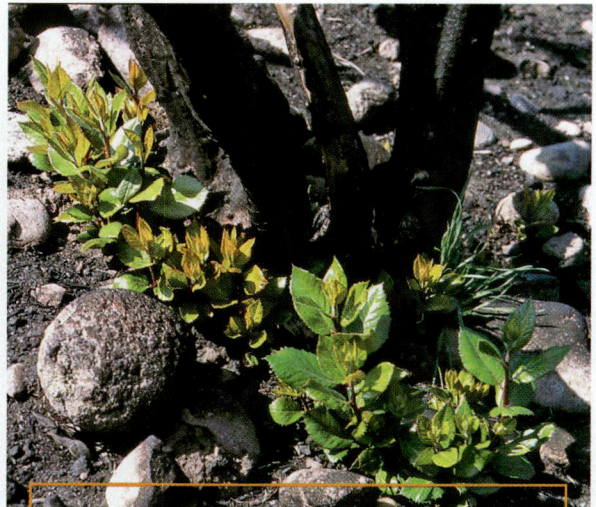

FIGURE 17.13 Effect of fire on community structure. Some woody shrubs, such as this toyon, resprout from their roots after a fire. In the absence of occasional fire, they are outcompeted and displaced by faster-growing but less fire-resistant species.

© Richard W. Halsey, California Chaparral Institute.

in what order. Modern ecologists recognize that three types of factors affect succession: (1) physical factors such as climate, (2) chance events such as the order in which pioneer species arrive, and (3) the frequency and extent of disturbances. Because the sequence of species arrivals and the frequency and extent of disturbances vary in unpredictable ways, it is difficult to predict exactly how the composition of any particular community will change in the future.

> Adapted to Disturbance In communities that are repeatedly subjected to a particular type of disturbance, individuals who withstand or benefit from that disturbance have a selective advantage. For example, some plants in areas subject to periodic fires produce seeds that germinate only after a fire has cleared away potential competitors. Other plants have an ability to resprout quickly after a fire (Figure 17.13). Because fire affects different species in different ways, the frequency of fire affects competitive interactions. For example, when naturally occurring fires are suppressed, plants adapted to periodic burning lose their competitive edge. The fire-adapted plants can be overgrown by plants that devote all of their energy to growing and reproducing, rather than investing in fire-related adaptations.

> Keystone Species A **keystone species** has a disproportionately large effect on a community relative to its abundance. Robert Paine coined the term to describe the results of his studies of a sea star (*Pisaster ochraceus*) common along rocky coastlines. To determine how this predator affected community structure, Paine compared the species richness of experimental plots, from which he had removed all sea stars, with that of control plots, in which he left the sea stars in place (Figure 17.14). Sea stars prey mainly on mussels, so when they were removed from experimental plots, mussels took over. The mussels crowded out seven other species of invertebrates and reduced the number of algal species. In control plots, species richness remained unaltered. Paine concluded that sea stars are a keystone species. They keep the number of species in the intertidal zone high by preventing mussels from dominating the habitat.

Keystone species need not be predators. For example, the large rodents called beavers can be a keystone species. A beaver cuts down trees by gnawing through their trunk, then uses felled trees to build a dam. Construction of a beaver dam creates a deep pool where a shallow stream would otherwise exist. By altering the physical conditions in a section of the stream, the beaver affects the types of fish and aquatic invertebrates that can live there.

FIGURE 17.14 Keystone species. The sea star *Pisaster* (shown *below*) is a keystone species along rocky shores. Removal of *Pisaster* from experimental plots caused a decline in species richness (*brown* dots and line). Control plots in which *Pisaster* was not removed had no comparable decline in species richness (*green* dots).

Credits: Left, © Nancy Sefton; right, © Cengage Learning.

A Kudzu native to Asia is overgrowing trees across the southeastern United States.

B Gypsy moths native to Europe and Asia feed on oaks through much of the United States.

C Nutrias native to South America are abundant in freshwater marshes and riversides of the Gulf States, the Chesapeake Bay region, and Oregon.

> **Exotic Species** Arrival of an **exotic species**—a species that was introduced to a new habitat and became established there—can also alter community structure. In its new home, an exotic species often has no coevolved parasites, pathogens, or predators to keep it in check, so its numbers can soar.

More than 4,500 exotic species have established themselves in the United States. The red imported fire ants described in Section 17.1 are one example. A look at three others will give you an idea of the variety of such species and the ways in which they affect native species.

One of the most notorious exotic species is a vine called kudzu. Native to Asia, it was deliberately introduced to the American Southeast as a food for grazers and to control erosion, but quickly became an invasive weed (Figure 17.15**A**).

Gypsy moths native to Europe and Asia entered the northeastern United States in the mid-1700s. Since then, they have extended their range into the Southeast, Midwest, and Canada. Gypsy moth caterpillars (Figure 17.15**B**) prefer to feed on oaks. Loss of leaves to gypsy moths weakens the trees, making them more susceptible to parasites and less efficient competitors.

Nutrias, large semiaquatic rodents from South America, were imported for their fur in the 1940s. Today, descendants of animals that either escaped or were intentionally released thrive in freshwater marshes and along rivers in twenty states. Their appetite for plants threatens native vegetation and crop plants, and their burrowing contributes to marsh erosion and damages levees, increasing the risk of flooding.

FIGURE 17.15 Three exotic species that have become pests in the United States. To learn about others, visit the National Invasive Species Information Center online at www .invasivespeciesinfo.gov.

Take-Home Message

What causes changes in community structure?

- In ecological succession, one array of species changes the habitat in a way that allows another array of species to take hold.
- Large and small disturbances shift community structure on an ongoing basis.
- A change in the presence or abundance of a keystone species has a great effect on other species in a habitat.
- An exotic species that leaves behind the predators, parasites, and competitors it evolved with can dramatically affect the structure of its adopted community.

exotic species A species that has been introduced to a new habitat and become established there.

keystone species A species that has a disproportionately large effect on community structure.

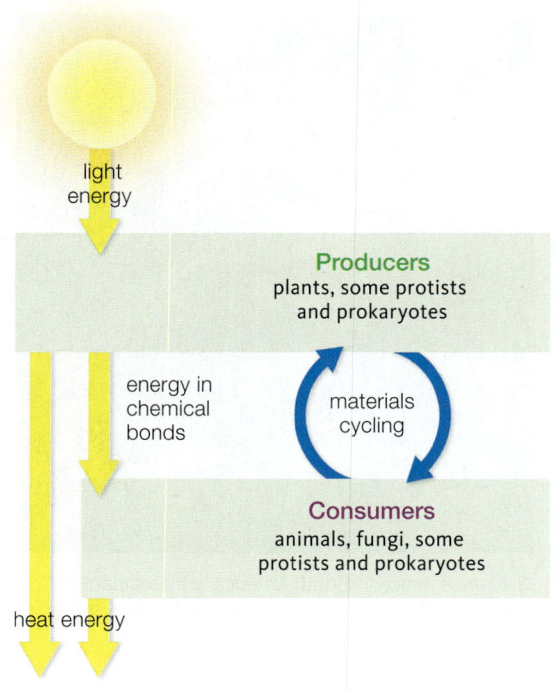

light energy

Producers
plants, some protists
and prokaryotes

energy in
chemical
bonds

materials
cycling

Consumers
animals, fungi, some
protists and prokaryotes

heat energy

FIGURE 17.16 Generalized model for the one-way flow of energy (*yellow* arrows) and the cycling of materials (*blue* arrows) in an ecosystem. Most producers convert sunlight energy to heat energy and energy in chemical bonds. Consumers use the chemical bond energy and also give off energy as heat. Eventually, all light energy that entered the ecosystem is converted to heat.

© Cengage Learning.

17.5 The Nature of Ecosystems

> Overview of the Participants Organisms of a community interact with their environment as an **ecosystem**. In all ecosystems, there is a one-way flow of energy and a cycling of essential materials (Figure 17.16). As explained in Section 1.3, an ecosystem's **producers** capture energy and use it to make their own food from inorganic materials in the environment. Usually the producer's energy source is sunlight and the producers are plants and photosynthetic bacteria and protists. An ecosystem's **consumers** obtain energy and carbon by feeding on tissues, wastes, and remains of producers and one another. Herbivores, predators, and parasites are consumers that feed on living organisms. **Detritivores** such as crabs and earthworms eat tiny bits of organic matter, or detritus. Finally, wastes and remains of organisms are broken down into inorganic building blocks by bacterial, protist, and fungal **decomposers**.

Light energy captured by producers is converted to bond energy in organic molecules, which is then released by metabolic reactions that give off heat as a by-product. This is a one-way process because organisms cannot convert heat back into chemical bond energy.

Unlike energy, with its one-way flow, nutrients cycle within an ecosystem. The cycle begins when producers take up hydrogen, oxygen, and carbon from inorganic sources such as the air and water. They also take up dissolved nitrogen, phosphorus, and other necessary minerals. Nutrients move from producers into the consumers who eat them. Decomposition returns nutrients to the environment, where producers take them up again.

> Food Chains and Webs All organisms of an ecosystem take part in a hierarchy of feeding relationships referred to as **trophic levels** ("troph" means nourishment). When one organism eats another, energy (in the form of chemical bonds) and nutrients are transferred from the eaten to the eater. All organisms at the same trophic level are the same number of transfers away from the energy input into that system.

A **food chain** is one sequence of steps by which energy captured by primary producers moves to higher trophic levels. Consider one food chain in a tallgrass prairie (Figure 17.17). The main producers in this ecosystem—grasses and other plants—are at the first trophic level. Energy flows from the plants to grasshoppers, to sparrows, and finally to hawks. Grasshoppers are primary consumers and are at the second trophic level. Sparrows that eat grasshoppers are second-level consumers and at the third trophic level. Hawks that eat sparrows are third-level consumers and at the fourth trophic level.

Food chains cross-connect with one another as a **food web**. Figure 17.18 shows some participants in an arctic food web. Nearly all food webs include two types of food chains. In grazing food chains, energy stored by producers

FIGURE 17.17
Animated! One food chain in a tallgrass prairie. Species at the first trophic level capture sunlight energy. Arrows represent the transfer of nutrients and energy from one trophic level to the next.

Credits: From left, © Van Vives;
© D. A. Rintoul; © D. A. Rintoul; © Lloyd
Spitalnik/ lloydspitalnikphotos.com.

First Trophic Level

Producer
Grass

Second Trophic Level

Primary Consumer
Grasshopper

Third Trophic Level

Second-Level Consumer
Sparrow

Fourth Trophic Level

Third-Level Consumer
Hawk

Higher Trophic Levels

A sampling of carnivores that feed on herbivores and one another

human (Inuk) arctic wolf arctic fox

gyrfalcon snowy owl ermine

Second Trophic Level

A sampling of primary consumers (herbivores) that eat plants

vole arctic hare lemming

mosquito flea

Parasitic consumers feed at more than one trophic level.

First Trophic Level

Examples of primary producers (plants)

grasses, sedges purple saxifrage arctic willow

Detritivores and decomposers (nematodes, annelids, saprobic insects, protists, fungi, bacteria)

FIGURE 17.18 Animated! Arctic food web. Arrows point from eaten to eater.

flows next to herbivores, which tend to be relatively large animals. In a detrital food chain, energy in producers flows to detritivores, which tend to be smaller animals, and to decomposers. In most land ecosystems, detrital food chains predominate. For example, in an arctic ecosystem, grazers such as voles, lemmings, and hares eat some plant parts. However, far more plant matter becomes detritus that sustains soil-dwelling insects and decomposers such as bacteria and fungi. Detrital food chains and grazing food chains interconnect as the ecosystem's overall food web.

consumers Organisms that eat organisms or their remains.

decomposers Consumers that feed on remains and break them into their inorganic building blocks.

detritivores Consumers that feed on small bits of organic material (detritus).

ecosystem A community and its environment.

food chain Sequence of steps by which energy moves from one trophic level to the next.

food web System of cross-connecting food chains.

producers Organisms that capture energy and make their own food from inorganic materials in the environment.

trophic level Position of an organism in a food chain.

> Primary Production and Inefficient Energy Transfers The flow of energy through an ecosystem begins with **primary production**: the capture and storage of energy by producers. Primary production, which is measured in terms of the amount of carbon taken up per unit area, varies seasonally and among habitats. On average, primary production is higher on land than it is in the oceans (Figure 17.19). However, because the oceans cover about 70 percent of Earth's surface, they contribute about half of Earth's total primary production.

An **energy pyramid** is a graphic representation of the proportion of the energy captured by producers that reaches higher trophic levels. Energy pyramids always have a large energy base, representing producers, and taper up. Figure 17.20 shows an energy pyramid for a freshwater ecosystem in Florida.

Only about 10 percent of the energy in tissues of organisms at one trophic level ends up in tissues of those at the next trophic level. Several factors limit the efficiency of transfers. All organisms lose energy as metabolic heat and this energy is not available to organisms at the next trophic level. Also, some energy gets stored in molecules that most consumers cannot break down. For example, most carnivores cannot access the energy tied up in bones, scales, hair, feathers, or fur. The inefficiency of energy transfers limits the possible length of food chains.

When people promote a vegetarian diet by touting the benefits of "eating lower on the food chain," they are referring to the inefficiency of energy transfers. A person eating a plant food involves only a single energy transfer. When a plant food is used to grow livestock, the animal uses some of the energy it obtains from plant food to sustain itself, loses some energy as heat, and invests some energy building inedible parts such as bones and hooves. Only a small percentage of the energy in the original plant food ends up as meat that a person can eat. Thus, feeding a population of meat-eaters requires far greater crop production than sustaining a population of vegetarians.

top carnivores	21
carnivores	383
herbivores	3,369
producers	20,810

FIGURE 17.20 Energy pyramid for a freshwater ecosystem. Numbers are the energy in kilocalories per square meter per year.

Figure It Out: What percentage of the sunlight energy captured by the producers was transferred to the herbivores that ate them?

Answer: 3,369/20,810 × 100 = 16 percent

© Cengage Learning.

Take-Home Message

How do organisms and their environment interact in ecosystems?

■ Materials cycle between organisms and their environment, but energy flow is one-way because energy in chemical bonds is converted to heat.

■ Energy and raw materials are taken up by producers, then flow to consumers.

■ Food chains and food webs describe routes by which energy and nutrients move from one trophic level to another.

■ Energy transfers between trophic levels are inefficient because energy is lost as metabolic heat, and some energy becomes tied up in materials that are not easily digested by consumers.

17.6 Nutrient Cycling in Ecosystems

In a **biogeochemical cycle**, ions or molecules of an essential substance flow among environmental reservoirs, and into and out of the world of life. Physical, chemical, and geological processes cause the slow movement of nutrients among environmental reservoirs. An ecosystem's producers take up various nutrients from the air, soil, and water. Consumers take in nutrients when they drink water and when they eat producers or one another.

We focus on four biogeochemical cycles that move important elements: the water cycle, phosphorus cycle, nitrogen cycle, and carbon cycle.

> **The Water Cycle** The **water cycle** moves water from oceans to the atmosphere, onto land and into freshwater ecosystems, and back to the oceans (Figure 17.21). Solar energy drives evaporation of water from the oceans and from freshwater reservoirs. Water that enters the lower atmosphere spends some time aloft as vapor, clouds, and ice crystals. By the process of precipitation, water falls from the atmosphere mainly as rain and snow.

Precipitation that falls on land mostly seeps into soil or joins surface runoff toward streams. Some soil water is taken up by plants, which then release most of it by transpiration—evaporation from their leaves.

Our planet has a lot of water, but 97 percent of it is salt water. Of the 3 percent that is fresh water, most is tied up in glaciers. **Groundwater**, another freshwater reservoir, includes water in the soil, and water stored in porous rock layers called **aquifers**. About half of the population of the United States relies on aquifers for drinking water. Surface water (water in streams, rivers, lakes, and freshwater marshes) constitutes less than 1 percent of Earth's fresh water.

> **The Phosphorus Cycle** The water cycle helps move minerals through ecosystems. Water that falls on land and makes its way back to the sea carries with it silt and dissolved minerals, including phosphorus.

Most of the phosphorus on Earth is bonded to oxygen as phosphate, an ionic compound found in rocks and sediments. There is no commonly occurring gaseous form of phosphorus, so the atmosphere is not one of its reservoirs.

Only 3 percent of Earth's water is fresh water, and most of that is in glacial ice.

aquifer Porous rock layer that holds some groundwater.

biogeochemical cycle A nutrient moves among environmental reservoirs and into and out of food webs.

energy pyramid Diagram that illustrates the energy flow in an ecosystem.

groundwater Water between soil particles and in aquifers.

primary production The energy captured by an ecosystem's producers.

water cycle Water moves from its main reservoir—the ocean—into the air, falls as rain and snow, and flows back to the ocean.

FIGURE 17.21 Animated! The water cycle. Water moves from the ocean into the atmosphere, onto land, and then back. The numbers *below* show the distribution of Earth's freshwater component among its environmental reservoirs.

Freshwater reservoir	Percent of fresh water
Polar ice, glaciers	68.7
Groundwater	30.1
Surface water	0.3
Other	0.9

© Cengage Learning.

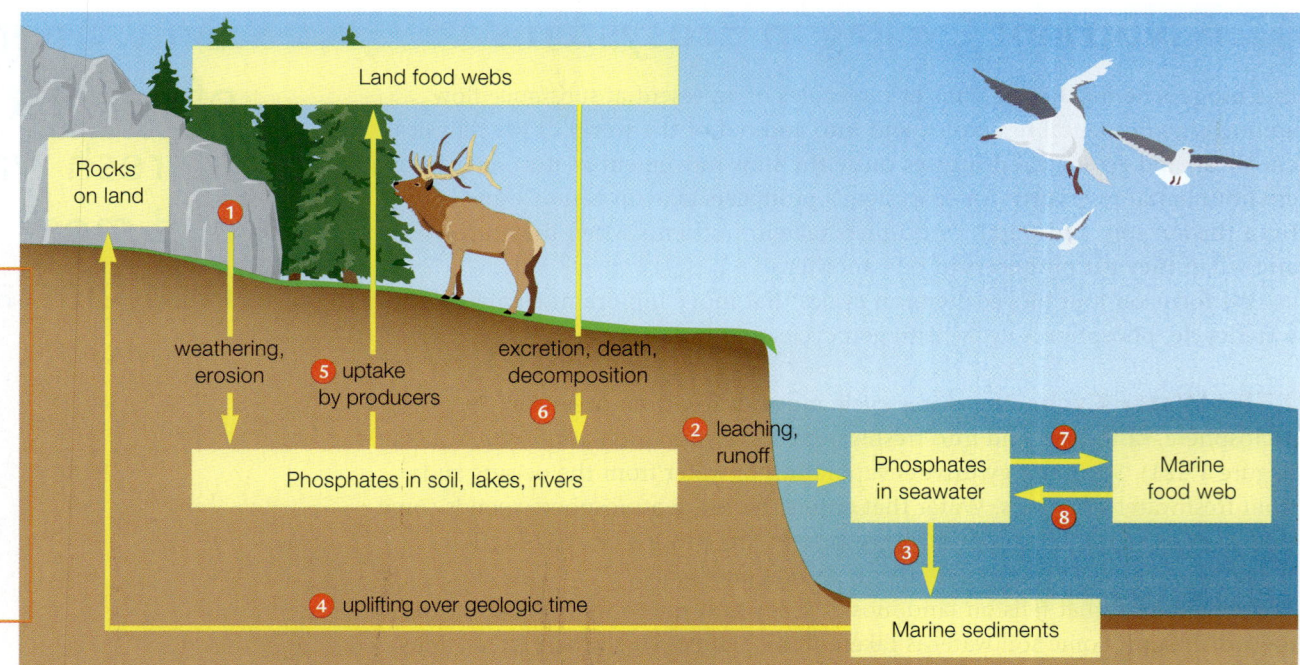

FIGURE 17.22 Animated! The phosphorus cycle. In this cycle, most of the phosphorus moves in the form of phosphate ions. Earth's main phosphorus reservoir is rocks and sediments.

© Cengage Learning.

Land food webs

Rocks on land ❶

weathering, erosion

❺ uptake by producers

excretion, death, decomposition

❻

❷ leaching, runoff

Phosphates in soil, lakes, rivers

Phosphates in seawater

❼

Marine food web

❽

❸

❹ uplifting over geologic time

Marine sediments

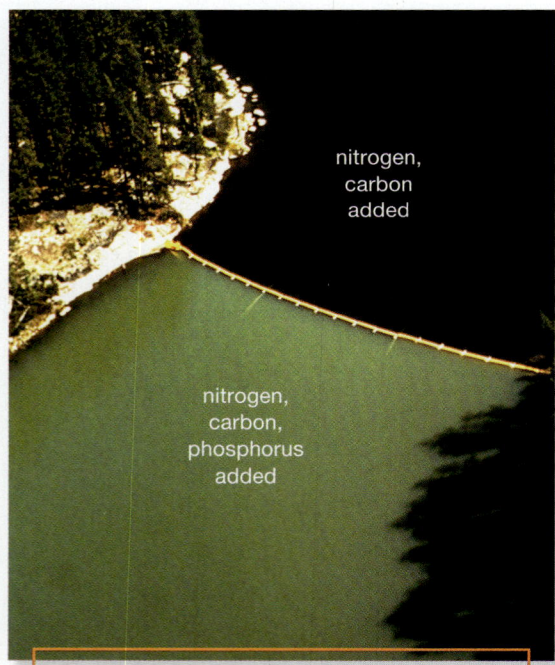

FIGURE 17.23 Results of a nutrient enrichment experiment. Researchers put a plastic curtain across a channel between two basins of a lake. They added nitrogen, carbon, and phosphorus on one side of the curtain (here, the *lower* part of the lake) and added nitrogen and carbon on the other side. Within months, the phosphorus-rich basin had a dense layer of single-celled algae covering its surface.

Fisheries & Oceans Canada, Experimental Lakes Area.

nitrogen, carbon added

nitrogen, carbon, phosphorus added

In the **phosphorus cycle**, phosphate moves among Earth's rocks, soil, and water, and into and out of food webs (Figure 17.22). In the environmental portion of the cycle, weathering and erosion move phosphate ions from rocks into soil, lakes, and rivers ❶. Leaching and runoff deliver phosphate ions to the ocean ❷, where most of the phosphorus comes out of solution and settles as deposits along the edges of continents ❸. Over millions of years, movements of Earth's crust can uplift parts of the seafloor onto land ❹, where weather releases phosphates from the rocks. Dissolved phosphates enter streams and rivers, and the environmental portion of the phosphorus cycle starts over again.

The biological portion of the phosphorus cycle begins when producers take up phosphorus. Roots of land plants take up dissolved phosphate from the soil water ❺. Land animals get phosphate by eating plants or one another. Phosphorus returns to the soil in wastes and remains ❻. In the seas, phosphorus enters food webs when producers take up dissolved phosphate from seawater ❼. As on land, wastes and remains replenish the supply ❽.

Lack of soil phosphates often limits plant growth, so fertilizers usually include phosphate. Phosphorus-rich droppings from seabird or bat colonies can be harvested as a natural fertilizer, but most commercial fertilizer contains phosphorus derived from rock that has been mined and then chemically treated.

Phosphate also serves as a limiting factor for aquatic producers, so adding phosphate to an aquatic habitat can cause a population explosion of algae (Figure 17.23). Such algal blooms (Section 13.6) threaten aquatic species. Humans encourage algal blooms by allowing phosphate-containing detergents, sewage, fertilizer runoff, and waste from stockyards to pollute aquatic environments.

〉 The Nitrogen Cycle Earth's atmosphere, which is about 80 percent gaseous nitrogen (N_2), is the largest nitrogen reservoir. In the **nitrogen cycle**, nitrogen moves among the atmosphere, through reservoirs in soil and water, and into and out of food webs (Figure 17.24).

Plants cannot use gaseous nitrogen because they do not have an enzyme that breaks the triple covalent bond that holds its two nitrogen atoms together. Some bacteria do have such an enzyme, and these organisms can carry out **nitrogen**

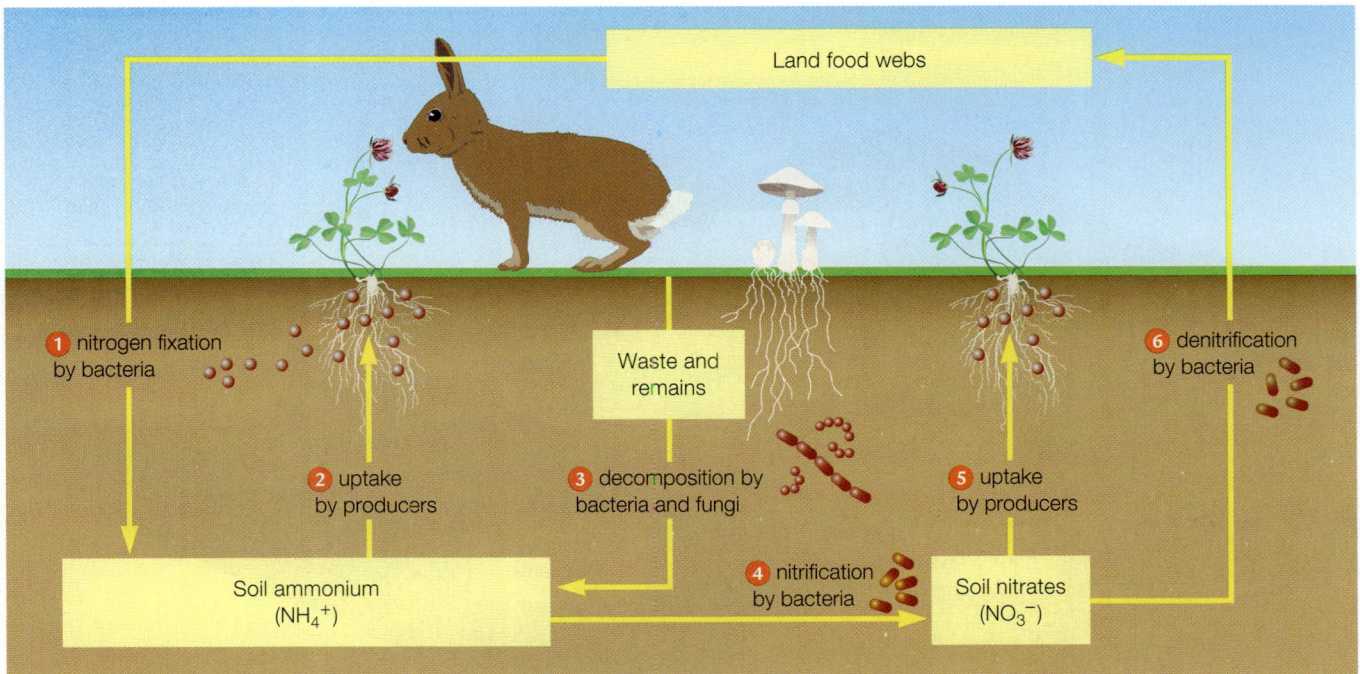

FIGURE 17.24 Animated! The nitrogen cycle. The main reservoir for nitrogen is the atmosphere. Activity of nitrogen-fixing bacteria converts gaseous nitrogen to forms that producers can use.

Figure It Out: What are the two forms of nitrogen that can be taken up by plants?

© Cengage Learning.

Answer: Ammonium and nitrate

fixation. They break the bonds in N_2, then use the nitrogen atoms to form ammonia, which dissolves and forms ammonium ions (NH_4^+) **1**. Plant roots take up ammonium from the soil **2** and use it in metabolic reactions. Consumers get nitrogen by eating plants or one another.

Bacterial and fungal decomposers also release ammonium when they break down the nitrogen-rich wastes and remains of organisms **3**. Other soil bacteria obtain energy by converting ammonium to nitrate (NO_3^-), a process called nitrification **4**. Like ammonium, nitrate can be taken up from the soil and used by producers **5**. Ecosystems lose nitrogen as some types of bacteria convert nitrate to gaseous forms that escape into the atmosphere **6**.

In the early 1900s, German scientists discovered a method of fixing atmospheric nitrogen and producing ammonium on an industrial scale. This process allowed the manufacture of synthetic nitrogen fertilizers that have boosted crop yields and help to feed the rapidly increasing human population. However, use of fertilizers, along with other human activities, has added large amounts of nitrogen-containing compounds to our water and air. Nitrates commonly pollute drinking water in agricultural areas, raising the risk of thyroid cancer. Nitrous oxide, a gas released by bacteria in overfertilized soil and by the burning of fossil fuel, is a greenhouse gas and contributes to destruction of the ozone layer (the layer filters out dangerous ultraviolet radiation from the sun).

❯ The Carbon Cycle Carbon occurs abundantly in the atmosphere, combined with oxygen as carbon dioxide (CO_2). In the **carbon cycle**, carbon moves among rocks, water, and the atmosphere, and into and out of food webs. After water, carbon is the most abundant substance in living organisms. All molecules of life (carbohydrates, fats, lipids, and proteins) have a carbon backbone.

On land, plants take up and use carbon dioxide in photosynthesis. Plants and most other land organisms release carbon dioxide back to the atmosphere when they carry out aerobic respiration. Bicarbonate ions (HCO_3^-) form when carbon dioxide dissolves in water. Aquatic producers take up bicarbonate and convert it to carbon dioxide for use in photosynthesis. As on land, most aquatic organisms carry out aerobic respiration and release carbon dioxide.

carbon cycle Movement of carbon among rocks, water, the atmosphere, and living organisms.
nitrogen cycle Movement of nitrogen among the atmosphere, soil, and water, and into and out of food webs.
nitrogen fixation Conversion of nitrogen gas to ammonia.
phosphorus cycle Movement of phosphorus among rocks, water, soil, and living organisms.

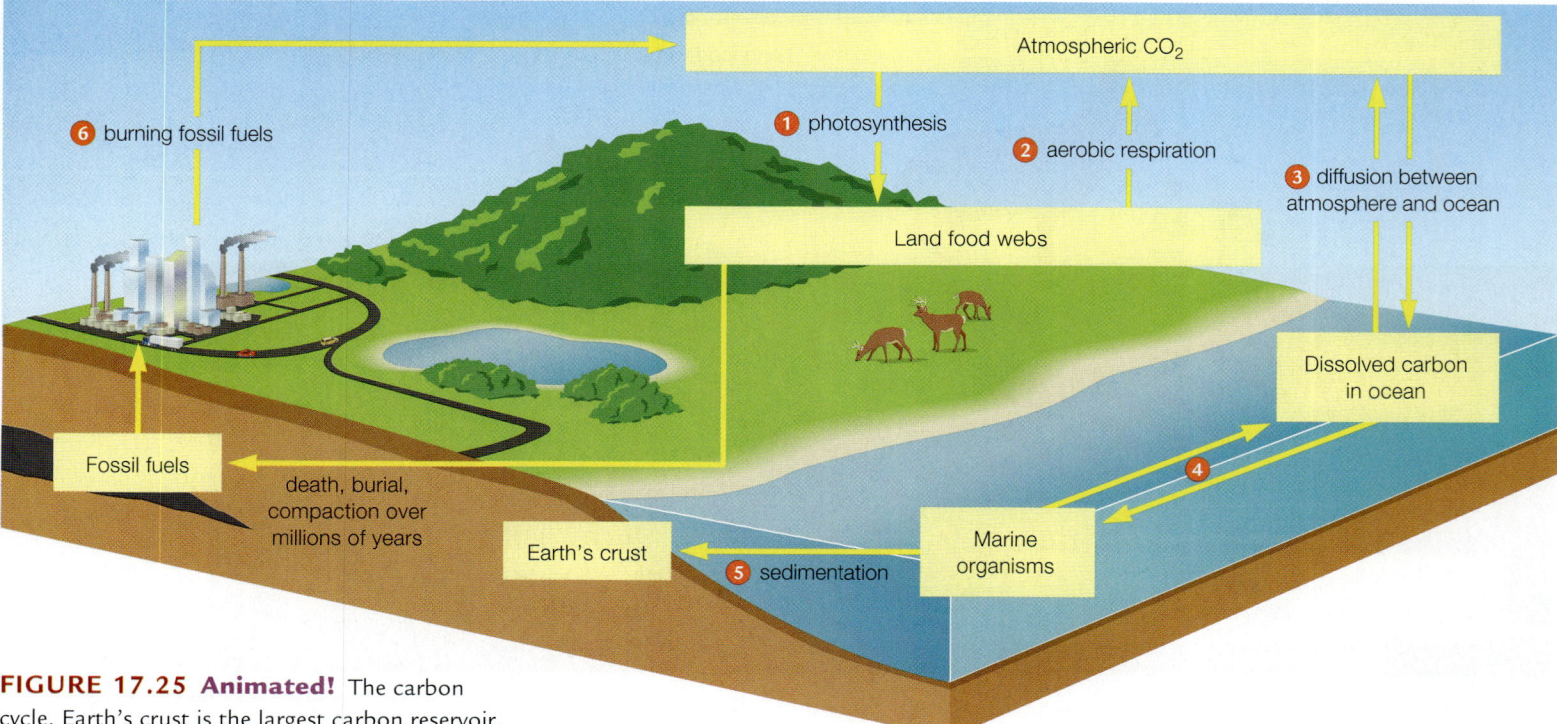

Atmospheric CO₂

6 burning fossil fuels

1 photosynthesis

2 aerobic respiration

3 diffusion between atmosphere and ocean

Land food webs

Dissolved carbon in ocean

Fossil fuels

death, burial, compaction over millions of years

Earth's crust

5 sedimentation

Marine organisms

4

FIGURE 17.25 Animated! The carbon cycle. Earth's crust is the largest carbon reservoir.

1 Carbon enters land food webs when plants take up carbon dioxide from the air and carry out photosynthesis.

2 Carbon returns to the atmosphere as carbon dioxide when plants and other land organisms carry out aerobic respiration.

3 Carbon diffuses between the atmosphere and the ocean. Carbon dioxide becomes bicarbonate when it dissolves in ocean water.

4 Marine producers take up bicarbonate for use in photosynthesis, and marine organisms release carbon dioxide produced by aerobic respiration.

5 Many marine organisms incorporate carbon into their shells. After they die, these shells become part of the sediments. Over time, these sediments become carbon-rich rocks such as limestone and chalk in Earth's crust.

6 Burning of fossil fuels derived from the ancient remains of plants adds additional carbon dioxide into the atmosphere.

© Cengage Learning.

global climate change Wide-ranging changes in rainfall patterns, average temperature, and other climate factors that result from rising concentrations of greenhouse gases.
greenhouse effect Warming of Earth's lower atmosphere and surface as a result of heat trapped by greenhouse gases.
greenhouse gas Atmospheric gas that helps keep heat from escaping into space and thus warms the Earth.

Sedimentary rocks such as limestone constitute Earth's largest carbon reservoir. These rocks formed over millions of years by the compaction of the carbon-rich shells of marine organisms. Plants do not take up dissolved carbon from the soil, so the carbon in these rocks is not readily accessible to organisms in land ecosystems. Fossil fuels are another reservoir of carbon that, under natural conditions, is largely unavailable to the world of life. Figure 17.25 shows the main carbon reservoirs and the processes that transfer carbon among them.

〉 The Greenhouse Effect and Global Climate Change By burning fossil fuels and wood, humans are putting extra carbon dioxide and nitrogen oxides into the air. At the same time, we are cutting forests, thus decreasing the global uptake of carbon dioxide by plants. The result of these activities is a well-documented change in the composition of the atmosphere.

These atmospheric changes can affect Earth's climate because carbon dioxide and nitrogen oxides are among the greenhouse gases. **Greenhouse gases** are atmospheric gases that slow the movement of heat from Earth to space (Figure 17.26). The process by which heat emitted by Earth's atmosphere warms its surface is called the **greenhouse effect**.

Without the greenhouse effect, Earth's surface would be cold and lifeless. But there can be too much of a good thing. Scientists estimate that atmospheric carbon dioxide is at its highest level in 15 millions years. As the concentrations of greenhouse gases in the atmosphere have risen, so has the average global temperature. In the past 100 years, Earth's average temperature has risen by about 0.74°C (1.3°F), and the rate of warming is accelerating.

A rise of a degree or two in average temperature may not seem like a big deal, but it is enough to increase the rate of glacial melting, raise sea level, alter wind patterns, shift the distribution of rainfall and snowfall, and increase the frequency and severity of hurricanes. Scientists refer to the many climate-related effects of the rise in greenhouse gases as **global climate change**. We discuss global climate change and its effects in detail in the next chapter.

FIGURE 17.26 The greenhouse effect.

1 Some light energy from the sun is reflected by Earth's atmosphere or surface.

2 More light energy reaches Earth's surface, and warms the surface.

3 Earth's warmed surface emits heat energy. Some of this energy escapes through the atmosphere into space. But some is absorbed and then emitted in all directions by greenhouse gases. The emitted heat warms Earth's surface and lower atmosphere.

NASA.

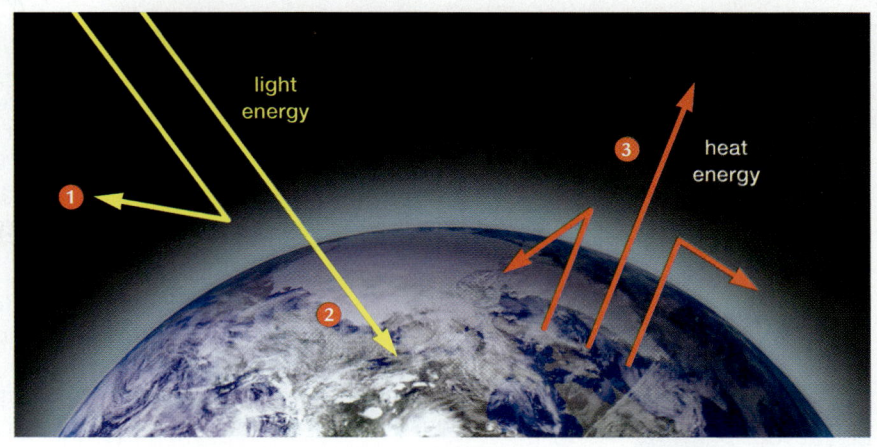

light energy

heat energy

Take-Home Message

How do nutrients cycle between organisms and their environment?

- Water moves on a global scale from the ocean (its main reservoir), through the atmosphere, onto land, then back to the ocean.

- Phosphorus cycles between its main reservoir—rocks and sediments—and soils and water. Phosphorus enters food webs when producers take up dissolved phosphates. The atmosphere does not play a significant role in this cycle.

- Nitrogen moves from its main reservoir—the atmosphere—into soils and water, and into and out of food webs. Bacteria play a pivotal role in the nitrogen cycle by producing forms of nitrogen that producers can take up and use.

- Carbon's main reservoir is rocks, but most carbon enters food webs when producers take up dissolved carbon from water or carbon dioxide from the air.

- Human activities cause nutrient imbalances in land and aquatic habitats. Our activities also add excess greenhouse gases such as carbon dioxide and nitrogen oxides to the air. The rise in greenhouse gases is affecting global climate.

WHERE YOU ARE GOING . . .

The next chapter explores the effects of human activities on species richness and nutrient cycles. Mutualistic relationships come up again in our discussions of plant nutrition (Section 27.5) and pollination (28.2).

17.7 Fighting the Foreign Fire Ants (revisited)

The species name for red imported fire ants is *Solenopsis invicta*. *Invicta* means "invincible" in Latin, and *S. invicta* lives up to its name. So far, chemical pesticides have done little to stop the spread of this exotic species. In fact, such chemicals may assist the nonnative invaders by killing off native ants that would otherwise compete with them.

A better defense against red imported fire ants may be tiny flies imported from Brazil for use as a biological control agent (Figure 17.27**A**). Phorid flies are parasitoids and *S. invicta* is their host. A female phorid fly lays an egg in the ant's body. The egg hatches into a larva that grows in and feeds on the ant's soft tissues. Eventually the larva enters the ant's head, causes the head to fall off (Figure 17.27**B**), then undergoes metamorphosis to an adult fly inside it.

Phorid flies are not expected to kill off all *S. invicta* in affected areas. Rather, the hope is that the flies will reduce the density of invading colonies. The phorid flies did not evolve with our native ants, so they do not harm them.

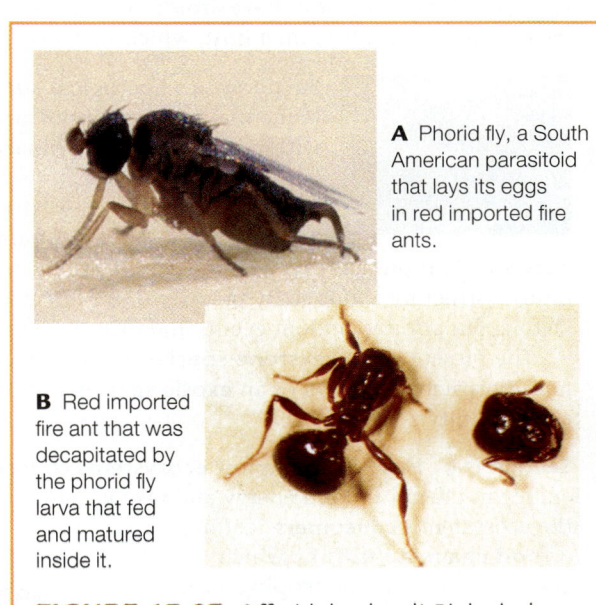

A Phorid fly, a South American parasitoid that lays its eggs in red imported fire ants.

B Red imported fire ant that was decapitated by the phorid fly larva that fed and matured inside it.

FIGURE 17.27 Off with her head! Biological control of red imported fire ants by a coevolved parasitoid fly.

Credits: (a) Photo by Scott Bauer/ USDA; (b) Photo by Sanford Porter, USDA/ARS.

Summary

Section 17.1 All species in a defined area are a **community**. Interactions among members of a community help keep populations in check. A species introduced to a new community without any natural enemies can increase in number and become a pest.

Section 17.2 Each species occupies a certain **habitat** and has a unique **niche**—the conditions and resources it requires, and the interactions it takes part in. The **species diversity** of a community is determined by nonbiological factors such as climate, as well as biological ones such as how species interact.

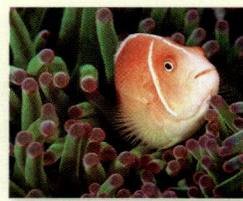

Section 17.3 Species interactions can result in **coevolution**. An association in which species live together is a **symbiosis**. In a **commensalism**, one species benefits and it neither helps nor harms the other. In a **mutualism**, two species exploit one another to their mutual benefit.

Interspecific competition harms both participants. **Competitive exclusion** occurs when species with identical resource needs share a habitat. **Resource partitioning** allows similar species to coexist.

Predation occurs when a free-living predator kills and eats its prey. In one type of **mimicry**, well-defended prey species have similar **warning coloration**. Less well-defended species also mimic well-defended ones. **Camouflage** hides both predators and prey.

Herbivory may or may not kill a plant. **Parasites** withdraw nutrients from a host, usually without killing it. **Brood parasites** lay eggs in another's nest. **Parasitoids** are insects whose larvae develop inside and feed on a host, which they eventually kill.

Section 17.4 **Ecological succession** is the sequential replacement of arrays of species within a community. **Primary succession** occurs in habitats without soil. **Secondary succession** takes place in disturbed habitats. The first species in a community are **pioneer species**. Their presence may help other potential colonists. Community structure is not easily predicted. It is affected by physical factors, but also by random events and disturbances such as fires.

The presence of a **keystone species** has a large effect on community structure. Arrival of an **exotic species** can drastically alter a community.

Section 17.5 **Producers** in most **ecosystems** convert sunlight energy to chemical bond energy and take up nutrients that they and the ecosystem's **consumers** require. A **food chain** is a path by which energy flows from one **trophic level** to another. Food chains intersect as **food webs**. In a typical land ecosystem, most energy in producers flows directly to **detritivores** and **decomposers**. Energy transfers are

inefficient (energy is lost as heat and tied up in inedible parts), so most ecosystems support no more than a few trophic levels.

The rate of **primary production**—the capture and storage of energy by producers—varies with climate, season, and other factors. **Energy pyramids** show how available energy decreases as it is transferred from one trophic level to the next.

Section 17.6 In a **biogeochemical cycle**, water or a nutrient moves through the environment, then through organisms, then back to an environmental reservoir.

In the **water cycle**, water moves from the ocean into the atmosphere, falls on land, and flows back to the ocean, its main reservoir. **Aquifers** and soil store **groundwater**, but most of Earth's fresh water is in the form of ice.

In the **phosphorus cycle**, living things take up dissolved forms of phosphorus released by Earth's rocks and sediments. No gaseous form of phosphorus plays a role in this cycle.

The atmosphere is the main reservoir in the **nitrogen cycle**. Bacteria and the fertilizer industry carry out **nitrogen fixation** (convert atmospheric nitrogen to ammonium that plants can take up). Bacteria and fungi that act as decomposers also release ammonium. Nitrogen fertilizers commonly pollute water in agricultural regions.

The global **carbon cycle** moves carbon from its reservoirs in rocks and seawater, through its gaseous form (CO_2) in the atmosphere, and through living organisms.

Carbon dioxide and nitrogen oxides are among the **greenhouse gases**. Through the **greenhouse effect**, they trap heat in Earth's atmosphere and make life possible. An increase in their concentration in the atmosphere is causing **global climate change**.

Self-Quiz   Answers in Appendix I

1. A species' habitat is like its address, and its _____ is like its occupation.
   a. richness   c. gene pool
   b. community   d. niche

2. Match the species interaction with a suitable description.
   ____ mutualism
   ____ competition
   ____ predation
   ____ parasitism
   ____ herbivory

   a. A snake kills and eats a mouse.
   b. A bee pollinates a flower while sipping floral nectar.
   c. An owl and a wood duck both need a tree cavity to nest.
   d. A mosquito sucks your blood.
   e. A goat grazes on grass.

3. With interspecific competition, selection favors individuals of both species who are most _____ the competing species.
   a. similar to   b. different from

4. Parasitoids are _____ that lay eggs in the bodies of their hosts.
   a. birds   b. reptiles   c. insects   d. fish

5. The establishment of a biological community on a newly formed volcanic island is an example of _____ .
   a. primary succession   c. competitive exclusion
   b. secondary succession   d. resource partitioning

Digging Into Data

Biological Control of Fire Ants

Ant-decapitating phorid flies are just one of the biological control agents used to battle imported fire ants. Researchers have also enlisted the help of a fungal parasite that infects the ants and slows production of their offspring.

To determine the effectiveness of these biological controls useful against imported red fire ants, USDA scientists treated infested areas with either traditional pesticides or pesticides plus biological controls (both phorid flies and the fungal parasite). The scientists left some plots untreated as controls. Figure 17.28 shows the results.

1. How did population size in the control plots change during the first four months of the study?

2. How did population size in the two types of treated plots change during this same interval?

3. If this study had ended after the first year, would you conclude that biological controls had a major effect?

4. How did the two types of treatment (pesticide alone versus pesticide plus biological controls) differ in their longer-term effects? Which is most effective?

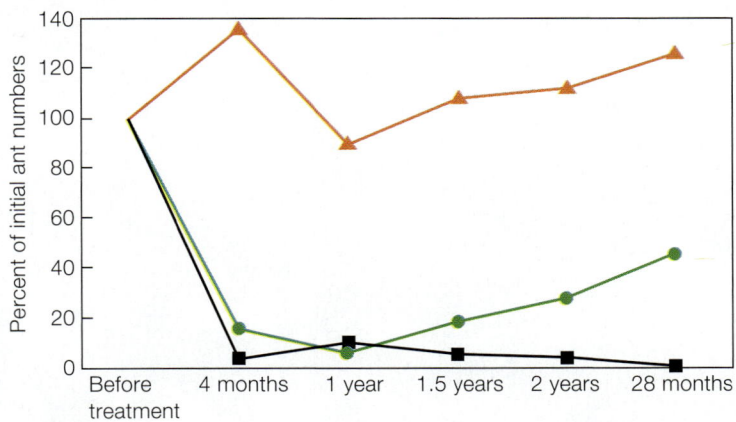

FIGURE 17.28 Effects of two methods of controlling red imported fire ants. The graph shows the numbers of red imported fire ants over a 28-month period. *Orange* triangles represent untreated control plots. *Green* circles are plots treated with pesticides alone. *Black* squares are plots treated with pesticide and biological control agents (phorid flies and a fungal parasite).

© Cengage Learning.

---

6. Match the terms with suitable descriptions.

_____ producer   a. steals parental care
_____ brood parasite   b. feeds on small bits of
_____ decomposer   organic matter
_____ detritivore   c. degrades organic wastes and
                            remains to inorganic forms
                            d. captures sunlight energy

7. In an energy pyramid diagram of a prairie food web, which of these organisms would be in the lowest (and largest) tier?
   a. grasses   c. grasshopper-eating sparrows
   b. grasshoppers   d. sparrow-eating hawks

8. Match each substance with its largest environmental reservoir. One reservoir choice will be used more than once.

_____ carbon   a. seawater
_____ water   b. rocks and sediments
_____ phosphorus   c. the atmosphere
_____ nitrogen

9. Earth's largest reservoir of fresh water is _____ .
   a. lakes   c. glacial ice
   b. groundwater   d. water in living organisms

10. _____ convert nitrogen gas to a form producers can take up.
   a. Fungi   c. Mammals
   b. Bacteria   d. Mosses

11. Land plants take up _____ for photosynthesis from the air.
   a. carbon dioxide   c. ammonium ions
   b. phosphate ions   d. nitrogen gas

12. Addition of _____ to water encourages algal blooms.
   a. carbon dioxide   c. salt
   b. phosphate ions   d. bicarbonate ions

13. A biological control agent is _____ a pest species.
   a. prey of   c. mutualistic with
   b. a descendant of   d. a natural enemy of

14. Greenhouse gases _____ .
   a. trap heat in the atmosphere
   b. are released by burning of fossil fuels
   c. may cause global climate change if they accumulate
   d. all of the above

15. A(n) _____ species is one that arrives early in succession.
   a. keystone   b. pioneer   c. commensal   d. exotic

Critical Thinking

1. With antibiotic resistance rising (Section 12.3), researchers are looking for ways to reduce use of these drugs. Some cattle once fed antibiotic-laced food now get feed that includes helpful bacteria that can live in the animal's gut. The idea is that if a large population of beneficial bacteria is in place, then harmful bacteria with the same resource needs are less likely to thrive. Explain why this idea makes sense in terms of species interactions.

2. Figure 17.8 shows a stingless fly that mimics a stinging wasp. Researchers have found that in such mimicry systems, mimics benefit most when they are rare relative to the well-protected model. Can you explain why?

3. Ectotherms such as invertebrates, fish, and amphibians convert more of the energy in the food they eat into body tissues than do endotherms such as birds and mammals. What allows ectotherms to make more efficient use of food energy?

The Biosphere 18
and Human Effects

. . . Where you Have Been

This chapter's topic is the biosphere, the highest level of organization in nature (Section 1.2). We revisit plant adaptations to drought (5.5), succession (17.4), primary production and trophic levels (17.5), the carbon cycle (17.6), deep-sea hydrothermal vents (13.2), and coral reefs (15.3).

18.1 A Long Reach

We began this book with the story of biologists who ventured into a remote forest in New Guinea and found many previously unknown species. At the far end of the globe, a U.S. submarine surfaced in Arctic waters and found polar bears hunting on the ice-covered sea (Figure 18.1). The bears were 435 kilometers (270 miles) from the North Pole and 805 kilometers (500 miles) from the nearest land.

FIGURE 18.1 Three polar bears investigate an American submarine that surfaced in ice-covered Arctic waters.

U.S. Navy photo by Chief Yeoman Alphanso Braggs.

Even seemingly remote regions are no longer beyond the reach of human explorers and human influence. Increasing levels of greenhouse gases are raising the temperature of Earth's atmosphere and seas. In the Arctic, the warming is causing sea ice to thin and to break up earlier in the spring. This raises the risk that polar bears hunting far from land will become stranded, unable to return to solid ground before the ice thaws.

Polar bears are top predators and their tissues contain a high amount of mercury and organic pesticides. These pollutants enter water and air in more temperate regions, then winds and ocean currents deliver them to polar realms. Pollutants also arrive when migratory birds that spend winters in more densely populated and polluted regions come to the Arctic in the spring. The birds release pollutants in their wastes and, if they die, in their remains.

In places less remote than the Arctic, human populations have a more direct effect. As we cover more and more of the world with our homes, factories, and farms, less habitat remains for other species. We also put species at risk by competing with them for resources, overharvesting them, and introducing nonnative competitors.

It would be presumptuous to think that humans alone have had a profound impact on the world of life. More than a billion years ago, the rise of oxygen-releasing photosynthetic cells changed the course of evolution by enriching the atmosphere with this gas (Section 5.6). Over life's long existence, the success of one species or group of organisms has often come at the expense of others. What is new is the increasing pace of change and the capacity of our own species to recognize and affect its role in this increase.

A century ago, Earth's physical and biological resources seemed inexhaustible. Now we know that many practices put into place when we were ignorant of how natural systems operate take a heavy toll on the biosphere. They threaten other species, ecosystem processes, and possibly our own long-term existence.

The **biosphere** is the sum of all places where we find life on Earth. With this chapter, we look at physical processes that affect the distribution of organisms through the biosphere. We also consider the ways that human activities are altering these processes and disrupting the world of life.

FIGURE 18.2 Variation in intensity of solar radiation with latitude. For simplicity, we depict two equal parcels of incoming radiation on an equinox, a day when incoming rays are perpendicular to Earth's axis.

Rays that fall on high latitudes **A** pass through more atmosphere (*blue*) than those that fall near the equator **B**. Compare the length of the *green* lines. Atmosphere is not to scale.

Also, energy in the rays that fall at the high latitude is spread over a greater area than energy that falls on the equator. Compare the length of the *red* lines. © Cengage Learning.

18.2 Factors Affecting Climate

⟩ Air Circulation Patterns **Climate** refers to average weather conditions, such as temperature, humidity, wind speed, cloud cover, and rainfall, over a long interval. Many factors influence a region's climate, which in turn affects the types of species that can live there.

A region's latitude (how far north or south it is) determines how much light energy it receives. On any given day, equatorial regions receive more sunlight than higher latitudes for two reasons (Figure 18.2). First, sunlight traveling to high latitudes passes through more atmosphere to reach Earth's surface than sunlight traveling to the equator. Fine particles of dust, water vapor, and greenhouse gases absorb some solar radiation or reflect it back into space, so less light energy reaches the poles. Second, the energy in any incoming parcel of sunlight is spread out over a smaller surface area at the equator than at the higher latitudes. As a result of these factors, Earth's surface warms more at the equator than at the poles.

Latitudinal differences in surface warming, and the resulting effects on air and water, give rise to global patterns of air circulation and rainfall (Figure 18.3). At the equator, intense sunlight warms the air and causes evaporation of water from the ocean. As the air heats up, it expands and rises ❶. The same effect causes a hot-air balloon to inflate and rise when the air inside it is heated. As the equatorial air mass rises, it flows north and south and begins to cool. Cool air can hold less moisture than warm air, so moisture leaves the air as rain that supports tropical rain forests.

By the time the air reaches about 30° north and south latitude, it has cooled and become dry. Being cool, the air sinks downward ❷. When it descends, it draws moisture from the soil. As a result, deserts often form at around 30° north and south latitude.

Air that continues flowing along Earth's surface toward the poles once again picks up heat and moisture. By a latitude of about 60° it has become warm and moist, and it rises, giving up moisture as rain ❸. In polar regions, cold air with little moisture descends ❹. Precipitation is sparse, and polar deserts form.

Landforms also affect rainfall. Moisture-laden winds give up water as rain as they rise over mountains. The dry region beyond the mountains is referred to as a **rain shadow**. Deserts often form downwind of coastal mountains.

⟩ Ocean Circulation Patterns The waters of Earth's oceans circulate on a global scale, and this movement influences climate. In the North Atlantic, arctic winds cool the ocean's upper layer of surface water, making it denser

❹ At the poles, cold, dry air sinks and moves toward lower latitudes.

❸ Air rises again at 60° north and south, where air flowing poleward meets air coming from the poles.

❷ At around 30° north and south latitude, the air—now cooler and dry—sinks.

❶ Warm, moist air rises at the equator. As the air flows north and south, it cools and loses moisture as rain.

FIGURE 18.3 Animated! Air circulation patterns that result from latitudinal differences in the amount of solar radiation reaching Earth.
© Cengage Learning.

biosphere All portions of the Earth where life exists.

climate Average weather conditions in a region over a long time period.

rain shadow Dry region on the downwind side of a coastal mountain range.

FIGURE 18.4 Global ocean circulation. The currents function like a giant conveyer belt that moves heat and nutrients.

After NASA graphic.

surface current, warm water

deep current, cold water

and causing it to sink (Figure 18.4). This cold, dense water flows southward in deep currents **1**. Wind-driven surface currents of warmer water move north to replace it **2**. This circulation of seawater in the North Atlantic transfers heat from the South Atlantic to the North Atlantic, thus keeping the eastern United States and Europe warmer than they would otherwise be. It is part of a larger current flow that distributes heat and nutrients throughout the world's oceans.

FIGURE 18.4 Global ocean circulation. The currents function like a giant conveyer belt that moves heat and nutrients.

After NASA graphic.

surface current, warm water

deep current, cold water

Take-Home Message

What causes winds and ocean currents that affect climate?

- Longitudinal differences in the amount of solar radiation reaching Earth produce global air circulation patterns.
- Sinking of seawater in cold regions and the wind-directed movement of warm water that replaces it create ocean currents that redistribute heat.
- The collective effects of air movements, ocean currents, and landforms determine regional temperature and moisture levels.

18.3 The Major Biomes

Climate differences allow different regions to support different types of plant life, which in turn support different kinds of animals. Scientists classify the world's land ecosystems into a variety of **biomes**, each characterized by its climate and main type of vegetation. Figure 18.5 shows the distribution of the major biomes. A biome typically includes multiple nonadjacent regions. A brief description of each of these biomes will illustrate how climate affects their distribution across the biosphere and how the differences among them contribute to the diversity of life on Earth.

biome Any of Earth's major land ecosystems, characterized by climate and main vegetation and found in several regions.

boreal forest Biome dominated by conifers that can withstand the cold winters at high northern latitudes.

temperate deciduous forest Biome dominated by broadleaf trees that grow in warm summers, then drop their leaves and go dormant during cold winters.

tropical rain forest Multilayered forest that occurs where warm temperatures and continual rains allow plant growth year-round. Most productive and species-rich biome.

❯ Forest Biomes Evergreen broadleaf (angiosperm) trees dominate the **tropical rain forests** of equatorial Asia, Africa, and South America (Figure 18.6**A**). Abundant rain, warm temperatures, and little variation in day length allow plants to grow year-round. As a result of continual growth, rain forests take up more carbon dioxide and release more oxygen per unit area than any other biome. Thus, they are sometimes described as Earth's "lungs."

Tropical rain forest is the most structurally complex biome, with many species of trees reaching 30 meters (100 feet) in height. Vines twine up the trees, and orchids and ferns grow on their trunks and branches. Little light reaches the forest floor, so few plants grow there. Constant warmth encourages rapid decomposition, and prevents leaf litter from accumulating. Thus, the soil is relatively nutrient-poor.

Per unit area, tropical rain forest holds more species of plants and animals than any other biome. It is the oldest biome, and its age is key to its diversity. Some modern rain forests have existed for more than 50 million years, which has provided ample time for many evolutionary branchings to occur.

Tropical regions where rains fall seasonally support tropical dry forests. The broadleaf trees that dominate such forests are smaller than rain forest trees and most lose their leaves and become dormant in the dry season.

Trees of **temperate deciduous forests** shed their leaves and become dormant in preparation for winter, when freezing temperatures do not favor growth (Figure 18.6**B**). In spring, deciduous trees flower and put out new leaves. At the same time, leaves that were shed and accumulated during the prior autumn decay to form a rich soil. During the growing season, a somewhat open canopy allows sunlight to reach the ground, where shorter understory plants flourish.

The most extensive biome is **boreal forest**, the conifer-dominated forest that sweeps across northern Asia, Europe, and North America (Figure 18.6**C**). It is also

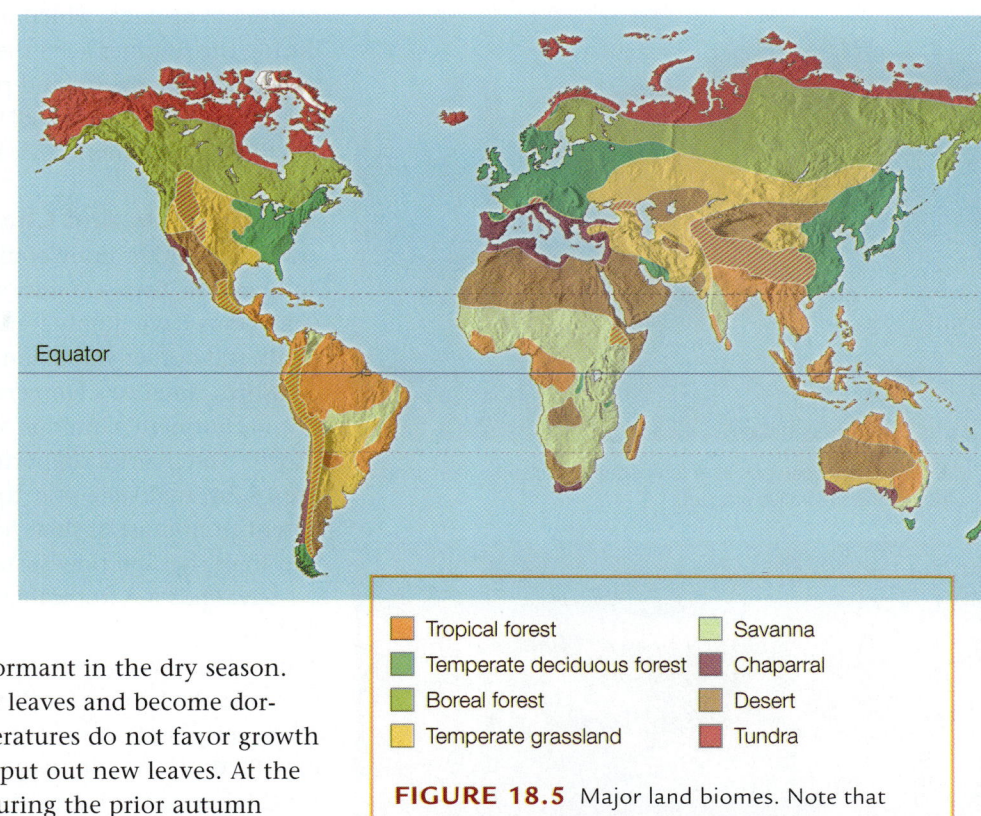

Equator

Tropical forest	Savanna
Temperate deciduous forest	Chaparral
Boreal forest	Desert
Temperate grassland	Tundra

FIGURE 18.5 Major land biomes. Note that most biomes encompass areas on more than one continent.

From RUSSELL/WOLFE/HERTZ/STARR. Biology, 2E. © 2011 Brooks/Cole, a part of Cengage Learning, Inc. Reproduced by permission.www.cengage.com/permissions.

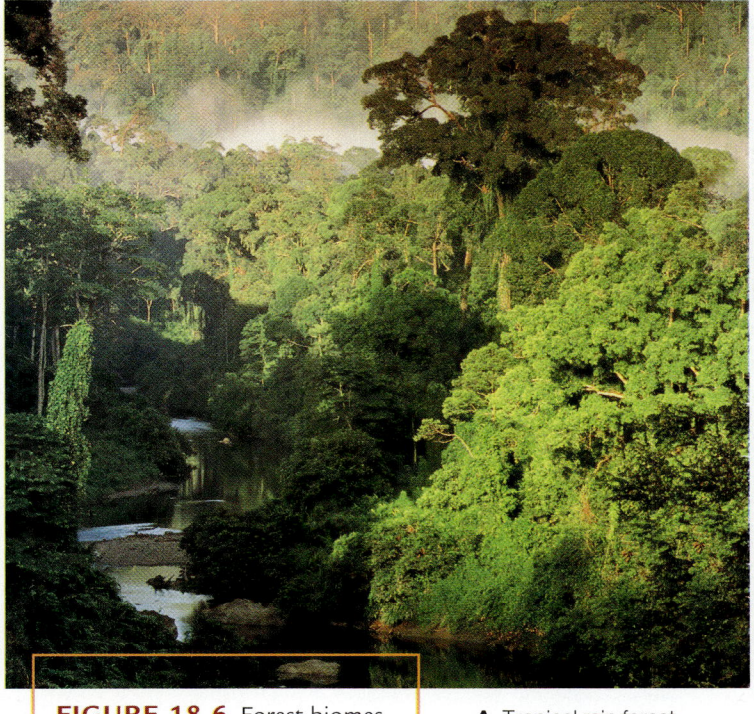

FIGURE 18.6 Forest biomes.

Credits: (a) © Frans Lanting; (b) © James Randklev/Corbis; (c) © Serg Zastavkin, Used under license from Shutterstock.com.

A Tropical rain forest in Southeast Asia.

B Temperate deciduous forest in New England in autumn.

C Boreal forest in Siberia during summer.

A North American prairie, a temperate grassland, where bison are among the native grazers.

B African savanna, a tropical grassland with scattered shrubs. It supports huge herds of grazing wildebeest.

C Chaparral in California. Shrubby drought-resistant plants with small, leathery leaves predominate.

FIGURE 18.7 Biomes dominated by plants adapted to occasional fire.

Credits: (a) © Danny Barron; (b) Jonathan Scott/ Planet Earth Pictures; (c) Jack Wilburn/ Animals Animals.

referred to as taiga, which is Russian for "swamp forest," because rains that fall during the summer keep the soil soggy. Winters are dry and cold. The conifers are mainly spruce, fir, and pine. The conical shape of these trees helps them shed snow, and their needlelike leaves help minimize evaporative water loss during the winter when they cannot take up water from the frozen ground.

> **Grasslands and Chaparral** Grasslands are dominated by perennial grasses and other nonwoody plants that tolerate grazing, intervals of drought, and periodic fires—conditions that prevent trees and shrubs from taking over. Grasslands have highly fertile soil. They typically occur in the middle of continents and border shrublands or desert.

North America's temperate grasslands, called **prairies**, once covered much of the continent's interior, where summers are hot and winters are cold and snowy. The prairies supported herds of elk, pronghorn antelope, and bison (Figure 18.7**A**) that were prey to wolves. Today, these predators and prey are largely absent from most of their former range. Nearly all prairies have been plowed, and their rich soil now sustains production of wheat and other crops.

Savannas are tropical grasslands with a few scattered shrubs and trees. They lie between the tropical forests and hot deserts of Africa, India, and Australia. Temperatures are warm year-round, and there is a distinct rainy season. Africa's savannas are famous for their abundant wildlife. Herbivores include giraffes, zebras, elephants, a variety of antelopes, and immense herds of wildebeests (Figure 18.7**B**). Lions and hyenas eat the grazers.

Chaparral is a biome dominated by drought-resistant, fire-adapted shrubs that have small, leathery leaves. It occurs along the western coast of continents, between 30 and 40 degrees north or south latitude. Mild winters bring a moderate amount of rain, and the summer is hot and dry. Chaparral is California's most extensive ecosystem (Figure 18.7**C**). It also occurs in regions bordering the Mediterranean, as well as in Chile, Australia, and South Africa.

> **Deserts** Low annual precipitation defines **desert**, a biome that covers about one-fifth of Earth's land surface. Many deserts are located at about 30° north and south latitude, where global air circulation patterns cause dry air to sink. Rain shadows also reduce rainfall. For example, the Himalayas prevent rain from falling in China's Gobi desert.

Lack of rainfall keeps the humidity in a desert low. With little water vapor to block rays, intense sunlight reaches and heats the ground. At night, the lack of insulating water vapor in the air allows the temperature to fall fast. As a result, deserts tend to have larger daily temperature shifts than other biomes.

Despite their harsh conditions, most deserts support some plant life (Figure 18.8**A**). Cacti are a family of plants native to Western Hemisphere deserts. All cacti are CAM plants (Section 5.5), which carry out gas exchange at night, when potential for evaporative water loss is lowest. When it does rain, cacti take up water and store it in their spongy tissues for use in drier times. Leaves modified as spines help cacti fend off animals that would like to tap this source of water.

Many deserts also contain woody shrubs such as mesquite and creosote. These plants have extensive root systems that take up the little water that is available. Mesquite roots can extend up to 60 meters (197 feet) beneath the soil surface. Desert shrubs often put out leaves only after a rain. Annual desert plants sprout and reproduce fast while soil remains moist after a rain.

> **Tundra** Tundra forms between the polar ice cap and the belts of boreal forests in the Northern Hemisphere, with the greatest area in northern Russia and Canada. Tundra is Earth's youngest biome, having appeared about 10,000 years

ago when glaciers retreated at the end of the last ice age. Emergence of land from beneath these glaciers opened the way to a process of primary succession that has produced the currently existing communities.

Snow blankets the arctic tundra for as many as nine months of the year. During the brief summer, plants grow fast under nearly continuous sunlight (Figure 18.8**B**). Lichens and shallow-rooted, low-growing plants provide the base for food webs. Many migratory birds nest in the tundra during the summer, when the air is thick with insects. Even in midsummer, only the surface layer of tundra soil thaws. Below that, a layer of permanently frozen soil called **permafrost**, can be as much as 500 meters thick (1,600 feet). Permafrost acts as a barrier that prevents drainage, so the soil above it remains perpetually waterlogged. Cool, anaerobic conditions slow decay, so organic remains accumulate, making the permafrost one of Earth's greatest stores of carbon.

Tropical rain forest is Earth's oldest biome, and tundra is its youngest.

Take-Home Message

How do environmental factors affect life in the major biomes?

- Near the equator, year-round rains and warmth support highly productive, species-rich tropical rain forests. At higher latitudes, the trees of temperate deciduous forests drop their leaves and become dormant during winter. At still-higher latitudes, conifers dominate Northern Hemisphere boreal forests.

- Plants of African savannas and temperate grasslands are adapted to grazing and periodic fire. Shrubby chaparral plants are also fire-adapted.

- Deserts receive little rain, so perennial desert plants have traits that reduce their water loss, allow them to store water, or tap into water deep in the soil. Annual desert plants complete their life cycle in a brief period after a rain.

- The most northerly and youngest biome is arctic tundra, where low plants are adapted to a short growing season and soil with a layer of permafrost.

chaparral Biome with cool wet winters, hot dry summers; dominant plants are shrubs with small, leathery leaves.

desert Biome with little precipitation; its perennial plants are adapted to withstand drought.

permafrost Layer of permanently frozen soil in the Arctic.

prairie Temperate grassland biome of North America. Its grasses and other plants are adapted to recover after grazing and the occasional fire.

savanna Tropical biome dominated by grasses and other plants adapted to grazing, as well as a scattering of shrubs.

tundra Northernmost biome, dominated by low plants that grow over a layer of permafrost.

A Mojave Desert after the winter rains. Annual plants sprout, flower, produce seeds, and die within weeks beneath slow-growing drought-adapted perennial cacti.

B Arctic tundra during the summer, when shrubby plants grow under nearly continuous sunlight. Permafrost underlies the upper layer of defrosted soil.

FIGURE 18.8 Desert and tundra, biomes with extreme climates.

Credits: (a) © George H. Huey/ Corbis; (b) © Darrell Gulin/ Corbis.

18.4 Aquatic Ecosystems

We can distinguish between types of aquatic ecosystems, just as we do biomes on land. Temperature, salinity, rate of water movement, and depth influence the composition of aquatic communities.

> **Freshwater Ecosystems** A lake is a body of standing fresh water. All but the shallowest lakes have zones that differ in their physical characteristics and species composition. Near shore, where sunlight penetrates all the way to the lake bottom, rooted aquatic plants and algae that attach to the bottom are primary producers. A lake's open waters include an upper well-lit zone and, if a lake is deep or cloudy, a zone where light does not penetrate. Producers in the well-lit water include photosynthetic protists and bacteria. In the deeper dark zone, consumers feed on organic debris that drifts down from above.

A lake undergoes succession, meaning the community of lake organisms changes over time (Section 17.4). A newly formed lake is deep, clear, and has few nutrients and a low primary productivity (Figure 18.9). As sediments accumulate, the lake becomes shallower. Nutrients accumulate and encourage growth of photosynthetic bacteria, diatoms, and other producers that cloud the water. These producers serve as food for tiny crustaceans, which are then eaten by fish.

Streams are flowing-water ecosystems. They typically originate from runoff or melting snow or ice. As they flow downslope, they grow and merge to form rivers. Properties of a stream or river vary along its length. The type of rocks a stream flows over can affect its solute concentration, as when limestone rocks dissolve and add calcium to the water. Shallow water that flows rapidly over rocks mixes with air and holds more oxygen than slower-moving, deeper water. Also, cold water holds more oxygen than warm water. As a result, different parts of a stream or river support species with different oxygen needs. For example, only cool, well-oxygenated water can support rainbow trout.

> **Marine Ecosystems** An **estuary** is a mostly enclosed coastal region where seawater mixes with nutrient-rich fresh water from rivers and streams. Water inflow continually replenishes nutrients, so estuaries have a high primary productivity. Photosynthetic bacteria and protists that live on mudflats often account for a large portion of an estuary's primary production. Plants adapted to withstand changes in water level and salinity also serve as producers.

Along ocean shores, organisms are adapted to the mechanical force of the waves and to tidal changes. Many species are underwater during high tide, then exposed to the air when the tide is low. Along rocky shores, where waves prevent detritus from piling up, algae that cling to rocks are the producers in grazing food chains. In contrast, waves continually rearrange loose sediments along sandy shores, and make it difficult for algae to take hold. Here, detrital food chains start with organic debris from land or offshore.

Warm, shallow, well-lit tropical seas hold **coral reefs**, formations made primarily of calcium carbonate secreted by generations of corals, which are invertebrate animals (Section 15.3). Like tropical rain forests, tropical coral reefs are home to an extraordinary assortment of species (Figure 18.10). The main producers in a coral reef community are photosynthetic dinoflagellates that live inside the reef-building corals. These protists provide their coral hosts with sugars. If a coral becomes stressed by an environmental change such as a shift in temperature, it expels its protist symbionts in a response called coral bleaching. If conditions improve fast, the symbiont population in the coral can recover. But when adverse conditions persist, the symbionts will not be restored, and the coral will die.

FIGURE 18.9 Low-nutrient lake. Crater Lake in Oregon formed when a collapsed volcano began to fill with snowmelt about 7,700 years ago. From a geologic standpoint, it is a young lake, and its clear water is a sign of its low primary productivity.

© Lindsay Douglas/ Shutterstock.

FIGURE 18.10 Coral reef in Fiji. Such reefs are the most productive and species-rich aquatic ecosystems. The backbone of the reef is a colony of coral polyps that have photosynthetic protists (dinoflagellates) living in their tissues.

© John Easley, www.johneasley.com.

A Computer model of seamounts off the coast of Alaska. Patton Seamount, at the rear, stands 3.6 kilometers (about 2 miles) tall, and peaks about 240 meters (800 feet) below the sea surface. The *inset* photo shows a flytrap anemone, one of many previously unknown species discovered at a seamount off the coast of California.

B Hydrothermal vent community. Bacteria and archaea that extract energy from minerals are the producers here. Consumers include crabs and giant tube worms (close-up in inset photo) that grow meters long without ever eating. The worms rely on bacteria that live in their tissues to produce their food.

FIGURE 18.11 Little-known ecosystems of the seafloor: seamounts and hydrothermal vents.

Credits: (a) NOAA; inset, Image courtesy of NOAA and MBARI; (b) NOAA/ Photo courtesy of Cindy Van Dover, Duke University Marine Lab, inset © Peter Batson/imagequestmarine.com.

In the brightly lit waters of the open ocean, photosynthetic microorganisms such as algae and bacteria are the primary producers, and grazing food chains predominate. Depending on the region, some light may penetrate as far as 1,000 meters (more than a half mile) beneath the sea surface. Below that, organisms live in continual darkness and detrital food chains are usually the rule.

On the seafloor, the greatest species richness occurs along the edges of continents and at **seamounts**, underwater mountains that stand 1,000 meters or more tall, but still lie below the sea surface (Figure 18.11**A**). Seamounts attract large numbers of fishes and are home to many marine invertebrates. Like islands, they harbor many species that evolved there and live nowhere else.

Hot, mineral-rich water spews out from the ocean floor at **hydrothermal vents**. When mineral-rich hot water mixes with cold seawater, the minerals settle out as extensive deposits. Bacteria and archaea that can extract energy from these deposits serve as primary producers for food webs that include invertebrates such as tube worms and crabs (Figure 18.11**B**). As explained in Section 13.2, one hypothesis holds that life originated near hydrothermal vents.

Take-Home Message

What factors shape aquatic ecosystems?

- Fast-flowing, cooler water holds more oxygen than warmer, slower-moving water, and the amount of available light decreases with the water's depth.

- In well-lit upper waters, producers such as aquatic plants, algae, and photosynthetic microbes are the base for grazing food chains. On coral reefs, the main producers are photosynthetic protists that live inside the coral's tissues.

- Detritus drifting down from above sustains most deep-water communities in lakes and oceans. However, hydrothermal vent communities on the ocean floor are sustained by energy that bacteria and archaea harvest from minerals.

coral reef In tropical sunlit seas, a formation composed of secretions of coral polyps that serves as home to many other species.

estuary A highly productive ecosystem where nutrient-rich water from a river mixes with seawater.

hydrothermal vent Place where hot, mineral-rich water streams out from an underwater opening in Earth's crust.

seamount An undersea mountain.

A Colorized photo of an ivory-billed woodpecker. Logging of old-growth swamp forests of the Southeast caused its decline.

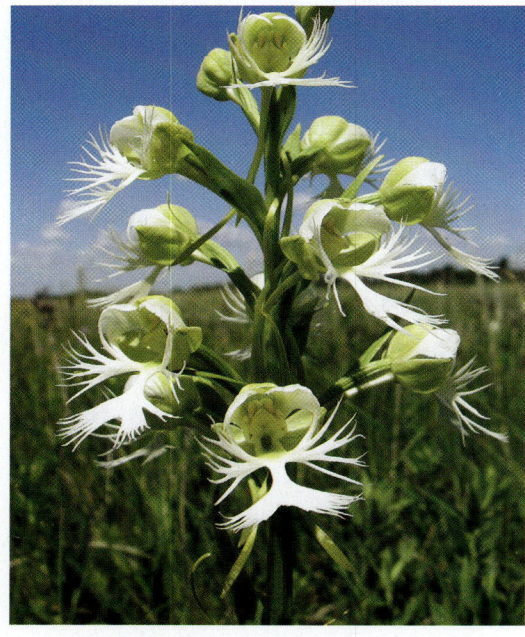

B Western prairie fringed orchid. It is threatened by conversion of prairie to farmland and rangeland.

C Texas blind salamander, from Edwards Aquifer. Extraction and pollution of the aquifer's fresh water threatens this salamander and 40 other species that live here and nowhere else.

FIGURE 18.12 Three North American species threatened by habitat loss or degradation.

Credits: (a) © George M. Sutton/ Cornell Lab of Ornithology; (b) © Ben Sullivan; (c) Joe Fries, U.S. Fish & Wildlife Service.

18.5 Human Effects on the Biosphere

The rising size of the human population and its increasing industrialization have far-reaching effects on the biosphere. We begin by discussing how human activities affect individual species, then turn to their wider impacts.

❯ Increased Species Extinctions Extinction, like speciation, is a natural process. Species arise and become extinct on an ongoing basis. The rate of extinction picks up dramatically during a mass extinction, when many kinds of organisms in many different habitats all become extinct in a relatively short period. We are currently in the midst of such an event. Unlike historical mass extinctions, this one cannot be blamed on some natural catastrophe such as an asteroid impact. Rather, this mass extinction is the outcome of the success of a single species—humans—and their effect on Earth.

Consider the plight of the ivory-billed woodpecker (Figure 18.12**A**). This spectacular bird evolved in the swamp forests of the American Southeast and it declined when these forests were cut for lumber. Until recently, the bird was thought to have become extinct in the 1940s. A possible 2004 sighting in Arkansas kicked off an extensive hunt for evidence of the bird. By the end of 2011, this search had produced some blurry photos, snippets of video, and a few recordings of what may or may not be ivory-bill calls and knocks. Definitive proof that the bird still lives remains elusive.

If the ivory-billed woodpecker survives, it is an **endangered species**, a species that faces extinction in all or part of its range. A **threatened species** is one that is likely to become endangered in the near future. Keep in mind that not all rare species are threatened or endangered. Some species have always been uncommon. If their numbers remain stable, they are not at risk of extinction. In the United States, the U.S. Fish and Wildlife Service determines which species will be listed as endangered or threatened and thus receive special protections.

Each species requires a specific type of habitat, and any loss, degradation, or fragmentation of that habitat reduces population numbers. An **endemic species**, one that remains confined to the area in which it evolved, is more likely to go extinct than a species with a more widespread distribution. Ivory-billed woodpeckers were endemic to swamp forests of the southeastern United States. Thus, logging of these forests harmed the only existing population of these birds.

Habitat loss affects nearly all of the more than 700 species of flowering plants currently listed as threatened or endangered in the United States. For example, widespread conversion of prairies to farmland and housing developments threaten the western prairie fringed orchid (Figure 18.12**B**).

Humans also degrade habitats in less direct ways. Edwards Aquifer in Texas consists of water-filled, underground limestone formations that supply drinking water to the city of San Antonio. Excessive withdrawals of water from this aquifer, along with pollution of the water that recharges it, endanger species that live in the aquifer's depths. The Texas blind salamander (Figure 18.12**C**) is one of these endangered species.

Deliberate or accidental species introductions degrade habitats too. Section 17.1 detailed the ways that red imported fire ants accidentally introduced to the United States from South America now threaten native species. Similarly, rats and domestic cats are decimating many of Hawaii's endemic bird species.

Overharvesting is another threat. You learned earlier about how overfishing caused a decline in the Atlantic codfish population (Section 16.4). The white abalone has suffered a similar fate. This marine mollusk (Figure 18.13) is native to kelp forests along the coast of California. During the 1970s, overharvesting reduced the species to about 1 percent of its original numbers. In 2001, the abalone became the first invertebrate to be listed as endangered by the U.S. Fish and Wildlife Service. Some white abalones remain in the wild, but population density is too low for effective reproduction. White abalones release eggs and sperm into the water, a mating strategy that works only when many animals live in close proximity. Individuals have been captured for a captive breeding program. If the program succeeds, their offspring can be reintroduced into the wild.

The extent of the species declines is staggering. The International Union for Conservation of Nature and Natural Resources (IUCN) reports that of the 48,677 species they assessed, 36 percent were threatened or endangered (Table 18.1). We do not know the level of threat for the vast majority of the approximately 1.8 million named species, or for the millions of species yet to be discovered.

FIGURE 18.13 White abalone. Overharvesting for food has driven this mollusk to near extinction.
John Butler, NOAA.

Table 18.1	Global Assessment of Threatened Species*		
	Described Species	Evaluated for Threats	Found to Be Threatened
Vertebrates			
Mammals	5,416	4,863	1,094
Birds	9,956	9,956	1,217
Reptiles	8,240	1,385	422
Amphibians	6,199	5,915	1,808
Fishes	30,000	3,119	1,201
Invertebrates			
Insects	950,000	1,255	623
Mollusks	81,000	2,212	978
Crustaceans	40,000	553	460
Corals	2,175	13	5
Others	130,200	83	42
Land Plants			
Mosses	15,000	92	79
Ferns and allies	13,025	211	139
Gymnosperms	980	909	321
Angiosperms	258,650	10,771	7,899
Protists			
Green algae	3,715	2	0
Red algae	5,956	58	9
Brown algae	2,849	15	6
Fungi			
Lichens	10,000	2	2
Mushrooms	16,000	1	1

* IUCN–WCU Red List, available online at www.iucnredlist.org.

endangered species A species that faces extinction in all or part of its range.

endemic species A species that evolved in one place and is found nowhere else.

threatened species A species likely to become endangered in the near future.

FIGURE 18.14 Deforestation. Ground view (*left*) and aerial view (*right*) of the Mato Grosso region of Brazil, where tropical forest is being cleared for pasture and plantations.

Credits: left, Courtesy of Guido van der Werf, Vrije Universiteit Amsterdam; right, NASA, courtesy the MODIS Rapid Response Team at Goddard Space Flight Center.

50 km

FIGURE 18.15 Dust blows from the expanding Sahara Desert into the Atlantic Ocean.

© Geoeye Satellite Image.

Atlantic Ocean   Africa

› Deforestation and Desertification

Deforestation, the removal of trees from an area, is an ongoing threat in many regions. Although the amount of forested land is currently stable or increasing in North America, Europe, and China, tropical forests continue to suffer heavy losses (Figure 18.14). Destruction of tropical rain forests is not merely a regional concern. As noted earlier, these forests have the highest primary production of any land biome, so their decline will significantly affect carbon storage and oxygen production. In addition, these forests harbor the greatest number of species of any biome. Thus, destruction of an area of tropical rain forest endangers far more species than destruction of a comparable area of temperate zone or boreal forest.

Poor agricultural practices can encourage soil erosion and lead to a rapid shift from grassland or woodland to desert, a process called **desertification**. During the mid-1930s, large portions of prairie on North America's southern Great Plains were plowed, exposing the soil to the force of the region's constant winds. Coupled with a drought, the result was an economic and ecological disaster. Winds carried more than a billion tons of topsoil aloft, turning the region into what came to be known as the Dust Bowl. A similar situation now exists in Africa, where expansion of the Sahara Desert causes clouds of dust to take flight over the Atlantic (Figure 18.15).

FIGURE 18.16 Animated! Acid rain.

Figure It Out: Is rain more acidic on the East Coast or the West Coast of the United States?   *Answer: The East Coast*

Credits: (a) © Cengage Learning; (b) Frederica Georgia/ Photo Researchers, Inc.

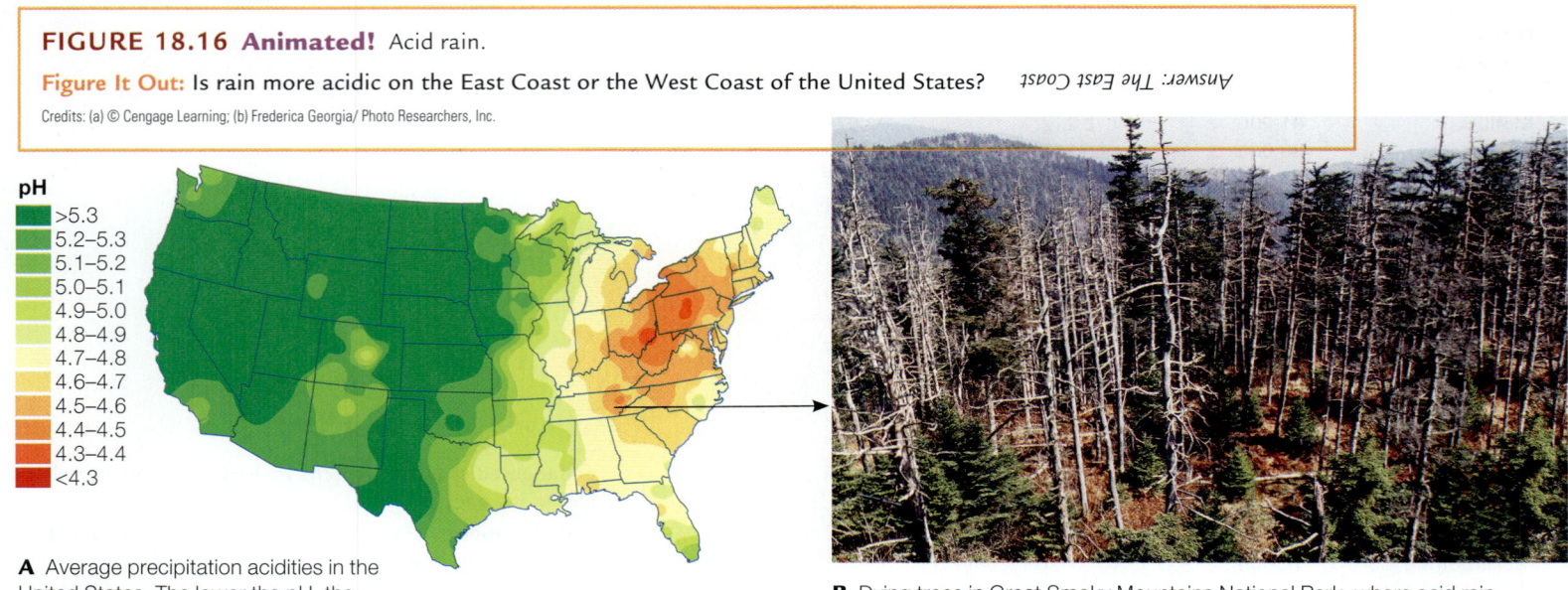

pH
- >5.3
- 5.2–5.3
- 5.1–5.2
- 5.0–5.1
- 4.9–5.0
- 4.8–4.9
- 4.7–4.8
- 4.6–4.7
- 4.5–4.6
- 4.4–4.5
- 4.3–4.4
- <4.3

A Average precipitation acidities in the United States. The lower the pH, the more acidic the rain.

B Dying trees in Great Smoky Mountains National Park, where acid rain harms leaves and causes loss of nutrients from soil.

FIGURE 18.17 Death by plastic. *Top*, Recently deceased Laysan albatross chick, dissected to reveal the contents of its gut. *Bottom*, Scientists found more than 300 pieces of plastic inside the bird. One of the pieces had punctured its gut wall, resulting in the chick's death. The chick was fed this junk by its parents, who gathered floating plastic from the ocean surface, mistaking it for food. Clair Fackler/ NOAA.

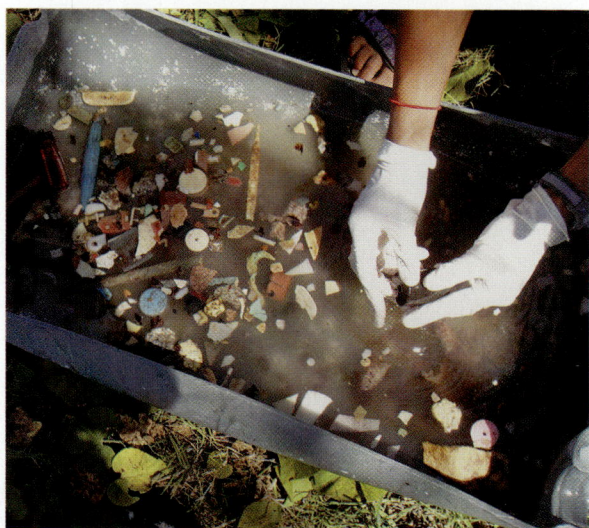

> **Acid Rain** A **pollutant** is a natural or man-made substance released into soil, air, or water in greater than natural amounts. It disrupts the physiological processes of organisms that evolved in its absence, or that are adapted to lower levels of it. **Acid rain** forms when air pollutants released by burning coal and other fossil fuels combine with water vapor and fall to Earth. The resulting rainfall can be ten times more acidic than normal (Figure 18.16**A**). Acid rain that falls onto or drains into waterways, ponds, and lakes harms aquatic organisms. When it falls on forests, it burns tree leaves and encourages loss of nutrient ions from the soil. As a result, trees become malnourished and more susceptible to disease. Effects are most pronounced at higher elevations where trees are frequently exposed to clouds of acidic droplets (Figure 18.16**B**).

> **Biological Accumulation and Magnification** Two processes can cause the concentration of a chemical pollutant in an organism's body to rise far above its concentration in the environment. First, some chemical pollutants taken up by organisms accumulate in their tissues. As a result, older individuals contain more of the pollutant than younger ones. Second, by the process of **biological magnification**, a chemical pollutant passes through food chains, becoming increasing concentrated in higher and higher trophic levels. For example, methylmercury released by coal-burning power plants is a toxic pollutant that can reach a very high concentration in large predatory fishes such as swordfish, bluefin tuna, and albacore (white) tuna. Methylmercury interferes with development of the nervous system, so pregnant women, nursing women, and children should avoid eating these fish.

> **The Trouble With Trash** Historically, humans buried unwanted material in the ground or dumped it out at sea. Trash was out of sight, and also out of mind. We now understand that chemicals seeping from buried trash can contaminate groundwater. The United States no longer dumps its trash at sea, but plastic and other garbage still enters our coastal waters. Foam cups and containers from fast-food outlets, plastic shopping bags, plastic water bottles, foam peanuts, and other material discarded as litter ends up in storm drains. From there it is carried to streams and rivers that can convey it to the sea. A seawater sample taken near the mouth of the San Gabriel River in Southern California had 128 times as much plastic as plankton by weight. Waste plastic that enters the ocean poses a threat to marine life. For example, seabirds often eat floating bits of plastic or feed the plastic to chicks, with deadly results (Figure 18.17).

> **Ozone Depletion and Pollution** Between 17 and 27 kilometers (10.5 and 17 miles) above sea level, the concentration of ozone gas (O_3) is so great that scientists refer to this region as the **ozone layer**. The ozone layer benefits us by absorbing most of the ultraviolet (UV) radiation in incoming sunlight. UV radiation, remember, can damage DNA and thus cause mutations (Section 6.4).

In the mid-1970s, scientists noticed that the ozone layer was thinning. Its thickness had always varied a bit with the season, but now there was steady

acid rain Rain containing sulfuric and/or nitric acid; forms when pollutants mix with water vapor in the atmosphere.

biological magnification A chemical pollutant becomes increasingly concentrated as it moves through a food chain.

deforestation Removal of all trees from a forested area.

desertification Conversion of grassland or woodlands to desertlike conditions.

ozone layer Region of upper atmosphere with a high ozone concentration; acts as a sunscreen against UV radiation.

pollutant A natural or man-made substance that is released into the environment in greater than natural amounts and that damages the health of organisms.

FIGURE 18.18 Ozone levels in Earth's upper atmosphere (stratosphere) in September 2007. *Purple* indicates the least ozone, with *blue*, *green*, and *yellow* indicating increasingly higher levels.

Check the current status of the ozone hole at NASA's website (http://ozonewatch.gsfc.nasa.gov/).

NASA.

decline from year to year. By the mid-1980s, the spring ozone thinning over Antarctica was so pronounced that people were referring to the low-ozone region as an "ozone hole" (Figure 18.18). Declining ozone quickly became an international concern. The ozone hole over the South Pole was a sign that the ozone layer was thinning worldwide.

Chlorofluorocarbons, or CFCs, were determined to be the main ozone destroyers. At the time, these odorless gases were widely used as propellants in aerosol cans, as coolants, and in solvents and plastic foam. In response to the potential threat posed by ozone thinning, countries worldwide agreed in 1987 to phase out the production of CFCs and other ozone-destroying chemicals. As a result of that agreement (the Montreal Protocol), concentrations of CFCs in the atmosphere are no longer rising dramatically. Unfortunately, CFCs break down quite slowly, so scientists expect them to remain at a level that significantly impairs the ozone layer for several decades.

Ozone gas is sometimes said to be "good up high, but bad nearby." The "good" ozone is high in Earth's atmosphere. In the lower atmosphere, where little ozone occurs naturally, it is considered a pollutant. It irritates human eyes and airways, and also slows plant growth. Ozone enters the lower atmosphere when sunlight encourages reactions among compounds released by burning or evaporating fossil fuels. Warm temperatures speed the reactions. To help reduce ozone pollution, avoid putting fossil fuels or their combustion products into the air at times that favor ozone production. On hot, sunny, still days, postpone filling a gas tank or using gas-powered appliances until evening, when there is less sunlight to power conversion of pollutants to ozone.

› Global Climate Change Earth's climate has varied greatly over its long history. There have been ice ages, when much of the planet was covered by glaciers, and unusually hot periods, when tropical plants and coral reefs thrived at what are now cool latitudes. Scientists can correlate past large-scale temperature changes with shifts in Earth's orbit, which varies in a regular fashion over 100,000 years, and Earth's tilt, which varies over 40,000 years. Changes in solar output and volcanic eruptions also affect Earth's temperature. However, these factors cannot explain the rise in average temperatures that we are currently experiencing. Scientists attribute this change to a human-induced increase in greenhouses gases such as carbon dioxide (Figure 18.19). These gases are released

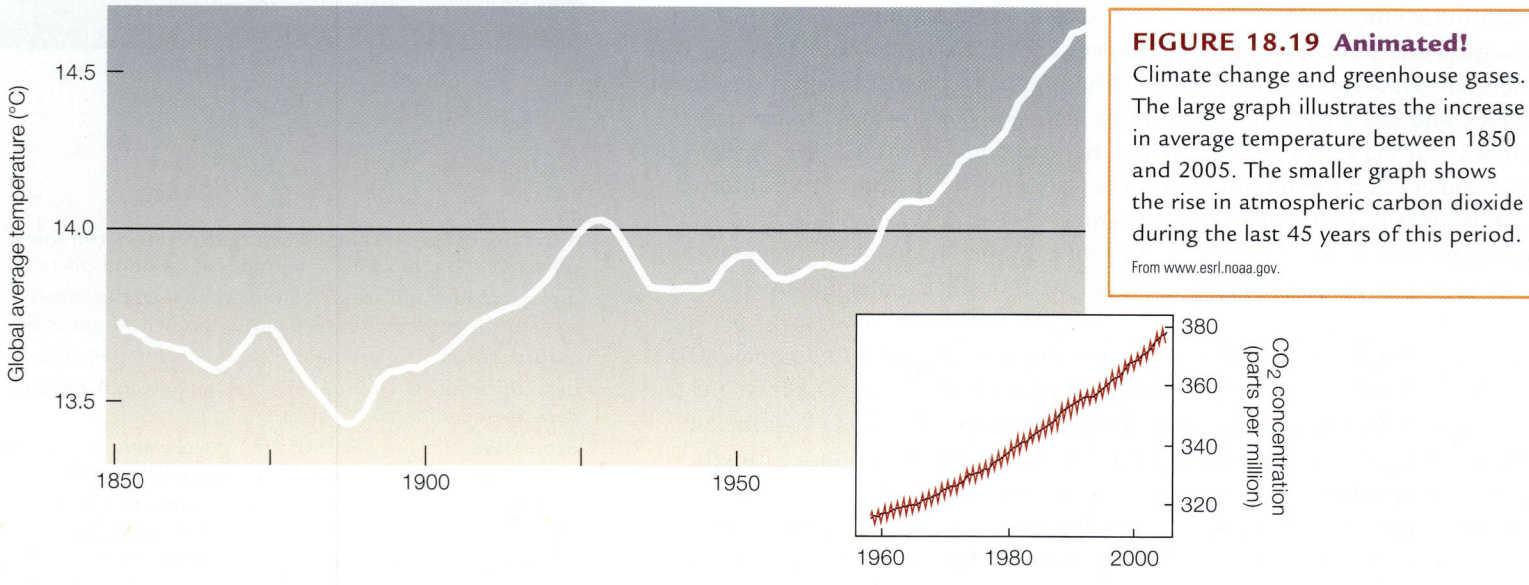

FIGURE 18.19 Animated!
Climate change and greenhouse gases. The large graph illustrates the increase in average temperature between 1850 and 2005. The smaller graph shows the rise in atmospheric carbon dioxide during the last 45 years of this period.

From www.esrl.noaa.gov.

A Muir Glacier 1941　　　　　　　　　　　　　　**B** Muir Glacier 2004

FIGURE 18.20 One sign of global climate change, retreating glaciers in Alaska.

Credits: (a) National Snow and Ice Data center, W. O. Field; (b) National Snow and Ice Data Center, B. F. Molnia.

by the burning of fossil fuels, a practice that has increased dramatically since the onset of the industrial revolution in the late 1700s. Greenhouse gases keep heat near Earth's surface from escaping into space (Section 17.6), so the higher their concentration, the more heat is trapped.

Warming temperature raises sea level because water expands as it is heated, and water released by melting glaciers (Figure 18.20) increases the volume of the sea. The temperature of the land and seas affects evaporation, winds, and currents. As a result, many weather patterns are expected to change as temperature rises. For example, warmer temperatures are correlated with extremes in rainfall patterns, with periods of drought interrupted by unusually heavy rains. Warmer seas are also expected to increase the intensity of hurricanes.

Climate change is already having widespread effects on biological systems. Temperature changes are cues for many temperate zone species, so warmer than normal springs cause deciduous trees to leaf out earlier, and spring-blooming flowers to blossom earlier. Animal migration times and breeding seasons are also shifting. Some species benefit from the warming, expanding their range to higher latitudes or elevations that were previously too cool to sustain them. Other species are harmed by the rising temperature. For example, warming of tropical waters stresses reef-building corals and has increased the frequency of coral bleaching events.

Take-Home Message

How do human activities affect the biosphere?

- We are increasing the rate of species extinctions by degrading, destroying, and fragmenting habitats, overharvesting species, and introducing exotic species.
- Human activities turn grasslands into deserts, strip woodlands of trees, and generate pollutants and trash that kill animals and damage ecosystems.
- Some pollutants have global effects, as when CFCs cause thinning of the ozone layer, and rising levels of greenhouse gases bring about climate change.

18.6 Maintaining Biodiversity

> **The Value of Biodiversity** Every nation enjoys several forms of wealth: material wealth, cultural wealth, and biological wealth, or biodiversity. We measure a region's **biodiversity** at three levels: the genetic diversity within its species, its species diversity, and its ecosystem diversity.

Why should we protect biodiversity? From a purely selfish standpoint, doing so is an investment in our future. Healthy ecosystems are essential to the survival of our own species. Other organisms produce the oxygen we breathe and the food we eat. They also remove waste carbon dioxide from the air and decompose and detoxify other wastes. Plants take up rain and hold soil in place, preventing erosion and reducing the risk of flooding.

Compounds discovered in wild species often serve as medicines. Two widely used chemotherapy drugs, vincristine and vinblastine, were extracted from the rosy periwinkle, a low-growing plant native to Madagascar's rain forests. Cone snails that live in tropical seas are the source of a highly potent pain reliever.

Wild relatives of crop plants serve as reservoirs of genetic diversity that plant breeders can draw upon to protect and improve crops. Wild plants often have genes that make them more resistant to disease or adverse conditions than their domesticated relatives. Plant breeders can use traditional cross-breeding methods or biotechnology to introduce genes from wild species into domesticated ones, thus creating improved varieties.

A decline in some species can warn us that the natural support system we depend on is in trouble. **Indicator species** are particularly sensitive to environmental change, and can be monitored as indicators of environmental health. For example, a decline in lichens tells us that air quality is deteriorating. The loss of mayflies from a stream tells us that the quality of water in a stream is declining.

In addition, there are ethical reasons to preserve biodiversity. As we have emphasized many times, all living species are the result of an ongoing evolutionary process that stretches back billions of years. Each species has a unique combination of traits. The extinction of a species removes its unique collection of traits from the world of life forever.

> **Conservation Biology** Biodiversity is currently in decline at all three levels and in regions worldwide. **Conservation biology** addresses these declines. The goals of this relatively new field of biology are to survey the range of biodiversity, and to find ways to maintain and use biodiversity in a manner that benefits human populations. The aim is to conserve as much biodiversity as possible by encouraging people to value it and use it in ways that do not destroy it.

With so many species and ecosystems at risk, conservation biologists must often make difficult choices about what areas to target for protection first. To prioritize their efforts, the biologists identify conservation hot spots, places that are richest in endemic species and under the greatest threat. The goal of prioritizing is to save representative examples of all of Earth's existing biomes. By focusing on hot spots, rather than individual species, scientists hope to maintain ecosystem processes that sustain biological diversity.

Conservation scientists of the World Wildlife Fund have defined 867 distinctive land ecoregions that they

The extinction of a species removes its unique collection of traits from the world of life forever.

FIGURE 18.21 The Klamath–Siskiyou forest, one of North America's conservation hot spots. Endangered northern spotted owls (*inset*) nest in old-growth trees of this forest.

Credits: David Patte, USFWS; inset, USFWS.

consider the top priority for conservation efforts. Each region has a large number of endemic species and is under threat. The Klamath–Siskiyou forest in southwestern Oregon and northwestern California is one hot spot (Figure 18.21). It is home to many rare conifers, and two endangered birds, the northern spotted owl and the marbled murrelet, nest in the forest's old growth. Endangered coho salmon breed in streams that run through it. Logging is the main threat to this region, but a newly introduced pathogen of conifers also poses a concern.

Protecting biological diversity can be a tricky proposition. Even in developed countries, people often oppose environmental protections because they fear such measures will have adverse economic consequences. However, taking care of the environment can make good economic sense. With a bit of planning, people can both preserve and profit from their biological wealth. For example, Costa Rica's Monteverde Cloud Forest Reserve protects more than 100 mammal species, 400 bird species, and 120 species of amphibians and reptiles. It is one of the few habitats left for the ocelot, puma, and jaguar (Figure 18.22). Each year, about 75,000 tourists visit the reserve, and the feeding, lodging, and guiding of these tourists provides much-needed employment to local people. These people also benefit from the ecological services that their forest continues to provide.

〉 Ecological Restoration

Sometimes, an ecosystem is so damaged, or there is so little of it left, that conservation alone is not enough to sustain biodiversity. **Ecological restoration** is work designed to bring about the renewal of a natural ecosystem that has been degraded or destroyed, fully or in part. Restoration work in Louisiana's coastal wetlands is an example. Louisiana contains more than 40 percent of the coastal wetlands in the United States. These marshes are an ecological and economic treasure, but they are troubled. Dams and levees built upstream of the marshes hold back sediments that would normally replenish sediments lost to the sea. Channels cut through the marshes for oil exploration and production have encouraged erosion, and the rising sea level threatens to flood existing plants. Since the 1940s, Louisiana has lost an area of marshland the size of Rhode Island. Restoration efforts now under way aim to reverse some of those losses and protect remaining marsh species (Figure 18.23).

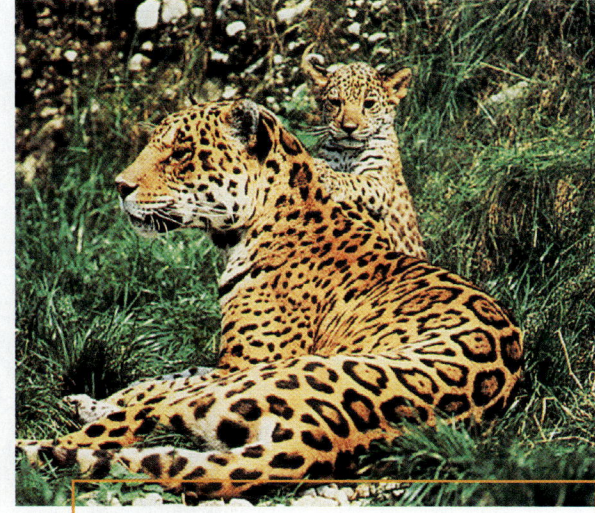

FIGURE 18.22 Jaguars, one of the many endangered species that live in Costa Rica's Monteverde Cloud Forest Reserve.

© Adolf Schmidecker/ FPG/ Getty Images.

biodiversity Of a region, the genetic diversity within its species, variety of species, and variety of ecosystems.

conservation biology Field of applied biology that surveys biodiversity and seeks ways to maintain and use it.

ecological restoration Actively altering an area in an effort to restore or create a functional ecosystem.

indicator species A species that is particularly sensitive to environmental changes and can be monitored to assess whether an ecosystem is threatened.

FIGURE 18.23 Ecological restoration in Louisiana's Sabine National Wildlife Refuge.

Credits: (a) Diane Borden-Bilot, U.S. Fish and Wildlife Service; (b) U.S. Fish and Wildlife Service.

A Where marsh has become open water, sediments are barged in and marsh grasses are planted on them.

B Restoration protects native species such as the roseate spoonbills that nest in the marsh during the summer.

FIGURE 18.24 Resource extraction. Bingham copper mine near Salt Lake City, Utah. The mine is 4 kilometers (2.5 miles) wide and 1,200 meters (0.75 miles) deep.

❯ **Living Sustainably** Ultimately, the health of our planet depends on our ability to recognize that the principles of energy flow and of resource limitation that govern the survival of all systems of life do not change. We must take note of these principles and find a way to live within our limits. The goal is living sustainably, which means meeting the needs of the present generation without reducing the ability of future generations to meet their own needs.

Promoting sustainable living begins with acknowledging the environmental consequences of our own lifestyle. People in industrial nations use enormous quantities of resources, and the extraction and delivery of these resources have negative effects on biodiversity. In the United States, the size of the average family has declined since the 1950s, while the size of the average home has doubled. All the materials used to build and furnish those larger homes come from the environment. For example, an average new home contains about 500 pounds of copper in its wiring and plumbing. Where does copper come from? Like most other mineral elements used in manufacturing, it is generally mined from the ground (Figure 18.24). Mining often generates pollution and produces ecological dead zones.

Nonrenewable mineral resources are also used in electronic devices such as phones, computers, televisions, and MP3 players. Constantly trading up to the newest device may be good for the ego and the economy, but it is bad for the environment. Reducing consumption by fixing existing products promotes sustainability, as does recycling. Recycling nonrenewable materials reduces the need for extraction of those resources from the environment, so it helps maintain healthy ecosystems.

Reducing energy use is another way to promote sustainability. Fossil fuels such as petroleum, natural gas, and coal supply most of the energy used by developed countries. You already know that use of these nonrenewable fuels contributes to global warming and acid rain. In addition, extraction and transportation of these fuels have negative impacts. Oil harms many species when it leaks from pipelines, ships, or wells. Renewable energy sources do not produce greenhouse gases, but they have their own drawbacks. For example, dams in rivers of the Pacific Northwest generate renewable hydroelectric power, but they may discourage endangered salmon from returning to streams above the dam to breed. Similarly, wind turbines can harm birds and bats. Panels used to collect solar energy are made using nonrenewable mineral resources, and manufacturing the panels generates pollutants.

In short, all commercially produced energy has some kind of negative environmental impact, so the best way to minimize that impact is to use less energy. Shop for energy-efficient appliances, use fluorescent bulbs instead of incandescent ones, and do not leave lights on in empty rooms. Walking, bicycling, and using public transportation are energy-efficient alternatives to driving. Shopping locally and purchasing locally produced goods also saves energy.

If you want to make a difference, learn about the threats to ecosystems in your own area. Are species threatened? If so, what are the threats? How can you help reduce those threats? Support efforts to preserve and restore local biodiversity. Many ecological restoration projects are supervised by trained biologists but carried out primarily through the efforts of volunteers (Figure 18.25).

Keep in mind that unthinking actions of billions of individuals are the greatest threat to biodiversity. Each of us may have little impact on our own, but our collective behavior, for good or for bad, will determine the future of the planet.

FIGURE 18.25 Volunteers restoring the Little Salmon River in Idaho so that salmon can migrate upstream to their breeding grounds.
Mountain Visions/NOAA.

Take-Home Message

What is biodiversity and how can we sustain it?

- Biodiversity is the genetic diversity of individuals of a species, the variety of species, and the variety of ecosystems. Worldwide, biodiversity is declining at all of these levels.

- Conservation biologists are working to identify threatened regions with high biodiversity and prioritize which receive protection.

- Ecological restoration is the process of re-creating or renewing a diverse natural ecosystem that has been destroyed or degraded.

- Individuals can help maintain biodiversity by using resources in a sustainable fashion.

WHERE YOU ARE GOING . . .

We return to the subject of pollutants in Section 25.1, when we look into the effect of agricultural chemicals on vertebrate endocrine systems and in Section 27.1, which describes how plants that can bioaccumulate pollutants are used to clean up soil.

18.7 A Long Reach (revisited)

The Arctic is not a separate continent, but rather a region that encompasses the northernmost parts of several continents. Eight countries, including the United States, Canada, and Russia, control parts of the Arctic and have rights to its oil, gas, and mineral deposits. Until recently, ice sheets covered the Arctic Ocean year-round, making it difficult for ships to move to and from the Arctic landmass. Those ice sheets are now shrinking (Figure 18.26). At the same time, ice on the Arctic landmass is melting. These changes will accelerate global warming and make it easier for people to remove minerals and fossil fuels from the Arctic. With the world supply of fuel and minerals dwindling, pressure to exploit Arctic resources is rising. However, conservationists warn that extracting these resources will harm Arctic species such as the polar bear that are already threatened by global climate change.

Arctic perennial sea ice 1979

Arctic perennial sea ice 2003

FIGURE 18.26 Decrease in Arctic perennial sea ice. Loss of ice accelerates global warming because ice reflects sunlight, whereas water and land absorb it, then radiate the absorbed energy as heat.
NASA.

Summary

Section 18.1 Humans and the chemical pollutants that their activities produce have reached even remote parts of the **biosphere**.

Section 18.2 **Climate** refers to average weather conditions over time. Variations in climate depend largely upon differences in the amount of solar radiation reaching different parts of Earth. The closer a region is to the equator, the more solar energy it receives. Warming of air at the equator is the start of global patterns of air circulation and ocean currents. Landforms also influence climate, as when coastal mountains cause a **rain shadow**.

 Section 18.3 **Biomes** are categories of major ecosystems on land. They are maintained largely by regional variations in climate and described mainly in terms of their dominant plant life.

Near the equator, high rainfall and mild temperatures support **tropical rain forests**, which are dominated by trees that remain leafy and active all year. This is Earth's most productive and oldest biome. In **temperate deciduous forests**, trees grow during warm summers, then lose their leaves and become dormant in winter. **Boreal forest**, dominated by conifers, is the most extensive biome. It is found only in the Northern Hemisphere where winters are cold and dry, and summers are cool and rainy.

Grasslands form at midlatitudes in the interior of continents and are dominated by plants adapted to grazing. **Prairie** is a North American grassland biome; **savanna** is an African one that also includes scattered shrubs. **Chaparral**, a biome dominated by shrubby plants with tough leaves, occurs in places with cool, wet winters and hot, dry summers. Like grasses, chaparral plants are adapted to periodic fires.

Deserts occur at latitudes about 30° north and south, where dry air descends and annual rainfall is sparse. Desert plants are adapted to withstand drought or complete their life cycle fast after a rain.

Tundra forms at high latitudes in the Northern Hemisphere. It is the youngest biome and lies over **permafrost**.

Section 18.4 Aquatic ecosystems have gradients of sunlight penetration, water temperature, salinity, and dissolved gases. Lakes are standing bodies of water. Different communities of organisms live at different depths and distances from the shore.

Streams and rivers are flowing bodies of water. Physical characteristics of a stream or river that vary along its length influence the types of organisms that live in it. A semi-enclosed area where nutrient-rich water from a river mixes with seawater is an **estuary**, a highly productive habitat. Seashores may be rocky or sandy. Grazing food chains based on algae form on rocky shores. Detrital food chains dominate sandy shores.

Coral reefs are species-rich ecosystems of warm, well-lit tropical waters. Photosynthetic protists that live in the coral's tissues are the main producers here.

The open ocean's upper waters hold photosynthetic organisms that form the basis for grazing food chains. Deeper water communities usually subsist on material that drifts down from above. However, bacteria and archaea that can obtain energy from minerals serve as the producers at **hydrothermal vent** ecosystems. **Seamounts** are undersea mountains that have a high species richness.

 Section 18.5 Humans are increasing the rate of extinctions by overharvesting, degrading and destroying habitats, and introducing exotic species. **Endangered species** are currently at risk of extinction; **threatened species** are likely to become endangered. **Endemic species** are more vulnerable to extinction than widely dispersed ones. Human activities can also threaten entire ecosystems, as when poor agricultural practices cause **desertification** or **deforestation** destroys a forest. **Pollutants** such as those that cause **acid rain** threaten forest ecosystems. **Biological magnification** occurs when a pollutant is passed along a food chain. Trash that gets into fresh water can make its way into the oceans, where it degrades marine ecosystems. Ozone is a pollutant near the ground, but depletion of the **ozone layer** is a global threat caused by use of chemicals called CFCs. Global agreement to phase out CFC use lessens this threat. Global climate change caused by rising concentrations of greenhouse gases also threatens the biosphere.

Section 18.6 **Biodiversity** includes the diversity of genes, species, and ecosystems. All three levels of biodiversity are declining. A decrease in biodiversity can harm humans. We rely on ecosystems to produce oxygen and decompose waste. We also benefit from many compounds produced by wild species, and by tapping their genetic diversity to enhance our crops. Loss of **indicator species** warns us that a habitat is being degraded.

Conservation biologists document the extent of biodiversity and look for ways to preserve it, while benefiting humans. They give priority to areas with high biodiversity that are most threatened. **Ecological restoration** is the work of actively renewing an ecosystem that has been damaged or destroyed. By living sustainably, we can maintain biodiversity and ensure that resources remain for future generations.

Self-Quiz   Answers in Appendix I

1. The most solar radiation reaches the ground at _____ .
   a. the equator   c. midlatitudes
   b. the North Pole   d. the South Pole

2. When air is heated, it _____ and can hold _____ water.
   a. sinks; less   c. rises; less
   b. sinks; more   d. rises; more

3. Most North American _____ has been converted to cropland.
   a. tundra   c. desert
   b. prairie   d. boreal forest

4. Plants in _____ are adapted to periodic fires.
   a. deserts   c. tropical rain forests
   b. boreal forests   d. chaparral

5. Permafrost underlies _____ .
   a. savanna   b. tundra   c. desert   d. prairie

6. The oldest and most productive biome is _____ .
   a. boreal forest   c. tropical rain forest
   b. tundra   d. desert

7. Bacteria and archaea that can obtain energy from minerals are the main producers at _____ .
    a. hydrothermal vents   c. coral reefs
    b. estuaries   d. seamounts

8. Which would hold more oxygen?
    a. a fast-moving, cool stream   b. a warm pond

9. Match the biome with the most suitable description.
    _____ tundra   a. fresh water and seawater mix
    _____ chaparral   b. low humidity, and little rainfall
    _____ desert   c. North American grassland
    _____ prairie   d. fire-adapted shrubs with tough leaves
    _____ estuary   e. low-growing plants over permafrost
    _____ boreal forest   f. most extensive biome
    _____ coral reef   g. main producers are protists
    _____ tropical rain   h. broadleaf trees are active all year
             forest

10. An _____ species has population levels so low it is at great risk of extinction in the near future.
    a. endemic   c. indicator
    b. endangered   d. exotic

11. An _____ species can be monitored to gauge the health of its environment.
    a. endemic   c. indicator
    b. endangered   d. exotic

12. The 1930s environmental disaster known as the Dust Bowl is an example of _____ .
    a. deforestation   c. ecological restoration
    b. desertification   d. species extinction

13. The ozone layer _____ .
    a. is getting thicker   c. is near the ground
    b. helps keep Earth warm   d. screens out UV radiation

14. Acid rain _____ .
    a. harms aquatic organisms   c. is a type of pollution
    b. kills trees   d. all of the above

15. Biological magnification causes pollutants to be high in _____ .
    a. producers   c. predators
    b. aquifers   d. the stratosphere

Critical Thinking

1. Like the salamander shown in Figure 18.12**C**, many animals that live in caves are blind but had sighted ancestors. By what process might a sighted animal species that colonized a cave lose its sight?

2. In one seaside community in New Jersey, the U.S. Fish and Wildlife Service suggested trapping and removing feral cats (domestic cats that live in the wild). The goal was to protect some endangered wild birds (plovers) that nested on the town's beaches. Many residents were angered by the proposal, arguing that the cats have as much right to be there as the birds. Do you agree? Why or why not?

3. Burning fossil fuel puts excess carbon dioxide into the atmosphere, but deforestation and desertification also affect carbon dioxide concentration. Explain how a global decrease in the amount of vegetation can bring about a rise in carbon dioxide.

Digging Into Data

Arctic PCB Pollution

Winds carry chemical contaminants produced and released at temperate latitudes to the Arctic, where the chemicals enter food webs. As a result of biological magnification, top carnivores in arctic food webs, including polar bears and people, end up with high concentrations of these chemicals in their body. Arctic people who eat a lot of local wildlife tend to have unusually high levels of pollutants called polychlorinated biphenyls, or PCBs. The Arctic Monitoring and Assessment Programme studies the effects of these industrial chemicals on the health and reproduction of Arctic people. Figure 18.27 shows the average PCB levels for women in native populations in the Russian Arctic and the sex ratios at birth for each PCB level.

1. Which sex was more common in offspring of women with less than 1 microgram per milliliter of PCB in serum?

2. At what PCB concentrations were women more likely to have daughters?

3. In some villages in Greenland, nearly all recent newborns are female. Would you expect PCB levels in those villages to be above or under 4 micrograms per milliliter?

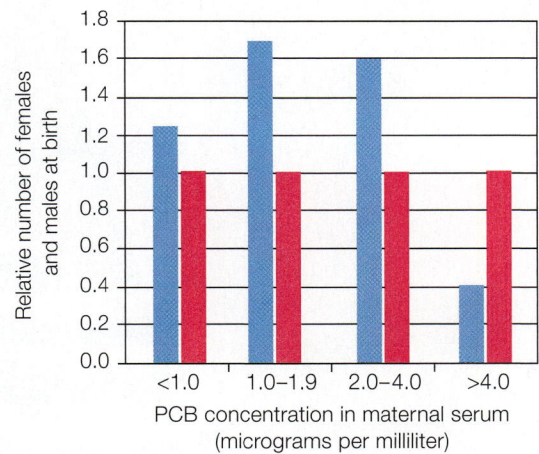

FIGURE 18.27 Effect of maternal PCB concentration on sex ratio of newborns in indigenous populations in the Russian Arctic. *Blue* bars indicate the relative number of males born per one female (*pink* bars).

© Cengage Learning.

4. Two arctic marine mammals that live in the same waters differ in the level of pollutants in their bodies. Bowhead whales have a lower pollutant load than ringed seals. What are some factors that might explain this difference?

5. The extent of damage done by acid rain depends to some extent on the type of rocks in the area where it falls. Acid rain that falls in a region with limestone or other calcium carbonate–rich rocks does less harm than acid rain that falls in a region without such rocks. How might the presence of limestone lessen the effects of an input of acid into an ecosystem? Hint: Review Section 2.5.

Animal Tissues and Organs

19

19.1 Stem Cell Studies

The new field of regenerative medicine aims to replace lost or diseased cells, tissues, or organs and so restore function. Use of stem cells may aid these efforts. As Figure 19.1 shows, **stem cells** are self-renewing cells that can divide and produce more stem cells ❶. In addition, stem cells can differentiate (Section 7.7), meaning they can develop into one of the specialized cell types that make up specific body parts ❷. In short, all cells in your body "stem" from stem cells.

Some cells such as blood cells and skin cells are continually replaced throughout life, with the replacements arising from adult stem cells. Each type of adult stem cell normally differentiates into only a limited number of cells. For example, stem cells in adult bone marrow can develop into a variety of blood cells, but not into nerve or muscle cells. Adults have few stem cells capable of becoming nerve cells or muscle cells. Thus, unlike skin and blood, nerves and muscles are not replaced if they get damaged or die. This is why a spinal cord injury can cause permanent paralysis. New nerves do not grow to replace damaged ones. Similarly, a heart remains forever weakened after a heart attack kills some of its muscle cells.

By contrast, a human embryonic stem cell is **pluripotent**, meaning it has the capacity to become any of the hundreds of cell types that make up a human body. Embryonic stem cells are formed during the early cell divisions, before the pinhead-sized embryo attaches to the wall of the uterus (womb). Of course, human embryonic stem cells used in research are not from embryos that formed inside a woman's body. Rather, they are derived from embryos produced by combining eggs and sperm in a laboratory. Once stem cells have been removed from a pre-implantation embryo, they can multiply in cell culture.

The first clinical trials of treatments involving human embryonic stem cells have just begun. Human embryonic stem cells must be induced to differentiate before they can be used in medical treatments. One clinical study currently under way in the United States uses retina cells derived from embryonic stem cells. (The retina is the light-detecting portion of the eye.) The laboratory-grown retina cells are transplanted into the eyes of people who have a degenerative disorder that causes blindness.

It is too early to know whether this or other proposed treatments involving human embryonic stem cells will be safe and effective. The human immune system may attack and destroy the transplanted cells, as it does other foreign cells. You will learn about immune responses in Chapter 22. If the transplanted cells do survive, other challenges remain. Cells of a human body do not operate independently. Rather, they attach to and interact with other cells, thus becoming organized into tissues and organs. It remains to be seen whether transplanted cells derived from embryos can make these necessary structural and functional attachments. Also, while a capacity for unlimited divisions is a good thing in embryonic cells, that same capacity could, if retained by transplanted cells, lead to tumor formation.

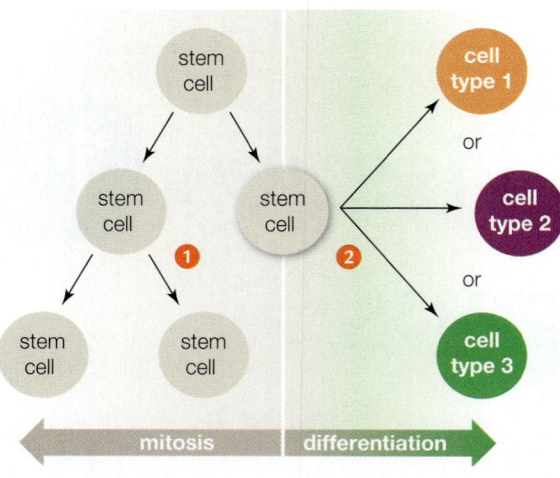

FIGURE 19.1 Characteristics of stem cells. Stem cells can divide to form new stem cells or differentiate to form specialized cell types.

© Cengage Learning.

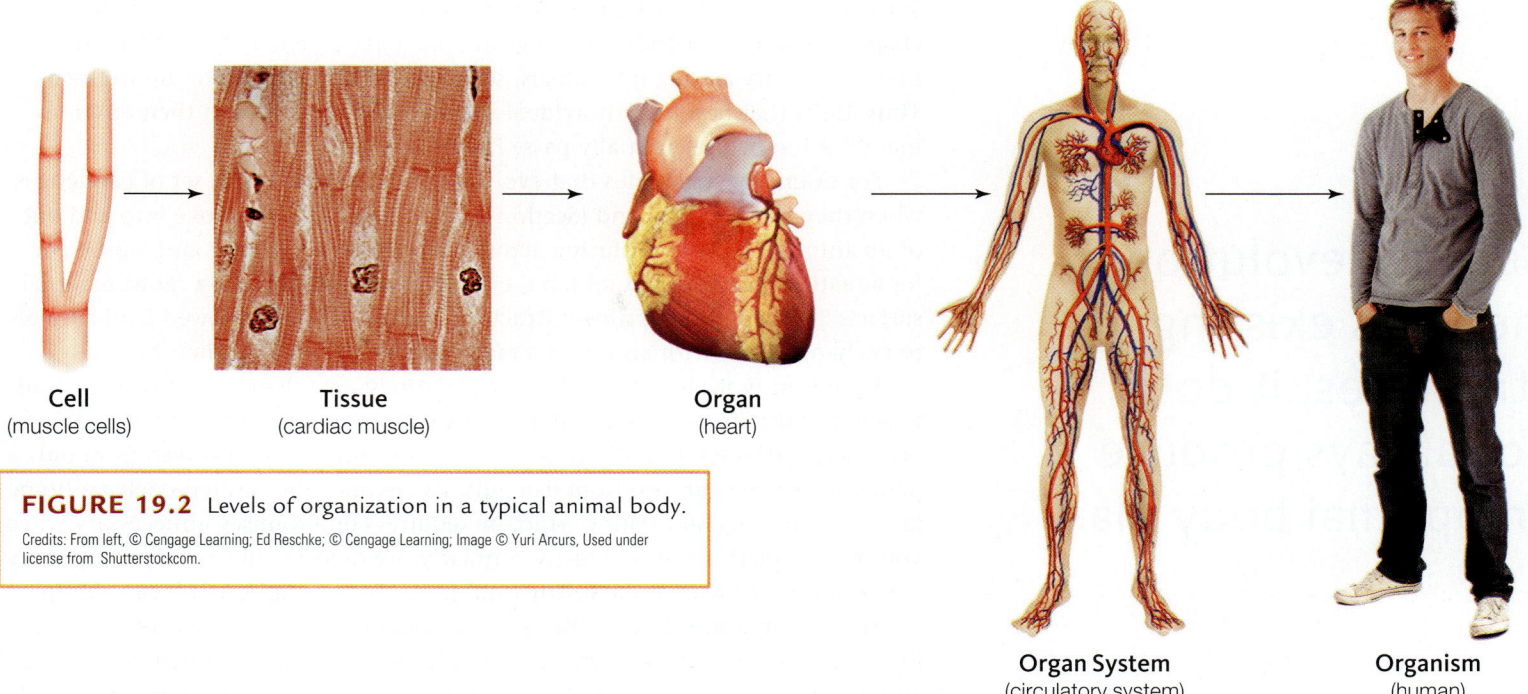

Cell
(muscle cells)

Tissue
(cardiac muscle)

Organ
(heart)

Organ System
(circulatory system)

Organism
(human)

FIGURE 19.2 Levels of organization in a typical animal body.

Credits: From left, © Cengage Learning; Ed Reschke; © Cengage Learning; Image © Yuri Arcurs, Used under license from Shutterstock.com.

19.2 Animal Structure and Function

❯ Organization and Integration All animals are multicelled, and nearly all have cells organized as tissues. A **tissue** consists of one or more cell types—and often an extracellular matrix—that collectively perform a specific task or tasks. Four types of tissue occur in all vertebrate bodies:

1. Epithelial tissues cover body surfaces and line internal cavities.
2. Connective tissues hold body parts together and provide structural support.
3. Muscle tissues move the body and its parts.
4. Nervous tissues detect stimuli and relay information.

Each tissue is characterized by the types of cells it includes and their proportions. For example, nervous tissue includes neurons, a type of cell not found in muscle tissue or epithelial tissue.

Some animal tissues become organized into organs. An **organ** is a structural unit of two or more tissues organized in a specific way and capable of carrying out specific tasks. For example, a human heart is an organ (Figure 19.2) that includes all four tissue types. The heart has a wall made up mostly of cardiac muscle tissue, it is enclosed by a sheath of connective tissue, and its internal chambers are lined with epithelial tissue. Nervous tissue delivers signals to and from the heart.

In **organ systems**, two or more organs and other components interact physically, chemically, or both in a common task. For example, the force generated by a beating heart moves blood through a system of blood vessels that extends throughout the body.

As an animal goes about its activities, it gains and loses solutes and water, and is exposed to temperature variations. Each living cell engages in metabolic activities that keep it alive. At the same time, organs and organ systems interact to keep solute concentrations and temperature within ranges that cells can tolerate. Such interactions keep conditions in the internal environment within tolerable limits, a process we call **homeostasis**.

homeostasis Process of maintaining favorable conditions inside the body.

organ Structural unit that is composed of two or more tissues and adapted to carry out a particular task.

organ system Organs that interact closely in some task.

pluripotent Capable of developing into any type of cell in a multicelled body.

stem cell A cell that can divide and create more stem cells or differentiate to become a specialized cell type.

tissue A collection of one or more specific cell types that are organized in a way that suits them to a task.

> **Evolution of Structure and Function** As you learned in earlier chapters, anatomical and physiological traits have a genetic basis. Many of these traits vary among individuals, so natural selection operates upon them. Thus, traits that best help individuals survive and reproduce in their environment tend to be preferentially passed on.

For example, vertebrates that evolved in water faced a new set of challenges when they moved onto land (Section 15.5). Gases can only move into and out of an animal's body by diffusing across a moist surface. That is not a problem for aquatic organisms, but on land, evaporation could dry out a moist external surface. The evolution of a new structural feature—lungs—allowed land animals to exchange gases with air across a moist surface deep inside their body.

Evolution typically does not produce entirely new tissues or organs. Instead, it modifies the structure and function of existing ones for new purposes, a process called exaptation (Section 12.6). For example, lungs evolved from outpouchings of the gut, not from fish gills. As anyone who has remodeled a home knows, modifying an existing structure requires compromises unnecessary when you build something entirely new. Similarly, we detect evidence of evolutionary compromises in many animal structural traits. For example, as a legacy of the lungs' ancient connection to the gut, the human throat opens to both the digestive tract and respiratory tract. As a result, food sometimes goes where air should and a person chokes. It would be safer if food and air entered the body through separate passages. However, because evolution modifies existing structures, it does not always produce an optimal body plan.

> Because evolution modifies existing structures, it does not always produce an optimal body plan.

Take-Home Message

How does the structure of an animal body relate to its function?

- All animals are multicelled. Most have cells organized as tissues, organs, and organ systems.

- Tissues, organs, and organ systems interact in homeostasis, the processes that keep conditions in the internal environment within the limits that cells tolerate.

- Anatomical and physiological traits can evolve by natural selection. Existing structures become modified over generations to better adapt their bearers to their environment.

- Evolutionary remodeling of existing structures for new purposes often results in body plans that are less than optimal.

19.3 Types of Animal Tissues

> **Epithelial Tissues** **Epithelial tissue**, or epithelium (plural, epithelia), is a sheetlike tissue in which cells attach to one another by tight junctions (Section 3.6). One side of an epithelium secretes a layer of proteins called the basement membrane that glues the epithelium to underlying connective tissue. The other side of the epithelium faces a body cavity or the external environment (Figure 19.3). Blood vessels do not run through epithelial tissue, so materials must move to and from epithelial cells by diffusing through the underlying connective tissue and basement membrane.

Most of what you see when you look in a mirror—your skin, hair, and nails—consists of epithelial tissue or structures derived from it. Epithelium even covers the outer surface of your eyeballs. On the inside, epithelium lines your body's many tubes and cavities, including blood vessels, airways,

FIGURE 19.3
Animated! General structure of an epithelium.
© Cengage Learning.

free surface of a simple epithelium

basement membrane (material secreted by epithelial cells)

underlying connective tissue

and the passages that make up the digestive, urinary, and reproductive tracts.

We describe an epithelial tissue in terms of the shape of the cells and number of cell layers. A simple epithelium is one cell thick. A stratified epithelium has multiple cell layers. Cells in squamous epithelium are flattened or platelike (Figure 19.4**A**). Cells of cuboidal epithelium are short cylinders that look like cubes when viewed in cross-section (Figure 19.4**B**). Cells in columnar epithelium are taller than they are wide (Figure 19.4**C**).

Different kinds of epithelia are adapted to different tasks. The thinnest type of epithelium, simple squamous epithelium, lines blood vessels and the inner surface of the lungs. It functions in exchange of materials. Gases and nutrients diffuse across it easily. In contrast, thicker stratified squamous epithelium has a protective function. This tissue makes up the outer layer of human skin.

Cells of cuboidal and columnar epithelium function in absorption and secretion of substances. In tissues such as the lining of the kidneys and small intestine, fingerlike projections called microvilli extend from the free surface of epithelial cells. Microvilli increase the surface area across which substances are absorbed. In tissues such as those of the upper airways and oviducts, the free epithelial surface is ciliated. The waving movement of cilia in the airways sweeps mucus with inhaled particles away from the lungs. Cilia in an oviduct move an egg toward the uterus.

Only epithelial tissue contains gland cells. Such cells secrete a substance that functions outside the cell. In most animals, secretory cells are clustered inside glands, which are organs that release substances onto the skin, or into a body cavity or a body fluid. **Exocrine glands** have ducts or tubes that deliver their secretions onto an internal or external surface. Exocrine secretions include mucus, saliva, tears, milk, digestive enzymes, and earwax. **Endocrine glands** have no ducts. They release signaling molecules called hormones into a body fluid. Most commonly, hormones enter the bloodstream, which then distributes them throughout the body.

Adults make few new muscle cells or nerve cells, but they do renew their epithelial cells. For example, each day you lose skin cells, and new ones replace them. An adult sheds about 0.7 kilogram (1.5 pounds) of skin each year. Similarly, the lining of your intestine is replaced every four to six days. All those cell divisions provide lots of opportunities for DNA replication errors that can lead to cancer. Thus, epithelium is the tissue most likely to become cancerous.

> **Connective Tissues** **Connective tissues** are the most abundant and widely distributed vertebrate tissues. They range from soft connective tissues to the more specialized types called bone tissue, cartilage, adipose tissue, and blood. Unlike epithelial cells, connective tissue cells are not tightly packed, but rather are spread out in a secreted extracellular matrix.

Two types of soft connective tissue (loose and dense) include the same components, but in different proportions. **Fibroblasts**, the main cell type in these tissues, make and secrete components of the extracellular matrix (Section 3.6). These components include polysaccharides and collagen, the body's most abundant protein. The extracellular matrix surrounds and supports the tissue's cells.

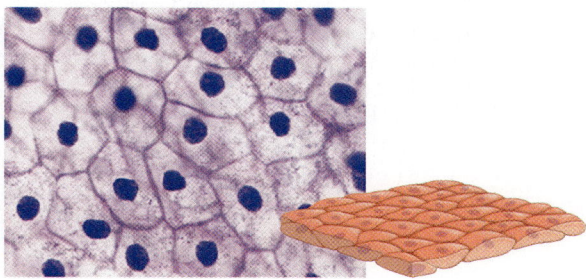

A Simple squamous epithelium
- Lines blood vessels, the heart, and air sacs of lungs
- Allows substances to cross by diffusion

B Simple cuboidal epithelium
- Lines kidney tubules, ducts of some glands, oviducts
- Absorbs, secretes, moves materials

mucus-secreting gland cell

C Simple columnar epithelium
- Lines some airways, parts of the gut
- Absorbs, secretes materials

FIGURE 19.4 Types of simple epithelia, with examples of their functions and locations.

Credits: photos from top, Ray Simmons/ Photo Researchers, Inc.; © Ed Reschki/ Peter Arnold, Inc.; © Don W. Fawcett; art, From RUSSELL/WOLFE/HERTZ/STARR. Biology, 1e. © 2008 Brooks/Cole, a part of Cengage Learning, Inc. Reproduced by permission. www.cengage.com/permissions.

connective tissue Animal tissue with an extensive extracellular matrix; provides structural and functional support.

endocrine gland Ductless gland that secretes hormones into a body fluid.

epithelial tissue Sheetlike animal tissue that covers outer body surfaces and lines internal tubes and cavities.

exocrine gland Gland that secretes milk, sweat, saliva, or some other substance through a duct.

fibroblast Main cell type in soft connective tissue; secretes collagen and other components of extracellular matrix.

Image © Stuart Miles, Used under license from Shutterstock.com.

A Loose connective tissue
• Underlies most epithelia
• Provides elastic support, stores fluid

B Dense, irregular connective tissue
• Deep in skin and around intestine, in kidney capsule
• Binds parts together, provides support, protection

C Dense, regular connective tissue
• Tendons and ligaments
• Stretchable attachment between parts

D Cartilage
• Internal framework of ears, nose, airways; covers ends of bones
• Supports soft tissues, cushions, reduces friction at joints

E Adipose tissue
• Under skin and around the heart and kidneys
• Stores energy-rich lipids, insulates, and cushions parts

F Bone tissue
• Bulk of most vertebrate skeletons
• Protects soft tissues, functions in movement, stores minerals, produces blood cells

G Blood
• Connective tissue with fluid matrix (plasma) and cellular components
• Transports substances, functions in body defenses, and helps maintain temperature

FIGURE 19.5 Animated! Micrographs of connective tissues, and descriptions of their functions in the human body.

Credits: (a) © John Cunningham/ Visuals Unlimited; (b,c) Ed Reschke; (d,g) Photo Researchers Inc.; (e) © University of Cincinnati, Raymond Walters College, Biology; (f) Michael Abbey/ Photo Researchers, Inc.; center, Image © Yuri Arcurs, Used under license from Shutterstock.com.

Loose connective tissue has fibroblasts and collagen fibers scattered widely in its matrix (Figure 19.5**A**). This tissue, the most common type in vertebrate bodies, holds organs and epithelia in place.

Two types of **dense connective tissue** both have an extracellular matrix full of fibroblasts and collagen fibers. In dense, irregular connective tissue, the fibers are oriented every which way, as in Figure 19.5**B**. This tissue makes up deep skin layers. It supports intestinal muscles and also forms capsules around organs that do not stretch, such as kidneys. Dense, regular connective tissue has fibroblasts in orderly rows between parallel, tightly packed bundles of fibers (Figure 19.5**C**). This organization helps keep the tissue from being torn when stretched. Tendons, which connect skeletal muscle to bone, and ligaments, which attach bones to one another, consist of dense, regular connective tissue.

〉 Specialized Connective Tissues All vertebrate skeletons include some **cartilage**, which has an extracellular matrix of collagen fibers and rubbery glycoproteins (Figure 19.5**D**). The skeleton of sharks and other cartilaginous fish consists entirely of cartilage. When you were an embryo, cartilage formed a model for your developing skeleton, then bone replaced most of it. Cartilage still serves as a scaffold for your outer ears, nose, and trachea (windpipe). Pads of cartilage between bones function like shock absorbers.

Blood vessels do not run through cartilage, so cartilage cells depend on diffusion of nutrients and oxygen from vessels in nearby tissues. Also, unlike cells of other connective tissues, cartilage cells seldom divide in adults, so torn cartilage does not repair itself.

Many cells store some fat, but only the cells of **adipose tissue** can bulge with so much fat that cellular components, including the nucleus, get pushed to

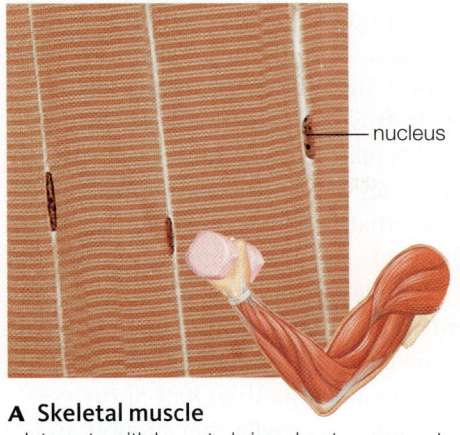

A Skeletal muscle
• Interacts with bone to bring about movement, maintain posture
• Reflex activated, but also voluntarily controlled

adjoining ends of abutting cells

B Cardiac muscle
• Occurs only in the heart wall
• Contraction is not under voluntary control

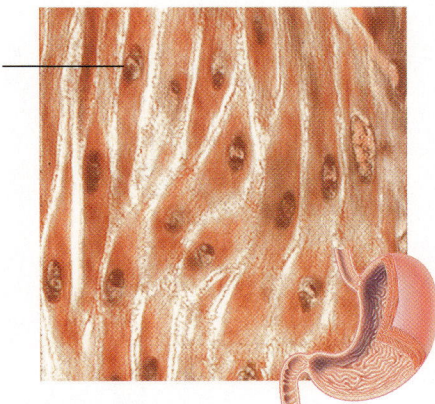

nucleus

C Smooth muscle
• Walls of digestive tract, arteries, reproductive tract, the bladder, other hollow organs
• Contraction is not under voluntary control

FIGURE 19.6 Animated! Micrographs of muscle tissues, and descriptions of their functions.

Credits: left (a,b) Ed Reschke; (c) Biophoto Associates/ Photo Researchers, Inc.; right (a–c) © Cengage Learning.

one side (Figure 19.5**E**). Small blood vessels that run through adipose tissue carry fats to and from cells. In addition to its energy-storage role, adipose tissue cushions body parts, and a layer of this tissue beneath the skin serves as insulation.

Bone tissue is a connective tissue in which a calcium-hardened extracellular matrix surrounds each cell (Figure 19.5**F**). This tissue is the main component of bones, the organs that interact with skeletal muscles to move a body. Bones also support and protect internal organs, and some produce blood cells.

Blood is considered a connective tissue because its cells and platelets descend from cells in bone (Figure 19.5**G**). Red blood cells transport oxygen. White blood cells defend the body against pathogens. Platelets are cell fragments that help blood clot. Plasma, the fluid portion of blood, consists mostly of water. It transports gases, proteins, nutrients, hormones, and other substances.

❯ Muscle Tissues Cells of muscle tissues contract in response to signals from nervous tissue. **Skeletal muscle tissue** attaches to bones and moves the body or its parts. It consists of cylindrical cells called muscle fibers that have multiple nuclei. The fibers parallel one another and have a striated (striped) appearance under the microscope (Figure 19.6**A**). Each fiber contains stacks of many units of contraction, each with the same array of protein filaments that function in muscle contraction. Skeletal muscles contract reflexively, but we can also deliberately cause them to contract when we want to move a body part. That is why skeletal muscles are sometimes described as "voluntary" muscles.

Only the heart wall contains **cardiac muscle tissue** (Figure 19.6**B**). Its cells contract as a unit in response to signals that flow through gap junctions (Section 3.6) between their abutting plasma membranes. Each branching cardiac cell has a single nucleus. It has far more mitochondria than other types of muscle cells, because the heart's constant contractions require a steady supply of ATP. Cardiac muscle tissue also is striated, although less conspicuously so. Cardiac muscle and smooth muscle are both "involuntary" muscle; we cannot make them contract.

We find layers of **smooth muscle tissue** (Figure 19.6**C**) in the wall of many soft internal organs, including the stomach, bladder, and uterus. Among other tasks, smooth muscle contractions propel material through the gut, adjust the diameter of blood vessels, and constrict the pupil of the eye. Smooth muscle cells are unbranched with tapered ends, and a single nucleus. Contractile units are not arranged in a repeating fashion, so smooth muscle is not striated. As in cardiac muscle, gap junctions relay signals between the cells.

adipose tissue Connective tissue with fat-storing cells.
blood Fluid connective tissue with cells that form inside bones.
bone tissue Connective tissue with cells surrounded by a mineral-hardened matrix of their own secretions.
cardiac muscle tissue Striated, involuntary muscle of the heart wall.
cartilage Connective tissue with cells surrounded by a rubbery matrix of their own secretions.
dense connective tissue Connective tissue with many fibroblasts and fibers in a random or a regular arrangement.
loose connective tissue Connective tissue with relatively few fibroblasts and fibers scattered in its matrix.
skeletal muscle tissue Striated, voluntary muscle that interacts with bone to move body parts.
smooth muscle tissue Involuntary muscle that lines blood vessels and hollow organs; not striated.

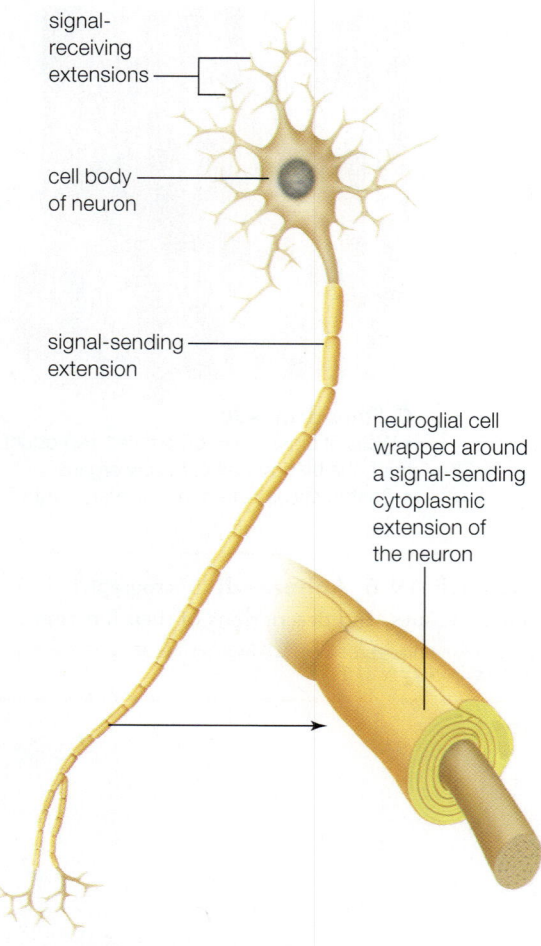

signal-receiving extensions

cell body of neuron

signal-sending extension

neuroglial cell wrapped around a signal-sending cytoplasmic extension of the neuron

FIGURE 19.7 Animated! Motor neuron and associated neuroglial cells.

© Cengage Learning.

> **Nervous Tissue** **Nervous tissue** allows a body to detect and respond to internal and external changes. This tissue detects stimuli, integrates information, and controls the actions of muscles and glands. Figure 19.7 shows a **neuron**, one kind of excitable cell in nervous tissue. In this context, "excitable" means that electrical signals can move along the cell's plasma membrane. Neurons also communicate with one another and with muscles or glands by secreting chemical messages. A neuron has a central cell body that holds the nucleus and other organelles. Thin cytoplasmic extensions project from the cell body. Some of these extensions are remarkably long. For example, cytoplasmic extensions that carry signals between a person's spinal cord and foot can be more than a meter in length. In a whale, the equivalent structure can be tens of meters long.

Neurons are not the only cells in nervous tissue. More than half of the volume of your nervous system consists of cells called neuroglia that support and protect the neurons.

Take-Home Message

What are the features of the four types of tissues?

- Sheetlike epithelial tissue covers the body surface and lines tubes and cavities. Tight junctions connect its cells, which have no matrix between them.

- Connective tissues consist of cells in a secreted extracellular matrix. Loose and dense connective tissues hold body parts in place. Adipose tissue stores fat. Bone and cartilage support body parts and function in movement. Blood is a transport medium with cellular components that were made inside bones.

- Muscle tissue consists of cells that contract. Skeletal muscle moves bones, and you can make it contract at will. Cardiac muscle (found in the wall of the heart) and smooth muscle (in walls of hollow organs) are involuntary, meaning you cannot control their contraction.

- Nervous tissue includes neurons. Neurons relay electrical signals along their plasma membrane and use chemicals to communicate with other cells.

cranial cavity

spinal cavity

thoracic cavity

diaphragm

abdominal cavity

pelvic cavity

FIGURE 19.8 Animated! Body cavities that hold many human organs.

© Cengage Learning.

19.4 Organs and Organ Systems

Tissues are organized into organs. The heart, stomach, and liver are organs, as are eyes, kidneys, and lungs. Most organs include all four types of tissues.

> **Organs in Body Cavities** Many human organs are located inside body cavities. Like other vertebrates, humans are bilateral and have a lined body cavity known as a coelom (Section 15.2). The diaphragm, a sheet of smooth muscle, divides the human coelom into an upper thoracic cavity and a lower cavity that has abdominal and pelvic regions (Figure 19.8). The heart and lungs are in the thoracic cavity. Digestive organs such as the stomach, intestines, and liver lie in the abdominal cavity. The bladder and reproductive organs are in the pelvic cavity. A cranial cavity in the head holds the brain, and a spinal cavity in the back holds the spinal cord. These last two cavities are not derived from the coelom.

> **Skin—Your Largest Organ** Your largest organ is not inside your body. It is the skin that covers your surface (Figure 19.9). The outermost region of skin, the **epidermis**, is a stratified squamous epithelium. In other words, it consists of many layers of squamous (flattened) cells ➊. Ongoing mitotic divisions of cells

epidermis
(stratified
squamous
epithelium)

dermis
mainly dense
connective tis-
sue

hypodermis
mainly adipose
tissue and loose
connective tissue

hair

duct of sweat gland

blood vessel

pressure-sensitive
nerve ending

smooth muscle

sweat gland

hair follicle

oil gland

FIGURE 19.9 Animated! Structure of human skin, your largest organ.

Figure It Out: Is a sweat gland an endocrine gland or an exocrine gland?

© Cengage Learning 2010.   *Answer: An exocrine gland*

in the deepest epidermal layer continually push cells toward the sheet's free surface. Abrasion and the pressure exerted by the growing cell mass beneath older cells flattens and kills them as they move upward. Dead cells at the skin surface get rubbed off or flake away.

The vast majority of epidermal cells are keratinocytes, cells that make filaments of the protein keratin. The presence of large amounts of keratin waterproofs the skin of reptiles, birds, and humans, and is an adaptation to life on land. The epidermis also contains melanocytes, cells that make the brownish protein melanin. Keratinocytes obtain melanin from melanocytes and store it in special organelles. Differences in the size and distribution of these organelles among human ethnic groups contribute to differences in their skin color.

The **dermis** beneath the epidermis consists mostly of dense connective tissue ❷. Nervous tissue threads through the dermis, as do capillaries (small blood vessels). Sweat glands in the dermis are made up of epidermal cells that migrated into the dermis during development. Remember, glands are epithelial tissue.

Epithelial tissue embedded in the dermis also forms hair follicles. The base of a hair follicle holds living hair cells, the fastest-dividing cells in the human body. As these cells divide, they push cells above them up, lengthening the hair. The part of a hair that extends beyond the skin surface is the keratin-rich remains of dead cells. A smooth muscle attaches to each hair and can reflexively pull the hair upright in response to cold or fright. Secretions from oil glands next to each hair follicle keep hair smooth and shiny. Like a sweat gland, an oil gland is epidermal tissue embedded in the dermis.

In most body regions, skin sits atop a layer of loose connective tissue and adipose tissue called the hypodermis ❸. This layer contains larger blood vessels that connect to the small ones running through the dermis.

Components of the skin work together to carry out a variety of tasks. Epidermis is a barrier that prevents pathogens from invading deeper tissues. Nerve endings in skin detect temperature, pressure, and touch and send signals to the spinal cord and brain. Skin also plays a role in temperature regulation.

dermis Deep layer of skin that consists of connective tissue with nerves and blood vessels running through it.

epidermis Outermost, epithelial skin layer.

nervous tissue Animal tissue composed of neurons and supporting cells; detects stimuli and controls responses to them.

neuron Main type of cell in nervous tissue; transmits electrical signals along its plasma membrane and communicates with other cells through chemical messages.

Integumentary System
Protects body from injury, dehydration, and some pathogens; controls its temperature; excretes certain wastes; receives some external stimuli.

Nervous System
Detects external and internal stimuli; controls and coordinates responses to stimuli; integrates all organ system activities.

Muscular System
Moves body and its internal parts; maintains posture; generates heat by increases in metabolic activity.

Skeletal System
Supports and protects body parts; provides muscle attachment sites; produces red blood cells; stores calcium, phosphorus.

Circulatory System
Rapidly transports many materials to and from cells; helps stabilize internal pH and temperature.

Endocrine System
Hormonally controls body functioning; with nervous system integrates short- and long-term activities. (Male testes added.)

Lymphatic System
Collects and returns some tissue fluid to the bloodstream; defends against infection and tissue damage.

Respiratory System
Rapidly delivers oxygen to the tissue fluid that bathes all living cells; removes carbon dioxide wastes of cells; helps regulate pH.

Digestive System
Ingests food and water; mechanically, chemically breaks down food and absorbs small molecules into internal environment; eliminates food residues.

Urinary System
Maintains the volume and solute composition of internal fluids; excretes excess fluid, solutes, and dissolved wastes.

Reproductive System
Male: Produces and transfers sperm to the female. Hormones of both systems also influence other organ systems. Female: Produces eggs; after fertilization, affords a protected, nutritive environment for the development of a new individual.

FIGURE 19.10 Animated! Vertebrate organ systems. © Cengage Learning.

> **Sun, Aging, and the Skin** When skin is exposed to sunlight, melanin production increases, producing a "suntan." The melanin has a protective function. It absorbs ultraviolet (UV) radiation that might otherwise damage underlying skin layers. A bit of UV radiation is a good thing because it stimulates melanocytes to make vitamin D, which is essential to good health. However, excessive UV exposure damages collagen and causes elastin fibers to clump. As a result, chronically tanned skin wrinkles and becomes leathery. UV also damages DNA in skin cells, increasing the risk of skin cancer (Section 6.4).

As we age, epidermal cells divide less often. Skin thins and becomes less elastic as collagen and elastin fibers become sparse. Glandular secretions that kept the skin soft and moist dwindle. Wrinkles deepen. Many people needlessly accelerate the aging process by excessive tanning or by smoking, which shrinks the skin's blood supply.

> **Organ Systems** Skin and structures derived from it such as hair, fur, hooves, claws, nails, and quills constitute a vertebrate's integumentary system, one of many vertebrate organ systems (Figure 19.10).

Organ systems work cooperatively to carry out reproduction and survival tasks. For example, several organ systems interact to provide cells with necessary substances and remove wastes (Figure 19.11). Food and water enter the body by way of the digestive system, which includes digestive organs such as the stomach and intestine, as well as organs that aid digestion such as the pancreas and gallbladder. The digestive system also eliminates undigested waste. Oxygen enters the body by way of the respiratory system, which includes lungs and airways that lead to them. The heart and blood vessels of the circulatory system deliver nutrients and oxygen to cells, and remove waste carbon dioxide and solutes from them. The circulatory system delivers the carbon dioxide to the respiratory system, which eliminates it. The circulatory system also moves excess water, salts, and soluble wastes to the urinary system. Organs of the urinary system include kidneys that filter wastes from the blood and make urine, and a bladder that stores urine until it can be eliminated.

Figure 19.11 does not include the nervous, endocrine, muscular, and skeletal systems, but these too help you obtain essential substances and eliminate wastes. For example, the nervous system detects changes in internal levels of water, solutes, and nutrients. Signals from the nervous and endocrine systems cause the urinary system to conserve or eliminate water. Signals from the nervous system also result in movement, as when your muscles and bones interact to lift you off the couch and move you toward the refrigerator or the toilet.

FIGURE 19.11 Flowchart of links between organ systems. This diagram shows some of the ways that organ systems interact to keep the body supplied with essential substances and eliminate unwanted wastes. Other organ systems that are not shown also take part in these tasks.

© Cengage Learning.

Take-Home Message

What are the roles of organs and organ systems?

- The cells and tissues of an organ are organized to collectively carry out one or more essential tasks. For example, the skin—an organ with all four tissue types—protects the body, helps make vitamin D, detects events in the outside environment, and helps maintain body temperature.

- Organ systems are composed of interacting organs. All vertebrates have the same types of organ systems.

- Survival and reproduction depend on interactions among all organ systems.

19.5 Control of Body Temperature

Homeostasis, again, is the process of keeping conditions in the body within limits cells can tolerate. In vertebrates, homeostasis involves interactions among sensory receptors, the brain, and muscle and glands. A sensory receptor is a cell or cell component that detects a specific stimulus. Sensory receptors involved in homeostasis function like internal watchmen that monitor the body for change. Information from sensory receptors throughout the body flows to the brain. The brain evaluates the incoming information, then signals muscles and glands to take any necessary actions to adjust conditions in the internal environment.

Homeostasis often involves **negative feedback**, a process in which a change causes a response that reverses the change. A heater with a thermostat provides a familiar nonbiological example of how negative feedback works. A person sets the thermostat to a desired temperature. When the temperature falls below this preset point, the heater turns on and emits heat. When the temperature rises to the desired level, the thermostat turns off the heater. Similarly, a negative feedback mechanism keeps your internal temperature near 37°C (98.6°F).

Recall that humans and other mammals are endotherms, which means "heated from within" (Section 15.6). Endotherms produce more metabolic heat and maintain a more stable body temperature than ectotherms such as lizards, whose temperature fluctuates with that of their environment (Figure 19.12).

When you exercise on a hot day, muscle activity generates heat, so your body's internal temperature rises. Sensory receptors in the skin detect the increase and signal the brain. The brain sends signals that bring about the body's response (Figure 19.13). Blood flow shifts, so more blood from the body's hot interior flows to the skin. The shift maximizes the amount of heat given off to the surrounding air. At the same time, sweat glands in the skin increase their output. Evaporation of sweat helps cool the body surface. You breathe faster and deeper, speeding the transfer of heat from the blood in your lungs to the air. Hormonal changes make you feel more sluggish. As your activity level slows and your rate of heat loss increases, your temperature falls.

Despite feedback mechanisms, hyperthermia—a dangerous rise in the body's temperature—kills about 175 Americans each year. Profuse sweating makes the

cooler---------------->warmer

FIGURE 19.12 Endotherm and ectotherm. Photo of a chameleon (an ectothermic lizard) being held by a human (an endotherm). It was taken using heat-sensitive film. The lizard blends into the background because it is about the same temperature as its surroundings. The human is warmer than its surroundings because he or she is producing a large amount of metabolic heat.
NASA/JPL-Caltech/L. Hermans.

FIGURE 19.13 Animated! Negative feedback control of body temperature.
Credits: left, © Photoroller/Shutterstock; right, © Cengage Learning.

Stimulus
Exertion on a hot day raises internal body temperature

Sensory Receptors
Receptors monitor internal temperature and signal the brain when it increases.

Brain
Brain receives signals from sensory receptors and signals muscles and glands.

Muscles and Glands

Skeletal muscles in the chest wall contract more frequently, increasing the rate of breathing.

Smooth muscle in blood vessels supplying the skin relax; more blood flows to skin, and more heat radiates to surrounding air.

Sweat glands secrete more sweat, which cools the body as it evaporates.

Endocrine glands that affect general activity levels slow secretion of hormones that stimulate activity.

Response
Body temperature declines

body lose water and salts, so it can change the concentration of body fluids. Blood flow to the gut and liver decreases. Starved of nutrients and oxygen, these organs release toxins that interfere with the function of the nervous system, as well as other organ systems. As a result, a person may become incapable of recognizing and responding to seemingly obvious signs that something is wrong.

To stay safe outside on a hot day, drink plenty of water and avoid excessive exercise. If you must exert yourself, take breaks to monitor how you feel. Wear light-colored clothing that allows evaporation. Avoid direct sunlight, wear a hat, and use sunscreen. Sunburn impairs the skin's ability to transfer heat to the air.

Receptors in the skin also notify the brain when conditions get chilly. The brain then diverts blood flow away from the skin, lessening movement of heat to the body surface, where it would be lost to the surrounding air. As another response, reflex contractions of smooth muscle in the skin cause hair to stand up, causing "goose bumps." Erect hair insulates better than hair that lies flat. With prolonged cold, the brain sends a signal to skeletal muscles, which respond by contracting ten to twenty times a second. This shivering response increases heat production by muscles, but it has a high energy cost.

Continued exposure to cold can cause hypothermia, in which the body's core temperature falls dangerously low. A decline to 35°C (95°F) alters brain functions. "Stumbles, mumbles, and fumbles" are the symptoms of early hypothermia. Severe hypothermia causes loss of consciousness, disrupts heart rhythm, and can be fatal.

negative feedback A change causes a response that reverses the change.

Take-Home Message

How does your body maintain its temperature?

- Sensory receptors detect a change in body temperature and notify the brain. Signals from the brain then cause muscles and glands to react to this change.
- Temperature regulation involves negative feedback, in which a change triggers a response that reverses the change.

19.6 Stem Cell Studies (revisited)

While some researchers are investigating ways to repair bodies using embryonic stem cells, others are looking into alternative sources of replacement cells. The ideal source of cells would be a patient's own body. Modifying a patient's own cells to provide replacements has two advantages. First, this method would not destroy human embryos. Second, it would minimize the likelihood of rejection of the replacement cells as foreign.

Scientists have recently developed several methods of inducing adult cells to return to an embryo-like pluripotent state. The cells can then be directed to differentiate into whatever type of adult cell a patient requires. Despite this progress, clinical tests of reprogrammed adult cells are probably years away. Before any treatment involving such cells can be tested in humans, it must be proven safe and effective in animal models.

Even so, the ability to induce pluripotency in adult cells of one type and convert them to another is already benefiting medical researchers. For example, it has allowed them to directly study the effects of Parkinson's disease. This disorder kills neurons deep within the brain, where it is difficult to study them. Researchers took skin cells from a patient with an inherited form of Parkinson's disease, made them pluripotent, then converted them to neurons. Growing and studying these neurons in the lab will provide insights into how genetic defects in these cells give rise to symptoms of Parkinson's disease.

where you are going . . .

Chapters 20–26 look at functions of vertebrate tissues and organs in more detail. Among other things, you will learn how epithelium functions in gas exchange (Section 21.7), nutrient absorption (23.3), and urine formation (23.6), and how hormones from endocrine glands affect other cells (Chapter 25). You will find out how muscles contract (20.4) and what causes the heart to beat (21.4). You will discover how blood cells transport gases (21.7) and protect against disease (22.2).

Summary

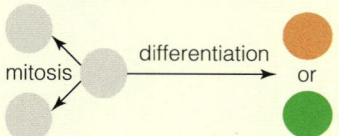

Section 19.1 **Stem cells** can divide to form more stem cells or differentiate to become specialized cells. All body parts develop from embryonic stem cells, which are **pluripotent**. Embryonic stem cells or adult stem cells made to behave like embryonic ones could one day help treat diseases and disorders.

Section 19.2 Animal bodies are structurally and functionally organized on several levels. A **tissue** is a group of cells and intercellular substances that perform a common task. Tissues make up **organs**, which interact as **organ systems**. Activities at all levels contribute to **homeostasis**, the process of maintaining conditions inside the body within tolerable limits.

Animal structure and function have a genetic basis and are shaped by natural selection. However, because evolution modifies existing structures, body parts often are less than optimal for a task.

Section 19.3 Most animals have four types of tissue. **Epithelial tissues** cover external body surfaces and line internal cavities and tubes. Epithelium has one free surface and one surface glued to an underlying tissue by its secretions. Tight junctions connect adjacent cells. **Endocrine glands** and **exocrine glands** are derived from epithelial tissue, as are hair and nails.

Connective tissues are the body's most abundant tissues. **Loose connective tissue**, which holds organs in place and underlies skin, and **dense connective tissue**, which makes up tendons and ligaments, have protein fibers, **fibroblasts**, and other cells in an extracellular matrix. Rubbery **cartilage**, fat-storing **adipose tissue**, mineralized **bone tissue**, and **blood** are specialized connective tissues.

Muscle tissues contract (shorten) when stimulated. **Skeletal muscle** moves body parts and is under voluntary control, whereas **smooth muscle** and **cardiac muscle** are not.

Neurons serve as communication lines in **nervous tissue**. They relay electrical signals along their plasma membrane and send and receive chemical signals. Cells called neuroglia support the neurons.

Section 19.4 An organ is a structural unit of different tissues combined in definite proportions and patterns that allow them to perform a common task. Skin is an organ with a deep **dermis** layer covered by **epidermis**. It functions in protection, temperature control, detection of shifts in the environment, vitamin D production, and defense. An organ system consists of two or more organs interacting in tasks that keep individual cells as well as the whole body functioning.

Section 19.5 Temperature regulation is an aspect of homeostasis. Human temperature regulation involves sensory receptors that detect changes, a brain that receives signals from receptors and coordinates responses, and muscles and glands that carry out responses. It is an example of **negative feedback** in which a change brings about a response that reverses the change.

Self-Quiz   Answers in Appendix I

1. _____ tissues are sheetlike with one free surface.
   a. Epithelial   c. Nervous
   b. Connective   d. Muscle

2. _____ connect(s) epithelium to underlying connective tissue.
   a. Gap junctions   c. Keratinocytes
   b. Basement membrane   d. Plasma membrane

3. Most animals have glands derived from _____ tissue.
   a. epithelial   c. muscle
   b. connective   d. nervous

4. Only _____ cells have cilia or microvilli at their surface.
   a. epithelial   c. muscle
   b. connective   d. nervous

5. The most abundant protein in the human body is _____ , made by fibroblasts.
   a. collagen   c. melanin
   b. keratin   d. hemoglobin

6. _____ consists mainly of plasma.
   a. Loose connective tissue   c. Cartilage
   b. Blood   d. Bone

7. Your body converts excess carbohydrates and proteins to fats that accumulate in _____ .
   a. fibroblasts   c. adipose tissue cells
   b. neurons   d. melanocytes

8. _____ tissues are the body's most abundant and widely distributed tissue.
   a. Epithelial   c. Nervous
   b. Connective   d. Muscle

9. Cells of _____ can shorten (contract).
   a. epithelial tissue   c. muscle tissue
   b. connective tissue   d. nervous tissue

10. _____ muscle tissue has a striped appearance and is under voluntary control.
   a. Skeletal   c. Cardiac
   b. Smooth   d. a and c

11. _____ detects and integrates information about changes and controls responses to those changes.
   a. Epithelial tissue   c. Muscle tissue
   b. Connective tissue   d. Nervous tissue

12. Thin cytoplasmic projections from cells called _____ carry signals between your spinal cord and your toes.
   a. neurons   c. fibroblasts
   b. neuroglial cells   d. keratinocytes

Digging Into Data

Growing Skin To Heal Wounds

Skin cells are relatively easy to grow in culture and are already in wide use for medical treatments. One product is produced using infant foreskins that were removed during routine circumcisions. The foreskin (a tissue that covers the tip of the penis) provides a rich source of keratinocytes and fibroblasts. These cells are grown in culture with other biological materials (Figure 19.14**A**).

The resulting product can be used to close chronic wounds such as those that occur as a result of diabetes. When placed over a wound, the cells release growth factors and other substances that encourage healing.

Diabetes is a disorder in which the blood sugar level is not properly controlled. Among other complications, this disorder reduces blood flow to the lower legs and feet. As a result, about 3 million diabetes patients have ulcers—open wounds that do not heal—on their feet. Each year, about 80,000 diabetics require amputations.

Figure 19.14**B** shows the results of a clinical experiment that tested the effect of a cultured skin product versus standard treatment for diabetic foot wounds. Patients were randomly assigned either to the experimental treatment group or to the control group. Their progress was monitored for 12 weeks.

1. What percentage of wounds had healed at 8 weeks when treated the standard way? When treated with cultured skin?

2. What percentage of wounds had healed at 12 weeks when treated the standard way? When treated with cultured skin?

3. How early was the healing difference between the control and treatment groups obvious?

FIGURE 19.14 Use of cultured skin cells.

Credits: (a) © Organogenesis, Inc., www.organo.com; (b) © Cengage Learning.

A Apligraf, a living cellular construct with a two-layered structure. The top layer is keratinocytes, and the lower layer is fibroblasts.

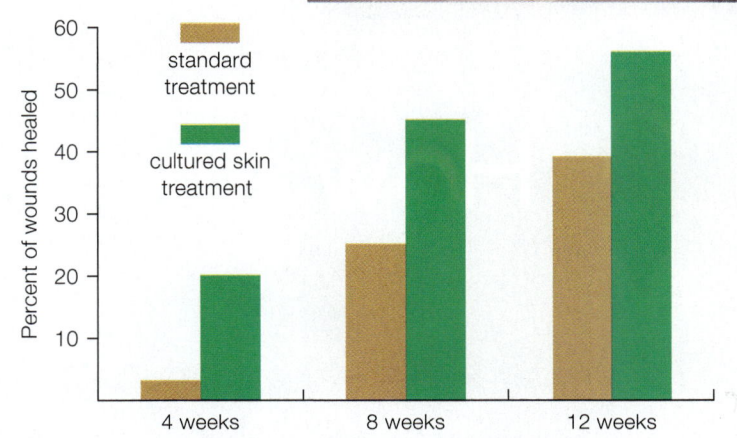

B Results of a multicenter study of the effects of standard treatment versus use of a cultured cell product for diabetic foot ulcers. Bars show the percentage of foot ulcers that had completely healed.

---

13. The functions of skin include _____ .
    a. defense against pathogens
    c. production of vitamin D
    b. helping to cool the body
    d. all of the above

14. When the level of sugar in your blood falls too low, your body senses this decline and converts glycogen to sugar, putting more sugar into your blood. This is an example of _____ .
    a. differentiation
    c. homeostasis
    b. negative feedback
    d. b and c

15. Match the terms with the most suitable description.
    _____ exocrine gland
    _____ endocrine gland
    _____ epidermis
    _____ dermis
    _____ smooth muscle
    _____ cardiac muscle
    _____ blood
    _____ melanin
    _____ collagen
    _____ cartilage

    a. outermost skin layer
    b. secretes through duct
    c. in heart only
    d. support in ears and nose
    e. contracts, not striated
    f. mainly loose connective tissue
    g. plasma, platelets, and cells
    h. ductless hormone secretor
    i. increases with sun exposure
    j. secreted by fibroblasts; most abundant protein in a human body

Critical Thinking

1. Leukemia is a cancer in which the body makes too many white blood cells. Affected people are sometimes treated with a bone marrow transplant. Radiation or chemotherapy drugs are used to kill the existing bone marrow, then a bone marrow transplant replaces it. Why would a bone marrow transplant affect the production of white blood cells?

2. Radiation and chemotherapy drugs preferentially kill cells that divide frequently, most notably cancer cells. These treatments also cause hair to fall out and disrupt gut function. Why?

3. Each level of biological organization has emergent properties that arise from the interaction of its component parts. For example, cells have a capacity for inheritance that the molecules making up the cell do not. Can you think of an emergent property of a tissue? Of an organ that contains that tissue?

4. The micrograph to the *right* shows cells from the lining of an airway leading to the lungs. The *gold* cells are ciliated and the *darker brown* ones secrete mucus. What type of tissue is this? How can you tell?

CNRI/ Photo Researchers, Inc.

20 How Animals Move

This chapter expands upon the introduction to connective and muscle tissues in Section 19.3. It draws on earlier discussions of motor proteins (3.5), energy-releasing pathways (5.6–5.7), knockout experiments (7.7), and muscular dystrophy (9.7).

20.1 Bulking Up Muscles

Like neurons, skeletal muscle cells seldom divide after birth, so muscles bulk up mainly by enlarging existing cells. Inside each muscle cell, protein filaments involved in muscle contraction are continually built up and broken down. Exercise tilts this process in favor of synthesis, so muscle cells get bigger and the muscle they are in gets stronger. Your body is like a machine that improves with use. The more you use your muscles, the stronger you become.

Hormones can also encourage increases in muscle mass. For example, the sex hormone testosterone causes muscle cells to build more proteins. Women make a little bit of testosterone, but men make much more, which is why men tend to be more muscular than women. Human growth hormone also stimulates synthesis of muscle proteins. Taking synthetic versions of these muscle-building, or "anabolic," hormones can bulk up muscles, but most sport organizations consider use of these drugs to be cheating.

Some people have a natural genetic advantage when it comes to putting on muscle. Liam, a young Michigan boy, is one of them (Figure 20.1A). His body does not respond normally to myostatin, a regulatory protein that slows synthesis of muscle proteins. Without myostatin's inhibitory effect, an individual has larger-than-normal muscles and unusual strength. By Liam's first birthday, he could do chin-ups. By age 3, he could lift 5-pound (2.3-kilogram) weights and had 40 percent more muscle than an average child his age.

Additional evidence that myostatin level affects athletic ability comes from a study of whippets, a type of dog bred for racing. Some whippets are homozygous for a mutation that prevents them from making myostatin. Such dogs, called bully whippets, are unusually heavily muscled compared to a normal whippet (Figure 20.1B). Bully whippets do not match the breeders' ideal of how a whippet should look, so they are not usually raced or bred. However, dogs that are heterozygous for the mutated allele do race. Compared to normal whippets, they make less myostatin, are more muscular, and are more likely to win races.

Several drug companies are working to develop chemicals that inhibit myostatin production or activity. The goal is to come up with treatments for muscle-wasting disorders such as muscular dystrophy. Will some athletes use such drugs to push their bodies beyond normal limits? No doubt they will. Nutritional supplements that purport to bind to myostatin and reduce its activity are already popular. In clinical tests, these supplements had no effect on strength, but hope for an extra edge keeps moving them off the shelf.

A Liam, a Michigan boy whose muscles do not respond to myostatin. As a result he is unusually strong for his age.

B Bully whippet (*above*). Such dogs are homozygous for a mutation that prevents myostatin production. Compare its bulky muscles to those of the typical whippet shown at the *right*.

FIGURE 20.1 Effects of disrupted myostatin function. Myostatin is a regulatory protein that inhibits synthesis of muscle proteins.

Credits: (a) The Muskegon Chronicle; (b) top, © Stuart Isett; bottom, © Lakatos Sandor/ Shutterstock.

FIGURE 20.2 Hydrostatic skeleton of an earthworm. Muscle contractions change the shape of an earthworm's body segments, just as squeezing a water-filled balloon alters its shape. In each body segment, muscles exert their force against the fluid trapped inside that segment. The fluid in all of the segments collectively constitutes a worm's hydrostatic skeleton.

Credits: (left) © Cengage Learning; right, © Johnna Goodyear/ Shutterstock.

20.2 Skeletal Systems

> **Types of Skeletons** Muscles bring about movement of a body or its parts by interacting with a skeleton. Soft-bodied invertebrates typically have a **hydrostatic skeleton**, an internal, fluid-filled chamber or chambers that muscles exert force against. For example, earthworms have a coelom divided into many fluid-filled chambers, one per segment (Section 15.3). Muscles alter the shape of a body segment by squeezing its fluid-filled chamber (Figure 20.2). By analogy, think about how squeezing a water-filled balloon changes the balloon's shape. Coordinated changes in the shape of the earthworm's many body segments allow it to move through the soil.

In animals with an **exoskeleton**, a cuticle, shell, or other secreted hard external body part receives the force of a muscle contraction. For example, muscles attached to a the cuticle of a fly's thorax change the shape of the thorax, thus causing the attached wings to flap up and down (Figure 20.3). The arthropod exoskeleton also protects the soft body tissues inside it. However, such a skeleton has some drawbacks. Because the exoskeleton consists of noncellular secreted material, it cannot grow with the animal. As an arthropod grows, it must periodically molt its old skeleton and replace it with a larger one. Repeatedly producing new exoskeleton uses resources that could otherwise be put to use in growth or reproduction. In addition, after an arthropod molts, it remains highly vulnerable to predators until its new exoskeleton hardens.

An **endoskeleton** is an internal framework of hardened elements to which muscles attach. Skeletal elements of an endoskeleton can grow with the animal. Echinoderms such as sea stars have an endoskeleton, as do all vertebrates. The term "vertebrate" refers to the **vertebral column**, or backbone, a skeletal feature common to all members of this group (Section 15.4). The vertebral column, along with the bones of the head and rib cage, constitute the axial skeleton. The appendicular skeleton consists of the pectoral (shoulder) girdle, the pelvic (hip) girdle, and limbs (or bony fins) attached to them (Figure 20.4).

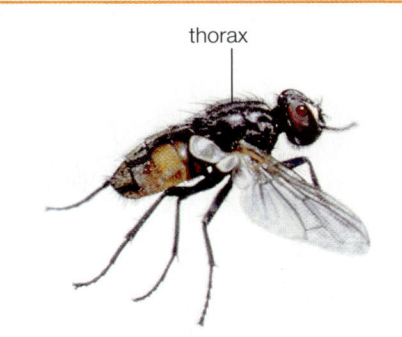

thorax

FIGURE 20.3 Fly exoskeleton. A fly's cuticle is an external skeleton. Muscles inside a fly's thorax pull on the cuticle of the thorax, thus changing its shape. The muscle-driven changes in thorax shape cause the fly's wings to pivot up and down.

Stephen Dalton/ Photo Researchers, Inc.

endoskeleton Hard internal parts that muscles attach to and move.
exoskeleton Hard external parts that muscles attach to and move.
hydrostatic skeleton Fluid-filled chamber or chambers that muscles act on, redistributing the fluid.
vertebral column Backbone.

FIGURE 20.4 Features of the vertebrate skeleton.

■ axial skeleton   ■ appendicular skeleton

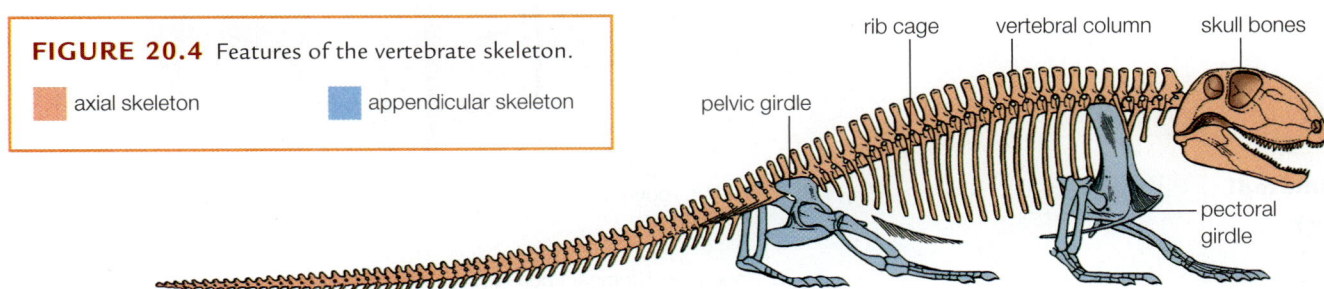

rib cage   vertebral column   skull bones

pelvic girdle

pectoral girdle

© Cengage Learning.

Skull
cranial bones
facial bones

Pectoral Girdle
clavicle (collarbone)
scapula (shoulder blade)

Rib Cage
sternum (breastbone)
ribs (12 pairs)

Upper Limb Bones
humerus (upper arm bone)
ulna (forearm bone)
radius (forearm bone)
carpals (wrist bones)
metacarpals (palm bones)
phalanges (finger bones)

Vertebral Column
vertebrae
intervertebral disk (cartilage)

Pelvic Girdle
(6 fused bones)

Lower Limb Bones
femur (thighbone)
patella (kneecap)
tibia (lower leg bone)
fibula (lower leg bone)
tarsals (ankle bones)
metatarsals (sole bones)
phalanges (toe bones)

A Bones of the human skeleton

nutrient canal
location of yellow marrow
compact bone tissue
spongy bone tissue

B Structure of a femur

spongy bone tissue
compact bone tissue
blood vessel
outer layer of dense connective tissue

C Spongy and compact bone tissue

> **The Human Skeleton** For a closer look at vertebrate skeletal features, consider the human skeleton (Figure 20.5**A**). The flattened cranial bones of the skull fit together to form a braincase that surrounds and protects the brain. The skull's facial bones include cheekbones and other bones around the eyes, the bone that forms the bridge of the nose, and bones of the jaw.

The vertebral column extends from the base of the skull to the pelvic girdle, and consists of 23 bones called **vertebrae**. Thick pads of cartilage, called **intervertebral disks**, separate adjoining vertebrae. The spinal cord runs through the backbone and connects with the brain by way of an opening at the base of the skull. Viewed from the side, our backbone has an S shape. This curvature keeps our head and torso centered over our feet and is an adaptation to upright walking (Section 15.7).

The rib cage attaches to the vertebral column. Both males and females have twelve pairs of ribs. Ribs and the breastbone, or sternum, form a protective cage around the heart and lungs.

The scapula (shoulder blade) and clavicle (collarbone) are bones of the human pectoral girdle. The upper arm has one bone, the humerus. The forearm has two bones, the radius and ulna. The wrist and palm each have multiple small bones, as does each finger.

The pelvic girdle protects internal organs and supports the weight of the upper body when you stand upright. It consists of two sets of fused bones, one set on each side of the body.

The largest bone in the body is the femur, or thighbone. It attaches to the lower leg bones, the tibia and fibula, at the knee. The knee is protected by the patella (kneecap). The smaller bone of the lower leg, the fibula, is not necessary for normal function. When a person has cancer of the jaw, reconstructive surgeons sometimes remove the cancerous jawbone and replace it with bone from the person's fibula. The ankle consists of multiple bones, as do the sole of the foot and each toe.

> **Bone Structure and Function** A bone is an organ. An outer sheath of connective tissue covers the bone, and nerves and blood vessels extend through tiny passageways in the bone's interior. Bone tissue itself consists of bone cells in a secreted extracellular matrix of collagen hardened with calcium and phosphorus.

FIGURE 20.5 Animated!
Human skeletal anatomy.
© Cengage Learning.

There are two types of bone. **Compact bone** makes up the outermost, weight-bearing part of a limb bone such as a femur (Figure 20.5**B**,**C**). Compact bone consists of many thin, concentric layers of extracellular matrix surrounding canals through which nerves and blood vessels that service living bone cells run. The shaft and ends of a femur contain **spongy bone**, which has internal spaces. **Red marrow** fills spaces in spongy bone of limbs and some other bones, such as the breastbone. Blood cells form mainly in red marrow. **Yellow marrow**, consisting mostly of adipose cells, fills the central cavity of adult limb bones.

Bones contain the body's main reserve of calcium and phosphorus. Ongoing mineral deposits to bone and removals from it help maintain the blood concentrations of calcium and phosphorus ions. Because calcium ions play a role in nerve cell function, muscle contraction, and other important processes, the calcium concentration in the blood is tightly controlled.

Some cells in bone continually secrete extracellular matrix and other bone cells break it down to release calcium. Until people reach their mid-twenties, the rate of matrix deposition exceeds the rate of matrix breakdown. As people age, matrix production slows and bone density declines. Osteoporosis is a disorder in which bone loss greatly outpaces bone formation, so bones become weaker and more likely to break. Osteoporosis most commonly occurs in postmenopausal women who no longer produce sex hormones that encourage bone deposition. However, about 20 percent of osteoporosis cases occur in men. To reduce risk of osteoporosis, ensure that your diet provides plenty of calcium and vitamin D, which the body needs to absorb calcium from food.

〉 Where Bones Meet—Skeletal Joints

A **joint** is an area where bones come together. Connective tissue holds bones securely in place at **fibrous joints** such as those that hold cranial bones together. Pads or disks of cartilage connect bones at **cartilaginous joints**. This flexible connection allows a bit of movement. Cartilaginous joints connect vertebrae to one another and connect some of the ribs to the sternum (breastbone). At **synovial joints**, the most common type of joint, the ends of the bones are covered with cartilage and enclosed in a fluid-filled capsule. Different kinds of synovial joints allow different kinds of movements. Ball-and-socket joints at the shoulders and hips allow rotational motion. At other joints, including some in the wrists and ankles, bones glide over one another. Joints at elbows and knees function like a hinged door, allowing the bones to move back and forth in one plane only.

Ligaments, which are cords of dense connective tissue, hold the bones of a synovial joint in place. For example, short cruciate ligaments cross one another in the center of the knee joint (Figure 20.6). A completely torn cruciate ligament may allow bones to shift, so the knee gives out when a person tries to stand. Such tears must be repaired surgically.

FIGURE 20.6 Anatomy of the knee, a hinge-type synovial joint. Bones are held in place by ligaments. The knee is also stabilized by wedges of cartilage called menisci (singular, meniscus). A torn ligament or damaged meniscus can impair knee function.

© Cengage Learning.

Take-Home Message

What are the features of animal skeletons?

■ Animal skeletons consist of hard structural elements or fluid-filled chambers that muscle contractions act upon.

■ Humans have a bony endoskeleton with some features such as an S-shaped backbone that are adaptations to upright posture and walking.

■ Bones are wrapped in connective tissue and have marrow in their interior. They are continually remodeled; minerals are removed and added as needed.

■ Bones meet at joints. At synovial joints, the most common type, bones are held in place by ligaments.

cartilaginous joint Joint where cartilage holds bones together and provides cushioning, as between vertebrae.

compact bone Dense bone with concentric layers of matrix.

fibrous joint Joint where dense connective tissue holds bones firmly in place.

intervertebral disk Cartilage disk between two vertebrae.

joint Region where bones come together.

ligament Strap of dense connective tissue that holds bones together at a joint.

red marrow Bone marrow that makes blood cells.

spongy bone Lightweight bone with many internal spaces.

synovial joint Joint such as the knee that is lubricated by fluid and allows movement of bones around the joint.

vertebrae Bones of the backbone (vertebral column).

yellow marrow Bone marrow that is mostly fat; fills cavity in most long bones.

20.3 How Bones and Muscles Interact

Muscles can only pull on bones. They cannot push.

Skeletal muscles are the functional partners of bones. A sheath of dense connective tissue encloses each muscle and extends beyond it as a cordlike or straplike **tendon**. Most often, one or more tendons attach each end of the muscle to a bone. Figure 20.7 shows the two muscles of the upper arm: the biceps and the triceps. Biceps means "two-headed" in Latin and refers to the fact that the upper portion of the biceps attaches to the scapula (shoulder blade) by way of two tendons ❶. At the opposite end of the muscle, a single tendon attaches the biceps to the radius in the forearm ❷.

Muscles can only exert force by pulling; they cannot push. Often two muscles work in opposition, meaning the action of one resists or reverses the action of another. For example, the triceps in the upper arm opposes the biceps.

When you pull your arm toward your shoulder, the triceps muscle relaxes as the biceps contracts. Contraction of the triceps coupled with relaxation of the biceps reverses this movement, extending the arm.

You can feel the biceps contract if you extend your arm outward, place your other hand over the biceps, then bend the arm at the elbow to bring your hand to your shoulder. Although the biceps shortens only about a centimeter, the hand moves through a much greater distance.

An elbow functions like a lever, a mechanism in which a rigid structure pivots about a fixed point. The bones are the rigid structures and the joint the fixed point. Use of a lever allows a small force, such as that exerted by a contracting biceps, to overcome a larger one, such as the gravitational force acting on the forearm.

Bear in mind that only skeletal muscle is the functional partner of bone. As Section 19.3 explained, smooth muscle is a component of some blood vessels and of the stomach, bladder, and other internal organs. Cardiac muscle is present only in the heart wall. We consider the functions of smooth muscle and cardiac muscle in later chapters.

The human body has close to 700 skeletal muscles, some near the surface, others deep inside the body wall. Collectively, skeletal muscles account for about 40 percent of the body weight of a young man of average fitness. Most skeletal muscles move bones, but some have other functions. Skeletal muscles that pull on facial skin change your facial expression. Others attach to and move the eyeball, or open and close eyelids. The tongue is a skeletal muscle, and sphincters of skeletal muscle allow voluntary control of defecation and urination.

tendons

biceps

radius

tendon

triceps

FIGURE 20.7 Animated! Muscles of the human upper arm.

❶ Two tendons attach the upper end of the biceps muscle to the scapula (not shown).

❷ A tendon at the lower end of the biceps attaches to the radius.

Contraction of the biceps pulls the forearm toward the shoulder. Contraction of the triceps muscle reverses this action. Thus, the biceps and triceps muscles work in opposition.

© Cengage Learning 2010.

Take-Home Message

How do skeletal muscles work together with bones?

- Tendons of dense connective tissue attach skeletal muscles to bones.
- When skeletal muscle contracts, it pulls on the bone to which it is attached.
- Skeletal muscles often work in opposition, with the action of one muscle opposing or reversing the action of another.

FIGURE 20.8 Animated! Structure of a vertebrate skeletal muscle, such as the quadriceps, the largest muscle in the thigh.

Credits: Left, © Frank B. Yuwono/ Shutterstock; Below, (a,b,c bottom) © Cengage Learning; (c top) © Don Fawcett/ Visuals Unlimited.

A Bundle of muscle fibers (cells) surrounded by connective tissue

B One myofibril, consisting of sarcomeres arranged end to end

sarcomere — sarcomere

Z line   Z line   Z line

C One sarcomere, with parallel arrays of thin myosin filaments and thick actin filaments

thick filament (myosin)

Z line   thin filaments (actin)   Z line

20.4 Skeletal Muscle Structure and Function

Skeletal muscles are the voluntary muscles that allow animals to move themselves or their parts. A skeletal muscle consists of connective tissue–wrapped bundles of cells called muscle fibers (Figure 20.8**A**). In addition to multiple nuclei and other organelles, a muscle fiber holds many **myofibrils**, threadlike, cross-banded structures that parallel the muscle's long axis (Figure 20.8**B**).

Each myofibril consists of many individual contractile units called **sarcomeres** arranged end to end. A sarcomere is tiny—shorter in length than a mitochondrion. When a myofibril is viewed under a microscope, dark-staining regions called Z lines define the ends of each sarcomere (Figure 20.8**C**). Between its Z lines, a sarcomere contains thin and thick protein filaments arrayed parallel to one another and to the muscle's long axis. Two coiled strands of the globular protein **actin** are the main components of each thin filament. **Myosin**, a motor protein (Section 3.5) with a club-shaped head, makes up thick filaments. The myosin heads of the thick filaments are just a few nanometers away from the thin actin filaments, which have sites that can bind myosin.

Muscle bundles, muscle fibers, myofibrils, and thick and thin filaments of a sarcomere all have the same parallel orientation. As a result, they all pull in the same direction when a muscle contracts.

actin Globular protein that makes up thin filaments in muscle fibers.
myofibrils Threadlike, cross-banded skeletal muscle components that consist of sarcomeres arranged end to end.
myosin Motor protein that makes up the thick filaments of muscle fibers.
sarcomere Contractile unit of skeletal and cardiac muscle.
tendon Strap of dense connective tissue that connects a skeletal muscle to bone.

relaxed sarcomere   contracted sarcomere

FIGURE 20.9 **Animated!** How a sarcomere changes during muscle contraction. Thick myosin filaments are fixed in place. The thin actin filaments are pulled from either side of the sarcomere toward its center, so they overlap more with thick filaments. Neither thick nor thin filaments change length.

© Cengage Learning.

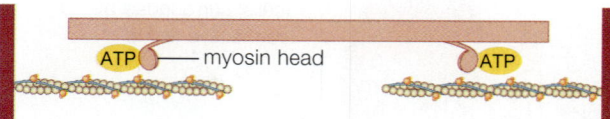

1 A relaxed sarcomere, in which myosin-binding sites on actin are blocked. Myosin heads have bound ATP and are in their low-energy state.

2 The myosin heads hydrolyze ATP to ADP and P_i. The heads are now in a high-energy state, ready to interact with actin, but the myosin-binding sites on actin are blocked.

3 When a nervous signal excites the muscle, myosin-binding sites open up. Myosin binds to actin, releasing bound phosphate and forming a cross-bridge between thin and thick filaments.

4 Release of phosphate triggers a power stroke. Myosin heads contract and pull the bound thin filaments inward.

5 When ATP binds to myosin, myosin releases its grip on actin and returns to its relaxed state, ready to act again.

FIGURE 20.10 **Animated!** Actin–myosin interaction during contraction. For simplicity we show only part of a sarcomere, with one thick filament and two thin ones.

© Cengage Learning.

> **How a Muscle Contracts** The **sliding-filament model** explains the process by which changes in a muscle's component sarcomeres cause it to contract. During muscle contraction, the length of thick and thin filaments does not change, and the thick filaments do not change position, but thin filaments slide inward (Figure 20.9). As the Z lines attached to the thin filaments are drawn closer together, the sarcomere shortens.

Let's take a closer look at the molecular basis for sarcomere contraction (Figure 20.10). When a sarcomere is relaxed, myosin cannot bind to actin because other components of the thin filament block the myosin-binding sites. The myosin can, however, bind ATP **1**. A myosin molecule with bound ATP is in a low-energy state. Converting ATP to ADP and phosphate (P_i) energizes the myosin head in a manner analogous to stretching a spring **2**. When a signal from a motor neuron excites a muscle, calcium ions are released from specialized endoplasmic reticulum in muscle fibers. The calcium causes proteins of the thin filaments to shift, allowing myosin heads to bind to actin. Binding of myosin creates a cross-bridge (an attachment) between the thin and thick filaments and causes release of P_i **3**. Release of P_i from the myosin head triggers the power stroke **4**. Like a stretched spring returning to its original shape, a myosin head snaps back toward the sarcomere center. As the head moves, it pulls the attached thin filament along and releases the bound ADP. As thin filaments are pulled along thick ones toward the sarcomere center, the sarcomere shortens.

A new molecule of ATP can now bind to the myosin head **5**. When it does, the cross-bridge between the thin and thick filaments breaks. The breaking of one cross-bridge does not allow a thin filament to slip backward because other cross-bridges hold the thin filament in place. During a contraction, each myosin head repeatedly binds, moves, and releases an adjacent thin filament.

> **Energy for Contraction** Muscle contraction is work and so it requires energy. A muscle fiber first uses its stored ATP, but it typically stores just enough ATP for a few seconds of contraction. Once that ATP is used up, the muscle fiber turns to its store of a small polypeptide called creatine phosphate (Figure 20.11). Phosphate transfers from creatine phosphate to ADP produce ATP, and keep contractions going for another 5 to 10 seconds. Some athletes take creatine supplements to increase the amount of creatine phosphate in muscles. Supplements can improve performance of tasks, such as weight lifting, that involve quick contraction, but excessive amounts may harm the kidneys.

We categorize muscle fibers based on how they produce most ATP to fuel their contractions. Red fibers have an abundance of large mitochondria and produce ATP mainly by aerobic respiration. They are colored bright red by myoglobin, an iron-containing protein that, like the blood

pathway 1
dephosphorylation of creatine phosphate

ADP + P$_i$

creatine

ATP

pathway 2
aerobic respiration

pathway 3
lactate fermentation

oxygen

glucose from bloodstream and from glycogen breakdown in cells

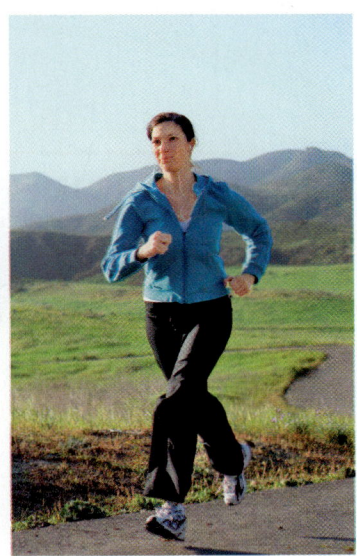

protein hemoglobin, reversibly binds oxygen. During periods of muscle activity, myoglobin releases bound oxygen, allowing aerobic respiration to continue even if blood flow is insufficient to meet the muscle's oxygen need. By contrast, white fibers have no myoglobin and fewer and smaller mitochondria. White fibers make ATP mainly by lactate fermentation (Section 5.7). The proportion of fiber types differs among muscles. Red fibers fatigue less easily than white fibers, so muscles involved in sustained activities tend to include more red fibers than those that typically contract only briefly. For example, muscles that move your arms and legs contain more red fibers than those that move your eyeballs.

For any given muscle, the relative proportions of the two fiber types vary among individuals. This variation arises as a result of both genetic and environmental factors. A predisposition to have more or less red fibers is inherited, but exercise can encourage conversion of white fibers to red ones.

Take-Home Message

How does skeletal muscle contract?

- A skeletal muscle shortens through combined decreases in the length of its numerous sarcomeres. Sarcomeres are the basic units of skeletal and cardiac muscle contraction.

- The parallel orientation of a skeletal muscle's components directs the force of contraction toward a bone.

- Interactions between myosin and actin filaments in the many sarcomeres of a muscle cell collectively bring about muscle contraction.

- Muscle contraction requires ATP, which can be provided by several different metabolic pathways.

20.5 Muscles and Health

› Effects of Exercise and of Inactivity When unrelenting stimulation keeps a skeletal muscle excited, muscle fatigue occurs, meaning the muscle's capacity to generate force declines despite ongoing stimulation. After a few minutes of rest, the fatigued muscle will contract again in response to stimulation.

Aerobic exercise—low intensity, but long duration—makes skeletal muscles more resistant to fatigue. It increases the blood supply to a muscle by encouraging the growth of new capillaries. It increases the number of mitochondria and

sliding-filament model Explanation of how interactions among actin and myosin filaments shorten a sarcomere and bring about muscle contraction.

the amount of myoglobin in existing red muscle fibers. As noted above, it also encourages conversion of white fibers to red ones.

Regular episodes of brief, intense exercise such as weight lifting result in increased synthesis of actin and myosin filaments. The increase helps a muscle exert more tension but does not improve endurance.

Exercise also benefits cardiovascular health by encouraging production of a muscle enzyme called lipoprotein lipase (LPL). This enzyme allows the muscle to take up fatty acids and triglycerides from the blood. Mutations that decrease LPL level are associated with a higher-than-normal blood lipid concentration and a heightened risk of cardiovascular disease. Muscle inactivity, as occurs during sitting, turns down LPL production with the same effect. Apparently, the sustained inactivity of leg muscles causes these muscles to transcribe less LPL mRNA. The inactivity-induced changes in LPL expression may contribute to the heightened risk of cardiovascular disease associated with prolonged television viewing (Figure 20.12). The inactivity-induced decrease in LPL appears to persist even if the sitter also gets regular exercise. To prevent the health problems associated with low LPL, avoid sitting for hours at a time. When you must sit, take periodic short breaks to stand up or walk. Any activity that requires your leg muscles to support your weight encourages the sustained production of LPL.

❯ Impaired Muscle Contraction Muscular dystrophies are genetic disorders in which skeletal muscles progressively weaken. With Duchenne muscular dystrophy, symptoms begin to appear in childhood. A mutation of a gene on the X chromosome causes this disorder. The affected gene encodes a protein (dystrophin) in the plasma membrane of muscle fibers. Having a defective form of this protein causes muscle fibers to break down (Figure 20.13).

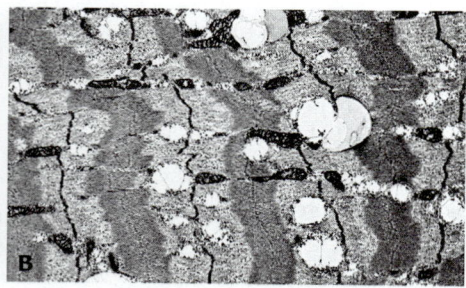

FIGURE 20.13 Electron micrographs of **A** normal skeletal muscle tissue and **B** muscle tissue of a person affected by muscular dystrophy.

Images courtesy of Department of Pathology, The University of Melbourne.

About 1 in 3,500 males are born with Duchenne muscular dystrophy. Like other X-linked disorders, it rarely causes symptoms in females, who nearly always have a normal version of the gene on their other X chromosome. Affected boys usually begin to weaken by the time they are three years old, and require a wheelchair in their teens. Eventually, skeletal muscles involved in respiration stop functioning, so breathing requires assistance of a ventilator. Affected individuals usually die before they reach their thirties.

When motor neurons cannot signal muscles to contract, or signaling is impaired, skeletal muscles weaken or become paralyzed. For example, muscle paralysis often occurs after poliovirus infects and kills motor neurons. Poliovirus most frequently infects children; those who survive an infection may be paralyzed or have a weakened voluntary muscle response as a result. Polio vaccines have been available since the 1950s, so the disease is on the decline. In the United States, it has been essentially eliminated. However, infections still occur in less developed countries. Also, some adults who had polio as children develop postpolio syndrome, the symptoms of which include fatigue and progressive muscle weakness.

Amyotrophic lateral sclerosis (ALS) also kills motor neurons. It is sometimes called Lou Gehrig's disease, after a famous baseball player whose career was cut short by the disease in the late 1930s. ALS usually causes death by respiratory failure within three to five years of diagnosis. Some inherited mutations cause ALS, but most cases arise in people with no family history of the disorder.

Take-Home Message

What factors affect whole muscle function?

- With continual stimulation, a muscle can become fatigued for a brief time, so that it no longer contracts.
- Exercise can reduce tendency of muscles to become fatigued and increase the size of muscle fibers.
- Muscular dystrophies are genetic disorders that destroy muscle. Muscle function can also be impaired by diseases or disorders that affect motor neurons.

20.6 Bulking Up Muscles (revisited)

Researchers have long been searching for ways to slow muscle loss resulting from muscular dystrophy, ALS, and even normal aging. Drugs that inhibit myostatin production or prevent myostatin from acting may help them reach this goal. One way to learn what effects such drugs might have is to study mice in which the myostatin gene was knocked out. The larger mouse in the photo at the *right* is a knockout mutant (Section 7.7). Its myostatin gene was disabled by genetic engineering, so it is larger and more muscular than the normal mouse beside it. The bad news is that such mice have unusually small, stiff, easily torn tendons. Thus, increased tendon injuries could be a side effect of myostatin-inhibiting drugs.

Se-Lin Lee, Johns Hopkins University School of Medicine.

WHERE YOU ARE GOING . . .

Our discussion of blood includes information about blood cells that form in bone and the role skeletal muscles play in moving blood through veins (Section 21.5). The skeletal muscle actions that bring about breathing are the topic of Section 21.8. We return to the dietary requirements for calcium and vitamin D in Section 23.4.

Section 20.1 Exercise makes muscles bigger, not by adding cells but by adding proteins to existing cells. Hormones and other molecules regulate this process.

Section 20.2 Muscles interact with skeletons. Earthworms and other soft-bodied invertebrates have a **hydrostatic skeleton** (fluid skeleton). Arthropods have an **exoskeleton** (external skeleton) and vertebrates have an **endoskeleton** (internal skeleton).

The human skeleton consists of a skull, a **vertebral column** (backbone), a rib cage, a pelvic girdle, a pectoral girdle, and paired limbs. The vertebral column consists of individual segments called **vertebrae**, with **intervertebral disks** between them. The spinal cord runs through the vertebral column and connects with the brain through a hole in the base of the skull. The shape of the human backbone is an evolutionary adaptation to upright walking.

Bones function in mineral storage, movement, and protection and support of soft organs. A long bone such as a femur (thighbone) has an outer layer of **compact bone** beneath a connective tissue sheath. Fatty **yellow marrow** fills most of the bone interior. **Spongy bone** with blood-producing **red marrow** fills the ends of the bone. Ongoing mineral deposits and removals help maintain blood levels of calcium and phosphorus, and also adjust bone strength. Bone density declines with age.

Bones meet at **joints**. **Fibrous joints** hold bones tightly in place and **cartilaginous joints** let them move a bit. **Synovial joints** allow the most motion. **Ligaments** connect bones at synovial joints.

Section 20.3 **Tendons** of connective tissue attach skeletal muscles to bones. When skeletal muscles contract, tendons transmit force that makes the bones move. Muscles can only pull on bones; they cannot push them. Many muscles work as opposing pairs.

Section 20.4 Skeletal muscles contract in response to signals from the nervous system. Internal organization of a skeletal muscle promotes a directional contraction. A skeletal muscle fiber contains many long **myofibrils**.

actin   myosin

Each myofibril consists of **sarcomeres**, units of muscle contraction, lined up end to end along its length. Each sarcomere has parallel arrays of thin filaments composed of **actin** and thick filaments composed of **myosin**. The **sliding-filament model** describes how ATP-driven sliding of myosin filaments past actin filaments shortens the sarcomere.

Muscle contraction requires an energy input. Muscle fibers make the ATP they require for contraction by dephosphorylation of creatine phosphate, aerobic respiration, and lactate fermentation. Aerobic respiration predominates in red muscle fibers, which have many mitochondria and contain the oxygen-storing protein myoglobin. Lactate fermentation predominates in white fibers.

Section 20.5 Exercise can increase the number of mitochondria and protein filaments in muscle fibers. Prolonged inactivity causes metabolic changes in muscle and these changes contribute

to an increased risk of cardiovascular disorders. Muscular dystrophies are genetic disorders that cause muscles to degenerate. Muscle function can also be impaired by diseases such as polio, which affect motor neurons (the neurons that signal muscles to contract).

Self-Quiz   Answers in Appendix I

1. A hydrostatic skeleton consists of _____ .
   a. a fluid in an enclosed space
   b. hardened plates at the surface of a body
   c. internal hard parts
   d. none of the above

2. Bones are _____ .
   a. mineral reservoirs
   b. skeletal muscle's partners
   c. sites where blood cells form (some bones only)
   d. all of the above

3. The human backbone _____ .
   a. consists of vertebrae and intervertebral cartilage disks
   b. is S-shaped
   c. encloses the spinal cord
   d. all of the above

4. Bones move when _____ muscles contract.
   a. cardiac
   b. skeletal
   c. smooth
   d. all of the above

5. A ligament connects _____ .
   a. bones at a joint
   b. a muscle to a bone
   c. a muscle to a tendon
   d. a tendon to bone

6. The knee is a _____ .
   a. hinge-type synovial joint
   b. fibrous joint
   c. cartilaginous joint
   d. ball-and-socket type of synovial joint

7. A vertebrate skeletal muscle _____ .
   a. contracts in response to signals from a motor neuron
   b. pushes against and moves a bone
   c. is an involuntary muscle
   d. none of the above

8. A sarcomere shortens when _____ .
   a. thick filaments shorten
   b. thin filaments shorten
   c. both thick and thin filaments shorten
   d. none of the above

9. Release of _____ from a muscle fiber's modified endoplasmic reticulum allows actin and myosin filaments to interact.
   a. myoglobin
   b. potassium ions
   c. calcium ions
   d. collagen

10. Binding of ATP to _____ activates it and prepares this protein to take part in muscle contraction.
    a. actin
    b. myosin
    c. collagen
    d. myostatin

11. ATP for muscle contraction can be formed by _____ .
    a. aerobic respiration
    b. lactate fermentation
    c. creatine phosphate breakdown
    d. all of the above

Digging Into Data

Building Stronger Bones

Tiffany (*right*) was born with multiple fractures in her arms and legs. By age six, she had undergone surgery to correct more than 200 bone fractures. Her fragile, easily broken bones are symptoms of osteogenesis imperfecta (OI), a genetic disorder caused by a mutation in a gene for collagen. As bones develop, collagen forms a scaffold for deposition of mineralized bone tissue. The scaffold forms improperly in children with OI. Figure 20.14 shows the results of an experimental test of a new drug. Bones of treated children, all less than two years old, were compared to bones of a control group of similarly affected, same-aged children who did not receive the drug.

1. An increase in vertebral area during the 12-month period of the study indicates bone growth. How many of the treated children showed such an increase?

2. How many of the untreated children showed an increase in vertebral area?

3. How did the rate of fractures in the two groups compare?

4. Do the results shown support the hypothesis that this drug can increase bone growth and reduce fractures in young children who have OI?

Treated child	Vertebral area in cm² (Initial)	(Final)	Fractures per year	Control child	Vertebral area in cm² (Initial)	(Final)	Fractures per year
1	14.7	16.7	1	1	18.2	13.7	4
2	15.5	16.9	1	2	16.5	12.9	7
3	6.7	16.5	6	3	16.4	11.3	8
4	7.3	11.8	0	4	13.5	7.7	5
5	13.6	14.6	6	5	16.2	16.1	8
6	9.3	15.6	1	6	18.9	17.0	6
7	15.3	15.9	0	Mean	16.6	13.1	6.3
8	9.9	13.0	4				
9	10.5	13.4	4				
Mean	11.4	14.9	2.6				

FIGURE 20.14 Results of a clinical trial of a drug treatment for osteogenesis imperfecta (OI), also known as brittle bone disease. The drug being tested is a compound that reduces the rate of bone breakdown. Nine children with OI received the drug. Six others were untreated controls. Surface area of specific vertebrae was measured before and after treatment. Fractures occurring during the 12 months of the trial were also recorded.

Credits: left, Courtesy of the family of Tiffany Manning; right, © Cengage Learning.

---

12. A red muscle fiber _____ .
    a. contains myoglobin
    b. produces ATP mainly by lactate fermentation
    c. has relatively few mitochondria
    d. fatigues faster than a white fiber

13. Red marrow _____ .
    a. is the outermost layer of long bones
    b. fills the shaft of most long bones
    c. consists mainly of fat
    d. produces blood cells

14. Match each bone with its description.
    _____ femur   a. part of skull
    _____ radius   b. thighbone
    _____ vertebra   c. segment of backbone
    _____ sternum   d. breastbone
    _____ cranial bones e. forearm bone

15. Match the words with their definitions.
    _____ myoglobin   a. genetic disorder of muscles
    _____ myostatin   b. makes ATP; requires oxygen
    _____ myosin   c. decline in bone density
    _____ collagen   d. binds and releases oxygen
    _____ osteoporosis   e. virus harms motor neurons
    _____ polio   f. slows muscle protein production
    _____ muscular dystrophy g. motor protein of a sarcomere
    _____ tendon   h. makes ATP without oxygen
    _____ triceps   i. main protein in bones
    _____ aerobic respiration j. opposes action of biceps
    _____ lactate fermentation k. connects a muscle to a bone

Critical Thinking

1. After death, a person no longer makes ATP, and calcium stored in the specialized endoplasmic reticulum of muscle fibers diffuses down its concentration gradient into the muscle cytoplasm. The result is rigor mortis—an unbreakable state of muscle contraction. Explain why the contraction occurs and why it is irreversible.

2. During strenuous activity, lactate (lactic acid) production can cause a burning sensation. However, once the activity stops, the lactate is quickly converted to pyruvate and used in ATP production. What metabolic pathway can make use of the pyruvate?

3. Zachary's younger brother Noah had Duchenne muscular dystrophy and died at the age of 16. Zachary is now 26 years old, healthy, and planning to start a family of his own. However, he worries that his sons might be at high risk for muscular dystrophy. His wife's family has no history of this genetic disorder. Review Sections 9.7 and 20.5 and decide whether Zachary's concerns are well founded.

4. Athletes tend to have stronger bones than nonathletes, but different sports strengthen different bones. Volleyball, basketball, gymnastics, and soccer all thicken the femur, whereas swimming, skating, and bicycling have little effect on this bone. What does this tell you about how exercise promotes increases in bone density?

5. Bully whippets are homozygous for a deletion of two base pairs in the myostatin gene. The deletion changes an mRNA codon in the middle of the myostatin mRNA from UGU to UGA. Use your knowledge of the genetic code (Section 7.4) to determine the effect of this mutation on the structure of the resulting protein.

Circulation and Respiration 21

This chapter expands on the discussion of circulatory systems in Section 15.2 and on descriptions of squamous epithelium, blood, and cardiac muscle (19.3). You will draw on your knowledge of hemoglobin (2.9), gap junctions (3.6), diffusion, osmosis, and membrane transport (4.5, 4.6), and sickle-cell anemia (7.6).

21.1 A Shocking Save

The heart is the body's most durable muscle. It begins to beat during the first month of human development, and keeps at it for a lifetime. A natural pacemaker in the heart wall generates electrical signals that stimulate contraction of heart muscle. This pacemaker sometimes malfunctions. When it does, electrical signaling becomes disrupted, the heart stops beating, and blood flow halts. Such an event, called a sudden cardiac arrest, occurs in more than 300,000 people per year in the United States. An inborn heart defect causes most cardiac arrests in people under age 36. Heart disease causes most in older people.

The chance of surviving sudden cardiac arrest rises by 50 percent when the affected person receives cardiopulmonary resuscitation (CPR) within four to six minutes of the event. A rescuer who carries out traditional CPR alternates mouth-to-mouth respiration that forces air into an affected person's lungs with chest compressions that keep the victim's blood moving. CPR can ensure that essential oxygen reaches the brain, but it cannot restart the heart. That requires a defibrillator, a device that resets the natural pacemaker by delivering an electric shock to the chest. You have probably seen this procedure depicted in hospital dramas.

Matt Nader (Figure 21.1) learned about the importance of CPR and defibrillation when he experienced sudden cardiac arrest during a high school football game. He came off the field after a play, sat on the bench to talk to his coach, and felt a burning pain in his chest. His vision suddenly blurred, then he passed out. Nader's parents, who are physicians, were watching the game and rushed from their seats. They quickly determined that Matt did not have a pulse and, as they examined him, he stopped breathing.

Matt's parents began CPR on their son. At the same time, someone ran to get the school's automated external defibrillator (AED), a device about the size of a laptop computer. The AED provides simple voice commands that direct the user to attach electrodes to a person in distress. It then determines whether the person has a heartbeat and, if required, shocks the heart.

In Nader's case, the AED restarted his heart, quite likely saving his life. Cardiologists determined that his sudden cardiac arrest had been caused by a genetic heart defect. To protect him, they implanted a small defibrillator inside his body. This device will provide a lifesaving shock if his heart stops again.

After his recovery, Matt Nader appeared before the Texas State Legislature to testify about his experience and advocate for wider availability of AEDs in schools. Thanks in part to his efforts, Texas has passed a law requiring that all high schools have an AED available at athletic events and practices.

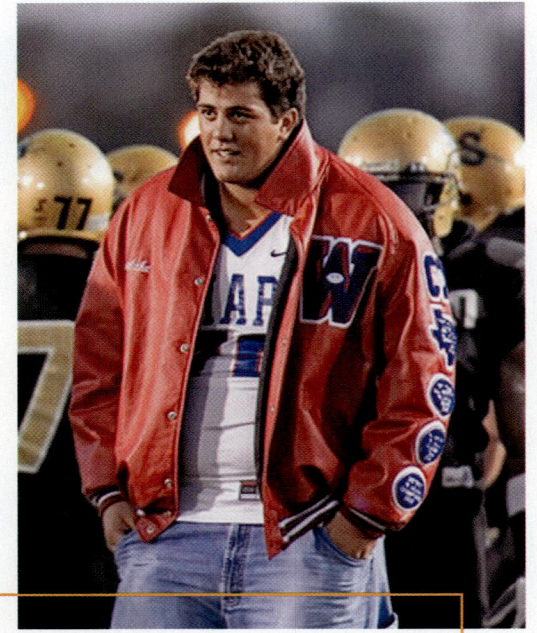

FIGURE 21.1 Surviving sudden cardiac arrest.

Matt Nader (*above*), a talented high school football player, found out he had a heart defect when his heart stopped during a game. CPR and quick defibrillation saved his life. Automated external defibrillators such as the one shown on the *right* can restart a heart and are designed to be simple enough to be used by a trained layperson.

Credits: (a) Courtesy of the family of Matt Nadar; (b) Courtesy of ZOLL Medical Corporation.

21.2 Moving Substances Through a Body

All animals must keep their cells supplied with nutrients and oxygen. Distributing these materials through the body is an important aspect of homeostasis, as is removing waste. In some invertebrates, including flatworms and sea anemones, solutes and gases simply diffuse through the **interstitial fluid**, the fluid between cells. As explained in Section 4.5, diffusion is the movement of a substance from an area where it is concentrated to one where it is less concentrated. Animals that rely on diffusion alone to move materials within their body tend to be small, with all cells close to the body surface and digestive cavity. Substances do not have to diffuse very far to reach all of the cells in such bodies.

Evolution of circulatory systems that could speed distribution of material through a larger, multi-layered body made possible more complex animal body plans. A circulatory system is an organ system that speeds the distribution of materials within an animal body. It includes one or more **hearts** (muscular pumps) that propel fluid through vessels that extend through the body.

> **Open and Closed Circulatory Systems** There are two types of circulatory systems. Arthropods and most mollusks have an **open circulatory system**, in which a heart or hearts pump fluid called hemolymph into vessels that open to internal cavities (Figure 21.2**A**). Hemolymph leaves the vessels and enters these cavities, where it mixes with interstitial fluid and makes direct exchanges with cells. The hemolymph is then sucked back into the heart through pores.

By contrast, annelids such as earthworms, cephalopod mollusks such as octopuses, and all vertebrates have a **closed circulatory system**, in which a heart or hearts pump blood through a continuous series of vessels (Figure 21.2**B**). A closed circulatory system distributes substances faster than an open one. It is "closed" because blood does not leave blood vessels to bathe tissues. Instead, transfers between blood and the cells of other tissues take place across the walls of **capillaries**, the smallest-diameter blood vessels.

capillaries Smallest-diameter blood vessels; site of exchanges of gases and other materials with the tissues.

closed circulatory system Circulatory system in which blood flows through a continuous network of vessels.

heart Muscular organ that pumps fluid through a body.

interstitial fluid Fluid between cells of a multicelled body.

open circulatory system Circulatory system in which the circulatory fluid (hemolymph) leaves vessels and flows among tissues.

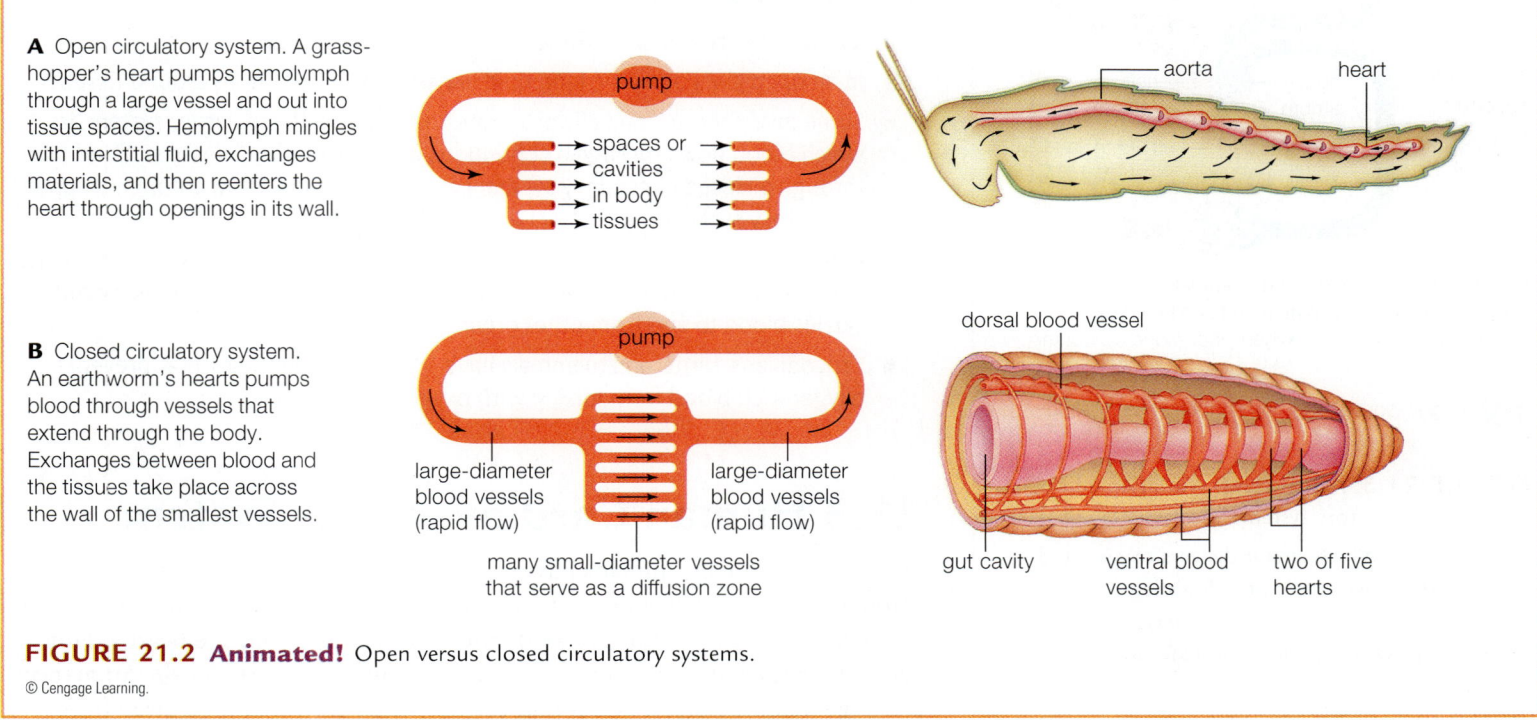

A Open circulatory system. A grasshopper's heart pumps hemolymph through a large vessel and out into tissue spaces. Hemolymph mingles with interstitial fluid, exchanges materials, and then reenters the heart through openings in its wall.

pump

→ spaces or →
→ cavities →
→ in body →
→ tissues →

aorta   heart

B Closed circulatory system. An earthworm's hearts pumps blood through vessels that extend through the body. Exchanges between blood and the tissues take place across the wall of the smallest vessels.

pump

large-diameter blood vessels (rapid flow)

large-diameter blood vessels (rapid flow)

many small-diameter vessels that serve as a diffusion zone

dorsal blood vessel

gut cavity   ventral blood vessels   two of five hearts

FIGURE 21.2 Animated! Open versus closed circulatory systems.

© Cengage Learning.

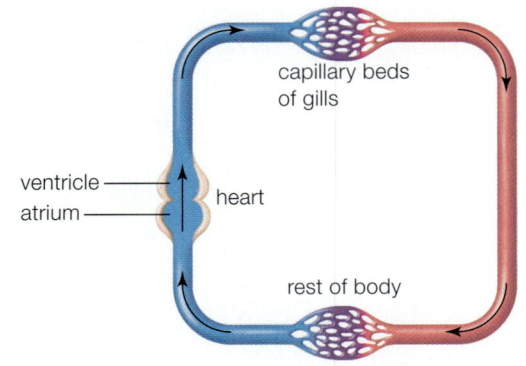

A In fish, the heart has one atrium and one ventricle. Force of the ventricle's contraction propels blood through the single circuit.

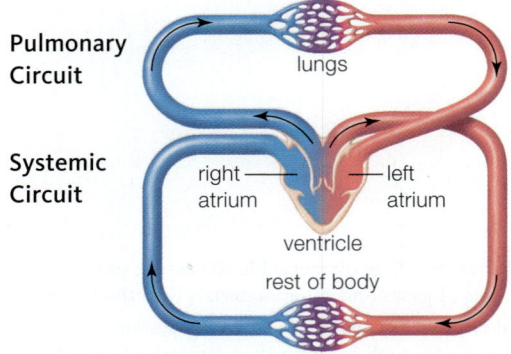

B In amphibians and most reptiles, the heart has three chambers: two atria and one ventricle. Blood flows in two partially separated circuits. Oxygenated and oxygen-poor blood mix a bit in the ventricle.

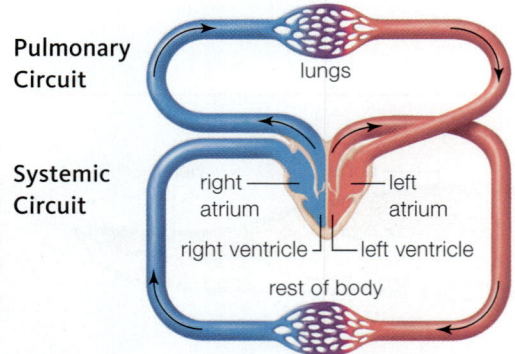

C In crocodilians, birds, and mammals, the heart has four chambers: two atria and two ventricles. Oxygenated blood and oxygen-poor blood do not mix.

oxygen-poor blood   oxygen-rich blood

FIGURE 21.3 Animated! Variation among vertebrate circulatory systems.

Figure It Out: Which circuit delivers oxygenated blood to the muscle cells of a dog's legs?

Answer: The systemic circuit. It services all nonlung parts of the body.

© Cengage Learning.

❯ **Evolution of Vertebrate Cardiovascular Systems** All vertebrates have a closed circulatory system, with a single heart and a network of blood vessels. These structures are collectively referred to as a cardiovascular system. The Greek word *kardia* means heart, and the Latin word *vasculum* means vessel.

The structure of the heart differs among vertebrate groups, as do the circuits through which blood flows. In most fishes, the heart has two main chambers, and blood flows in a single circuit (Figure 21.3**A**). One chamber of the heart, an **atrium** (plural, atria), receives blood. From there, the blood enters a **ventricle**, a chamber that pumps blood out of the heart. The pressure exerted by the contracting ventricle drives the blood through a series of vessels, into networks of capillaries inside each gill, similar networks in body tissues and organs, and finally back to the heart. Capillaries dissipate the pressure from the ventricle's contraction, so the blood is not under much pressure when it leaves the gill capillaries, and even less as it travels back to the heart.

Adapting to life on land involved modifications of respiratory and circulatory systems. Amphibians and most reptiles have a three-chamber heart, with two atria emptying into one ventricle (Figure 21.3**B**). The heart of these animals propels blood through two circuits, increasing the speed of blood flow. The force of one contraction drives blood through the **pulmonary circuit**, to the lungs and then back to the heart. (The Latin word *pulmo* means lung.) A second contraction sends the newly oxygenated blood through the **systemic circuit**, which runs through capillaries in body tissues before returning to the heart.

In a three-chamber heart, oxygenated and oxygen-poor blood mix in the single ventricle. Oxygen delivery improved with evolution of a four-chamber heart, which has two atria and two ventricles (Figure 21.3**C**). Crocodilians, birds, and mammals have a four-chamber heart. With two fully separated circuits, only oxygen-rich blood flows to tissues, and blood pressure can be regulated independently in each circuit. Strong contraction of the heart's left ventricle keeps blood moving fast through the long systemic circuit. Less forceful contraction of the right ventricle protects delicate lung capillaries.

Take-Home Message

What are features of animal circulatory systems?

- In an open circulatory system, fluid leaves vessels and seeps around tissues.
- Some invertebrates and all vertebrates have a closed circulatory system, in which blood flows through a continuous network of blood vessels. Blood flow through a closed circulatory system is faster than blood flow through an open circulatory system.
- In fishes, blood flows in one circuit. In other vertebrates, it flow in two circuits. The pulmonary circuit carries blood to and from lungs. The systemic circuit carries blood to and from other organs.
- Crocodilians, birds, and mammals have a four-chamber heart that prevents oxygen-rich blood from mixing with oxygen-poor blood.

21.3 Human Cardiovascular System

Like other mammals, humans have a four-chamber heart that pumps blood through two circuits (Figure 21.4). The short pulmonary circuit (*gold* boxes) carries blood to and from lungs. Blood enters this circuit when the heart's right ventricle pumps oxygen-poor blood into two pulmonary arteries ❶. An **artery** is a large-diameter blood vessel that carries swiftly moving blood away from

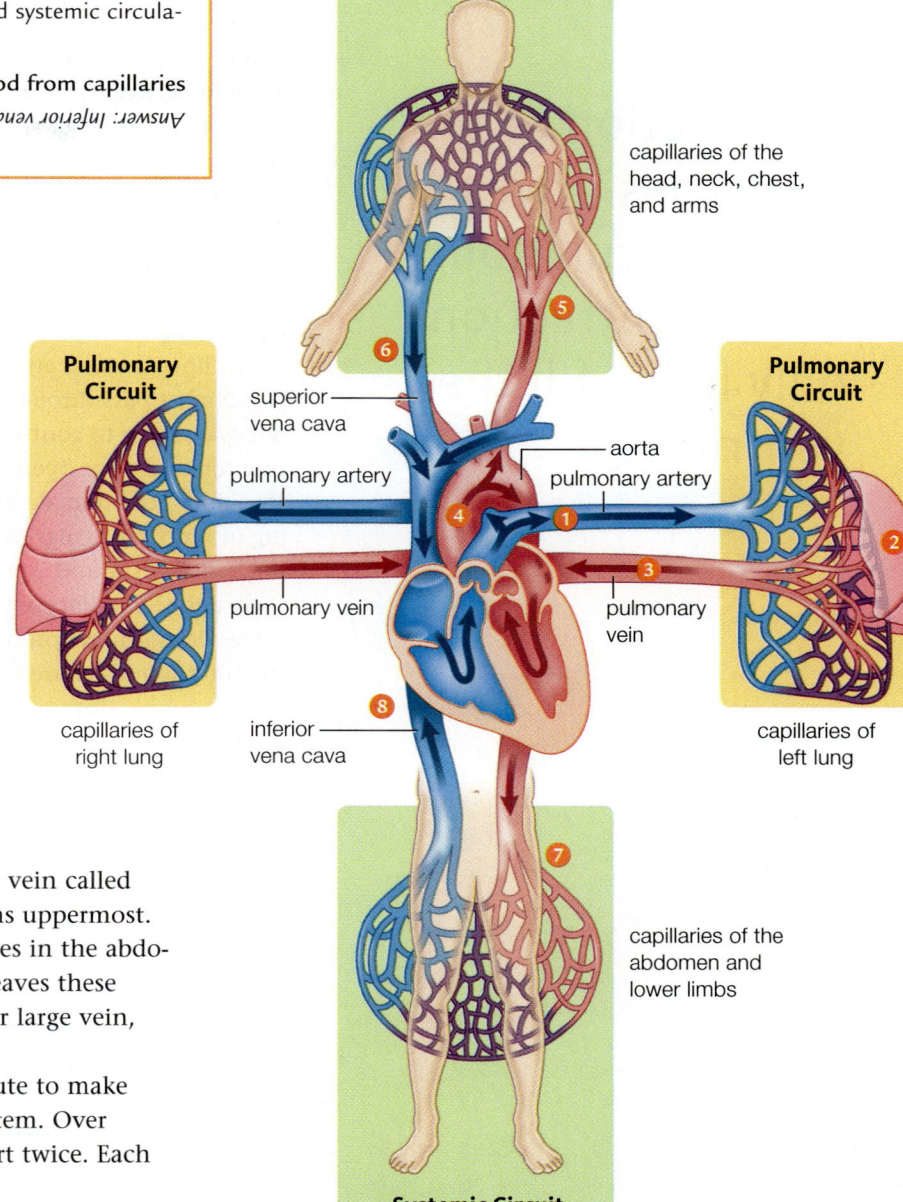

Systemic Circuit

capillaries of the head, neck, chest, and arms

Pulmonary Circuit

superior vena cava

pulmonary artery

aorta
pulmonary artery

pulmonary vein

Pulmonary Circuit

capillaries of left lung

capillaries of right lung

inferior vena cava

capillaries of the abdomen and lower limbs

Systemic Circuit

the heart. Each pulmonary artery delivers blood to one lung. As blood flows through pulmonary capillaries (capillaries of the lung) ❷, it picks up oxygen and releases carbon dioxide. The newly oxygenated blood then returns to the heart through pulmonary veins ❸ that empty into the left atrium. A **vein** is a large vessel that carries blood back toward the heart.

The oxygenated blood next travels through the longer systemic circuit. The heart's left ventricle pumps blood into the body's largest artery, the aorta ❹. Some vessels that branch from the aorta deliver blood to the capillaries of the upper body ❺. Blood gives up oxygen and picks up carbon dioxide as it flows through these capillaries, then returns to the heart's right atrium, through a large vein called the superior vena cava ❻. In this context, *superior* means uppermost. Other branches from the aorta deliver blood to capillaries in the abdomen and lower limbs ❼. The oxygen-poor blood that leaves these capillaries returns to the heart's right atrium via another large vein, called the inferior (meaning lower) vena cava ❽.

In a person at rest, it takes a blood cell about a minute to make its way through the two loops of the cardiovascular system. Over the course of its travels, the cell passes through the heart twice. Each time, contraction of a ventricle speeds it on its way.

Take-Home Message

What are the components of a human cardiovascular system and how do they interact?

- The human cardiovascular system has a four-chamber heart that pumps blood through a network of blood vessels in two circuits.

- Oxygen-poor blood pumped out of the heart's right ventricle travels through the pulmonary circuit. Pulmonary arteries carry the blood to the lungs, then pulmonary veins return the now-oxygenated blood to the left atrium.

- Oxygen-rich blood pumped out of the left ventricle travels through the systemic circuit. The aorta gives off branches that supply capillary beds of the body. The superior vena cava and inferior vena cava return oxygen-poor blood to the right atrium.

artery Large-diameter blood vessel that carries blood away from the heart.

atrium Heart chamber that pumps blood into a ventricle.

pulmonary circuit Circuit through which blood flows from the heart to the lungs and back.

systemic circuit Circuit through which blood flows from the heart to the body tissues and back.

vein Large-diameter vessel that returns blood to the heart.

ventricle Heart chamber that pumps blood out of the heart and into an artery.

21.4 The Human Heart

The heart lies inside the thoracic cavity, between the lungs (Figure 21.5**A**). A double-layered sac of tough connective tissue (pericardium) protects and anchors it. Fluid between the sac's two layers provides lubrication for the heart's continual motions. Inside the sac, the heart wall consists mostly of cardiac muscle cells and its chambers are lined with epithelium (Section 19.3).

The two sides of the human heart are separated by a thick septum (Figure 21.5**B**). Each side of the heart has two chambers: an atrium that receives blood from veins, and a ventricle that pumps blood into arteries. The superior vena cava and inferior vena cava deliver oxygen-poor blood from the body to the right atrium. The pulmonary veins deliver oxygen-rich blood to the left atrium.

To flow from an atrium into a ventricle, or from a ventricle into an artery, blood must pass through a heart valve. Pressure-sensitive valves function like one-way doors to control the flow of blood through the heart. High fluid pressure forces a valve open, and valves snap shut when pressure declines. The "lup-dup" sound made by a beating heart arises from the closing of valves. Closing of atrioventricular (AV) valves between the atria and the ventricle produces the first heart sound. Simultaneous closing of valves at the entrances to the aorta and the pulmonary artery produces the second heart sound. When heart valves do not function normally, movement of blood through them makes a swishing sound referred to as a heart murmur.

〉 The Cardiac Cycle With each heartbeat, the heart's chambers go through a sequence of relaxation and contraction known as the **cardiac cycle** (Figure 21.6). As we begin our observation of the cardiac cycle, both the atria and ventricles are relaxed. The pressure exerted by blood entering the atria has forced

Contraction of the ventricles drives blood circulation. Atrial contraction only helps fill the ventricles.

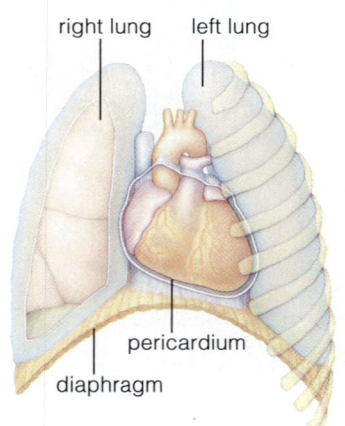

right lung   left lung

pericardium

diaphragm

A The heart sits between the lungs.

FIGURE 21.5
Animated! Location and structure of the human heart. Arrows indicate the path of oxygenated blood (*red*) and oxygen-poor blood (*blue*).

© Cengage Learning.

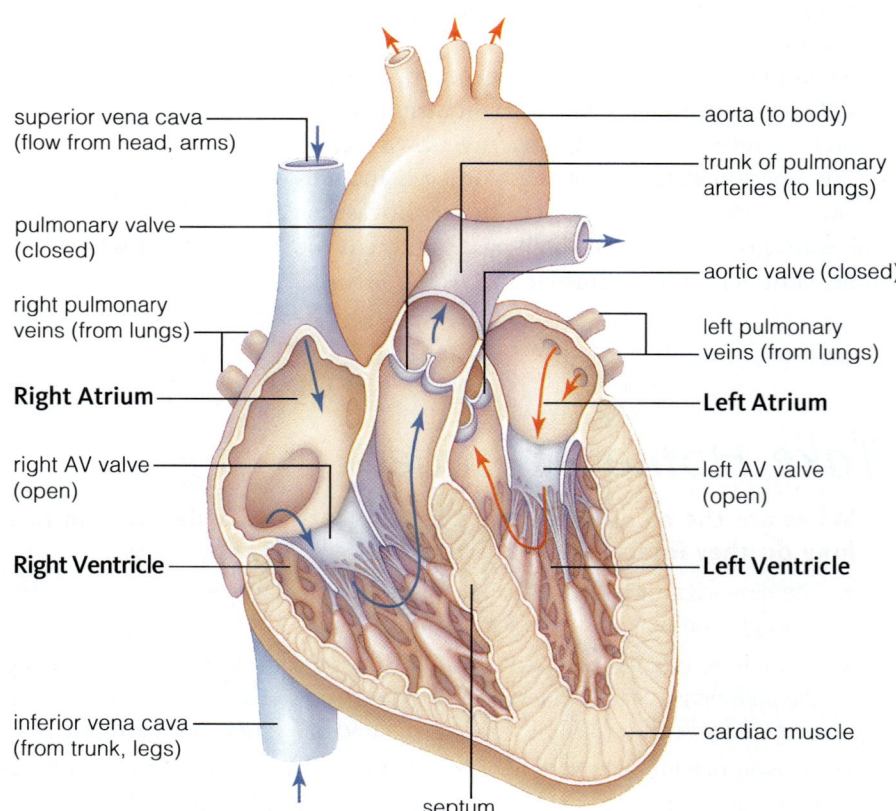

superior vena cava (flow from head, arms)

aorta (to body)

trunk of pulmonary arteries (to lungs)

pulmonary valve (closed)

aortic valve (closed)

right pulmonary veins (from lungs)

left pulmonary veins (from lungs)

Right Atrium

Left Atrium

right AV valve (open)

left AV valve (open)

Right Ventricle

Left Ventricle

inferior vena cava (from trunk, legs)

cardiac muscle

septum

B Cutaway view of the heart.

the AV valves open, and blood is filling the ventricles ❶. Contraction of the atria forces more blood into the ventricles ❷. Next, the ventricles begin contracting. The resulting rise in fluid pressure inside the ventricles causes the AV valves to shut. Pressure in the ventricles continues to rise until the aortic and pulmonary valves open, and blood rushes out of the ventricles and into the aorta and pulmonary arteries ❸. As fluid pressure in the ventricles declines, aortic and pulmonary valves close. At this point, the atria are starting to fill again ❹.

Contraction of ventricles is the driving force for blood circulation. Atrial contraction only helps fill ventricles. The structure of the cardiac chambers reflects this difference in function. Atria need only to generate enough force to squeeze blood into the ventricles, so they have relatively thin walls. Ventricle walls are much thicker. Contraction of muscle in the ventricle walls has to be strong enough to create a pressure wave that propels blood through an entire circuit. The left ventricle, which pumps blood throughout the entire body, has thicker walls than the right ventricle, which pumps blood only to the lungs and back.

❶ Relaxed atria fill. Fluid pressure opens AV valves and blood flows into the relaxed ventricles.

❷ Atrial contraction squeezes more blood into the still-relaxed ventricles.

❸ Ventricles start to contract and the rising pressure pushes the AV valves shut. A further rise in pressure causes the aortic and pulmonary valves to open.

❹ As blood flows into the arteries, pressure in the ventricles declines and the aortic and pulmonary valves close.

FIGURE 21.6 Animated! The cardiac cycle. **Figure It Out: Which valves open in response to rising atrial pressure?** *Answer: The AV valves*

© Cengage Learning.

❯ **Setting the Pace of Contractions** The sinoatrial (SA) node, a clump of cells in the wall of the right atrium, serves as the **cardiac pacemaker** (Figure 21.7). In a healthy person at rest, the SA node sends out a muscle-exciting electrical signal about seventy times a minute. Gap junctions (Section 3.6) between adjacent muscle cells allow electrical signals generated by the SA node to spread across the atria, causing the atria to contract. At the same time, special noncontractile muscle fibers convey the excitatory signal to the atrioventricular (AV) node. This clump of cells is the only electric bridge to the ventricles, meaning the signal to contract can only reach the ventricles by way of the AV node. The time it takes for a signal to reach and cross this bridge allows blood arriving from the atria to fill the ventricles before the ventricles contract. From the AV node, the excitatory signal travels along conducting fibers in the septum between the heart's left and right halves. The fibers extend to the heart's lowest point and up the ventricle walls. As the excitatory signal spreads via gap junctions, ventricles contract from the bottom up, with a wringing motion.

As noted in Section 21.1, malfunctions of a heart's natural pacemaker can be dangerous or deadly. Implanting a battery-operated electrical pacemaker can regulate the heart rate of a person whose heart would otherwise beat too slow or too fast. A heart that stops periodically can be treated by an implantable defibrillator. The defibrillator administers an electrical shock to a stopped heart, and thus allows the SA node to reestablish a normal rhythm.

SA node (cardiac pacemaker)

AV node

fibers that relay signals

FIGURE 21.7 Animated! Components of the heart's signaling system. Signals from the SA node travel to the AV node and then on to the ventricles along junctional fibers.

© Cengage Learning.

Take-Home Message

How does the human heart function?

- The four-chamber heart is a muscular pump. Contraction of the thick-walled ventricles drives blood circulation. Atrial contraction fills the ventricles.

- The SA node is the cardiac pacemaker. Its spontaneous signals make cardiac muscle fibers of the heart wall contract in a coordinated fashion.

cardiac cycle Sequence of contraction and relaxation of heart chambers that occurs with each heartbeat.

cardiac pacemaker Group of heart cells (SA node) that emits rhythmic signals calling for muscle contraction.

plasma

blood cells →

red blood cell   white blood cell   platelet

FIGURE 21.8 Components of blood.

Credits: left, © Cengage Learning; right, National Cancer Institute/ Photo Researchers, Inc.

FIGURE 21.9 Animated! Blood clotting. Cells and platelets become trapped by threads of fibrin derived from a plasma protein (fibrinogen).

Professor P. Motta/ Department of Anatomy/ University La Sapienca, Rome/ Photo Researchers, Inc.

arteriole Blood vessel that delivers blood from an artery to capillaries; site of adjustments to blood distribution.

blood pressure Pressure exerted by blood against the walls of blood vessels.

diastolic pressure Blood pressure when ventricles are relaxed.

lymph Fluid that has left capillaries and entered lymph vessels.

plasma Protein-rich fluid portion of blood.

platelet Cell fragment that functions in blood clotting.

red blood cell Hemoglobin-filled, oxygen-transporting cell.

systolic pressure Blood pressure when the ventricles are contracting.

venule Blood vessel that connects a capillary to a vein.

white blood cell Blood cell that helps defend the body against pathogens and cleans up cellular debris.

21.5 Blood and Blood Vessels

❯ Components and Functions of Blood An average adult has about 4.5 liters (nearly 5 quarts) of blood. The blood consists of plasma, red blood cells, white blood cells, and platelets (Figure 21.8). All blood cells and platelets arise from stem cells in red bone marrow.

Plasma, the fluid portion of blood, constitutes 50 to 60 percent of the total blood volume. It serves as the transport medium for blood cells and platelets and contains dissolved nutrients, wastes, signaling molecules, and hundreds of plasma proteins. Some plasma proteins transport lipids and fat-soluble vitamins. Others help blood clot or act against pathogens.

Disk-shaped erythrocytes, or **red blood cells**, are the most numerous cells in blood. As many as 30 trillion circulate in your body. Their red color comes from hemoglobin, an iron-containing pigment that can reversibly bind oxygen. As red blood cells develop in bone marrow, hemoglobin accumulates inside of them. Before they enter the circulation, they expel their nucleus and other organelles. Mature red blood cells circulate for about four months.

Your 10 billion leukocytes, or **white blood cells**, play a variety of roles in housekeeping and defense. Some types patrol tissues. Others accumulate in lymph nodes. White blood cells engulf and digest cellular debris such as aged red blood cells. They also defend the body against viruses, bacteria, and other pathogens. The next chapter discusses their role in defense in more detail.

Platelets are bits of cytoplasm wrapped in plasma membrane. Each platelet lasts only five to nine days, but hundreds of thousands circulate in the blood. If an injury occurs, platelets release substances that initiate blood clotting. When small vessels are cut or torn, platelets clump together and temporarily fill the breach. They release substances that attract more platelets. During clot formation, enzymes convert fibrinogen, a soluble plasma protein, to insoluble threads of fibrin. Fibrin forms a mesh that traps cells and platelets (Figure 21.9). The entire mass is called a blood clot. Eventually, the clot retracts and forms a compact patch that seals the breach in the blood vessel.

❯ High-Pressure Flow in Arteries Blood pumped out of ventricles enters arteries, which have a muscular wall reinforced with elastic tissue (Figure 21.10**A**). The elasticity of arteries helps keep blood moving, even between contractions. When a ventricle contracts, it forces blood into the arteries, causing their elastic walls to bulge. When the ventricle relaxes, artery walls rebound inward and squeeze the blood inside them a bit farther away from the heart. The bulging of an artery with each ventricular contraction is referred to as the pulse.

Blood pressure, the pressure exerted by blood against the walls of vessels that enclose it, is highest in arteries and declines as blood proceeds through a circuit. Blood pressure is usually measured in the brachial artery of the upper arm. Two pressures are recorded. **Systolic pressure**, the highest pressure of a cardiac cycle, occurs as the contracting ventricles force blood into the arteries. **Diastolic pressure**, the lowest blood pressure of a cardiac cycle, occurs when the ventricles are fully relaxed. Blood pressure is typically written as systolic value/diastolic value. It is measured in millimeters of mercury (mm Hg), a standard unit for measuring pressure. Normal blood pressure is about 120/80 mm Hg. In conversation, you would say this is a pressure of "120 over 80."

❯ Adjusting Resistance at Arterioles Your body can alter the distribution of blood flow by adjusting the diameter of **arterioles**, vessels that receive blood from an artery and deliver it to capillaries (Figure 21.10**B**). Nervous or endocrine signals cause smooth muscle ringing an arteriole to constrict, so that the vessel narrows and less blood flows through it. Alternatively, signals can cause the smooth muscle to relax, so that the vessel dilates and blood flow increases. For example, after you eat, arterioles that supply blood to capillaries in your gut dilate, and arterioles supplying blood to capillaries that service skeletal muscles in your legs narrow. Conversely, when you run, arterioles servicing your legs dilate and those supplying the gut constrict.

> **Exchanges at Capillaries** The speed of blood flow drops dramatically when blood enters a network of capillaries, where blood exchanges fluids and solutes with cells (Figure 21.10**C**). Every cell is near at least one capillary. This proximity is essential because diffusion cannot move substances over long distances fast enough to keep cells alive.

Materials move between capillaries and body cells in several ways. A capillary wall is of a single layer of squamous epithelium. In most tissues, capillaries are leaky, with narrow spaces between the cells. At the arterial end of a capillary, pressure exerted by the beating heart forces plasma fluid rich in oxygen (O_2) and nutrients through these spaces into the surrounding interstitial fluid (Figure 21.11 **1**). Plasma proteins are too big to exit through the spaces between cells, so they remain in the vessel. Along the length of the capillary, oxygen diffuses from the blood into the interstitial fluid, while nutrients such as glucose are transported in the same direction **2**. Carbon dioxide (CO_2) diffuses into the capillary and other metabolic wastes are transported into it **3**. Near the venous end of the capillary bed, where blood pressure is lower, water moves by osmosis from the interstitial fluid into the protein-rich plasma **4**. With all the leaking and reentry of fluid, there is a small *net* outward flow from a capillary bed. The extra fluid enters vessels of the lymphatic system and becomes **lymph** **5**. Lymph vessels feed into lymphatic ducts that return fluid to veins near the heart.

> **Back to the Heart** Blood from capillary beds enters **venules**, or "little veins," which merge into veins (Figure 21.10**D,E**). One-way, pressure-sensitive valves prevent blood in veins from flowing backwards. The flexible vein wall can bulge greatly under pressure, so the veins can serve as a blood reservoir. In a person standing at rest, veins can hold up to 70 percent of the total blood volume.

FIGURE 21.10 Animated! Types of blood vessels and relationship among them.

From Russell/Wolfe/Hertz/Starr. Biology, 1e. © 2008 Brooks/Cole, a part of Cengage Learning, Inc.

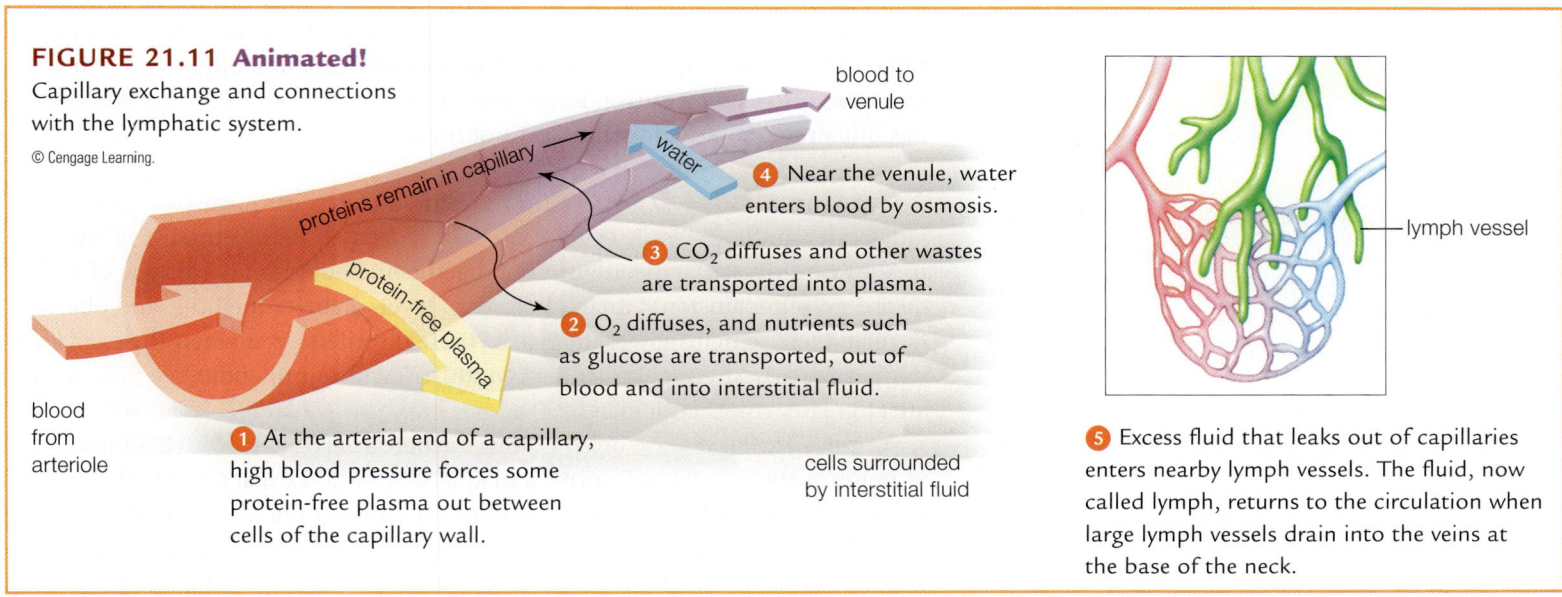

FIGURE 21.11 Animated!
Capillary exchange and connections with the lymphatic system.

© Cengage Learning.

4 Near the venule, water enters blood by osmosis.

3 CO_2 diffuses and other wastes are transported into plasma.

2 O_2 diffuses, and nutrients such as glucose are transported, out of blood and into interstitial fluid.

1 At the arterial end of a capillary, high blood pressure forces some protein-free plasma out between cells of the capillary wall.

cells surrounded by interstitial fluid

5 Excess fluid that leaks out of capillaries enters nearby lymph vessels. The fluid, now called lymph, returns to the circulation when large lymph vessels drain into the veins at the base of the neck.

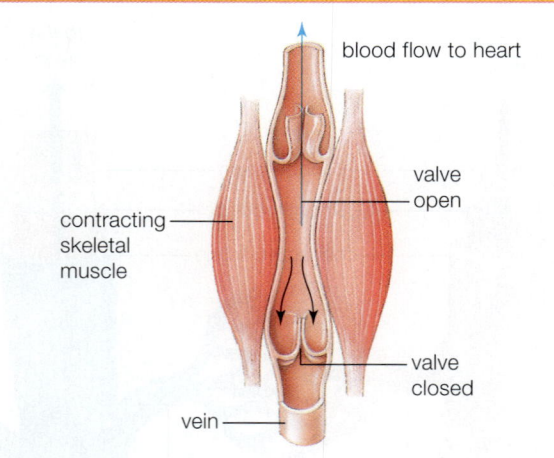

FIGURE 21.12 Skeletal muscle's effect on a vein. When a skeletal muscle next to a vein contracts, the muscle's increased bulk pushes on the vein, forcing blood inside it through a one-way valve and in the direction of the heart. © Cengage Learning.

Like arteries and arterioles, veins have some smooth muscle in their wall. When blood needs to circulate faster, as during exercise, the muscle in vein walls contracts. The resulting increase in blood pressure drives more blood back to the heart. In addition, contractions of skeletal muscles help keep the blood moving. When these muscles contract, they bulge and press against nearby veins. The additional pressure on the blood in the veins helps move it toward the heart (Figure 21.12).

Take-Home Message

How do blood and blood vessels function?

- Blood consists mainly of plasma, a protein-rich fluid with solutes and dissolved gases. Red blood cells distribute oxygen. White blood cells defend the body. Platelets are cell fragments that help blood clot after an injury.

- Blood flows into arteries under high pressure. Elastic arterial walls help propel blood forward when ventricles are relaxed. Adjusting arteriole diameter alters how much blood flows to different body regions.

- Exchanges between the blood and cells occur at capillary beds. Venules and veins return blood to the heart.

21.6 Blood and Cardiovascular Disorders

❭ Blood Disorders When red blood cells are in short supply or abnormal, oxygen delivery to cells falters, causing shortness of breath, fatigue, and chills. This condition is called anemia, and it has many causes. Sickle-cell anemia is a genetic disorder. It arises from a mutation that causes hemoglobin to change shape at a low oxygen concentration (Section 7.6). A low-iron diet can cause anemia because the lack of raw material slows production of iron-containing hemoglobin that fills red blood cells. Chronic blood loss, as from heavy menstrual periods or a bleeding ulcer, also lowers the supply of red blood cells.

Problems can also arise when blood has an impaired or excessive tendency to clot. Hemophilia is a genetic disorder that impairs clotting (Section 9.7). It most commonly occurs because an enzyme that crosslinks strands of fibrin into a stable mesh is defective. Other disorders cause spontaneous clot formation in vessels. Such clots can stay in the vessel where they formed, or break loose and travel in the blood. In either case, the clot can clog a vessel and halt blood flow. A clot that blocks a blood vessel in the brain can cause a stroke, in which interrupted blood flow impairs brain cell function. Bursting of a blood vessel in the brain can also cause a stroke. Either way, a person who survives a stroke often has impairments caused by the death of brain cells.

❭ Cardiovascular Disease In the United States, cardiovascular disorders kill about a million people every year. Tobacco smoking tops the list of risk factors for cardiovascular disorders. Other factors include a family history of such disorders, hypertension, a high cholesterol level, diabetes mellitus, and obesity. Regular exercise helps lower the risk of these disorders even when the exercise is not particularly strenuous. Gender is another factor; until about age fifty, males are at greater risk than females.

Cardiovascular disease often involves **atherosclerosis**, in which a buildup of lipids in the arterial wall forms a plaque that narrows space inside the vessel (Figure 21.13). Cholesterol plays a role in this "hardening of the arteries." Most of us absorb more cholesterol than we need from the food in our gut. Genetics affects how different people's bodies deal with the excess. Most cholesterol in

A Normal artery.

B Artery narrowed by an atherosclerotic plaque and a clot that has stuck to it.

FIGURE 21.13 Atherosclerosis. Credits: (a) Ed Reschke; (b) Biophoto Associates/ Photo Researchers, Inc.

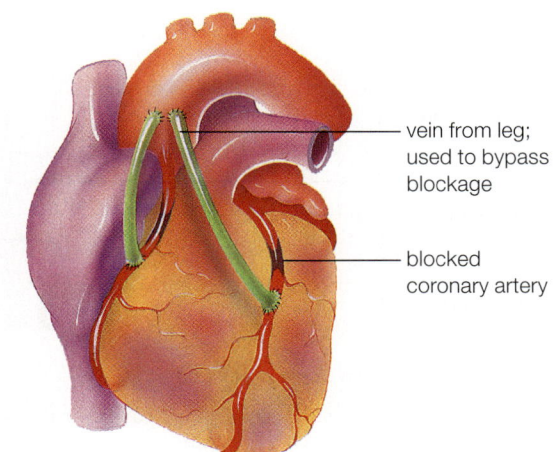

A Coronary bypass surgery. Veins from another part of the body are used to divert blood past the blockages. This illustration shows a "double bypass," in which veins (*green*) from elsewhere in the body are placed to divert blood around two blocked coronary arteries.

B Balloon angioplasty and the placement of a stent. After a balloonlike device is inflated in an artery to open it and flatten the plaque, a tube of metal (the stent) is inserted and left in place to keep the artery open.

FIGURE 21.14 Two methods of treating a blocked coronary (heart) artery. Such blockages are the main cause of heart attacks in older adults.

© Cengage Learning.

the blood is bound to plasma proteins in clumps called low-density lipoproteins, or LDLs, which most kinds of cells take up. A smaller amount of cholesterol is bound to plasma proteins as high-density lipoproteins, or HDLs. The liver takes up HDLs and uses them to make bile, a substance that is secreted into the gut and eventually leaves the body in feces. With a high concentration of LDLs or a low concentration of HDLs, lipids build up on artery walls, rather than being eliminated by the digestive tract. The site of lipid accumulation becomes inflamed, and fibrous connective tissue is laid down. The resulting atherosclerotic plaque narrows the artery, slowing flow through it.

Narrowing of the small arteries that deliver blood to the heart tissue can lead to a heart attack, in which impaired blood flow in coronary arteries kills cardiac muscle cells. A person has about a 30 percent chance of dying with a first heart attack. Heart cells do not divide after birth, so the cells killed in a heart attack cannot be replaced. Thus, heart attack survivors have a weakened heart.

Several methods are used to treat clogged coronary arteries. With coronary bypass surgery, a bit of blood vessel from elsewhere in the body is stitched to the aorta and to the coronary artery below a clogged region (Figure 21.14**A**). In laser angioplasty, laser beams vaporize plaques. In balloon angioplasty, doctors inflate a small balloon in a blocked artery to flatten the plaques. A wire mesh tube called a stent is then inserted to help keep the vessel open (Figure 21.14**B**).

Hypertension refers to chronically high blood pressure (above 140/90). It is sometimes described as a silent killer, because many affected people are unaware they have it. In some people, salt intake causes water retention that raises blood pressure, but the cause of hypertension is usually unknown. Regardless of its cause, hypertension makes the heart work harder than normal, causing it to become enlarged and less efficient. High blood pressure also increases risk of atherosclerosis and damages small blood vessels throughout the body. In the kidney, such damage is a leading cause of kidney failure.

Take-Home Message

How do circulatory disorders affect health?

- Anemia causes fatigue and weakness by reducing oxygen delivery to tissues.
- A clot or an atherosclerotic plaque can clog a blood vessel and slow or halt blood flow, thus causing a stroke or heart attack.
- Hypertension puts a strain on the heart and damages small blood vessels such as those in the kidney.

atherosclerosis Narrowing of an artery's interior due to lipid deposition and inflammation.
hypertension Chronically high blood pressure.

21.7 Animal Respiration

In Chapter 5 you learned about aerobic respiration, an energy-releasing pathway that requires oxygen (O_2) and produces carbon dioxide (CO_2). Here we focus on the physiological processes of **respiration**, which supply cells with oxygen from the environment, and deliver waste carbon dioxide to that environment.

> **Two Sites of Gas Exchange** Gases enter and leave an animal's body across a thin, moist layer of epithelial cells called the **respiratory surface** (Figure 21.15). A typical respiratory surface is one or two cell layers thick. It must be thin because gases diffuse quickly only over very short distances. It is kept moist because gases cross a cell membrane quickly only if they first dissolve in fluid. Once dissolved, gases diffuse across the lipid bilayer of plasma membranes. The larger the area of a respiratory surface, the greater the speed of gas exchange.

A second exchange of gases takes place internally, at the plasma membrane of body cells. Here, oxygen diffuses from interstitial fluid into a cell, and carbon dioxide diffuses in the opposite direction. In most invertebrates and all vertebrates, a circulatory system enhances the movement of gases between these two sites of gas exchange.

> **Respiratory Systems** Some invertebrates that live in aquatic or continually damp land environments have no special respiratory organs. They exchange gases across their entire body surface. In sea anemones and flatworms, gases diffuse from water into the body across the outer surface and the surface of their internal gastrovascular cavity. Similarly, an earthworm exchanges gases with the air all along the length of its body.

FIGURE 21.15 Two sites of gas exchange.

1 Cells of a respiratory surface exchange gases with the external environment and with fluid inside the body.

2 Other body cells exchange gases with the internal environment (interstitial fluid).

In some invertebrates, gases simply diffuse between the two sites. In animals with a circulatory system, blood transports gases rapidly between the two exchange sites.

© Cengage Learning.

External environment (air or water) — *Internal environment (interstitial fluid)* — O_2 — CO_2 — cells of the respiratory surface **1** — other body cells **2**

FIGURE 21.16 Sea slug with feathery external gills.

© Hal Beral VWPics/ SuperStock.

gills

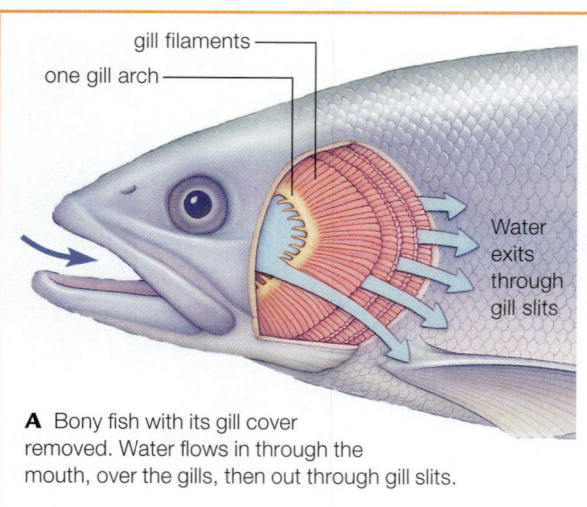

gill filaments — one gill arch

Water exits through gill slits

A Bony fish with its gill cover removed. Water flows in through the mouth, over the gills, then out through gill slits.

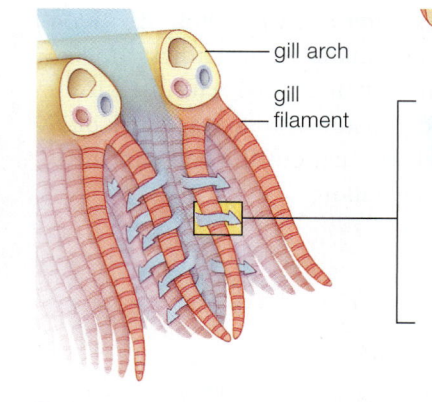

gill arch — gill filament

B Each gill has bony gill arches with many thin gill filaments attached.

respiratory surface — fold with a capillary bed inside — direction of blood flow — oxygen-poor blood from deep in body — oxygenated blood back toward body — water flow

C Flow of blood and water in opposite directions maximizes diffusion of oxygen from the water into the blood.

FIGURE 21.17 Animated! Fish respiration. © Cengage Learning.

Other aquatic animals have **gills**, filamentous respiratory organs that increase the surface area available for gas exchange. As hemolymph or blood passes through the gills, it exchanges gases with water that surrounds them. Some sea slugs have external gills (Figure 21.16), as do all larval and some adult amphibians. Lobsters and clams have internal gills.

Fishes have a pair of internal gills: rows of slits at the back of the mouth that extend to the body's surface (Figure 21.17**A**). Water flows into the mouth and pharynx, then over arrays of gill filaments (Figure 21.17**B**). Each filament has blood vessels that carry blood into and out of it (Figure 21.17**C**). Water flows first over the blood vessel that moves blood from the gill back to the rest of the body. Blood in this vessel has less oxygen than the water, so oxygen diffuses into the blood. Next, the water flows over a vessel moving oxygen-poor blood from the body into the gill. Although the water already gave up some oxygen, it still contains more than the blood in this vessel. So a second helping of oxygen diffuses into the blood.

The most diverse air-breathing land invertebrates are insects. They have a hard exoskeleton that helps conserve water but also prevents gas exchange across the body surface. Insects have a **tracheal system** that consists of repeatedly branching, air-filled tubes that begin at small openings across the body surface (Figure 21.18). Gas exchange occurs at the tips of the finest branches, where gases dissolve in fluid and diffuse into cells. The circulatory system does not play a role in most insect respiration.

Lungs are saclike internal respiratory organs. More than 450 million years ago, lungs evolved from outpouchings in the gut of some fishes. In oxygen-poor water, an ability to gulp air and exchange gases across a respiratory surface in lungs gave these fish an advantage. Later, an ability to breathe air allowed amphibians that descended from these fish to live on land.

Just about all land vertebrates have lungs. Frogs have small lungs that function in oxygen uptake, but most carbon dioxide diffuses outward across their skin. Toads, whose skin in usually dry, have larger lungs. In reptiles, birds, and mammals, the skin is waterproof, so the only respiratory surface is inside the lungs. Birds have a unique system of air sacs connected to their lungs (Figure 21.19). The air sacs are an adaptation to the high oxygen demands of flight. They do not function directly in gas exchange, but rather ensure that there is a constant flow of fresh air across the respiratory surface. As you will learn in the section that follows, each mammalian inhalation mixes fresh air with some oxygen-depleted air that remained in the lungs after the prior exhalation.

trachea (tube inside body)

spiracle (opening to body surface)

FIGURE 21.18 Tracheal tube respiratory system of an insect.

© Cengage Learning.

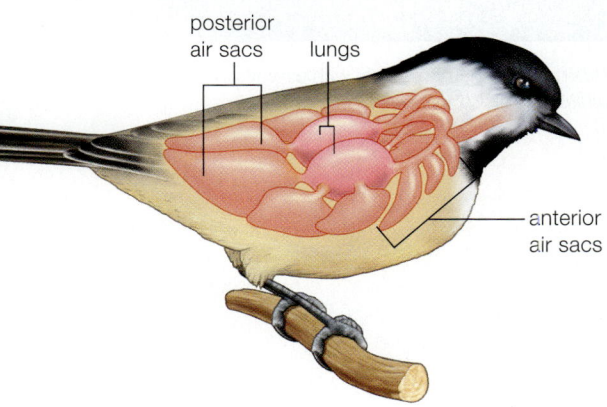

posterior air sacs

lungs

anterior air sacs

FIGURE 21.19 Animated!
Bird respiratory system with air sacs that keep air moving through lungs.

From Russell/Wolfe/Hertz/Starr. Biology, 1e. © 2008 Brooks/Cole, a part of Cengage Learning, Inc.

Take-Home Message

What structural traits facilitate animal respiration?

- Respiration is a physiological process by which animals obtain oxygen and get rid of waste carbon dioxide by diffusion across a moist respiratory surface.

- In some invertebrates, the body surface serves as the respiratory surface.

- Gills increase the surface area for gas exchange in some invertebrates and vertebrates. In fish gills, opposing flow of water and blood increases the efficiency of gas exchange.

- A system of tracheal tubes delivers air deep inside an insect's body.

- Amphibians have small lungs but also exchange gases across their skin.

- Reptiles, birds, and mammals rely on lungs for respiration. Air sacs keep air flowing continually through a bird's lungs.

gills Folds or body extensions that increase the surface area for respiration.

lungs Internal saclike organs that serve as the respiratory surface in most land vertebrates and some fish.

respiration Physiological process by which gases enter and leave an animal body.

respiratory surface Moist surface across which gases are exchanged between animal cells and their environment.

tracheal system Tubes that deliver air from body surface to tissues of insects.

nasal cavity

pharynx (throat)

epiglottis

larynx (voice box)

trachea (windpipe)

left lung

bronchus

bronchiole

diaphragm

intercostal muscle

alveolar sac (sectioned)

alveoli

Resin cast of human lungs, with airways in *white* and blood vessels in *red*.

FIGURE 21.20

Animated! Human respiratory organs and associated structures. The diaphragm and intercostal muscles (muscles between the ribs) are muscles of respiration. The epiglottis is a flap of tissue that folds over the trachea and prevents food from entering it when you swallow.

Credits: left, Martin Dohrn, Royal College of Surgeons/ Photo Researchers, Inc.; right, © Cengage Learning.

21.8 Human Respiratory Function

> From Airways to Alveoli Take a deep breath. Now look at Figure 21.20 to get an idea of where the air traveled in your respiratory system. If you are healthy and sitting quietly, air probably entered through your nose, rather than your mouth. Air from the nostrils enters the nasal cavity, where it is warmed and moistened. The moist, warm air then flows into the **pharynx**, or throat.

Air continues to the **larynx**, a short airway also known as the voice box because it contains a pair of vocal cords (Figure 21.21). Each vocal cord consists of skeletal muscle covered by mucus-secreting epithelium. Contraction of the vocal cords changes the size of the **glottis**, the gap between them. Adjusting the size of the glottis, tension on vocal cords, and speed of air movement past them allows you to make different sounds.

When you breathe without speaking, the glottis is wide and the **epiglottis**, a tissue flap at the entrance to the larynx, points upward. With the epiglottis in its upward position, air can enter the **trachea**, or windpipe. When you swallow, the epiglottis flops down and covers the larynx entrance, so food and fluids enter the esophagus. The esophagus connects the pharynx to the stomach.

The trachea branches into two airways, one to each lung. Each airway is a **bronchus** (plural, bronchi). Human lungs are cone-shaped organs in the thoracic cavity, one on each side of the heart. The rib cage encloses and protects the lungs. A two-layer-thick pleural membrane covers each lung's outer surface and lines the inner thoracic cavity wall.

Once inside a lung, air moves through finer and finer branchings of a "bronchial tree." The branches are called **bronchioles**. At the tips of the finest bronchioles are respiratory **alveoli** (singular, alveolus), little air sacs where gases

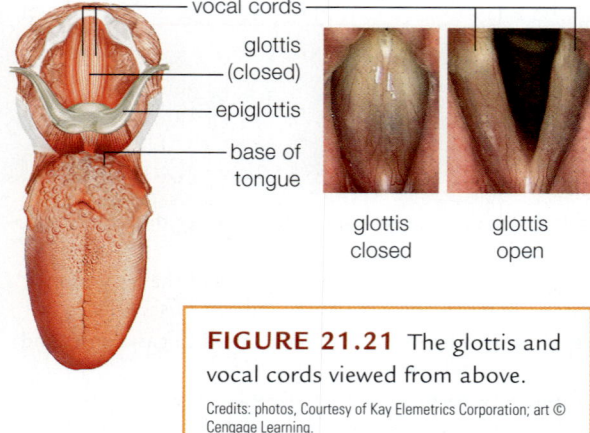

vocal cords

glottis (closed)

epiglottis

base of tongue

glottis closed

glottis open

FIGURE 21.21 The glottis and vocal cords viewed from above.

Credits: photos, Courtesy of Kay Elemetrics Corporation; art © Cengage Learning.

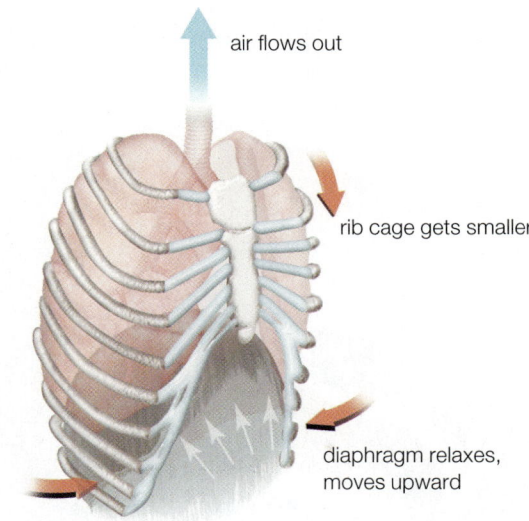

air flows in

rib cage expands

diaphragm contracts and flattens downward

A Inhalation

air flows out

rib cage gets smaller

diaphragm relaxes, moves upward

B Exhalation

are exchanged. Air in alveoli exchanges gases with blood flowing through pulmonary capillaries. Blood oxygenated in these capillaries returns to the heart, which then pumps it to the body's tissues.

❯ How You Breathe A broad sheet of smooth muscle beneath the lungs, the **diaphragm**, partitions the coelom into a thoracic cavity and an abdominal cavity. Of all smooth muscle, only the diaphragm can be controlled voluntarily. You can make it contract by deliberately inhaling. Movements of the diaphragm and **intercostal muscles**, the skeletal muscles between the ribs, allow you to breathe.

A **respiratory cycle** includes one breath in (inhalation) and one breath out (exhalation). Inhalation is always active; muscle contractions drive it. Changes in the volume of the lungs and thoracic cavity cause the movement of air.

When you inhale, the diaphragm contracts and moves downward. At the same time, external intercostal muscles between the ribs contract, moving the rib cage upward and outward (Figure 21.22**A**). The thoracic cavity expands, and air is pulled into the lungs.

Exhalation is usually passive. When muscles that caused inhalation relax, the lungs passively recoil and lung volume decreases. This decrease in volume compresses alveolar sacs, and pushes air out of the lungs (Figure 21.22**B**).

Exhalation becomes active when you exercise vigorously or consciously attempt to expel more air. During active exhalation, internal intercostal muscles contract, pulling the thoracic wall inward and downward. At the same time, muscles of the abdominal wall contract. Abdominal pressure increases and pushes the diaphragm upward. As a result, the volume of the thoracic cavity decreases more than normal, and a bit more air is forced outward.

You do not have to think about breathing. Neurons in a part of your brain stem act as a pacemaker for respiration. When you rest, these neurons send out signals 10 to 14 times per minute. Nerves carry these signals to the diaphragm and intercostal muscles, causing contractions that result in inhalation. When you are more active, muscle cells increase their rate of aerobic respiration and produce more CO_2. This CO_2 enters blood, causing changes that are detected by receptors in arteries and the brain. In a homeostatic response to these changes, the brain alters the breathing pattern, so you breathe faster and more deeply.

alveoli Tiny, thin-walled air sacs that are the site of gas exchange in the lung.
bronchiole Small airway leading to alveoli.
bronchus Airway connecting the trachea to a lung.
diaphragm Dome-shaped muscle at base of thoracic cavity that alters the size of this cavity during breathing.
epiglottis Tissue flap that folds down to prevent food from entering the airways when you swallow.
glottis Opening formed when the vocal cords relax.
intercostal muscles Muscles between the ribs; help alter the size of the thoracic cavity during breathing.
larynx Short airway containing the vocal cords (voice box).
pharynx Throat; opens to airways and digestive tract.
respiratory cycle One inhalation and one exhalation.
trachea Major airway leading to the lungs; windpipe.

cells of alveolar wall

cells of capillary wall

O₂

CO₂

fused basement membranes of both epithelial cell layers

B

A

FIGURE 21.23 The site of gas exchange.
A Cutaway view of a cluster of alveoli. Each alveolus is filled with air and surrounded by a network of pulmonary capillaries.
B The respiratory surface consists of the wall of the alveolus, the wall of the capillary, and their fused basement membranes. Gases diffuse across this three-layered boundary.

© Cengage Learning.

FIGURE 21.24 Animated! Hemoglobin. This protein consists of four globin molecules (*blue* and *green*). Each polypeptide associates with an iron-containing heme group (*red*) that can reversibly bind a molecule of O_2.

© Cengage Learning.

› **Exchanges at the Respiratory Membrane** Inhaled air exchanges gases with the blood at the alveoli. Alveolus means cavity in Latin, and each alveolus is a tiny sphere with a central air-filled cavity (Figure 21.23**A**). Alveolar epithelium contains squamous cells and secretory cells. The secretory cells release a substance (called a surfactant) that lubricates alveolar walls, preventing them from sticking together when the alveolus has to inflate. Pulmonary capillaries run across the outer surface of each alveolus. The basement membranes secreted by the alveolar and capillary cells fuse together as the respiratory membrane (Figure 21.23**B**). Gases diffuse quickly across this thin membrane between the air inside alveoli and the blood flowing through a capillary.

› **Transport of Gases** Some oxygen can dissolve in plasma, but not enough to meet the body's needs. In all vertebrates and many invertebrates, hemoglobin increases blood's oxygen-carrying capacity. Nearly all the oxygen you inhale becomes bound to hemoglobin in your red blood cells. Hemoglobin is a protein that consists of four polypeptide chains (globin), each with an associated iron-containing cofactor called a heme (Figure 21.24). Oxygen binds reversibly to the iron in these hemes. Hemoglobin tends to let go of oxygen in regions where the oxygen concentration is low, the temperature is warm, the pH is low, and the concentration of carbon dioxide is high. Such conditions occur in contracting muscles and other metabolically active tissues. After giving up oxygen to tissues, red blood cells return to the pulmonary capillaries to pick up more.

Carbon dioxide diffuses into the blood from body tissues, where its concentration exceeds that of the blood. On their return trip from the tissues to the lungs, red blood cells carry a small amount of carbon dioxide bound to hemoglobin. However most carbon dioxide is transported to the lungs as bicarbonate ions (HCO_3^-). Enzymes inside red blood cells carry out a reaction that combines carbon dioxide with water to form these ions. Once formed, the bicarbonate diffuses out of the red blood cell and travels to the lungs in the plasma. In the pulmonary capillaries, bicarbonate is converted back to carbon dioxide. The carbon dioxide diffuses into the air in alveoli and is exhaled.

› **Respiratory Disorders and the Effects of Smoking** Bronchitis is a common respiratory disorder in which the mucus-producing lining of the bronchi becomes inflamed. Inflamed epithelial cells secrete extra mucus that triggers coughing. Bacteria can colonize the mucus, leading to more inflammation, more mucus, and more coughing. Bronchitis often arises after an upper respiratory infection. Smoking irritates airways and can cause chronic bronchitis.

With asthma, an inhaled allergen or irritant triggers inflammation and constriction of airways, making breathing difficult. A tendency to have asthma is inherited, but avoiding potential irritants such as cigarette smoke and air pollutants can reduce the frequency of attacks. An acute asthma attack is treated with inhaled drugs that cause smooth muscle ringing the airways to relax.

With emphysema, tissue-destroying bacterial enzymes digest the thin, elastic alveolar wall. As these walls disappear, the area of the respiratory surface declines. Over time, the lungs become distended and inelastic, leaving the person constantly feeling short of breath. Some people inherit a genetic predisposition to emphysema. They do not have a functioning gene for an enzyme that inhibits bacterial attacks on alveoli. However, tobacco smoking is by far the main risk factor for emphysema (Figure 21.25).

Each year, about 5 million people die of disorders brought on by smoking tobacco. In addition to emphysema, smoking brings on cancers, heart attacks, and strokes. As G. H. Brundtland, the former director of the World Health Organization, once put it, tobacco remains the only legal consumer product that kills half of its regular users.

Take-Home Message

How does the human respiratory system function?

- Air flows to and from the lungs through a system of airways. Air enters through the nose or mouth, and flows through the pharynx, larynx (voice box), and trachea to the two bronchi that carry air into the lungs.

- Inhalation is always active. Contraction of the diaphragm and muscles of the rib cage increase the volume of the thoracic cavity and lungs, so air is sucked into the lungs.

- Exhalation is usually passive. As muscles of respiration relax, the volume of the thoracic cavity and lungs decrease, pushing air out of the lungs.

- Gas exchange takes place in alveoli. Oxygen diffuses from air inside alveoli into pulmonary capillaries. Carbon dioxide diffuses in the opposite direction.

- Blood carries gases between the lungs and body tissues. Most of the oxygen in blood is bound to hemoglobin in red blood cells. Most carbon dioxide is converted to bicarbonate, which travels in the plasma.

FIGURE 21.25 The effect of smoking on lungs. **A** Normal human lungs. **B** Lungs from a smoker with emphysema. © O. Aurbach/ Visuals Unlimited.

21.9 A Shocking Save (revisited)

Most cardiac arrests do not occur in a hospital, so the presence of a bystander willing to carry out CPR or to use an AED often means the difference between life and death. Sadly, although most cardiac arrests are witnessed, only about 15 percent of victims get CPR before trained personnel arrive.

Most people have an understandable reluctance to engage in mouth-to-mouth contact with a stranger. A new procedure called CCR (cardiocerebral resuscitation) relies on chest compressions alone to move air into and out of the lungs. Proponents of CCR argue that as long as a person's airway is clear, chest compressions will move enough air into and out of a victim's lungs to oxygenate the blood flowing to the person's heart and brain. Early studies indicate that CCR may be as good as or even better than traditional CPR at saving most people who experience a sudden cardiac arrest.

Another promising development is the rising availability of AEDs. In 2010, approximately 200,000 AEDs were purchased for installation in public places. That's the good news. The bad news is that most people still do not know what an AED is, where they are available, or how simple it is to use one. The photo at the *upper right* shows the symbol for an AED in public place.

Sign in an airport indicating where an AED is stored.

WHERE YOU ARE GOING . . .

White blood cells and the lymphatic system play an integral role in immunity, the subject of Chapter 22. The role of the blood and lymphatic system in the distribution of nutrients is covered in Section 23.3; filtration of the blood by kidneys, in Section 23.6. You'll learn about blood loss during menstruation in Section 26.3, and how blood inflates the penis during an erection in Section 26.4. Section 26.6 explains how gases are exchanged between fetal and maternal bloodstreams.

Summary

Section 21.1 When the heart stops pumping, blood flow halts and brain cells begin to die. CPR can keep some oxygenated blood moving to cells, but it cannot restart a heart. Reestablishing the normal rhythm requires a shock from a defibrillator.

Section 21.2 In some invertebrates, substances simply move to and from cells by diffusion through the **interstitial fluid**. Most animals have a circulatory system, in which a **heart** pumps blood through blood vessels. In an **open circulatory system**, hemolymph leaves vessels and flows among tissues. In a **closed circulatory system**, blood stays inside vessels and exchanges take place at **capillaries**. A fish heart has one **atrium** and one **ventricle**, so blood flows in a single circuit. In other vertebrates, a **pulmonary circuit** services the lungs and a **systemic circuit** extends through the rest of the body.

Sections 21.3, 21.4 Each half of the human heart has two chambers: an upper atrium and a lower ventricle, with valves between them. During the **cardiac cycle**, electrical signals from the **cardiac pacemaker** (the SA node) trigger contraction of the atria, then the ventricles. The sounds we associate with a heartbeat are caused by closing of heart valves. Contraction of the ventricles drives blood flow away from the heart through **arteries**. Blood returns to the atria through **veins**.

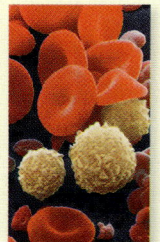

Sections 21.5, 21.6 Blood consists of plasma, blood cells, and platelets. **Plasma** is water with dissolved ions and molecules. **Red blood cells** are packed with hemoglobin and function mainly in the transport of oxygen. Many kinds of **white blood cells** function in housekeeping and defense. **Platelets** release substances that initiate blood clotting. All blood cells and platelets arise from stem cells in bone marrow.

Blood pressure, the pressure exerted by blood on the walls of blood vessels, is usually measured in arteries. It is recorded as **systolic pressure** over **diastolic pressure**. Chronic high blood pressure, or **hypertension**, threatens health, as does **athero-sclerosis**, which narrows arteries. **Arterioles** adjust the distribution of blood through the body. Capillaries are diffusion zones for exchanges between blood and interstitial fluid. **Venules** carry blood to veins, which convey blood back to the heart and also function as a blood volume reservoir. Valves keep blood from moving backward in veins. **Lymph** is a mix of interstitial fluid and protein-free plasma that escaped from capillaries. Lymph vessels return lymph to veins.

Section 21.7 **Respiration** is a physiological process by which gases enter and leave an animal body across a **respiratory surface**. In small, flattened animals, the body surface can serve as the respiratory surface. Sea slugs have external **gills** and fish have internal ones. Insects have a **tracheal system** that conveys air deep inside the body. Reptiles, birds, and mammals rely on their **lungs** for gas exchange. In birds, air sacs connected to lungs keep air flowing continually across the respiratory surface.

Section 21.8 In humans, air flows through the nasal cavities, the **pharynx**, the **larynx**, **trachea**, **bronchi**, and **bronchioles**, which end at **alveoli** deep inside the lungs. When you swallow, the **epiglottis** covers the **glottis** so food does not enter the larynx.

A **respiratory cycle** is one inhalation and one exhalation. Inhalation requires energy. Contraction of the **diaphragm** and **intercostal muscles** between the ribs expand the chest cavity and pull air into the lungs. These events are reversed during exhalation, which is usually passive.

The walls of alveoli and of pulmonary capillaries combine as the respiratory membrane. Oxygen follows its concentration gradient from alveolar air spaces into the pulmonary capillaries, then into red blood cells, where it binds reversibly with hemoglobin. In capillary beds in the tissues, hemoglobin releases oxygen, which diffuses across interstitial fluid into cells. Carbon dioxide diffuses from cells into the interstitial fluid, then into the blood where it combines with water to form bicarbonate. In the lungs, bicarbonate is converted back to carbon dioxide and water. The carbon dioxide diffuses from the blood into air in alveoli, then is exhaled.

Bronchitis and emphysema are respiratory disorders that can be caused or worsened by smoking. Worldwide, smoking is a leading cause of death.

Self-Quiz   Answers in Appendix I

1. All vertebrates have _____ .
   a. an open circulatory system   c. a four-chamber heart
   b. a closed circulatory system d. both b and c

2. Signals initiated by the _____ cause the heart to contract.
   a. AV node   c. pericardium
   b. SA node   d. brain

3. Blood pressure is highest in the _____ and lowest in the _____ .
   a. arteries; veins   c. veins; arteries
   b. arterioles; venules   d. capillaries; arterioles

4. Contraction of _____ is the main force driving the flow of blood away from the heart.
   a. the atria   c. the ventricles
   b. arterioles   d. skeletal muscle

5. At rest, the largest volume of blood is in the _____ .
   a. arteries   c. veins
   b. capillaries   d. arterioles

6. In the blood, most oxygen is transported _____ .
   a. in red blood cells   c. bound to hemoglobin
   b. in white blood cells   d. both a and c

7. The _____ circuit carries blood from the heart to the lungs, then back to the heart.
   a. pulmonary   b. systemic

8. The heart chamber with the thickest wall pumps blood into the _____ .
   a. aorta   c. pulmonary vein
   b. pulmonary artery   d. superior vena cava

Digging Into Data

Risks of Radon

Radon is a colorless, odorless gas emitted by many rocks and soils. It is formed by the radioactive decay of uranium and is itself radioactive. There is some radon in the air almost everywhere, but routinely inhaling a lot of it raises the risk of lung cancer. Radon also seems to increase cancer risk far more in smokers than in nonsmokers. Figure 21.26 is an estimate of how radon in homes affects risk of lung cancer mortality. Note that these data show only the death risk for radon-induced cancers. Smokers are also at risk from lung cancers that are caused by tobacco.

1. If 1,000 smokers were exposed to a radon level of 1.3 pCi/L over a lifetime (the average indoor radon level), how many would die of a radon-induced lung cancer?

2. How high would the radon level have to be to cause approximately the same number of cancers among 1,000 nonsmokers?

3. The risk of dying in a car crash is about 7 out of 1,000. Is a smoker in a home with an average radon level (1.3 pCi/L) more likely to die in a car crash or of radon-induced cancer?

Radon level (pCi/L)	Risk of cancer death from lifetime radon exposure	
	Never smoked	Current smokers
20	36 out of 1,000	260 out of 1,000
10	18 out of 1,000	150 out of 1,000
8	15 out of 1,000	120 out of 1,000
4	7 out of 1,000	62 out of 1,000
2	4 out of 1,000	32 out of 1,000
1.3	2 out of 1,000	20 out of 1,000
0.4	>1 out of 1,000	6 out of 1,000

FIGURE 21.26 Estimated risk of lung cancer death as a result of lifetime radon exposure. Radon levels are measured in picocuries per liter (pCi/L). The Environmental Protection Agency considers a radon level above 4 pCi/L unsafe. To learn about testing for radon and what to do about a high radon level, visit the EPA's radon information site at www.epa.gov/radon.

© Cengage Learning.

9. In human lungs, gas exchange occurs at the _____ .
   a. bronchi
   b. pericardium
   c. alveoli
   d. epiglottis

10. When you breathe quietly, inhalation is _____ and exhalation is _____ .
   a. passive; passive
   b. active; active
   c. passive; active
   d. active; passive

11. During inhalation, _____ .
   a. the thoracic cavity expands
   b. the glottis closes
   c. the diaphragm relaxes
   d. all of the above

12. The diaphragm is _____ muscle.
   a. smooth   b. skeletal   c. cardiac

13. Match each disorder with its description.
   ____ anemia
   ____ atherosclerosis
   ____ hypertension
   ____ emphysema
   ____ hemophilia
   ____ cardiac arrest
   ____ asthma
   ____ bronchitis

   a. heart stops beating
   b. alveoli walls break down
   c. airways constrict
   d. impaired blood clotting
   e. chronic high blood pressure
   f. inflammation of bronchi
   g. too few red blood cells
   h. narrows arteries

14. Match the words with their descriptions.
   ____ plasma
   ____ alveolus
   ____ hemoglobin
   ____ veins
   ____ SA node
   ____ trachea
   ____ glottis
   ____ ventricle
   ____ atrium

   a. receives blood from veins
   b. fluid component of blood
   c. cardiac pacemaker
   d. gap between vocal cords
   e. site of gas exchange
   f. drives blood flow from heart
   g. windpipe
   h. blood volume reservoir
   i. reversibly binds oxygen

15. Arrange these structures in the order in which inhaled air flows through them.
   ____ larynx
   ____ pharynx
   ____ trachea
   ____ bronchus
   ____ bronchiole

Critical Thinking

1. Sitting motionless for long periods, as on airline flights, allows blood to pool and clots to form in legs. Long-distance flights cause clot formation in about 1 percent of air travelers, and the risk is the same regardless of whether a person is in a first-class seat or an economy seat. Physicians recommend that air travelers drink plenty of fluids and periodically get up and walk around the cabin. Given what you know about blood flow in the veins, explain why these precautions can lower the risk of clot formation.

2. Blood entering systemic capillaries has the same oxygen and carbon dioxide concentration as the blood in pulmonary veins. Blood in systemic veins has the same oxygen and carbon dioxide concentration as the blood in pulmonary arteries. Explain the reason for each of these similarities.

3. High blood pressure can cause edema, a condition in which excess fluid pools in the tissues, causing them to appear swollen. Which type of blood vessel do you think the fluid escapes from? Explain your reasoning.

4. Carbon monoxide is a colorless, odorless gas produced by combustion, including burning of tobacco. If inhaled, it binds tightly to the oxygen-binding sites on hemoglobin. Why does exposure to low levels of carbon monoxide produce symptoms like those of anemia?

22 Immunity

. . . WHere YOU HaVe BeeN

In this chapter, we discuss bacteria (Sections 3.4, 13.5) and viruses (13.4) as human pathogens. Phagocytosis (4.6) by white blood cells (21.5) plays a large role in immune responses. The chapter also returns to eukaryotic cell structure (3.5, 3.6), DNA hybridization (10.2) and gene therapy (10.5), AIDS (13.1), vaccines (13.4, 16.5), epithelial tissue (19.3) and skin (19.4), the endocrine and circulatory systems (19.4, 19.5, 21.3), and cardiovascular disorders (21.6).

22.1 Frankie's Last Wish

In October of 2000, Frankie McCullough had known for a few months that something was not quite right. She had not had an annual checkup in many years; after all, she was only 31 and had been healthy her whole life. It never occurred to her to doubt her own invincibility until the moment she saw the doctor's face change as he examined her cervix.

The cervix is the lowest part of the uterus, or womb. Cervical cells can become cancerous, but the process is usually slow. The cells pass through several precancerous stages that are detectable by routine Pap tests. Precancerous and even early-stage cancerous cells can be removed from the cervix before they spread to other parts of the body. However, plenty of women like Frankie do not take advantage of regular exams. They do not see a gynecologist unless they have pain or bleeding, which can be symptoms of advanced cervical cancer. Even with treatment, 85 percent of the women who end up with this type of cancer will die within 5 years. Thousands of women die of cervical cancer each year in the United States, and many more die in places where routine gynecological testing is not common.

Healthy cervical cells are transformed into cancerous ones by infection with human papillomavirus (HPV), a virus that infects skin and mucous membranes. A few of the 100 or so different strains of HPV cause warts on the hands or feet, or in the mouth. About 30 others that infect the genital area sometimes cause genital warts, but usually there are no symptoms of infection. Genital HPV is spread very easily by sexual contact: A woman has a 50% chance of being infected with genital HPV within three years of becoming sexually active.

A genital HPV infection usually goes away on its own, but not always. A persistent infection with one of about 10 strains is the main risk factor for cervical cancer (Figure 22.1). Types 16 and 18 are particularly dangerous: One of the two is found in more than 70 percent of all cervical cancers.

In 2006, the U.S. Food and Drug Administration (FDA) approved Gardasil, a vaccine that protects the recipient against infection by four types of genital HPV, including types 16 and 18. The vaccine also prevents cervical cancer caused by these HPV strains. It is most effective in girls who have not yet become sexually active, because they are least likely to have already become infected with HPV.

The HPV vaccine came too late for Frankie McCullough. Despite radiation treatments and chemotherapy, her cervical cancer spread quickly. She died in 2001, leaving a wish for other people: awareness. "If there is one thing I could tell a young woman to convince them to have a yearly exam, it would be not to assume that your youth will protect you. Cancer does not discriminate; it will attack at random, and early detection is the answer." Almost all women newly diagnosed with invasive cervical cancer have not had a Pap test in five years.

FIGURE 22.1 HPV and cervical cancer.

Above, a Pap test reveals HPV-infected cervical cells among normal ones. Infected cells have enlarged, often multiple nuclei surrounded by a clear area. These changes sometimes lead to cervical cancer, which is treatable if detected early enough.

Inset, Frankie McCullough (waving), who died of cervical cancer at age 32.

Credits: above, Biomedical Imaging Unit, Southampton General Hospital/ Photo Researchers, Inc.; inset, In memory of Frankie McCullough.

22.2 Integrated Responses to Threats

> **Evolution of the Body's Defenses** You continually cross paths with a tremendous array of viruses, bacteria, fungi, parasitic worms, and other agents of disease, but you need not lose sleep over this. Humans coevolved with these pathogens, so you have defenses that protect your body from them. The evolution of **immunity**, an organism's capacity to resist and combat infection, began well before multicelled eukaryotes evolved from free-living cells. Mutations in the genes for membrane proteins introduced new molecular patterns that were unique in cells of a given type. As multicellularity evolved, so did mechanisms of identifying the patterns as self, or belonging to one's own body.

By about 1 billion years ago, nonself recognition had also evolved. Cells of all modern multicelled eukaryotes bear a set of receptors that collectively can recognize around 1,000 different nonself cues, which are called pathogen-associated molecular patterns (PAMPs). As their name suggests, PAMPs occur mainly on or in pathogens. They include proteins that make up bacterial flagella and pili (Section 3.4), some components of bacterial cell walls, double-stranded RNA unique to some viruses, and so on. A PAMP is an example of **antigen**, which is any molecule or particle recognized by the body as nonself. Binding of a cell's receptors to antigen triggers a set of immediate, general defense responses. In mammals, for example, the binding activates **complement**: a set of proteins that circulate in inactive form throughout the body in blood and tissue fluids. Activated complement proteins can destroy invading cells or mark them for uptake by phagocytic cells (those that engulf material by phagocytosis).

Pattern receptors and the responses they trigger are part of **innate immunity**, a set of immediate, general defenses that help protect multicelled organisms from infection. Vertebrates have an additional set of defenses carried out by interacting cells, tissues, and proteins. This **adaptive immunity** tailors immune defenses to a vast array of specific pathogens that an individual may encounter during its lifetime. Table 22.1 compares adaptive and innate immunity.

> **Three Lines of Defense** The mechanisms of adaptive immunity evolved within the context of innate immunity. The two systems were once thought to operate independently of each other, but we now know they function together. We describe both systems together in terms of three lines of defense. The first line of defense includes the physical, chemical, and mechanical barriers that usually keep pathogens on the outside of the body (Figure 22.2). Innate immunity, the second line of defense, begins after a tissue is damaged, or after antigen is detected inside the body. Its general response mechanisms quickly rid the body of many invaders. An innate immune response triggers adaptive immunity, the third line of defense. White blood cells divide to form huge populations that target a specific antigen and destroy anything bearing it. Some white blood cells that form during an adaptive response persist after infection ends. If the same antigen returns, these memory cells mount a secondary response.

> **The Defenders** White blood cells participate in all immune responses. Many kinds circulate through the body in blood and lymph; others populate the lymph nodes, spleen, and other tissues. All communicate with one another by secreting and responding to chemical signaling molecules. These molecules, which include proteins and polypeptides called cytokines, allow cells throughout the body to coordinate their activities during an immune response.

Table 22.1	Innate and Adaptive Immunity Compared	
	Innate Immunity	Adaptive Immunity
Response time	Immediate	About a week
How antigen is detected	Fixed set of receptors for pathogen-associated molecular patterns (PAMPs)	Antigen receptors produced by gene recombinations
Specificity	About 1,000 PAMPs	Billions of antigens
Persistence	None	Long-term

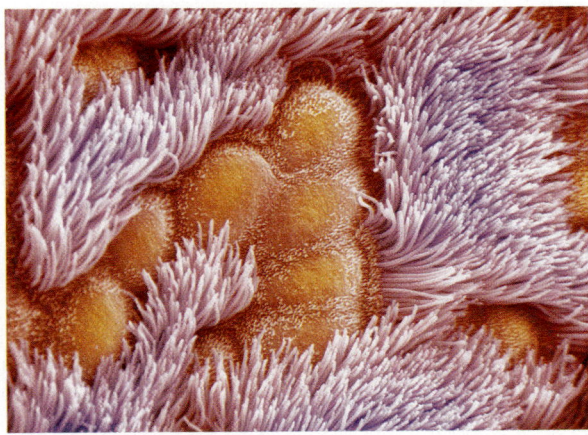

FIGURE 22.2 A physical barrier to infection.

Mucus and the sweeping action of cilia keep pathogens from getting a foothold in airways to the lungs. Bacteria and other particles get stuck in mucus secreted by goblet cells (*gold*). Cilia (*pink*) on other cells sweep the mucus to the throat for disposal.

Dr. Richard Kessel and Dr. Randy Kardon/ Tissues & Organs/ Visuals Unlimited.

adaptive immunity In vertebrates, set of immune defenses that can be tailored to specific pathogens encountered by an organism during its lifetime.

antigen A molecule or particle that the immune system recognizes as nonself. Triggers an immune response.

complement A set of proteins that circulate in inactive form in blood, and when activated play a role in immune responses.

immunity The body's ability to resist and fight infections.

innate immunity In all multicelled organisms, set of immediate, general defenses against infection.

neutrophil   monocyte   basophil   lymphocyte   eosinophil

FIGURE 22.3 Animated! White blood cells that circulate in the blood. Staining reveals details such as cytoplasmic granules that contain enzymes, toxins, and signaling molecules.

© Antonio Zamora, www.scientificpsychic.com.

Different types of white blood cells are specialized for specific tasks. Those that are phagocytic (Section 4.6) engulf and digest pathogens, dead cells, or other particles. Neutrophils, which circulate in blood (Figure 22.3), are the most abundant phagocytes. Phagocytic **macrophages** that migrate through tissues and tissue fluids develop from monocytes, which patrol the blood. **Dendritic cells** are phagocytes that alert the adaptive immune system to the presence of antigen in solid tissues.

Many white blood cells have secretory vesicles called granules, which can contain cytokines, local signaling molecules, destructive enzymes, and toxins such as hydrogen peroxide. In response to a trigger such as antigen binding, these cells degranulate (they release the contents of their granules). Neutrophils have granules, as do eosinophils that target parasites too big for phagocytosis. Basophils and mast cells degranulate in response to injury or antigen. Mast cells, unlike most other white blood cells, stay anchored in tissues. These cells are often closely associated with nerves, and they also degranulate in response to signaling molecules secreted by cells of the endocrine and nervous systems.

Lymphocytes are a special category of white blood cells that have a central role in adaptive immunity. There are several kinds of **T cells** (T lymphocytes), including cytotoxic T cells that can kill infected or cancerous body cells. **B cells** (B lymphocytes) make antibodies. A lymphocyte called an NK cell (natural killer cell) kills cancerous body cells undetectable by cytotoxic T cells. Collectively, lymphocytes have the capacity to target billions of specific antigens.

Take-Home Message

What is immunity?

- Innate immunity of multicelled organisms comprises a system of immediate, general defenses against a fixed number of antigens.

- Vertebrate adaptive immunity is a system of defenses that can specifically target billions of different antigens.

- White blood cells are central to both systems; signaling molecules such as cytokines integrate their activities.

22.3 Surface Barriers

Your skin is in constant contact with the external environment, so it picks up many microorganisms. It normally teems with about 200 different kinds of yeast, protozoa, and bacteria. If you showered today, there are probably thousands of microorganisms on every square inch of your external surfaces. If you did not, there may be billions. They tend to flourish in warm, moist areas, such as between the toes. Huge populations inhabit cavities and tubes that open on the body's surface, including the eyes, nose, mouth, and anal and genital openings.

Microorganisms that typically live on human surfaces, including the interior tubes and cavities of the digestive and respiratory tracts, are called **normal flora** (Figure 22.4). Our surfaces provide them with a stable environment and nutrients. In return, their populations deter more dangerous species from colonizing (and penetrating) body surfaces. Normal flora in the digestive tract help us

B cell B lymphocyte. Lymphocyte that can make antibodies.

dendritic cell Phagocytic white blood cell that alerts the immune system to the presence of antigen in solid tissues.

lysozyme Antibacterial enzyme that occurs in body secretions such as mucus.

macrophage Phagocytic white blood cell that patrols tissues and tissue fluids.

normal flora Microorganisms that typically live on human surfaces, including the interior tubes and cavities of the digestive and respiratory tracts.

T cell T lymphocyte. Lymphocyte central to adaptive immunity; some kinds target infected or cancerous body cells.

digest food, and also they make essential nutrients such as vitamins K and B_{12}.

Normal flora are helpful only on body surfaces; they can cause or worsen many conditions when they invade tissues. Serious illnesses associated with normal flora include pneumonia; ulcers; colitis; whooping cough; meningitis; abscesses of the lung and brain; and cancers of the colon, stomach, and intestine. The bacterial agent of tetanus, *Clostridium tetani*, is considered a normal inhabitant of human intestines. The bacteria responsible for diphtheria, *Corynebacterium diphtheriae*, was normal skin flora before widespread use of the vaccine eradicated the disease. *Staphylococcus aureus*, a resident of human skin and linings of the mouth, nose, throat, and intestines (Figure 22.4**B**), is also a leading cause of human bacterial disease. Antibiotic-resistant strains of *S. aureus* are now widespread. A particularly dangerous kind, MRSA (methicillin-resistant *S. aureus*), is resistant to a wide range of antibiotics. MRSA is now a permanent resident of hospitals around the world.

In contrast to body surfaces, the blood and tissue fluids of healthy people are typically sterile (free of microorganisms). Surface barriers usually prevent normal flora from entering the body's internal environment. The tough outer layer of vertebrate skin (Section 19.3) is an example. Microorganisms flourish on skin's waterproof, oily surface, but they rarely penetrate its thick layer of dead cells (Figure 22.5).

The thinner epithelial tissues that line the body's interior tubes and cavities also have surface barriers. Figure 22.4**B** shows how sticky mucus secreted by cells of these linings can trap microorganisms. The mucus contains **lysozyme**, an enzyme that kills bacteria. In the sinuses and respiratory tract, the coordinated beating of cilia sweeps trapped microorganisms away before they have a chance to breach the delicate walls of these structures.

Your mouth is a particularly inviting habitat for microorganisms because it offers plenty of nutrients, warmth, moisture, and surfaces for colonization. Accordingly, it harbors huge populations of normal flora that can resist lysozyme in saliva. Microorganisms that get swallowed are typically killed by gastric fluid, a potent brew of protein-digesting enzymes and acid in the stomach. Any that survive passage to the small intestine are usually killed by bile salts. The hardy ones that reach the large intestine must compete with about 500 resident species that are specialized to live there and have already established large populations. Any that displace normal flora are typically flushed out by diarrhea.

Lactic acid produced by *Lactobacillus* helps keep the vaginal pH outside the range of tolerance of most fungi and other bacteria. Urination's flushing action usually prevents pathogens from colonizing the urinary tract.

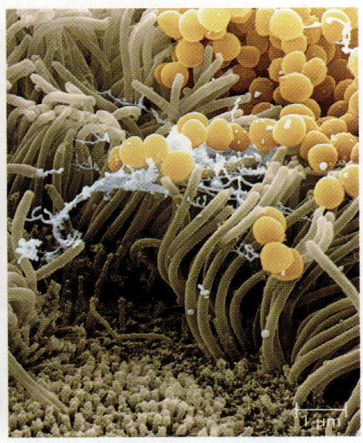

A *Staphylococcus epidermidis*, a common bacterial inhabitant of human skin.

B *Staphylococcus aureus* bacteria (*yellow*) stuck in mucus secreted by human nasal epithelial cells.

FIGURE 22.4 Examples of normal flora.

Credits: (a) © David Scharf, 1999. All rights reserved; (b) Juergen Berger/ Photo Researchers, Inc.

skin surface

dead cells

living cells

FIGURE 22.5 One surface barrier to infection: the epidermis of human skin. Its thick, waterproof layer of dead cells usually keeps normal skin flora from penetrating internal tissues.

Veronika Burmeister/ Visuals Unlimited, Inc.

Take-Home Message

What prevents ever-present microorganisms from entering the body's internal environment?

- Surface barriers usually keep microorganisms that contact or inhabit body surfaces from invading the internal environment.
- Skin's tough epidermis is one barrier. Mucus, lysozyme, and often the sweeping action of cilia protect the linings of internal tubes and cavities.
- Resident populations of normal flora deter more dangerous microorganisms from colonizing the body's internal and external surfaces.

membrane attack complex

lipid bilayer

pore

A Complement. Activated complement can assemble into membrane attack complexes that insert themselves into a lipid bilayer (*top*). The resulting pores (*bottom*) cause the cell to burst.

B Phagocytosis. This macrophage is in the process of engulfing tuberculosis bacteria (*red*).

C Neutrophil nets. Two *Klebsiella* bacteria (*purple*) are ensnared in material that has been ejected by a neutrophil in lung tissue.

FIGURE 22.6 Animated! A few of the defense mechanisms of innate immunity.

Credits: (a) top, © Cengage Learning; bottom, Robert R. Dourmashkin, courtesy of Clinical Research Centre, Harrow, England; (b) Photo Researchers Inc.; (c) © 2010, Papayannopoulos et al. Originally published in *J. Cell Biol.* 191:677-691. doi: 10.1083/jcb.201006052 (Image by Volker Brinkman and Abdul Hakkim).

22.4 The Innate Immune Response

What happens if a pathogen slips by surface defenses and enters the body's internal environment? A second line of defense—the fast-acting, general mechanisms of innate immunity—can keep an invading pathogen from establishing a population in body tissues or fluids.

❯ Complement About 30 different kinds of complement proteins circulate through the blood and interstitial fluid of a vertebrate body. Some of these proteins become activated when they encounter cytoplasmic or mitochondrial proteins leaking out of damaged body cells. Others are activated by antigen, or antibody bound to antigen (we return to antibodies in Section 22.5).

Activated complement proteins activate other complement proteins, which activate other complement proteins, and so on. These cascading reactions quickly produce huge amounts of activated complement. The proteins diffuse into surrounding tissues to form a concentration gradient of activated complement around an affected site.

Some of the activated complement proteins bind to pathogens, forming a coating that enhances uptake by phagocytic white blood cells. Others assemble into complexes that can insert themselves into a lipid bilayer of a foreign cell or ailing body cell, opening large channels in the plasma membrane that cause the cell to burst (Figure 22.6**A**).

Healthy body cells continuously produce proteins that inactivate complement, a negative feedback mechanism that prevents a complement activation cascade from spreading too far into healthy tissue. Microorganisms do not make these inhibitory proteins, so they are preferentially singled out for destruction.

❯ Phagocytosis Dendritic cells, macrophages, and neutrophils bear receptors for activated complement, and these phagocytes can follow a complement gradient back to an affected site. Dendritic cells are concentrated in the lining of respiratory airways and other tissues that contact the external environment. Phagocytosis by these cells plays a critical role in protecting the lungs from pathogens and other harmful particles. However, the main function of dendritic cells is to present antigen to T cells (more on how this works in Section 22.5).

Macrophages in interstitial fluid engulf essentially everything except undamaged body cells (Figure 22.6**B**), but they are more than just scavengers. Upon engulfing antigen, a macrophage secretes cytokines that alert the adaptive immune system to the presence of the threat.

A neutrophil that engulfs a microorganism releases the contents of its granules. Enzymes and toxins released into extracellular fluid destroy all cells in the vicinity, sick and healthy alike. Recently, researchers discovered that neutrophils literally explode in response to a certain combination of signaling molecules and complement, in the process ejecting their nuclear DNA and associated proteins along with the contents of their granules. The mixture solidifies into a net that traps pathogens in the vicinity of the secreted antimicrobial compounds (Figure 22.6**C**). Neutrophil nets are very effective at killing bacteria.

❯ Inflammation Tissue damage or infection triggers **inflammation**, a fast, local response that simultaneously destroys affected tissues and jump-starts the healing process (Figure 22.7). Inflammation begins when a white blood cell releases the contents of its granules. Degranulation can occur in response to a number of stimuli, including pattern receptor binding to antigen, for example on the surface of a bacterium ❶. Activated complement proteins also trigger degranulation when they bind to receptors on a white blood cell's plasma membrane.

mast cell — bacterium — complement-coated bacterium

local signaling molecules

macrophage

blood vessel

FIGURE 22.7 Animated! Example of inflammation as a response to bacterial infection.

❶ Pattern receptors on mast cells in the tissue recognize and bind to bacterial antigen.

❷ The mast cells release signaling molecules (*blue* dots) that cause arterioles to widen. The resulting increase in blood flow reddens and warms the tissue.

❸ The signaling molecules also increase capillary permeability, which allows phagocytic white blood cells to squeeze through the vessel walls into the tissue. Plasma proteins that leak out of the capillaries cause the tissue to swell with fluid.

❹ Bacterial antigens activate complement (*purple* dots). Activated complement binds to the bacteria.

❺ Phagocytic white blood cells in the tissue recognize and engulf the complement-coated bacteria.

© Cengage Learning 2010.

Among the substances released by the degranulating cells are local signaling molecules called prostaglandins and histamines ❷. The release of these substances has two effects. First, nearby arterioles widen. As a result, blood flow to the area increases, speeding the arrival of more phagocytic white blood cells ❸. Second, spaces open up between cells that make up nearby capillary walls, so phagocytes in the bloodstream can more easily exit the blood vessels ❹. By the time this occurs, any invading cells have become coated with activated complement, which makes them easy targets for the phagocytes ❺.

Symptoms of inflammation include redness and warmth that result from increased blood flow to the affected area. The increased permeability of capillary walls allows more plasma proteins to escape into interstitial fluid. The fluid becomes hypertonic with respect to blood, and water follows by osmosis. The tissue swells with excess fluid, putting pressure on nerves and thus causing pain.

Inflammation continues as long as its triggers do. Once these stimuli subside, for example after invading bacteria have been cleared from an infected tissue, macrophages begin to produce compounds that suppress inflammation and promote tissue repair. If the stimulus persists, inflammation becomes chronic. Chronic inflammation is not a normal condition and does not benefit the body. Rather, it causes or contributes to many diseases, including asthma, Crohn's disease, rheumatoid arthritis, atherosclerosis, diabetes, and cancer.

❯ **Fever** Fever is a temporary rise in body temperature above the normal 37°C (98.6°F) that often occurs in response to infection. The hypothalamus regulates body temperature (Section 19.5). Some cytokines stimulate brain cells to release prostaglandins, which cause the hypothalamus to raise the body's internal temperature set point. As long as the temperature of the body is below the new set point, the hypothalamus sends out signals that cause blood vessels in the skin to constrict, which reduces heat loss from the skin. The signals also trigger reflexive movements called shivering, or "chills," that increase the metabolic heat output of muscles. Both responses raise the body's internal temperature.

Fever enhances immune defenses by increasing the rate of enzyme activity, thus speeding up metabolism, tissue repair, and formation and activity of phagocytic cells. In addition, many pathogens multiply more slowly at the higher temperature, so white blood cells can get a head start in the proliferation race against them.

fever An internally induced rise in core body temperature above the normal set point as a response to infection or tissue damage.

inflammation A local response to tissue damage or infection; characterized by redness, warmth, swelling, and pain.

A The bacterial cause of acne, *Propionibacterium acnes*.

B *Left*, bacteria (*Streptococcus mutans*, *gold*, and *Porphyromonas gingivalis*, *green*), and yeast (*red*) in dental plaque. *Right*, micrograph of toothbrush bristles scrubbing plaque on a tooth surface.

FIGURE 22.8 Normal flora associated with common health problems.

Credits: (a) Kwangshin Kim/ Photo Researchers, Inc.; (b) left, © Dennis Kunkel Microscopy, Inc./ Phototake; right, www.zahnarztstuttgart.com.

A fever is a sign that the body is fighting something, so it should never be ignored. However, a fever of 40.6°C (105°F) or less does not necessarily require treatment in an otherwise healthy adult. Body temperature usually will not rise above that value, but if it does, immediate hospitalization is recommended. Brain damage or death can occur if the body's temperature reaches 42°C (107.6°F).

❯ Examples of Innate Responses

Acne is one effect of an innate response. This skin condition is caused in part by a major constituent of normal flora, *Propionibacterium acnes* (Figure 22.8**A**), a bacterium that feeds on sebum. Sebum is a greasy mixture of fats, waxes, and glycerides that lubricates hair and skin. During puberty, higher levels of steroid hormones trigger an increase of sebum secretion into hair follicles. The excess sebum combines with dead, shed skin cells and blocks the follicle openings. *P. acnes* bacteria can survive on the surface of the skin, but greatly prefer anaerobic habitats such as the interior of blocked hair follicles. There, they multiply to tremendous numbers. Secretions of the flourishing populations of bacteria leak into internal tissues of the follicles and initiate inflammation. The resulting pustules are called acne.

Another common innate response occurs in the mouth, which is normally inhabited by at least 400 species of microorganisms. These normal flora can accumulate in **dental plaque**, a thick biofilm of various bacteria and occasional archaea, their extracellular products, and saliva glycoproteins that sticks tenaciously to teeth (Figure 22.6**B**). Plaque is harmful. Some of the organisms in it are fermenters, and the lactic acid they produce dissolves minerals in teeth and results in cavities. Other metabolic products of these organisms damage tight junctions (Section 3.6) that normally seal gum tissue to teeth. Deep pockets open up between the weakened gum tissue and teeth, and a nasty collection of anaerobic bacteria and archaea accumulates in them. Their secretions trigger inflammation of the surrounding gum tissue, a condition called periodontitis. All species of oral bacteria associated with periodontitis are also found in atherosclerotic plaque (Section 21.6), and they contribute to the inflammation that fuels coronary artery disease.

Take-Home Message

What happens when antigen is detected inside the body?

- Antigen or tissue damage activates complement. Activated complement recruits phagocytic white blood cells, coats pathogens in an affected area, and destroys foreign or ailing cells.

- Phagocytic white blood cells release cytokines after engulfing antigen-bearing particles such as complement-coated microorganisms. The cytokines trigger fever and alert the immune system to the presence of the threat.

- Inflammation occurs when granular white blood cells release local signaling molecules that increase blood flow and attract phagocytes to a site of tissue damage or infection.

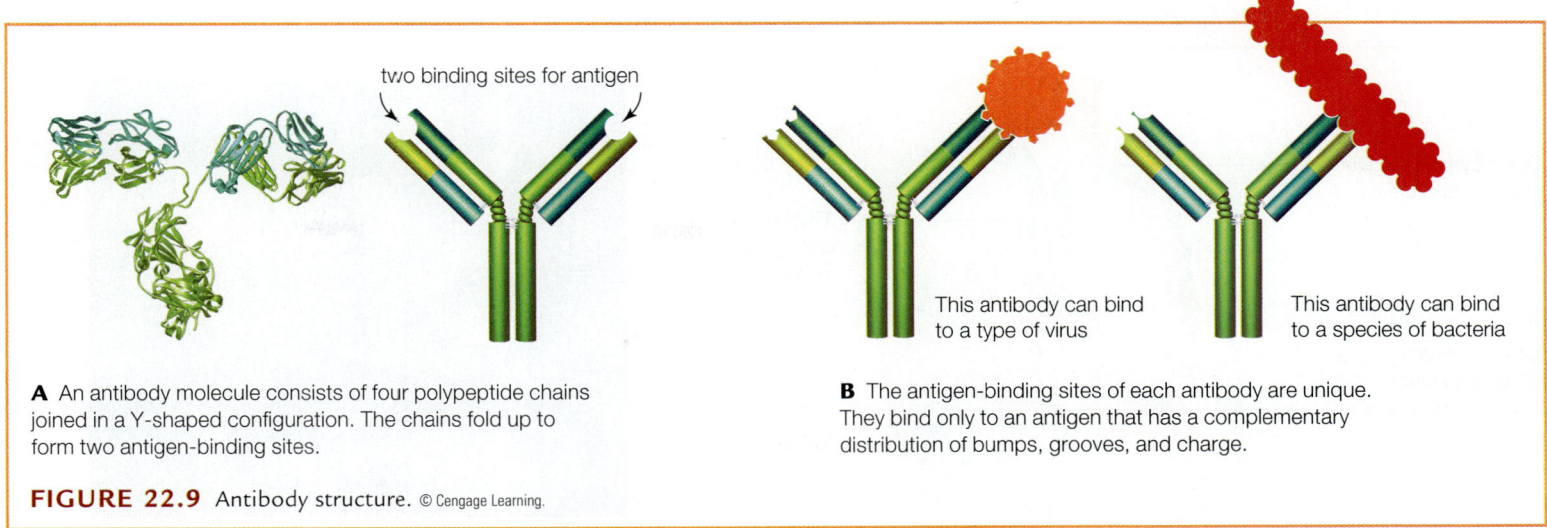

A An antibody molecule consists of four polypeptide chains joined in a Y-shaped configuration. The chains fold up to form two antigen-binding sites.

two binding sites for antigen

This antibody can bind to a type of virus

This antibody can bind to a species of bacteria

B The antigen-binding sites of each antibody are unique. They bind only to an antigen that has a complementary distribution of bumps, grooves, and charge.

FIGURE 22.9 Antibody structure. © Cengage Learning.

22.5 Antigen Receptors

If innate immune mechanisms do not quickly rid the body of an invading pathogen, an infection may become established in internal tissues. By that time, long-lasting mechanisms of adaptive immunity have already begun to target the invaders specifically. These mechanisms are triggered by white blood cells that detect antigen via antigen receptors. Plasma membrane proteins that recognize PAMPs are one type of antigen receptor. Antibodies are another. **Antibodies** are Y-shaped proteins made and secreted only by B cells. Many antibodies circulate in blood, and they can enter interstitial fluid during inflammation, but they do not kill pathogens directly. Instead, they activate complement and facilitate phagocytosis. Antibodies can prevent pathogens from attaching to body cells, and they can also neutralize some toxic molecules.

Each antibody molecule can bind to a specific antigen. The tips of the "Y" are two antigen-binding sites with a specific distribution of bumps, grooves, and charge (Figure 22.9**A**). These binding sites are the receptor part of an antibody: They can bind only to antigen with a complementary distribution of bumps, grooves, and charge (Figure 22.9**B**).

B cell receptors are antibodies that are not secreted; they remain attached to the B cell's plasma membrane. Your T cells bear special antigen receptors called **T cell receptors**, or TCRs. Part of a TCR recognizes antigen as nonself. Another part recognizes certain proteins in the plasma membrane of body cells as self. These self-proteins are called **MHC markers** (*right*), after the genes that encode them. MHC genes have thousands of alleles, so the cells of even closely related individuals rarely bear the same MHC markers.

an MHC marker
© Cengage Learning.

❯ Antigen Receptor Diversity Humans can make billions of unique B and T cell receptors. This diversity arises because the genes that encode these receptors occur in several segments on different chromosomes, and there are several different versions of each segment. The segments become spliced together during B and T cell differentiation, but which version of each segment gets spliced into the antigen receptor gene of a particular cell is random. As a B or T cell differentiates, it ends up with one of about 2.5 billion different combinations of gene segments.

antibody Y-shaped antigen receptor made only by B cells.
B cell receptor Antigen receptor on the surface of a B cell; an antibody that has not been released from the B cell's plasma membrane.
dental plaque On teeth, a thick biofilm composed of bacteria, their extracellular products, and saliva proteins.
MHC markers Self-proteins on the surface of human body cells.
T cell receptor (TCR) Antigen receptor on the surface of a T cell.

Right Lymphatic Duct
Collects and drains
lymph from right upper
portion of the body

**Left Lymphatic
(Thoracic) Duct**
Collects and drains
lymph from most of
the body

Lymph Vessels
Return excess interstitial
fluid and reclaimable
solutes to the blood

Bone Marrow
Marrow in some
bones produces
blood cells

Thymus Gland
Site where
T cells mature

Spleen
Antibodies are
produced here.
Also a disposal
site for old red
blood cells and
other debris

Lymph Nodes
Filter bacteria and
other agents of
disease from lymph

arrays of resident
lymphocytes

lymph
enters
node

lymph exits node

A Major components of the human lymphatic system and their functions.

B The lymph node. Lymph is filtered through at least one
node before it merges with the bloodstream. The fluorescence
micrograph shows T cells (*blue*) that are passing through
a lymph node and interacting with populations of resident
B cells (*green*) and antigen-presenting dendritic cells (*red*).

FIGURE 22.10 Animated! Battlegrounds of the adaptive immune system.

Credits: (a, b bottom) © Cengage Learning; (b top) Image courtesy of Dr. Fabien Garcon, The Babraham Institute.

B cell

B cell receptor
© Cengage Learning.

Like all other blood cells, lymphocytes form in bone marrow. A new B cell
is already making receptors before it even leaves the marrow. The base of each
receptor is embedded in the lipid bilayer of the cell's plasma membrane, and
the two arms of the "Y" project into the extracellular environment. A mature
B cell bristles with more than 100,000 receptors (*left*). T cells also form in bone
marrow, but they mature in the thymus gland, which is part of the endocrine
system (Section 19.4 and Figure 22.10**A**). There, they encounter hormones that
stimulate them to make receptors.

❭ Antigen Processing and Presentation A new lymphocyte is "naive,"
which means that no antigen has bound to its receptors yet. B cell receptors
can bind directly to antigen, but T cell receptors cannot. T cell receptors recognize and bind only to antigen that has been processed by an antigen-presenting
cell. Macrophages, B cells, and dendritic cells do the processing and presenting
(Figure 22.11). First, one of these cells engulfs something bearing antigen ❶. A
vesicle that contains the antigen-bearing particle forms in the cell's cytoplasm
and fuses with a lysosome ❷. Remember from Section 3.5 that lysosomes are
vesicles filled with powerful digestive enzymes. These enzymes now proceed to

Golgi body

nucleus

lysosome

MHC molecule

① ② ③ ④

FIGURE 22.11 Antigen processing: What happens when a B cell, macrophage, or dendritic cell engulfs an antigenic particle—in this case, a bacterium.

Figure It Out: From what organelles do the lysosomes bud?

Answer: Golgi bodies

© Cengage Learning.

① An endocytic vesicle forms around a bacterium as it is engulfed by a phagocytic cell.

② The vesicle fuses with a lysosome, which contains enzymes and MHC molecules.

③ Lysosomal enzymes digest the bacterium to molecular bits. The bits bind to MHC molecules.

④ The vesicle fuses with the cell's plasma membrane by exocytosis. When it does, the antigen–MHC complex becomes displayed on the cell's surface.

break down the ingested particle into molecular bits. Lysosomes also contain MHC markers that bind to some of the antigen-bearing bits **③**. The resulting antigen–MHC complexes become displayed at the cell's surface when the vesicles fuse with (and become part of) the plasma membrane **④**. The display of MHC markers paired with antigen fragments serves as a call to arms for T cells.

Antigen-presenting cells interact with T cells in lymph nodes. Every hour, about 500 different naive T cells pass through each node, inspecting antigen presented by arrays of B cells, dendritic cells, and macrophages (Figure 22.10**B**). Some of these phagocytic cells are permanent residents of the node, stationed there to capture and process antigenic particles in lymph (Section 21.5). Other phagocytes migrate to a lymph node and take up residence there after engulfing and processing antigen in other parts of the body. In both cases, the phagocytes present antigen to the naive T cells passing through the node. As you will see shortly, T cells with receptors that recognize and bind to antigen presented by a phagocyte (Figure 22.12) initiate an adaptive response.

During an infection, the lymph nodes swell because T cells accumulate inside them. When you are ill, you may notice your swollen lymph nodes as tender lumps under the jaw or elsewhere in your body.

Take-Home Message

What are antigen receptors?

- Each B or T cell makes antigen receptors that can bind a specific antigen. Humans are capable of producing billions of unique antigen receptors.

- Antibodies are antigen receptors made only by B cells. Antibodies released into the circulatory system activate complement and facilitate phagocytosis. B cell receptors are antibodies that are not secreted; they remain attached to the B cell's plasma membrane.

- T cell receptors can recognize their specific antigen only in conjunction with MHC markers displayed on the surface of an antigen-presenting cell.

lipid bilayer of T cell

T cell receptor

antigen bit

MHC molecule

lipid bilayer of antigen-presenting cell

FIGURE 22.12 T cell receptors recognize and bind antigen–MHC complexes on the surface of a phagocyte. The binding provokes an adaptive immune response targeting the antigen.

© Cengage Learning.

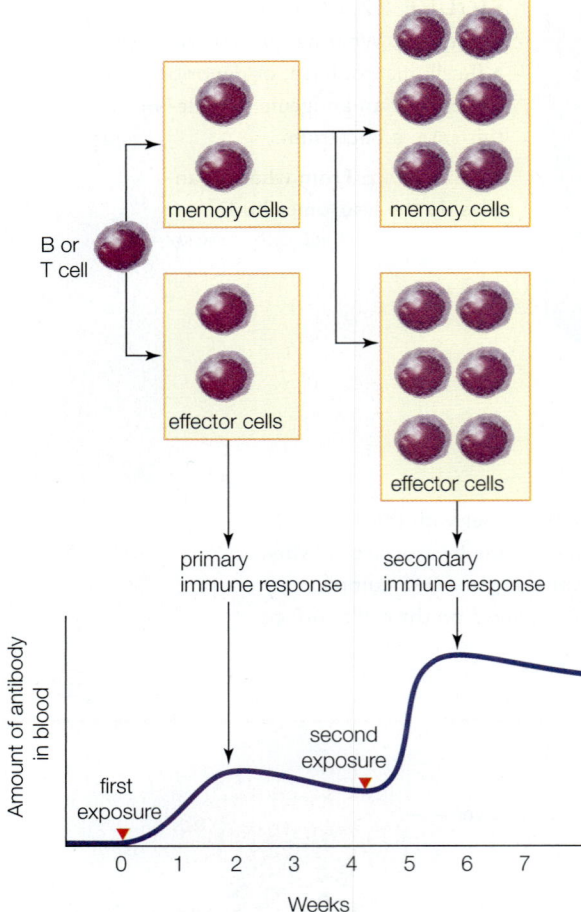

FIGURE 22.13 Animated! Primary and secondary immune responses.

A first exposure to an antigen triggers a primary response in which effector cells fight the infection.

Memory cells also form in a primary response but are set aside, sometimes for decades. If the antigen returns at a later time, the memory cells initiate a faster, stronger secondary response.

© Cengage Learning.

22.6 Adaptive Immune Responses

Like a boxer's one-two punch, adaptive immunity has two separate arms: two types of responses that work together to eliminate diverse threats. Why two arms? Not all threats present themselves in the same way. For example, bacteria, fungi, or toxins can circulate in blood or interstitial fluid. These threats are intercepted quickly by B cells and other phagocytes that interact in an **antibody-mediated immune response**. In this response, B cells are triggered to make antibodies specific to antigen detected in extracellular fluid. However, an antibody-mediated immune response is not the most effective way of countering other threats. For example, viruses, bacteria, fungi, and protists that reproduce inside body cells may be vulnerable to an antibody-mediated response only when they slip out of one cell to infect another. Such intracellular pathogens are targeted primarily by the **cell-mediated immune response**, in which cytotoxic T cells and NK cells detect and destroy infected or cancerous body cells.

In a typical adaptive immune response, a naive T cell recognizes and binds to an antigen–MHC complex displayed on the surface of an antigen-presenting cell. The encounter triggers the formation of **effector cells**, which are B and T cells that act at once in a primary immune response. Some **memory cells** also form, and these long-lived B and T cells are reserved for possible future encounters with the same antigen. Memory cells can persist for decades after the initial infection ends. If the same antigen enters the body at a later time, these memory cells carry out a faster, stronger secondary response (Figure 22.13).

The tide of battle turns when the effector cells have destroyed most of the antigen-bearing agents. With less antigen present, fewer lymphocytes form. Antibodies in blood and interstitial fluid glue antigen-bearing particles and debris into clumps that can be quickly cleared by the liver and spleen. Complement proteins assist in the cleanup. The immune response subsides after the antigen-bearing particles have been cleared from the body.

In both antibody-mediated and cell-mediated responses, the actions of effector and memory cells are responsible for the four defining characteristics of adaptive immunity:

Self/nonself recognition, based on the ability of T cell receptors to recognize self (in the form of MHC markers), and that of all antigen receptors to recognize nonself (in the form of antigen).

Specificity, which means that adaptive immune responses are tailored to combat specific antigens.

Diversity, which refers to the diversity of antigen receptors on a body's collection of lymphocytes.

Memory, the capacity of the adaptive immune system to "remember" an antigen (via memory cells).

❯ Example of an Antibody-Mediated Response Suppose that you accidentally nick your finger. Being opportunists, some *Staphylococcus aureus* bacteria on your skin immediately enter the cut, invading your internal environment. Complement in interstitial fluid quickly attaches to carbohydrates in the bacterial cell walls, and complement activation cascades begin.

Within an hour, complement-coated bacteria tumbling along in lymph vessels reach a lymph node in your elbow. There, they filter past an army of naive B cells (Figure 22.14). One of the naive B cells residing in that lymph node makes antigen receptors that recognize a carbohydrate in *S. aureus* cell walls ❶. Via those receptors, the B cell binds to one of the bacteria. The complement coating stimulates the B cell to engulf and digest the bacterium.

FIGURE 22.14 Animated! An antibody-mediated immune response.

1 The B cell receptors on a naive B cell bind to an antigen on the surface of a bacterium (*red*). The bacterium's complement coating (*purple* dots) triggers the B cell to engulf it. Fragments of the bacterium bound to MHC markers become displayed at the surface of the B cell.

2 A dendritic cell engulfs the same kind of bacterium that the B cell encountered. Fragments of the bacterium bound to MHC markers become displayed at the surface of the dendritic cell.

3 Antigen–MHC complexes displayed by the dendritic cell are recognized by TCRs on a naive helper T cell. The two cells interact, and then the T cell begins to divide. Its descendants differentiate into effector helper T cells and memory helper T cells.

4 TCRs on one of the effector helper T cells recognize and bind to the antigen–MHC complexes on the B cell. Binding causes the T cell to secrete cytokines (*blue* dots).

5 The cytokines induce the B cell to begin mitosis. Its many descendants differentiate into effector B cells and memory B cells.

6 The effector B cells begin making and secreting huge numbers of antibodies, all of which recognize the same antigen as the original B cell receptor. The new antibodies circulate throughout the body and bind to any remaining bacteria.

© Cengage Learning.

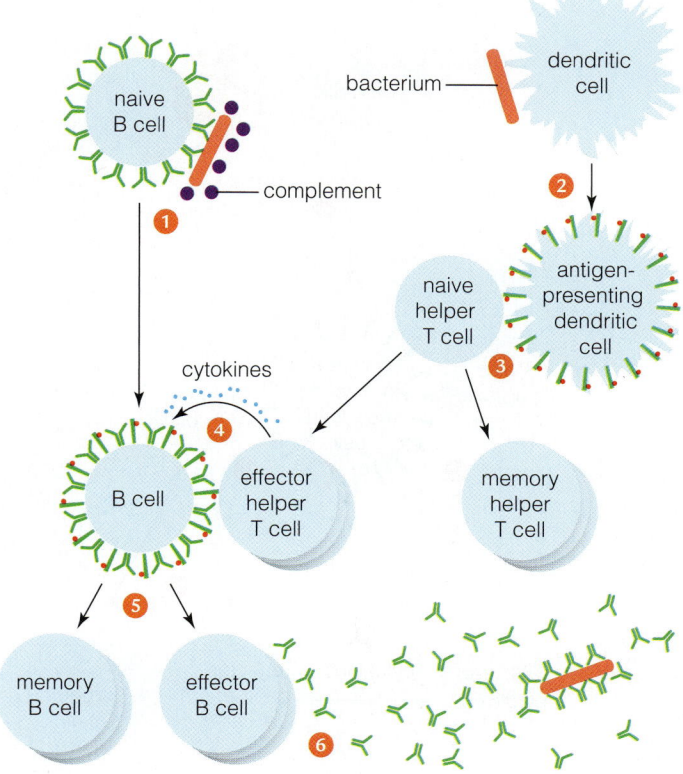

Meanwhile, more *S. aureus* cells have been secreting metabolic products into interstitial fluid around your cut. The secretions are attracting phagocytes. A dendritic cell engulfs several bacteria, then migrates to the lymph node in your elbow. By the time it gets there, it has digested the bacteria and is displaying their fragments as antigens bound to MHC markers on its surface **2**.

Within a couple of hours, one of your T cells has recognized and bound to the *S. aureus* antigen on the dendritic cell **3**. This T cell is called a helper T cell because it helps other lymphocytes produce antibodies and kill pathogens. The helper T cell and the dendritic cell interact in the lymph node for about 24 hours, then disengage. The helper T cell returns to the circulatory system and begins to divide, and a gigantic population of identical helper T cells forms. These clones differentiate into effector and memory cells, each of which has receptors that recognize the same *S. aureus* antigen.

Let's go back to that B cell in the lymph node. By now, it is displaying bits of the digested *S. aureus* bacterium bound to MHC molecules on its plasma membrane. The new helper T cells recognize the antigen–MHC complexes displayed by the B cell. One of these helper T cells binds to the B cell. Like long-lost friends, the two cells stay together for a while and communicate **4**. One of the messages that is communicated consists of cytokines secreted by the helper T cell. The cytokines stimulate the B cell to begin mitosis after the two cells disengage. The B cell gives rise to a huge population of descendant cells that differentiate into effector and memory cells **5**. The effector B cells start releasing antibodies **6**. These antibodies are a secreted version of the original B cell's receptors, so they recognize the same *S. aureus* antigen.

Tremendous numbers of antibodies now circulate throughout the body and attach themselves to any *S. aureus* cells. An antibody coating prevents the bacteria from attaching to body cells and brings them to the attention of phagocytic cells for quick disposal. Antibodies also glue the foreign cells together into clumps that are quickly removed from the circulatory system.

antibody-mediated immune response Immune response in which antibodies are produced in response to antigen.

cell-mediated immune response Immune response involving cytotoxic T cells and NK cells that destroy infected or cancerous body cells.

effector cell Antigen-sensitized B cell or T cell that forms in an immune response and acts immediately.

memory cell Long-lived, antigen-sensitized B or T cell that can act in a secondary immune response.

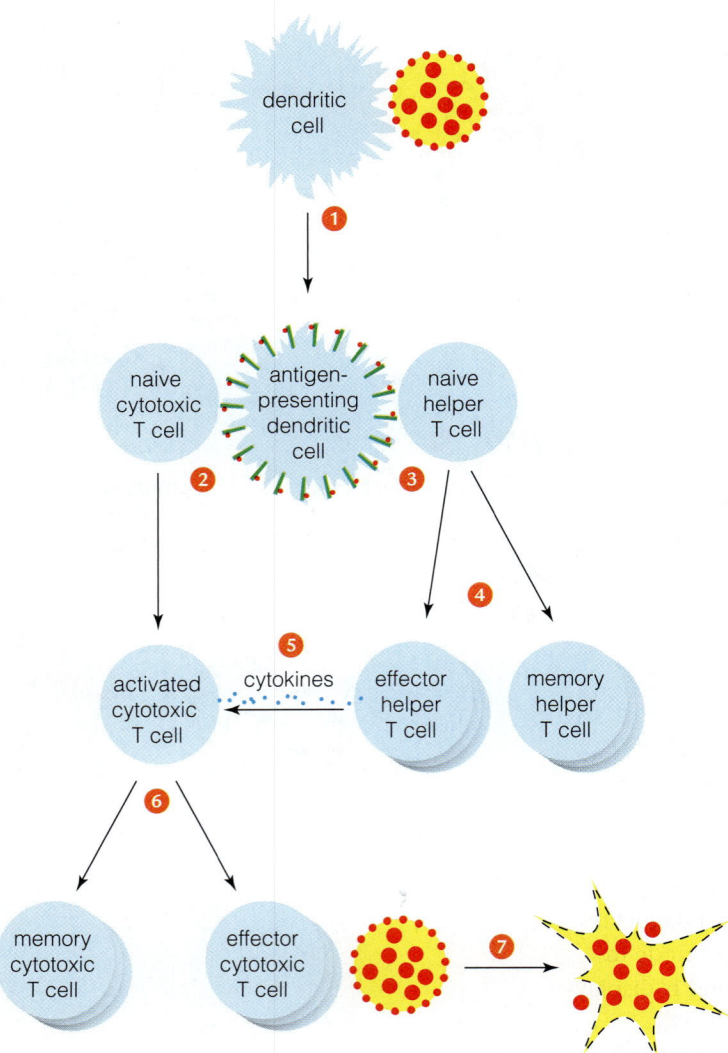

› A Tale of T Cells: An Example of a Cell-Mediated Response

Ailing body cells typically display antigens that are not found on healthy cells. For example, cancer cells display altered body proteins, and body cells infected with intracellular pathogens display polypeptides of the infecting agent. Both types of cell are detected and killed in a cell-mediated response.

A cell-mediated response often starts in interstitial fluid during inflammation, when a dendritic cell engulfs and digests a sick body cell or the remains of one (Figure 22.15). The dendritic cell begins to display antigen that was part of the sick cell as it migrates to the spleen or a lymph node **1**. There, the dendritic cell presents its antigen–MHC complexes to naive cytotoxic T cells **2** and naive helper T cells **3**. Some of the naive cells have T cell receptors that recognize and bind to the complexes on the dendritic cell. The binding activates the T cells.

Activated helper T cells begin to divide repeatedly, and their many descendants differentiate into effector and memory helper T cells **4**. The new effector helper T cells recognize and bind to antigen–MHC complexes displayed by macrophages. Interacting with a helper T cell causes a macrophage to increase its production of pathogen-busting enzymes and toxins, and also to secrete more phagocyte-attracting cytokines.

Effector helper T cells also secrete cytokines **5**. Cytotoxic T cells that have been activated by interacting with an antigen-presenting cell recognize these cytokines as a signal to divide repeatedly, and their many descendants differentiate into effector and memory cytotoxic T cells ❻. All of the new cells recognize and bind the same antigen—the one displayed by that first ailing cell.

The effector cytotoxic T cells circulate throughout blood and interstitial fluid, and bind to any other body cell displaying the original antigen together with MHC markers ❼. When a cytotoxic T cell binds to an ailing body cell (Figure 22.16), it releases protein-digesting enzymes and small molecules called perforins. Perforins assemble into complexes that, like membrane attack complexes, insert themselves into the ailing cell's plasma membrane to form large channels. The channels allow the enzymes to enter the body cell, which then bursts or commits suicide.

In order to kill a body cell, cytotoxic T cells must recognize the MHC molecules on the surface of the cell. However, some infections or cancer can alter a body cell so much that part or all of its MHC markers are missing. NK cells are crucial for eliminating such cells from the body. Unlike cytotoxic T cells, NK cells can kill body cells that lack MHC markers. NK cells recognize certain proteins displayed by body cells that are under stress. Stressed body cells with normal MHC markers are not killed; only those with altered or missing MHC markers are destroyed.

As occurs in an antibody-mediated response, memory cells form in a primary cell-mediated response. These long-lasting cells do not act immediately. If the antigen returns at a later time, the memory cells will mount a faster, stronger secondary response.

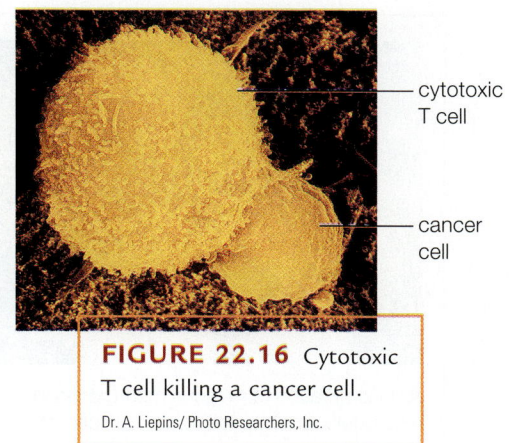

FIGURE 22.16 Cytotoxic T cell killing a cancer cell.

Dr. A. Liepins/ Photo Researchers, Inc.

Take-Home Message

What happens during an adaptive immune response?

- Vertebrate adaptive immunity has four defining characteristics: self/nonself recognition, specificity, diversity, and memory.

- During an antibody-mediated response, B cells produce antibodies that bind to an antigen in blood or interstitial fluid. Antibody binding can prevent a pathogen from entering body cells, neutralize a toxin, and facilitate the elimination of both from the body.

- During a cell-mediated immune response, cytotoxic T cells and NK cells kill infected body cells or those that have been altered by cancer.

- Memory cells that form during adaptive responses are set aside, sometimes for decades. If the antigen returns at a later time, these cells initiate a faster, stronger secondary response.

Even a small failure in immune function can have a major effect on health.

22.7 Immunity Gone Wrong

Despite built-in quality controls and redundancies in immune system functions, immunity does not always work as well as it should. Its complexity is part of the problem, because there are simply more opportunities for failure to occur in systems with many components. Even a small failure in immune function can have a major effect on health.

〉 Allergies In millions of people, exposure to a normally harmless substance stimulates an immune response. Any substance that is ordinarily harmless yet provokes such responses is an **allergen**. Sensitivity to an allergen is called an **allergy**. Drugs, foods, pollen, dust mite feces, fungal spores, and venom from bees, wasps, and other insects are among the most common allergens.

allergen A normally harmless substance that provokes an immune response in some people.
allergy Sensitivity to an allergen.

A Inflammation of mucous membranes in respiratory airways due to a pollen allergy causes hay fever.

B Direct contact with an allergen can cause a rash—an itchy, irritated patch of skin. In this case, the offending allergen was nickel, a metal, in a ring.

C Anaphylactic shock is a potentially lethal systemic response to an allergen.

FIGURE 22.17 Allergies.

Credits: (a) © Colin Hawkins/ Cultura/ Getty Images; (b) Biophoto Associates/ Photo Researchers, Inc.; (c) Hayley Witherell.

With a typical allergy, a first exposure to an allergen stimulates B cells to make and secrete antibodies that become anchored to mast cells and basophils. With later exposures, antigen binds to the anchored antibody. Binding triggers the anchoring cell to release the contents of its granules, and inflammation is initiated. If allergen is detected by mast cells in the lining of the respiratory tract, a copious amount of mucus is secreted and the airways constrict; sneezing, stuffed-up sinuses, and a drippy nose result (Figure 22.17**A**). Antihistamines relieve these symptoms by dampening the effects of histamines released by the degranulating cells. Other drugs can inhibit mast cell degranulation, thus preventing histamine release.

Skin rashes and other contact allergies do not involve antibodies; they are caused by a cell-mediated response to an allergen (Figure 22.17**B**).

> **Overly Vigorous Responses** Immune defenses that eliminate a threat can also damage host tissues. Thus, multiple mechanisms that limit immune responses are always in play. For example, one complement protein can become activated even in the absence of infection or tissue damage. Without the fail-safe mechanism of inhibitory molecules that deactivate this protein, complement cascades would occur constantly, with disastrous results to body tissues.

Acute illnesses arise when mechanisms that limit immune responses fail. Exposure to an allergen sometimes causes a severe, whole-body allergic reaction called anaphylactic shock. Huge amounts of cytokines and inflammatory molecules such as histamines and prostaglandins are released all at once in all parts of the body. Too much fluid leaks from blood into tissues, so blood pressure drops too much (a reaction called shock) and tissues swell dramatically. The swelling tissue constricts airways and may block them. Anaphylactic shock is rare but life-threatening and requires immediate treatment (Figure 22.17**C**). It may occur at any time, upon exposure to even a tiny amount of allergen. Risks include any prior allergic reaction.

Other types of hyperactive immune responses are not as well understood. Severe episodes of asthma or septic shock occur when too many neutrophils degranulate at once. In a "cytokine storm," too many white blood cells release cytokines at the same time. The cytokine overdose triggers immediate, widespread inflammation that can cause organ failure, with potentially fatal results. Cytokine storm triggered by infection with H5N1 influenza virus (Section 13.4) is a reason that this strain of bird flu has an unusually high mortality rate.

> **Autoimmune Disorders** People usually do not make antibodies to molecules that occur on their own healthy body cells, in part because the thymus has a built-in quality control mechanism that weeds out T cells with defective receptors. If this mechanism fails, mature lymphocytes that do not discriminate between self and nonself may be produced. Such lymphocytes can mount an **autoimmune response**, which is an immune response that targets one's own tissues. Autoimmunity is beneficial when a cell-mediated response targets cancer cells, but in most cases it is not.

A neurological disorder, multiple sclerosis, occurs when self-reactive T cells attack nerves. Symptoms range from weakness and loss of balance to paralysis and blindness. Specific alleles for MHC markers increase susceptibility, but a bacterial or viral infection may trigger the disorder.

> **Immunodeficiency and AIDS** Impaired immune function is dangerous and sometimes lethal. Immune deficiencies render individuals vulnerable to infections by opportunistic agents that are typically harmless to those in good health. Primary immune deficiencies, which are present at birth, are the out-

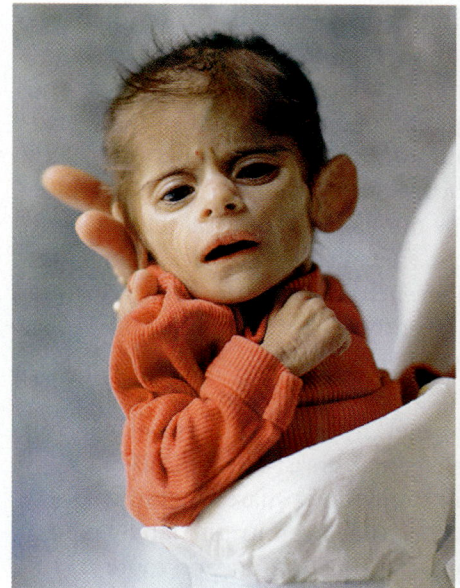

A This baby contracted AIDS from his mother's breast milk.

B Lesions of Kaposi's sarcoma, a cancer that is a common symptom of HIV infection in older AIDS patients.

C HIV (*red*) budding from the surface of a T cell. The virus infects the very cells that would otherwise protect the body from infection.

FIGURE 22.18 Three views of AIDS.

Credits: (a) © Peter Turnley/ Corbis; NIBSC/ Photo Researchers, Inc.; (b) © Zeva Oelbaum/ Peter Arnold, Inc.; (c) © R. Dourmashkin/ Wellcome Images.

come of mutations. Severe combined immunodeficiencies (SCIDs, Section 10.5) are examples. Secondary immune deficiency is the loss of immune function after exposure to an outside agent, such as a virus. **AIDS** (acquired immune deficiency syndrome) is the most common secondary immune deficiency. AIDS is a collection of disorders that occur as a result of infection with HIV, the human immunodeficiency virus (Section 13.4). This virus cripples the immune system, so it makes the body very susceptible to infections by other pathogens and to rare forms of cancer. Worldwide, approximately 33.3 million individuals are currently infected with HIV (Figure 22.18**A** and Table 22.2).

A person recently infected with HIV appears to be in good health, perhaps fighting a cold or "the flu." But symptoms eventually emerge that foreshadow AIDS: fever, many enlarged lymph nodes, chronic fatigue and weight loss, and drenching night sweats. Then, infections caused by normally harmless microorganisms strike. Yeast infections of the mouth, esophagus, and vagina often occur, as well as a form of pneumonia caused by the fungus *Pneumocystis jiroveci*. Gastrointestinal inflammation due to infection by a yeast or virus causes diarrhea. Colored lesions that erupt are evidence of Kaposi's sarcoma, a type of connective tissue cancer that is common among AIDS patients but rare among the general population (Figure 22.18**B**). Other cancers are also common, as are infections by cancer-causing viruses such as Epstein–Barr virus. These medical problems are unusual in people with healthy immune systems.

HIV mainly infects macrophages, dendritic cells, and helper T cells (Figure 22.18**C**). When virus particles enter the body, dendritic cells engulf them. The dendritic cells then migrate to lymph nodes, where they present processed HIV antigen to naive T cells. An army of HIV-neutralizing antibodies and HIV-specific cytotoxic T cells forms. For years or even decades, these responses keep the level of HIV low. During this stage, infected people often have no symptoms of AIDS, but they can pass the virus on because it persists in a few helper T cells, in a few lymph nodes. Eventually, the level of antibodies in the blood plummets,

Table 22.2	Global HIV and AIDS Cases	
Region	AIDS Cases	% of Adults Infected
Sub-Saharan Africa	22,500,000	5.0
Caribbean Islands	240,000	1.0
Central Asia/East Europe	1,400,000	0.8
Latin America	1,400,000	0.5
North America	1,500,000	0.5
South/Southeast Asia	4,100,000	0.3
Australia/New Zealand	57,000	0.3
Western/Central Europe	820,000	0.2
Middle East/North Africa	460,000	0.2
East Asia	770,000	0.1
Approx. worldwide total	33,247,000	0.8

Source: Joint United Nations Programme HIV/AIDS, 2010 report

AIDS Acquired immune deficiency syndrome. A collection of diseases that develops after a virus (HIV) weakens the immune system.

autoimmune response Immune response that targets one's own tissues.

and the production of T cells slows. Why this happens is still a major topic of research, but its effect is certain: The immune system becomes progressively less effective at fighting the virus. The number of virus particles rises, and more and more helper T cells become infected. Lymph nodes begin to swell with infected T cells. Eventually, secondary infections and tumors kill the patient.

Drugs that slow the disease's progress target processes unique to viral replication. A three-drug "cocktail" is currently the most successful AIDS therapy, and has changed the typical course of the disease from a short-term death sentence to a long-term, often manageable illness. Preventive use of a drug that consists of two reverse transcriptase inhibitors has recently been shown to greatly reduce the HIV infection rate in high-risk populations.

HIV is not transmitted by casual contact; most infections are the result of having unprotected sex with an infected partner. Infected mothers can transmit HIV to a child during pregnancy, labor, delivery, or breast-feeding. HIV also travels in tiny amounts of infected blood in the syringes shared by intravenous drug abusers, or by hospital patients in less developed countries. Many people have become infected via blood transfusions, but this transmission route is becoming rarer as most blood is now tested prior to use for transfusions.

Most AIDS tests check blood, saliva, or urine for antibodies that bind to HIV antigens. These antibodies are detectable in 99 percent of infected people within three months of exposure to the virus.

Take-Home Message

What happens when the immune system does not function as well as it should?

- Normally harmless substances may induce an immune response in some people. Sensitivity to such allergens is called an allergy.
- Misdirected or compromised immunity, which sometimes occurs as a result of mutation or environmental factors, can have severe or lethal outcomes.
- AIDS is a secondary immune deficiency caused by HIV infection. HIV infects lymphocytes and so cripples the human immune system.

22.8 Vaccines

Immunization refers to processes designed to induce immunity. In active immunization, a preparation that contains antigen—a **vaccine**—is administered orally or injected. The first immunization elicits a primary immune response, just as an infection would. A second immunization, or booster, elicits a secondary immune response for enhanced protection. In passive immunization, a person receives antibodies purified from the blood of another individual. Passive immunization offers immediate benefit for someone who has been exposed to a lethal agent such as tetanus, rabies, Ebola virus, or a venom or toxin. Because the antibodies were not made by the recipient's lymphocytes, effector and memory cells do not form, so benefits last only as long as the injected antibodies do.

The first vaccine was developed in the late 1700s, a result of desperate attempts to survive devastating smallpox epidemics (Figure 22.19). Edward Jenner, an English physician, injected liquid from a cowpox sore into the arm of a healthy boy. Six weeks later, Jenner injected the boy with liquid from a smallpox sore. Luckily, the boy did not get smallpox. He was immune—protected from infection. Jenner named his procedure "vaccination," after the Latin word for cowpox (*vaccinia*). Though it was still controversial, the use of Jenner's vaccine

When enough individuals refuse vaccination, outbreaks of preventable and sometimes fatal diseases occur.

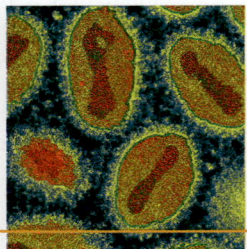

FIGURE 22.19
Young survivor and the cause of her disease, smallpox viruses. Worldwide vaccinations have eradicated naturally occurring cases of smallpox.

Credits: left, James Hicks, Centers for Disease Control and Prevention; right, Eye of Science/ Photo Researchers, Inc.

spread quickly through Europe, then to the rest of the world. The last known case of naturally occurring smallpox was in 1977.

We now know that the cowpox virus is an effective vaccine for smallpox because the antibodies it elicits also recognize smallpox virus antigens. Our knowledge of how the immune system works has allowed us to develop many other vaccines that are part of public programs worldwide (Table 22.3). These vaccines have overwhelmingly reduced suffering and deaths from many infectious diseases, but public confidence is a necessary part of their success. When enough individuals refuse vaccination, outbreaks of preventable and sometimes fatal diseases occur.

We still have no vaccine effective against HIV infection. Antibodies produced during a normal adaptive response exert selective pressure on HIV virus, which has a very high mutation rate, so a large number of variants of the HIV virus exist. A successful HIV vaccine must efficiently recognize and neutralize all variants, including those that have not yet arisen. Immunization with live, weakened HIV virus is an effective vaccine in chimpanzees, but the risk of infection from the vaccination itself far outweighs its potential benefits in humans. Other types of HIV vaccines are notoriously ineffective.

One promising strategy involves reverse engineering HIV antibodies isolated from people with AIDS. Some of these antibodies can neutralize one or two strains of virus, but most are utterly ineffective. Researchers recently isolated three antibodies with a much greater ability to neutralize HIV than other antibodies studied to date; one of them is effective against almost all known strains of HIV. All three antibodies came from a man that has lived with AIDS for more than twenty years. His antibodies are being studied in painstaking detail in order to discover the exact parts of the virus they recognize. Those parts of the virus are being used to create new vaccines. Genes encoding the three antibodies are also being inserted into viral vectors for use in gene therapy (Section 10.5), the idea being that the vector will deliver the genes into body cells, which will then start producing antibodies.

Table 22.3	Recommended Immunization Schedule
Vaccine	**Age of Vaccination**
Hepatitis B	Birth
Hepatitis B boosters	1–2 months and 6–18 months
Rotavirus	2, 4, and 6 months
DTP: diphtheria, tetanus, and pertussis (whooping cough)	2, 4, and 6 months
DTP boosters	15–18 months, 4–6 years, and 11–12 years
HiB (*Haemophilus influenzae*)	2, 4, and 6 months
HiB booster	12–15 months
Pneumococcal	2, 4, and 6 months
Pneumococcal booster	12–15 months
Inactivated poliovirus	2 and 4 months
Inactivated poliovirus boosters	6–18 months and 4–6 years
Influenza	Yearly, 6 months and older
MMR (measles, mumps, rubella)	12–15 months
MMR booster	4–6 years
Varicella (chicken pox)	12–15 months
Varicella booster	4–6 years
Hepatitis A series (2 doses)	12–23 months
HPV series (3 doses)	11–12 years
Meningococcal	11–12 years

Source: Centers for Disease Control and Prevention (CDC), 2011

Take-Home Message

How do vaccines work?

- Administering a vaccine elicits an immune response in the recipient that protects against future encounters with a disease-causing agent.
- Worldwide vaccination programs have greatly reduced suffering and deaths from many diseases. An effective vaccination program requires widespread participation.

22.9 Frankie's Last Wish (revisited)

The Gardasil HPV vaccine consists of viral proteins that self-assemble into virus-like particles (VLPs). These proteins are produced by a recombinant yeast, *Saccharomyces cerevisiae*. The yeast carries genes for one surface protein from each of four strains of HPV, so the VLPs carry no viral DNA. Thus, the VLPs are not infectious, but the antigenic proteins they consist of elicit an immune response at least as strong as infection with HPV virus.

WHERE YOU ARE GOING . . .

Chemical signaling returns in context of the nervous system in Chapter 24. Celiac disease (Section 23.4), multiple sclerosis (24.2), and most thyroid disorders (25.4) are autoimmune diseases. Section 26.5 discusses AIDS and other sexually transmitted diseases.

immunization Any procedure designed to promote immunity to a specific disease.

vaccine A preparation introduced into the body in order to elicit immunity to a specific antigen.

Summary

Section 22.1 Screenings, treatments, and vaccines for diseases such as cervical cancer are a direct outcome of our increasing understanding of the way the human body interacts with pathogens.

Section 22.2 The body's ability to resist and fight infections is called **immunity**. A microorganism or other antigen-bearing agent that breaches surface barriers triggers **innate immunity**, a set of general defenses that can prevent pathogens from becoming established in the body. **Adaptive immunity**, which can specifically target billions of different **antigens**, follows. **Complement** and signaling molecules such as cytokines help coordinate the activities of white blood cells such as phagocytic **dendritic cells** and **macrophages**. Lymphocytes (**B cells**, **T cells**, and NK cells) are white blood cells with special roles in immune responses.

Section 22.3 Most **normal flora** that colonize body surfaces (including the linings of tubes and cavities) do not cause disease unless they penetrate inner tissues. Vertebrates fend off pathogens at body surfaces with physical, mechanical, and chemical barriers (including **lysozyme**).

Section 22.4 An innate immune response can be triggered by activated complement or by phagocytic white blood cells that engulf a pathogen or other antigen-bearing particle. Complement activated by the presence of antigen or tissue damage attracts phagocytes, coats pathogens, and forms membrane attack complexes that induce cell death.

Inflammation begins when white blood cells in infected or damaged tissue release signaling molecules that attract phagocytes and increase blood flow to the affected area. Plasma proteins that leak from capillaries cause swelling and pain. Prolonged exposure to an inflammation-provoking stimulus can cause chronic inflammation, a condition associated with many diseases. Normal flora that cause **dental plaque** can end up in atherosclerotic plaque, where they fuel inflammation associated with cardiovascular disease. **Fever** increases the metabolic rate and slows pathogen replication.

Section 22.5 **T cell receptors** are the basis of self/nonself discrimination; they recognize antigen displayed with **MHC markers** by an antigen-presenting cell. **B cell receptors** are **antibodies** that have not been released from a B cell. Both kinds of antigen receptors collectively have the ability to recognize billions of specific antigens, a diversity that arises from random splicing of antigen receptor genes. Phagocytic white blood cells that engulf, process, and present antigen to T cells are central to all adaptive immune responses.

Section 22.6 Four main characteristics of adaptive immunity are self/nonself recognition, sensitivity to specific threats, potential to intercept a tremendous number of diverse pathogens, and memory. Two arms of adaptive immunity work together to rid the body of a specific pathogen. Antibodies that recognize a specific antigen are produced by B cells during an **antibody-mediated immune response**. Antibody binding tags antigen-bearing particles for phagocytosis. Cytotoxic T cells that form in a **cell-mediated immune response** kill body cells that have been altered by infection or cancer. NK cells can kill ailing body cells that are missed by cytotoxic T cells.

In both types of responses, many **effector cells** form and target the antigen-bearing particles. **Memory cells** that also form are reserved for a later encounter with an antigen, in which case they trigger a faster, stronger secondary response.

Section 22.7 **Allergens** are normally harmless substances that induce immune responses; hypersensitivity to an allergen is called **allergy**. A malfunction in the immune system or its checks and balances can cause dangerous acute illnesses, or chronic and sometimes deadly diseases. Immune deficiency is a reduced capacity to mount an immune response. In an **autoimmune response**, a body's own cells are inappropriately recognized as foreign and attacked.

AIDS is caused by the human immunodeficiency virus (HIV). An initial adaptive immune response to infection rids the body of most—but not all—of the virus. The virus infects helper T cells, so it eventually cripples the adaptive immune system.

Section 22.8 **Immunization** with **vaccines** designed to elicit immunity to specific diseases saves millions of lives each year as part of worldwide health programs. Effective vaccination programs require widespread participation.

Self-Quiz   Answers in Appendix I

1. Which of the following is not among the first line of immune defenses against infection?
   a. skin
   b. acidic gastric fluid
   c. lysozyme in saliva
   d. resident bacterial populations
   e. complement activation
   f. flushing action of diarrhea

2. Which of the following is not part of an innate immune response?
   a. phagocytic cells
   b. fever
   c. histamines
   d. cytokines
   e. inflammation
   f. complement activation
   g. presenting antigen
   h. all take part

3. Which of the following is not part of an adaptive immune response?
   a. phagocytic cells
   b. antigen-presenting cells
   c. histamines
   d. cytokines
   e. antigen receptors
   f. complement activation
   g. antibodies
   h. all take part

4. Activated complement proteins _____ .
   a. form pore complexes
   b. promote inflammation
   c. attract phagocytes
   d. all of the above

5. _____ trigger immune responses.
   a. Cytokines
   b. Lysozymes
   c. Antibodies
   d. Antigens
   e. Histamines
   f. all of the above

6. Name a characteristic of innate immunity.

7. Antibodies are _____ .
   a. antigen receptors
   b. made only by B cells
   c. proteins
   d. all of the above

8. Name a characteristic of adaptive immunity.

9. A dendritic cell engulfs a bacterium, then presents bacterial bits on its surface along with a(n) _____ .
   - a. MHC marker
   - b. antibody
   - c. T cell receptor
   - d. antigen

10. Antibody-mediated responses are most effective against _____ .
   - a. intracellular pathogens
   - b. extracellular pathogens
   - c. cancerous cells
   - d. both a and b
   - e. both b and c
   - f. a, b, and c

11. Cell-mediated responses are most effective against _____ .
   - a. intracellular pathogens
   - b. extracellular pathogens
   - c. cancerous cells
   - d. both a and b
   - e. both a and c
   - f. a, b, and c

12. _____ are targets of cytotoxic T cells.
   - a. Extracellular virus particles in blood
   - b. Virus-infected body cells or tumor cells
   - c. Parasitic worms in the liver
   - d. Bacterial cells in pus
   - e. Pollen grains in nasal mucus

13. Allergies occur when the body responds to _____ .
   - a. pathogens
   - b. toxins
   - c. normally harmless substances
   - d. all of the above

14. Match the immune cell with its main function.
   - _____ dendritic cell
   - _____ B cell
   - _____ helper T cell
   - _____ NK cell
   - _____ cytotoxic T cell
   - a. kills virus-infected cells
   - b. antigen-presenter
   - c. activates other lymphocytes
   - d. kills cancer cells
   - e. makes antibodies

15. Match the immunity concepts.
   - _____ anaphylactic shock
   - _____ immune memory
   - _____ autoimmunity
   - _____ inflammation
   - _____ immune deficiency
   - _____ antigen receptor
   - _____ antigen processing
   - a. recognizes antigen
   - b. inadequate immune response
   - c. general defense mechanism
   - d. immune response against one's own body
   - e. secondary response
   - f. systemic allergic reaction
   - g. presenting antigen together with MHC molecules

Critical Thinking

1. A tissue or organ or transplanted from one human into another is immediately attacked by the recipient's immune system. Which type of lymphocyte participates in this transplant rejection?

2. Before each flu season, you get a flu shot, which consists of a vaccine against several strains of influenza virus. This year, you get "the flu" anyway. What happened? (There are at least three explanations.)

3. Elena developed chicken pox when she was in first grade. Later in life, when her children developed chicken pox, she remained healthy even though she was exposed to countless virus particles daily. Explain why.

4. Antibodies used in passive immunization are effective, but only in the immediate term. Antibodies produced by one's own immune system can last up to about six months in the bloodstream, but those delivered in passive immunization often last for less than a week. Why the difference?

Digging Into Data

Cervical Cancer Incidence in HPV-Positive Women

In 2003, Michelle Khan and her coworkers published their findings on a 10-year study in which they followed cervical cancer incidence and HPV status in 20,514 women. All women who participated in the study were free of cervical cancer when the test began. Pap tests were taken at regular intervals, and the researchers used a DNA probe hybridization test (Section 10.2) to detect specific types of HPV in the women's cervical cells.

The results are shown in Figure 22.20 as a graph of the incidence rate of cervical cancer by HPV type. HPV-positive women are often infected by more than one type, so the data were sorted into groups based on the women's HPV status ranked by type: either positive for HPV16; or negative for HPV16 and positive for HPV18; or negative for HPV16 and 18 and positive for any other cancer-causing HPV; or negative for all cancer-causing HPV.

1. At 110 months into the study, what percentage of women who were not infected with any type of cancer-causing HPV had cervical cancer? What percentage of women who were infected with HPV16 also had cervical cancer?

2. In which group would women infected with both HPV16 and HPV18 fall?

3. Is it possible to estimate from this graph the overall risk of cervical cancer that is associated with infection of cancer-causing HPV of any type?

4. Do these data support the conclusion that being infected with HPV16 or HPV18 raises the risk of cervical cancer?

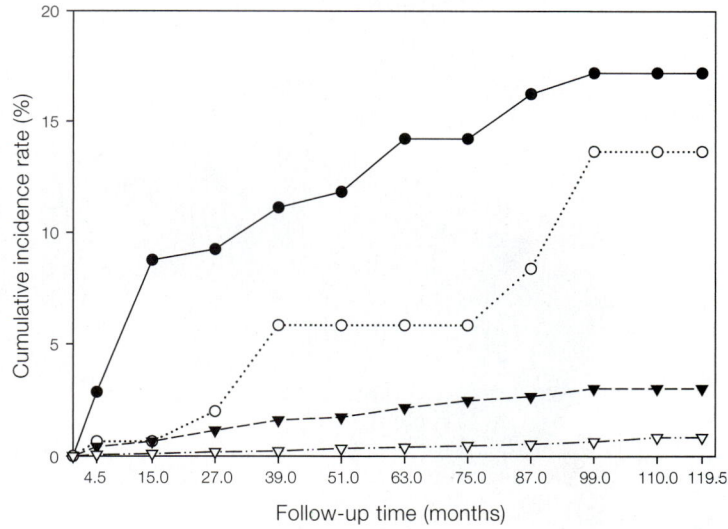

FIGURE 22.20 Cumulative incidence rate of cervical cancer correlated with HPV status in 20,514 women aged 16 years and older.

The data were grouped as follows:
- ● Closed circles: HPV16 positive
- ○ Open circles: HPV16 negative and HPV18 positive
- ▼ Closed triangles: All other cancer-causing HPV types combined
- ▽ Open triangles: no cancer-causing HPV type detected

© Cengage Learning.

Digestion and Excretion

23

23.1 Regulating Appetite

Food supplies your body with raw materials and fuel. When the food you eat contains more energy than you need at the time, you store the excess in the bonds of organic compounds—mainly in fat molecules deposited in adipose tissue. For most of our species' history, an ability to store energy as fat provided an evolutionary advantage. Putting on fat when food is abundant increases the likelihood of survival if food later becomes scarce. However, most people in the United States now have more than enough food all of the time. As a result, about two-thirds of adults are overweight or obese. Their excessive weight is a matter of public health concern because, compared to people at a healthy weight, obese people have a higher risk of type 2 diabetes, cardiovascular disease, some cancers, and other disorders.

Why do some people stay slim with no apparent effort, whereas others diet constantly but remain heavy? Genetics certainly plays a role. In terms of weight class (thin, average, overweight, or obese), people raised by adoptive parents resemble their biological parents more than their adoptive ones. Geneticists have also pinpointed some genes that affect weight. The first to be discovered, the *ob* gene, encodes leptin, a hormone made by adipose cells. Leptin acts in the brain and suppresses appetite. Mice that do not have a functional *ob* gene overeat and become fat (Figure 23.1). Similarly, people who cannot make leptin are obese. When leptin-deficient mice or humans are injected with leptin, they eat less and slim down.

FIGURE 23.1 Two normal mice (*left*) and a mouse that cannot make leptin (*right*).

Courtesy of Dr. Jeffrey M. Friedman, Rockefeller University.

Human leptin deficiency is extremely rare, but variations in another obesity-related gene, *fto*, are common. About 16 percent of people of European ancestry are homozygous for an *fto* allele that predisposes them to obesity. Compared to people with two low-risk alleles, they are almost twice as likely to be obese. The *fto* gene is expressed most strongly in the brain and it influences appetite. People who have the high-risk allele tend to eat more food before they feel full.

Genetics can explain why one person is somewhat more likely than another to be overweight, but it cannot explain a national trend toward weight gain. Since 1980, the proportion of adults in the United States who are obese has doubled, and the proportion of children who are obese has tripled. There is no reason to think that the proportion of alleles that predispose people to obesity has increased in this time. Decreasing exercise and increasing food intake are likely causes of obesity's rise.

This discussion of food intake leads us into the world of nutrition, which encompasses all the processes by which we ingest and digest food, then absorb the released nutrients for use as energy sources and building blocks for cells.

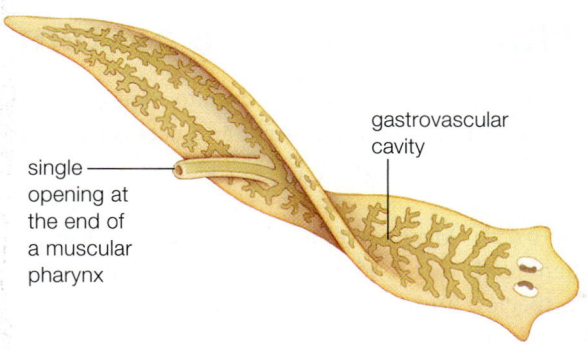

A Saclike gastrovascular cavity of a planarian (a flatworm).

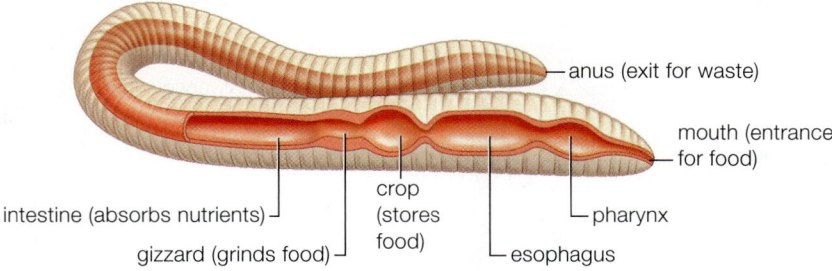

B Tubular digestive tract of an earthworm (an annelid).

23.2 Animal Digestive Systems

In sponges, individual cells engulf and break down food, then expel wastes. However, most animals have a digestive system that carries out these tasks:

1. **Ingestion**: Taking food into the chamber where digestion occurs. This chamber is an extension of the external environment, to which it opens. Although body tissues enclose this chamber, it is not part of the body's internal environment. Thus, in animals with a digestive system, digestion is mainly extracellular.

2. **Digestion**: Breaking food down into components that can be absorbed into the body. Mechanical digestion smashes food into smaller and smaller fragments. Chemical digestion breaks down large molecules in food into their component subunits.

3. **Nutrient absorption**: Movement of nutrient molecules into cells that line the digestive chamber.

4. **Elimination**: Expelling leftover material that was not digested and absorbed.

Flatworms and cnidarians have the simplest type of digestive system. Their saclike **gastrovascular cavity** has a single opening through which food enters and wastes leave. In flatworms, this opening is at the tip of a muscular pharynx (Figure 23.2**A**). Because of the two-way traffic in a gastrovascular cavity, each load of food must be broken down, its nutrients absorbed, and any wastes eliminated before a new load of food can enter.

Most invertebrates and all vertebrates have a **complete digestive tract**, a tube that has a mouth at one end and an **anus** at the other (Figure 23.2**B**). Material moves in one direction, from the mouth to the anus. Along the way, it passes through regions specialized for food storage, food breakdown, nutrient absorption, and waste elimination. Unlike a gastrovascular cavity, a tubular gut can carry out all these tasks simultaneously. An earthworm can take in new food even if food it ate earlier has not yet completed its journey through the digestive tract. The body plan of animals with a complete digestive tract is sometimes described as a "tube within a tube." The body wall forms the outer tube and the tubular digestive tract lies within it, inside the coelom (Figure 23.2**C**).

Take-Home Message

Where do animals digest their food?

■ In some invertebrates digestion occurs in a saclike gastrovascular cavity.

■ Other invertebrates and all vertebrates have a more efficient complete digestive tract, which is a tube that has specialized regions along its length.

C Schematic diagram of the tube-within-a-tube body plan of an animal with a complete digestive tract.

FIGURE 23.2 Animated! Two types of animal digestive systems. In both, breakdown of food occurs in a space that is an extension of the external environment, rather than a part of the body's internal environment.

Credits: (a,c) © Cengage Learning; (b) From RUSSELL/WOLFE/HERTZ/STARR. Biology, 2E. © 2011, Brooks/Cole, a part of Cengage Learning, Inc. Reproduced by permission. www.cengage.com/permissions.

anus Opening through which wastes leave a complete digestive tract.

complete digestive tract Tubular gut with two openings.

digestion Breakdown of food into smaller bits and then into its component molecules.

elimination Expulsion of unabsorbed material from the digestive tract.

gastrovascular cavity Saclike gut with one opening.

ingestion Taking food into the digestive system.

nutrient absorption Movement of nutrient molecules from a digestive cavity into the body proper.

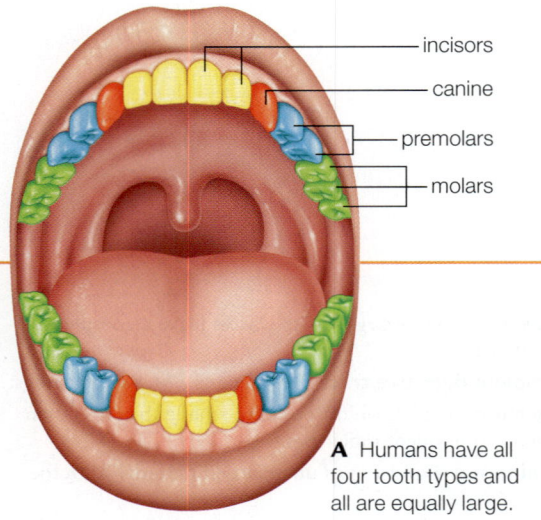

Digestive Tract

Mouth

Pharynx (throat)

Esophagus

Stomach

Small
intestine

Large
intestine
(colon)

Rectum

Anus

Accessory Organs

Salivary glands

Liver

Gallbladder

Pancreas

FIGURE 23.3 **Animated!** The human digestive system.

© Cengage Learning.

23.3 Human Digestive System

Figure 23.3 shows components of the human digestive system, which includes a tubular digestive tract and accessory organs that secrete substances into it.

〉 In the Mouth Mechanical digestion begins in the mouth, where teeth tear food apart and smash it into smaller pieces. As a species, humans are omnivores, meaning our diet includes both plant and animal material, and the structure of our teeth adapts us to this varied diet. We have all four types of mammalian teeth, and all are equally well developed (Figure 23.4**A**). By comparison, carnivorous (meat-eating) mammals typically have enlarged canine teeth for piercing and tearing flesh (Figure 23.4**B**) and herbivores (plant eaters) tend to have reduced canine teeth (Figure 23.4**C**). Premolars of carnivores have a narrow, bladelike surface that helps them shear through meat. By contrast, those of herbivores and humans are broad and flat, the better for grinding plant material.

Movements of the tongue help mix food with saliva from salivary glands. The tongue, a bundle of membrane-covered skeletal muscle, attaches to the floor of the mouth. Salivary glands open into the mouth beneath the tongue and on the inner surface of the cheeks next to the upper molars. Salivary amylase, an enzyme in saliva, begins the process of chemical digestion by breaking large polymers of starch into smaller ones such as disaccharides (two-sugar units). This reaction, like other digestive reactions, is an example of hydrolysis (Section 2.6). Saliva also contains glycoproteins that combine with water to form mucus. The presence of mucus helps small food bits stick together in easy-to-swallow clumps.

〉 Swallowing—Down the Tube The human pharynx, or throat, is the entrance to both the digestive and respiratory tracts (Section 21.8). When the tongue pushes food to the back of the throat, the food triggers a swallowing reflex. As you swallow, the flaplike epiglottis flops down and vocal cords constrict, blocking the route between the pharynx and larynx. This reflex directs food into the **esophagus**, the muscular tube that connects the pharynx to the stomach.

If food or drink accidentally enters the larynx, a reflexive cough generally expels it. However, sometimes food gets through the larynx and obstructs the

incisors

canine

premolars

molars

A Humans have all four tooth types and all are equally large.

B Carnivores have enlarged canine teeth and sharp premolars.

C Herbivores have reduced canines, and large flat premolars and molars.

FIGURE 23.4 Variations in mammalian teeth.

Credits: (a) © Cengage Learning; (b,c) From RUSSELL/WOLFE/HERTZ/STARR. Biology, 2E. © 2011, Brooks/Cole, a part of Cengage Learning, Inc. Reproduced by permission. www.cengage.com/permissions.

A   **B**

trachea. When this happens, a person chokes. A choking person cannot cough, speak, or breathe. To assist someone who is choking, the American Red Cross suggests first administering five sharp slaps to the victim's back between the shoulder blades, followed by five abdominal thrusts (Figure 23.5).

When swallowing proceeds normally and food enters the esophagus, waves of smooth muscle contraction, called peristalsis, move food to the stomach. The gastroesophageal sphincter connects the lower end of the esophagus with the stomach. (Gastro- or gastric refers to the stomach). A **sphincter** is a ring of muscle that controls passage of material through a tube or a body opening. The gastroesophageal sphincter opens to allow food into the stomach, then closes to prevent stomach acid from splashing into the esophagus. In some people, this sphincter does not shut properly, and acid reflux occurs; acidic fluids irritate the esophagus, causing the pain commonly called "heartburn." Occasional heartburn poses no threat to health. However, chronic heartburn, which can raise the risk of esophageal cancer, should be investigated by a physician.

⟩ The Stomach The **stomach**, a muscular, stretchable sac, has three functions. First, it mixes and stores ingested food. Second, it secretes substances that begin to break down food, especially proteins. Third, it helps control the passage of food into the small intestine.

The stomach has a wall of smooth muscle layers and a lining of glandular epithelium (Figure 23.6). When the stomach is empty, this lining has numerous folds. The folds disappear as the stomach fills with food. Arrival of food in the stomach also stimulates smooth muscle contractions and causes cells in the stomach lining to secrete **gastric fluid**. This fluid includes a strong acid, enzymes, and mucus. Stomach contractions mix food and gastric fluid together to form a semiliquid mass called **chyme**.

Protein digestion begins in the stomach. The acidity of gastric fluid denatures (unfolds) proteins in chyme. Then, an enzyme in gastric fluid (pepsin) cuts the unfolded proteins into polypeptides—another hydrolysis reaction. Why doesn't this process also break down proteins in the stomach wall? Mucus secreted by the stomach lining protects the underlying tissue.

When something disrupts the protective mucus layer, gastric fluid and enzymes can erode the stomach lining, causing an ulcer. Most ulcers occur after certain bacteria (*Helicobacter pylori*) make their way through gastric mucus and infect cells of the stomach lining. Antibiotics can halt the bacterial infection and allow the ulcer to heal. Continual use of aspirin, ibuprofen, and other nonsteroidal anti-inflammatory drugs (NSAIDs) can also cause ulcers, because one of their side effects is to decrease the production of mucus by the stomach lining.

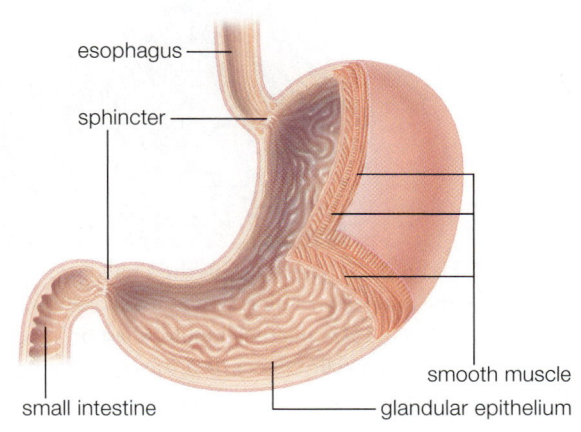

esophagus

sphincter

smooth muscle

small intestine

glandular epithelium

FIGURE 23.6 Animated! Stomach structure and function. The innermost layer is glandular epithelium that secretes gastric fluid. Three smooth muscle layers differ in their orientation, and their contractions cause a wringing action that mixes and moves food. Sphincters at either end of the stomach can close to keep acidic chyme inside the stomach.

After A. Vander et al., *Human Physiology: Mechanisms of Body Function*, Fifth Edition, McGraw-Hill, 1990.

chyme Mix of food and gastric fluid.

esophagus Muscular tube that connects the pharynx (throat) to the stomach.

gastric fluid Fluid secreted by the stomach lining; contains enzymes, acid, and mucus.

sphincter Ring of muscle that controls passage through a tubular organ or body opening.

stomach Muscular organ that mixes food with gastric fluid secreted by its lining.

① The highly folded interior of the small intestine.

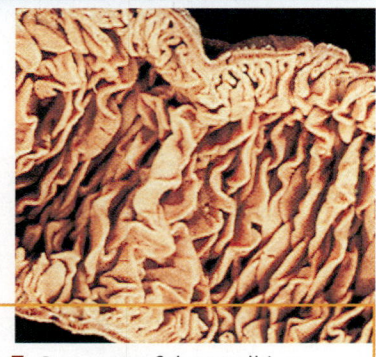

FIGURE 23.7 Structure of the small intestine, the main site of digestion and absorption.

Figure It Out: Are microvilli multicelled or smaller than a cell? *Answer: Smaller than a cell*

Credits: (1) top, © Cengage Learning, bottom, Microslide courtesy Mark Nielsen, University of Utah; (2–4) After Sherwood and others.

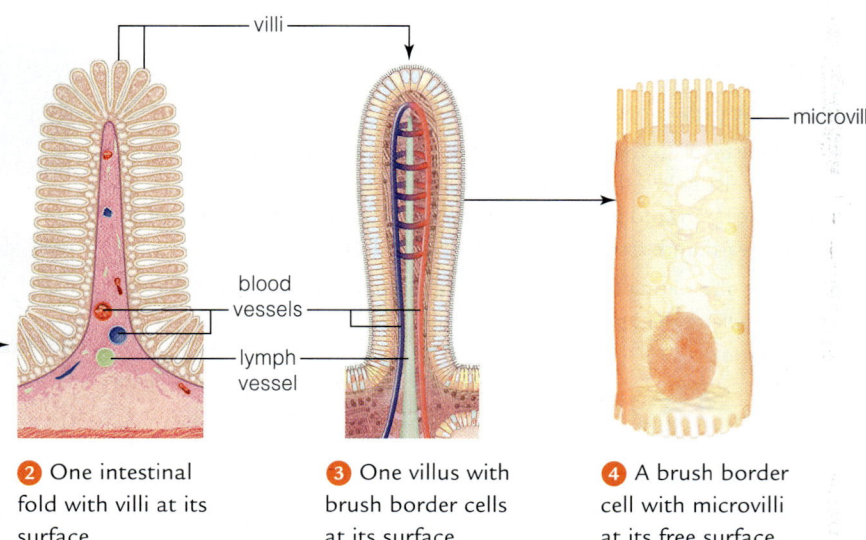

villi

blood vessels

lymph vessel

microvilli

② One intestinal fold with villi at its surface.

③ One villus with brush border cells at its surface.

④ A brush border cell with microvilli at its free surface.

bile Mix of salts, pigments, and cholesterol produced by the liver; aids in fat digestion.

gallbladder Organ that receives bile from the liver and expels it into the small intestine.

liver Organ that produces bile, stores glycogen, and detoxifies many substances.

microvilli Thin projections that increase the surface area of brush border cells.

pancreas Organ that secretes digestive enzymes into the small intestine and hormones into the blood.

small intestine Longest portion of the digestive tract, and the site of most digestion and absorption.

villi Multicelled projections from the lining of the small intestine.

› **Digestion in the Small Intestine** Muscle contractions force chyme through a sphincter into the **small intestine**. The small intestine is "small" only in its diameter, which is about 2.5 cm (1 inch). It is the longest segment of the gut. Uncoiled, it would extend for about 5 to 7 meters (16 to 23 feet).

Most digestion and absorption takes place in the small intestine, and its structure reflects its function (Figure 23.7). The interior of the small intestine, like that of the stomach, is highly folded ①. Unlike the interior folds of the stomach, intestinal folds are permanent. Each fold has many multicelled projections called **villi** (singular, villus), each about a millimeter long ②. The millions of villi that project from the intestinal lining give it a velvety appearance. Cells at the surface of each villus have even tinier projections called **microvilli** (singular, microvillus) ③. The 1,700 or so microvilli at the surface of a cell make its outer edge look like a brush ④, so these cells are sometimes called brush border cells. Collectively, the many folds and projections of the small intestinal lining increase its surface area several hundredfold.

Chemical digestion of carbohydrates and proteins is completed in the small intestine, with help from the neighboring pancreas and liver (Figure 23.8**A**). The **pancreas** is a large gland that secretes enzymes and a buffer into the small intestine. Pancreatic and small intestinal enzymes break carbohydrates down into monosaccharides (simple sugars) and polypeptides into amino acids.

Fat digestion occurs entirely in the small intestine. It requires enzymes and **bile**, a mix of salts, pigments, and cholesterol. Bile is produced in the **liver**, a large organ that also stores glycogen and detoxifies substances. Between meals, the main bile duct from the liver to the small intestine is closed, and bile enters the **gallbladder**, which stores it. Eating a fatty meal stimulates the gallbladder to contract and squirt bile into the small intestine. Bile enhances fat digestion by helping to keep fat droplets from clumping together, a process called emulsification. Compared to bigger globules, tiny droplets present a greater surface area to fat-digesting pancreatic and intestinal enzymes. These enzymes break lipids down into fatty acids and glycerol.

Sometimes components of bile accumulate as hard pellets called gallstones (Figure 23.8**B**). Most gallstones do no harm, but some can be dangerous because they block or become lodged in a duct. In this case, the gallstones or the entire gallbladder can be removed surgically.

› **Absorption in the Small Intestine** Figure 23.9 summarizes the sites where nutrients are digested and the absorbable products of this process. Absorp-

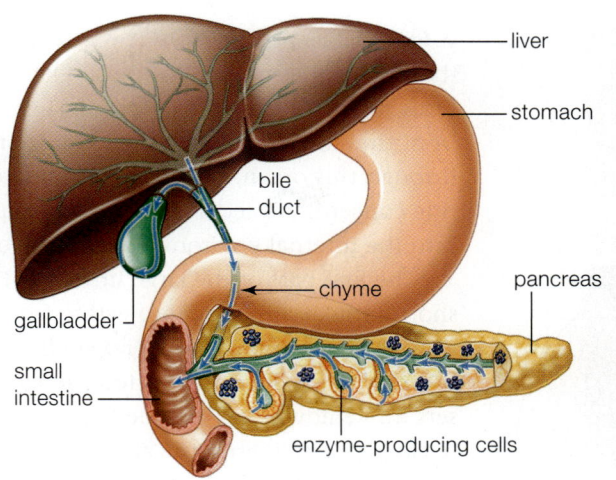

FIGURE 23.8 Organs that empty into the small intestine.

Credits: (a) From RUSSELL/WOLFE/HERTZ/STARR. Biology, 2E. © 2011, Brooks/Cole, a part of Cengage Learning, Inc. Reproduced by permission. www.cengage.com/permissions; (b) © C. James Webb/ Phototake. All rights reserved.

A Chyme from the stomach joins bile and pancreatic enzymes, which enter through ducts.

B Gallstone cut in half. Such stones can cause problems if they block the bile duct.

tion is the passage of solutes and water across the lining of the digestive tract and into the body proper. The process is facilitated by the large surface area of the small intestine, and by rings of smooth muscle in the intestinal wall. The repeated contraction and relaxation of these rings sloshes chyme back and forth, forcing it against the absorptive lining.

Products of carbohydrate digestion (monsaccharides) and protein digestion (amino acids) move through brush border cells, diffuse through the interstitial fluid inside the villus, then enter a blood capillary (Figure 23.10). Products of fat digestion (fatty acids and glycerol) also cross the brush border cell and enter the interstitial fluid. However, these molecules are taken up by lymph vessels, which eventually carry them to the blood.

All blood from capillaries in the small intestine flows through a vein (the hepatic portal vein) to the liver. Thus, blood with substances absorbed from the small intestine goes through the liver before returning to the heart and flowing to the rest of the body. The liver helps protect the body by detoxifying or removing from the blood dangerous substances that were ingested or formed by digestion. Chapter 4 explained the role of the liver in detoxifying alcohol.

	Carbohydrates	Proteins	Lipids
Mouth	Salivary amylase begins digestion		
Stomach		Acid, pepsin begin digestion	
Small intestine	Pancreatic and intestinal enzymes complete digestion	Pancreatic and intestinal enzymes complete digestion	Pancreatic and intestinal enzymes carry out digestion with the assistance of bile
Absorbable products	Monosaccharides (simple sugars)	Amino acids	Fatty acids, glycerol

FIGURE 23.9 Nutrient breakdown locations and products. © Cengage Learning.

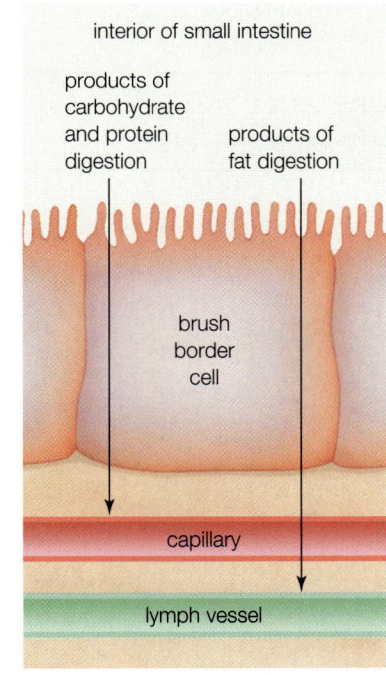

FIGURE 23.10 Animated!
Absorption. Nutrients move across brush border cells into a villus, then enter the blood or lymph. © Cengage Learning.

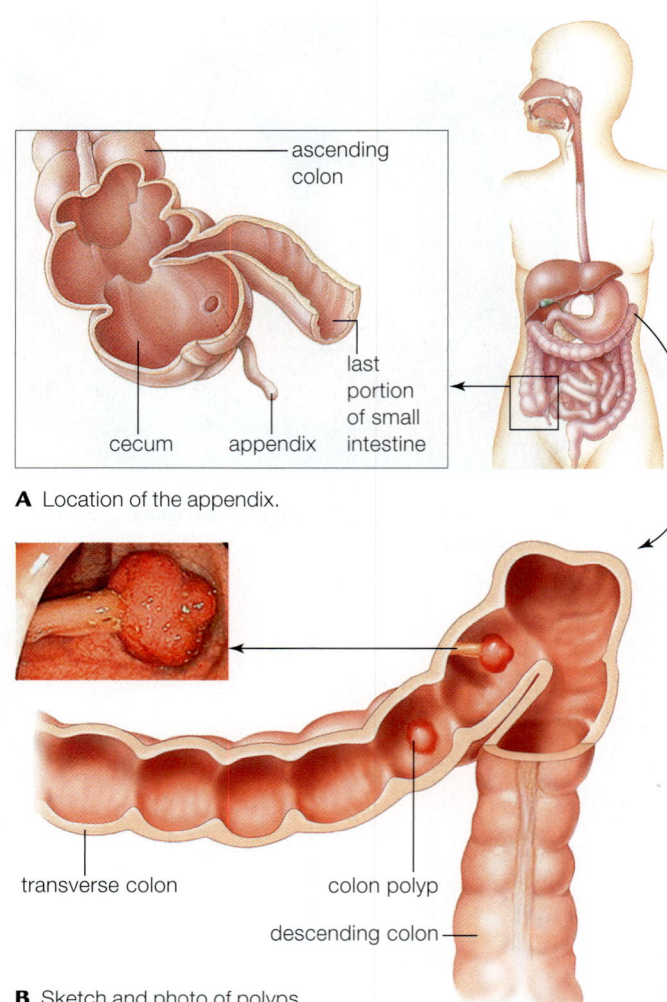

A Location of the appendix.

B Sketch and photo of polyps in the colon.

ascending colon

last portion of small intestine

cecum   appendix

transverse colon   colon polyp

descending colon

FIGURE 23.11 The large intestine.

Credits: (a) © Cengage Learning; (b) National Cancer Institute.

> **Concentrating and Eliminating Wastes** Not everything that enters the small intestine can be or should be absorbed. Contractions propel indigestible material, shed mucosal cells, and some water from the small intestine into the **large intestine**. As the wastes travel through this organ, they become compacted as **feces**. The large intestine concentrates feces by actively pumping sodium ions across its wall, into the internal environment. Water follows by osmosis.

The first part of the large intestine, the cup-shaped cecum, has a short, tubular **appendix** projecting from it (Figure 23.11**A**). Inflammation of the appendix, called appendicitis, requires prompt treatment. It often occurs after a bit of feces lodges in the appendix and infection sets in. Removing an inflamed appendix prevents it from bursting and releasing bacteria into the abdominal cavity. Such ruptures can cause a life-threatening infection.

Beyond the cecum, the portion of the large intestine known as the **colon** ascends the wall of the abdominal cavity, extends across that cavity, and descends. It connects to the **rectum**, the region of the large intestine that stores feces until it is expelled.

Contraction of smooth muscle in the colon wall mixes the colon's contents and propels them toward the rectum. Materials move more slowly through the colon than through other regions of the digestive tract and the colon has a moderate pH. These conditions encourage the growth of bacteria such as *Escherichia coli*, which produce vitamin B_{12} that we absorb across our colon lining.

After a meal, signals from autonomic nerves cause much of the colon to contract forcefully and propel feces to the rectum. The resulting stretching of the rectum wall activates a defecation reflex that expels feces. We can override this reflex by contracting a skeletal muscle sphincter at the anus. Healthy adults typically defecate about once a day, on average.

Some people are genetically predisposed to develop colon polyps, which are small growths on the colon wall (Figure 23.11**B**). Most polyps are benign, but some can become cancerous. If detected in time, colon cancer is highly curable. Blood in feces and dramatic changes in bowel habits may be symptoms of colon cancer and should be reported to a doctor. Also, anyone over the age of 50 should have a periodic colonoscopy, a procedure in which clinicians use a camera to examine the interior of the colon for polyps or cancer.

Take-Home Message

What happens to food that enters the human digestive tract?

- Mechanical and chemical breakdown of food begins in the mouth. Saliva contains an enzyme that begins carbohydrate digestion.

- Gastric fluid contains acid and digestive enzymes that begin the process of protein digestion in the stomach.

- In the small intestine, intestinal and pancreatic enzymes complete digestion of proteins and carbohydrates, and bile from the gallbladder aids in fat digestion.

- Most nutrients released by digestion are absorbed across the richly folded lining of the small intestine, which has a vast surface area.

- The large intestine absorbs water and mineral ions from its contents and compacts undigested residues. The resulting feces are stored in the rectum before being expelled through the anus.

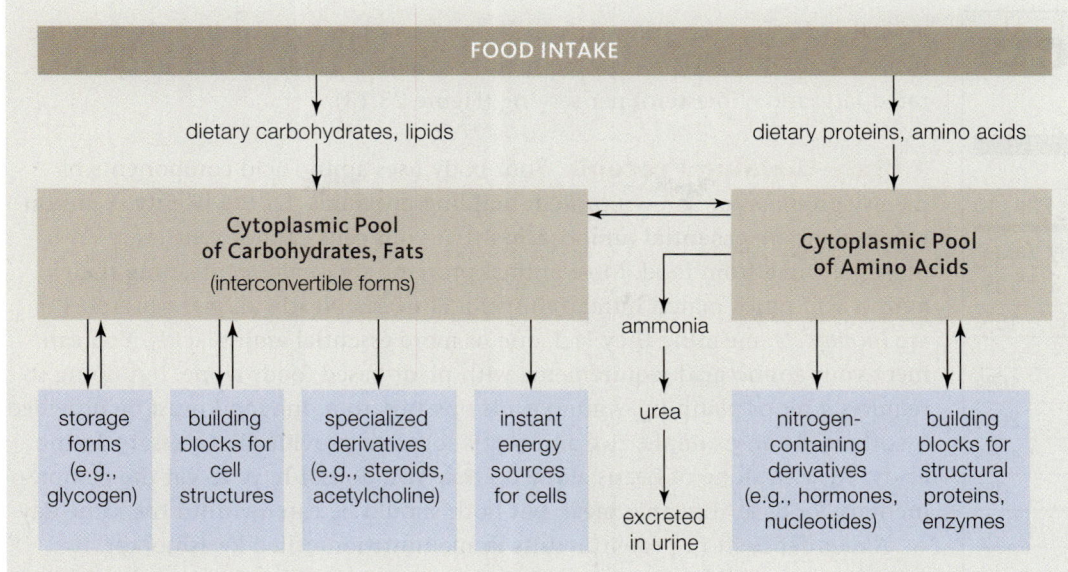

FOOD INTAKE

dietary carbohydrates, lipids → Cytoplasmic Pool of Carbohydrates, Fats (interconvertible forms)

dietary proteins, amino acids → Cytoplasmic Pool of Amino Acids

ammonia

urea

excreted in urine

storage forms (e.g., glycogen)

building blocks for cell structures

specialized derivatives (e.g., steroids, acetylcholine)

instant energy sources for cells

nitrogen-containing derivatives (e.g., hormones, nucleotides)

building blocks for structural proteins, enzymes

FIGURE 23.12 Fate of nutrients absorbed into the body from the gut.

Figure It Out: The liver converts ammonia, a toxic product of digestion, to urea that is excreted in the urine. What type of nutrient (carbohydrate, lipid, or protein) is the source of the ammonia?

© Cengage Learning.   *Answer: Protein*

23.4 Human Nutrition

Absorbed nutrients can be burned as fuel in energy-releasing pathways, stored, or used as building blocks. Section 5.8 described how carbohydrates, fats, and proteins are converted to intermediates that enter the ATP-producing pathway of aerobic respiration. Figure 23.12 summarizes the major routes by which the body uses organic compounds from food.

❯ Energy-Rich Carbohydrates Fresh fruits, whole grains, and vegetables, especially legumes such as peas and beans, contain abundant complex carbohydrates (Section 2.7). The body breaks down the starch in these foods to glucose, your primary source of energy. When glucose absorption exceeds the body's immediate needs, the excess is stored. Liver and muscle cells store excess glucose as glycogen. The liver and adipose cells also use glucose to build fats.

Carbohydrates are also the body's source of dietary fiber. There are two types of fiber. Soluble fiber consists of polysaccharides that form a gel when mixed with water. Eating foods high in soluble fiber helps lower one's cholesterol level and reduces the risk of heart disease. Insoluble fiber such as cellulose does not dissolve and passes through the human digestive tract more or less intact. Eating insoluble fiber helps prevent constipation. Processed carbohydrates such as white flour, refined sugar, and corn syrup are sometimes said to provide only "empty calories" because these foods provide little vitamins or fiber.

You may have noticed some grain-based foods labeled as "gluten-free." Gluten is a protein found in wheat and many other grains. An estimated 1 percent of the population has celiac disease, a genetic disorder in which gluten causes an autoimmune reaction that harms the small intestine's villi and interferes with nutrient absorption. The disease is treated by eliminating gluten from the diet.

❯ Good Fat, Bad Fat Fats serve as an energy source and as building blocks of cell membranes. They also help you take up fat-soluble vitamins. Your body can assemble most fats you need from carbohydrates. Fats the body cannot make, such as linoleic acid, are **essential fatty acids**. Vegetable oils are a good source of these fatty acids.

Butter, cheese, and fatty meats are rich in saturated fats and cholesterol. Overconsumption of these foods increases the risk for heart disease and stroke,

> A diet high in soluble fiber can help lower the risk of heart disease.

appendix Tubular projection from the first part of the large intestine.

colon Longest portion of the large intestine.

essential fatty acids Fatty acids that the body cannot make and must obtain from the diet.

feces Unabsorbed food material and cellular waste that is produced by digestion.

large intestine Organ that concentrates and stores waste, then delivers it to the anus for excretion.

rectum Portion of the large intestine that stores feces until they are expelled.

Nutrition Facts

Serving Size 1 cup (228g)
Servings Per Container 2

Amount Per Serving

Calories 250	Calories from Fat 110

	% Daily Value*
Total Fat 12g	18%
Saturated Fat 3g	15%
Trans Fat 1.5g	
Cholesterol 30mg	10%
Sodium 470mg	20%
Total Carbohydrate 31g	10%
Dietary Fiber 0g	0%
Sugars 5g	
Protein 5g	

Vitamin A	4%
Vitamin C	2%
Calcium	20%
Iron	4%

*Percent Daily Values are based on a 2,000 calorie diet. Your Daily Values may be higher or lower depending on your calorie needs:

		Calories:	2,000	2,500
Total Fat	Less than		65g	80g
Sat Fat	Less than		20g	25g
Cholesterol	Less than		300mg	300mg
Sodium	Less than		2,400mg	2,400mg
Total Carbohydrate			300g	375g
Dietary Fiber			25g	30g

as well as for cancers. *Trans* fatty acids, or "*trans* fats," are similarly bad for the heart. All food labels are now required to show the amounts of *trans* fats, saturated fats, and cholesterol per serving (Figure 23.13).

> **Body-Building Proteins** Your body uses amino acid components of dietary proteins for its own protein-building programs. Of the twenty common types, eight are **essential amino acids** that your cells cannot synthesize. You must get these from food. Most animal proteins are *complete*, meaning their amino acid ratios match human nutritional needs. Nearly all plant proteins are *incomplete*, meaning they lack one or more essential amino acids. You can meet your amino acid requirements with plant-based foods alone, but doing so requires a bit of planning. Amino acids missing from one food must be provided by others. As an example, rice and beans together provide all necessary amino acids, but rice alone or beans alone do not. You do not have to eat the complementary foods at the same meal, but both should be eaten within the same day.

A diet deficient in protein results in malnutrition called kwashiorkor. It mainly affects children in developing countries. Affected individuals have brittle hair, muscle loss, and fatigue. Their liver enlarges and their abdomen swells with fluid. Long-term effects included stunted growth and impaired intelligence.

FIGURE 23.13 Food label. Information on such labels can be used to ensure that you obtain the nutrients you need without exceeding recommended limits on less healthy substances such as salt and *trans* fats.

Figure It Out: What fraction of the fat in a serving of this macaroni and cheese product comes from the least healthy forms of fat (saturated fat and *trans* fat)?

Answer: Of the total fat content in a serving (12 grams), 3 g are saturated fat and 1.5 g are trans fat. Thus 4.5 g, or 1/3 the fat, is from unhealthy sources.

Credits: left, © Food Collection/ SuperStock; right, USDA.

Table 23.1	Sources and Functions of Major Vitamins	
Vitamin	**Dietary Sources**	**Main Functions**
Fat-soluble vitamins		
A	Yellow/orange fruits, green vegetables, milk, egg yolk, fish, liver	Used in synthesis of visual pigments, bone, teeth; maintains the skin
D	Fish liver oils, egg yolk, fortified milk; also synthesized in skin	Promotes bone health; enhances calcium absorption
E	Whole grains, dark green vegetables, vegetable oils	Counters effects of free radicals; helps maintain cell membranes; blocks breakdown of vitamins A, C in gut
K	Green leafy vegetables, cabbage	Acts in blood clotting; ATP formation via electron transfer chains
Water-soluble vitamins		
B$_1$ (thiamine)	Legumes, whole grains, green leafy vegetables, meats, eggs	Connective tissue formation; folate utilization; coenzyme action
B$_2$ (riboflavin)	Whole grains, poultry, fish, egg white, milk	Coenzyme in carbohydrate and amino acid metabolism
B$_3$ (niacin)	Green leafy vegetables, potatoes, peanuts, poultry, fish, meats	Coenzyme in carbohydrate metabolism
B$_6$	Spinach, tomatoes, potatoes, meats	Coenzyme in amino acid metabolism
Pantothenic acid	Meats, yeast, egg yolk	Coenzyme in glucose metabolism, fatty acid and steroid synthesis
Folate (folic acid)	Dark green vegetables, whole grains, yeast, lean meats; made by bacteria in large intestine	Coenzyme in nucleic acid and amino acid metabolism; required for normal nervous system development of embryo
B$_{12}$	Poultry, fish, red meat, dairy foods (not butter)	Coenzyme in nucleic acid metabolism
Biotin	Legumes, egg yolk	Coenzyme in fat, glycogen formation, and amino acid metabolism
C (ascorbic acid)	Fruits and vegetables, especially citrus, berries, cantaloupe, green pepper, cabbage, broccoli	Collagen synthesis; used in carbohydrate metabolism; structural role in bone, teeth, cartilage; may counter free radicals

> ## Vitamins and Minerals

Vitamins are organic substances that are required in small amounts for normal metabolism (Table 23.1). Vitamins A, E, D, and K are fat-soluble vitamins. Such vitamins are not harmed by heating, and are abundant in both cooked and fresh foods. Fat-soluble vitamins can be stored in the body's own fat. Because the body stores these vitamins, they do not need to be replenished daily and deficiencies rarely occur. In fact, taking megadoses of fat-soluble vitamins can be harmful. For example, taking an excessive amount of vitamin A during pregnancy can cause birth defects. Some studies have found that taking large amounts of vitamin E can actually increase the risk of death.

Water-soluble vitamins such as vitamin C and the B vitamins are not stored by the body. It is difficult to accumulate too much of these vitamins, because any excess is eliminated in the urine. Water-soluble vitamins must be replenished each day to prevent a deficiency. These vitamins are destroyed by heat, so uncooked foods are the best source. Many of the B vitamins are coenzymes. For example, the coenzyme NADH, which plays an essential role in aerobic respiration (Section 5.6), is derived from vitamin B_3 (niacin).

Minerals are inorganic substances needed in small amounts for normal growth and metabolism. For example, all cells use iron in electron transfer chains, and red blood cells use it to make hemoglobin. You can obtain iron from meat, and many cereals are now fortified with iron. As another example, calcium is necessary for normal function of nerves and muscles, and to build strong bones. Dairy products and tinned fish such as sardines are good sources of calcium. Some cereal products are also fortified with calcium.

> ## USDA Dietary Recommendations

The United States Department of Agriculture (USDA) reviews research on human nutrition and issues a set of dietary guidelines. The guidelines are intended to summarize nutrition-related discoveries and explain how dietary choices affect the risk of chronic health problems. At the USDA website (www.ChooseMyPlate.gov), an interactive program based on these guidelines generates dietary recommendations specific for a person's age, sex, height, weight, and activity level (Figure 23.14). Current guidelines recommend minimizing one's intake of refined grains, saturated fats, *trans* fatty acids, sugar or other high-calorie sweeteners, and salt. They also recommend eating more vegetables and fruits with a high potassium and fiber content, more whole grains, and more fat-free or low-fat milk products. People who lack the enzymes that break down the milk sugar lactose and thus are lactose-intolerant, are advised to choose alternative high-calcium foods.

> ## Overweight and Obesity

Excess weight is a pressing public health issue. Since the 1970s, the percentage of overweight children and adolescents has doubled. One recent study concluded that if current trends continue, weight-related deaths could decrease the life expectancy in the United States by as much as 5 years over the next decade. Is your current weight healthy? One way to find out is to calculate your **body mass index** (BMI), using this formula:

$$BMI = \frac{\text{weight (pounds)} \times 700}{\text{height (inches)}^2}$$

A BMI of 25 to 29.9 is defined as overweight, and a BMI of 30 or more is defined as obese. Having a BMI higher than 25 carries an increased risk of premature death due to weight-related illness.

It is difficult to lower weight-related risks with diet alone. Eat less, and the body slows its metabolic rate to conserve energy. To maintain your weight, you must balance energy intake with energy used for metabolic activities. The energy

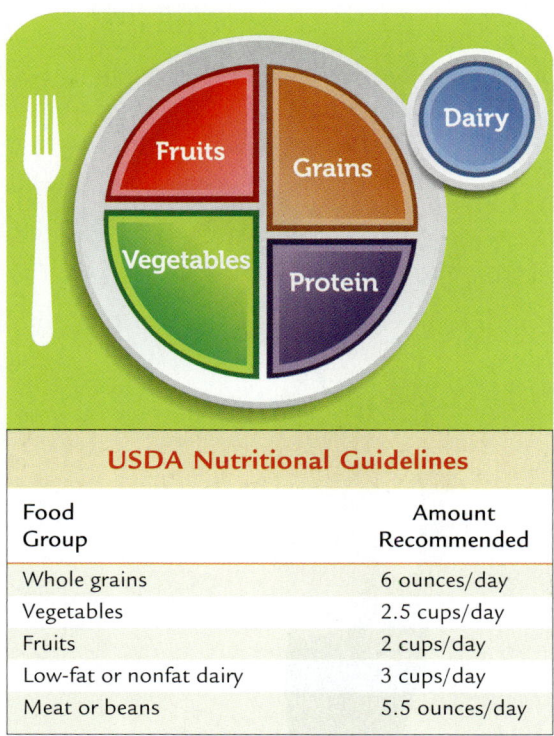

USDA Nutritional Guidelines

Food Group	Amount Recommended
Whole grains	6 ounces/day
Vegetables	2.5 cups/day
Fruits	2 cups/day
Low-fat or nonfat dairy	3 cups/day
Meat or beans	5.5 ounces/day

FIGURE 23.14 Example of nutritional guidelines from the United States Department of Agriculture (USDA). These recommendations are for females between ages ten and thirty who get less than 30 minutes of vigorous exercise daily. Portions add up to a 2,000-kilocalorie daily intake. USDA.

body mass index Formula used to determine a healthy weight based on height.

essential amino acid Amino acid that the body cannot make and must obtain from food.

mineral Inorganic substance that is required in small amounts for normal metabolism.

vitamin Organic substance required in small amounts for normal metabolism.

stored in food is measured as kilocalories, or Calories (with a big C). For a rough estimate of how many kilocalories you should take in each day to maintain your preferred weight, multiply that weight (in pounds) by 10 if you are not active physically, 15 if moderately active, and 20 if highly active. Next, subtract one of the following amounts from the multiplication value:

Age:	25–34	Subtract:	0
	35–44		100
	45–54		200
	55–64		300
	Over 65		400

Age is a factor because resting metabolic rate drops as a person gets older, so less food is required. So, if you are 25, are highly active, and want to maintain a weight of 120 pounds, you need to eat $120 \times 20 = 2,400$ kilocalories.

❭ Eating Disorders Eating too little can be as dangerous as eating too much, or more so. Anorexia nervosa is an eating disorder in which a person who has access to food does not eat enough to keep her weight within 15 percent of normal. Young women in developed countries are disproportionately affected, and societal pressures to be thin play a role in encouraging weight loss. However, how a person responds to such pressures is determined in part by genetics. Scientists have pinpointed several genes that increase risk of anorexia.

Anorexia damages organ systems throughout the body. Starved of essential calcium, the body breaks down bone. Inadequate iron intake causes anemia. Heart muscle weakens and heart rhythms can become disrupted. Most deaths related to anorexia occur as a result of sudden cardiac arrest.

Bulimia nervosa, another eating disorder, is also most common among young women. Bulimics tend to be near normal weight, but perceive themselves as heavy. They "binge and purge," eating far too much, then inducing vomiting. Induced vomiting bathes the esophagus and teeth in gastric fluid, making teeth pitted and brittle, and increasing the risk of esophageal cancer. Loss of gastric fluid to vomiting can also alter blood pH, which can disrupt normal breathing, alter the heart rhythm, and lead to convulsions.

Both anorexia and bulimia are treated with counseling and medical attention to the damage done by the disorder. Many patients respond favorably to treatment and are able to maintain a normal weight.

Credit: © Dana Heinemann/ Shutterstock.

Take-Home Message

How do we use the nutrients in our food?

- Carbohydrates are broken down to glucose, the body's main energy source. Foods rich in complex carbohydrates also supply fiber and vitamins.
- Fats are burned for energy and used as building materials.
- Proteins are the source of amino acids that the body uses to build its own proteins.
- Vitamins are organic substances required in small amounts for metabolism. Fat-soluble vitamins can be stored in the body. Water-soluble vitamins must be replenished daily.
- Minerals are inorganic substances required in small amounts for essential metabolic tasks.
- Maintaining a healthy weight requires balancing energy intake (calories in food) with energy expenditures (calories burned by metabolism and activity).

23.5 Regulating Internal Fluids

> **Fluid Homeostasis** By weight, all organisms consist mostly of water with salts and other solutes. Fluid outside cells—the extracellular fluid—serves as the body's internal environment. In humans and other vertebrates, extracellular fluid consists mostly of interstitial fluid (the fluid in spaces between cells) and plasma (the fluid portion of the blood). Keeping the solute composition and volume of the extracellular fluid within the range that living cells can tolerate is an essential aspect of homeostasis. Water and solute gains need to be balanced by water and solute losses. Eating and drinking add water and nutrients to the extracellular fluid. Metabolism adds wastes such as urea, a nitrogen-containing by-product of protein digestion. In all vertebrates, a urinary system rids the body of unwanted solutes and excess water.

> **The Human Urinary System** The human **urinary system** consists of two kidneys, two ureters, a bladder, and a urethra (Figure 23.15). **Kidneys** are bean-shaped organs about as large as an adult's fist, enclosed in a protective outer capsule of connective tissue. The kidneys filter water, mineral ions, nitrogen-rich wastes, and other substances from blood. Then they adjust the composition of the filtrate and return nearly all water and nonwaste solutes to blood. **Urine**, the fluid that remains, consists of urea and other wastes, water, and excess solutes.

Urine flows from each kidney into a **ureter**, a tube that empties into the **urinary bladder**. This hollow muscular organ stores urine. An urge to urinate arises when the stretch receptors in the bladder are stimulated. Smooth muscle in the bladder wall contracts in a reflex response, and urine flows into the **urethra**, through which it exits the body. After the age of 2 or 3 years, most people can control when they urinate by controlling a sphincter of skeletal muscle that encircles the urethra. Only when they deliberately relax this sphincter, does the force of bladder contraction propel urine out of the body.

kidney (one of a pair)
Constantly filters water and all solutes except proteins from blood; reclaims water and solutes as the body requires, and excretes the remainder as urine

ureter (one of a pair)
Channel for urine flow from a kidney to the urinary bladder

urinary bladder
Stretchable container for temporarily storing urine

urethra
Channel for urine flow between the urinary bladder and body surface

POSTERIOR
right kidney | vertebral column | left kidney
peritoneum | abdominal cavity
ANTERIOR

FIGURE 23.15 Animated! Organs of the human urinary system and their functions. These organs are located *outside* the peritoneum, the lining of the abdominal cavity.
© Cengage Learning.

Take-Home Message

What is the function of a urinary system?

- A urinary system regulates the composition and volume of extracellular fluid. Kidneys filter the blood and form urine.
- Urine from kidneys flows through ureters to a bladder, where it is stored until it flows out of the body through the urethra.

kidney Organ that filters blood and forms urine.
ureter Tube that carries urine from a kidney to the bladder.
urethra Tube through which urine from the bladder flows out of the body.
urinary bladder Hollow, muscular organ that stores urine.
urinary system Organ system that filters blood, and forms, stores, and expels urine.
urine Mix of water and soluble wastes formed and excreted by the urinary system.

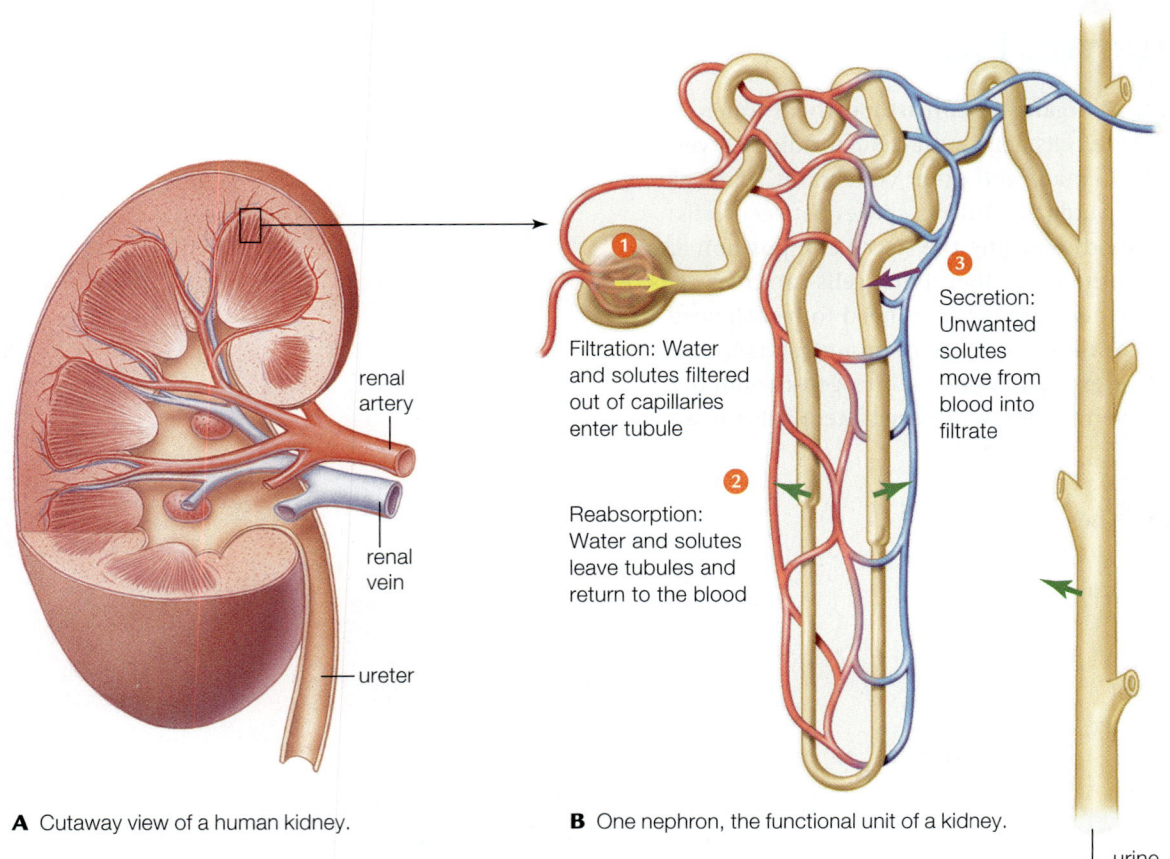

FIGURE 23.16 Animated!
Urine formation in the kidney.

Nephron tubules interact with capillaries to form urine by three processes.

1 Filtration (*yellow* arrow): Blood pressure forces fluid out of capillaries and into the entrance to a kidney tubule, forming the filtrate.

2 Reabsorption (*green* arrows): Water and solutes leave kidney tubules and return to blood of adjacent capillaries.

3 Secretion (*purple* arrow): Solutes move from the blood in capillaries into adjacent tubules.

© Cengage Learning.

renal artery

renal vein

— ureter

Filtration: Water and solutes filtered out of capillaries enter tubule

Reabsorption: Water and solutes leave tubules and return to the blood

Secretion: Unwanted solutes move from blood into filtrate

A Cutaway view of a human kidney.

B One nephron, the functional unit of a kidney.

urine

23.6 Kidney Function

> **How Urine Forms** A kidney is a blood-cleansing organ. A renal artery carries blood to be cleansed to each kidney, and a renal vein transports cleansed blood away from it (Figure 23.16**A**). The Latin word *renal* means "relating to kidneys." Inside a kidney, blood flows into arterioles. The arterioles branch into capillaries that associate with kidney tubules as a **nephron**.

Nephrons cleanse the blood and form urine by three processes: filtration, reabsorption, and secretion (Figure 23.16**B**). Each nephron starts in the outer region of the kidney, where the wall of a kidney tubule cups around a cluster of capillaries. These capillaries are structurally specialized for filtering; they are much leakier than capillaries in most parts of the body. **Filtration** occurs when blood pressure forces water and solutes out through spaces between cells in the walls of the capillaries **1**. Proteins and blood cells are too big to get out through these spaces, so they remain in the blood.

The solute-rich filtrate forced out of the capillaries enters the first part of the kidney tubule, where **reabsorption** begins. By this process, water and essential solutes that left the blood return to it as the filtrate flows through tubules **2**. About 99 percent of the water returns to the blood, along with all of the glucose and amino acids. Wastes such as urea and the breakdown product of hemoglobin remain in the filtrate. Degraded hemoglobin is what gives your urine its yellow color.

As the filtrate flows through the tubule, unwanted solutes from the blood are moved into it, a process called **tubular secretion 3**. Secretion is essential to regulating the pH of the internal environment. Enzymes can function only within a limited pH range. An excess of hydrogen ions (H^+) can disrupt many essential

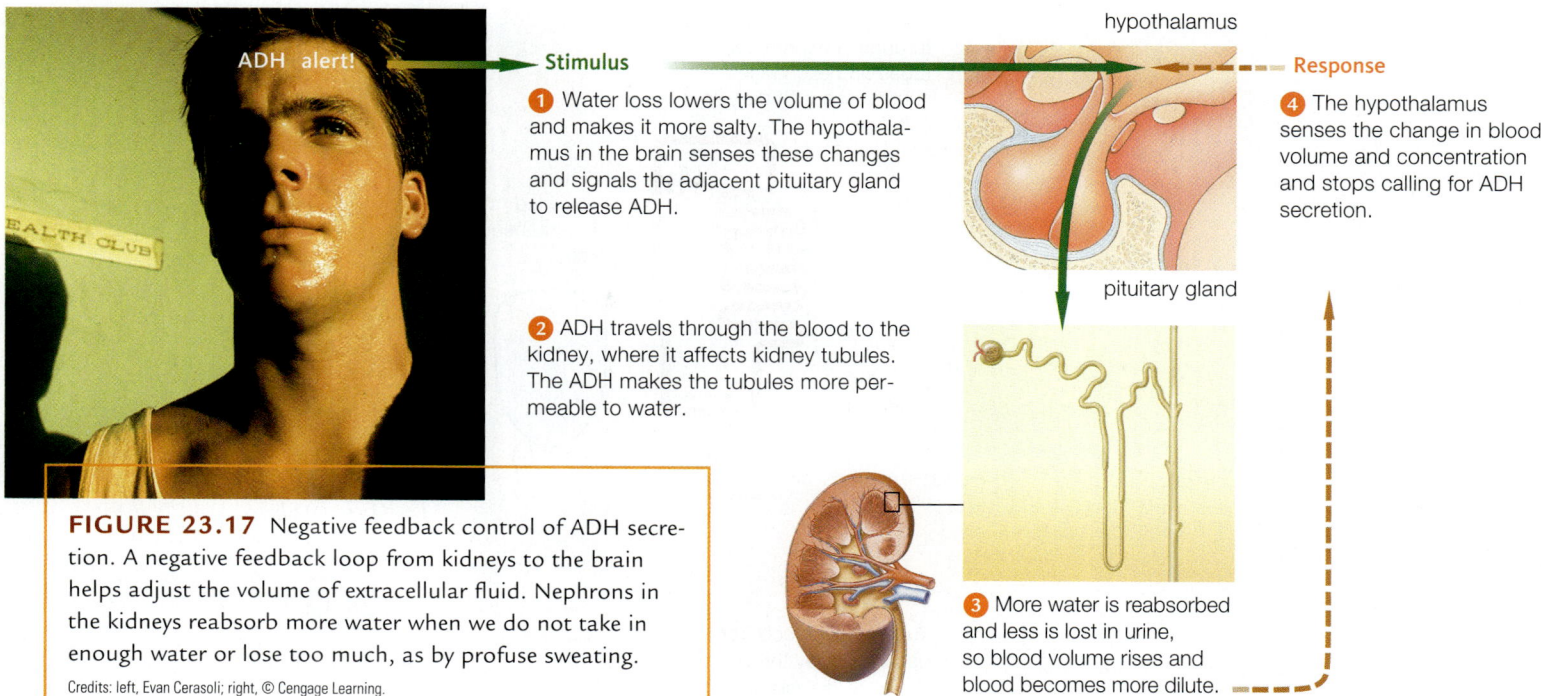

ADH alert!

Stimulus

① Water loss lowers the volume of blood and makes it more salty. The hypothalamus in the brain senses these changes and signals the adjacent pituitary gland to release ADH.

② ADH travels through the blood to the kidney, where it affects kidney tubules. The ADH makes the tubules more permeable to water.

hypothalamus

Response

④ The hypothalamus senses the change in blood volume and concentration and stops calling for ADH secretion.

pituitary gland

③ More water is reabsorbed and less is lost in urine, so blood volume rises and blood becomes more dilute.

FIGURE 23.17 Negative feedback control of ADH secretion. A negative feedback loop from kidneys to the brain helps adjust the volume of extracellular fluid. Nephrons in the kidneys reabsorb more water when we do not take in enough water or lose too much, as by profuse sweating.

Credits: left, Evan Cerasoli; right, © Cengage Learning.

metabolic processes. Kidneys help maintain the body's acid–base balance by regulating how much H^+ is secreted into the urine.

> **Feedback Control of Urine Formation** The concentration of urine can vary. Sip soda all day and your urine will be dilute and light in color. Spend 8 hours or so without drinking, as when you sleep, and your urine becomes concentrated and darker in color. Your urine also becomes more concentrated when you lose a lot of water by sweating.

Antidiuretic hormone (ADH) is a signaling molecule that increases water reabsorption, thus making the urine more concentrated (Figure 23.17). When you lose water, the resulting change in the volume and concentration of body fluid is detected by the hypothalamus, a brain region that is a control center for many homeostatic responses ①. The hypothalamus stimulates the pituitary gland to secrete ADH. The ADH travels through the blood to cells of the kidney tubules, where it causes the tubules to become more permeable to water ②. More water is reabsorbed, so less departs in the urine. Increased water reabsorption returns the body fluid's salt concentration and volume back to normal ③. The hypothalamus senses the change and stops calling for ADH secretion ④.

Alcohol inhibits ADH secretion. As a result, more fluid remains in the filtrate and the drinker needs to urinate more frequently. The excessive urination can lead to dehydration, which contributes to the symptoms of a hangover.

> **Impaired Kidney Function** Sometimes the substances dissolved in urine come out of solution and form hard kidney stones inside the kidney. About one million Americans receive treatment for kidney stones each year, with men four times as likely as women to be affected.

Most kidney stones do not cause symptoms, but some enter a ureter and cause extreme pain. When the ureter contracts around the stone, the ureter wall can be damaged, making the urine bloody. Usually the fluid pressure exerted on the stone by the urine pent up behind it forces the stone through the urinary tract and out of the body. If the kidney stone does not pass after a reasonable

antidiuretic hormone Hormone produced by the pituitary gland; increases water reabsorption by the kidney.

filtration Blood pressure forces water and small solutes, but not blood cells or proteins, out across the walls of capillaries.

nephron Kidney tubule and associated capillaries; filters blood and forms urine.

reabsorption Water and solutes enter capillaries.

tubular secretion Substances are moved out of capillaries and into kidney tubules.

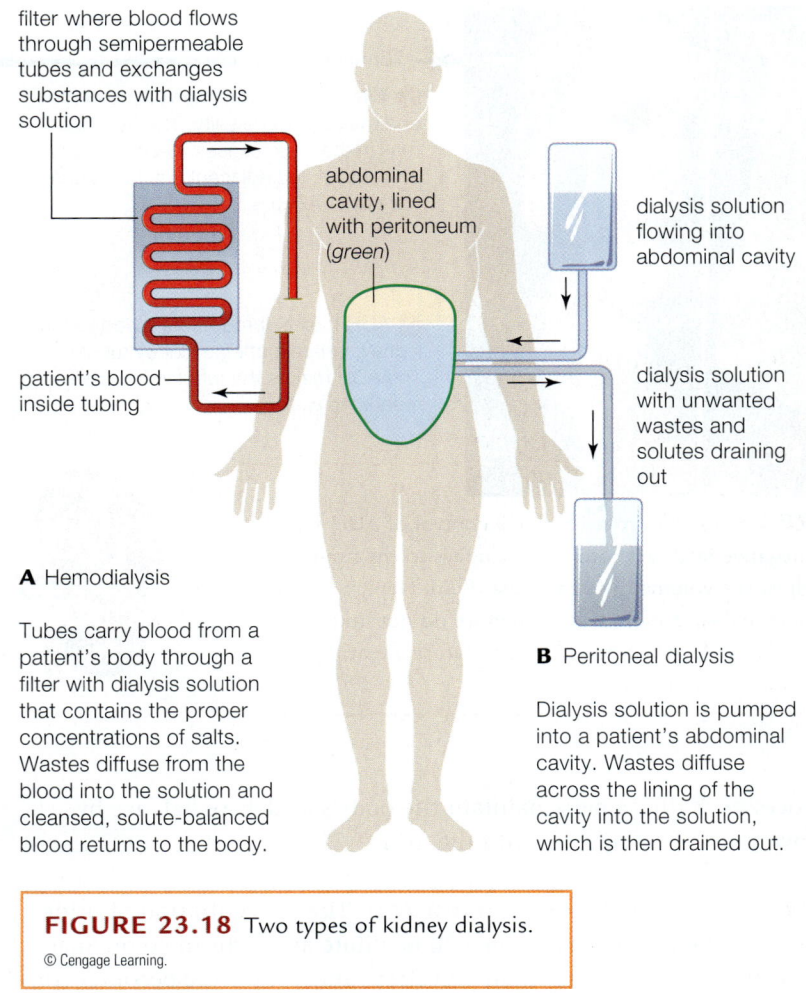

filter where blood flows through semipermeable tubes and exchanges substances with dialysis solution

abdominal cavity, lined with peritoneum (green)

dialysis solution flowing into abdominal cavity

patient's blood inside tubing

dialysis solution with unwanted wastes and solutes draining out

A Hemodialysis

Tubes carry blood from a patient's body through a filter with dialysis solution that contains the proper concentrations of salts. Wastes diffuse from the blood into the solution and cleansed, solute-balanced blood returns to the body.

B Peritoneal dialysis

Dialysis solution is pumped into a patient's abdominal cavity. Wastes diffuse across the lining of the cavity into the solution, which is then drained out.

FIGURE 23.18 Two types of kidney dialysis.
© Cengage Learning.

time, causes constant pain, or blocks flow of urine, it can be removed surgically or broken up by shock waves administered from outside the body.

The best way to prevent kidney stones is to drink lots of fluids. A diet high in protein can raise the risk of stone formation, so no one who has impaired kidney function or a history of kidney stones should begin a high-protein diet without discussing it with a doctor.

Kidney function is measured as the rate of filtration. When the filtration rate falls by half, a person is in kidney failure. The vast majority of kidney failure occurs as a complication of diabetes mellitus or high blood pressure. These disorders damage small blood vessels, including capillaries that filter blood into nephrons. Kidney failure can be fatal because wastes build up in the blood and interstitial fluid. The pH of body fluids rise, and changes in concentrations of other ions interfere with metabolism.

Kidney dialysis is used to restore proper solute balance in a person with kidney failure. "Dialysis" refers to the exchange of solutes between two solutions that are separated by a semipermeable membrane. With hemodialysis, a dialysis machine is connected to a patient's blood vessel (Figure 23.18**A**). The machine pumps the patient's blood through semipermeable tubes submerged in a warm solution of salts, glucose, and other substances. As the blood flows through the tubes, wastes dissolved in the blood diffuse out and solute concentrations return to normal levels. Cleansed, solute-balanced blood is returned to the patient's body. Typically, a person with kidney failure has hemodialysis three times a week at a dialysis center. Each treatment takes several hours. Peritoneal dialysis can be done at home. Each night, dialysis solution is pumped into a patient's

kidney dialysis Procedure used to cleanse blood and restore proper solute concentrations in a person with impaired kidney function.

abdominal cavity (Figure 23.18**B**). Wastes diffuse across the peritoneal lining into the fluid, which is drained out the next morning. With this method of dialysis, the peritoneal lining functions as the semipermeable membrane.

When kidney failure is permanent, dialysis must be continued for the rest of a person's life, or until a kidney becomes available for transplant. The National Kidney Foundation estimates that every day, 17 people die of kidney failure while waiting for a transplant. Most kidneys used as transplants come from people who had arranged to be organ donors after their death. However, an increasing number of kidneys are removed from a living donor (Figure 23.19). A kidney transplant from a living donor has a better chance of success than one from a deceased person. One kidney is adequate to maintain good health, so the risks to a living donor are mainly related to the surgery—unless the donor's remaining kidney fails.

The benefits of organs from living donors, a lack of donated organs, and high dialysis costs have led some to suggest that people should be allowed to sell a kidney. Critics argue that it is unethical to tempt people to risk their health for money. Section 10.4 describes another potential alternative. Someday, genetically modified pigs may provide organs for transplant.

Take-Home Message

How does urine form?

- Filtration forces water and small solutes into the kidney tubules. The driving force for filtration is blood pressure.

- Most of the filtered water and solutes are reabsorbed into capillaries around the tubules.

- Wastes that are not absorbed end up in the urine, along with any excess solutes secreted into the filtrate.

- The hormone ADH is secreted by the brain and acts in the kidney to promote water reabsorption, thus concentrating the urine.

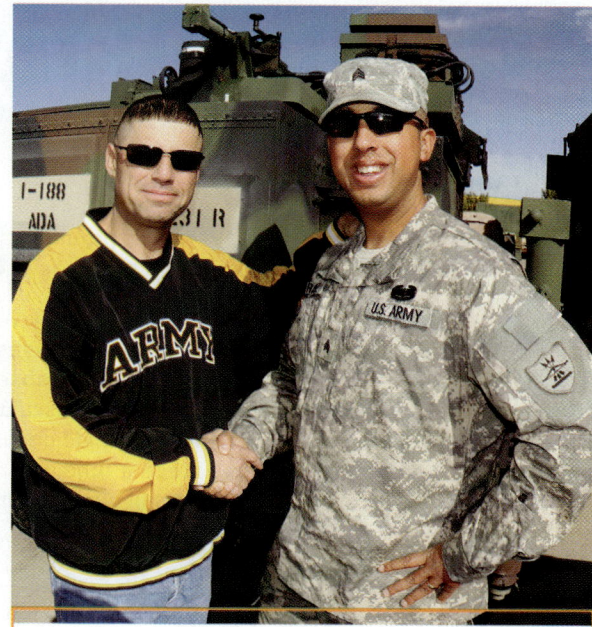

FIGURE 23.19 Living donor kidney transplantation. Spc. John Chase (*left*) with Sgt. Francisco Raat (*right*), who donated a kidney to him. The two National Guardsmen had served together in Afghanistan before Spc. Chase went into kidney failure.

US Army, Senior Master Sgt. David H. Lipp

23.7 Regulating Appetite (revisited)

Leptin decreases appetite. Another hormone, ghrelin, increases it. Ghrelin is secreted by cells in the stomach lining, as well as cells in the brain. This secretion increases when the stomach is empty, and declines after a big meal.

In one study of ghrelin's effects, a group of obese volunteers were placed on a low-fat, low-calorie diet for six months. They lost weight, but the concentration of ghrelin in their bloodstream climbed dramatically. As a result, they were hungrier than ever!

Gastric bypass surgery may be more effective than standard weight loss methods in part because, unlike dieting, it lowers ghrelin levels. The surgery is designed for extremely obese people and it effectively reduces the size of the stomach and small intestine. After surgery, decreased stomach size and ghrelin secretion makes people feel full faster and less hungry between meals. In addition, shortening the length of small intestine decreases absorption so fewer calories are absorbed from the food that is eaten. The results can be dramatic, but gastric bypass surgery also has some major drawbacks. There are the risks of anesthesia and the possibility of complications. Also, after surgery, the reduced capacity of the shortened small intestine for absorption increases the risk for vitamin and mineral deficiencies.

WHERE YOU ARE GOING . . .

Section 24.4 describes how the nervous system controls digestion and urination. The mechanism of action of hormones such as leptin is covered in Section 25.2, and the brain regions that make and secrete ADH are covered in Section 25.3. Section 26.6 mentions how urine tests for pregnancy work and explains how a developing fetus obtains essential nutrients and disposes of wastes.

Summary

Section 23.1 The body's fat cells produce leptin, a molecule that suppresses appetite. Leptin deficiency causes obesity in mice, but is rare in humans. Another molecule, ghrelin, increases appetite.

Section 23.2 Most animals have a digestive system that carries out **ingestion** and **digestion** of food, **nutrient absorption**, and **elimination** of digestive waste. The **gastrovascular cavity** of flatworms and cnidarians is saclike, with a single opening. Most animals have a **complete digestive tract**: a tube with two openings (mouth and **anus**), and specialized areas between them.

Section 23.3 Like other vertebrates, humans have a complete digestive tract. Chewing mixes food with saliva. An enzyme in saliva starts the process of starch digestion. The swallowing reflex moves food from the throat into the **esophagus**. Food continues through a **sphincter** into the **stomach**, a muscular sac with a lining that secretes **gastric fluid**. Protein digestion begins in the stomach.

The **small intestine** receives **chyme** (food mixed with gastric fluid) from the stomach, enzymes from the **pancreas**, and **bile** from the **gallbladder**. Bile, which assists in fat digestion, is made by the **liver**. Digestion of all nutrients is completed in the small intestine.

Most nutrients and fluid are absorbed into the internal environment across the wall of the small intestine. Intestinal folds are covered with multicelled projections (**villi**). Brush border cells at the villus surface have membrane extensions called **microvilli**. All of the folding and extensions greatly increase the surface area for absorption.

The **large intestine** consists mainly of the **colon**, which absorbs minerals and water, and concentrates undigested residues as **feces**. The **rectum** stores feces until they exit through the anus. The **appendix** is a small projection near the start of the large intestine.

Section 23.4 The body burns carbohydrates as fuel and whole grains provide fiber. Fats are another source of energy. **Essential fatty acids** must be obtained from the diet, as must **essential amino acids**. **Vitamins** and **minerals** are necessary in small amounts for normal metabolism. To maintain body weight, energy intake must balance energy output. Having a high **body mass index** (**BMI**) raises the risk of health problems.

Section 23.5 An animal body must maintain the volume and composition of its fluid components within a narrow range. The human **urinary system** interacts with other organ systems to balance the intake and output of solutes and water. The urinary system consists of a pair of blood-filtering **kidneys** that produce **urine**. The urine drains into a pair of **ureters**, is stored in a **urinary bladder**, and leaves the body through the **urethra**.

Section 23.6 Kidneys contain an enormous number of nephrons, or blood-filtering units. A **nephron** has a cup-shaped entrance around one set of blood capillaries and tubular regions that associate with another set of blood capillaries. Three processes contribute to urine formation by the nephrons:

Filtration. Blood pressure drives water and most solutes except proteins out of the capillaries at the start of a nephron. These enter the nephron's tubular parts.

Reabsorption. Water and solutes to be conserved move out of the tubular parts, then into the capillaries that thread around the nephron. A small volume of water and solutes remains inside the nephron as the forming urine.

ADH (**antidiuretic hormone**) is a hormone that adjusts urine concentration by promoting water reabsorption. As a result, more water returns to the blood rather than leaving in the urine.

Tubular secretion. H⁺ and some other substances move from capillaries into the nephron for excretion.

Good kidney function is essential to life. **Kidney dialysis** or a transplant is an option for those with permanent kidney damage.

Self-Quiz   Answers in Appendix I

1. A digestive system functions in _____ .
   a. secreting enzymes
   b. absorbing nutrients
   c. eliminating wastes
   d. all of the above

2. Protein digestion begins in the _____ .
   a. mouth
   b. stomach
   c. small intestine
   d. colon

3. Digestion is completed and most nutrients are absorbed in the _____ .
   a. mouth
   b. stomach
   c. small intestine
   d. colon

4. Bile has roles in _____ digestion and absorption.
   a. carbohydrate
   b. fat
   c. protein
   d. amino acid

5. Bacteria that produce vitamin B₁₂ live in the _____ .
   a. stomach
   b. small intestine
   c. large intestine
   d. esophagus

6. Match each organ with a digestive function.
   ____ gallbladder
   ____ large intestine
   ____ liver
   ____ small intestine
   ____ stomach
   ____ pancreas
   a. makes bile, stores glycogen
   b. compacts undigested residues
   c. secretes enzymes, buffer
   d. absorbs most nutrients
   e. secretes gastric fluid
   f. stores, secretes bile

7. Essential fatty acids are _____ .
   a. *trans* fats
   b. saturated fats
   c. vitamins
   d. not made by the body

8. Microvilli _____ .
   a. increase the surface area of the stomach
   b. are the main type of cell in the lining of the small intestine
   c. project from the surface of cells in the small intestine
   d. produce bile

9. Iron is an example of a(n) _____ .
   a. vitamin
   b. mineral
   c. essential fatty acid
   d. essential amino acid

10. Urea forms as a breakdown product of _____ .
    a. nucleic acids
    b. simple sugars
    c. saturated fats
    d. proteins
    e. complex carbohydrates
    f. a through d

11. Filtration moves _____ and other substances into kidney tubules.
    a. water
    b. fiber
    c. proteins
    d. red blood cells

12. Water loss triggers a(n) _____ in ADH secretion.
    a. increase
    b. decrease

13. Kidneys return water and small solutes to the blood by the process of _____ .
    a. filtration
    b. reabsorption
    c. tubular secretion
    d. both a and b

14. Kidneys adjust the blood acidity by increasing or decreasing the _____ of H^+.
    a. filtration
    b. reabsorption
    c. secretion
    d. both a and b

15. Match each structure with a function.
    _____ ureter
    _____ bladder
    _____ urethra
    _____ nephron
    _____ pituitary gland

    a. stores urine
    b. delivers urine to body surface
    c. carries urine from kidney to bladder
    d. secretes ADH
    e. many inside a kidney

Critical Thinking

1. Starch and sugar have the same number of calories per gram. However, not all vegetables are equally calorie dense. For example, a serving of boiled sweet potato provides about 1.2 calories per gram, while a serving of kale yields only 0.3 calories per gram. What could account for the difference in the calories your body obtains from these two foods?

2. Oats and beans have a large amount of soluble fiber. A diet high in soluble fiber has been found to lower cholesterol, but it also can increase flatulence (gas). The gas is produced when bacteria living in the digestive tract break down the soluble fiber. In what part of the digestive tract do you think the most gas is produced?

3. Some people who have gallstones experience pain after they eat, with fatty meals causing the greatest discomfort. Why do such meals trigger gallbladder pain?

4. Diabetes insipidus is a medical disorder in which a person produces an unusually large amount of highly dilute urine. Some cases are caused by a gene mutation, but in most people diabetes insipidus arises after a head injury. Explain how an injury to the head could affect kidney function.

5. Marine mammals and desert rodents both have highly efficient kidneys that produce only a tiny amount of very concentrated urine. What selective pressure shaped this trait in both animals?

Digging Into Data

Pesticides and Organic Food

Products labeled as "organic" fill an increasing amount of space on supermarket shelves. What does this label mean? A food that carries the USDA's organic label must be produced without pesticides such as malathion and chlorpyrifos, which farmers typically use on conventionally grown fruits, vegetables, and many grains.

Does eating organic food significantly affect the level of pesticide residues in a child's body? Chensheng Lu of Emory University used urine testing to find out (Figure 23.20). For fifteen days, the urine of twenty-three children (aged 3 to 11) was monitored for breakdown products of pesticides. During the first five days, children ate their standard diet of conventionally grown foods. For the next five days, they ate organic versions of the same types of foods and drinks. For the final five days, the children returned to their standard diet.

1. During which phase of the experiment did the children's urine contain the lowest level of the malathion metabolite?

2. During which phase of the experiment was the maximum level of the chlorpyrifos metabolite detected?

3. Did switching to an organic diet lower the amount of pesticide residues excreted by the children?

4. Even in the conventional diet phases of this experiment, the highest pesticide metabolite levels detected were far below those known to be harmful. Given these data, would you spend more to buy organically grown foods?

Study Phase	No. of Samples	Malathion Metabolite		Chlorpyrifos Metabolite	
		Mean (µg/liter)	Maximum (µg/liter)	Mean (µg/liter)	Maximum (µg/liter)
1. Conventional	87	2.9	96.5	7.2	31.1
2. Organic	116	0.3	7.4	1.7	17.1
3. Conventional	156	4.4	263.1	5.8	25.3

FIGURE 23.20 *Above*, levels of metabolites (breakdown products) of malathion and chlorpyrifos detected in the urine of children taking part in a study of the effects of an organic diet. The difference in the mean level of metabolites in the organic and inorganic phases of the study was statistically significant. *Right*, the USDA organic food label.

Credits: above, © Cengage Learning; right, USDA.

Neural Control and the Senses 24

. . . WHERE YOU HAVE BEEN

The nervous signals discussed in this chapter involve receptor proteins (Section 3.3) and transport proteins (4.6). The chapter also refers back to discussions of PET scans (2.2), tumors (8.5), and animal body plans (15.2), and provides an example of morphological convergence (11.6).

24.1 In Pursuit of Ecstasy

Taking Ecstasy, an illegal drug, can make you feel socially accepted, less anxious, and more aware of your surroundings and of sensory stimuli. It can also leave you dying in a hospital, bleeding from all body openings as your temperature skyrockets. It can send your family and friends spiraling into horror and disbelief as they watch you stop breathing. Lorna Spinks ended life that way at age 19. Her anguished parents released the photographs in Figure 24.1 because they wanted others to know what their daughter did not: Ecstasy can kill.

Ecstasy is a psychoactive drug, meaning it alters psychological functions. Its active ingredient, MDMA, is similar in structure to methamphetamine:

methamphetamine   Ecstasy (MDMA)

Both drugs interfere with the function of neurons (the signaling cells of the nervous system) by disrupting their ability to send, receive, respond to, and clear away chemical signals. MDMA binds to active transporters in the membranes of brain neurons. The binding results in unnaturally high concentrations of a chemical signaling molecule that influences mood and memory. The resulting overabundance of this molecule in the brain is the source of Ecstasy's psychoactive effects: increased energy, empathy, and euphoria, and decreased anxiety.

However, MDMA's effects extend beyond the brain. Under the drug's influence, the body behaves as it would in an emergency. The heart races, blood pressure increases, jaw muscles clench, and pupils widen. Saliva production slows, so the mouth feels dry. Drinking water to alleviate this feeling can cause problems because MDMA also inhibits urine formation and urination. When you drink a large amount of fluid without getting rid of the excess, your extracellular fluid becomes too dilute—a condition that can cause convulsions. Alternatively, drinking too little and being physically active while on MDMA can cause a life-threatening rise in body temperature. Lorna Spinks died because MDMA caused her body temperature to soar, shutting down her organ systems.

Fortunately, Ecstasy use rarely results in life-threatening complications. When such complications do occur, rapid medical treatment is essential. If you suspect someone is having a bad reaction to Ecstasy or any other psychoactive drug, get medical help fast and be honest about the cause of the problem. Immediate, informed medical action may save a life.

FIGURE 24.1 Photos of Lorna Spinks alive (*top*), and just after her death (*bottom*). She died after taking two Ecstasy tablets.

Credits: top, © Press Association Images; bottom, © SWNS.com.

© Cengage Learning.

A Sensory neurons become excited when their receptor endings detect a specific stimulus, such as light or touch. They send chemical signals to interneurons or motor neurons.

B Interneurons receive chemical signals from sensory neurons (as shown) or other interneurons. They send chemical signals to other interneurons or to motor neurons (as shown).

C Motor neurons receive chemical signals from sensory neurons or interneurons (as shown). They send chemical signals to the muscle or gland that they control.

FIGURE 24.2 The three types of neurons. Arrows indicate direction in which signals flow. © Cengage Learning.

24.2 Neurons—The Great Communicators

❭ Neurons and Supporting Cells Of all multicelled organisms, animals respond fastest to external stimuli. **Neurons**, the excitable cells that make up the communication lines of the nervous system, facilitate these responses. In most animals, information flows among three types of neurons: sensory neurons, interneurons, and motor neurons (Figure 24.2). **Sensory neurons** are activated by specific stimuli such as light, touch, or heat. They signal interneurons or motor neurons. **Interneurons** receive signals from sensory neurons and other interneurons, and send signals of their own to motor neurons. **Motor neurons** control muscles and glands.

The structure of each type of neuron adapts it to a specific function. All neurons have a cell body, a region that contains the nucleus and most other organelles. All also have a single **axon**, a cytoplasmic extension that transmits electrical signals along its length and releases chemical signals at its terminal endings. In most sensory neurons, one end of the axon has receptor endings that detect some stimulus (Figure 24.2**A**). The other end has axon terminals that send signals to interneurons or motor neurons. In addition to their axon, interneurons and motor neurons have branched cytoplasmic extensions called dendrites. The **dendrites** receive chemical signals sent by other neurons, and convert them into electrical signals (Figure 24.2**B**,**C**).

Neuroglial cells, or **neuroglia**, act as a framework that holds neurons in place (*glia* means glue in Latin). They also help neurons function. For example, some types of neuroglial cells wrap around axons (Figure 24.2**D**). These neuroglia produce an insulating material, **myelin**, that functions like insulation on a wire. A myelin coating increases the speed at which signals travel along an axon.

The importance of myelin becomes clear when we consider the symptoms of multiple sclerosis (MS), an autoimmune disease in which the immune system attacks and destroys the myelin of axons in the brain and spinal cord. Affected people suffer from muscle weakness, fatigue, impaired coordination, numbness, and vision problems. Symptoms of MS typically begin between ages 20 and 40, with women affected twice as often as men. There is no cure, but medical treatment can slow progression of the disease.

❭ Resting Potential The ability of neurons to communicate with one another and with other cells arises from properties of the neuron plasma membrane. A neuron's plasma membrane, like that of other cells, is composed mainly of a lipid bilayer that is impermeable to ions and large molecules.

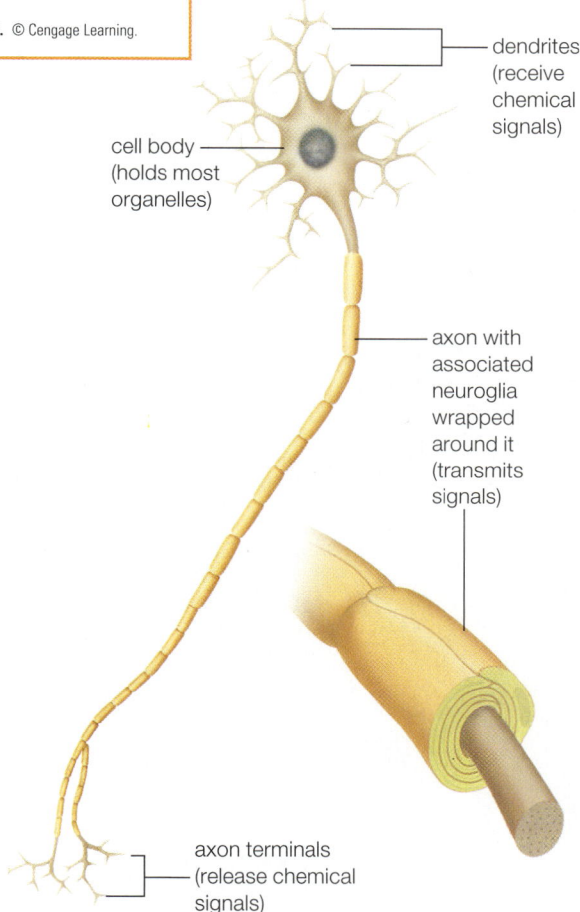

D Closer look at a motor neuron and associated neuroglia.

axon Cytoplasmic extension of a neuron that transmits electrical signals along its length and releases chemical signals at its endings.

dendrite Neuron's signal-receiving cytoplasmic extension.

interneurons Neurons that receive signals from, and send signals to, other neurons; they integrate information.

motor neurons Neurons that control muscles and glands.

myelin Material made by some neuroglia, insulates axons.

neuroglia Cells that support neurons.

neurons Excitable cells that make up the communication lines of nervous systems.

sensory neurons Neurons that detect specific stimuli.

Transport proteins control the movement of ions across the membrane. Also, like other cells, neurons have electrical and concentration gradients across their plasma membrane. Their cytoplasm has more negatively charged components than the interstitial fluid outside the cell.

Negatively charged proteins in the cytoplasm contribute to the gradient across the membrane. Being large and charged, these molecules cannot diffuse across the lipid bilayer of the cell membrane. The distributions of positively charged potassium ions (K^+) and positively charged sodium ions (Na^+) also play a role in the electrical gradient. There are more sodium ions in the fluid just outside a cell than there are inside it. The reverse is true for potassium ions. We can illustrate these ion concentration gradients as follows (the larger text represents higher concentrations; the green ball represents proteins):

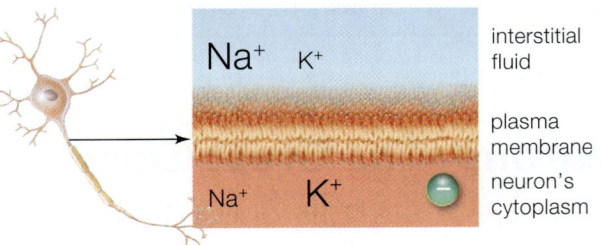

As in a battery, the separation in charge constitutes potential energy. Potential energy across a membrane is called membrane potential. The membrane potential of a neuron that is not being stimulated is its **resting potential**.

❯ The Action Potential Neurons are said to be "excitable" cells, meaning they undergo an action potential when properly stimulated. An **action potential** is a brief reversal of the electric gradient across the plasma membrane. It results from the opening and closing of voltage-gated channel proteins. Such transport proteins have a gate that opens at a particular voltage, or membrane potential (Figure 24.3).

An action potential begins in a trigger zone, the part of the axon adjacent to the cell body. When a neuron is at rest, this region's gated channels for sodium and potassium are closed, and there are more negatively charged ions inside the axon than outside it ❶.

A stimulus such as a signal from another neuron shifts the membrane potential of the resting neuron. If the stimulus is large enough, the trigger zone reaches **threshold potential**. This is the membrane potential at which gated sodium channels in the trigger zone open and an action potential begins.

When gated sodium channels open, they allow sodium ions to follow their concentration gradient and diffuse from interstitial fluid into the neuron ❷. As sodium ions flow into the axon's cytoplasm, the cytoplasm becomes more positively charged, and more sodium channels open. More sodium ions flow in, causing even more sodium channels to open:

This accelerating inward flow of sodium ions is an example of positive feedback, in which a response amplifies the conditions that set it in motion.

As a result of the inward flow of sodium ions, the membrane potential reverses—within a millisecond, the fluid outside of the cell becomes more negatively charged than the fluid inside. This brief reversal of potential across the membrane is the action potential.

action potential Brief reversal of the charge difference across a neuron membrane.
resting potential Membrane potential of a neuron at rest.
threshold potential Membrane potential at which voltage-gated sodium channels in a neuron axon open, causing an action potential.

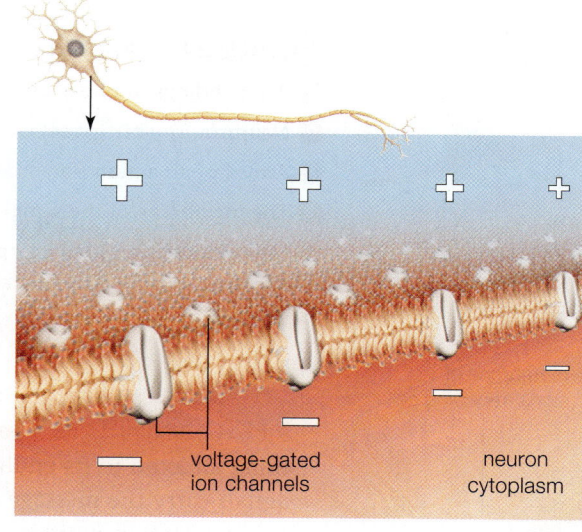

❶ A neuron's trigger zone at resting potential. *White* pluses and minuses indicate that the cytoplasm's charge is negative relative to that of interstitial fluid, and all voltage-gated channels are closed. There are more sodium ions in the interstitial fluid just outside of the cell than there are inside of it. The reverse is true for potassium ions.

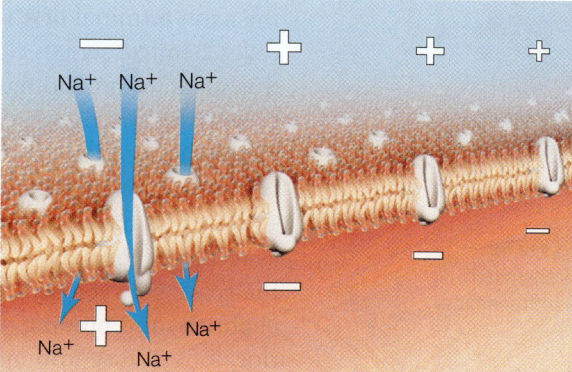

❷ If the membrane potential reaches threshold, gated sodium channels in the membrane open. Sodium ions (Na⁺) flow through the channels and down their concentration gradient, from interstitial fluid into the cytoplasm. The influx of sodium ions reverses the charge across the membrane. This reversal is called an action potential.

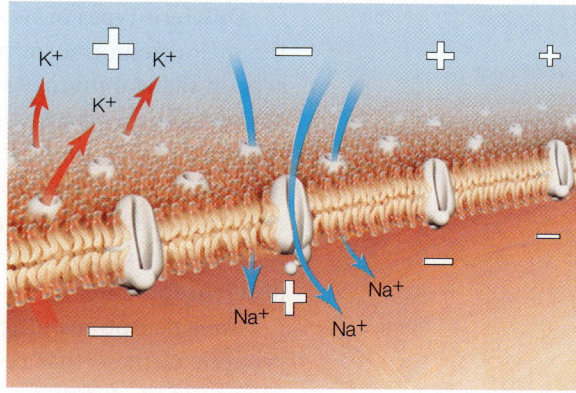

❸ The reversal in membrane potential causes gated sodium channels to close, and gated potassium channels to open. Potassium ions (K⁺) flow through the channels from the cytoplasm to interstitial fluid. At the same time, sodium ions that diffuse away from the site of an action potential trigger a new action potential in the adjacent patch of membrane.

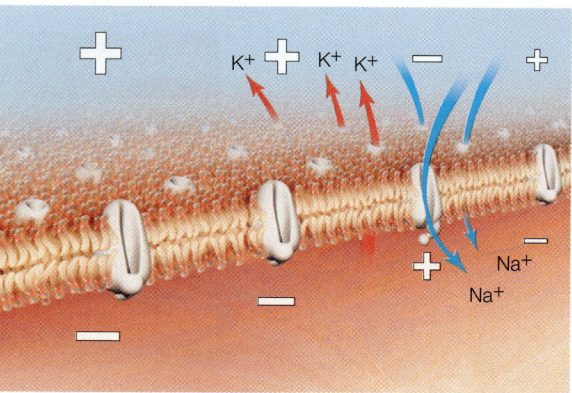

❹ The decline in voltage at the site of the original action potential causes gated potassium channels to shut.

After an action potential occurs in a region, gated sodium channels briefly become unable to open. As a result, an action potential moves in one direction only, toward the axon terminals.

When the reversal in voltage across the membrane reaches a certain point, it triggers gated sodium channels to close and gated potassium channels to open ❸. Potassium ions follow their concentration gradient and diffuse out of the cell through the now-open channels. As a result, the axon's cytoplasm once again becomes more negatively charged than the interstitial fluid. This decline in membrane potential causes the gates on potassium channels to swing shut ❹.

An action potential is sometimes described as an all-or-nothing event because once the threshold potential is reached, an action potential always occurs and all action potentials are the same size. The peak voltage achieved during an action potential is always the same—the voltage always rises to the point where it triggers the closing of gated sodium channels and the opening of gated potassium channels. These events halt the voltage rise and bring the action potential in that region of the membrane to an end.

An action potential affects each region of the membrane only briefly as it moves along the axon from the trigger zone toward the axon terminals. After voltage-gated sodium channels in one region of an axon open, some of the sodium ions that rush inward diffuse into adjoining regions, where they lower the membrane potential to threshold level and set in motion a new action potential.

An action potential cannot move backward, because after a sodium channel's gate closes, it cannot open for a brief period. However, sodium gates in regions farther along the axon can and do swing open as these regions reach threshold. As gates of sodium channels open in one region of the axon after another, the action potential moves steadily toward the axon terminals. The signal moves without weakening because, as previously noted, all action potentials are the same size. The voltage is reversed to the same extent in each region that reaches threshold.

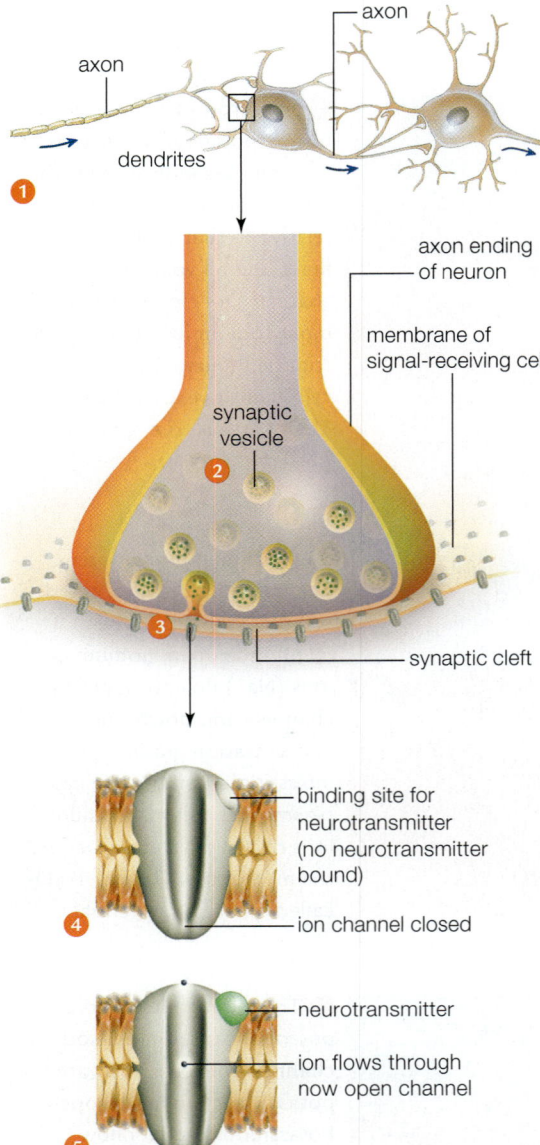

axon

axon

dendrites

1

axon ending
of neuron

membrane of
signal-receiving cell

synaptic
vesicle

2

3

synaptic cleft

binding site for
neurotransmitter
(no neurotransmitter
bound)

4

ion channel closed

neurotransmitter

ion flows through
now open channel

5

FIGURE 24.4 Animated! Chemical synapse.

1 Axon endings of one neuron send signals to another neuron at a chemical synapse.

2 Neurotransmitter (*green*) is stored in vesicles inside an axon ending.

3 Arrival of an action potential at the axon ending causes neurotransmitter release.

4 The membrane of the signal-receiving cell has receptors for neurotransmitter.

5 Neurotransmitter binding opens a channel through the receptor. The opening allows ions to flow into the signal-receiving cell.

© Cengage Learning.

> ❯ **How Messages Flow From Cell to Cell** Action potentials can travel along an axon to its endings, but they cannot jump from cell to cell. Chemical signals carry the message between cells (Figure 24.4). A **chemical synapse** is the communication point between two neurons, or between a neuron and a muscle or gland **1**. At the synapse, only a narrow space (the synaptic cleft) separates the axon terminal of a signal-sending neuron from an adjacent signal-receiving cell. Chemical signals called **neurotransmitters** convey information across this space. Axon endings store neurotransmitter in synaptic vesicles in their cytoplasm **2**. The arrival of an action potential at an axon's endings causes exocytosis (Section 4.6). The synaptic vesicles move to the plasma membrane and fuse with it. As fusion occurs, neurotransmitter enters the synaptic cleft **3**.

The plasma membrane of a signal-receiving cell includes proteins that serve as receptors for neurotransmitters **4**. Neurotransmitter molecules diffuse across the space between cells and bind to these receptors. Some receptors are ion channels, and binding of a neurotransmitter opens a tunnel through their interior. Ions diffuse into or out of the signal-receiving cell through the tunnel **5**. Arrival of these ions alters the membrane potential of the signal-receiving cell, pushing it either closer to its threshold potential or farther away from it.

Different types of neurons make different neurotransmitters. Table 24.1 lists the major neurotransmitters and their effects. In vertebrates, acetylcholine (ACh) acts on skeletal muscle, smooth muscle, the heart, many glands, and the brain. When motor neurons release acetylcholine into synapses with skeletal muscle cells, the muscle undergoes an action potential. The muscle then contracts. ACh is also released at synapses with heart muscle. Here it inhibits contraction. How can the same neurotransmitter have different effects? Skeletal muscle cells and heart cells have different types of ACh receptors. Both bind ACh, but each lets a different ion enter the muscle cell after ACh binds.

After neurotransmitter molecules have acted, they must be removed from synaptic clefts so that new signals can be sent. Some neurotransmitter molecules simply diffuse away. Membrane transport proteins pump others back

Table 24.1	Major Neurotransmitters and Their Effects
Neurotransmitter	Examples of Effects
Acetylcholine (ACh)	Induces skeletal muscle contraction, slows cardiac muscle contraction rate, affects mood and memory
Epinephrine and norepinephrine	Speed heart rate; dilate the pupils and the airways to lungs; slow gut contractions; increase anxiety
Dopamine	Dampens excitatory effects of other neurotransmitters; has roles in memory, learning, fine motor control
Serotonin	Elevates mood; role in memory
Endorphins	Dull pain, elevate mood, and promote feelings of well-being

into the neuron or into neighboring neuroglial cells. Enzymes secreted into the cleft break down neurotransmitter that remains in it. The importance of such enzymes is illustrated by the effects of nerve gases, which inactivate the enzyme that breaks down ACh. Exposure to a nerve gas causes ACh to accumulate in synaptic clefts, and the excess neurotransmitter results in confusion, headaches, and muscle paralysis. If nerve gas dosage is high enough, death follows.

> **Disrupted Signaling** Some people have neurological disorders caused by disrupted signaling at synapses. Others deliberately disrupt synaptic function by using psychoactive drugs.

Alzheimer's disease, the leading cause of dementia (loss of the ability to think), involves damage to brain neurons that release ACh. It begins with forgetfulness. As the disease progresses, a person becomes increasingly confused, cannot communicate, and eventually becomes incapable of living independently. Drugs that inhibit the enzymatic breakdown of ACh can slow the mental decline in some people.

Several neurotransmitters affect mood. An abnormally low level of one or more neurotransmitters can result in depression—a persistent feeling of sadness and inability to experience pleasure. The most widely prescribed antidepressants, including Prozac and Paxil, increase the level of serotonin by blocking proteins that pump it back into axon terminals.

Parkinson's disease results from a shortage of the neurotransmitter dopamine in the brain region that governs motor control (Figure 24.5). Tremors are an early symptom. Later, the sense of balance becomes impaired, and voluntary movement, including speech, becomes difficult. Because the symptoms arise from a dopamine shortage, patients can be treated with a drug (levodopa) that the body converts to dopamine.

Decreased dopamine in a different part of the brain can cause attention deficit hyperactivity disorder (ADHD). Affected people have trouble concentrating and controlling impulses. Drugs such as Ritalin reduce symptoms of ADHD by increasing the dopamine level in brain regions that control motivation.

All addictive drugs stimulate release of dopamine, which normally functions in reward-based learning. In nearly all animals, dopamine release provides pleasurable feedback when the animal engages in behavior that enhances survival or reproduction. This response is adaptive, because it helps animals learn to repeat beneficial behaviors. Drugs that cause dopamine release tap into this ancient learning pathway. Users inadvertently teach themselves that taking the drug is essential to their well-being.

Analgesics are drugs that relieve pain. Narcotic analgesics such as morphine, codeine, heroin, fentanyl, and oxycodone mimic natural neurotransmitters (endorphins) that act in the brain to dull pain and elevate mood. Narcotic analgesics cause a rush of euphoria and are highly addictive. Ketamine and PCP (phencyclidine) are pain killers that work by a different mechanism. They result in an out-of-body experience and numb the extremities by slowing removal of neurotransmitter from synapses.

Depressants such as alcohol and barbiturates slow motor responses by inhibiting ACh output. Alcohol also stimulates the release of endorphins, so users typically experience a brief euphoria followed by depression.

Stimulants make users feel alert but also anxious, and they can interfere with fine motor control. Nicotine is a stimulant that, among other effects, blocks receptors for ACh in the brain. Cocaine and amphetamines are also stimulants. Ecstasy is a type of amphetamine, as is methamphetamine.

Hallucinogens distort sensory perception and bring on a dreamlike state. LSD (lysergic acid diethylamide) resembles serotonin and binds to receptors for

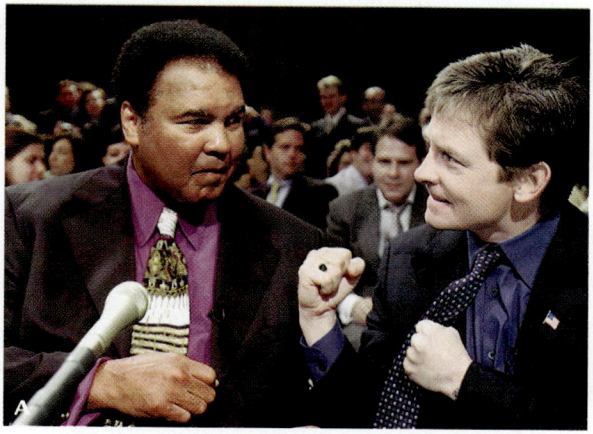

A Parkinson's disease affects about half a million people in the United States, including former heavyweight champion Muhammad Ali and actor Michael J. Fox.

Normal   With Alzheimer's

B Brain PET scans. *Red* and *yellow* indicate regions with high metabolic activity in dopamine-secreting neurons.

FIGURE 24.5 Parkinson's disease.

Credits: (a) AP Images; (b) From Neuro Via Clinical Research Program, Minneapolis VA Medical Center.

chemical synapse Region where a neuron's axon terminal transmits signals to another cell.

neurotransmitter Chemical signal released by a neuron's axon terminals.

it. LSD is not addictive. However, LSD users can be injured or killed because they do not perceive and respond to hazards such as oncoming cars. Flashbacks, which are brief distortions of perception, may occur years after the last intake of LSD. Two related drugs, mescaline and psilocybin, have weaker effects.

Marijuana comes from *Cannabis* plants. Smoking a lot of marijuana can cause hallucinations. More often, users become relaxed, sleepy, uncoordinated, and inattentive. The active ingredient, THC (delta-9-tetrahydrocannabinol), alters brain levels of dopamine, serotonin, norepinephrine, and GABA. Chronic use can impair short-term memory and decision-making ability.

Take-Home Message

What are neurons and how do they transmit signals?

- Neurons are the cells that make up the communication lines of animal nervous systems.

- In response to a stimulus, electrical signals called action potentials move along a neuron's axon to the axon terminals.

- The arrival of an action potential triggers axon terminals to release signaling molecules called neurotransmitters. The neurotransmitter molecules diffuse across a narrow space and bind to receptors on an adjacent cell. The binding may have excitatory or inhibitory effects on the signal-receiving cell.

- Alzheimer's disease and Parkinson's disease result from impaired neuron signaling. Psychoactive drugs exert their effects by altering neuron signaling.

24.3 Animal Nervous Systems

> **Invertebrate Nervous Systems** Cnidarians such as sea anemones and jellies are among the simplest animals that have neurons. These radial, aquatic animals have a **nerve net**, which is a mesh of interconnected neurons. Information can flow in any direction among cells of the nerve net, and there is no centralized, controlling organ that functions like a brain. By causing cells in the body wall to contract, the nerve net can alter the size of the animal's mouth, change the shape of the body, or shift the position of tentacles.

Most animals have a bilaterally symmetrical body (Section 15.2). Evolution of bilateral body plans was accompanied by a concentration of sensory neurons and interneurons in a head.

For example, a planarian's head has a pair of ganglia (Figure 24.6**A**). A **ganglion** (plural, ganglia) is a cluster of neuron cell bodies that functions as an integrating center. A planarian's ganglia receive signals from eyespots and chemical-detecting cells on its head. The ganglia also connect to a pair of nerve cords that run the length of the body. Nerves cross the body between the cords, giving the nervous system a ladderlike appearance. A **nerve** is a bundle of axons wrapped in connective tissue.

Crayfish and other arthropods have paired nerve cords that connect to a simple brain (Figure 24.6**B**). In addition, a pair of ganglia in each body segment provides local control over the muscles in that segment.

> **The Vertebrate Nervous System** The nervous system of vertebrates has two divisions. Nearly all interneurons are located in the **central nervous system**—the brain and spinal cord. Nerves that extend through the rest of the body make up the **peripheral nervous system**. Figure 24.7 illustrates the location of the human brain ❶, spinal cord ❷, and some peripheral nerves. Each

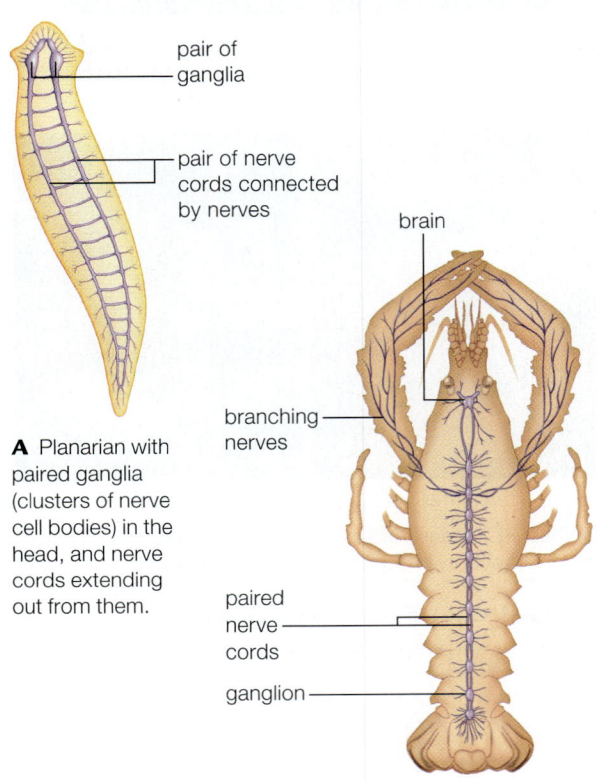

pair of ganglia

pair of nerve cords connected by nerves

brain

branching nerves

A Planarian with paired ganglia (clusters of nerve cell bodies) in the head, and nerve cords extending out from them.

paired nerve cords

ganglion

B A crayfish has a brain, and paired nerve cords with ganglia in each of its segments.

FIGURE 24.6 Animated! Invertebrate nervous systems. © Cengage Learning.

central nervous system Brain and spinal cord.
ganglion Cluster of nerve cell bodies.
nerve Axons bundled in connective tissues.
nerve net Mesh of interacting neurons with no central control, as in the nervous system of jellies.
peripheral nervous system Nerves that extend through the body and carry signals to and from the central nervous system.

FIGURE 24.7 Animated! Components of the human nervous system. The brain ❶ and spinal cord ❷ make up the central nervous system. The peripheral nervous system includes spinal nerves, cranial nerves, and their branches, which extend through the rest of the body. Peripheral nerves such as the sciatic nerve ❸ carry signals to and from the central nervous system.

© Cengage Learning.

peripheral nerve has both sensory and motor components. Sensory axons convey signals from sensory receptors to the central nervous system. Motor axons carry signals away from the central regions to muscles. For example, a sciatic nerve extends from the spinal cord down each leg ❸. It is composed of the bundled extensions of hundreds of neurons. Some axons in this bundle carry signals to skeletal muscles that move the leg, or to smooth muscle that adjusts the blood flow through the leg's arterioles. Other axons in the bundle convey messages from sensory neurons in the leg's muscles, joints, and skin.

Figure 24.8 shows the structure of a vertebrate nerve. Each nerve consists of bundles of axons enclosed in connective tissue. As previously noted, each axon is largely encased by neuroglial cells that wrap around it to form a myelin sheath. The sheath speeds transmission of action potentials by preventing diffusion of sodium ions out of the axon into interstitial fluid. As a result, the ions diffuse more rapidly along the length of the axon. Action potentials occur in unsheathed areas, or nodes, of the axon, which are the regions between the neuroglia.

outer connective tissue of one nerve

many neurons bundled together inside a connective tissue sheath

axon of one neuron

myelin sheath

axon

unsheathed node

FIGURE 24.8 Animated! Structure of a peripheral nerve. The myelin sheath around axons in the nerve increases the speed at which action potentials travel. Ions cannot enter or leave the axon in the sheathed region, so they diffuse more quickly inside the axon. At each unsheathed node, arrival of sodium that diffused from the previous node triggers an action potential. Thus an action potential in a myelinated axon "jumps" from node to node.

© Cengage Learning.

Take-Home Message

How are animal nervous systems organized?

- The simplest nervous system is a nerve net, a mesh of interconnecting neurons with no central integrating region. Jellies have this type of system.

- Most animals have a bilaterally symmetrical body with an integrating center at their head end. Nerves that run through the body carry signals to and from this center.

- Vertebrates have a central nervous system (brain and spinal cord) and a peripheral nervous system (nerves that extend through the body and carry signals to and from the central nervous system).

- Myelin sheaths around axons in vertebrate nerves speed information flow.

24.4 The Peripheral Nervous System

The vertebrate peripheral nervous system has two types of nerves (Figure 24.9). Nerves of the **somatic nervous system** relay commands to skeletal muscles. This is the only part of the nervous system that is normally under voluntary control. The somatic nervous system also relays information from sensory receptors in the skin and joints to the central nervous system. You can feel a pinch or wiggle your toes because of signals that travel along somatic nerves.

Nerves of the **autonomic nervous system** relay signals to smooth and cardiac muscle, and to glands. They also relay signals about internal conditions to the central nervous system. Signals traveling along autonomic nerves keep you breathing, adjust your heart rate, and inform your brain of your blood pressure.

There are two divisions of the autonomic system: sympathetic and parasympathetic. Both divisions service most organs and work antagonistically, meaning that the signals from one division oppose signals from the other (Figure 24.10). **Sympathetic neurons** are most active in times of excitement or danger. Their axon terminals release norepinephrine. **Parasympathetic neurons** are most active in times of relaxation. Release of ACh from their axon terminals promotes housekeeping tasks such as digestion and urine formation.

When something startles or scares you, parasympathetic neurons reduce their output and sympathetic ones increase theirs. The increased sympathetic signals raise your heart rate and blood pressure, make you sweat more and breathe faster, and induce adrenal glands to secrete epinephrine (also known as adrenaline). The signals put you in a state of intense arousal, so you are ready to fight or make a fast getaway. This reaction is commonly described as the "fight or flight" response.

Opposing sympathetic and parasympathetic signals govern most organs. For instance, both act on the smooth muscle cells in the wall of the digestive tract. As sympathetic neurons release norepinephrine at synapses with these cells, parasympathetic neurons release ACh at other synapses with the same cells. One signal tells the muscle to slow its contractions, and the other calls for increased activity. Integration of these conflicting commands by the signal-receiving muscle cells determines the muscle's response.

Peripheral Nervous System

Somatic Nervous System
Controls skeletal muscle
Relays signals from joints, skin

Autonomic Nervous System
Controls internal organs, glands
Relays signals about internal organs

Parasympathetic division
Encourages housekeeping tasks

Sympathetic division
Prepares the body for "fight or flight"

FIGURE 24.9 Functional divisions of the peripheral nervous system. This system consists of spinal nerves, cranial nerves, and their branches, which extend through the body. © Cengage Learning.

You can feel a pinch or wiggle your toes because of signals that travel along somatic nerves.

Take-Home Message

How does the peripheral nervous system function?

- The peripheral nervous system consists of nerves that extend through the body and connect with the central nervous system.
- Neurons of the somatic part of this system control skeletal muscle and convey information about the external environment to the central nervous system.
- The autonomic system carries information about internal conditions to the brain and sends signals to smooth muscle, cardiac muscle, and glands. Most organs receive both sympathetic and parasympathetic stimulation.
- Sympathetic stimulation prepares a body for "fight or flight." Parasympathetic stimulation encourages housekeeping tasks.

Sympathetic Effects	Organ	Parasympathetic Effects
Widens pupils	Eyes	Narrows pupils
Increases salivation	Salivary glands	Decreases salivation
Increases heart rate	Heart	Decreases heart rate
Widens airways	Airways	Constricts airways
Slows secretions and movements	Stomach	Increases secretions and movements
Slows secretions to digestive tract	Liver, pancreas	Increases secretions to digestive tract
Increases secretion	Adrenal gland	Decreases secretion
Slows secretions and movements	Small intestine, large intestine	Increases secretions and movements
Inhibits urination	Bladder	Stimulates urination
Promotes ejaculation	Genitals	Promotes erection, lubrication

(most ganglia near spinal cord)

optic nerve · midbrain · medulla oblongata · vagus nerve · cervical nerves (8 pairs) · thoracic nerves (12 pairs) · lumbar nerves (5 pairs) · sacral nerves (5 pairs) · pelvic nerve · (all ganglia in walls of organs)

FIGURE 24.10 Animated! Autonomic nerves and their effects. Autonomic signals travel to organs by a two-neuron path. The first neuron has its cell body in the brain or spinal region (indicated in *red*). This neuron synapses on a second neuron at a ganglion. Ganglia of sympathetic neurons are close to the spinal cord. Parasympathetic ganglia are in or near the organs they affect. Axons of these second neurons then synapse with the organ.

Figure It Out: What effect does stimulation of the vagus nerve (a parasympathetic nerve) have on heartbeat?

© Cengage Learning.

Answer: It slows the heartbeat.

24.5 The Central Nervous System

The brain and spinal cord are the organs of the central nervous system (CNS). Three membranes called the **meninges** enclose and protect these organs, which are bathed in clear **cerebrospinal fluid**. This fluid forms when water and solutes seep out of capillaries in the brain. The composition of the cerebrospinal fluid is controlled by a **blood–brain barrier** that prevents most unwanted substances from entering the fluid. The barrier consists of the wall of blood capillaries and some supporting neuroglial cells. Sometimes bacteria or viruses breach the blood–brain barrier and enter the meninges and cerebrospinal fluid. The resulting infection, called meningitis, can be life-threatening.

Two visibly different types of tissue occur in both the brain and spinal cord. White matter consists of bundles of myelin-sheathed axons. In the CNS, such bundles are called tracts, rather than nerves. The tracts carry information from one part of the central nervous system to another. Gray matter consists of cell bodies, dendrites, and supporting neuroglial cells.

autonomic nervous system Set of nerves that relay signals to and from internal organs and to glands.

blood–brain barrier Protective mechanism that prevents unwanted substances from entering cerebrospinal fluid.

cerebrospinal fluid Fluid around brain and spinal cord.

meninges Membranes that surround and protect the brain and spinal cord.

parasympathetic neurons Neurons of the autonomic system that encourage housekeeping tasks.

somatic nervous system Set of nerves that control skeletal muscle and relay signals from joints and skin.

sympathetic neurons Neurons of the autonomic system that prepare the body for danger or excitement.

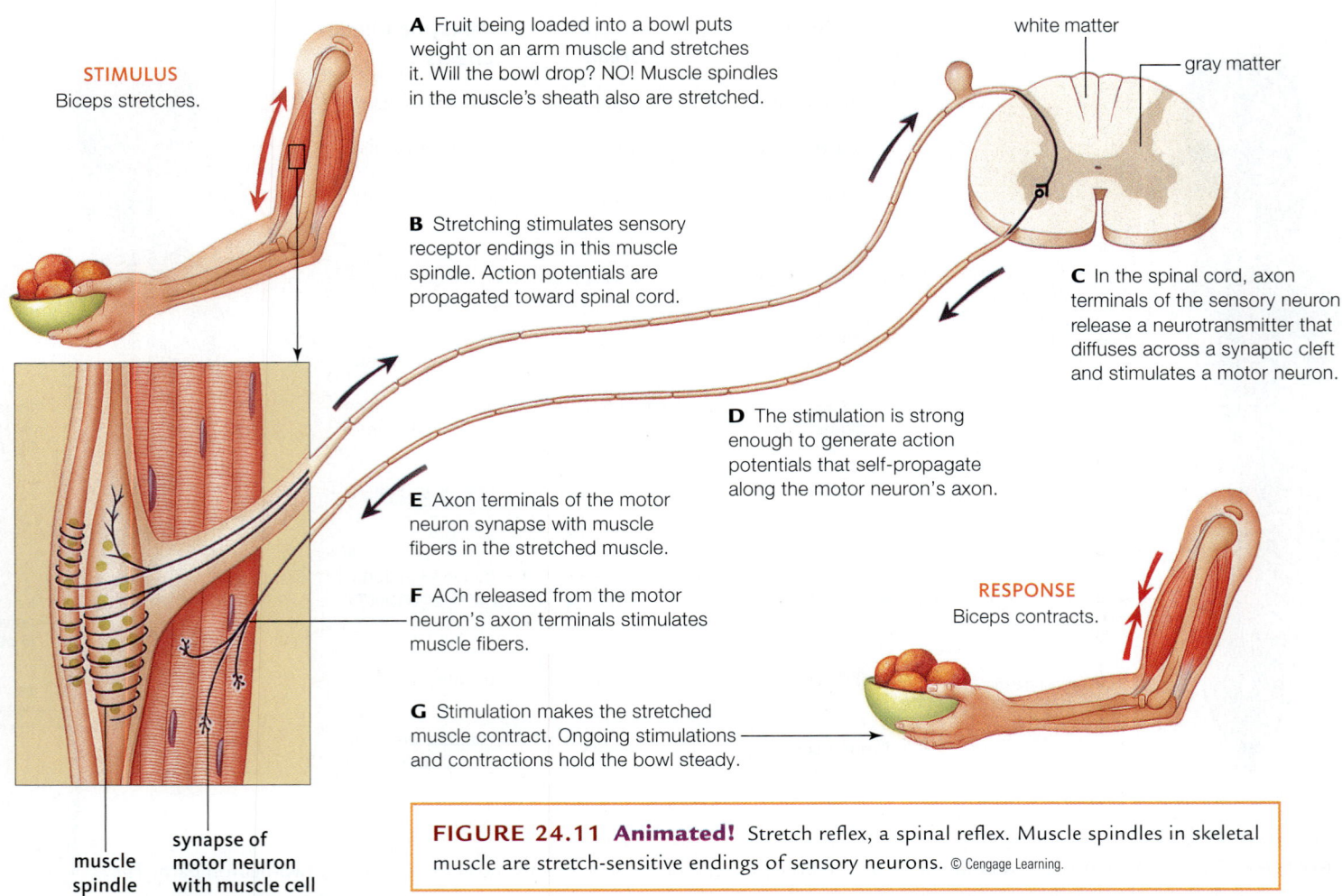

STIMULUS
Biceps stretches.

A Fruit being loaded into a bowl puts weight on an arm muscle and stretches it. Will the bowl drop? NO! Muscle spindles in the muscle's sheath also are stretched.

B Stretching stimulates sensory receptor endings in this muscle spindle. Action potentials are propagated toward spinal cord.

white matter

gray matter

C In the spinal cord, axon terminals of the sensory neuron release a neurotransmitter that diffuses across a synaptic cleft and stimulates a motor neuron.

D The stimulation is strong enough to generate action potentials that self-propagate along the motor neuron's axon.

E Axon terminals of the motor neuron synapse with muscle fibers in the stretched muscle.

F ACh released from the motor neuron's axon terminals stimulates muscle fibers.

RESPONSE
Biceps contracts.

G Stimulation makes the stretched muscle contract. Ongoing stimulations and contractions hold the bowl steady.

muscle spindle

synapse of motor neuron with muscle cell

FIGURE 24.11 Animated! Stretch reflex, a spinal reflex. Muscle spindles in skeletal muscle are stretch-sensitive endings of sensory neurons. © Cengage Learning.

❭ **The Spinal Cord** Your **spinal cord** is about as thick as your thumb. It runs through the vertebral column and connects peripheral nerves with the brain. An injury that disrupts the signal flow through the spinal cord can cause a permanent loss of sensation and paralysis. Symptoms depend on where the cord is damaged. Nerves carrying signals to and from the upper body originate higher in the cord than nerves that govern the lower body. An injury to the lower region of the cord often paralyzes the legs. An injury to the highest cord regions can paralyze all limbs, as well as muscles used in breathing. More than 240,000 Americans now live with a spinal cord injury.

In addition to transmitting signals to and from the brain, the spinal cord functions in some reflex pathways. A **reflex** is an automatic response to a stimulus—it is a movement or other action that does not require thought. The stretch reflex, illustrated in Figure 24.11, is an example. This reflex causes a muscle to contract after gravity or some other force stretches it. For example, stretching of the biceps causes action potentials in sensory receptors called muscle spindles. The action potentials travel along an axon to the spinal cord, where they synapse with motor neurons that control the biceps. Signals from the sensory neurons cause action potentials in the motor neurons. In response, the biceps contracts and steadies the arm.

❭ **Regions of the Human Brain** The average human brain weighs 1,240 grams, or 3 pounds. It contains about 100 billion interneurons, and neuroglia make up more than half of its volume. During development, the brain becomes organized as three functional regions: the hindbrain, midbrain, and forebrain corpus callosum   hypothalamus   thalamus   cerebrum

midbrain

cerebellum

pons

medulla oblongata

Forebrain	
Cerebrum	Localizes, processes sensory inputs; initiates, controls skeletal muscle activity; governs memory, emotions, abstract thought
Thalamus	Relays sensory signals to and from cerebral cortex; has a role in memory
Hypothalamus	With pituitary gland, functions in homeostatic control. Adjusts volume, composition, temperature of internal environment; governs behaviors that ensure homeostasis (e.g., thirst, hunger)
Midbrain	Relays sensory input to the forebrain
Hindbrain	
Pons	Bridges cerebrum and cerebellum, also connects spinal cord with forebrain. With the medulla oblongata, controls rate and depth of respiration
Cerebellum	Coordinates motor activity for moving limbs and maintaining posture, and for spatial orientation
Medulla oblongata	Relays signals between spinal cord and pons; functions in reflexes that affect heart rate, blood vessel diameter, and respiratory rate. Also involved in vomiting, coughing, other reflexive functions

FIGURE 24.12 Human brain. *Top*, right half of a brain, with major structures labeled. *Right*, three main brain regions, major structures in each region, and their functions.

C. Yokochi and J. Rohen, *Photographic Anatomy of the Human Body*, 2nd Ed., Igaku-Shoin, Ltd., 1979.

(Figure 24.12). The hindbrain sits atop the spinal cord. The portion just above the cord, the medulla oblongata, influences the strength of heartbeats and the rhythm of breathing. It also controls reflexes such as swallowing, coughing, vomiting, and sneezing. Above the medulla oblongata lies the pons, which also affects breathing. Pons means "bridge," and tracts extend through the pons to the midbrain.

The **cerebellum** lies at the back of the brain and is about the size of a plum. It is densely packed with neurons, having more than all other brain regions combined. The cerebellum controls posture and coordinates voluntary movements. Excessive alcohol consumption disrupts coordination by impairing neurons in the cerebellum. That is why police often ask a person suspected of driving while drunk to walk a straight line.

In humans, the midbrain is the smallest of the three brain regions. It plays an important role in reward-based learning. The pons, medulla, and midbrain are collectively referred to as the brain stem.

The forebrain contains the **cerebrum**, the largest part of the human brain. A fissure divides the cerebrum into right and left hemispheres. A thick band of tissue, called the corpus callosum, connects the hemispheres. Each hemisphere has an outer layer of gray matter called the cerebral cortex. Our large cerebral cortex is the part of the brain responsible for our unique capacities such as language and abstract thought.

Most sensory signals destined for the cerebrum pass through the adjacent thalamus, which sorts them and sends them to the proper region of the cerebral cortex. The **hypothalamus** ("under the thalamus") is the center for homeostatic control. It receives signals about the state of the body and regulates thirst, appetite, sex drive, and body temperature. It is also an endocrine gland that interacts with the adjacent pituitary gland.

cerebellum Hindbrain region that coordinates voluntary movements.
cerebrum Forebrain region that controls higher functions.
hypothalamus Homeostatic control center in forebrain.
reflex Automatic response to a stimulus.
spinal cord Portion of central nervous system that connects peripheral nerves with the brain.

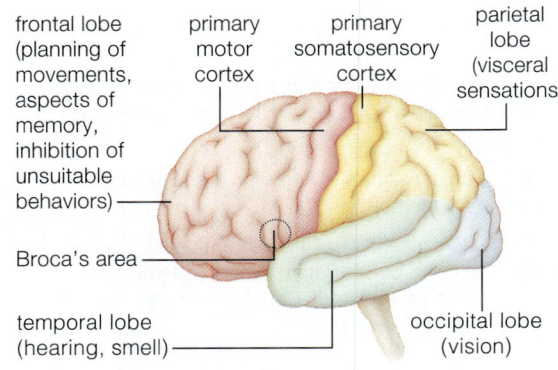

frontal lobe
(planning of
movements,
aspects of
memory,
inhibition of
unsuitable
behaviors)

primary
motor
cortex

primary
somatosensory
cortex

parietal
lobe
(visceral
sensations)

Broca's area

temporal lobe
(hearing, smell)

occipital lobe
(vision)

FIGURE 24.13 Lobes of the cerebrum and the location of some association and sensory areas of the cerebral cortex. Broca's area in the frontal lobe governs speech production.

© Cengage Learning.

❯ A Closer Look at the Cerebral Cortex The cerebral cortex is a 2-millimeter-thick, highly folded outer layer of gray matter. Prominent folds in the cortex are used as landmarks to define the cerebrum's frontal, parietal, temporal, and occipital lobes (Figure 24.13).

Much of each frontal lobe is devoted to areas that integrate information and bring about conscious actions. During the 1950s, more than 20,000 people had their frontal lobes deliberately damaged when they underwent frontal lobotomy. Doctors carried out this procedure to treat mental illness, personality disorders, and even severe headaches. Frontal lobotomies sometimes made patients calmer, but the procedure also permanently blunted emotions and impaired their ability to plan, to concentrate, and to behave appropriately in social situations.

The primary motor cortex near the rear of each frontal lobe controls skeletal muscles. Each hemisphere controls and receives signals from the opposite side of the body. For example, signals to move your right arm originate in the motor cortex of your left hemisphere. The primary somatosensory cortex of the parietal lobe receives sensory input from the skin and joints. When someone taps you on your left shoulder, signals that arrive in the primary somatosensory cortex of your right parietal lobe alert you to the tap. Another sensory area in the parietal lobe receives signals about taste. In the occipital lobe, the primary visual cortex integrates incoming signals from both eyes. The perception of sounds and odors arises in the primary sensory areas of the temporal lobe.

The two hemispheres differ somewhat in their function. About 90 percent of people are right-handed, and their left hemisphere is more active in controlling movement and in language. Keep in mind that the abilities of each hemisphere are flexible. When a stroke or injury damages tissue on one side of the brain, the other hemisphere often can take on new tasks. People can even function with a single hemisphere.

❯ Brain Tumors Neurons do not generally divide, so they do not give rise to tumors. However, uncontrolled division of cells that give rise to neuroglial cells sometimes results in a tumor. Tumors can also arise from epithelial cells in the meninges or endocrine glands of the brain, such as the pituitary. In addition, brain tumors can result when cancerous cells arrive from elsewhere in the body. Most tumors that originate in the brain are not cancer. However, even a benign (noncancerous) tumor can pose a serious threat. Benign tumors do not spread to new regions, but growth of a tumor within the confined space of the skull can put pressure on surrounding nervous tissue and damage neurons.

Take-Home Message

What are the functions of the spinal cord and brain?

- The spinal cord relays information between peripheral nerves and the brain. It also has a role in some reflexes.
- The brain stem (hindbrain and midbrain) regulates breathing and adjusts heart rate. It also controls reflexes such as swallowing and coughing. The brain stem's cerebellum coordinates voluntary movements.
- The hypothalamus deep inside the forebrain has important roles in thirst, temperature regulation, and other responses related to homeostasis.
- The cerebrum makes up the bulk of the forebrain. Its outer layer, the cerebral cortex, controls voluntary activity, sensory perception, abstract thought, and language and speech.
- The two cerebral hemispheres differ somewhat in their function. In most people, the left hemisphere dominates.

24.6 The Senses

〉 Sensory Receptors Sensory receptors at the endings of sensory neurons function as the sentries of a nervous system. These specialized structures monitor conditions and detect changes in the internal and external environment. **Thermoreceptors** respond to heat or to cold. **Mechanoreceptors** detect changes in pressure, position, or acceleration. Pain receptors (nociceptors) detect injury. **Chemoreceptors** detect substances dissolved in the fluid around them. **Photoreceptors** include light-sensitive pigments that respond to light energy.

Animals monitor the environment in different ways depending on the kinds and numbers of sensory receptors they have. Many animals have evolved sensory capacities we lack. Pythons have thermoreceptors that allow them to detect warm-blooded prey (Figure 24.14**A**). Bats have mechanoreceptors that detect ultrasound (high-pitched sound) beyond our range of hearing (Figure 24.14**B**).

All sensory receptors convert stimulus energy into action potentials. Given that action potentials are "all-or-nothing" events that do not vary in size, how does the brain know how strong a stimulus is? Three mechanisms provide this information. First, an animal's brain is prewired to interpret action potentials of certain nerves in certain ways. For example, the brain always interprets signals from the optic nerve as "light."

Second, as stimulus strength increases, so does the rate at which a sensory receptor has action potentials. The same receptor detects the sound of a polite whisper or an exuberant whoop. The brain interprets the difference from variations in the rate of action potentials.

Third, a stronger stimulus can recruit more sensory receptors than a weaker stimulus. Gently touch a small patch of skin on one of your arms and you activate a few receptors. Press harder and you activate more. The brain interprets the heightened activity as an increase in stimulus intensity.

In some cases, the rate of action potentials in a sensory neuron falls or stops even though a stimulus continues at constant strength. A reduced response to an ongoing stimulus is called sensory adaptation. Some mechanoreceptors in skin adapt quickly to sustained stimulation. For example, soon after you put on clothing, receptors adapt and stop notifying you of its presence.

〉 Somatic and Visceral Sensations Sensory neurons responsible for somatic sensations are located in skin, muscle, tendons, and joints. **Somatic sensations** are localized to a specific body part. **Visceral sensations** arise from neurons in the walls of soft internal organs, and are often difficult to pinpoint. It is easy to determine exactly where someone is touching you, but less easy to say exactly where you feel a stomachache.

Many receptors detect touch, pressure, cold, warmth, and pain near the body surface. Regions with the most sensory receptors, such as the fingertips, are the most sensitive. Less sensitive regions, such as the back of the hand, have far fewer receptors.

Mechanoreceptors near skeletal muscle, joints, and ligaments detect limb motions. Muscle spindles that trigger the stretch reflex described in Figure 24.11 are an example. The response of these receptors depends on how much and how fast a muscle is stretched.

Pain is the perception of a tissue injury. Signals from pain receptors in skin, skeletal muscles, joints, and tendons give rise to somatic pain. Visceral pain is associated with organs inside body cavities. It occurs as a response to a smooth muscle spasm, inadequate blood flow to an internal organ, over-stretching of a hollow organ, and other abnormal conditions. Injured or distressed body cells release chemicals that stimulate nearby pain receptors. Signals from the pain

A Thermoreceptors in pits above and below a python's mouth allow it to detect the body heat of its prey.

B Mechanoreceptors inside a bat's ear allow the animal to detect high-pitched, or ultrasonic, pressure waves.

FIGURE 24.14 Examples of sensory receptors.

Credits: (a) © Eric A. Newman; (b) Merlin D Tuttle, Bat Conservation International.

chemoreceptor Sensory receptor that responds to the binding of a particular chemical.
mechanoreceptor Sensory receptor that responds to pressure, position, or acceleration.
photoreceptor Light-sensitive sensory receptor.
somatic sensations Sensations that arise when sensory neurons in skin and near joints and muscle are activated.
thermoreceptor Temperature-sensitive sensory receptor.
visceral sensations Sensations that arise when sensory neurons in soft internal organs are activated.

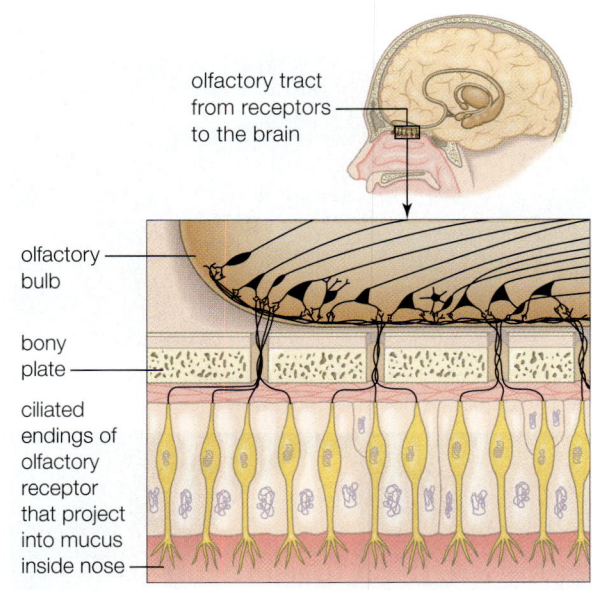

olfactory tract from receptors to the brain

olfactory bulb

bony plate

ciliated endings of olfactory receptor that project into mucus inside nose

FIGURE 24.15 Animated! Pathway from sensory endings of olfactory receptors in the human nose to the cerebral cortex. Receptor axons pass through holes in a bony plate between the lining of the nasal cavities and the brain.

© Cengage Learning.

receptors then travel along the axons of sensory neurons to the spinal cord. Here, the sensory axons synapse with the spinal interneurons that relay signals about pain to the brain.

Pain is an adaptive mechanism that alerts an animal to tissue damage. The animal's response to pain gives the damaged part time to heal. However, incapacitating pain can prevent an animal from responding to threats. Therefore, when a vertebrate is injured, its brain releases some natural pain relievers called **endorphins**. Endorphins inhibit flow of pain-related signals to the brain, making pain less incapacitating. Synthetic opiates such as morphine mimic the effect of endorphins. Aspirin reduces pain in a different way. It slows production of prostaglandins, local signaling molecules that increase the sensitivity of pain receptors. Cells release prostaglandins in response to tissue damage.

〉 The Chemical Senses—Smell and Taste

Smell (olfaction) and taste are chemical senses. Stimuli are detected by chemoreceptors that send signals when a specific chemical dissolved in the surrounding fluid binds to them. Binding of a chemical to an olfactory receptor causes action potentials to flow along axons to one of the brain's two olfactory bulbs. In these small brain structures, axons synapse with groups of cells that sort out components of a scent. From there, information flows along the olfactory tract to the cerebrum, where it is further processed (Figure 24.15).

Many animals use olfactory cues to find food and avoid predators. Many also communicate by use of pheromones. **Pheromones** are signaling molecules that are secreted by one individual and affect another member of the same species. For example, olfactory receptors on antennae of a male silk moth allow him to locate a pheromone-secreting female more than a kilometer upwind.

In reptiles and most mammals, a cluster of sensory cells forms a vomeronasal organ that detects pheromones. In primates, including humans, a reduced version of this organ is located in the nasal septum separating the two nostrils. Whether the human vomeronasal organ is functional, and what role, if any, it plays in human behavior, remains unclear.

Depending on the animal, chemoreceptors involved in taste can be on antennae, legs or tentacles, or inside the mouth. Humans have about 10,000 sensory organs called taste buds on the surface of the mouth, throat, and especially the upper part of the tongue (Figure 24.16). Each taste bud contains taste receptor cells and neurons. Receptor-covered microvilli of the taste receptor cells extend out through a pore, and thus come in contact with food molecules in saliva.

We perceive many different tastes, but all arise from a combination of five primary sensations: sweet (elicited by simple sugars), sour (acids), salty (NaCl or other salts), bitter (alkaloids), and umami (amino acids such as glutamate, which provides the savory taste typical of aged cheese and meat). Monosodium glutamate (MSG), a commonly used artificial flavor enhancer, stimulates the receptors responsible for the sensation of umami.

Each taste receptor cell is most sensitive to one of the five primary tastes. Taste buds in all regions of the tongue include all five types of taste receptor cells. Thus, contrary to popular belief, the ability to detect a specific taste such as bitterness does not map to just one specific region of the tongue.

taste bud

hairlike ending of taste receptor

sensory nerve

section through circular papilla

FIGURE 24.16 Animated! Taste receptors in the human tongue. The structures called circular papillae enclose epithelial tissue that contains taste buds. A human tongue has approximately 10,000 of these sensory organs, each of which has as many as 150 chemoreceptors.

© Cengage Learning.

› Detecting Light Vision involves detection of light by photoreceptors, and processing of information from those receptors to form a mental image of objects in the environment. Some invertebrates such as earthworms have photoreceptors at their body surface. These animals use light as a cue to orient movements. However, they do not form a mental image of their surroundings as animals with eyes do.

Eyes are sensory organs with a dense array of photoreceptors. The most effective eyes have a **lens**, a transparent structure that bends light rays passing through it so the light falls on photoreceptors. Insects have compound eyes with many separate light-detecting units, each with its own lens (Figure 24.17). Compound eyes cannot detect visual details, but they are highly sensitive to movement. Cephalopod mollusks such as squids and octopuses have the most complex eyes of any invertebrate. Their camera eyes have an adjustable opening that allows light to enter a dark chamber. Each eye's single lens focuses incoming light onto a **retina**, a tissue densely packed with photoreceptors. The retina of a camera eye is analogous to the light-sensitive film in a film camera.

Vertebrates also have camera eyes, and because they are not closely related to cephalopod mollusks, camera eyes presumably evolved independently in the two lineages. This is an example of morphological convergence (Section 11.6).

FIGURE 24.17 Compound eyes of a fly.
Ablestock.com/ photos.com.

› The Human Eye A human eyeball sits inside a protective, cuplike, bony cavity called the orbit. Skeletal muscles that run from the rear of the eye to bones of the orbit move the eyeball.

Eyelids, eyelashes, and tears protect the delicate eye tissues. Periodic blinking is a reflex that spreads a film of tears over the eyeball's exposed surface. A protective mucous membrane, the conjunctiva, lines the inner surface of the eyelids and folds back to cover most of the eye's outer surface. Conjunctivitis, also called pinkeye, is an inflammation of this membrane caused by a viral or bacterial infection.

The eyeball is spherical and has a three-layered structure (Figure 24.18). A **cornea** that covers the front of each eye consists of a transparent protein called crystallin. A dense, white, fibrous sclera covers the rest of the eye's outer surface.

The eye's middle layer includes the choroid, iris, and ciliary body. The blood vessel–rich choroid is darkened by the brownish pigment melanin. This dark layer contributes to clear vision by reducing light reflection within the eyeball. Attached to the choroid and suspended behind the cornea is a muscular, doughnut-shaped **iris**. It too contains melanin. Whether your eyes are blue, brown, or green depends on the amount of melanin in your iris.

Light enters the eye's interior through the **pupil**, an opening at the center of the iris. Muscles of the iris adjust pupil diameter in response to light conditions. In bright light the pupil contracts, so less light gets in. In low light the pupil widens, so more light enters the eye. Sympathetic stimulation also widens the pupil, presumably allowing a better look at the source of danger or excitement.

A ciliary body of muscle, fibers, and secretory cells attaches to the choroid and holds the lens in place behind the pupil. The stretchable, transparent lens is about 1 centimeter (1/2 inch) across and shaped like a bulging disk.

The eye has two internal chambers. The ciliary body produces the fluid that fills the anterior chamber. Called aqueous humor, this fluid bathes the iris and lens. A jellylike vitreous body fills the larger chamber behind the lens. The innermost layer of the eye, the retina, lines the back of this chamber. The retina contains the light-detecting photoreceptors.

FIGURE 24.18 Animated! The human eye.
© Cengage Learning.

cornea Outermost layer at the front of the eye; bends light.
endorphins Natural pain-reliever molecules.
iris Ring of smooth muscle with pupil at its center; adjusts how much light enters the eye.
lens Structure that focuses light on an eye's photoreceptors.
pheromones Signaling molecules that affect another member of the same species.
pupil Opening through which light enters the eye.
retina Layer of eye that contains photoreceptors.

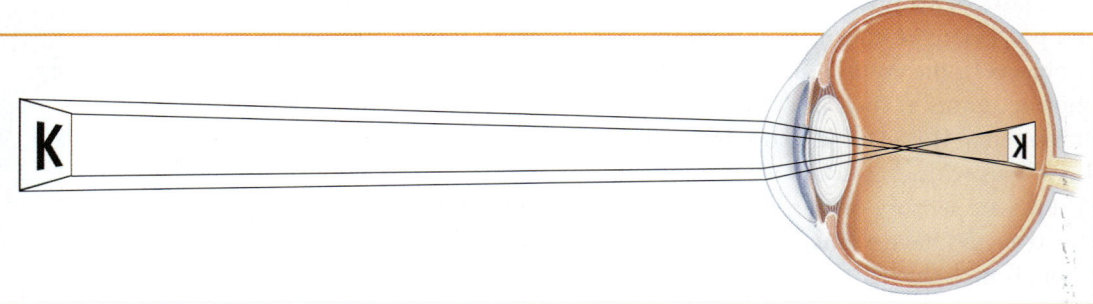

FIGURE 24.19 Pattern of retinal stimulation in the human eye. Light rays that fall on the retina produce a pattern that is both upside down and inverted left to right.

Cengage Learning.

The cornea and lens bend light rays coming from different points so they all converge at the back of the eye, on the retina. The image formed on the retina is an upside-down mirror image of the real world (Figure 24.19). However, the brain interprets this image so you perceive the world in its correct orientation.

When you see an object, you are perceiving light rays that are reflected from that object. The properties of light reflected from near and distant objects hit the eye at different angles, but adjustments of the lens ensure that all light rays become focused on the retina. Lens adjustments are carried out by a ciliary muscle that encircles the lens and attaches to it. When you focus on something close up, the ciliary muscle in each eye contracts, causing the lens to bulge outward. As a result, rays of light from the nearby object are bent so they are focused onto the retina (Figure 24.20**A**). When an object is farther away, light rays do not have to be bent as much to be focused on the retina. Ciliary muscles relax a bit, allowing the lens to flatten (Figure 24.20**B**).

Continuous viewing of a close object such as a computer screen or book keeps the ciliary muscles contracted. To reduce eyestrain, take breaks to focus on more distant objects. This will allow your ciliary muscles to relax.

About 150 million Americans have disorders that impair their eyes' ability to focus. With astigmatism, an unevenly curved cornea does not properly focus incoming light on the lens. Nearsightedness occurs when the distance from the front to the back of the eye is longer than normal or when ciliary muscles react too strongly. In either case, distance vision is poor because rays from distant objects converge in front of the retina instead of on it. Close vision is normal. In farsightedness, the distance from front to back of the eye is unusually short, ciliary muscles are weak, or the lens is not flexible enough to stretch easily. As a result, light rays from nearby objects get focused behind the retina. The lens typically loses its flexibility as a person ages. That is why most people who are over age forty have impaired close vision and require reading glasses.

Changes in the structure of proteins in the lens can result in a cataract, a clouding of the lens. Excessive exposure to ultraviolet radiation, smoking, use of steroids, and some diseases such as diabetes promote cataract formation. Typically, both eyes are affected. At first, a cataract scatters light and blurs vision. Eventually, the lens may become opaque, causing blindness. Cataract surgery restores normal vision by replacing a clouded lens.

> **At the Retina** As previously noted, the cornea and lens bend light rays so they fall on the retina. Figure 24.21**A,B** shows how a physician uses a lighted magnifying instrument to examine the retina inside the eyeball, and what she sees. The **fovea**, the area of the retina that is richest in photoreceptors, appears as a reddish spot in an area relatively free of blood vessels. With normal vision, most light rays are focused on the fovea.

The retina consists of several cell layers (Figure 24.21**C**). Rod cells and cone cells are the photoreceptors (Figure 24.21**D**). They lie beneath several layers of

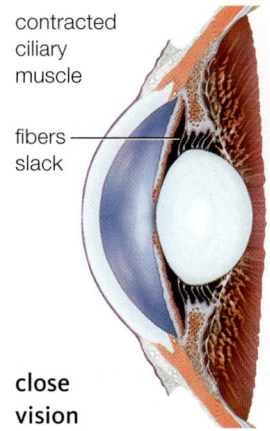
contracted ciliary muscle

fibers slack

close vision

A Contraction of ciliary muscle allows fibers to slacken so the lens rounds up and focuses light from a close object on the retina.

relaxed ciliary muscle

fibers taut

distance vision

B Relaxation of ciliary muscle pulls fibers taut so the lens is stretched into a flat shape that focuses light from a distant object on the retina.

FIGURE 24.20 How the eye focuses. The lens is encircled by ciliary muscle. Elastic fibers attach the muscle to the lens. The shape of the lens is adjusted by contracting or relaxing the ciliary muscle, increasing or decreasing the tension on the fibers, and changing the shape of the lens.

Bo Veisland/ Photo Researchers, Inc.

A A doctor uses a lighted magnifying instrument to look through the pupil and examine the retina.

start of the optic nerve

fovea

B What the doctor sees. Photoreceptors are most concentrated in the fovea. The start of the optic nerve causes a blind spot.

interneurons that process visual signals. **Rod cells** detect very dim light and respond to changes in light intensity that indicate motion. The sense of color and of acute daytime vision starts with light absorption by **cone cells**. There are three types, each having a different pigment. One cone cell pigment absorbs mainly red light, another absorbs mainly blue, and a third absorbs green. Normal human color vision requires all three kinds of cones. Individuals who lack one or more cone types are color blind (Section 9.7). They often have trouble distinguishing red from green in dim light, and some cannot do so even in bright light.

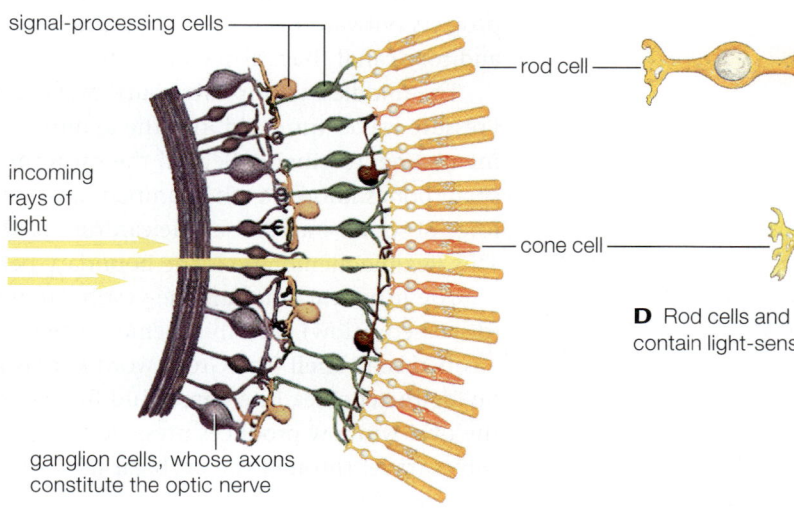

signal-processing cells

incoming rays of light

rod cell

cone cell

stacked pigment-containing discs

D Rod cells and cone cells are photoreceptors that contain light-sensitive pigment.

ganglion cells, whose axons constitute the optic nerve

C Cell layers of the retina.

   Signal integration and processing start in the retina. Signals flow from the rods and cones to adjacent cell layers that process the information and send signals to ganglion cells. Bundled axons of ganglion cells leave the retina as the beginning of the optic nerve. The part of the retina where the optic nerve exits lacks photoreceptors. It cannot respond to light and thus is a called "blind spot." We all have a blind spot in each eye, but usually do not notice it because information that is missed by one eye is provided by the other.

❯ Hearing Sounds are waves of compressed air—one form of mechanical energy. When you clap your hands or shout, you create pressure waves that move through the air. The frequency of those waves, the number per second, determines a sound's pitch. The amplitude (height) of the waves determines the loudness of the sound, which we measure in decibels. The human ear can detect a one-decibel difference between sounds. An increase of 10 on this scale is an increase of tenfold in loudness. Normal conversation is about 60 decibels, a food blender at high speed is about 90 decibels, and a chain saw is about 100 decibels. Music at a rock concert is about 120 decibels, as is the sound heard through the earbuds of an iPod or similar device cranked up to its maximum volume.

cone cell Photoreceptor that provides sharp vision and allows detection of color.
fovea Region of vertebrate eye with densest concentration of photoreceptors.
rod cell Photoreceptor active in dim light; provides coarse perception of image and detects motion.

inner ear
vestibular apparatus, cochlea

outer ear
pinna, auditory canal

middle ear
eardrum, ear bones

❶ The outer ear's flap and canal collect sound waves.

middle ear bones:
stirrup
anvil
hammer

oval window (behind stirrup)

auditory nerve

auditory canal

eardrum

round window

cochlea

❷ The eardrum and middle ear bones amplify sound.

Figure 24.22 shows the three regions of the human ear. The outer ear gathers sounds. Its pinna, a skin-covered, sound-collecting, folded flap of cartilage, projects outward from the side of the head ❶. The outer ear also includes the auditory canal that carries sound to the middle ear.

The middle ear amplifies sound waves and transmits them to the inner ear ❷. Pressure waves funneled into the auditory canal cause the **eardrum**, a thin membrane, to vibrate. Behind the eardrum is an air-filled cavity with a set of small bones known as the hammer, anvil, and stirrup. These bones transmit the force of sound waves from the eardrum to the smaller surface of the oval window. This flexible membrane is the boundary between the middle ear and inner ear.

The inner ear contains the vestibular apparatus, which functions in balance (discussed below), and the cochlea. The pea-sized, fluid-filled **cochlea** resembles a coiled snail shell (the Greek word *koklias* means snail). Internal membranes divide the cochlea into three fluid-filled ducts ❸. Pressure from the stirrup on the oval window produces pressure waves in the fluid in these ducts. As these waves travel through the cochlear fluid, they cause the membranous walls of the ducts to vibrate.

The **organ of Corti**, the organ responsible for hearing, sits on a membrane (the basilar membrane) in one of the ducts ❹. It contains arrays of mechanoreceptors called hair cells. Specialized cilia extend from the hair cells into an overlying membrane. When pressure waves move the membranes, the hair cells bend and undergo action potentials. These signals then travel along an auditory nerve into the brain.

The brain determines the volume and pitch of a sound by assessing how many action potentials flow along the optic nerve, and where in the cochlea those potentials come from. The greater the volume of the sound, the more the hair cells bend. The pitch of a sound determines where hair cells bend the most. Higher-pitched sounds set up vibrations near the base of the coil, whereas lower-pitched sounds cause vibrations deeper inside the coil.

❯ **Sense of Balance** The **vestibular apparatus** of the inner ear is an organ of equilibrium with three fluid-filled semicircular canals (Figure 24.23). Rotations of the head, stops, or starts stimulate mechanoreceptors inside this apparatus. The brain integrates the signals from organs of equilibrium in both ears, with the input from receptors in the skin, joints, and tendons to determine your body's position and motion.

Mismatched signals can cause motion sickness. On a curvy road, passengers in a car experience changes in acceleration and direction that indicate "motion" to their vestibular apparatus. At the same time, signals from their eyes about

cochlea Coiled structure in the inner ear that holds the sound-detecting organ of Corti.
eardrum Membrane of middle ear that vibrates in response to sound waves.
organ of Corti Sound-detecting organ in the cochlea.
vestibular apparatus In vertebrates, and organ of equilibrium in the inner ear.

3 One coil of the cochlea in cross-section. The organ of Corti detects pressure waves in fluid-filled ducts inside the cochlea.

4 Pressure waves cause the basilar membrane beneath the organ of Corti to move upward. The movement pushes hair cells against an overlying membrane. The resulting action potentials travel along the auditory nerve to the brain.

objects inside the car tell their brain that the body is at rest. Driving minimizes motion sickness because a driver focuses on sights rushing past outside the car, making visual signals more consistent with vestibular signals.

Take-Home Message

How do sensory organs function?

- Different animals have different types of sensory receptors, and so have different sensory experiences.
- Receptors throughout the body give rise to somatic and visceral sensations.
- Smell and taste are chemical senses. They involve chemoreceptors that are activated by binding of a particular chemical.
- Human vision occurs when the various components of an eye focus light onto photoreceptors on the retina at the back of the eye.
- Hearing is the detection of pressure waves. Components of the ear collect, amplify, and sort out these waves, which are detected by mechanoreceptors. Mechanoreceptors in the inner ear also have a role in the sense of balance.

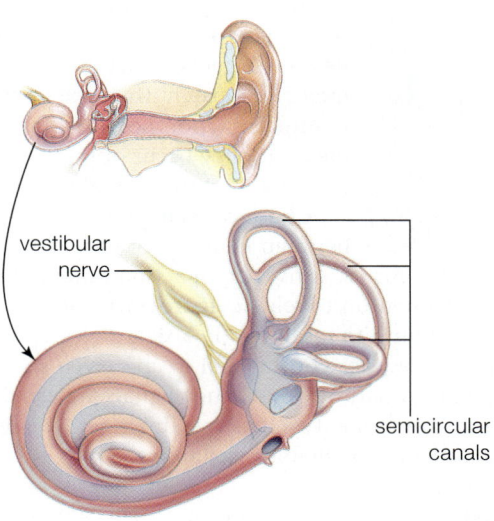

FIGURE 24.23 The vestibular apparatus and the adjacent cochlea.

© Cengage Learning.

24.7 **In Pursuit of Ecstasy** (revisited)

Now that you know a bit more about how the brain functions, take a moment to reconsider the effects of MDMA, the active ingredient in Ecstasy. MDMA binds to and interferes with the action of transport proteins that pump serotonin from the synaptic cleft back into axon terminals. The resulting excess of serotonin acts on postsynaptic cells and produces the pleasurable sensations that users of Ecstasy seek. However, in the aftermath of drug use, serotonin-releasing neurons are left depleted of this essential neurotransmitter. Low serotonin levels are associated with depression, and transient depression is a well-documented after-effect of Ecstasy use. Depletion of serotonin may also be responsible for MDMA's detrimental effect on memory.

WHeRe You aRe GOInG . . .

Section 25.3 describes the endocrine functions of the hypothalamus and pituitary gland, and Section 25.6 explains how stress elicits a fight–flight type response. Section 26.6 explains how drinking while pregnant can have permanent effects on prenatal development of a child's nervous system.

Summary

Section 24.1 Ecstasy is an illegal drug that enhances mood and energizes a person through its effects on the brain. However, it can also cause memory impairment, panic attacks, and even death.

Section 24.2 **Neurons** are electrically excitable cells that signal other cells by means of chemical messages. **Sensory neurons** detect stimuli. **Interneurons** relay signals between neurons. **Motor neurons** signal effectors (muscles and glands). **Neuroglia** support and assist neurons, as by insulating **axons** with a covering of **myelin**.

Signals reaching a neuron's **dendrites** can disturb the ion distribution across the plasma membrane. Stimulation may cause a cell that was at **resting potential** to reach its **threshold potential**. The resulting **action potential** travels along the neuron's axon.

An action potential is an abrupt, brief reversal in the voltage across a neuron's plasma membrane. The reversal triggers an action potential at an adjacent membrane patch, then the next, and on to axon terminals. Action potentials are "all or nothing," meaning they occur only if a cell reaches threshold potential and they are always the same size.

Chemical synapses allow information to move between neurons, or between a motor neuron and a cell it controls. Arrival of an action potential triggers axon terminals of a neuron to release **neurotransmitters** that can bind to receptors on a signal-receiving cell. The response of that cell to any given signal depends in part on what other signals are arriving at the same time. Psychoactive drugs act by altering signaling at chemical synapses.

Section 24.3 The simplest nervous system is a **nerve net**, a mesh of neurons with no central command. Most animals have a bilateral nervous system with a cluster of **ganglia** or a brain at the head end. The vertebrate nervous system is functionally divided into the **central nervous system** (brain and spinal cord) and the **peripheral nervous system** (nerves that connect the spinal cord and brain with the body).

Nerves are bundles of axons inside connective tissue sheaths. Most vertebrate axons are enclosed in insulating myelin that speeds signal transmission.

Sections 24.4, 24.5 Nerves of the peripheral nervous system extend from cell bodies in the brain or spinal cord out through the body. They relay signals in both directions. The **somatic nervous system** sends signals to skeletal muscles and receives them from receptors in joints and the skin. The **autonomic nervous system** connects to internal organs. Its two divisions often work in opposition. During a fight–flight situation, **sympathetic neurons** predominate. When you are in a relaxed state, **parasympathetic neurons** predominate.

The organs of the central nervous system are enclosed in protective membranes (**meninges**) and surrounded by **cerebrospinal fluid**. The **blood–brain barrier** keeps unwanted substances from entering this fluid.

The **spinal cord** relays signals from the body to the brain. It also is the integrating center for reflexes that do not involve the brain. A **reflex** is an automatic response to a stimulus.

The brain's hindbrain and midbrain constitute the brain stem, which regulates breathing and reflexes such as swallowing and coughing. The brain stem also includes the **cerebellum**, which coordinates voluntary movements. The forebrain contains the **hypothalamus**, which regulates functions related to homeostasis. The bulk of the forebrain is the **cerebrum**. Its thin surface layer of gray matter, the cerebral cortex, governs complex functions such as language and abstract thought. It has specific areas that receive different types of sensory input or control voluntary movements.

Section 24.6 Sensory receptors respond to a specific stimulus by undergoing action potentials, as when **thermoreceptors** respond to heat or cold. The brain evaluates information from sensory receptors based on which nerve delivers them, the rate of action potentials, and the number of axons firing in a given interval.

The **somatic sensations** such as touch arise from sensory receptors located in skin, or near muscles or joints. **Visceral sensations** arise from receptors near organs in body cavities. Pain is a sensation that arises when tissues are damaged, and **endorphins** can reduce it.

Chemoreceptors involved in taste and smell detect specific chemicals. Human taste receptors are concentrated in taste buds on the tongue and walls of the mouth. Olfactory receptors line human nasal passages. Many animals also have chemoreceptors that detect social signaling molecules called **pheromones**.

In a human eye, light passes through the **cornea** at the eye's surface and enters its interior through the **pupil**, whose diameter is controlled by the **iris**. The **lens** focuses light onto **photoreceptors** (**rod cells** and **cone cells**) in the **fovea** of the **retina**. Other cells of the retina integrate and process signals, which travel along optic nerves to the brain for final processing and interpretation.

The human outer ear collects sound waves. Sound waves vibrate the **eardrum**, and middle ear bones transmit these vibrations to the **cochlea** of the inner ear. The **organ of Corti** inside the cochlea contains **mechanoreceptors** that, when excited, send signals to the brain. The **vestibular apparatus** in the inner ear detects movement and changes in the body's position.

Self-Quiz   Answers in Appendix I

1. The _____ of a neuron releases neurotransmitter.
   a. axon terminals   c. dendrites
   b. cell body   d. all of the above

2. _____ occur mainly in the brain and spinal cord.
   a. Sensory neurons   c. Interneurons
   b. Motor neurons   d. b and c

3. A(n) _____ has one axon and no dendrites.
   a. sensory neuron   c. interneuron
   b. motor neuron   d. neuroglial cell

4. When you sit quietly, output from the _____ system prevails.
   a. sympathetic   c. both
   b. parasympathetic   d. neither

5. An action potential occurs when _____ .
   a. a neuron reaches threshold potential
   b. gated sodium channels close
   c. sodium–potassium pumps stop working
   d. gated potassium channels open

6. Skeletal muscles are controlled by the _____ system.
   a. sympathetic   c. somatic
   b. parasympathetic   d. both a and b

7. A _____ is an automatic response that does not require thought.

8. The two halves of the cerebrum _____ .
   a. have identical functions
   b. are part of the autonomic nervous system
   c. are connected by the corpus callosum
   d. consist mainly of motor neurons

9. The blood–brain barrier controls what enters _____ .
   a. blood   c. peripheral nerves
   b. cerebrospinal fluid   d. both a and b

10. Which is a somatic sensation?
   a. hearing   c. touch   e. both a and c
   b. smell   d. taste   f. all of the above

11. _____ is a reduced response to an ongoing stimulus.
   a. Propagation   c. Sensory adaptation
   b. Perception   d. Synaptic integration

12. Chemoreceptors play a role in the sense of _____ .
   a. hearing   b. smell   c. vision   d. pain

13. In a vertebrate eye, photoreceptors are in the _____ .
   a. conjunctiva   c. lens
   b. cornea   d. retina

14. Color blindness arises when _____ are missing or defective.
   a. hair cells   c. cone cells
   b. rod cells   d. neuroglial cells

15. Match each structure with its function.
   _____ rod cell   a. detects pheromones
   _____ cochlea   b. connects to spinal cord
   _____ cerebellum   c. sorts out pressure waves
   _____ brain stem   d. protects brain and spinal
   _____ cerebral cortex   cord from some toxins
   _____ taste bud   e. speeds signal transmission
   _____ myelin   f. contains chemoreceptors
   _____ neurotransmitter g. secreted at synapse
   _____ blood–brain   h. governs higher thought
          barrier   i. coordinates voluntary moves
   _____ vomeronasal organ j. detects light

Digging Into Data

Prenatal Effects of Ecstasy

Animal studies are often used to assess effects of prenatal exposure to illicit drugs. For example, Jack Lipton used rats to study the behavioral effect of prenatal exposure to MDMA, the active ingredient in Ecstasy. He injected female rats with either MDMA or saline solution when they were 14 to 20 days pregnant. This is the period when their offspring's brains were forming. When those offspring were 21 days old, Lipton tested their response to a new environment. He placed each young rat in a new cage and used a photobeam system to record how much each rat moved around before settling down. Figure 24.24 shows his results.

1. Which rats moved around most (caused the most photobeam breaks) during the first 5 minutes in a new cage, those prenatally exposed to MDMA or the controls?

2. How many photobeam breaks did the MDMA-exposed rats make during their second 5 minutes in the new cage?

3. Which rats moved around the most during the last 5 minutes of the study?

4. Does this study support the hypothesis that MDMA affects a developing rat's brain?

FIGURE 24.24 Effect of prenatal exposure to MDMA on activity levels of 21-day-old rats placed in a new cage. Movements were detected when the rat interrupted a photobeam. Rats were monitored at 5-minute intervals for a total of 20 minutes. *Blue* bars are results for rats whose mothers received saline, *red* bars are rats whose mothers received MDMA. © Cengage Learning.

Critical Thinking

1. The blood–brain barrier in human newborns is not yet fully developed. Explain why this makes careful monitoring of diet and environmental chemical exposure particularly important.

2. Axons of some human motor neurons extend from the base of the spinal cord to the big toe. In a giraffe, the longest axons are several meters long. What are some functional challenges involved in development and maintenance of such lengthy cellular extensions?

3. After a leg injury, pain makes a person avoid putting too much weight on the affected leg. Shielding the injury gives it time to heal. An injured insect shows no such shielding response. Some have cited the lack of such a response as evidence that insects do not feel pain. But insects do produce substances similar to our natural painkillers. Is the presence of these compounds in insects sufficient evidence to conclude that they feel pain?

25 Endocrine Control

... WHERE YOU HAVE BEEN

This chapter focuses on endocrine glands (Section 19.3). Knowing about properties of steroids (2.8), proteins (2.9), and the plasma membrane (3.3), and the function of promoters (7.3), will help you understand hormone action. We will return to glucose metabolism (5.8), genetic engineering (10.4), and urine formation (23.6).

25.1 Hormones in the Balance

We live in a world awash in synthetic chemicals. We drink from plastic bottles, wear synthetic fabrics, slather ourselves with synthetic skin products, and dose our food with synthetic pesticides. Enormous numbers of man-made compounds are used to manufacture our computers and other electronic gadgets. What do we know about the safety of all these substances? We have learned by sad experience that some synthetic chemicals harm the environment and threaten human health. For example, DDT (a pesticide) and PCBs (used to make electronic products, caulking, and solvents) are endocrine disruptors. Endocrine disruptors are molecules that interfere with the action of **animal hormones**, the signaling molecules secreted into blood by endocrine glands. As a result of this interference, hormone disruptors can impair animal growth, development, and reproduction. DDT was banned in 1972 and PCBs in 1979. However, because both were widely used for years and are highly stable, they still persist in the environment.

Many other endocrine disruptors remain in use. Atrazine, a chemical that has been in wide use as an herbicide for more than forty years, is among them. In the United States, 76 million pounds of atrazine are sprayed each year, mostly to kill weeds in cornfields (Figure 25.1**A**). Atrazine-contaminated runoff from fields seeps into soil and water. Atrazine breaks down within a year, but until then it accumulates in ponds, streams, wells.

Living in atrazine-contaminated water can disrupt hormone function in aquatic animals and diminish their reproductive capacity. Exposing genetically male frog tadpoles to atrazine alters their sexual development. As adults, such frogs have a lower level of the male sex hormone testosterone, smaller testes, and a reduced sperm count compared to frogs raised in atrazine-free water. Atrazine even causes some males to develop ovaries in addition to testes, or in place of them (Figure 25.1**B**).

Researchers have also shown that atrazine alters reproduction and development in freshwater fish. One group found that exposing fish embryos to atrazine levels comparable to those in runoff from atrazine-treated fields doubles the proportion of embryos that develop into females. Another group showed that exposing adult fish to atrazine during their breeding period can lower their reproductive capacity. Female minnows that breed in atrazine-contaminated water produce fewer eggs than those that breed in pure fresh water.

Atrazine is usually applied to fields in spring, so the time of peak atrazine runoff often occurs when fish and frogs are mating and when their young are developing. Thus, atrazine's effects could help push some endangered fish and frogs toward extinction.

FIGURE 25.1 Benefits and costs of herbicide applications. **A** Atrazine keeps cornfields free of weeds, reducing the need for tilling that can cause soil erosion.
B Atrazine also interferes with endocrine function in aquatic animals. This photo shows two frogs mating. The one on the bottom in this photo is a genetic male that has been turned into an egg-laying female by exposure to atrazine.

Credits: (a) © David Ryan/ SuperStock; (b) Courtesy of Prof. Tyrone Hayes, UC Berkeley.

Pineal gland
- Melatonin (affects sleep–wake cycles, onset of puberty)

Parathyroid glands (not shown, on rear of thyroid)
- Parathyroid hormone (raises blood calcium level)

Thyroid gland
- Thyroid hormone (metabolic effects)
- Calcitonin (lowers blood calcium level)

Thymus gland
- Thymosins (help T cells mature)

Adrenal glands
Adrenal cortex
- Cortisol (affects metabolism, immune response)
- Aldosterone (acts in kidneys)
Adrenal medulla
- Epinephrine, norepinephrine (cause fight–flight response)

Hypothalamus
- Hormones that regulate pituitary's anterior lobe
- Antidiuretic hormone (ADH), oxytocin (stored and released by pituitary)

Pituitary gland
Anterior lobe makes and secretes:
- Adrenocorticotropic hormone (ACTH; stimulates adrenal gland)
- Thyroid-stimulating hormone (TSH; stimulates thyroid gland)
- Luteinizing hormone (LH; stimulates ovaries and testes)
- Follicle-stimulating hormone (FSH; stimulates ovaries and testes)
- Prolactin (stimulates mammary glands)
- Growth hormone (affects overall growth)
Posterior lobe secretes:
- Antidiuretic hormone (ADH; acts in kidneys to promote water reabsorption)
- Oxytocin (causes contraction of smooth muscle of reproductive tract and milk ducts)

Pancreas
- Insulin (lowers blood glucose)
- Glucagon (raises blood glucose)

Gonads (ovaries or testes)
- Estrogens, progesterone, testosterone (affect sex organs and influence secondary sexual traits)

25.2 Hormones and the Endocrine System

Endocrine glands and cells secrete animal hormones into the blood, which carries the hormones to target cells in the body. Like neurotransmitters, hormones affect cells by binding to receptor proteins. We define a hormone's target as a cell that has receptors for that hormone.

In humans and other vertebrates, the many sources of hormones interact as an **endocrine system**. Figure 25.2 provides an overview of the main components of the human endocrine system. Other vertebrates have a similar system. The endocrine system and nervous system are so closely linked that scientists sometimes refer to them collectively as the neuroendocrine system. Both endocrine glands and neurons receive signals from the hypothalamus in the forebrain (Section 24.5). Most organs respond to hormones as well as signals from the nervous system.

Invertebrates also produce hormones, although they do not have the same endocrine organs as vertebrates. For example, an insect must molt its cuticle periodically as it grows, and hormones control the molting process. Chemicals that disrupt the action of these hormones are used as insecticides.

FIGURE 25.2 Animated! Overview of major glands of the human endocrine system and primary effects of their main hormones. The endocrine system also includes endocrine cells in many organs, such as the heart, kidneys, stomach, liver, small intestine, and skin. Section 23.1 described the effects of leptin and other hormones that affect body weight. © Cengage Learning 2010.

animal hormone Signaling molecule secreted by an endocrine gland or cell.
endocrine system Hormone-producing glands and secretory cells of a vertebrate body.

Table 25.1	Types and Examples of Hormones
Steroids	Testosterone, estrogens, progesterone, aldosterone, cortisol
Amines	Melatonin, epinephrine, thyroid hormone
Peptides	Glucagon, oxytocin, antidiuretic hormone, calcitonin, parathyroid hormone
Proteins	Growth hormone, insulin, prolactin, follicle-stimulating hormone, luteinizing hormone

> **Types of Hormones** There are two main categories of hormones: steroid hormones, and hormones derived from amino acids. **Steroid hormones** are lipids derived from cholesterol. The sex hormones testosterone and estrogen are examples (Table 25.1). Being lipids, these hormones can diffuse through a cell's plasma and nuclear membranes (Figure 25.3**A**). The hormones can then bind to receptors inside the cell, forming a hormone–receptor complex. This complex binds to a promoter region of DNA (Section 7.3), where it either inhibits or enhances gene transcription. Some steroid hormones also produce effects by binding to receptors at the cell membrane.

In some types of breast cancer, binding of the steroid hormone estrogen to receptors on the surface and inside the cancer cells encourages cell division. The drug tamoxifen binds to the estrogen receptors of these cells, prevents estrogen from binding to these receptors, and so slows the growth of the cancer.

Hormones derived from amino acids include amines (modified amino acids), peptides (short amino acid chains), and proteins. Peptide and protein hormones cannot diffuse across a lipid bilayer. Instead, they bind to receptors at a target cell's surface. Often the binding starts a cascade of enzyme reactions. For example, receptors for the peptide hormone glucagon span the plasma membrane of liver cells (Figure 25.3**B**). Binding of glucagon to these receptors activates an enzyme that converts ATP to a compound called cyclic AMP (cAMP). The cAMP serves as a **second messenger**, meaning it relays signals from outside the cell to target molecules inside the cell. Inside liver cells, cAMP activates enzymes that activate other enzymes, and so on. The series of reactions eventually activates enzymes that break glycogen into its glucose subunits.

> **The Importance of Receptors** Blood flow distributes a hormone throughout the body, but the hormone only affects cells that have receptors for it. All hormone receptors are proteins, so mutations can interfere with their function. Consider that development of male genitals in an XY embryo requires the action of the hormone testosterone. XY individuals who have total androgen insensitivity syndrome secrete testosterone, but a mutation alters receptors for it. Without functional receptors, it is as if testosterone is not present. The embryo forms testes, but they do not descend into the scrotum, and genitals appear female. Such individuals are often not diagnosed until their teens. Being genetically male, they do not have female reproductive organs and do not menstruate.

Variations in receptor structure affect how tissues respond to a hormone. The receptor for a hormone in one tissue may cause a different cellular response than a somewhat different receptor for the same hormone in another tissue. Most tissues have receptors for many hormones. The response called up by one hormone may oppose or reinforce that of another. For example, a skeletal muscle cell has receptors for glucagon, insulin, cortisol, epinephrine, estrogen, testosterone, growth hormone, somatostatin, and thyroid hormone, as well as others. The level of all of these hormones influences what happens in the cell.

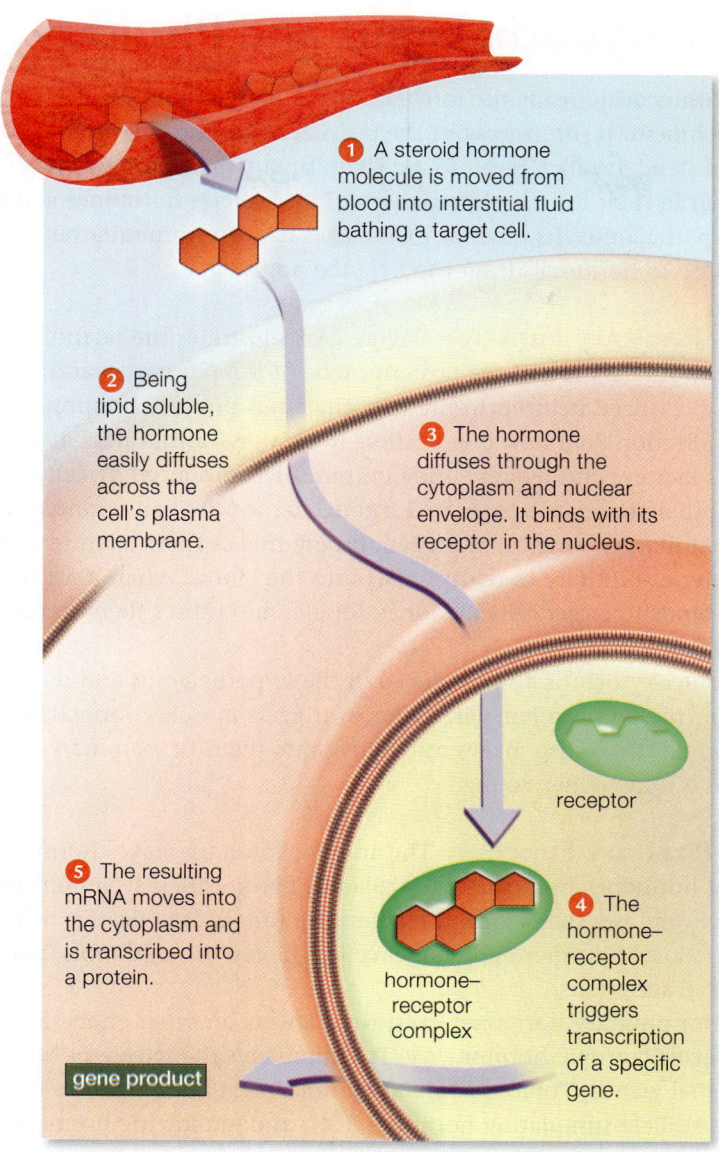

1 A steroid hormone molecule is moved from blood into interstitial fluid bathing a target cell.

2 Being lipid soluble, the hormone easily diffuses across the cell's plasma membrane.

3 The hormone diffuses through the cytoplasm and nuclear envelope. It binds with its receptor in the nucleus.

5 The resulting mRNA moves into the cytoplasm and is transcribed into a protein.

receptor

4 The hormone–receptor complex triggers transcription of a specific gene.

hormone–receptor complex

gene product

A Example of steroid hormone action inside a target cell.

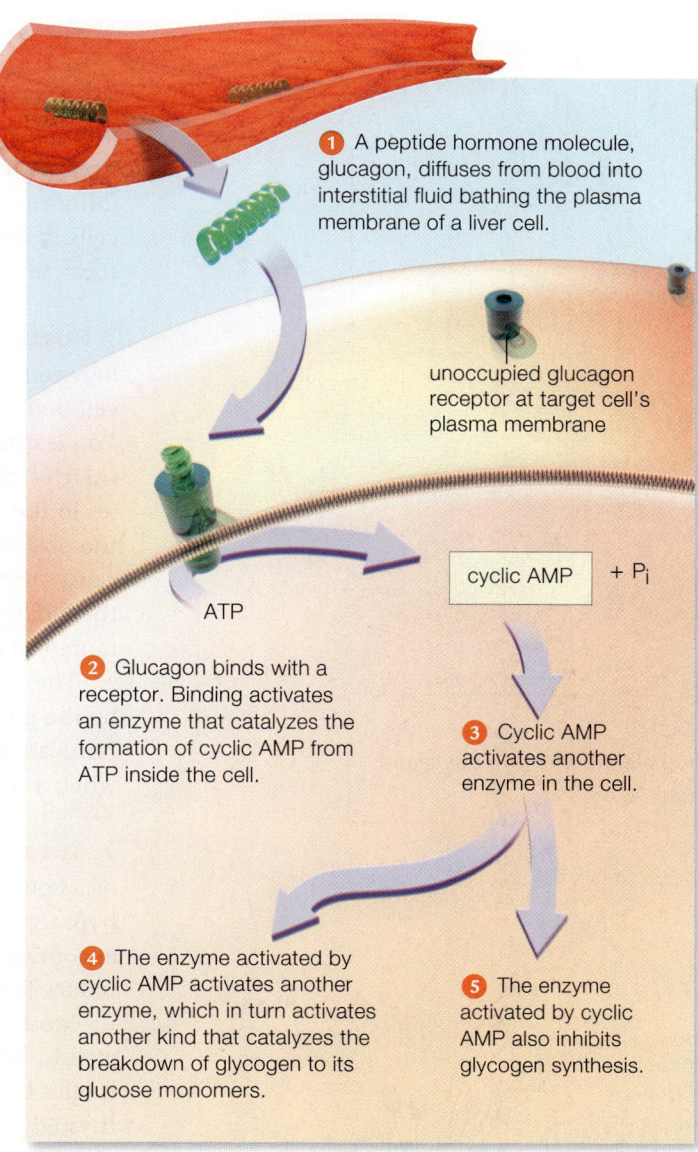

1 A peptide hormone molecule, glucagon, diffuses from blood into interstitial fluid bathing the plasma membrane of a liver cell.

unoccupied glucagon receptor at target cell's plasma membrane

ATP

cyclic AMP + P_i

2 Glucagon binds with a receptor. Binding activates an enzyme that catalyzes the formation of cyclic AMP from ATP inside the cell.

3 Cyclic AMP activates another enzyme in the cell.

4 The enzyme activated by cyclic AMP activates another enzyme, which in turn activates another kind that catalyzes the breakdown of glycogen to its glucose monomers.

5 The enzyme activated by cyclic AMP also inhibits glycogen synthesis.

B Example of peptide hormone action.

Take-Home Message

How do hormones influence cells?

- Components of the endocrine system produce and secrete hormones. Blood distributes a hormone through the body, but only cells with a receptor protein for that hormone are affected by it. Hormone molecules reversibly bind with a target cell's receptors.

- Steroid hormones are derived from cholesterol. They can diffuse across cell membranes and bind to intracellular receptors. Many receptor–hormone complexes bind to DNA, and influence protein synthesis. Some steroid hormones also can bind to receptors at a cell's surface.

- Peptide hormones and protein hormones consist of amino acids. These hormones cannot enter cells, so they bind to receptors in the plasma membrane. Binding often sets in motion a chain of enzyme activations inside the cell.

- Different types of cells have different receptors for the same hormone. The effect of a hormone varies with the properties of the receptor it binds to.

- Most cells have receptors for, and are influenced by, many different hormones.

FIGURE 25.3 The two main mechanisms of hormone action.
Figure It Out: Where are receptors for a peptide hormone located?
© Cengage Learning. *Answer: At the cell surface*

second messenger Molecule that forms inside a cell when a hormone binds at the cell surface; sets in motion reactions that alter enzyme activity inside the cell.
steroid hormone Hormone derived from cholesterol.

25.3 The Hypothalamus and Pituitary

The **hypothalamus** deep inside the forebrain is the body's main regulator of the internal environment. It interacts with the pea-sized **pituitary gland** that connects to it by a slender stalk (Figure 25.4). The pituitary gland has two lobes that differ somewhat in their function. Its posterior lobe secretes hormones made by cells in the hypothalamus. Its anterior lobe makes its own hormones but releases them in response to hormones from the hypothalamus.

❯ Posterior Pituitary Function Figure 25.5 illustrates the relationship between the hypothalamus and the posterior lobe of the pituitary gland. The cell bodies of specialized neurons in the hypothalamus produce hormones ❶. You learned in Section 23.6 how one of these hormones—antidiuretic hormone (ADH)—affects kidneys and reduces urine output. ADH produced by cell bodies in the hypothalamus gets transported through axons to axon terminals in the posterior pituitary ❷. Arrival of an action potential at these axon terminals causes the posterior pituitary to release ADH into the blood. When ADH reaches the kidney, it binds to target cells in kidney tubules and causes them to reabsorb more water ❸.

The hormone oxytocin is also produced in the hypothalamus and released by the posterior pituitary. In females, oxytocin triggers muscle contractions that bring about childbirth. It also makes milk move into ducts of mammary glands when a female is nursing her young.

❯ Anterior Pituitary Function The anterior pituitary makes hormones of its own, but hormones from the hypothalamus control their secretion. Most hypothalamic hormones that act on the anterior pituitary are releasers, which encourage secretion of hormones by target cells. The rest are inhibitors, which reduce target cell secretions.

Most anterior pituitary hormones exert their effects on other endocrine glands. Adrenocorticotropic hormone (ACTH) stimulates the release of hormones by adrenal glands. Thyroid-stimulating hormone (TSH) regulates the thyroid gland. Follicle-stimulating hormone (FSH) and luteinizing hormone (LH) affect sex hormone secretion and production of gametes by gonads—a male's testes or a female's ovaries.

The anterior pituitary hormone prolactin targets mammary glands, the exocrine glands that secrete milk. Prolactin stimulates and sustains milk production after a woman has given birth.

Growth hormone (**GH**) is an anterior pituitary hormone with widespread effects throughout the body. It encourages the growth of bone and soft tissues in the young, and it influences metabolism in adults.

❯ Growth Disorders Normally, a surge of GH production during teenage years results in a growth spurt, then GH production declines with age. Disorders that alter the amount of GH secreted or the timing of its secretion have dramatic effects on body size and form. Pituitary dwarfism occurs when the body produces too little GH or receptors do not respond to it properly during childhood.

FIGURE 25.4 Location of the hypothalamus and pituitary gland.
© Cengage Learning.

hypothalamus

anterior lobe of pituitary

posterior lobe of pituitary

cell bodies of neurons in the hypothalamus

❶ Cell bodies of neurons in the hypothalamus make ADH.

axons

❷ ADH is moved through axons into posterior pituitary.

anterior pituitary

posterior pituitary

❸ ADH released in the posterior pituitary enters the blood, and binds to target cells in kidney.

ADH

FIGURE 25.5 Production and secretion of ADH. The hormone is produced in the hypothalamus, but stored in and released from the posterior pituitary gland.
© Cengage Learning.

A Insufficient growth hormone slowed the growth of Jyoti Amge of Nagpur, India, shown here with her same-age classmates. Now over age 18 and no longer growing, she stands a bit more than 2 feet (61 centimeters tall).

B Excess growth hormone during childhood made Bao Xishun one of the world's tallest men. He stands 2.36 meters (7 feet 9 inches) tall.

Affected individuals are short but normally proportioned (Figure 25.6**A**). Pituitary dwarfism can be inherited, or it can result from a pituitary tumor or injury. Injections of recombinant human growth hormone (rhGH) made through genetic engineering (Section 10.4) increase the growth rate of children who have a naturally low GH level. However, such treatment is controversial. Many people object to the idea that short stature is a defect to be "cured."

Excessive GH secretion during childhood causes gigantism. Affected people have a normally proportioned body, but are unusually large (Figure 25.6**B**). Excess production of GH during adulthood causes acromegaly, a condition in which bones and cartilage thicken. Facial bones and bones of the hands and feet are the most obviously affected (Figure 25.6**C**). The Greek word *acro* means extremities, and *megas* means large. Both gigantism and acromegaly most often arise as the result of a benign pituitary tumor.

Injections of rhGH have been touted as a way to slow normal aging or boost athletic performance. However, such uses are not approved by regulatory agencies, have not been shown effective in clinical trials, and can have negative side effects, including increased risk of high blood pressure and diabetes.

C A woman before and after she became affected by acromegaly. Excess growth hormone secretion thickens fingers, enlarges ears, lips, and nose, and makes the brow and chin protrude.

FIGURE 25.6 Growth hormone disorders.

Credits: (a) BARM/ Fame Pictures; (b) China Daily/ Reuters; (c) Courtesy of Dr. William H. Daughaday, Washington University School of Medicine, from A. I. Mendelhoff and D. E. Smith, eds., *American Journal of Medicine*, 1956, 20:133.

Take-Home Message

How do the hypothalamus and pituitary function?

- The hypothalamus produces two hormones, antidiuretic hormone (ADH) and oxytocin, that are secreted by the pituitary gland's posterior lobe. ADH promotes water reabsorption by kidneys. Oxytocin causes contraction of muscle during childbirth and nursing.

- The hypothalamus also makes releasers and inhibitors that control secretions by the pituitary's anterior lobe. Hormones secreted by the anterior pituitary govern secretions of the thyroid gland, adrenal glands, gonads, and mammary glands. The anterior pituitary also secretes growth hormone, which has effects on growth and development of tissues throughout the body.

growth hormone (GH) Anterior pituitary hormone that promotes growth and development throughout the body.
hypothalamus Forebrain region that controls processes related to homeostasis and has endocrine functions.
pituitary gland Pea-sized endocrine gland in the forebrain that interacts closely with the adjacent hypothalamus.

Thyroid hormone increases metabolic rate, so producing too little of it can cause weight gain.

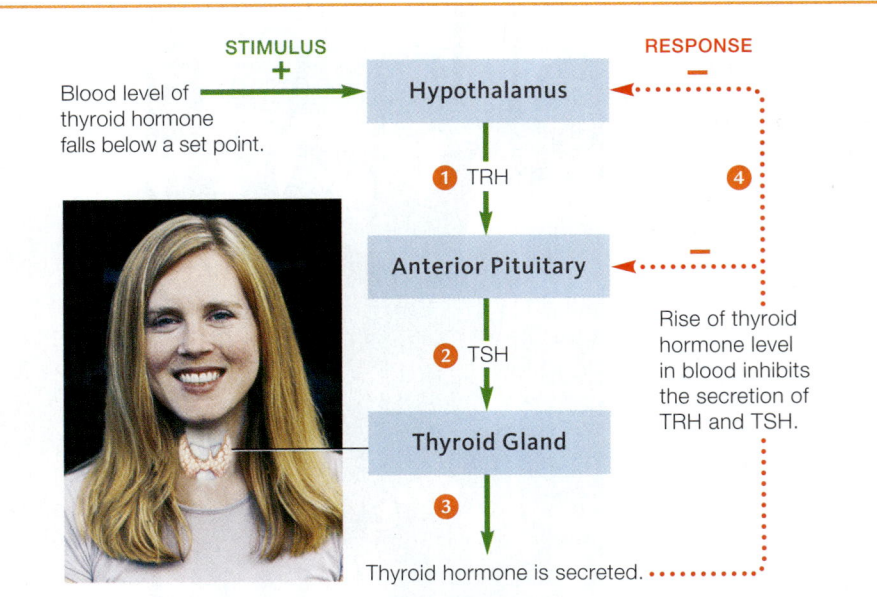

FIGURE 25.7 The negative feedback loop that governs thyroid function.
Figure It Out: What effect does a high concentration of thyroid hormone have on the hypothalamus? *Answer: It inhibits secretion of thyroid-releasing hormone (TRH).*
Credits: left, Gary Head; right, © Cengage Learning.

FIGURE 25.8 A goiter (enlarged thyroid). In this case, it was caused by an iodine deficiency.
Scott Camazine/ Photo Researchers, Inc.

25.4 Thyroid and Parathyroid Glands

❯ **The Thyroid Gland** The **thyroid gland** lies at the base of the neck, in front of the trachea. In children and adolescents, it secretes calcitonin, a hormone that encourages bone to take up calcium. Normal human adults produce little calcitonin. However, in some other species, this hormone plays a major role in controlling blood calcium level.

The thyroid gland also secretes two molecules (triiodothyronine and thyroxine) that we refer to collectively as thyroid hormone. Thyroid hormone acts throughout the body to increase the resting metabolic rate, the rate at which the body uses energy to maintain itself. People who have a high metabolic rate release more energy as heat than those with a lower metabolic rate, and they can eat more food without gaining weight.

The anterior pituitary gland and hypothalamus regulate thyroid hormone secretion by way of a negative feedback loop (Figure 25.7). A low blood concentration of thyroid hormone causes the hypothalamus to secrete thyroid-releasing hormone (TRH) ❶. This releasing hormone causes the anterior pituitary to secrete thyroid-stimulating hormone (TSH) ❷. TSH in turn stimulates the secretion of thyroid hormone ❸. When the blood level of thyroid hormone rises, secretion of TRH and TSH declines ❹.

Thyroid hormone contains the mineral iodine, so a lack of iodine in the diet can cause thyroid hormone deficiency, or hypothyroidism. When iodine is in short supply, signals calling for thyroid hormone secretion cannot be turned off by the normal feedback mechanism. The ongoing stimulation of the thyroid causes it to enlarge. A visibly enlarged thyroid is called a goiter (Figure 25.8). A thyroid hormone deficiency also causes fatigue, confusion, increased sensitivity to cold temperature, and weight gain.

In the United States, where consumption of iodized salt has made dietary iodine deficiency rare, most thyroid disorders arise as a result of an autoimmune disorder. Depending on the type of cells the immune system attacks, the amount of thyroid hormone can increase or decrease. In either case, the thyroid may become enlarged. An excess of thyroid hormone causes nervousness and irritability, a chronic fever, and weight loss. Often, altered metabolism causes tissues behind the eyeball to swell, causing eyes to bulge in their sockets.

Thyroid hormone also plays a role in normal development of the nervous system development. If a mother lacks iodine during her pregnancy or has an immune disorder that lowers her production of thyroid hormone, development of her child's nervous system may be impaired, resulting in lowered intelligence. A low level of thyroid hormone during infancy or early childhood also stunts growth and impairs mental ability.

Exposure to a variety of pollutants can disrupt thyroid hormone function. For example, exposure to nitrate derived from synthetic fertilizers, which frequently pollutes drinking water in agricultural regions, raises the risk of thyroid cancer. Nitrate exerts its negative effects by binding to membrane transport proteins in place of iodine, and thus preventing uptake of this essential mineral. Some insecticides, herbicides, fungicides, and other synthetic chemicals also increase the risk of thyroid disorders.

> **Parathyroid Glands** **Parathyroid glands** are the main regulators of the blood's calcium level. There are four of these glands, each about the size of a grain of rice, on the thyroid's rear surface. When the calcium level in the blood declines, the glands release parathyroid hormone (PTH), which targets cells in bones and the kidneys. PTH increases the breakdown of bone, thus putting calcium ions into the bloodstream. In the kidneys, PTH increases calcium reabsorption, so less calcium is lost in the urine. It also increases production of an enzyme that activates vitamin D. This vitamin helps the intestine take up calcium from food.

Rickets is an nutritional disorder that most often occurs when children age 2 to 6 year lack vitamin D. The deficiency in this vitamin impairs the child's ability to take up calcium from the gut and use it to build new healthy bone. The low blood calcium level also encourages secretion of PTH, which in turn encourages the breakdown of existing bone. Long bones, which are the fastest growing, are most affected. Typical symptoms of rickets include bowed legs and pelvic deformities (Figure 25.9).

In adults, a benign parathyroid tumor sometimes causes excessive PTH secretion that leads to osteoporosis. As with rickets, the high PTH causes breakdown of existing bone and adds calcium to the blood. The elevated level of blood calcium raises the risk of kidney stones.

FIGURE 25.9 Characteristic bowed legs of a child with rickets. This nutritional disorder arises when a lack of vitamin D inhibits calcium uptake and leads to excessive PTH secretion. The resulting withdrawal of calcium from leg bones softens them, causing them to bend beneath the child's weight.

Biophoto Associates/ Photo Researchers, Inc.

Take-Home Message

What are the functions of the thyroid and parathyroid glands?

- The thyroid gland regulates the metabolic rate of cells throughout the body. Normal thyroid function requires a diet with adequate iodine.

- Thyroid hormone also plays a role in nervous system development. Low thyroid hormone during prenatal development or early childhood impairs intelligence.

- The parathyroid glands control the level of calcium in the blood. Parathyroid hormone increases breakdown of bone and lessens calcium output in urine, thus increasing the blood calcium level.

parathyroid glands Four small endocrine glands in the neck whose hormone product increases the level of calcium in blood.

thyroid gland Endocrine gland at the base of the neck; produces thyroid hormone, which increases the metabolic rate and is essential for normal brain development.

25.5 The Pancreas

The pancreas sits in the abdominal cavity, just behind the stomach (Figure 25.10). We saw in Section 23.3 that the pancreas secretes digestive enzymes into the small intestine, but it also secretes hormones into the bloodstream. The bulk of the organ produces the digestive enzymes. The hormone-producing cells are grouped in clusters called pancreatic islets.

❯ Controlling Blood Glucose Food intake and intracellular demands for glucose vary and both have the potential to shift the amount of glucose in the blood beyond its ideal range. However, opposing effects of two pancreatic hormones, glucagon and insulin, help smooth out fluctuations in blood glucose level. **Glucagon** is a hormone secreted by some cells of the pancreatic islets. It targets cells in the liver, where it causes the breakdown of glycogen into glucose. Thus, glucagon raises the level of glucose in blood. **Insulin** is a hormone secreted by other cells of the pancreatic islets. It causes its main targets—liver, fat, and skeletal muscle cells—to take up more glucose. Thus, insulin lowers the level of glucose in the blood.

When blood glucose level rises above a set point, the pancreas secretes less glucagon and more insulin (Figure 25.10). As glucose is taken up and stored inside cells, blood glucose declines.

In contrast, a decline in blood glucose below the set point increases glucagon secretion and slows insulin secretion. The resulting release of glucose from the liver causes the blood glucose level to rise.

FIGURE 25.10 Animated! Maintaining blood glucose level. *Above,* location of the pancreas. *Right,* how cells that secrete insulin and glucagon respond to a change in the level of glucose circulating in blood. These two hormones work antagonistically to maintain the glucose level in its normal range.

❶ *After* a meal, glucose enters blood faster than cells can take it up, so its concentration in blood increases.

❷ In the pancreas, the increase stops cells from secreting glucagon.

❸ It also stimulates other cells to secrete insulin.

❹ In response to insulin, adipose and muscle cells take up and store glucose; cells in the liver and muscle make more glycogen.

❺ As a result, insulin *lowers* the blood level of glucose.

❻ *Between* meals, the glucose level in blood declines.

❼ The decrease causes glucagon secretion.

❽ It also slows insulin secretion.

❾ In the liver, glucagon causes cells to break glycogen down into glucose, which enters the blood.

❿ As a result, glucagon raises the blood glucose level.

© Cengage Learning.

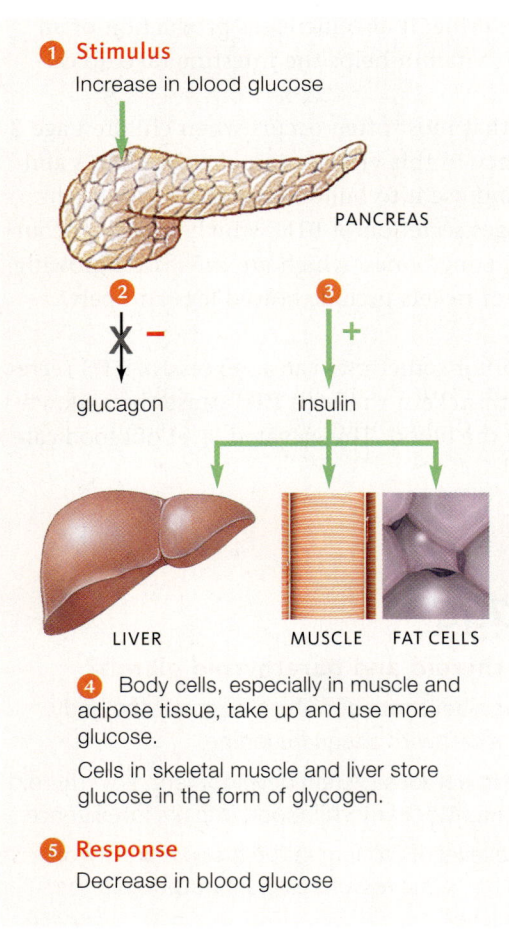

❶ **Stimulus**
Increase in blood glucose

PANCREAS

❷ − glucagon

❸ + insulin

LIVER · MUSCLE · FAT CELLS

❹ Body cells, especially in muscle and adipose tissue, take up and use more glucose.

Cells in skeletal muscle and liver store glucose in the form of glycogen.

❺ **Response**
Decrease in blood glucose

❻ **Stimulus**
Decrease in blood glucose

❼ + glucagon

❽ − insulin

❾ Cells in liver break down glycogen faster. The released glucose monomers enter blood.

❿ **Response**
Increase in blood glucose

Table 25.2	Some Complications of Diabetes
Eyes	Changes in lens shape and vision; damage to blood vessels in retina; blindness
Skin	Increased susceptibility to bacterial and fungal infections; patches of discoloration; thickening of skin on the back of hands
Digestive system	Gum disease; delayed stomach emptying that causes heartburn, nausea, vomiting
Kidneys	Increased risk of kidney disease and kidney failure
Heart and blood vessels	Increased risk of heart attack, stroke, high blood pressure, and atherosclerosis
Hands and feet	Impaired sensations of pain; formation of calluses, foot ulcers; poor circulation in feet especially may lead to tissue death that can only be treated by amputation

> **Diabetes Mellitus** Diabetes mellitus is an endocrine disorder whose name loosely translates as "passing honey-sweet water." Diabetics produce sweet urine because their liver, fat, and muscle cells do not take up and store glucose as they should. The result is high blood sugar, or hyperglycemia.

Cells that do not take up enough glucose must break down proteins and fats for energy (Section 5.8). Fat and protein breakdown produces harmful waste products, so relying heavily on them for energy causes an excess of these substances to accumulate. At the same time, high blood sugar causes some cells to overdose on glucose and produce other harmful substances. Accumulation of harmful molecules causes complications associated with diabetes (Table 25.2).

There are two main types of diabetes. Type 1 diabetes occurs when an immune reaction, genetic defect, tumor, or toxin destroys insulin-producing cells of pancreatic islets so little or no insulin is produced. Type 1 diabetes is also known as juvenile-onset diabetes because symptoms usually appear in childhood and adolescence. Affected people must monitor the amount of sugar in their blood and supply themselves with insulin via injections. New devices called insulin pumps can be programmed to deliver the amounts of insulin required to smooth out blood sugar fluctuations (Figure 25.11).

Type 1 diabetes accounts for only 5 to 10 percent of diabetes cases. In the more common type 2 diabetes, the insulin level is normal, but target cells stop responding to it. As a result, blood sugar levels remain high. It typically occurs in middle age, as insulin production declines. Genetics is a factor, but obesity increases the risk. Diet, exercise, and oral medication can control most cases of type 2 diabetes, but some affected people eventually need insulin injections.

The rate of type 2 diabetes is soaring worldwide. By one estimate, more than 150 million people are now affected. A trend toward Western diets and sedentary lifestyles contributes to the increasing risk for the disease. Because of the many negative effects of diabetes, the increasing number of cases is currently a pressing concern of public health officials.

FIGURE 25.11 A diabetic with an insulin pump. The device is programmed to deliver insulin through a hollow tube that projects through the skin and into the body. The pump helps smooth out fluctuations in blood sugar, thus lowering risk of complications that arise from excessively low or high blood sugar.

Credits: © Elizabeth Musar; inset, © Manny Hernandez/ Diabetes Hands Foundation, www.tudiabetes.com.

Take-Home Message

How do pancreatic hormones regulate blood glucose?

- The pancreas secretes two hormones that work in opposition to control the level of glucose in the blood.
- Insulin secreted in response to high blood glucose encourages target cells to take up and store glucose.
- Glucagon secreted in response to low blood glucose causes target cells to break down glycogen to glucose.

glucagon Pancreatic hormone that causes liver cells to break down glycogen and release glucose, thus raising the blood glucose level.

insulin Pancreatic hormone that causes cells to take up glucose from the blood, thus lowering blood glucose level.

25.6 The Adrenal Glands

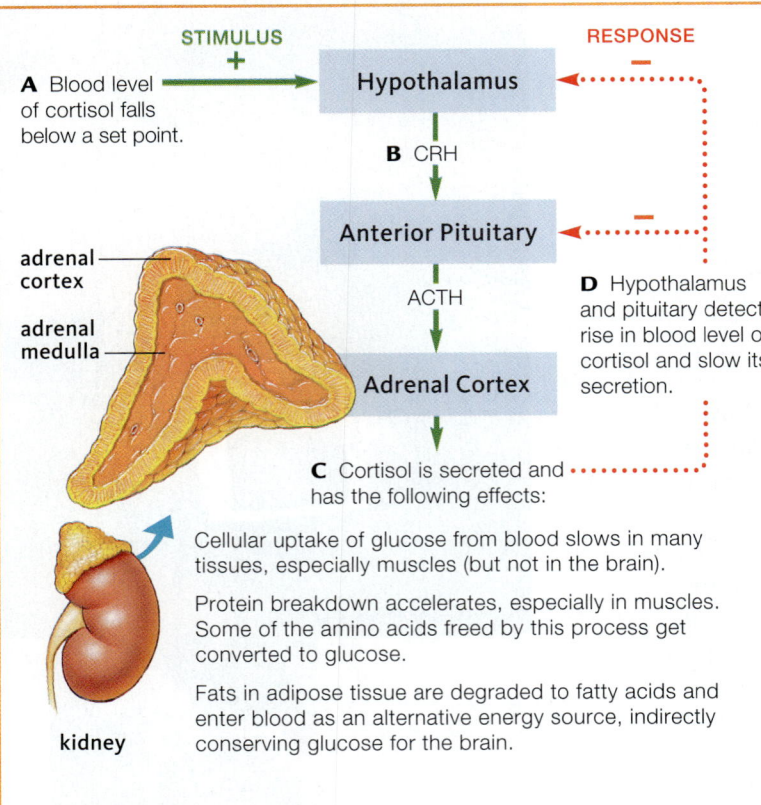

STIMULUS
+

RESPONSE
−

A Blood level of cortisol falls below a set point.

Hypothalamus

B CRH

Anterior Pituitary

ACTH

D Hypothalamus and pituitary detect rise in blood level of cortisol and slow its secretion.

adrenal cortex

adrenal medulla

Adrenal Cortex

C Cortisol is secreted and has the following effects:

Cellular uptake of glucose from blood slows in many tissues, especially muscles (but not in the brain).

Protein breakdown accelerates, especially in muscles. Some of the amino acids freed by this process get converted to glucose.

Fats in adipose tissue are degraded to fatty acids and enter blood as an alternative energy source, indirectly conserving glucose for the brain.

kidney

FIGURE 25.12 Structure and function of the adrenal gland. A negative feedback loop governs cortisol secretion. However, the nervous system can override the feedback controls during times of stress or danger. © Cengage Learning.

On top of each kidney is a grape-sized **adrenal gland**. (In Latin *ad–* means near, and *renal* refers to the kidney.) Each adrenal gland has two regions: an outer **adrenal cortex**, and an inner **adrenal medulla**. The two regions of the gland are controlled by different mechanisms, and they secrete different hormones.

The adrenal cortex secretes aldosterone and cortisol. Aldosterone promotes reabsorption of sodium and water in the kidney. Like ADH, aldosterone concentrates the urine. The hormone **cortisol** has wide-ranging effects on metabolism and immunity. Cortisol induces liver cells to break down their stored glycogen, and suppresses uptake of glucose by other cells. It prompts adipose cells to degrade fats, and skeletal muscles to degrade proteins. The breakdown products of fats and proteins serve as alternative energy sources (Section 5.8).

Under most circumstances, negative feedback loops with the anterior pituitary and hypothalamus maintain blood cortisol levels. Figure 25.12 shows what happens when the level of cortisol in blood decreases below its set point. This decline triggers secretion of corticotropin-releasing hormone (CRH) by the hypothalamus. CRH causes the anterior pituitary to secrete ACTH (adrenocorticotropin). ACTH in turn causes the adrenal cortex to secrete cortisol.

Cortisol secretion rises dramatically with injury, illness, anxiety, or starvation. Under these circumstances, the nervous system overrides the feedback loop that stabilizes the cortisol level, and the cortisol level in the blood soars. At the same time, the adrenal medulla increases its output. The adrenal medulla contains specialized neurons of the sympathetic division (Section 24.4). Like other sympathetic neurons, those in the adrenal medulla release norepinephrine and epinephrine. However, in this case, the norepinephrine and epinephrine do not act as neurotransmitters at a synapse. Instead, they enter the blood and function as hormones. Epinephrine and norepinephrine released into the blood travel to organs, where they have the same effect as direct stimulation by a sympathetic nerve. Remember that sympathetic stimulation plays a role in the fight–flight response. Epinephrine and norepinephrine dilate the pupils, increase the respiratory rate, and make the heart beat faster. They prepare the body to deal with an exciting or dangerous situation.

The increased secretion of cortisol, epinephrine, and norepinephrine during stressful times helps the body deal with an immediate threat by diverting resources from maintenance tasks. In the short term, a fear or stress response can be highly adaptive. When you are threatened with starvation or being pursued by a predator, short-term survival is of paramount importance. However, a chronically high cortisol level can cause health problems such as Cushing's syndrome. Affected people have a puffy, rounded "moon face" and tend to put on fat around their torso. Blood pressure and blood glucose become unusually high. Impaired immunity makes infections common. Skin thins, bone density declines, and muscles shrink. Wounds may be slow to heal. Women's menstrual cycles are erratic or nonexistent, and men may be impotent. Cushing's syndrome is usually caused by use of synthetic steroids such as cortisone shots and prednisone. These drugs are given to relieve pain, inflammation, and other ail-

adrenal cortex Outer portion of adrenal gland; secretes aldosterone and cortisol.
adrenal gland Endocrine gland located above kidney; secretes hormones with roles in urine formation and stress responses.
adrenal medulla Inner portion of the adrenal gland; secretes epinephrine and norepinephrine.
cortisol Adrenal cortex hormone that influences metabolism and immunity; secretions rise with stress.
estrogens Sex hormones produced by a female's ovaries.
gonads Ovaries and testes; organs that produce gametes and secrete sex hormones.
progesterone Sex hormone produced by a female's ovaries.
puberty Period when reproductive organs mature and begin to function.
testosterone Sex hormone produced by a male's testes.

ments. The body converts them to cortisol. A tumor of the pituitary or adrenal gland can also cause the oversecretion of cortisol, thus leading to Cushing's syndrome (Figure 25.13).

Because cortisol levels rise with starvation or improper nutrition, anorexics, bulimics, and people on fad diets tend to have an elevated cortisol level. So do people who are facing anxiety-provoking situations. As a group, poor people have more health problems than those who are well off—even controlling for differences, such as variations in diet and access to health care. By one hypothesis, stress-related high cortisol levels may be an important link between poverty and poor health.

An abnormally low cortisol level, referred to as Addison's disease, is also unhealthy. It can be caused by some infections such as tuberculosis. However, in developed countries, it usually results from autoimmune attacks on the adrenal glands. President John F. Kennedy had the autoimmune form of Addison's disease. Symptoms often include fatigue, weakness, depression, weight loss, and darkening of the skin. If cortisol declines too much, blood sugar and blood pressure can fall to life-threatening levels. People with Addison's disease are treated with a synthetic form of cortisol.

Woman with elevated cortisol levels as a result of an adrenal gland tumor. She has the characteristic puffy moon face.

The same woman after removal of the tumor lowered her cortisol to normal levels.

FIGURE 25.13 Cushing's syndrome.

Permission obtained from Blackwell Publishing © Holt RIG and Hanley NA (2006) *Essential Endocrinology & Diabetes*, edn 5.

Take-Home Message

What are the functions of the adrenal glands?

- The adrenal cortex secretes aldosterone and cortisol. Aldosterone makes the urine more concentrated. Cortisol affects metabolism and the stress response.

- The adrenal medulla releases epinephrine and norepinephrine, which prepare the body for excitement or danger.

25.7 Hormones and Reproductive Behavior

> The Gonads The **gonads** are primary reproductive organs: a male's testes (singular, testis) and a female's ovaries. The testes and ovaries synthesize the same steroid sex hormones—estrogens, progesterone, and testosterone—in different proportions. Ovaries produce mostly estrogen and progesterone. Testes produce mainly testosterone.

Sex hormone production steps up during **puberty**, the period during which the body's reproductive organs and structures mature and begin to function. In females, estrogen secretion results in development of female secondary sexual traits such as enlarged breasts and wider hips. **Estrogens** and **progesterone** also regulate maturation of eggs and prepare the body for pregnancy.

In males, **testosterone** production rises at puberty. The hormone causes development of male secondary sexual traits such as facial hair and regulates production of sperm. The small amount of estrogen produced by the testes may play an essential role in sperm development.

The hypothalamus and anterior pituitary control the secretion of sex hormones (Figure 25.14). In both males and females, the hypothalamus produces GnRH (gonadotropin-releasing hormone). This releaser causes the anterior pituitary to secrete follicle-stimulating hormone (FSH) and luteinizing hormone (LH). FSH and LH cause the gonads to secrete sex hormones. We will return to the subject of sex hormones and their role in reproduction in Chapter 26.

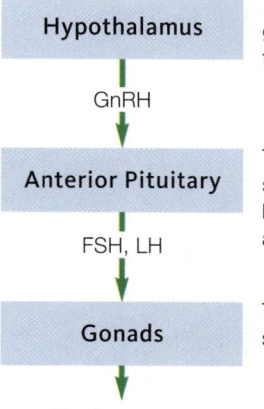

The hypothalamus produces gonadotropin-releasing hormone that acts on the anterior pituitary.

Hypothalamus

GnRH

The anterior pituitary makes follicle-stimulating hormone and luteinizing hormone, both of which target cells in a male's testes or a female's ovaries.

Anterior Pituitary

FSH, LH

The testes or ovaries produce sex hormones.

Gonads

Sex hormones

FIGURE 25.14 Control of sex hormone secretion.

© Cengage Learning.

pineal gland

FIGURE 25.15 Location of the human pineal gland. Exposing the retina to light inhibits secretion of melatonin by this gland. Thus, melatonin secretion typically peaks at about 2 A.M., while we are asleep.

Credits: left, © Cengage Learning; right, © iStockphoto.com/ Diane Diederich.

Routinely working night shifts or having poor sleep habits disrupts melatonin secretion and raises the risk of cancer.

⟩ The Pineal Gland The small, pinecone-shaped **pineal gland** lies deep inside the brain (Figure 25.15). It secretes **melatonin**, but these secretions halt when the retina detects light and signals flow along the optic nerve to the brain. Light varies with time of day and day length varies seasonally, so melatonin secretion rises and falls in daily and seasonal cycles.

Fluctuations in melatonin level influence seasonal behavior in many species. For example, in some male songbirds, long winter nights lead to an increase in melatonin secretion. The rise in melatonin indirectly prevents singing and other courtship behaviors by slowing testosterone secretion. As long as the levels of testosterone stay low, the bird will not sing. In the spring, there are fewer hours of darkness and the melatonin level in the blood declines. With the inhibitory effects of melatonin lifted, the bird's testes secrete additional testosterone. The result is the chorus of birdsong we enjoy in the spring.

Melatonin also affects sex hormone production in humans. A drop in the blood level of melatonin is correlated with the onset of puberty and may trigger it. Also, certain endocrine disorders that increase the production of melatonin delay puberty. Others that lower melatonin production accelerate it.

Melatonin also regulates human sleep–wake cycles, in part by affecting body temperature. Just after sunrise, melatonin secretion decreases, body temperature rises, and we awaken. At night, melatonin secretion causes a decline in body temperature and we become sleepy. Because cycles of melatonin secretion that affect sleep–wake rhythms are set by exposure to light, travelers who fly across many time zones are advised to spend some time in the sun. The exposure to light can help them reset their internal clock and minimize the effects of jet lag.

Some people in latitudes where there are big seasonal shifts in day length have seasonal affective disorder (SAD), commonly called the "winter blues." During winter months, people with SAD tend to feel lethargic and depressed. Researchers have found that affected people secrete more melatonin in winter than in summer, whereas most people secrete the same amount year-round. Exposure to bright artificial light in the early morning is used to treat SAD.

Melatonin has a protective effect against some cancers. It directly inhibits division of cancer cells in animals. It also suppresses production of sex hormones, which encourage the growth of certain cancers. Routinely working night shifts or having poor sleep habits disrupts its secretion and raises the risk of cancer. For example, one study showed that female night-shift nurses have lower

© Steve Byland/ Shutterstock.

melatonin Hormone produced by the pineal gland; affects onset of puberty, sleep–wake cycles.
pineal gland Endocrine gland in the forebrain that secretes melatonin; secretion declines when the eye is exposed to light.

melatonin levels and a higher risk of breast cancer than their dayshift counter-parts. Other studies suggest that a woman's risk of breast cancer decreases with the length of her average night's sleep. In men, night-shift work increases the risk of prostate cancer.

Take-Home Message

What are the endocrine roles of the gonads and the pineal gland?

- A female's ovaries and a male's testes are gonads that make sex hormones as well as gametes. Males make mostly testosterone. Females make mostly estrogens and progesterone.

- In both sexes, sex hormone secretion is regulated by hormones secreted by the pituitary gland, in response to a releaser from the hypothalamus.

- The pineal gland deep inside the brain produces melatonin, which influences sleep–wake cycles and onset of puberty. Melatonin also has a protective effect against some cancers.

25.8 Hormones in the Balance (revisited)

When human cells in culture are exposed to the herbicide atrazine, they increase the rate at which they convert testosterone to estrogen. The resulting rise in estrogen and decrease in testosterone may explain the feminizing effects of this herbicide on aquatic animals. Endocrine disruption may also occur in people who live in agricultural regions, where atrazine commonly taints drinking water. In women, atrazine pollution of drinking water is associated with an increased risk of irregular menstrual cycles.

Other commonly used industrial chemicals called phthalates (pronounced THAL-aytes) also interfere with sex hormone function. Phthalates are widely used in soft plastics, cleansers, and personal care products. Concerns about their effects led the U.S. Consumer Product Safety Commission to ask manufactur-ers to stop using phthalates in rattles and toys designed for teething infants in 1999. The following year, Osvaldo Rosario reported that a high concentration of phthalates in the blood was correlated with premature breast development in young Puerto Rican girls. Figure 25.16 shows an affected girl whose breasts began to develop before she reached the age of two. Phthalates cause this effect by acting like estrogens: They stimulate the development of female secondary sexual traits such as enlarged breasts.

In males, an increased exposure to phthalates is associated with suppressed testosterone secretion. Adult men who have a high blood phthalate concentra-tion tend to have lower sperm quality than those with a lower phthalate level. Phthalates can also disrupt male prenatal development. A woman's chances of giving birth to a son with somewhat feminized genitals, smaller genitals, or an undescended testicle increases with the level of phthalates in her blood.

Pediatrician Sheela Sathyanarayan found that the more scented powders, shampoos, and lotions a mother used on her infant, the higher the level of phthalates in the child's urine. Her findings led the American Academy of Pedi-atrics to recommend that parents choose unscented products for use on infants and young children. Similarly, many obstetricians recommend that pregnant women use personal care products that are labeled as organic, fragrance-free, or free of phthalates. Phthalates do not have to be listed as ingredients if they are a component of a fragrance.

WHERE YOU ARE GOING . . .

Sex hormones play a central role in our animal reproduction and development, the topics of Chapter 26. Plant hormones are the focus of Section 28.6.

FIGURE 25.16 Breast development in a 23-month-old girl. Exposure to phthalates may increase risk of early breast development. Dr. Carlos J. Bourdony.

Summary

Section 25.1 Synthetic chemicals such as phthalates and the weed-killer atrazine are endocrine disruptors, meaning they interfere with the function of some **animal hormones**.

Section 25.2 All vertebrates have an **endocrine system** consisting of endocrine glands and cells that secrete hormones. In most cases, hormones travel through the bloodstream to target cells elsewhere in the body. A cell is a target of a hormone only if it has receptors that can bind that hormone.

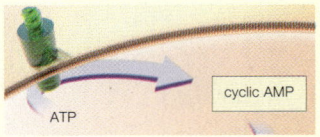

Steroid hormones are lipid soluble; they can enter a target cell and interact directly with DNA. Peptide and protein hormones bind to membrane receptors at the cell surface. Binding commonly leads to formation of a **second messenger** molecule. The second messenger causes a series of enzyme activations in the cytoplasm.

Section 25.3 The **hypothalamus**, a forebrain region, is structurally and functionally linked with the adjacent **pituitary gland**.

The posterior pituitary releases antidiuretic hormone (ADH) and oxytocin made by cells in the hypothalamus. Oxytocin causes contraction of smooth muscle in milk ducts and the uterus. ADH acts on kidney tubules and promotes concentration of the urine.

The anterior pituitary makes and secretes its own hormones: Adrenocorticotropin tells the adrenal cortex to secrete cortisol; thyroid-stimulating hormone calls for thyroid hormone secretion; and follicle-stimulating hormone and luteinizing hormone stimulate hormone production by male and female gonads and have roles in gamete formation. Prolactin stimulates milk production by the mammary glands. **Growth hormone** has growth-promoting effects on cells throughout the body.

Section 25.4 A feedback loop to the anterior pituitary and hypothalamus governs the **thyroid gland** in the base of the neck. Iodine is required for synthesis of thyroid hormone, which increases the metabolic rate and plays an important role in development of the nervous system.

The four **parathyroid glands** located at the rear of the thyroid gland are the main regulators of calcium levels in the blood. They release parathyroid hormone in response to low calcium levels. Parathyroid hormone acts on cells in bone, the kidneys, and the intestine to raise calcium levels in the blood.

Section 25.5 **Insulin** and **glucagon** are pancreatic hormones secreted in response to shifts in the blood level of glucose. Insulin stimulates glucose uptake by muscle and liver cells and thus lowers the blood glucose level. Glucagon stimulates the release of glucose, which increases blood levels. With the metabolic disorder diabetes mellitus, blood glucose remains chronically high because of either insufficient insulin or receptors that do not respond to insulin.

Section 25.6 There is an **adrenal gland** atop each kidney. The **adrenal cortex** secretes aldosterone (which targets the kidney) and **cortisol** (the stress hormone). Cortisol secretion is governed by a negative feedback loop to the anterior pituitary gland and hypothalamus. In times of stress, the nervous system overrides feedback controls over cortisol secretion so cortisol level rises.

Norepinephrine and epinephrine released by neurons of the **adrenal medulla** influence organs the same way that sympathetic stimulation does. They cause a fight–flight response.

Section 25.7 **Gonads** (ovaries and testes) secrete **estrogens**, **progesterone**, and **testosterone**. These sex hormones are steroids that act in reproduction and in development of secondary sexual traits.

Melatonin secretion by the vertebrate **pineal gland** varies seasonally because exposure to light suppresses it. In humans, it affects the daily sleep–wake cycle and the onset of **puberty**. Disrupted melatonin secretion raises the risk of some cancers.

Self-Quiz   Answers in Appendix I

1. _____ , signaling molecules released by endocrine cells and glands, enter the blood and are distributed through the body.
   a. Hormones
   b. Neurotransmitters
   c. Endocrine disruptors
   d. all of the above

2. Antidiuretic hormone and oxytocin are hormones produced in the hypothalamus but released from the _____ .
   a. pineal gland
   b. pituitary gland
   c. pancreas
   d. thalamus

3. Protein hormones typically bind to receptors _____ .
   a. in the DNA
   b. in the cytoplasm
   c. at the plasma membrane
   d. both a and c

4. Match each pituitary hormone with its target.
   _____ antidiuretic hormone   a. gonads (ovaries, testes)
   _____ oxytocin   b. mammary glands, uterus
   _____ luteinizing hormone   c. kidneys
   _____ growth hormone   d. most body cells

5. Overproduction of _____ causes acromegaly and gigantism.
   a. melatonin
   b. insulin
   c. growth hormone
   d. cortisol

6. The _____ regulate(s) calcium levels in the blood.
   a. hypothalamus
   b. pancreas
   c. pineal gland
   d. parathyroid glands

7. _____ lowers blood sugar level; _____ raises it.
   a. Glucagon; insulin
   b. Insulin; glucagon
   c. Melatonin; insulin
   d. Cortisol; glucagon

8. A rise in the concentration of thyroid hormone in the blood slows the release of thyroid hormone. This is an example of _____ .
   a. positive feedback
   b. negative feedback

9. The _____ produces digestive enzymes and hormones.
   a. hypothalamus
   b. pancreas
   c. pineal gland
   d. parathyroid gland

10. Thyroid hormone _____ .
    a. requires vitamin D for activity
    b. influences development of the nervous system
    c. is secreted in response to signals from the adrenal gland
    d. all of the above

11. True or false? Only women make follicle-stimulating hormone and luteinizing hormone.

12. A person with an overly active thyroid gland is more likely to be unusually _____ .
    a. heavy   c. cold
    b. anxious   d. both a and b

13. During stressful situations, the adrenal glands increase their output of _____ .
    a. cortisol   c. norepinephrine
    b. epinephrine   d. all of the above

14. The male sex hormone testosterone is secreted in response to secretion of hormones by the _____ .
    a. testes   c. pituitary gland
    b. ovaries   d. pancreas

15. Match the hormone source listed at left with the most suitable description at right.
    _____ adrenal cortex
    _____ thyroid gland
    _____ pineal gland
    _____ parathyroid glands
    _____ pancreatic islets
    _____ posterior pituitary
    _____ hypothalamus
    _____ testes

    a. make gametes and hormones
    b. major control center
    c. influence blood calcium level
    d. stress increases secretions
    e. light inhibits secretion
    f. hormones require iodine
    g. regulate blood sugar level
    h. secretes hormones made by cell bodies in the hypothalamus

Critical Thinking

1. A woman affected by type 1 (insulin-dependent) diabetes miscalculates and injects herself with too much insulin. She soon begins to feel confused and sluggish. She calls for medical assistance and injects herself with the glucagon her doctor prescribed for such an emergency. When an ambulance arrives, she is given dextrose (a sugar) intravenously. How did the excess insulin cause her symptoms? How did the glucagon injection and the intravenous sugar help reverse them?

2. A tumor in the pituitary gland can cause a woman to produce milk even when she is not pregnant. In which lobe of the pituitary would such a tumor be located, and what hormone would it affect?

3. Women who are completely blind tend to undergo puberty at an earlier age than sighted women. They also are less likely to have breast cancer. By one hypothesis, blindness encourages early puberty and discourages breast cancer by its effect on melatonin secretion. Based on this information, would you expect the average melatonin level in blind women to be higher or lower than that of sighted women? Explain your answer.

4. Sex hormone secretion is governed by a negative feedback loop to the hypothalamus and pituitary, similar to that for thyroid hormone or cortisol. Because of this, a veterinarian can use a blood test

Digging Into Data

Sperm Counts Down on the Farm

Evidence that contamination of water by agricultural chemicals affects the reproductive function of some aquatic animals has raised concerns that these chemicals may also affect human health or reproduction. To look into possible reproductive effects of chemical exposure, epidemiologist Shanna Swan and her colleagues studied sperm collected from men in four cities in the United States that vary in their proximity to agricultural fields (Figure 25.17). All of the men were partners of women who had become pregnant and were visiting a prenatal clinic, so all were fertile. Of the four cities, Columbia, Missouri, is located in the county with the most farmlands. New York City in New York is in an area with no agriculture.

1. In which cities did researchers record the highest and lowest sperm counts?

2. In which cities did samples show the highest and lowest sperm motility (ability to move)?

3. Aging, smoking, and sexually transmitted diseases adversely affect sperm. Could differences in any of these variables explain the regional differences in sperm count?

4. Do these data support the hypothesis that living near farmlands can adversely affect male reproductive function?

	Location of clinic			
	Columbia, Missouri	Los Angeles, California	Minneapolis, Minnesota	New York, New York
Average age	30.7	29.8	32.2	36.1
Percent nonsmokers	79.5	70.5	85.8	81.6
Percent with history of STD	11.4	12.9	13.6	15.8
Sperm count (million/ml)	58.7	80.8	98.6	102.9
Percent motile sperm	48.2	54.5	52.1	56.4

FIGURE 25.17 Data from a study of sperm collected from male partners of pregnant women who visited prenatal health clinics in one of four cities. STD stands for sexually transmitted disease. © Cengage Learning.

to find out whether a female dog has been neutered. Dogs who still have their ovaries have a lower blood level of luteinizing hormone (LH) than dogs who have been neutered. Explain why removing a dog's ovaries would result in an elevated level of LH.

5. The chemical bisphenol A (BPA) is used to make plastic bottles and to line cans. Evidence from animal experiments and epidemiological studies indicates that the risk of obesity rises with increased BPA intake. How might endocrine-disrupting effects of such a chemical influence body weight?

Reproduction and Development 26

This chapter returns to sexual reproduction and meiosis (Sections 8.6 and 8.7), cell differentiation (7.7), prenatal testing (9.9), sex hormones (25.7), and the viruses HIV (13.1, 22.7) and HPV (22.1).

A Fertility specialist injecting a single sperm into an egg. The video screen shows a magnified view of the egg.

26.1 Assisted Reproduction

A couple who cannot conceive naturally or a woman who wishes to use sperm from a nonpartner donor may now choose to conceive through *in vitro* fertilization (IVF), an assisted reproductive technique that combines an egg and sperm outside the body. To obtain eggs for IVF, a woman is given hormones that encourage many eggs to mature at once. The mature eggs are removed from her ovaries via a hollow needle. They are then combined with sperm for fertilization (Figure 26.1A). After fertilization, each zygote undergoes mitotic divisions, forming a microscopic ball of cells that can be placed in a woman's womb to develop to term.

Louise Brown was the first child to be born after conception by IVF. Her birth in 1978 shocked scientists and the public alike. Many were appalled by the idea of what the media referred to as "test tube babies." Scientists expressed concern that this unnatural procedure would produce children with psychological and genetic defects. Ethicists and religious leaders warned about the dire societal implications of manipulating human embryos.

Despite these initial reservations, IVF has become widely accepted and practiced. Worldwide, the technique has resulted in the birth of more than 3 million children. The first test tube babies are now adults and have begun having children of their own. Louise Brown is among them. In 2007, at age 28, she conceived naturally and gave birth to a healthy son (Figure 26.1B).

Research into IVF has also opened the way to a variety of assisted reproductive technologies. If a man makes sperm, but does not ejaculate them or does not ejaculate a large enough number to allow fertilization by normal means, intracytoplasmic sperm injection can put his sperm inside his partner's egg. A woman who does not make viable eggs but who wishes to carry a child can be implanted with an early embryo derived from an egg donor. A woman who produces viable eggs but cannot, or does not want to, carry them herself can have her egg fertilized by IVF. The resulting blastocyst can then be implanted in a surrogate mother. Eggs, sperm, and embryos can also be frozen for years before being used. With all IVF procedures, prospective parents can screen embryos for genetic defects before they are implanted (Section 9.9).

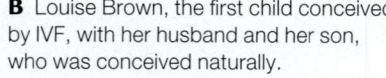

B Louise Brown, the first child conceived by IVF, with her husband and her son, who was conceived naturally.

FIGURE 26.1 *In vitro* fertilization.

Credits: (a) © Heidi Specht, West Virginia University; (b) © Craig Hibbert/ Mail on Sunday/ Solo Syndication.

26.2 Modes of Reproduction

> **Asexual Reproduction** With **asexual reproduction**, a single individual produces offspring that are genetically identical to the parent and one another. As a result, the parent has all its genes represented in each offspring. Asexual reproduction can be advantageous in a stable environment where gene combinations that make the parent successful can be expected to do the same for offspring. However, being locked into a particular gene combination can be a disadvantage in a changing environment.

Many invertebrates reproduce asexually. Some do this by fragmentation—a piece breaks off and grows into a new individual. In others, new individuals bud from existing ones (Figure 26.2**A**). In some insects, fish, amphibians, birds, and lizards, new individuals can develop from unfertilized eggs by a process known as **parthenogenesis**. No mammals are known to reproduce asexually.

> **Sexual Reproduction** With **sexual reproduction**, two parents produce gametes that combine at fertilization. As a result of the gene shuffling during meiosis and fertilization, each of the resulting offspring has a unique combination of paternal and maternal genes (Section 8.7). Thus, by reproducing sexually, a parent increases the likelihood that some of its offspring will inherit a combination of genetic traits that suits them to a changed environment.

Some species switch between sexual and asexual reproduction. Plant-sucking insects called aphids provide one example. During the summer, a female aphid settles on a plant and gives birth to many wingless daughters that develop inside the mother's body from unfertilized eggs. An aphid's daughters live on the same plant as their mother, or on an adjacent one, so they all experience more or less similar conditions. In autumn, males are produced and sexual reproduction yields genetically variable females. These females spend the winter in a resting state. In the spring, they seek out new plants on which to raise the new generation of genetically identical daughters.

> **Variations on Sexual Reproduction** Sexually reproducing animals that can produce both eggs and sperm are called **hermaphrodites**. Simultaneous hermaphrodites produce eggs and sperm at the same time. Tapeworms and some roundworms are simultaneous hermaphrodites that can fertilize themselves. Earthworms and slugs are simultaneous hermaphrodites too, but they require a partner (Figure 26.2**B**). During mating, each slug both donates sperm to and receives sperm from its partner. Some other mollusks and certain fishes are sequential hermaphrodites, meaning they can switch from one sex to another during the course of a lifetime. More typically, vertebrates have separate sexes that remain fixed for life, with each individual making either eggs or sperm.

Fertilization may be internal or external. With external fertilization, large numbers of gametes are released into the water. Sperm swim to eggs, and development typically occurs in the environment. Fertilization is external in most aquatic invertebrates and bony fishes. With internal fertilization, sperm are deposited directly into a female's reproductive tract (Figure 26.2**C**). Fertilization is internal in cartilaginous fishes, some bony fishes, and most land animals, including insects, slugs, and the amniotes (reptiles, birds, and mammals).

Eggs that have been fertilized internally may be released to develop in the external environment, as is the case in birds, or they may develop inside their mother's body, as occurs in cartilaginous fishes, some lizards and snakes, and most mammals. Depending on the group, young that develop inside their mother may be nourished by nutrient reserves within the egg, by nutrients that diffuse from their mother, or by some combination of the two.

A Asexual reproduction in a hydra. A new individual (*left*) is budding from the body of its parent.

B Sexual reproduction in banana slugs. Each slug both donates sperm to and receives sperm from its partner.

C Sexual reproduction in elephants. The male is inserting his penis into the female. Eggs will be fertilized and the offspring will develop inside the mother's body, nourished by nutrients delivered by her bloodstream.

FIGURE 26.2 Animal reproduction.

Credits: (a) Biophoto Associates/ Photo Researchers, Inc.; (b) Courtesy of Christine Evers; (c) © Gabriela Staebler Wildlife Photography.

asexual reproduction Reproductive mode by which offspring arise from a single parent only.

hermaphrodite Individual animal that produces both eggs and sperm.

in vitro **fertilization** Procedure in which eggs are removed from a woman's ovary and fertilized outside her body.

parthenogenesis Individuals develop from unfertilized eggs.

sexual reproduction Reproductive mode by which two parents produce offspring having a combination of their genes.

A Eggs form and mature in female reproductive organs. Sperm form and mature in male reproductive organs.

Gamete Formation

B A sperm penetrates an egg. Their nuclei fuse. A zygote has formed.

Fertilization

C Mitotic cell divisions form a ball of cells, a blastula. Each cell gets regionally different parts of the egg cytoplasm.

Cleavage

D A gastrula, an early embryo that has primary tissue layers, forms by cell divisions, cell migrations, and rearrangements.

Gastrulation

E Details of the body plan fill in as different cell types interact and form tissues and organs in predictable patterns.

Organ Formation

F Organs grow in size, take on mature form, and gradually assume specialized functions.

Growth, Tissue Specialization

FIGURE 26.3 Overview of reproductive and developmental processes that occur in animals with tissues and organs. © Cengage Learning.

apoptosis Programmed cell death.

blastula Hollow ball of cells that forms early in animal development.

cleavage Mitotic division of cells in an early embryo.

differentiation Process by which cells become specialized; occurs as different cell lineages express different subsets of their genes.

ectoderm Outermost tissue layer of an animal embryo.

endoderm Innermost tissue layer of an animal embryo.

fertilization Egg and sperm combine and form a zygote.

gastrula Three-layered body formed when an animal blastula undergoes gastrulation during early development.

gastrulation Animal developmental process by which cell movements produce a three-layered gastrula.

mesoderm Middle tissue layer of a three-layered animal embryo.

yolk Nutritious material in many animal eggs.

zygote Cell formed by fusion of gametes; first cell of a new individual.

❯ Stages in Reproduction and Development Animals develop through a series of stages. Figure 26.3 shows the stages of reproduction and development typical of animals that have tissues and organs.

Reproduction begins with gamete formation. Gametes (eggs or sperm) form by meiosis of germ cells in primary reproductive organs (gonads). A sperm consists of paternal DNA and a bit of cellular equipment that helps it reach and penetrate an egg. It has little cytoplasm. An egg is much larger than a sperm and more complex internally. As it matures, enzymes, mRNA transcripts, and other materials become stored in its cytoplasm. Most animal eggs also include **yolk**, a mixture of lipids and proteins that nourishes a developing individual.

Fertilization begins when a sperm penetrates an egg. It ends when sperm and egg nuclei fuse, forming a **zygote**, the first cell of the new individual. A zygote's cytoplasm is almost entirely derived from the egg.

Developmental processes transform a single-celled zygote into an adult animal. Figure 26.4 shows one example of vertebrate development. After fertilization, development begins with **cleavage**, during which mitotic cell divisions increase the number of cells without increasing a zygote's original volume ❶. Materials are not randomly distributed in the egg cytoplasm, so cleavage distributes different mRNAs to different cells. Which of these "maternal messages" are inherited helps determine the fate of a cell lineage. For example, cells of one lineage alone might receive maternal mRNA that turns on a master gene that directs development in a particular way (Section 7.7).

Cleavage ends with the formation of the **blastula**, a hollow ball of cells around a fluid-filled center ❷. After the blastula forms, cell divisions slow and the blastula undergoes **gastrulation**, a process of structural reorganization during which some cells on the blastula's surface move inward. These cellular rearrangements produce a **gastrula**, an embryonic form with three primary tissue layers (also called germ layers) ❸. Cellular descendants of these tissues give rise to all tissues and organs of the adult.

Ectoderm, the outermost primary tissue, emerges first. Its cellular descendants become nervous tissue and the skin's epidermis. The innermost layer of **endoderm** is the source of the gut's inner lining and organs derived from it. In most animals, **mesoderm** forms between the outer and inner primary tissue layers. It gives rise to muscles, most of the skeleton, the circulatory, reproductive, and excretory systems, and connective tissues of the gut and skin.

Because all cells of an embryo descended from the same zygote, they all have the same number and kinds of genes. How then, do different kinds of cells arise? During development, different cell lineages begin to express some groups of genes and not others. As Section 7.7 explained, selective gene expression begins the process of **differentiation**, by which cell lineages become specialized. For example, all cells in your body express the genes for enzymes that carry out glycolysis. However, only cells in the lens of your eye turned on the gene for crystallin, a transparent protein.

Organ formation begins as different tissues arise and become organized in predictable patterns ❹. Growth and tissue specialization is the final stage of animal development ❺. This stage usually continues into early adulthood. Tissues and organs develop in a genetically programmed, orderly sequence. A body takes on its characteristic form as cells of various lineages divide, migrate, alter their size and shape, and die at specific times and in specific locations. The tail of a larval frog (a tadpole) disappears during metamorphosis as cells of the tail commit suicide on cue, a process known as **apoptosis**. As you will see, a similar process of cell suicide shapes a human hand in the early embryo. The hand starts out as a paddlelike structure in which webs of flesh connect the digit. Death of cells in the webbing sculpts a five-fingered hand.

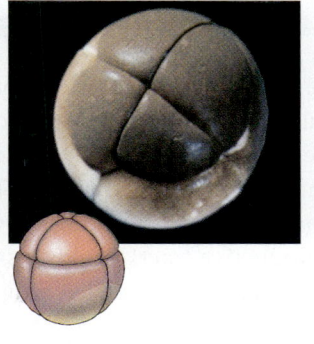

gray
crescent

blastocoel

blastula

1 Here we show the first three divisions of cleavage, a process that carves up the zygote's cytoplasm. Cleavage results in a blastula, a ball of cells with a fluid-filled cavity.

2 Cleavage is over when the blastula forms.

  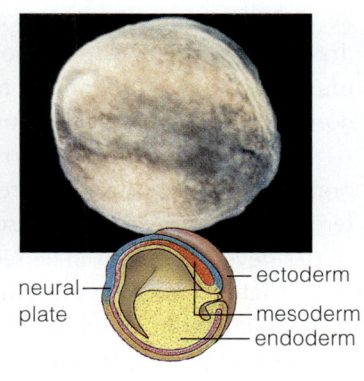

ectoderm
dorsal lip
future gut cavity

yolk
plug

neural plate
ectoderm
mesoderm
endoderm

neural tube

notochord
gut cavity

3 The blastula becomes a three-layered gastrula by a process called gastrulation. At the dorsal lip, a fold of ectoderm above the first opening that appears in the blastula, cells migrate inward and start rearranging themselves.

4 Organs begin to form as a primitive gut cavity opens up. A neural tube, then a notochord and other organs form from the primary tissue layers.

Tadpole, a swimming larva with segmented muscles and a notochord extending into a tail.

Limbs grow and the tail is absorbed during metamorphosis to the adult form.

Sexually mature, four-legged adult leopard frog.

5 The frog's body form changes as it grows and its tissues specialize. The embryo becomes a tadpole, which metamorphoses into an adult.

FIGURE 26.4 Animated!
Stages in the development of a vertebrate, the leopard frog.

Credits: (1–4) photos, Carolina Biological Supply Company; art, © Cengage Learning; (5) left, center, © David M. Dennis/ Tom Stack & Associates, Inc.; right, © John Shaw/ Tom Stack & Associates.

Take-Home Message

How do animals reproduce and develop?

- Some animals reproduce asexually, producing offspring identical to one another and to their parents. Most animals reproduce sexually, and so produce genetically variable offspring.

- In sexually reproducing animals, gametes combine and form a zygote. Cleavage (a series of mitotic divisions) produces a blastula, a hollow ball of cells.

- Gastrulation, a series of cellular rearrangements, yields a three-layered gastrula. Cells of the gastrula differentiate to form specialized tissues.

- Cell migrations and programmed cell death are among the processes that shape organs and give rise to the final body form.

26.3 The Human Reproductive System

We turn now to details of reproduction and development in our own species. Human gametes form by meiosis inside pairs of primary sex organs, or gonads. Sperm form inside male gonads, the **testes** (singular, testis). Eggs form inside female gonads, the **ovaries**. The male reproductive tract also has components that store sperm and that move them from the male's body into the female reproductive tract. A female's reproductive system has components that receive sperm and sustain developing offspring.

❯ **Male Reproductive Anatomy** Both male and female gonads form deep inside the body. Before a male is born, his testes descend into the **scrotum**, a pouch of skin-covered smooth muscle suspended below the pelvic girdle (Figure 26.5). When a man becomes cold or frightened, reflexive muscle contractions draw the testes closer to his body. When he feels warm, relaxation of smooth muscle in the scrotum allows his testes to hang lower, so the sperm-making cells do not overheat. These cells function best just below normal body temperature.

When a male enters puberty (the stage of development when reproductive organs mature), his testes increase their production of the sex hormone testosterone (Section 25.7). They also begin making sperm. The increased testosterone causes development of male secondary sexual traits such as thickened vocal cords that deepen the voice; increased growth of hair on the face, chest, armpits, and pubic region; and increased muscle mass.

Sperm that form in a testis are pushed by cilia into the epididymis (plural, epididymides), a coiled duct perched on a testis. The Greek *epi–* means upon and *didymos* means twins. In this context, the "twins" refers to the two testes. Secretions from the epididymis wall nourish the sperm and help them mature.

A Frontal view, showing position of reproductive organs relative to pelvic girdle and bladder.

bladder

FIGURE 26.5 Animated! Male reproductive anatomy.

Figure It Out: Which glandular structure do sperm first travel past during ejaculation: a seminal vesicle, the bulbourethral gland, or the prostate gland?

© Cengage Learning.   Answer: A seminal vesicle

Prostate Gland
An exocrine gland that contributes some fluid to the semen

urinary bladder

Urethra
Duct with dual functions: channel for ejaculation of sperm during sexual arousal and for excretion of urine at other times

cylinders of spongy tissue that swell with blood during an erection

urethra

Penis
Male organ of sexual intercourse

Testis
One of a pair of gonads, packed with small, sperm-producing tubes (seminiferous tubules) and cells that secrete testosterone and other sex hormones

scrotum

anus

Ejaculatory Duct
One pair of ducts that carry sperm to the penis

Seminal Vesicle
One of a pair of exocrine glands that contributes fructose-rich fluid to semen

Bulbourethral Gland
One of a pair of exocrine glands that secrete mucus

Vas Deferens
One of a pair of ducts that carry sperm to the penis

Epididymis
One of a pair of ducts in which sperm mature and are stored

B Cutaway side view, with cross-section of penis.

The last region of each epididymis stores mature sperm and is continuous with the first portion of a **vas deferens** (plural, vasa deferentia). In Latin, *vas* means vessel, and *deferens*, to carry away. A vas deferens is a duct that carries sperm away from an epididymis, and to a short ejaculatory duct. Ejaculatory ducts deliver sperm to the urethra, the duct that extends through a male's penis to an opening at the body surface.

The penis is the male organ of intercourse. It has a rounded head (the glans) at the end of a narrower shaft. Nerve endings in the glans make it highly sensitive to touch. Normally, when a man is not sexually excited, a retractable tube of skin called the foreskin covers the glans of his penis. In many cultures, males undergo circumcision, an elective surgical procedure that removes the foreskin.

Beneath the skin of the penis, connective tissue encloses three elongated cylinders of spongy tissue. When a male becomes sexually excited, signals from the nervous system cause blood to flow into the spongy tissue faster than it flows out. As fluid pressure rises, the normally limp penis becomes erect.

Sperm stored in the epididymides continue their journey toward the body surface only when a male reaches the peak of sexual excitement and ejaculates. During **ejaculation**, smooth muscle in the walls of the epididymides and vasa deferentia undergoes rhythmic contractions that propel sperm and accessory gland secretions out of the body as a thick, white fluid called semen. **Semen** is a complex mixture of sperm, proteins, nutrients, ions, and signaling molecules. Sperm account for less than 5 percent of its volume. The other 95 percent is accessory gland secretions. **Seminal vesicles**, which are exocrine glands near the base of the bladder, secrete fructose-rich fluid into the vasa deferentia. Sperm use the fructose (a sugar) as their energy source. Another exocrine gland, the **prostate gland,** is the other major contributor to semen volume. It encircles the urethra and secretes fluid into it. Prostate gland secretions help raise the pH of the female reproductive tract, so it is more hospitable to sperm.

A young, healthy man has a prostate gland about the size of a walnut, but it may enlarge with inflammation or age. Because the urethra runs through the prostate gland, prostate enlargement can narrow this duct and cause difficulty urinating. When this happens, medication, laser treatments, or surgery can alleviate symptoms. An enlarged prostate can also be a sign of cancer. Doctors diagnose prostate cancer with blood tests and by physical examination. The blood test detects compounds made only by prostate cancers. The physical exam involves inserting a finger into the rectum and feeling the prostate for any bumps or size change that could indicate the presence of cancer. Most prostate cancers grow relatively slowly and those in older men often do not require treatment. However, some prostate cancers grow quickly and can spread.

Testicular cancer is the most common cancer among men aged 15 to 34. To detect it, a man should examine his testes once a month, after a warm shower or bath, to look and feel for lumps, enlargement, or hardening.

> ### How Sperm Form
Although smaller than a golf ball, a testis holds 125 meters of coiled seminiferous tubules (Figure 26.6**A**). Diploid male germ cells line the inner wall of each tubule (Figure 26.6**B**). These cells divide repeatedly by mitosis, and their offspring differentiate into primary spermatocytes that undergo meiosis to form immature sperm. Large, elongated nurse cells inside seminiferous tubules surround and support the sperm as they develop.

A mature sperm is a haploid, flagellated cell (Figure 26.6**C**). The flagellum allows the sperm to swim toward an egg. Mitochondria in an adjacent midpiece supply the energy required for movement. The "head" of the sperm contains DNA and has an enzyme-containing cap. These enzymes help the sperm penetrate an egg during fertilization.

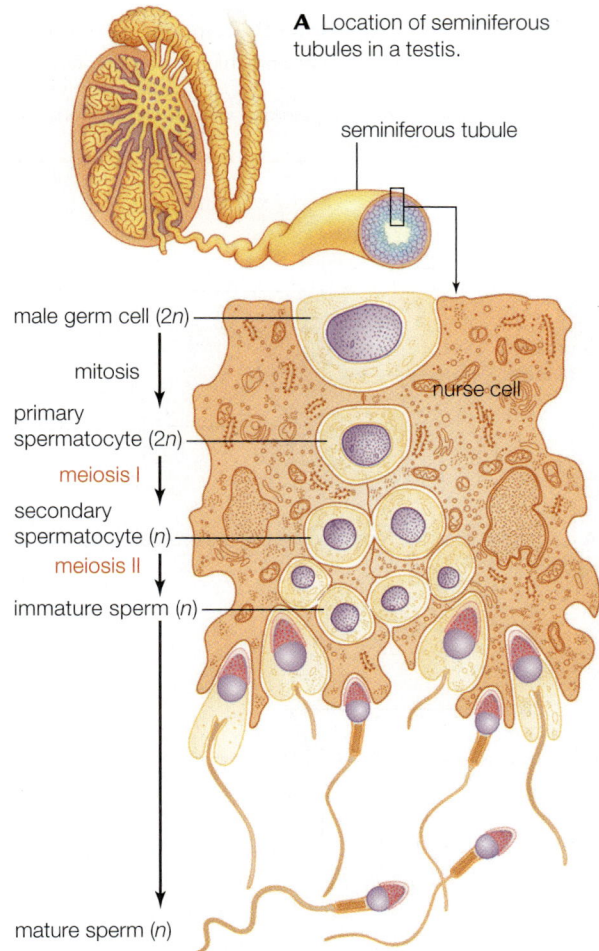

A Location of seminiferous tubules in a testis.

seminiferous tubule

male germ cell (2*n*)

mitosis

nurse cell

primary spermatocyte (2*n*)

meiosis I

secondary spermatocyte (*n*)

meiosis II

immature sperm (*n*)

mature sperm (*n*)

B Sperm formation inside a tubule.

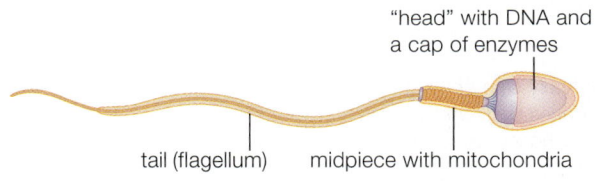

"head" with DNA and a cap of enzymes

tail (flagellum)   midpiece with mitochondria

C Structure of a mature sperm.

> **FIGURE 26.6 Animated!** How sperm form.
> © Cengage Learning.

ejaculation Smooth muscle contractions that propel semen through the penis and out of the body.

ovary Organ in which eggs form.

prostate gland Exocrine gland that encircles a male's urethra; its secretions contribute to semen.

scrotum Pouch that holds the testes.

semen Sperm mixed with fluid secreted by exocrine glands.

seminal vesicles Exocrine glands that secrete fluid into the vasa deferentia; main source of semen volume.

testes In animals, organs that produce sperm.

vas deferens Main duct that conveys mature sperm toward the urethra.

FIGURE 26.7 Animated! Components of the female reproductive system and their functions.

Credits: (a,c) From RUSSELL/WOLFE/HERTZ/STARR. Biology, 1E. © 2008 Brooks/Cole, a part of Cengage Learning, Inc. Reproduced by permission. www.cengage.com/permissions; (b) © Cengage Learning.

A Location of female reproductive organs, frontal view.

B External sex organs, collectively referred to as the vulva.

The same hypothalamic and pituitary hormones control gamete formation in both sexes. Gonadotropin-releasing hormone (GnRH) secreted by the hypothalamus stimulates anterior pituitary cells to secrete luteinizing hormone (LH) and follicle-stimulating hormone (FSH). In males, LH and FSH target cells inside the testes and encourage testosterone secretion and sperm production.

❯ Female Reproductive Anatomy

A female's almond-sized ovaries lie deep inside her body (Figure 26.7**A,C**). A ciliated **oviduct** connects each ovary to the **uterus**, a hollow, pear-shaped organ above the urinary bladder. The uterus is commonly called the womb. If fertilization occurs, an embryo forms and develops inside the uterus. A thick layer of smooth muscle makes up most of the uterine wall. The uterine lining (the endometrium) consists of glandular epithelium, connective tissues, and blood vessels. The lowest portion of the uterus narrows, forming the **cervix**, which connects to the vagina.

The **vagina** is a muscular tube that extends from the cervix to the body's surface. It functions as the female organ of intercourse and also as the birth canal. Typically, when a girl is born, a membrane called the hymen partially covers the external entrance to her vagina. The hymen thins as a girl approaches puberty. It is usually stretched open during the first episode of intercourse, if it has not previously been broken by other physical activity.

The vulva is the externally visible portion of the female reproductive system (Figure 26.7**B**). Two pairs of skin folds enclose vaginal and urethral openings. Fatty tissue fills the outer folds (the labia majora). Thin inner folds (the labia minora) have a rich blood supply and swell during sexual arousal. The tip of an erectile organ called the clitoris lies at the anterior junction of the labia minora. The clitoris and penis develop from the same embryonic tissue. Both have an abundance of highly sensitive touch receptors, and both swell with blood and become erect during sexual arousal. Only a tiny portion of the clitoris is external and visible. The bulk of the organ consists of erectile tissue that wraps around the walls of the vagina.

Ovary One of two female gonads. Makes eggs and secretes female sex hormones (estrogens and progesterone).

Oviduct One of a pair of ducts through which oocytes are propelled from an ovary to the uterus; usual site of fertilization.

Uterus Womb, chamber in which an embryo develops. Includes myometrium (smooth muscle layer) and endometrium (epithelial lining). Narrowed lower portion (the cervix) secretes mucus into the vagina.

Vagina Organ of sexual intercourse: birth canal.

Clitoris Highly sensitive erectile organ. Only the tip is externally visible; bulk of the organ extends internally on either side of the vagina.

Labium minus One of a pair of inner skin folds (the labia minora).

Labium majus One of a pair of fatty outer skin folds (the labia majora).

C Frontal view, showing position of reproductive organs.

oocyte

ovary

follicle

surgical instrument

corpus luteum

ovary

secreted protein — primary oocyte

follicle cells

ovarian follicle

first polar body

secondary oocyte

ovulation

corpus luteum

FIGURE 26.8 Animated! The ovarian cycle.

Credits: left, © Cengage Learning; (4) Courtesy of Dr. M. Jacques Donnez and Dr. Jean-Christophe.Lousse, Catholic University of Louvain, Belgium. From *Fertility and Sterility*, "Laparoscopic observation of spontaneous human ovulation," Volume 90, Issue 3, Pages 833–834, September 2008; (5) Courtesy of Ed Uthman.

1 An ovarian follicle begins to mature. The primary oocyte enlarges and secretes proteins, and follicle cells around it divide.

2 A fluid-filled cavity begins to form in the follicle's cell layer.

3 Primary oocyte completes meiosis I and undergoes unequal cytoplasmic division to form a secondary oocyte and the first polar body.

4 Ovulation. The follicle wall ruptures, releasing the secondary oocyte and surrounding cells. The photo (*above left*) was taken by a surgeon who was performing a procedure on a patient and noticed that she was ovulating.

5 Follicle cells left behind after ovulation develop into a corpus luteum that secretes hormones. The photo (*above right*) shows a normal human ovary removed from a woman who had cancer of the uterus.

6 If pregnancy does not occur, the corpus luteum degenerates.

> **Egg Formation and the Ovarian Cycle** Unlike male germ cells, female germ cells do not divide after birth. A girl is born with about 2 million primary oocytes in her ovaries. An **oocyte** is a general term for an immature egg. A primary oocyte is an immature egg that has entered meiosis but stopped in prophase I. When a female reaches puberty, her primary oocytes begin to mature, one at a time, in an approximately 28-day ovarian cycle.

Figure 26.8 shows how an oocyte matures during the course of this cycle. A primary oocyte and the cells around it constitute an **ovarian follicle** 1. In the first part of the ovarian cycle, cells around the oocyte divide repeatedly as the oocyte enlarges and secretes a layer of proteins. As the follicle matures, a fluid-filled cavity opens around the oocyte 2. Often more than one follicle begins to develop, but typically only one becomes fully mature. In that follicle, the primary oocyte completes meiosis I and undergoes unequal cytoplasmic division. The result is a large secondary oocyte and a tiny polar body 3. The polar body has no reproductive function and will later disintegrate. The secondary oocyte goes on to enter meiosis II, but halts in metaphase II.

About two weeks after the onset of the cycle, the follicle wall ruptures and **ovulation** occurs: The secondary oocyte and polar body are ejected into the adjacent oviduct 4. Once in the oviduct, an oocyte must meet up with sperm within 12 to 24 hours for fertilization to occur. It will not complete meiosis until it has been penetrated by a sperm.

Meanwhile, back in the ovary, the cells of the ruptured follicle develop into a hormone-secreting structure called the **corpus luteum** 5. (The name means "yellow body" in Latin and refers to its yellowish color.) If pregnancy does not occur, the corpus luteum breaks down 6. Once the corpus luteum is gone, a new follicle can begin to mature.

cervix Narrowed region of uterus; connects to vagina.
corpus luteum Hormone-secreting structure that forms from follicle cells after ovulation.
oocyte Immature egg.
ovarian follicle Immature egg and surrounding cells.
oviduct Ciliated tube connecting an ovary to the uterus.
ovulation Release of a secondary oocyte from an ovary.
uterus Muscular chamber where offspring develop; womb.
vagina Female organ of intercourse and birth canal.

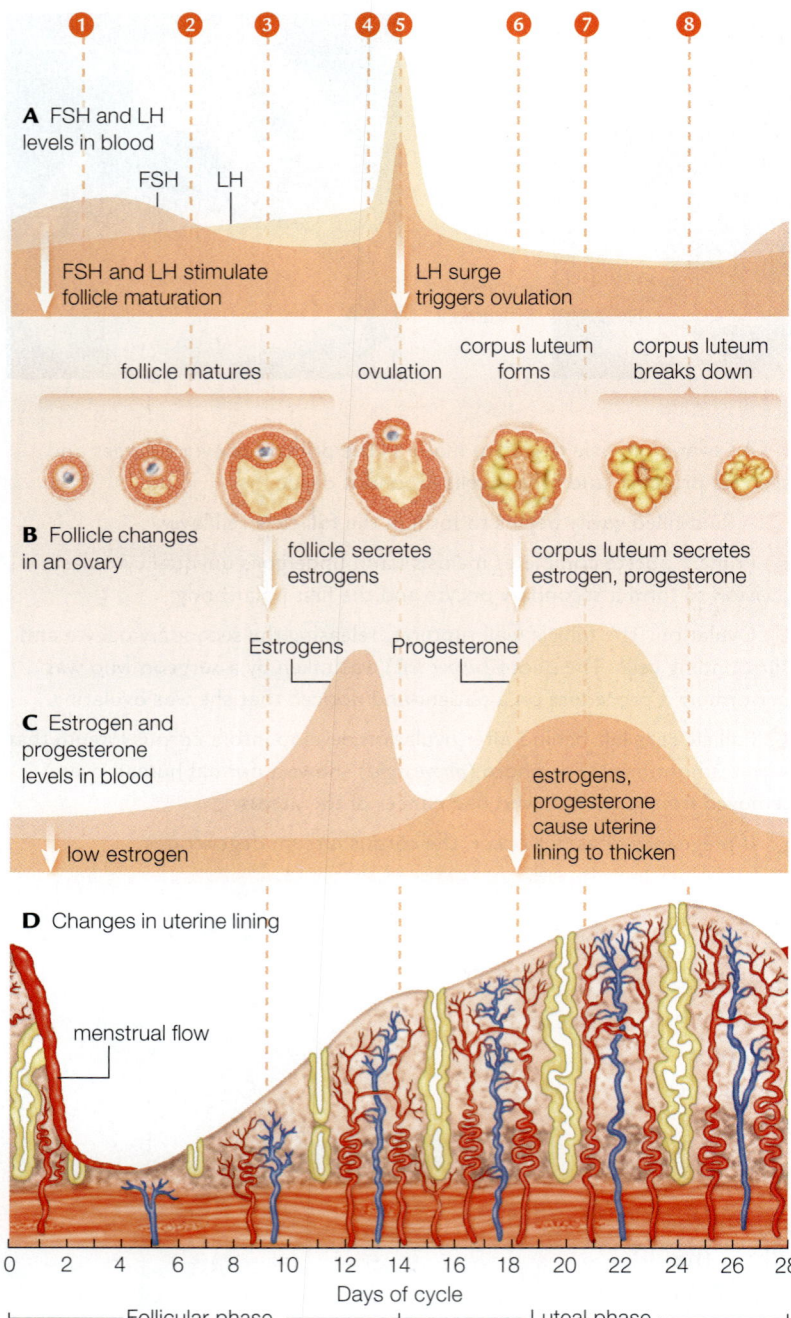

A FSH and LH levels in blood

FSH   LH

FSH and LH stimulate follicle maturation

LH surge triggers ovulation

follicle matures   ovulation   corpus luteum forms   corpus luteum breaks down

B Follicle changes in an ovary

follicle secretes estrogens

corpus luteum secretes estrogen, progesterone

Estrogens   Progesterone

C Estrogen and progesterone levels in blood

low estrogen

estrogens, progesterone cause uterine lining to thicken

D Changes in uterine lining

menstrual flow

0 2 4 6 8 10 12 14 16 18 20 22 24 26 28
Days of cycle

|— Follicular phase —|— Luteal phase —|

FIGURE 26.9 Animated! Changes in a human ovary and uterus correlated with changing hormone levels. Onset of menstrual flow is considered Day 1 of the approximately 28-day menstrual cycle.

A,B Prompted by GnRH from the hypothalamus, the anterior pituitary secretes FSH and LH, which stimulate a follicle to grow and an oocyte to mature in an ovary. A midcycle surge of LH triggers ovulation and the formation of a corpus luteum. A decline in FSH after ovulation stops more follicles from maturing.

C,D Early on, estrogen from a maturing follicle encourages repair and rebuilding of uterine lining. After ovulation, the corpus luteum secretes some estrogen and more progesterone that primes the uterus for pregnancy. If pregnancy occurs, the corpus luteum will persist, and its secretions will stimulate the maintenance of the uterine lining.

© Cengage Learning.

❯ Hormones and the Menstrual Cycle Cyclic events in the ovaries occur in synchrony with cyclic changes in the uterus. We refer to the approximately monthly changes in the uterus as the **menstrual cycle**. The first day of the menstrual cycle is marked by the onset of **menstruation**: the flow of bits of uterine lining and a small amount of blood from the uterus, through the cervix, and out of the vagina.

Figure 26.9 shows how hormone levels and the thickness of the uterine lining change during the menstrual cycle. It also shows how those events are tied to the ovarian cycle.

Like testes, ovaries are under the control of GnRH. At the onset of the menstrual cycle, GnRH from the hypothalamus causes cells in anterior pituitary to increase their secretion of FSH and LH ❶. As its name implies, follicle-stimulating hormone stimulates an ovarian follicle to begin maturing ❷. As a follicle matures, cells surrounding the oocyte secrete estrogens (a type of female sex hormone). Rising estrogens ❸ encourage the uterine lining to thicken.

The pituitary detects the rising level of estrogens in the blood and responds with an outpouring of LH ❹. The LH surge encourages the primary oocyte to complete meiosis I and undergo cytoplasmic division. The LH surge also causes the follicle to swell and burst. Thus, a midcycle surge of LH is the trigger for ovulation ❺.

Immediately after ovulation, the estrogen level declines until the corpus luteum forms. The corpus luteum secretes some estrogen and a lot of progesterone (another type of female sex hormone) ❻. Estrogens and progesterone cause the uterine lining to thicken and encourage blood vessels to grow through it. The uterus is now ready for a pregnancy ❼.

If a pregnancy does not occur, the corpus luteum persists for about twelve days. The estrogen and progesterone it secretes keep the pituitary from secreting FSH, so no other follicles start to mature. When the corpus luteum begins to break down, estrogen and progesterone levels fall ❽. The hypothalamus senses this decline, and stimulates the pituitary to begin secreting FSH and LH once again. In the uterus, the decline in estrogens and progesterone causes thickened lining to break down, and menstruation begins.

Many women occasionally experience discomfort a week or so before they menstruate. Body tissues swell because premenstrual changes influence secretion of aldosterone, a hormone that stimulates reabsorption of sodium and, indirectly, water (Section 25.6). Breasts may become tender because hormonal changes cause the milk ducts inside them to enlarge. Cycle-induced hormonal changes can also cause depression, irritability, or anxiety, and headaches. Regular recurrence of these symptoms is known as premenstrual syndrome (PMS). During menstruation, local signaling molecules stimulate contractions of smooth muscle in the uterine wall. Many women barely notice these contractions, but others experience them as menstrual cramps (Figure 26.10).

Menstruation usually lasts for five to seven days. Unusually heavy or persistent menstrual flow can be caused by fibroids, which are benign tumors of the uterus. A woman who needs to change pads or tampons on an hourly basis should discuss this condition with her doctor. Fibroids can be surgically removed, usually without impairing fertility.

Some women have irregular menstrual cycles. Excessive exercise or malnutrition can halt the cycles, but they usually resume when physical activity lessens or diet improves. Menstrual cycles halt permanently and fertility ends when a woman completes **menopause**, usually at about age 50. Hormonal changes related to menopause cause many women to have hot flashes. A woman gets abruptly and uncomfortably hot, flushed, and sweaty as blood surges to her skin. Hormone replacement therapy (taking estrogen and in some cases progesterone) can relieve symptoms, but may pose some health risks.

Vaginal bleeding between menstrual periods and bleeding after menopause require medical attention. Such bleeding can be a sign of uterine cancer, the most common cancer of the female reproductive tract. Ovarian cancer is less common, but causes more deaths. It does not alter menstrual cycles, and so often goes undetected until it has enlarged or spread. There are no early symptoms of cervical cancer either, but this type of cancer can be detected by a Pap smear (a procedure in which cervical cells are sampled, stained, and examined under a microscope for changes indicating cancer).

Take-Home Message

What are the roles of human reproductive organs?

- A male's testes produce sperm and secrete the sex hormone testosterone. Sperm and secretions from accessory glands form semen. Semen is propelled through a series of ducts and leaves the body through an opening in the penis.

- A woman's ovaries make eggs and secrete sex hormones (estrogens and progesterone). Oviducts connect the ovaries to the uterus, where offspring develop. The vagina is the organ of intercourse and also the birth canal.

- In men, germ cells divide continually to produce sperm. By contrast, a woman is born with all the immature eggs she will ever have.

- From puberty until menopause, a woman's eggs mature—one at a time—in an approximately monthly cycle. During each cycle, the uterine lining thickens in preparation for pregnancy. If pregnancy does not occur, the woman sheds the uterine lining (menstruates), and the cycle begins again.

26.4 How Pregnancy Happens

❭ Sexual Intercourse For males, intercourse requires an **erection**. When a male is not sexually aroused, his penis remains limp because the arterioles that transport blood into the large cylinders of spongy tissue that make up the bulk of the penis remain constricted. With arousal, signals from the parasympathetic nervous system induce the vessels supplying this spongy tissue to widen. Inward flow of blood now exceeds outward flow, and the increase in fluid pressure expands the internal chambers. As a result, the penis enlarges and stiffens, so it can be inserted into a female's vagina.

The ability to achieve and sustain an erection peaks during the late teens. As a man ages, he may have episodes of erectile dysfunction. With this disorder, the penis cannot stiffen enough for intercourse. The National Institutes of Health estimates that erectile dysfunction affects about 30 million men in the United States alone. Smokers and men with circulatory problems are most often affected. Viagra and similar drugs prescribed for erectile dysfunction cause blood

FIGURE 26.10 Menstrual cramps. During menstruation, local signaling molecules called prostaglandins cause contraction of uterine smooth muscle. The contractions help expel bits of shed uterine lining, but many women find them painful. Aspirin and nonsteroidal anti-inflammatory drugs such as ibuprofen can reduce menstrual cramps by decreasing the production of prostaglandins.

© Dragos Iliescu/ Shutterstock.

erection Penis becomes erect as a result of increased fluid pressure inside its spongy tissue.

menopause Of a human female, the end of fertility and of menstrual cycles.

menstrual cycle Approximately monthly cycle in which the uterine lining thickens in preparation for pregnancy, then is shed if pregnancy does not occur.

menstruation Flow of blood and bits of shed uterine lining out through the vagina.

Fertilization

Ovulation

- oviduct
- ovary
- uterus
- opening of cervix
- vagina

sperm enter vagina

- follicle cell
- egg nucleus
- secreted protein

1

2

3 nuclei fuse

fusion of sperm nucleus with egg nucleus

4

FIGURE 26.11 Animated! Fertilization.

1 Sperm surround a secondary oocyte and release enzymes that digest their way through the protein layer around the oocyte.

2 When a sperm enters the oocyte, the protein layer around the oocyte changes so other sperm cannot enter. Entry of a sperm also stimulates completion of meiosis II in the oocyte's nucleus.

3 The sperm tail degenerates; its nucleus enlarges and fuses with the oocyte nucleus.

4 At nuclear fusion, a zygote forms and fertilization is completed.

© Cengage Learning.

vessels that carry blood into the penis to widen and deliver more blood. Such drugs can cause headaches and (rarely) sudden hearing loss. They also interact with other drugs, and should never be taken without a prescription.

In a sexually excited female, the vaginal wall, labia, and clitoris swell with blood. Her cervix and glands near the entrance to the vagina increase their secretion of mucus.

During intercourse, mechanical stimulation of the male's penis and the female's clitoris and vaginal wall may lead to **orgasm**, or sexual climax. During orgasm, neurotransmitters flood the brain, causing sensations of release and pleasure. In females, muscles of the vaginal wall and pelvis contract rhythmically. In males, similar reflexive contractions force sperm and contents of seminal vesicles and the prostate gland into the urethra. These substances mix, forming semen, which is ejaculated in the vagina.

❯ Fertilization On average, an ejaculation will put 150 million to 350 million sperm into the vagina. Within thirty minutes some have made their way into the female's oviducts. Just a few hundred will survive the long journey to the upper region of the oviduct, where eggs usually are fertilized (Figure 26.11).

A secondary oocyte released at ovulation has a wrapping of follicle cells over a layer of proteins **1**. Sperm make their way between follicle cells to the protein layer. Binding of receptors in the plasma membrane of a sperm's head to egg proteins triggers release of protein-digesting enzymes from the cap on the sperm's head. The collective effect of enzyme from many sperm clears a passage to the oocyte's plasma membrane. Receptors in this membrane bind a sperm's plasma membrane, and the two membranes fuse, allowing the sperm into the oocyte. Usually only one sperm enters because its entry triggers changes to the protein coat that prevent other sperm from binding. Once inside the oocyte, a sperm breaks down, but its nucleus remains intact. Sperm entry prompts the secondary oocyte to complete meiosis II, forming a mature egg, or **ovum** (plural, ova), along with a second polar body **2**. Egg and sperm nuclei fuse **3**. The result is a diploid zygote, the first cell of a new individual **4**.

Sometimes two eggs mature and are released at the same time, and each is fertilized by a different sperm. The result is fraternal twins. Such twins are no more similar than any other siblings.

Take-Home Message

How does a pregnancy occur?

- In preparation for intercourse, a male's penis becomes inflated by the inflow of blood and a female's reproductive tract becomes lubricated with mucus.

- During orgasm, smooth muscles of the reproductive tract undergo rhythmic contractions. In males, ejaculation forces sperm out of the penis.

- Fertilization usually occurs in the oviduct. Usually an oocyte changes after one sperm penetrates it, preventing other sperm from binding. The result of fertilization is a zygote, the first cell of the new individual.

26.5 Factors Affecting Fertility

〉 Contraception Methods used to prevent pregnancy are termed **contraception**. Table 26.1 lists common options and compares their effectiveness. The most effective option is abstinence—no sex—which is 100 percent effective. However, it can take great self-discipline. Rhythm methods are forms of abstinence; a woman simply avoids sex during her fertile period. She calculates when she is ovulating by recording how long her menstrual cycles last, checking her temperature each morning, monitoring the thickness of her cervical mucus, or by checking some combination of these indicators.

Withdrawal, or removing the penis from the vagina before ejaculation, requires great willpower. It also may be ineffective because sperm can be in pre-ejaculation fluids. Similarly, rinsing out the vagina (douching) immediately after intercourse is unreliable. Typically, within seconds of ejaculation some sperm swim through the cervix and into the uterus where douche cannot reach.

Surgical methods are highly effective, but are meant to make a person permanently sterile. Men may opt for a vasectomy. A doctor makes a small incision into the scrotum, then cuts and ties off each vas deferens. A tubal ligation blocks or cuts a woman's oviducts.

Other fertility control methods use physical and chemical barriers to stop sperm from reaching an egg. Spermicidal foams and jellies poison sperm. They are not always reliable, but their use with a condom or diaphragm reduces the chance of pregnancy. A diaphragm is a flexible, dome-shaped device that is positioned inside the vagina so it covers the cervix. A diaphragm is relatively effective if it is first fitted by a doctor and used correctly with a spermicide. A cervical cap is a similar but smaller device. Condoms are thin, tight-fitting sheaths worn over the penis during intercourse. Latex ones have the advantage of also protecting against sexually transmitted diseases (STDs). However, all condoms can tear or leak.

An intrauterine device, or IUD, is inserted into the uterus by a physician. Some IUDs thicken cervical mucus so sperm cannot swim through it. Others shed copper, which keeps an early embryo from implanting in the uterus.

The birth control pill is the most common fertility control method in developed countries. "The Pill" is a mix of synthetic estrogens and progesterone-like hormones that prevents both maturation of oocytes and ovulation. When taken consistently, it is highly effective. Its use lowers risk of ovarian and uterine cancer but raises risk of breast, cervical, and liver cancer. A birth control patch is a small, flat adhesive patch applied to skin. It delivers the same mixture of hormones as an oral contraceptive, and it blocks ovulation the same way.

Both birth control pills and the birth control patch raise the risk of stroke and other serious cardiovascular disorders. The risk of these complications is greatest in older women and women who smoke.

Hormone injections or implants prevent ovulation. Each injection acts for several months, whereas the implant lasts for three years. Both methods are quite effective, but may cause sporadic, heavy bleeding.

> Typically, some sperm swim through the cervix within seconds of ejaculation.

Table 26.1	Common Methods of Contraception	
Method	**Description**	**Pregnancy Rate***
Abstinence	No intercourse	0% per year
Rhythm method	Avoid intercourse in female's fertile period	25% per year
Withdrawal	End intercourse before male ejaculates	27% per year
Douche	Wash semen from vagina after intercourse	60% per year
Vasectomy	Cut or close off male's vasa deferentia	>1% per year
Tubal ligation	Cut or close off female's oviducts	>1% per year
Condom	Enclose penis, block sperm entry to vagina	15% per year
Diaphragm, cervical cap	Cover cervix, block sperm entry to uterus	16% per year
Spermicides	Kill sperm	29% per year
Intrauterine device	Prevent sperm entry to uterus or prevent implantation of embryo	>1% per year
Oral contraceptives	Prevent ovulation	>1% per year
Hormone patches, implants, or injections	Prevent ovulation	>1% per year
Emergency contraception pill	Prevent ovulation	15–25% per use**

* Percent of users who become pregnant with consistent, correct usage.

** Not meant for regular use. 15–25 percent of users become pregnant despite use of pill.

contraception Methods used to prevent pregnancy.
orgasm Sexual climax during which neurotransmitter release causes smooth muscle contractions accompanied by a feeling of release and pleasure.
ovum Mature egg.

FIGURE 26.12 Tubal pregnancy. Such pregnancies occur when an embryo implants in an oviduct, rather than the uterus. The embryo cannot develop to term and its presence can rupture the oviduct, causing bleeding, infection, and even maternal death.

Dr. E. Walker/ Photo Researchers, Inc.

1 cm

Table 26.2	Common STDS	
STD	Cases per year in the U.S.	Pathogen
Trichomoniasis	7,400,000	Protist
HPV infection	6,000,000	Virus
Chlamydia	3,000,000	Bacteria
Genital herpes	1,000,000	Virus
Gonorrhea	300,000	Bacteria
HIV infection	40,000	Virus
Syphilis	14,000	Bacteria

If a woman has unprotected sex or a condom breaks, she can turn to emergency contraception. Some so-called "morning-after pills" are available without a prescription to women over age 17. The most widely used emergency contraceptive, Plan B, delivers a large dose of synthetic progesterone that prevents ovulation and interferes with fertilization. It is most effective when taken immediately after intercourse, but has some effect up to three days later. Morning-after pills are not meant for regular use. They cause nausea, vomiting, abdominal pain, headache, and dizziness.

› Abortion An induced abortion is the deliberate dislodging and removal of an embryo or a fetus from the uterus. In the United States, about half of unplanned pregnancies end in induced abortion. The drug mifepristone (RU-486) together with a prostaglandin can be used to terminate a pregnancy during the first nine weeks of pregnancy. The two chemicals interfere with progesterone receptors in the uterus. Without progesterone's effects, the uterine lining cannot be maintained, and neither can pregnancy.

› Infertility About 10 percent of couples in the United States are infertile, meaning they do not have a successful pregnancy despite a year of trying. Risk of infertility increases with age and with obesity. In about two-thirds of couples, the woman is infertile. Hormonal disorders can prevent a woman from having regular menstrual periods or keep follicles from maturing normally. Endometriosis is another common cause of female infertility. With this disorder, tissue that normally lines the uterus grows elsewhere in the pelvis. Like normal uterine tissue, the misplaced tissue proliferates and bleeds in response to changing hormone levels. The result is pain and scarring that can distort the oviducts or scar them. Endometriosis can also make intercourse painful, thus lessening the frequency of intercourse and opportunities for conception.

Oviducts scarred as a result of endometriosis or a sexually transmitted disease can cause an embryo to implant in an oviduct (Figure 26.12). The resulting tubal pregnancy cannot develop to term and can threaten the life of the mother. Infertility also results when an embryo can reach the uterus, but fibroids or other uterine problems interfere with its ability to implant there.

Male infertility arises when a man makes few or abnormal sperm, or when sperm are not ejaculated properly. Hormonal disorders that lower testosterone can prevent sperm maturation. Genetic disorders can interfere with sperm motility. It only takes one sperm to fertilize an egg, but it takes the binding of many to release enough enzyme to clear a way to the egg plasma membrane. This is why a man who has healthy sperm but a low sperm count can be functionally infertile. In males, as in females, sexually transmitted disease can scar ducts of the reproductive system, causing blockages that keep sperm from leaving the body. In some men, the valve between the bladder and the urethra does not function properly, so sperm end up in the bladder rather than being ejaculated.

› Sexually Transmitted Disease Each year, pathogens that cause sexually transmitted diseases, or STDs, infect many millions of Americans (Table 26.2). Women are more easily infected than men, and suffer from more complications. For example, pelvic inflammatory disease (PID), a secondary outcome of bacterial STDs, scars the female reproductive tract and can cause infertility, chronic pain, and tubal pregnancies. Having an STD while pregnant increases the risk of miscarriage, premature delivery, and birth defects.

The most common STD in the United States, trichomoniasis, is caused by a flagellated protozoan (Section 13.6). Infected women produce a yellowish discharge and have a sore, itchy vagina. Men are usually symptom-free. In both sexes, untreated infections cause infertility. A single dose of an antiprotozoal drug can quickly cure the infection.

Chlamydia and gonorrhea are bacterial STDs that often occur together. Both can cause a discharge from the penis or vagina and pain during urination. However, these symptoms occur in only about half of infected women and even fewer infected men. If untreated, both diseases can lead to infertility. Both can also be passed from mother to child during delivery, resulting in eye infections and other problems in the newborn (Figure 26.13**A**). An untreated gonorrhea infection can also harm joints and the heart.

The earliest symptom of syphilis, another bacterial STD, is flattened, painless ulcers called chancres (pronounced "shankers") at the site of initial infection. If the infection goes untreated, bacteria spread and more widely distributed chancres appear (Figure 26.13**B**). A long-term infection can destroy organs throughout the body, causing joint pain, mental illness, blindness, and death.

Treatment with antibiotics can cure most bacterial STDS, but antibiotic resistance is on the rise. For example, a strain of gonorrhea recently discovered in Japan is resistant to all currently used antibiotics. Many bacteria strains that cause STDs resist commonly used drugs such as penicillin and tetracycline.

Some types of human papillomavirus (HPV) cause bumplike growths called warts on the hands or feet. Sexually transmitted strains of the virus can become established on the genitals (Figure 26.13**C**) or wherever else sexual behavior introduces them. In addition to warts, some sexually transmitted types of HPV cause cancer. HPV infection causes nearly all cervical cancers in women (Section 22.1) and most anal cancers in homosexual men. Oral sex can introduce HPV into the mouth and raise the risk of cancers of the mouth and throat. A recently approved vaccine can prevent HPV infection in both males and females, if given before viral exposure. No current treatment can cure an HPV infection, but the resulting warts can be removed by surgery, or by burning or freezing them.

The herpes simplex virus 2 causes genital herpes. The initial exposure to the virus commonly results in small sores at the site of infection. After these sores heal, the infection periodically becomes reactivated, causing tingling or itching, which may or may not be accompanied by visible sores. Antiviral drugs can promote healing of sores caused by an active infection and lessen the likelihood of viral reactivation, but they cannot eliminate the virus entirely.

Infection by HIV (human immunodeficiency virus) can cause AIDS (acquired immune deficiency syndrome), in which the immune system weakens and infectious organisms take hold. Section 13.4 described the HIV virus and Section 22.7 explained how its effects on the immune system lead to AIDS. To reduce the risk of HIV transmission during vaginal or anal sex, doctors recommend use of a latex condom and a lubricant. The lubricant helps prevent small abrasions that could allow the virus to enter the body. If you think you could have been exposed to HIV, get tested as soon as possible. Early treatment may prevent development of AIDS.

A Chlamydia-inflamed eyes of an infant who was infected by its mother during delivery.

B Chancres caused by a syphilis infection that has spread throughout the body.

C Genital warts (raised white areas) on the labia of a woman infected by HPV.

FIGURE 26.13 Consequences of STDs. Use of a latex condom minimizes the risk of transmission of pathogens that can cause STDs.

Credits: (a) Western Ophthalmic Hospital/ Photo Researchers, Inc.; (b) CNRI/ Photo Researchers, Inc; (c) CDC/ Joe Millar.

Take-Home Message

What factors affect human fertility?

- Contraceptives can prevent ovulation or fertilization.

- Infertility affects women more than men. Female infertility arises from hormonal disorders, blocked reproductive ducts, or uterine problems. Infertile men may have a low sperm count, blocked reproductive ducts, or problems ejaculating.

- Sexually transmitted diseases can scar the reproductive tract of both males and females, increasing the risk of infertility.

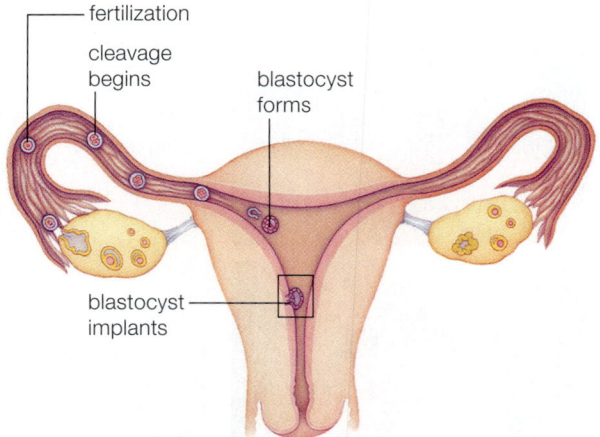

fertilization
cleavage begins
blastocyst forms
blastocyst implants

FIGURE 26.14 Animated! Fertilization to implantation. *Above*, where events occur. *Below*, steps in early development. © Cengage Learning.

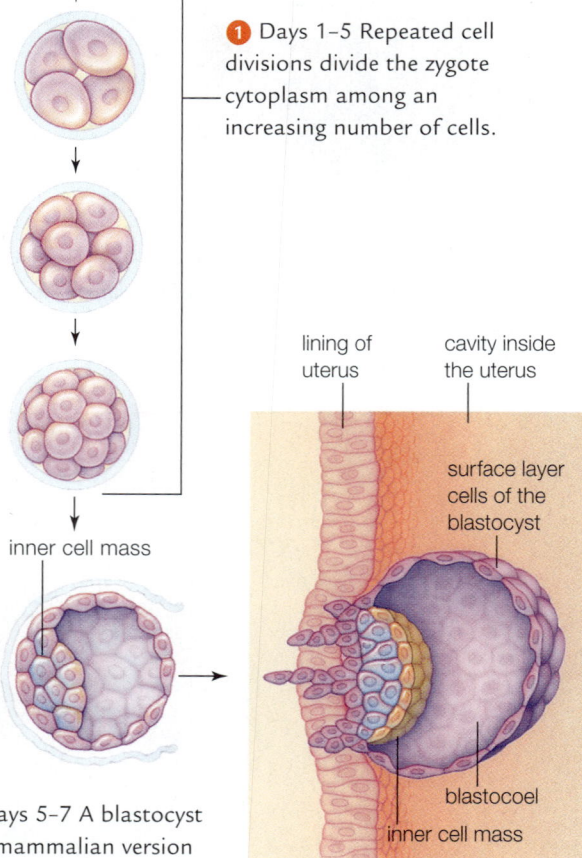

❶ Days 1–5 Repeated cell divisions divide the zygote cytoplasm among an increasing number of cells.

inner cell mass

❷ Days 5–7 A blastocyst (the mammalian version of a blastula) forms, expands, and sheds its protein coat. Cells of the inner cell mass are destined to become the embryo proper.

26.6 Human Prenatal Development

Development that occurs before birth is referred to as prenatal development. In humans, prenatal development typically lasts for 38 weeks after fertilization. During that time, cell divisions transform a single-celled zygote into a newborn with about a trillion cells of many different types.

❭ Cleavage and Implantation For fertilization to occur naturally, a human egg must meet up with sperm in the upper portion of an oviduct (Figure 26.14). Cleavage begins within a day, as cilia propel the zygote through the oviduct. The zygote divides by mitosis to form two cells, which become four, which become eight, and so on ❶. Sometimes a cluster of four or eight cells splits in two and each cluster develops independently, resulting in identical twins. More typically, all of the cells adhere tightly to one another as the divisions continue.

By about 5 days after fertilization, the dividing cluster of cells has reached the uterus and formed a blastocyst, the mammalian version of the blastula ❷. The blastocyst consists of an outer layer of cells, a cavity filled with their fluid secretions, and an inner cell mass. Of 200 to 250 cells in the human blastocyst, only the thirty or so in the inner cell mass give rise to the embryo. The other cells give rise to the extraembryonic membranes.

Before a blastocyst can implant, it must escape from the protein coat that surrounds it. Once this coat has ruptured, the blastocyst attaches to the lining of the uterus and begins burrowing into it ❸.

As implantation continues, the inner cell mass develops into an embryonic disk with two flattened layers of cells ❹. At the same time, the extraembryonic membranes begin to form. A membrane called the **amnion** encloses a fluid-filled amniotic cavity between the embryonic disk and the blastocyst surface. Fluid in this cavity (amniotic fluid) acts as a buoyant cradle in which an embryo grows and moves freely, protected from temperature changes and mechanical impacts. As the amnion forms, other cells move around the inner wall of the blastocyst, forming the lining of a yolk sac. In reptiles and birds, this sac holds yolk that nourishes the developing embryo. In humans, it has no nutritional role. Some

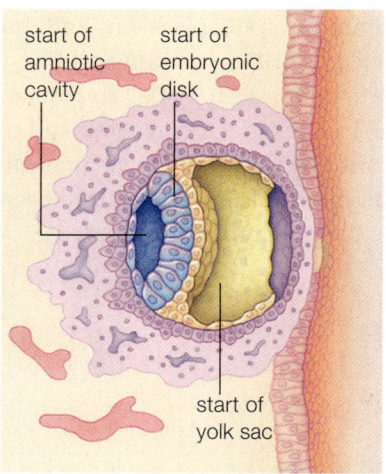

lining of uterus
cavity inside the uterus
surface layer cells of the blastocyst
blastocoel
inner cell mass

❸ Days 8–9 Implantation begins. The blastocyst attaches to the lining of the uterus and begins to burrow into it.

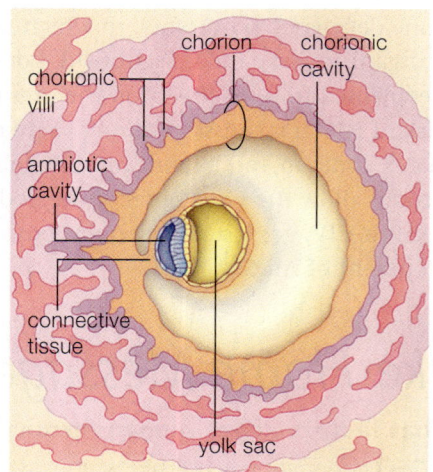

start of amniotic cavity
start of embryonic disk
start of yolk sac

❹ Days 10–11 The inner cell mass becomes a two-layered embryonic disk. Cell migrations begin to form extra-embryonic membranes.

actual size

chorion
chorionic cavity
chorionic villi
amniotic cavity
connective tissue
yolk sac

❺ Day 14 Projections from the chorion (chorionic villi) begin to grow into blood-filled spaces in the endometrium. The amniotic cavity has filled with fluid.

actual size

cells that form in a human yolk sac give rise to the embryo's first blood cells. Others migrate to the gonads and become germ cells.

As implantation continues, spaces in the uterine tissue around the blastocyst become filled with blood seeping in from ruptured capillaries. An extraembryonic membrane called the **chorion** develops with many tiny fingerlike projections (chorionic villi) that extend into blood-filled maternal tissues ❺. The chorion will become part of the **placenta**, an organ that functions in exchanges of materials between a mother and her developing child.

Implantation of a blastocyst in the uterus prevents menstruation because the chorion secretes **human chorionic gonadotropin** (**HCG**). This is the first hormone that a new human produces and it prevents degeneration of the corpus luteum. By the beginning of the third week, HCG can be detected in a mother's blood or urine. At-home pregnancy tests have a "dip stick" with a region that changes color when exposed to urine containing HCG.

> **Embryonic and Fetal Development** Gastrulation occurs about two weeks after fertilization, when cells from the outer layer of the embryonic disk migrate inward and form a three-layered embryo. After gastrulation, two neural folds appear on the surface of the embryonic disk (Figure 26.15). The folds will merge into a neural tube that develops into the spinal cord and brain.

By the end of the third week, somites begin to appear on either side of the neural tube. These paired segments of mesoderm will develop into the bones and skeletal muscles of the head and trunk (Section 11.6).

The heart begins to beat a bit more than three weeks after fertilization. At first it is a linear tube of contractile cells that resembles a fish heart. The sound of its beating is still too faint to hear, even with a stethoscope. As development continues, this tubular heart bends back on itself, forming a heart with three chambers like that of an amphibian, and finally a heart with four chambers.

Bands of tissue called pharyngeal arches form at the onset of the fourth week. These will later contribute to the pharynx, larynx, and the face, neck, mouth, and nose. In fishes, pharyngeal arches develop into gills, but a human embryo never has gills. It receives oxygen by way of the placenta.

At-home pregnancy tests check urine for the presence of HCG, a hormone produced by the early embryo.

paired neural folds

Day 16 Gastrulation has occurred and the neural tube is forming by the merging of two neural folds.

future brain

somites

Days 18–23 Bumps of mesoderm (somites) form. They will develop into muscle and bone of the head and trunk.

pharyngeal arches

Days 24–25 Pharyngeal arches appear. They will become structures in the head and neck.

FIGURE 26.15 Organ formation begins. © Cengage Learning.

amnion Extraembryonic membrane that encloses an amniote embryo and the amniotic fluid.

chorion Outermost extraembryonic membrane of amniotes; major component of the placenta in placental mammals.

human chorionic gonadotropin (**HCG**) Hormone produced by the chorion and later by the placenta; encourages maintenance of the uterus, thus sustaining a pregnancy.

placenta Of placental mammals, an organ composed of maternal and embryonic tissues that allows the exchange of substances between a developing individual and its mother.

Week 4

Weeks 5–6

yolk sac
connecting stalk
embryo

forebrain

future lens

pharyngeal arches

developing heart

upper limb bud

somites

neural tube forming

lower limb bud

tail

actual length

head growth exceeds growth of other regions

retinal pigment

future external ear

upper limb differentiation (hand plates develop, then digital rays of future fingers; wrist, elbow start forming)

umbilical cord formation between weeks 4 and 8 (amnion expands, forms tube that encloses the connecting stalk and a duct for blood vessels)

foot plate

actual length

FIGURE 26.16 Human embryonic and fetal development.

Figure It Out: Is a human embryo ever as big as your fist?

Answer: No. By the time the developing individual is as big as a fist, it is a fetus.

When the fourth week ends, the embryo is 500 times its starting size, but it remains less than 1 centimeter long (Figure 26.16). A conspicuous tail accounts for about a sixth of the body's length. The pace of growth now slows as details of the embryo's organs begin to fill in. As development continues, limbs form. Suicide of cells connecting the digits sculpts individual toes and fingers from the previously paddlelike feet and hands.

By the end of the eighth week, all organ systems have formed, cell suicide has usually eliminated the tail, and we define the individual as a human **fetus**. The fetal heartbeat can be detected at about five months, and a mother usually begins to feel reflexive fetal movements about five to six months after fertilization. Soft, fuzzy hair, the lanugo, covers the fetus. The fetal skin is wrinkled,

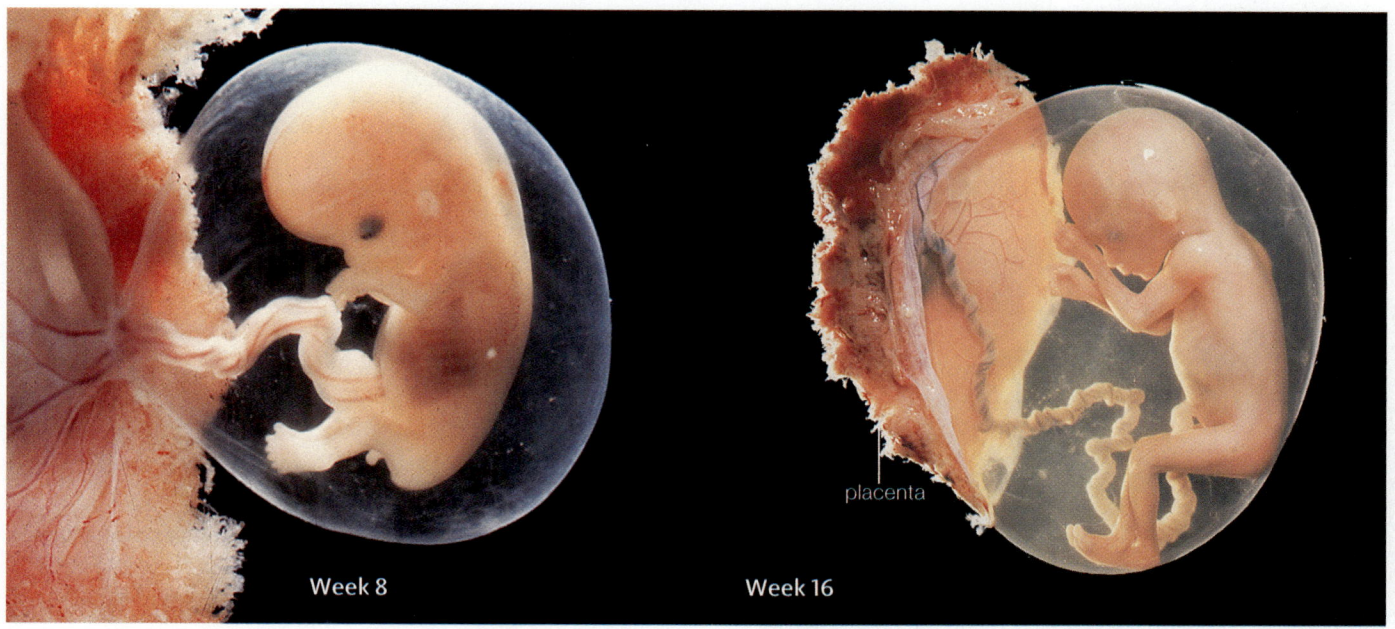

Week 8   Week 16

placenta

final week of embryonic period; embryo looks distinctly human compared to other vertebrate embryos

upper and lower limbs well formed; fingers and then toes have separated

primordial tissues of all internal, external structures now developed

tail has become stubby

actual length

Length: 16 centimeters
          (6.4 inches)
Weight: 200 grams
          (7 ounces)

WEEK 29
Length: 27.5 centimeters
          (11 inches)
Weight: 1,300 grams
          (46 ounces)

WEEK 38 (full term)
Length: 50 centimeters
          (20 inches)
Weight: 3,400 grams
          (7.5 pounds)

During fetal period, length measurement extends from crown to heel (for embryos, it is the longest measurable dimension, as from crown to rump).

reddish, and protected from abrasion by a thickened, cheeselike coating. Eyelids form and the eyes open in the seventh month.

The optimal birthing time is about 38 weeks after fertilization. Technological advances have allowed us to keep premature infants alive at increasingly early stages. As of 2011, the earliest a child has been born and survived is 21 weeks. Survival of children born before 28 weeks (7 months) still remains uncertain, mainly because the lungs have not fully developed. The prospects for survival start to drop after this. By 36 weeks, the survival rate is 95 percent. However, a fetus born between 36 and 38 weeks still has some trouble breathing and maintaining a normal body temperature. Premature birth also increases the risk of a variety of birth defects.

fetus Developing human between 9 weeks after fertilization and birth.

amniotic fluid

umbilical cord

fetal blood vessels inside umbilical cord

pool of maternal blood

placenta

chorionic villus, with fetal blood vessels inside it

maternal blood in uterine lining

FIGURE 26.17 Animated! The life-support system of a developing human. *Left*, artist's depiction of the view inside the uterus, showing a fetus connected by an umbilical cord to the pancake-shaped placenta. *Right*, the placenta consists of maternal and fetal tissue. Fetal blood flowing in vessels of chorionic villi exchanges substances by diffusion with maternal blood around the villi. However, the bloodstreams do not mix.

© Cengage Learning.

› **Functions of the Placenta** All exchanges of materials between an embryo and its mother take place by way of the placenta, a pancake-shaped, blood-engorged organ made of uterine lining and extraembryonic membranes (Figure 26.17).

The placenta begins forming early in pregnancy. By the third week, maternal blood has begun to pool in spaces in the lining of the uterus tissue. Chorionic villi—the tiny fingerlike projections from the chorion—grow into the pools of maternal blood. Embryonic blood vessels extend through the umbilical cord to the placenta, and into the chorionic villi, where they are surrounded by the pooled maternal blood. The maternal and embryonic bloodstreams never mix. Instead, substances move between maternal and embryonic blood by diffusing across the walls of the embryonic vessels in the chorionic villi. Oxygen and nutrients diffuse from maternal blood into embryonic vessels in the villi. Wastes diffuse the other way, and the mother's body disposes of them.

The placenta also has a hormonal role. From the third month on, it produces large amounts of HCG, progesterone, and estrogens. These hormones encourage the ongoing maintenance of the uterine lining.

› **Maternal Effects on Prenatal Development** An embryo or fetus depends on its mother to supply nutrients. It is also exposed to pathogens and toxins that enter the mother's body.

A pregnant woman must eat enough to gain 20 to 25 pounds, on average. If she does not, her newborn may be seriously underweight. A low birth weight is correlated with a higher risk of postdelivery complications and impaired brain function. Supplements of some minerals and vitamins can help ensure that development proceeds normally. For example, taking B-complex vitamins during early pregnancy helps reduce the risk of severe neural tube defects in the embryo. Folate (folic acid) is especially important in this regard.

Some pathogens can cross the placenta. Diseases are especially dangerous in the first six weeks after fertilization, when organ systems are still forming. For example, if a pregnant woman gets rubella (German measles) in this critical period, there is a 50 percent chance that some of her child's organs will not form properly. If she is infected while embryonic ears are forming, her newborn may be deaf. If she is infected from the fourth month of pregnancy onward, the disease will have no notable effect. A woman can avoid risks associated with rubella by getting vaccinated against the virus before pregnancy.

Another pathogen that can affect development is a protist that sometimes lurks in garden soil, cat feces, and undercooked meat. It causes toxoplasmosis. The disease often does not cause symptoms, so a pregnant woman may become infected and not realize it. If the parasite crosses the placenta, it can infect her child and lead to developmental problems, a miscarriage, or stillbirth. To minimize the risk, pregnant women should eat well-cooked meat and avoid cat feces. Let someone else change that litter box.

Alcohol passes across the placenta, so when a pregnant woman drinks, her embryo or fetus feels the effects. Alcohol exposure can cause fetal alcohol syndrome, or FAS. A small head and brain, facial abnormalities, slow growth, mental impairment, heart problems, and poor coordination characterize affected children (Figure 26.18). The damage is permanent. Children affected by FAS never catch up, physically or mentally. As a result, most doctors now advise women who are pregnant or attempting to become pregnant to avoid alcohol.

Caffeine interferes with nervous system development in animals, and may also harm human embryos. A recent study showed that women who took in 200 milligrams a day of caffeine (the amount in one and a half cups of coffee) had twice as many miscarriages as those who avoided caffeine. The study's authors advise pregnant women to choose decaffeinated or caffeine-free beverages.

Smoking or exposure to secondhand smoke increases the risk of miscarriage and adversely affects fetal growth and development. Carbon monoxide in the smoke can outcompete oxygen for the binding sites on hemoglobin, so the embryo or fetus of a smoker gets less oxygen than that of a nonsmoker. In addition, levels of the addictive stimulant nicotine in amniotic fluid can be even higher than those in the mother's blood.

Some medications cause birth defects. Isotretinoin (Accutane), a highly effective treatment for severe acne, is often prescribed for young women. If taken early in a pregnancy, it can cause heart problems or facial and cranial deformities in the embryo. Some antidepressants also increase the risk of birth defects. Paroxetine (Paxil) and related drugs inhibit the reuptake of serotonin. Use of these drugs during early pregnancy increases the likelihood of heart malformations. Taking them later in pregnancy increases risk that an infant will have fatal heart and lung disorders.

Highly addictive drugs such as methamphetamine and cocaine also cross the placenta. They increase the risk of a prenatal stroke and decrease birth weight.

FIGURE 26.18 Child with fetal alcohol syndrome. Obvious symptoms are low and prominently positioned ears, improperly formed cheekbones, and an abnormally wide, smooth upper lip. Growth-related complications, heart problems, and nervous system abnormalities are also common. © George Steinmetz.

Take-Home Message

How does a human develop?

- Cleavage produces a blastocyst, which implants in the uterus. Projections from its extraembryonic membranes invade maternal tissue and form a placenta that permits exchanges between a mother and embryo without intermingling their blood. Pathogens and toxins, as well as nutrients, can cross the placenta.

- Gastrulation occurs about two weeks after fertilization, and produces an embryo with three tissue layers. Chordate features such as a neural tube, notochord, and pharyngeal arches form later in the embryonic period.

- The fetal period begins at the end of the eighth week. Heartbeats are detected at about five months, and limb movements are felt at five to six months.

- Lungs are not fully formed until about 28 weeks (7 months), so births before this point are highly risky. A fetus born before 22 weeks will not survive.

A Fetus positioned for childbirth; its head is against the mother's cervix, which is dilating.

B Muscle contractions stimulated by oxytocin force the fetus out through the vagina.

C The placenta detaches from the wall of the uterus and is expelled.

FIGURE 26.19 Labor and childbirth.

© Cengage Learning.

26.7 From Birth Onward

〉 The Process of Birth As a woman's pregnancy nears full term, her body prepares for the upcoming birth. Her cervix, which has remained firm and has helped keep the fetus from slipping out of the uterus prematurely, becomes thinner, softer, and more flexible. This prepares it to stretch enough to allow the passage of the fetus.

The process of birth is known as **labor**, or parturition. Often a woman finds she is about to begin labor when "her water breaks." This phrase describes the rupturing of the amnion, which allows a large amount of amniotic fluid to drain out of the vagina. During labor, the cervix dilates and the fetus moves through it into the vagina, and then out of its mother's body (Figure 26.19).

The hormone oxytocin stimulates smooth muscle contractions during labor. As the fetus nears full term, it usually "drops," or shifts down, so the head touches the cervix. Receptors in the cervix sense mechanical pressure and signal the hypothalamus, which calls for oxytocin secretion by the posterior pituitary.

The binding of oxytocin to smooth muscle of the uterus increases the strength of contractions, which causes more mechanical pressure. This leads to more oxytocin secretion, and so on. The result is a positive feedback cycle— stretching of the cervix triggers more oxytocin secretion, which results in more stretching. This continues until the fetus has been expelled and there is no longer any mechanical pressure on the cervix. Synthetic oxytocin is sometimes given intravenously to induce or increase contractions.

Strong contractions help detach the placenta from the uterus and expel it as the afterbirth. Contractions also help to stop the bleeding where the placenta was attached to the wall of the uterus, by causing blood vessels at this ruptured attachment site to constrict. The umbilical cord is cut and tied off. A few days after shriveling up, the cord's stump has become the navel.

〉 Nourishing the Newborn When a woman is not pregnant, her breasts consist mostly of adipose tissue (Figure 26.20**A**). Her milk ducts and mammary glands are small and inactive. If she becomes pregnant, prolactin secreted by her anterior pituitary triggers growth of mammary glands in preparation for production of milk to nourish her newborn (Figure 26.20**B**).

For the first few days after a woman gives birth, her mammary glands produce a clear fluid rich in proteins and lactose (milk sugar). Then, milk production begins in earnest. In addition to lactose and proteins, milk contains easily digested fats, essential vitamins and minerals, and enzymes that assist in digestion. Substances in milk encourage growth of beneficial bacteria and kill harmful ones. Milk also contains maternal antibodies that coat the lining of the newborn's throat and gut lining, lessening the risk of dangerous infections.

Breast-feeding benefits a mother as well as her child. Oxytocin released in response to the nipple stimulation forces milk into milk ducts, and it causes uterine contractions that help restore the uterus to its pre-pregnancy state. Hormonal effects of breast-feeding slow the return of menstrual cycles, thus acting as a natural contraceptive. Producing milk takes a lot of energy, so breast-feeding helps a woman lose weight gained during her pregnancy. In addition, breast-feeding lowers a woman's risk of cancers of the reproductive tract and protects against inherited forms of breast cancer that affect premenopausal women.

Women who are breast-feeding should keep in mind that toxins that enter their body may end up in their milk. Exposure to nicotine in milk interferes with an infant's ability to sleep, as does exposure to alcohol. Alcohol in milk may also interfere with a child's early motor development. Some viral pathogens, including HIV, can also be transmitted in breast milk.

labor Expulsion of a placental mammal from its mother's uterus by muscle contractions.

Postnatal Development Like many animals, humans change in size and proportion until reaching sexual maturity. Postnatal growth is most rapid between years thirteen and nineteen. Sex hormone secretions increase and bring about development of secondary sexual traits as well as sexual maturity. Bones do not become fully mature until adulthood.

Take-Home Message

What happens during and after birth?

- Hormonal changes prepare a woman's body for labor and nursing.
- During labor, hormone-stimulated contraction of uterine smooth muscle forces the fetus out of the body.
- Milk secreted by mammary glands provides nutrition and also has substances that protect the newborn from infection.
- Growth and development continue after birth. Bones are not fully mature until early adulthood.

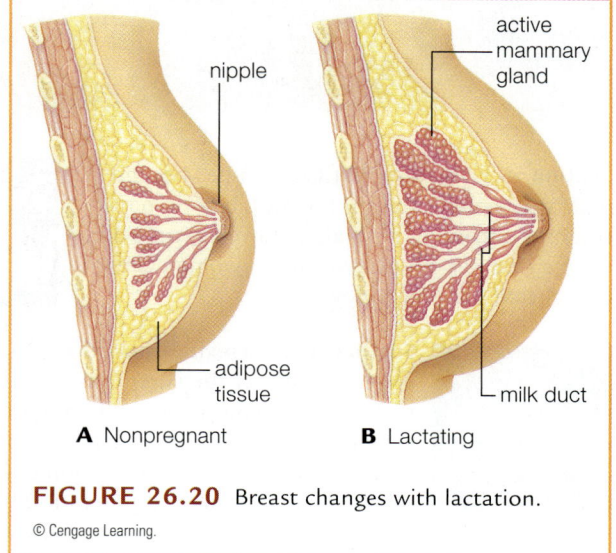

FIGURE 26.20 Breast changes with lactation.

© Cengage Learning.

26.8 Assisted Reproduction (revisited)

One negative effect of IVF has been an increase in the number of multiple births. Embryos conceived by IVF frequently fail to take hold and develop, so doctors often place several in a woman's womb to increase the odds that one will survive. If all the embryos do develop, the woman faces the risks associated with a multiple pregnancy. Compared to a woman carrying a single child, she is more likely to undergo miscarriage, stillbirth, or premature birth, and the children that she carries are at an increased risk of low birth weight and a variety of birth defects.

Professional organizations have urged fertility doctors to limit the number of embryos that they place in a woman's uterus, but this advice sometimes goes unheeded. In 2009, Nadya Suleman's doctor put twelve embryos conceived by IVF into her uterus. Suleman subsequently gave birth to octuplets (Figure 26.21). The babies were nine weeks premature and all were underweight. Suleman already had six other children, also conceived by IVF, so she is now the mother of fourteen. The doctor who placed 12 embryos in Suleman's uterus has since had his medical license revoked for transferring such an excessive number of embryos.

Often an IVF procedure produces many more embryos than parents want to use, raising the question of what to do with those left over. Typically, any extra embryos are placed in frozen storage. For example, Suleman still had 29 embryos stored away after her last IVF procedure. Some parents have donated embryos to would-be mothers who carried them to term. Over a thousand children have been born as a result of such procedures. Other parents have donated embryos to researchers as a source of material for stem cell studies. Still others have allowed the fertility clinic to destroy their embryos. In the United States, an estimated half a million embryos remain frozen.

WHERE YOU ARE GOING . . .

This chapter concludes our discussion of animal physiology. The next two chapters cover plant function, reproduction, and development.

FIGURE 26.21 Nadya Suleman with the eight infants she carried to term after implantation of twelve embryos conceived by IVF.

© TTA Media/ Splash News.

Summary

Section 26.1 ***In vitro* fertilization** has become a widely accepted solution for couples who have trouble conceiving naturally. However, inserting a large number of embryos into a woman can complicate her pregnancy.

Section 26.2 **Asexual reproduction** produces genetic copies of the parent, as when some animals produce young by **parthenogenesis. Sexual reproduction** produces variable offspring, which may be advantageous in environments where conditions change from one generation to the next.

Most animals reproduce sexually and have separate sexes, but some are **hermaphrodites** that produce both eggs and sperm. With external fertilization, gametes are released into water. Most animals on land have internal fertilization, meaning gametes meet inside a female's body.

Sexual reproduction begins with the formation of gametes—sperm and eggs—that combine at **fertilization** to form a **zygote**. Eggs often contain nutritious **yolk**. During **cleavage**, mitosis increases the number of cells but not the original volume of the zygote. Cleavage yields a **blastula**, a hollow, fluid-filled cluster of cells (*right*). During **gastrulation**, cells arrange themselves as a **gastrula** with three tissue layers: **ectoderm**, **mesoderm**, and **endoderm**. Organs form as cells in these layers **differentiate**. Development involves cell migrations, cell shape changes, and death of cells on cue (**apoptosis**). In the final stage, organs enlarge and develop specialized properties.

Section 26.3 **Testes** are primary male reproductive organs and are contained within the **scrotum**. They produce sperm and the sex hormone testosterone. Accessory glands such as the **seminal vesicles** and the **prostate gland** produce the other components of **semen**. These components are added to the sperm as they pass through the **vasa deferentia**, which carry them to the urethra. Semen leaves the body when a man has an **ejaculation**.

Ovaries are the female's primary reproductive organs. They produce eggs and sex hormones: estrogens and progesterone. Pituitary hormones (LH, FSH) control production of both sperm and eggs.

Males produce sperm continually, but a female is born with all the immature eggs (**oocytes**) she will ever have. After puberty, these eggs mature one at a time during an approximately monthly **menstrual cycle**. During this cycle, FSH encourages maturation of an **ovarian follicle**. A surge of LH near the midpoint of the cycle triggers **ovulation**, which sends an oocyte down an **oviduct** and on its way toward the **uterus**. The **cervix** is the portion of the uterus that connects to the **vagina**.

After ovulation, the remains of the ovarian follicle become a hormone-secreting **corpus luteum**. Hormones from the maturing follicle and the corpus luteum encourage thickening of the uterine lining. If pregnancy does not occur, the corpus luteum breaks down, and the uterine lining is shed during **menstruation. Menstrual cycles** cease permanently when a woman enters **menopause**.

Section 26.4 Intercourse requires that a male achieve an **erection**. A sexual climax, or **orgasm**, produces pleasurable sensations and contractions of reproductive ducts. Millions of sperm are typically ejaculated, but usually only one penetrates the secondary oocyte. When it does, the oocyte completes meiosis II, becoming a mature **ovum**. The ovum and sperm unite to form the new zygote.

Section 26.5 **Contraception** refers to methods to prevent pregnancy. Humans can choose to prevent pregnancy by abstaining from sex entirely; or to reduce the chance of pregnancy by abstaining during a woman's fertile period, surgically severing reproductive ducts, placing a physical or chemical barrier at the entrance to the uterus, or administering synthetic female sex hormones to prevent ovulation. An abortion ends an existing pregnancy.

Infertility can arise as a result of hormone disorders that affect sperm or egg production, blocked reproductive ducts, lack of sperm motility, or uterine problems that interfere with embryonic implantation or development.

Sexually transmitted diseases (STDs) are caused by protozoal, bacterial, and viral pathogens and are spread by unsafe sex. Viral STDs are incurable. Bacterial and protozoal STDs can be cured by medications, although antibiotic resistance is increasing. If untreated, STDs can cause sterility some can be fatal.

Section 26.6 After fertilization, a blastocyst forms by cleavage and burrows into the wall of the uterus. It begins producing a hormone (**human chorionic gonadotropin**) that prevents ovulation. An extraembryonic membrane called the **amnion** encloses the developing embryo and the amniotic fluid. Another membrane, the **chorion,** combines with maternal tissue to form a **placenta** that passes oxygen and nutrients between mother and embryo. The placenta can also deliver toxins.

Gastrulation occurs after the embryo has implanted. By the end of the eighth week, the embryo has lost its tail and become a **fetus**, and all organ systems have formed.

Section 26.7 Hormones prepare a woman's body for pregnancy, **labor**, and nursing. During labor, uterine contractions expel the fetus and placenta. Milk secreted by mammary glands supplies nutrients and also contains antibodies that help a newborn resist infection. Development and growth continue until adulthood.

Self-Quiz   Answers in Appendix I

1. A(n) _____ produces both eggs and sperm.
   a. hermaphrodite   c. ovary
   b. asexual organism   d. testis

2. The end product of cleavage is a(n) _____ .
   a. blastula   b. gastrula   c. embryo   d. fetus

3. A(n) _____ is a hollow ball of cells that undergoes gastrulation.
   a. gastrula   c. embryonic disk
   b. blastula   d. inner cell mass

4. Meiotic divisions of germ cells in the _____ give rise to sperm.
   a. seminiferous tubules   c. penis
   b. prostate gland   d. seminal vesicles

5. A male has an erection when _____ .
   a. muscles running the length of the penis contract
   b. nurse cells release a surge of testosterone
   c. the posterior pituitary releases oxytocin
   d. spongy tissue inside the penis fills with blood

6. In humans, fertilization usually occurs in _____ .
   a. the uterus   b. the vagina   c. an oviduct   d. an ovary

7. During a menstrual cycle, a midcycle surge of _____ secreted by
   the pituitary triggers ovulation.
   a. estrogen   b. progesterone   c. LH   d. FSH

8. After ovulation, the corpus luteum secretes _____ .
   a. LH   c. progesterone
   b. FSH   d. prolactin

9. A _____ implants in the lining of the human uterus.
   a. zygote   b. blastocyst   c. gastrula   d. fetus

10. Number these events in human development in the correct order.
    _____ gastrulation
    _____ blastocyst forms
    _____ zygote forms
    _____ tail disappears
    _____ neural tube forms
    _____ heart begins beating

11. Which of the following STDs are caused by bacteria?
    a. chlamydia   d. trichomoniasis
    b. gonorrhea   e. a and b
    c. genital warts   f. all of the above

12. Human milk contains _____ .
    a. antibodies   c. easily digested fats and proteins
    b. lactose   d. all of the above

13. During labor, secretion of oxytocin stimulates _____ .
    a. uterine contractions   c. production of milk proteins
    b. softening of the cervix d. ovulation

14. Match each human reproductive structure with its description.
    _____ testis   a. maternal and fetal tissues
    _____ vas deferens   b. adds fluid to sperm in semen
    _____ placenta   c. produces testosterone
    _____ vagina   d. produces estrogen and
    _____ ovary   progesterone
    _____ oviduct   e. conveys egg to uterus
    _____ prostate gland   f. secretes milk
    _____ mammary gland   g. birth canal
                                h. conveys sperm toward urethra

Digging Into Data

Multiple Births and Birth Defects

People considering fertility treatments should be aware that such treatments raise the risk of multiple births, and that multiple pregnancies are associated with an increased risk of some birth defects.

Figure 26.22 shows the results of Yiwei Tang's study of birth defects reported in Florida from 1996 to 2000. Tang compared the incidence of various defects among single and multiple births. She calculated the relative risk for each type of defect based on type of birth, and corrected for other differences that might increase risk such as maternal age, income, race, and medical care during pregnancy. A relative risk of less than 1 means a defect occurs less often with multiple births than single births. A relative risk greater than 1 means that multiples are more likely to have a defect.

1. What type of birth defect was most common in newborns of the single-birth group?

2. Was that type of defect more or less common among newborns of the multiple-birth group?

3. Tang found that multiples have more than twice the risk of single newborns for one type of defect. Which type?

4. Does a multiple pregnancy increase the relative risk of chromosomal defects in offspring?

	Prevalence of Defect		Relative Risk
	Multiples	Singles	
Total birth defects	358.50	250.54	1.46
Central nervous system defects	40.75	18.89	2.23
Chromosomal defects	15.51	14.20	0.93
Gastrointestinal defects	28.13	23.44	1.27
Genital/urinary defects	72.85	58.16	1.31
Heart defects	189.71	113.89	1.65
Musculoskeletal defects	20.92	25.87	0.92
Fetal alcohol syndrome	4.33	3.63	1.03
Oral defects	19.84	15.48	1.29

FIGURE 26.22 Prevalence, per 10,000 live births, of various types of birth defects among multiple and single births. Relative risk for each defect is given after researchers adjusted for mother's age, race, previous adverse pregnancy experience, education, Medicaid participation during pregnancy, and the infant's sex and number of siblings.

© Cengage Learning.

Critical Thinking

1. Signals from parasympathetic nerves are necessary for a male to have an erection. Explain why anxiety can impede erections.

2. Of all birds, the flightless kiwi has the highest proportion of yolk in its eggs and the longest incubation time. Explain why birds with the longest incubation times have the yolkiest eggs.

3. Fraternal twins are nonidentical siblings that form when two eggs are released and fertilized at the same time. Explain why an increased level of FSH raises the likelihood of fraternal twins.

4. A nursing mother who has an alcoholic drink will secrete alcohol into her milk for two to three hours afterward. Alcohol also decreases her oxytocin production. What effect will the decreased oxytocin have on the woman's ability to nurse her infant?

Plant Form and Function

27

27.1 Leafy Cleanup Crews

A J-Field during its use as a testing site for weapons.

B J-Field today. Poplar trees are helping to remove toxic waste left behind in the soil.

FIGURE 27.1 Phytoremediation.

Credits: (a) OPSEC Control Number #4 077-A-4; (b) Billy Wrobel, 2004.

From World War I until the 1970s, the United States Army tested and disposed of weapons at J-Field, Aberdeen Proving Ground in Maryland (Figure 27.1**A**). Obsolete chemical weapons and explosives were burned in open pits, together with plastics, solvents, and other wastes. Lead, arsenic, mercury, and other metals heavily contaminated the soil and groundwater. So did highly toxic organic compounds, including TCE (trichloroethylene). TCE damages the nervous system, lungs, and liver, and exposure to large amounts of it can be fatal. Today, the toxic groundwater is seeping toward nearby marshes and the Chesapeake Bay.

To protect the bay and clean up the soil, the Army and the Environmental Protection Agency turned to phytoremediation: the use of plants to take up and concentrate or degrade environmental contaminants. They planted poplar trees that cleanse groundwater by taking up TCE and other organic pollutants from it (Figure 27.1**B**).

Like other vascular plants, poplar trees take up soil water through their roots. Along with the water come nutrients and chemical contaminants, including TCE. Although TCE is toxic to animals, it does not harm the trees. The poplars break down some of it, but they release most of the toxin into the atmosphere. Airborne TCE is the lesser of two evils: It breaks down much more quickly in air than it does in groundwater.

In other types of phytoremediation, groundwater contaminants accumulate in the tissues of the plants, which are then harvested for safer disposal elsewhere.

Analyses at J-Field since the poplars were planted show that the phytoremediation strategy is working splendidly. By 2001, the trees had removed 60 pounds of TCE from the soil and groundwater. In addition, they had helped foster communities of microorganisms that were also breaking down toxins. Researchers estimate that the trees will have removed the majority of contaminants from J-Field by 2030.

Phytoremediation uses the ability of plants to take up toxic chemicals dissolved in soil water. Plants have this capacity not for our benefit, but rather to meet their own needs for water and nutrients. Scientists who know about plant physiology are finding ways to put plants to work as leafy cleanup crews. Compared to other methods of cleaning up toxic waste sites, phytoremediation is usually less expensive—and it is more appealing to neighbors.

27.2 Organization of the Plant Body

Like cells of most other multicelled organisms, those in plants are organized as tissues, organs, and organ systems. The plant body has two organ systems: roots and shoots (Figure 27.2). Most shoots are above the ground and most roots are below it, but there are many exceptions. Roots are specialized to absorb water and dissolved minerals, and they often serve to anchor the plant. They are as essential as a plant's shoots, and often at least as extensive. A shoot system includes stems, leaves, and reproductive organs such as flowers. Stems provide a structural framework for the plant's growth. Leaves are specialized to intercept sunlight for photosynthetic production of sugars.

❯ Types of Plant Tissues

Roots, stems, leaves, flowers—all plant organs consist of simple and complex tissues. Simple tissues consist primarily of one type of cell. Complex tissues have two or more cell types.

© Ross E. Koning, plantphys.info.

Parenchyma is a simple tissue that makes up most of the soft internal parts of a plant. The shape of parenchyma cells varies with their function, but all have a thin, flexible cell wall (*left*). These cells are alive in mature tissue and can continue to divide, so they play an important role in wound repair.

Parenchyma has various roles depending on its location. In leaves and soft stems, for example, chloroplast-containing parenchyma (*left*) is photosynthetic; parenchyma cells in stems and roots store starch, proteins, water, and oils in their central vacuoles. Parenchyma also has specialized functions such as structural support, nectar secretion, and gas exchange in leaves and stems.

© ISM/ Phototake.

© Ross E. Koning, plantphys.info.

Collenchyma is a simple, stretchable tissue that supports rapidly growing plant parts such as young stems and leaf stalks. It consists of collenchyma cells, which are elongated and alive at maturity. A complex polysaccharide called pectin imparts flexibility to these cells' primary wall, which is thickened unevenly where three or more of the cells abut (*inset*).

© Ross E. Koning, plantphys.info.

Variably shaped cells of **sclerenchyma** are dead at maturity, but the thick, lignin-containing cell walls that remain after the cells die help this simple tissue resist compression. Fibers (*left*) and sclereids are typical sclerenchyma cells. Fibers are long, tapered cells; they occur in bundles that structurally support vascular tissues in stems and leaves. The bundles flex and twist, but resist stretching. We use fibers of some plants in cloth, rope, paper, and other commercial products. Sclereids strengthen hard seed coats such as peach pits, and they make pear flesh gritty.

© Donald L. Rubbelke/ Lakeland Community College.

Dermal tissues cover and protect exposed surfaces. The first dermal tissue to form on a plant is **epidermis**, which in most species consists of one layer of cells on the plant's outer surface (*inset*). Epidermal cells secrete waxy substances on their outward-facing cell walls. The deposits form a waterproof cuticle that helps the plant conserve water and repel pathogens. The epidermis of leaves and young stems includes specialized cells and, often, hairs and other epidermal cell outgrowths. Paired epidermal cells form stomata (Section 5.5). Plants control the diffusion of water vapor and gases across epidermis by opening and closing

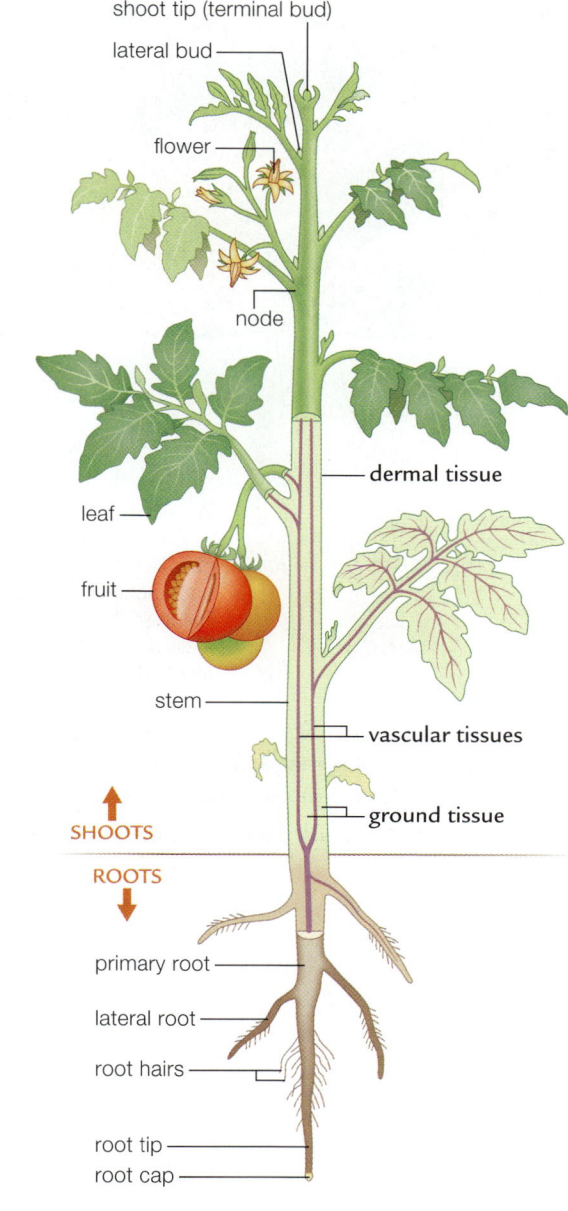

FIGURE 27.2 Animated! Body plan of a tomato plant.

© Cengage Learning.

---

collenchyma Simple plant tissue composed of living cells with unevenly thickened walls; provides flexible support.

dermal tissues Tissues that cover and protect the plant body.

epidermis Outermost tissue layer of a young plant.

parenchyma Simple plant tissue that consists of living parenchyma cells; makes up most of a plant's internal parts.

sclerenchyma Simple plant tissue that is dead at maturity; its lignin-reinforced cell walls structurally support plant parts.

B *Above*, components of phloem.

Living sieve-tube members connect end-to-end, forming tubes of phloem.

Companion cells provide metabolic support for sieve tube members and load sugars into them.

A *Left*, components of xylem. In mature tissue, vessel members and tracheids are dead, but their pitted cell walls remain and serve as conduits for water.

sieve plate

companion cell

sieve-tube member

stack of vessel members

pit

tracheid

FIGURE 27.3 Animated! Components of vascular tissues. Parenchyma cells and fibers of sclerenchyma are also visible in the micrograph.

Figure It Out: What are the green structures inside the parenchyma cells?

Answer: Chloroplasts

Credits: (a) top, Andrew Syred/ Photo Researchers, Inc.; (a) bottom, (b) © Cengage Learning.

these gaps. In older stems and roots, a complex dermal tissue called periderm replaces epidermis.

© ISM/ Phototake.

Pipelines of complex **vascular tissues** (*left*) distribute water and nutrients through all parts of the plant body. The two types of vascular tissue, xylem and phloem, are typically bundled together with sclerenchyma fibers in young plant parts. Both vascular tissues are composed of elongated conducting tubes.

Xylem, which conducts water and mineral ions, consists of two types of cells, tracheids and vessel members (Figure 27.3**A**). Both types of cells are dead in mature tissue, but their stiff, waterproof walls remain. These interconnected walls form tubes that conduct water and also lend structural support to the plant. Water can move laterally between the tubes as well as upward through them.

Phloem, which conducts sugars and other organic solutes, consists of two types of cells, sieve-tube members and companion cells (Figure 27.3**B**). Both types of cells are alive in mature tissue. Sieve-tube members connect end to end at sieve plates, forming sieve tubes that can conduct sugars between all parts of the plant. Each sieve-tube member has an associated **companion cell**, a parenchyma cell that provides it with metabolic support and transfers sugars into it.

© age footstock/ SuperStock.

Ground tissue, which is defined as everything other than dermal and vascular tissue, accounts for the bulk of a plant. It includes the tissues specialized for photosynthesis and for storage (the *inset* shows starch-packed organelles in the ground tissue of a potato). Ground tissue consists mostly of parenchyma, but can also include

companion cell In phloem, parenchyma cell that loads sugars into sieve tubes.

cotyledon Seed leaf; part of a flowering plant embryo.

ground tissue Tissue that makes up the bulk of the plant body; all plant tissue other than dermal and vascular tissue.

phloem Complex vascular tissue of plants; distributes sugars through its sieve tubes.

vascular tissues Tissues that distribute water and nutrients through a plant body.

xylem Complex vascular tissue of plants; its tracheids and vessel members distribute water and mineral ions.

FIGURE 27.4 Locations of some tissues in a stem of buttercup (a eudicot).
© Ross E. Koning, plantphys.info.

vascular tissues

xylem phloem

ground tissue

sclerenchyma (fibers)   parenchyma   collenchyma

epidermis (dermal tissue)

Table 27.1	Overview of Flowering Plant Tissues	
Tissue Type	**Main Components**	**Main Functions**
Simple Tissues		
Parenchyma	Parenchyma cells	Photosynthesis, storage, secretion, tissue repair
Collenchyma	Collenchyma cells	Pliable structural support
Sclerenchyma	Fibers or sclereids	Structural support
Complex Tissues		
Dermal		
Epidermis	Epidermal cells, their secretions and outgrowths	Secretion of cuticle; protection; control of gas exchange and water loss
Periderm	Cork cambium; cork cells; parenchyma	Forms protective cover on older stems, roots
Vascular		
Xylem	Tracheids, vessel members; parenchyma cells; sclerenchyma cells	Water-conducting tubes; structural support
Phloem	Sieve-tube members, parenchyma cells; sclerenchyma cells	Sugar-conducting tubes and their supporting cells

Eudicots / Monocots

In seeds, two cotyledons (seed leaves of embryo)

In seeds, one cotyledon (seed leaf of embryo)

Flower parts in multiples of four or five

Flower parts in multiples of three

Leaf veins usually in a netlike array

Leaf veins usually parallel with one another

Pollen grains have three pores or furrows

Pollen grains have one pore or furrow

Vascular bundles organized in a ring in ground tissue of stem

Vascular bundles throughout ground tissue of stem

FIGURE 27.5 Animated! Some structural differences between eudicots and monocots.

Credits: Clockwise from top left, © Catalin Petolea/ Shutterstock; © Dr. Morley Read/ Shutterstock; ©ArjaKo's/ Shutterstock; Gary Head; Courtesy of Janet Wilmhurst, Landcare Research, New Zealand; © Frans Holthuysen, Making the invisible visible, Electron Microscopist, Phillips Research; Courtesy of Dr. Thomas L Rost; © gresei/ Shutterstock.

other simple tissues. Parenchyma cells that contain chloroplasts compose mesophyll, the only ground tissue that is photosynthetic.

❯ Eudicots and Monocots With more than 260,000 species (and counting), flowering plants dominate the plant kingdom. Magnoliids, eudicots, and monocots (Section 14.7) are the major angiosperm groups. In this chapter, we focus on eudicots and monocots. Eudicots include flowering shrubs and trees, vines, and many nonwoody plants such as tomatoes and dandelions. Lilies, orchids, grasses, and palms are examples of monocots.

Table 27.1 lists the main types of flowering plant tissues, and Figure 27.4 shows some of their locations in a eudicot stem. Eudicots and monocots have the same types of tissues organized in the same systems, but the two lineages differ in many aspects of their tissue organization and hence in their structure. The names of the groups refer to one difference: the number of seed leaves, or **cotyledons**, in their embryos. The plant embryo in a monocot seed has a single cotyledon; the embryo in a eudicot seed has two. Figure 27.5 shows some additional differences.

Take-Home Message

What is the basic structure of a flowering plant?

- Plants typically have aboveground shoots and stems, and belowground roots. All plant parts consist of ground, vascular, and dermal tissues.

- Ground tissues make up most of a plant. Vascular tissues that thread through ground tissue distribute water and solutes. Dermal tissue covers and protects plant surfaces.

- Eudicots and monocots have the same types of tissues, but differ somewhat in their pattern of tissue organization.

27.3 Stems, Roots, and Leaves

A A cross-section of a corn stem shows how vascular bundles in monocot stems are dispersed throughout the ground tissue.

B A cross-section of a buttercup stem shows the cylindrical arrangement of vascular bundles characteristic of eudicot stems. This cylinder divides the stem's ground tissue into regions of cortex and pith.

FIGURE 27.6 Comparing the structure of a primary shoot from **A** a monocot and **B** a eudicot. Cells appear different colors in the sections because they have been stained with different dyes.

› **Stems** Stems are shoots that form the basic structure of a flowering plant, providing support for flowers and keeping leaves positioned for photosynthesis. They can grow above or below the soil, and in many species they are specialized for storage or reproduction. New shoots develop at **nodes**, which are regions of a stem where leaves form.

Inside most stems, the cells of primary xylem and phloem are organized into long, multistranded cords called **vascular bundles**. The main function of these bundles is to conduct water, ions, and nutrients between different parts of the plant. Some components of the bundles—sclerenchyma fibers and the lignin-reinforced walls of tracheids—also play an important role in supporting upright stems.

Vascular bundles extend through the ground tissue of all stems and leaves, but the arrangement of the bundles in these plant parts differs between monocots and eudicots. The vascular bundles of monocot stems are distributed throughout the ground tissue (Figure 27.6**A**). By contrast, the vascular bundles of most eudicot stems form a cylinder that runs parallel with the long axis of the shoot (Figure 27.6**B**). This cylinder divides the ground tissue into regions called cortex and pith. Cortex comprises all ground tissue between the cylinder and epidermis; pith is all ground tissue inside the cylinder.

Specialized stems allow some plants to store nutrients, to reproduce asexually, or both. Stolons are stems that branch from the main stem of the plant and grow horizontally along the ground or just under its surface. Stolons may look like roots, but they have nodes (roots do not have nodes). Roots and leafy shoots that sprout from the nodes develop into new plants. Sto-

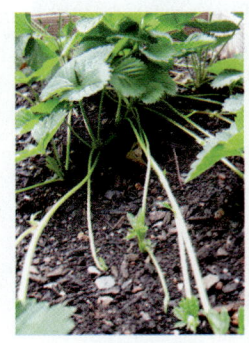

A New plants sprout from runners (stolons) of a strawberry plant.

B The main stems of turmeric plants are underground rhizomes.

C Scales surround the stem at the center of an onion, which is a bulb.

D The stem of arrowroot, or taro, is a corm.

E Potatoes are tubers that grow on stolons. The "eyes" are the nodes of the tubers.

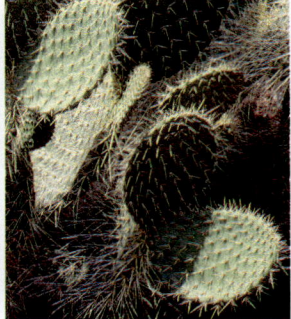

F Prickly pear and many other succulent plants have fleshy (sometimes spiky) cladodes.

FIGURE 27.7 Some examples of specialized stem structure.

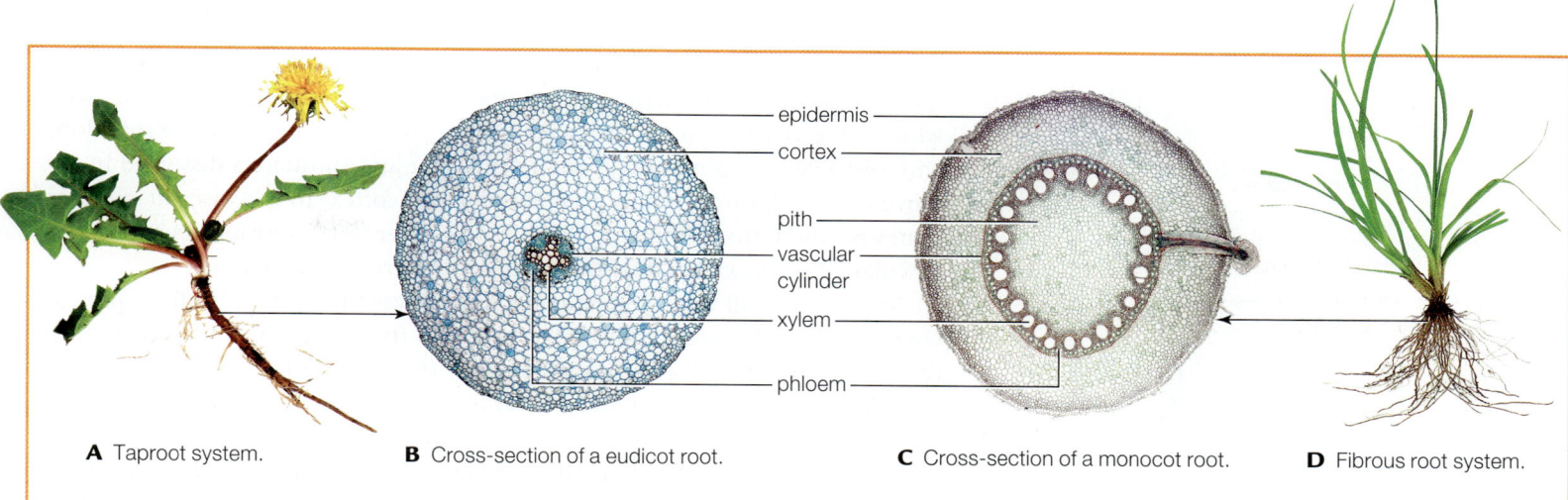

| A Taproot system. | B Cross-section of a eudicot root. | C Cross-section of a monocot root. | D Fibrous root system. |

FIGURE 27.8 Animated! Structure of young roots. A taproot system is characteristic of eudicots; a fibrous root system, of monocots.

Credits: (a) ©Tom Biegalski/ Shutterstock; (b) © Brad Mogen/ Visuals Unlimited; (c) Michael Clayton/ University of Wisconsin, Department of Botany; (d) © design56/ Shutterstock.

lons are commonly called runners because in many plants they "run" along the surface of the soil (Figure 27.7**A**).

Some grasses have rhizomes, which are fleshy stems that typically grow under the soil and parallel to its surface. A rhizome is the main stem of the plant, and it also serves as the plant's primary storage tissue. Shoots that sprout from nodes grow aboveground for photosynthesis and flowering (Figure 27.7**B**).

A bulb is a short section of underground stem encased by overlapping layers of thickened, modified leaves called scales (Figure 27.7**C**). The scales develop from the base of the bulb, as do roots. Bulb scales contain starch and other substances that a plant holds in reserve for times when conditions in the environment are unfavorable for growth. When favorable conditions return, the plant uses these stored substances to sustain rapid growth. The dry, paperlike outer scale of many bulbs serves as a protective covering.

A corm is a thickened underground stem that stores nutrients (Figure 27.7**D**). Like a bulb, a corm can grow roots from its base. Unlike a bulb, a corm is solid rather than layered, and it has nodes from which new plants develop. Tubers are thickened portions of underground stolons; they are the plant's primary storage tissue. Tubers are like corms in that they have nodes from which new shoots and roots sprout, but no roots or shoots grow from their base (Figure 27.7**E**).

Cacti and other succulents have cladodes, which are flattened, photosynthetic stems that store water. New plants form at the nodes. The cladodes of some plants appear leaflike, but most are unmistakably fleshy (Figure 27.7**F**).

〉 Roots Roots take up water and mineral ions, and transport these substances to other parts of the plant. The first structure to emerge from a seed is a root. In most eudicots, this primary root thickens as it grows longer, and it usually develops many lateral branchings. The primary root together with its lateral branchings constitutes a taproot system (Figure 27.8**A**). A taproot is generally larger in diameter than its branchings. By contrast, the primary root that emerges from a monocot seed is quickly replaced with a mat of adventitious roots, which sprout from the base of the stem. All of the roots in the resulting fibrous root system are about equal in diameter—there is no dominant, central root as in a taproot system (Figure 27.8**D**).

In a typical eudicot root, most of the ground tissue lies outside a central column of vascular tissue called the **vascular cylinder** (Figure 27.8**B**). In a typical monocot root, the vascular cylinder divides ground tissue into regions of cortex and pith (Figure 27.8**C**).

node A region of stem where leaves form.
vascular bundle Multistranded cord of xylem and phloem in a stem or leaf.
vascular cylinder A root's central column of vascular tissue.

water flow

vascular cylinder

root hair

primary phloem
primary xylem
endodermis

cortex

epidermis

FIGURE 27.9 Animated!
Internal structure of a eudicot
root. © Cengage Learning.

The outermost layer of a root is a sheet of epidermis. In a young root, many specialized cells of this dermal tissue send out fine extensions called **root hairs**, which collectively increase the root's surface area for taking up soil water along with dissolved oxygen and mineral ions. Water and substances dissolved in it move from soil, through a root's epidermis and cortex, to the vascular cylinder. Osmosis drives this movement (Section 4.5). After water and solutes enter the vascular cylinder, xylem distributes them to the rest of the plant.

Most of the soil water that enters a root moves through cell walls of parenchyma cells in the cortex. Soil water can reach the vascular cylinder by diffusing through these cell walls, but it cannot enter the cylinder the same way. The outer boundary of the vascular cylinder is a sheet of **endodermis**, a single layer of cells that separates the root cortex from the conductive tissues inside the vascular cylinder (Figure 27.9). The endodermal cells that make up endodermis secrete a waxy substance into their walls wherever they abut. This substance forms a waterproof band that prevents water from diffusing through endodermal cell walls into the vascular cylinder. As a result, water and substances dissolved in it can enter a vascular cylinder only by passing through the cytoplasm of an endodermal cell. The presence of endodermis allows a plant to regulate which solutes enter xylem to be distributed to the rest of the plant body.

> **Leaves** Most leaves are thin, with a high surface-to-volume ratio. Monocot leaves tend to be long and narrow. The base of the leaf wraps around the stem, forming a sheath around it (Figure 27.10**A**). In eudicots, a short stalk called a petiole attaches the leaf to a stem (Figure 27.10**B**). Eudicot leaves vary widely in their structure and some are highly specialized (Figure 27.10**C**).

In most flowering plants, leaves are the main organs of photosynthesis. Leaves also function in gas exchange, and they are the major site of evaporative water loss. The outermost tissue of a leaf is a sheet of epidermis one cell thick. This surface tissue may be smooth, sticky, or slimy, with hairs, scales, spikes, hooks, and other specializations (Figure 27.10**D**). Cells of the epidermis secrete a translucent, waxy cuticle that slows water loss (Figure 27.11 ❶). Typically, a

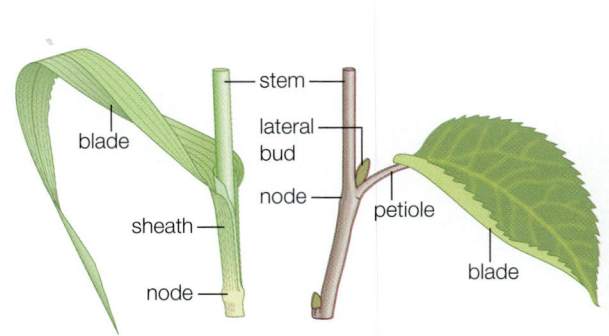

stem

lateral
bud

node

petiole

blade

blade

sheath

node

A A monocot leaf   **B** A simple eudicot leaf

FIGURE 27.10 Leaf structures.

Credits: (a,b) © Cengage Learning; (c) Perennou Nuridsany/ Photo Researchers, Inc.;
(d) © Dan Legere Photography/ www.danlegere.daportfolio.com.

C Example of specialized leaf form. Carnivorous plants of the genus *Nepenthes* grow in nitrogen-poor soil. They secrete acids and protein-digesting enzymes into fluid in a cup-shaped modified leaf. The enzymes release nitrogen from small prey, such as insects, frogs, and rats, that are attracted to odors from the fluid and then drown in it.

D Example of specialized epidermal surface structure: trichomes, which are hairs or other fine outgrowths of epidermal cells. Glandular trichomes such as the ones on this marijuana leaf secrete substances that deter herbivory. Marijuana trichomes secrete a chemical (tetrahydrocannabinol, or THC), that has a psychoactive effect in humans.

① epidermis

③ palisade mesophyll

④ spongy mesophyll

xylem

⑤ vascular tissue

phloem

② stomata

FIGURE 27.11 Animated! Internal structure of a eudicot leaf.

① The upper leaf surface is epidermis with a secreted layer of cuticle.

② A leaf's lower surface, which is also covered with cuticle, usually has more stomata than the upper surface.

③ The bulk of the leaf is mesophyll, which is photosynthetic parenchyma. In many leaves, mesophyll occurs in two forms: elongated palisade mesophyll attached to the upper epidermis, with **④** spongy mesophyll below it.

⑤ Vascular bundles of xylem (*blue*) and phloem (*pink*) form the leaf's veins.

© Cengage Learning.

leaf's upper surface, which receives the most direct sunlight, secretes a thicker cuticle than the lower surface, which tends to be shaded. The lower surface usually has more stomata **②**.

The bulk of a leaf consists of mesophyll, which is photosynthetic parenchyma. Mesophyll in eudicot leaves, which are typically oriented perpendicular to the sun, is arranged in two layers. Elongated cells of palisade mesophyll lie just beneath the upper epidermis **③**. These cells have more chloroplasts than cells of the spongy mesophyll layer below. Cells of spongy mesophyll are more rounded than those of palisade mesophyll, and have larger air spaces between them **④**. Blades of grass and other monocot leaves that grow vertically can intercept light from all directions. The mesophyll in such leaves is typically not divided into distinct layers.

Leaf veins are vascular bundles strengthened with fibers **⑤**. Inside each vein, strands of xylem transport water and dissolved ions to the mesophyll, while strands of phloem transport products of photosynthesis (sugars) away from it. In most eudicot leaves, large veins branch into a network of minor veins; in most monocots, all veins are similar in length and run parallel with the leaf's long axis (examples are shown in Figure 27.5).

Take-Home Message

How are roots and shoots organized?

- The arrangement of vascular tissue inside roots and shoots differs between eudicots and monocots.
- Stems form the basic structural framework of the plant. Many plants have modified stems that function in storage and reproduction.
- Roots absorb water and dissolved minerals from soil. Inside each root is a vascular cylinder, with long strands of xylem and phloem.
- Leaves are the main site of photosynthesis, and their structure varies widely. Components include mesophyll (photosynthetic cells), veins (bundles of vascular tissue), and cuticle-secreting epidermis.

endodermis In plant roots, a layer of cells that separates vascular cylinder from cortex.

root hairs Fine, absorptive extensions of an epidermal cell in young roots.

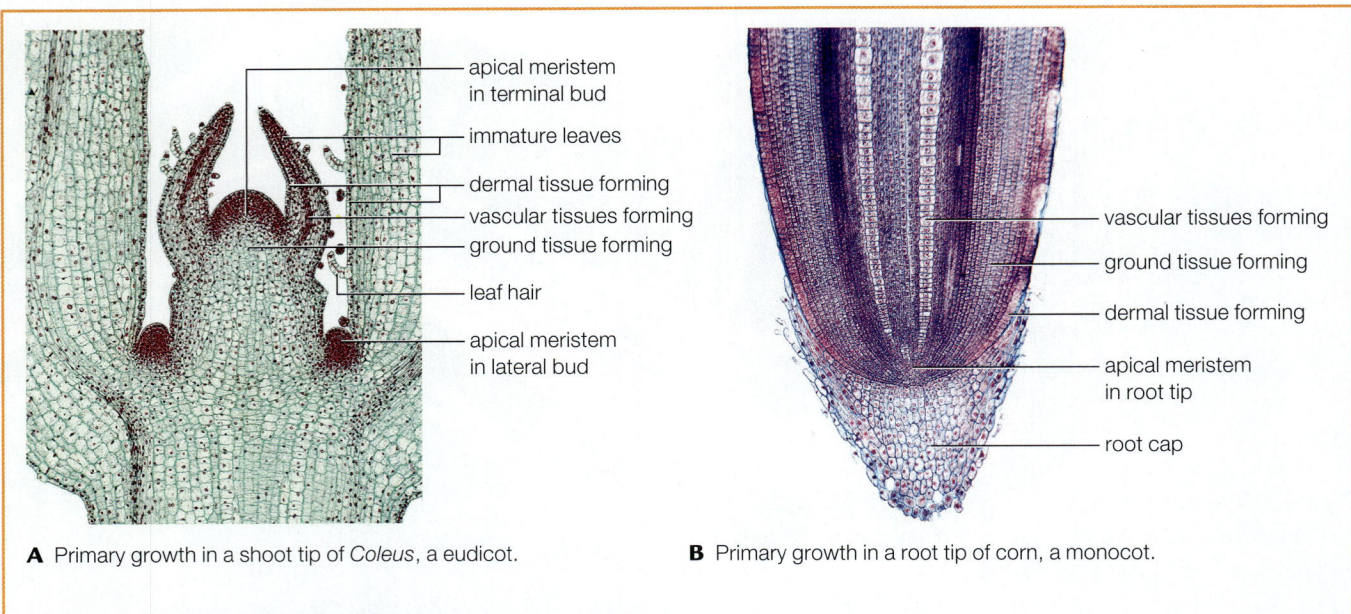

A Primary growth in a shoot tip of *Coleus*, a eudicot.

B Primary growth in a root tip of corn, a monocot.

FIGURE 27.12 Zones of primary growth in roots and shoots.

Credits: (a) © Dale M. Benham, Ph.D., Nebraska Wesleyan University; (b) Biodisc/ Visuals Unlimited.

27.4 Plant Growth

All plant tissues arise from the activity of **meristems**, which are regions of undifferentiated cells that can divide rapidly. Cells just under a meristem differentiate and give rise to shoots, stems, and roots as they elongate and mature.

New plant parts lengthen by activity at **apical meristems** in shoot tips and root tips. The lengthening of young shoots and roots is called **primary growth**. Some eudicots also undergo **secondary growth**, which means that their stems and roots thicken over time. The thickening originates at **lateral meristems**, which are cylindrical layers of meristem that run lengthwise through stems and roots.

❯ Primary Growth During a plant's primary growth, its stems mainly lengthen (as opposed to thickening), and it produces soft organs such as leaves. In a shoot, primary growth occurs at the terminal bud, which is the shoot's tip (Figure 27.12**A**). A mass of apical meristem lies just below the surface of the bud. Cells at the center of the mass are completely undifferentiated, and they divide continually during the growing season. The divisions push some of the cells toward the edge of the mass. Cells that are displaced away from the center of the meristem begin to differentiate. The shoot grows longer mainly because differentiating cells under the mass of meristem elongate as they take on their specialized functions. The apical meristem remains at the tip of the lengthening shoot, continually renewing itself and also producing cells that divide and differentiate below it.

Masses of tissue that bulge out near the sides of the apical meristem develop into leaves. As the stem lengthens, more leaves form and mature in tiers, one after the next, and lateral buds form at nodes. Depending on hormonal signals, divisions of apical meristem cells in a lateral bud can give rise to either a branch, a leaf, or a flower.

As in stems, most of the primary growth in roots originates at apical meristems. Cells in the apical meristem of a root tip divide repeatedly, and some

apical meristem Meristem in the tip of a shoot or root; gives rise to primary growth.

bark In woody plants, all living and dead tissue that lies outside of the vascular cambium.

cork Component of periderm; waterproofs, insulates, and protects the surfaces of woody stems and roots.

cork cambium Lateral meristem that gives rise to periderm.

lateral meristem Vascular cambium or cork cambium; cylindrical sheet of meristem that runs lengthwise through shoots and roots; gives rise to secondary growth.

meristem Zone of undifferentiated plant cells that can divide rapidly.

periderm Plant dermal tissue that replaces epidermis during secondary growth of woody stems and roots.

primary growth Lengthening of young shoots and roots; originates at apical meristems.

secondary growth Thickening of older stems and roots; originates at lateral meristems.

vascular cambium Lateral meristem that produces secondary xylem and phloem.

wood Accumulated secondary xylem.

differentiate to form the specialized tissues of a root (Figure 27.12**B**). Some of these cells form a root cap, a dome-shaped mass of cells that protects the soft, young root as it grows through soil. Cells of the root cap are shed continually as the root lengthens.

) Secondary Growth Plants thicken and become woody during secondary growth (**wood** is accumulated secondary xylem). Woody eudicots have two types of lateral meristems (Figure 27.13**A**), vascular cambium and cork cambium (*cambium* is the Latin word for change). **Vascular cambium** produces secondary vascular tissue: secondary xylem on its inner surface, and secondary phloem on its outer surface (Figure 27.13**B**). As the core of xylem thickens, it also displaces the vascular cambium toward the surface of the stem or root. The displaced cells of the vascular cambium divide in a widening circle, so the tissue's cylindrical form is maintained (Figure 27.13**C**).

As seasons pass, the expanding inner core of xylem puts increasing pressure on the stem or root surface. In time, the pressure ruptures the cortex and the outer secondary phloem. Then, another lateral meristem, the **cork cambium**, forms and gives rise to periderm. **Periderm** is a dermal tissue that consists of parenchyma and cork, as well as the cork cambium that produces them. The **cork** component of periderm has densely packed rows of dead cells, the walls of which are thickened with a waxy substance. Cork protects, insulates, and waterproofs the stem or root surface. What we call **bark** is secondary phloem and periderm—all of the living and dead tissues that lie outside the ring of vascular cambium.

A In eudicots, secondary growth (thickening of older stems and roots) occurs at two lateral meristems, vascular cambium and cork cambium. Vascular cambium gives rise to secondary xylem and phloem; cork cambium, to periderm.

vascular cambium
cork cambium
cortex
pith

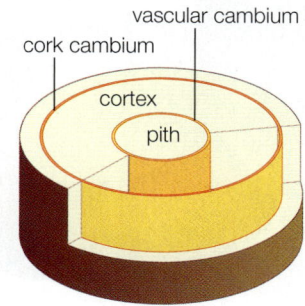

primary phloem
vascular cambium
primary xylem

secondary phloem
vascular cambium
secondary xylem
cork cambium

C Pattern of division and differentiation of vascular cambium cells (*orange*). With each division, one descendant cell differentiates and the other remains undifferentiated.

Cells that differentiate on the cortex (outer) side of the vascular cambium become part of phloem (P_1, P_2, P_3).

Cells that differentiate on the pith (inner) side of the vascular cambium become part of xylem (X_1, X_2, X_3).

X_1 X_1
X_2 X_2
X_3 X_3
P_2 P_2
P_1 P_1
P_3 P_3
P_2 P_2
P_1 P_1

X_1 X_1
X_2 X_2
P_1 P_1

division
differentiation
division
differentiation
division
differentiation
division
differentiation

B In spring, primary growth resumes at terminal and lateral buds. Secondary growth occurs at vascular cambium. Divisions of meristem cells in the vascular cambium expand the inner core of xylem, displacing the vascular cambium (*orange*) toward the surface of the stem or root.

FIGURE 27.13 Animated!
Secondary growth in a eudicot stem.
Figure It Out: Which meristem is the source of wood?

Answer: Vascular cambium

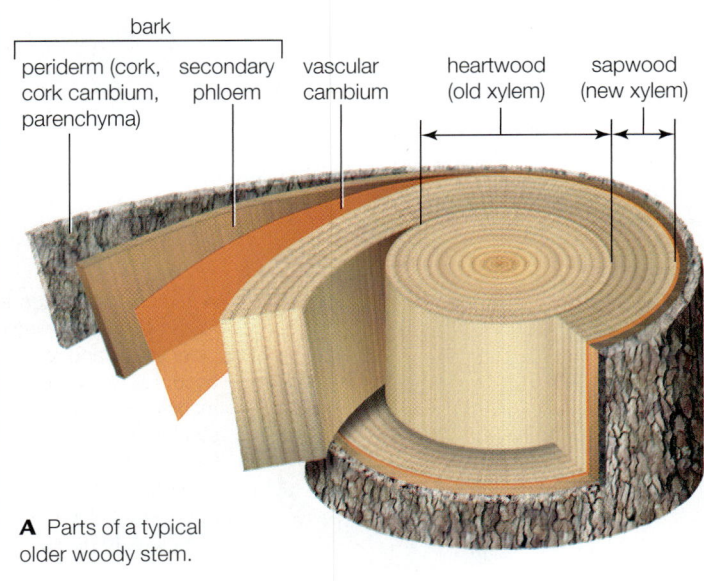

A Parts of a typical older woody stem.

vessel in xylem   direction of growth →

early | late | early | late | early | late | early | late | early

B Early and late wood in an ash tree. The early wood formed during warm, wet springs. The late wood formed during warm, dry summers. Each band of early and late wood is one growth ring, or "tree ring."

1587–1589   1606–1612

C The relative thicknesses of a tree's rings can be used to estimate average annual rainfall long before records of climate were kept. This is a section of a bald cypress tree that was living near English colonists when they first settled in North America. Narrower annual rings mark years of severe drought.

FIGURE 27.14 Structure of woody stems.

Rings visible in the heartwood and sapwood are regions of early and late wood. In most temperate zone trees, one ring forms each year.

Credits: (a) © Cengage Learning; (b) © Peter Gasson, Royal Botanic Gardens, Kew; (c) © David W. Stahle, Department of Geosciences, University of Arkansas.

Wood's structure and function change with age (Figure 27.14**A**). Over time, the secondary xylem at the center of an older woody stem becomes so clogged by tannins, gums, and oils that it can no longer transport fluid through the stem. This heartwood is dense and strong, so it provides excellent structural support. Sapwood, the region of still-functional secondary xylem, lies between heartwood and the vascular cambium. During the spring in temperate zones, sugar-rich fluid (sap) travels through sapwood, from tree roots to buds. New Englanders collect maple tree sap to make maple syrup.

In trees native to regions where growth slows or stops during the winter, vascular cambium gives rise to large-diameter, thin-walled xylem cells (early wood) during spring. Late wood, with small-diameter, thick-walled xylem cells, forms during summer. Thus, a cross-section reveals rings that consist of alternating bands of early and late wood (Figure 27.14**B**). In tropical regions where weather does not vary seasonally, trees grow at the same rate all year and do not have these growth rings.

Tree rings can be used to estimate average annual rainfall; to date archaeological ruins; to gather evidence of wildfires, floods, landslides, and glacier movements; and to study the ecology and effects of parasitic insect populations. How? Some tree species, such as redwoods and bristlecone pines, add wood over centuries, one ring per year. Count an old tree's rings, and you have an idea of its age. If you know the year in which the tree was cut, you can find out when a particular ring formed by counting the rings backward from the outer edge. Thickness and other features of the rings offer clues about the environmental conditions that prevailed during the years they formed (Figure 27.14**C**).

Take-Home Message

How does growth occur in plants?

- Primary growth (lengthening) occurs at apical meristems in shoot and root tips.

- Secondary growth, which thickens the older roots and shoots, occurs at lateral meristems (vascular cambium and cork cambium) in eudicots.

- Vascular cambium gives rise to secondary xylem and phloem. Cork cambium gives rise to periderm, a dermal tissue.

- Bark consists of all living and dead tissue that lie outside the vascular cambium.

27.5 Plant Nutrition

Sixteen elements are required for growth and maintenance of a plant body. Carbon, oxygen, and hydrogen atoms are abundantly available in carbon dioxide and water. Plants obtain the other nutrients when their roots take up minerals dissolved in soil water.

Carbon, oxygen, and hydrogen are macronutrients, which means they are required in amounts above 0.5 percent of the plant's dry weight. Other macronutrients include nitrogen, phosphorus, and sulfur (which are components of proteins and nucleic acids in all organisms), and potassium and calcium (which affect processes such as cell signaling). Magnesium, a component of chlorophyll, is another macronutrient.

Seven elements (chlorine, iron, boron, manganese, zinc, copper, and molybdenum) are micronutrients, meaning they make up traces—typically a few parts per million—of the plant's dry weight. Among other roles, these elements serve as enzyme cofactors (Section 4.4).

Soils vary in their nutrient content and other properties, so some are better for plant growth than others. Most plants do best in soil that contains some decaying organic matter, or **humus**. Humus provides essential minerals and helps soil retain water. The proportions of mineral particles (sand, silt, and clay) are also important. The biggest sand grains are about one millimeter in diameter, visible with the naked eye. Silt particles are hundreds or thousands of times smaller than sand grains, so they are too small to see. Clay particles are even smaller. Clay particles enhance the nutritive value of soils because they have a negative charge that attracts positively charged mineral ions in soil water. Thus, clay-rich soil retains dissolved nutrients that would otherwise trickle past roots too quickly to be absorbed. Sand and silt do not bind mineral ions as well as clay, but they are also necessary for plant growth. Water and air occupy spaces between the particles and organic bits in soil. Without enough sand and silt to intervene between tiny particles of clay, soil packs so tightly that it excludes air. Without air, root cells cannot secure enough oxygen for aerobic respiration.

> **Root Adaptations for Nutrient Uptake** Structures such as hairs adapt roots to absorb water and nutrients. Most flowering plants also take part in mutualisms that provide nutritional benefit (Section 14.9).

A **mycorrhiza** (plural, mycorrhizae) is a mutualism between a root and a fungus that grows on or in it (Figure 27.15**A**). Filaments of the fungus (hyphae) form a velvety cloak around roots, or penetrate the root cells. Collectively, the hyphae have a large surface area, so they absorb mineral ions from a larger volume of soil than roots alone. The fungus absorbs some sugars and nitrogen-rich compounds from root cells. In return, the root cells get some scarce minerals that the fungus is better able to absorb.

Some types of soil bacteria form mutualisms with clover, peas, and other legumes. Like all other plants, legumes require a lot of nitrogen. Nitrogen gas is abundant in air, but plants do not have an enzyme that can break it apart. Nitrogen-fixing bacteria in soil do (Section 17.6). These bacteria infect roots and thus become symbionts in localized swellings called **root nodules** (Figure 27.15**B**). The bacteria share their fixed nitrogen with the plant, thus providing an advantage in areas with nitrogen-poor soil. In return, the plant provides these anaerobic bacteria with an oxygen-free environment, and shares its photosynthetically produced sugars with them.

A Filaments (hyphae) of a mycorrhizal fungus associate with a plant root. The filaments deliver mineral nutrients to the root and receive sugars from it.

B Bacteria that live in root nodules on this soybean plant fix nitrogen from air, and share it with the plant.

FIGURE 27.15 Mutualisms that enhance a plant's ability to take up nutrients.

Credits: (a) Photo courtesy of Iowa State University Plant and Insect Diagnostic Clinic, (b) © Wally Eberhart/ Visuals Unlimited.

Take-Home Message

How do plants obtain essential nutrients?

- Plants require sixteen elemental nutrients. Oxygen, hydrogen, and carbon are abundant in water and air; the others come from mineral ions taken up in soil water.

- Soils vary in their nutrient content. Humus (decaying organic material) and clay help soils retain mineral ions.

- Mycorrhizae and root nodules enhance a plant's ability to take up nutrients.

humus Decaying organic matter in soil.
mycorrhiza Mutualism between a fungus and a root.
root nodules Of some plant roots, swellings that contain nitrogen-fixing bacteria.

cuticle   epidermal cell

stoma

guard cells

FIGURE 27.16 Cuticle and stoma, water-conserving structures in a section of pincushion leaf. In this species, guard cells are recessed beneath a ledge of cuticle. The ledge creates a small cup that traps moist air above the stoma.

Dr. Keith Wheeler/ Photo Researhers, Inc.

20 μm   guard cells

A This stoma is open. When guard cells swell with water, they bend so a gap (the stoma) opens up between them. The gap allows the plant to exchange gases with air.

B This stoma is closed. When the guard cells lose water, they collapse against each other so the gap between them closes. A closed stoma limits water loss, but it also limits gas exchange.

FIGURE 27.17 Stomata in action. Whether a stoma is open or closed depends on how much water is plumping up guard cells. The round structures inside the cells are chloroplasts. Guard cells are the only type of plant epidermal cell with these organelles.

Courtesy of E. Raveh.

27.6 Fluid Movement in the Plant Body

❭ Controlling Water Loss In land plants, at least 90 percent of the water transported from roots to leaves is lost by evaporation. Only about 2 percent is used in metabolism, but that amount must be maintained or photosynthesis, growth, membrane functions, and other processes will shut down. A plant that is running low on water cannot move around to seek out more, as most animals can. A cuticle and stomata help the plant conserve the water it already holds in its tissues (Figure 27.16).

A cuticle is impermeable to water, so it restricts water loss. It is also impermeable to gases, so it restricts gas exchanges between the plant and the air. Gas exchange is important because the concentrations of carbon dioxide and oxygen in the plant's internal air spaces affect the rate of critical metabolic pathways such as photosynthesis and aerobic respiration. If a plant were impermeable to gases, it could not take in carbon dioxide to run photosynthesis. Neither could it sustain aerobic respiration for very long, because this pathway uses oxygen.

Stomata help a plant balance its need to conserve water with the requirement to exchange gases with the air. A pair of specialized epidermal cells defines each stoma. When these two **guard cells** swell with water, they bend slightly so a gap forms between them (Figure 27.17**A**). The gap is the stoma. When the guard cells lose water, they collapse against one another, so the gap between them closes (Figure 27.17**B**).

When stomata are closed, the plant conserves water; when they are open, gases flow into and out of the plant's interior tissues. Carbon dioxide for photosynthesis enters the plant, then diffuses through air in the spaces between mesophyll cells. Oxygen released by photosynthesis diffuses in the opposite direction. Water vapor usually diffuses out of stomata when they are open, because the interior of a leaf or stem usually contains more water than the environment does. Indentations or other structures that reduce air movement next

cohesion–tension theory Explanation of how water moves through vascular plants: Transpiration creates a tension that pulls a cohesive column of water through xylem, from roots to shoots.

guard cell One of a pair of cells that define a stoma across the epidermis of a leaf or stem.

transpiration Evaporation of water from aboveground plant parts.

to an open stoma decrease the rate of evaporation from it. The stoma shown in Figure 27.16 has a ledge of cuticle over it for this reason.

Stomata open or close based on an integration of cues that include humidity, the level of carbon dioxide inside the leaf, light intensity, and hormonal signals from other parts of the plant. These cues trigger changes in osmotic pressure (Section 4.5) that open or close the guard cells. For example, when the sun comes up in the morning, the light causes guard cells to begin pumping potassium ions into their cytoplasm. Water follows the ions by osmosis, plumping the guard cells so the gap between them opens. Carbon dioxide from the air diffuses through the open stomata into the plant's tissues, and photosynthesis begins.

› Water Moves Through Xylem

Water that enters a root travels to the rest of the plant inside continuous pipelines of xylem. These pipelines consist of the walls of dead vessel members and tracheids. Dead tracheids and vessel members cannot be expending any energy to pump water upward against gravity, so how does water in xylem move all the way from roots to leaves that may be more than 100 meters (330 feet) above the soil?

The movement of water in vascular plants is driven by two features of water: evaporation and cohesion. By the **cohesion–tension theory**, water in xylem is pulled upward by air's drying power, which creates a continuous negative pressure called tension. The tension extends from leaves at the tips of shoots to roots that may be hundreds of feet below them (Figure 27.18).

Most of the water a plant takes up is lost by evaporation, typically from open stomata on the plant's leaves and stems ❶. The evaporation of water from aboveground plant parts, particularly at stomata, is called **transpiration**. Transpiration's effect on water inside a plant is a bit like what happens when you suck a drink through a straw. Transpiration puts negative pressure (pulls) on continuous columns of water in xylem. A column of water resists breaking into droplets as it moves through a narrow conduit such as a straw or a xylem tube. Why? Water molecules are connected by hydrogen bonds (Section 2.4), so a pull on one tugs all of them. Thus, the negative pressure created by transpiration (tension) pulls on the entire column of water that fills a xylem tube ❷. Because of water's cohesion, the tension extends from leaves that may be hundreds of feet in the air, down through stems, into young roots where water is being absorbed from soil ❸.

❶ Water evaporates from aboveground plant parts (a process called transpiration), mainly through open stomata.

❷ The evaporation exerts tension (pulls) on the narrow columns of water that fill xylem tubes. The tension extends from leaves to roots because liquid water has cohesion. Hydrogen bonds among water molecules collectively impart cohesion to liquid water.

❸ As long as evaporation continues, the tension it creates drives the uptake of more water molecules from soil.

FIGURE 27.18 Animated! Cohesion–tension theory of water transport in vascular plants.

© Cengage Learning.

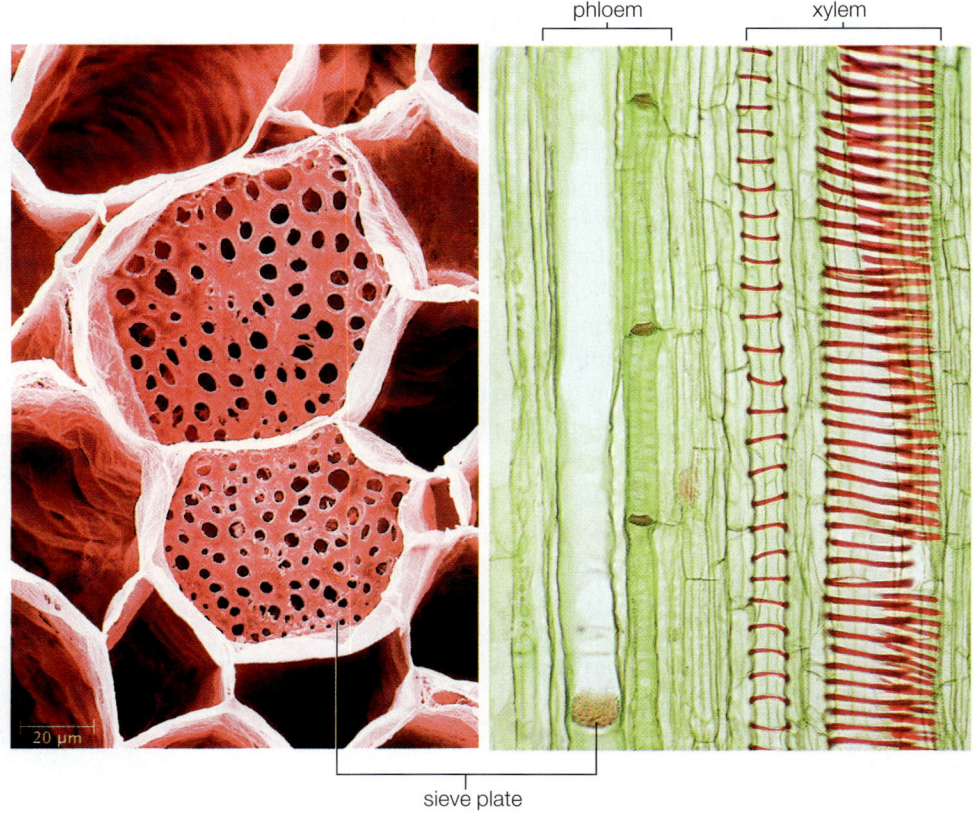

phloem | xylem

20 µm

sieve plate

> **FIGURE 27.19** Sieve plates.

The scanning electron micrograph on the *left* shows sieve plates on the ends of two side-by-side sieve-tube members. The light micrograph on the *right* shows sieve plates in two columns of phloem. *Red* stripes are rings of lignin in xylem vessel members. The spacing of the rings differs because older vessels stretch lengthwise as cells in newer tissue enlarge.

Credits: left, © J.C. Revy/ ISM/ Phototake; right, © M.I. Walker/ Wellcome Images.

pressure flow theory Explanation of how organic compounds move through vascular plants: Fluid flow through sieve tubes is driven by a difference in turgor between source and sink regions.

sink Region of a plant where sugars are being used or stored.

source Region of a plant where sugars are being produced or released from storage.

translocation Movement of organic molecules through phloem.

> **Sugars Flow Through Phloem** Unlike a conducting tube of xylem, a sieve tube in phloem consists of living cells—a stack of sieve-tube members. Sieve-tube members abut at their perforated end walls, which are called sieve plates after their appearance (Figure 27.19). As a sieve-tube member cell matures, it loses most of its organelles, including the nucleus, Golgi bodies, and almost all of its ribosomes. How can it stay alive without these organelles? Each sieve-tube member has an associated companion cell that retains its nucleus and other components at maturity. Many plasmodesmata connect the cytoplasm of the two cells, so the companion cell can provide all of the metabolic functions necessary to sustain its paired sieve-tube member.

A plant produces sugar molecules during photosynthesis. Some of these molecules are used by the cells that make them. The remainder move into sieve tubes, where they are conducted to other parts of the plant. The movement of sugars and other organic molecules through phloem is called **translocation**. Inside sieve tubes, sugars flow from a **source** (a region of the plant where they are being produced or released from storage) to a **sink** (a region where they are being broken down for energy, remodeled into other compounds, or stored for later use). Photosynthetic tissues are typical source regions; developing shoots and fruits are typical sink regions. However, source and sink regions can change. For example, in some plants, sugars move from shoots to roots during autumn in preparation for a seasonal period of winter dormancy. As dormancy ends in spring, sugars move in the opposite direction, from roots to budding shoots.

Why do sugars flow from a source to a sink? A pressure gradient between the two regions drives the movement (Figure 27.20). The gradient is set up at source regions, where sugars move into companion cells, then into sieve tubes ❶. The sugars can be pumped into the sieve tube by active transport proteins, or they can diffuse from the companion cell into the tube through plasmodesmata. Either way, sugar loading at a source region increases the solute concentration of a sieve tube's cytoplasm so that it becomes hypertonic with respect to that of the surrounding cells. Water follows the sugars by osmosis, moving into the sieve tube from the surrounding cells ❷. The cell walls of mature sieve-tube members cannot expand very much, so the influx of water raises the osmotic pressure (turgor) inside the tube.

Pressure inside a sieve tube can be very high—up to five times higher than that of an automobile tire. The high fluid pressure pushes the sugar-rich cytoplasm inside a sieve tube toward sink regions, where the osmotic pressure is lower ❸. The pressure inside a sieve tube decreases at sink regions because sugars are moving from the sieve tube into sink cells. Water follows, again by osmosis ❹, so turgor inside sieve-tube members decreases in these regions.

The explanation of how osmotic pressure pushes sugar-rich fluid inside a sieve tube from source to sink is called the **pressure flow theory**.

Take-Home Message

How do fluids move through a plant body?

- A plant loses water mainly by transpiration through open stomata. Closed stomata minimize water loss, but also prevent gas exchange with air.

- Transpiration puts water in xylem into a continuous state of negative tension from roots to leaves. The tension pulls columns of water in xylem upward through the plant.

- At source regions, sugars move into companion cells, then into sieve tubes. Sugars exit sieve tubes at sink regions. A pressure gradient pushes sugar-laden cytoplasm inside a sieve tube from a source to a sink.

27.7 Leafy Cleanup Crews (revisited)

With elemental pollutants such as lead or mercury, the best phytoremediation strategies use plants that take up toxins and store them in aboveground tissues. The toxin-laden plant parts can then be harvested for safe disposal. Researchers have genetically modified some plants to enhance their absorptive and storage capacity. In the case of organic toxins such as TCE, the best phytoremediation strategies use plants with biochemical pathways that break down the compounds to less toxic molecules. Researchers are beefing up these pathways by transferring genes from bacteria or animals into plants, and by enhancing expression of genes that encode molecular participants in the plants' own detoxification pathways.

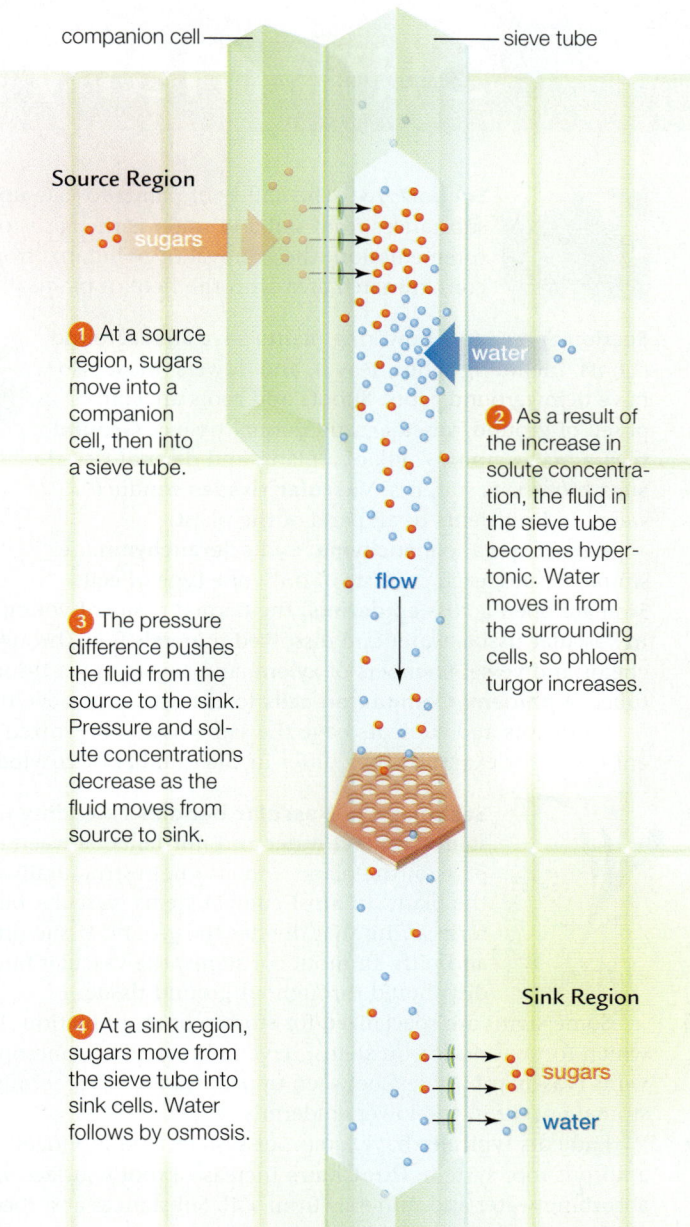

companion cell — — sieve tube

Source Region

→ sugars

❶ At a source region, sugars move into a companion cell, then into a sieve tube.

water

❷ As a result of the increase in solute concentration, the fluid in the sieve tube becomes hypertonic. Water moves in from the surrounding cells, so phloem turgor increases.

flow

❸ The pressure difference pushes the fluid from the source to the sink. Pressure and solute concentrations decrease as the fluid moves from source to sink.

Sink Region

❹ At a sink region, sugars move from the sieve tube into sink cells. Water follows by osmosis.

→ sugars

→ water

FIGURE 27.20 Translocation of organic compounds in phloem from source (shaded *pink*) to sink (shaded *yellow*).

Water molecules (*blue* balls) move from cell to cell by diffusing through plasma membranes. Sugar molecules (*red* balls) are actively transported across plasma membranes, and they also diffuse through plasmodesmata.

© Cengage Learning.

WHERE YOU ARE GOING . . .

The next chapter continues our exploration of plant form and function, with a focus on how plants reproduce and develop.

© OPSEC Control Number #4 077-A-4.

Summary

Section 27.1 The ability of plants to take up substances from soil and water is the basis for phytoremediation, the removal of pollutants from a contaminated area with the help of plants.

Section 27.2 Most flowering plants have aboveground shoots, including stems, leaves, and flowers. Most also have belowground roots. Shoots and roots are composed of ground, vascular, and dermal tissues. **Ground tissue** makes up the bulk of a plant, and **dermal tissues** protect its surfaces. **Vascular tissues** conduct water and nutrients to all parts of the plant.

Parenchyma, **collenchyma**, and **sclerenchyma** are simple tissues; each consists of only one type of cell. Stomata open across **epidermis**, the dermal tissue of young plants. In vascular tissue, water and dissolved minerals flow through tracheids and vessel members of **xylem**, and sugars travel through sieve tubes of **phloem**. **Companion cells** load sugars into sieve tubes.

Monocots and eudicots have the same tissues organized in different ways. For example, they differ in the number of **cotyledons**.

Section 27.3 **Vascular bundles** extending through stems conduct water and nutrients between different parts of the plant, and also help structurally support the plant. In most eudicot stems, vascular bundles form a ring that divides the ground tissue into cortex and pith. In monocot stems, the vascular bundles are distributed throughout ground tissue.

Some stems are specialized for storage or reproduction. Leaves, which form at **nodes** on stems, have photosynthetic mesophyll and veins (vascular bundles) between upper and lower epidermis. Most stomata occur in the lower epidermis.

Eudicots typically have a taproot system, and monocots have a fibrous root system. **Root hairs** increase a root's surface area for absorbing water and minerals from soil. Substances absorbed by a root must cross **endodermis** to enter the **vascular cylinder** and be distributed to the rest of the plant.

Section 27.4 All plant tissues originate at **meristems**, which are regions of undifferentiated cells that retain their ability to divide. **Primary growth** (lengthening) arises at **apical meristems** in the tips of young shoots and roots. **Secondary growth** (thickening) arises at **lateral meristems** (**vascular cambium** and **cork cambium**) in older stems and roots. Cell divisions in vascular cambium add secondary xylem (**wood**) and secondary phloem. Cell divisions in cork cambium give rise to **periderm**, which includes **cork**. **Bark** is periderm and secondary phloem of a woody stem.

Section 27.5 Plant growth requires steady sources of elemental nutrients. Oxygen, carbon, and hydrogen atoms are abundant in air and water; nitrogen, phosphorus, sulfur, and other elements are available in soil water. The availability of water and mineral ions in a particular soil depends on its proportions of sand, silt, and clay, and also on its **humus** content.

Mycorrhizae and nitrogen-fixing bacteria in **root nodules** enhance nutrient uptake in many plants.

Section 27.6 A cuticle helps a plant conserve water; stomata help it balance water conservation with gas exchange required for photosynthesis and aerobic respiration. A stoma is defined by a pair of **guard cells**. Environmental signals trigger guard cells to pump ions into or out of their cytoplasm; water follows the ions by osmosis. Water moving into guard cells plumps them, so the stoma opens. Water diffusing out of the cells causes them to collapse against each other, so the stoma closes.

The **cohesion–tension theory** explains how water moves through plants: **Transpiration** (evaporation from plant parts) pulls water upward inside xylem from roots to leaves. The cohesion of water molecules keeps the columns of fluid continuous inside the narrow xylem vessels.

Sugars move through a plant by **translocation** in sieve tubes, which consist of stacked sieve-tube members separated by perforated sieve plates. By the **pressure flow theory**, the movement of sugar-rich fluid through a sieve tube is driven by a pressure gradient between **source** and **sink** regions.

Self-Quiz   Answers in Appendix I

1. Roots and shoots lengthen through activity at _____ .
   a. apical meristems   c. vascular cambium
   b. lateral meristems   d. cork cambium

2. _____ conducts water and mineral ions; _____ conducts sugars.
   a. Phloem; xylem   b. Xylem; phloem

3. Mesophyll consists of _____ .
   a. waxes and cutin   c. parenchyma cells
   b. lignified cell walls   d. cork but not bark

4. Which of the following cell types are alive in mature tissue?
   a. companion cells   c. tracheids
   b. sieve-tube members   d. vessel members

5. Xylem and phloem are _____ tissues.
   a. ground   b. vascular   c. dermal   d. both b and c

6. A cylindrical arrangement of vascular bundles is characteristic of _____ .
   a. monocot stems   c. monocot roots
   b. eudicot stems   d. eudicot roots

7. The activity of lateral meristems _____ older roots and stems.
   a. lengthens   b. thickens   c. both a and b

8. Plants obtain carbon from _____ .
   a. soil water   b. air   c. both a and b

9. Plants obtain phosphorus from _____ .
   a. soil water   b. air   c. both a and b

10. Decomposing matter in soil is called _____ .
    a. clay   c. silt
    b. humus   d. sand

11. Water evaporation from plant parts is called _____ .
    a. translocation
    c. transpiration
    b. cohesion
    d. tension

12. When guard cells swell, _____ .
    a. transpiration ceases
    c. stomata open
    b. sugars enter phloem
    d. root cells die

13. Tree rings occur when _____ .
    a. there are droughts during growth seasons
    b. environmental conditions influence xylem cell size
    c. heartwood alternates with sapwood
    d. periderm replaces epidermis

14. Water transport from roots to leaves occurs by _____ .
    a. a pressure gradient inside sieve tubes
    b. different solutes at source and sink regions
    c. the pumping force of xylem vessels
    d. transpiration, tension, and cohesion of water

15. Sugar transport from leaves to roots occurs by _____ .
    a. a pressure gradient inside sieve tubes
    b. different solutes at source and sink regions
    c. the pumping force of xylem vessels
    d. transpiration, tension, and cohesion of water

16. Match the plant parts with the suitable description.
    _____ apical meristem   a. mass of secondary xylem
    _____ lateral meristem   b. source of primary growth
    _____ cotyledon   c. vascular bundle
    _____ leaf vein   d. source of secondary growth
    _____ vascular cylinder   e. one in monocot seeds
    _____ wood   f. central column in roots

Critical Thinking

1. Suppose you hammer a nail into an oak tree trunk five feet above the ground. If you return fifteen years later, after the tree has grown 20 feet taller, how high will the nail be?

2. Iron and manganese are components of chlorophyll. How does a plant obtain these minerals? What effect would you expect a deficiency in one of these minerals to have on plant growth?

3. Farmers sometimes plant clovers or other legumes in fields during seasons when they are not growing crops. When a crop-planting season approaches, the "cover crop" is plowed under. Explain how this procedure enhances soil fertility.

4. Allen is studying the rate of water uptake by tomato plant roots. He notices that several environmental factors, including wind and relative humidity, affect the rate. Explain how they might do so.

5. Was the micrograph shown at *right* taken from a section of stem or a root? Monocot or eudicot?

© Mike Clayton/ University of Wisconsin Department of Botany.

Digging Into Data

TCE Uptake by Transgenic Plants

Plants used for phytoremediation take up organic pollutants from the soil or air, then transport the chemicals to plant tissues, where they are stored or broken down. Researchers are now designing transgenic plants that have enhanced ability to take up or break down toxins.

In 2007, Sharon Doty and her colleagues published the results of their efforts to design plants useful for phytoremediation of soil and air containing organic solvents. The researchers used *Agrobacterium tumefaciens* (Section 10.4) to deliver a mammalian gene into poplar plants. The gene encodes cytochrome P450, an enzyme involved in the breakdown of a range of organic molecules, including solvents such as TCE. The results of one of the researchers' tests on these transgenic plants are shown in Figure 27.21.

1. How many transgenic plants did the researchers test?

2. In which group did the researchers see the slowest rate of TCE uptake? The fastest?

3. On day 6, what was the difference between the TCE content of air around transgenic plants and that around vector control plants?

4. Assuming no other experiments were done, what two explanations are there for the results of this experiment? What other control might the researchers have used?

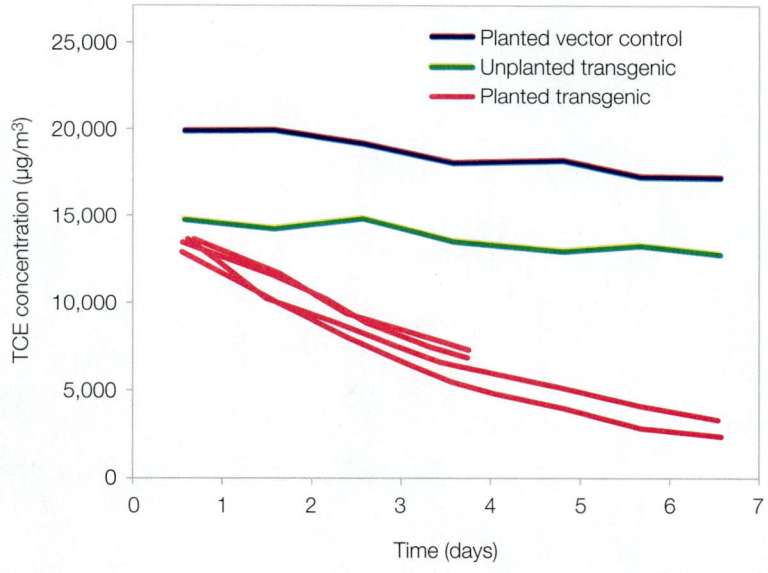

FIGURE 27.21 TCE uptake by transgenic poplar plants.

Individual potted plants were kept in separate sealed containers with an initial 15,000 micrograms of TCE (trichloroethylene) per cubic meter of air. Samples of the air in the containers were taken daily and measured for TCE content.

Controls included a tree transgenic for a *Ti* plasmid with no cytochrome P450 in it (vector control), and a bare-root transgenic tree (one that was not planted in soil).

© Cengage Learning.

Plant Reproduction and Development

28

... WHERE YOU HAVE BEEN

This chapter continues our discussions of plant structure (Chapter 27), evolution (Sections 9.8 and 12.5), and life cycles (14.7). A review of gene expression and control (7.2, 7.7) will help you understand how plant hormones work. You will also revisit carbohydrates (2.6), receptor proteins (3.3), aerobic respiration (5.6), gametes (8.7), inheritance (9.3), feedback mechanisms (19.5), and estrogens in mammals (26.3).

28.1 Plight of the Honeybee

In the fall of 2006, commercial beekeepers worldwide began to notice something was amiss in some of their honeybee hives. All of the adult worker bees were missing from the hives, but there were no bee corpses to be found. The bees had suddenly (and quite unusually) abandoned their queen and many living larvae. The few young workers that remained were reluctant to feed on abundant pollen and honey stored in their own hive, or in other abandoned hives.

A third of the colonies did not survive the following winter. By spring, the phenomenon had a name: colony collapse disorder (CCD). Since then, honeybee populations have continued to decline. As of 2011, about thirty percent of bee colonies in the United States and twenty percent of those in Europe have been affected.

CCD is a problem that impacts more than just bees and honey lovers. Nearly all of our food crops are the fruits of angiosperms (Section 14.7). Angiosperms, like other seed-bearing plants, produce pollen. Many angiosperm species rely on honeybees and other **pollinators** to carry pollen from one plant to another (Figure 28.1**A**). Fruits usually do not form from unpollinated flowers, and in some species they only form when flowers receive pollen from a different plant of the same species. Even species with flowers that can self-pollinate tend to make bigger fruits and more of them when they are cross-pollinated (Figure 28.1**B**).

Many types of insects pollinate plants, but honeybees are especially efficient pollinators of a wide variety of species. They are also the only ones that tolerate living in man-made hives that are loaded onto trucks and transported to wherever crops require pollination. Loss of their portable pollination service is a huge threat to our agricultural economy. Currently, about half of the world's leading crops—about $212 billion worth—depend on pollination, mainly by bees.

CCD is currently in the spotlight because it directly affects our food supply. However, other pollinator populations are also dwindling. Flowering plants rose to dominance in part because they coevolved with animal pollinators. Most flowers are specialized to attract and be pollinated by a specific type (or even a specific species) of animal. Those adaptations become handicaps if coevolved pollinator populations decline. Wild animal species that depend on the plants for fruits and seeds will also be affected. Recognizing the prevalence and importance of these interactions is our first step toward finding workable ways to protect them.

A Honeybees are efficient pollinators of a variety of flowering plants, including berries.

B Raspberry flowers can pollinate themselves, but the fruit that forms from a self-fertilized flower is of lower quality than that of a cross-pollinated flower. The two berries on the *left* formed from self-pollinated flowers. The one on the *right* formed from an insect-pollinated flower.

FIGURE 28.1 Importance of insect pollinators.

Credits: (a) © Alan McConnaughey, www.flickr.com/photos/engrpiman; (b) Courtesy of James H. Cane, USDA-ARS Bee Biology and Systematics Lab, Utah State University, Logan, UT.

FIGURE 28.2 Animated! Life cycle of a typical flowering plant.

Credits: left, © Cengage Learning; right, © Martin Ruegner/ Photographer's Choic RF/ Getty Images.

FIGURE 28.3 Animated! Anatomy of a typical flower. Male gametophytes are produced by stamens; female gametophytes, by carpels. The photo (*above left*) shows flowers of a cherry tree.

© Cengage Learning.

28.2 Sexual Reproduction in Flowering Plants

Remember from Section 14.2 that the life cycle of flowering plants is dominated by the sporophyte, a diploid spore-producing body (Figure 28.2). Flowers are the reproductive structures of sporophytes. Spores that form by meiosis inside flowers develop into haploid gametophytes, which produce gametes (Section 8.7).

〉 Anatomy of Flowers A flower forms at the tip of a special reproductive shoot (Figure 28.3). Its parts are modified leaves that develop from a thickened region of stem called a receptacle ❶. A typical flower has four spirals or rings (whorls) of modified leaves. The outermost whorl is a calyx, which is a ring of leaflike sepals ❷. The sepals of most flowers are photosynthetic and inconspicuous; they serve to protect the flower's more delicate parts. Just inside the calyx is the corolla (from the Latin *corona*, or crown), a whorl of petals ❸. Petals are usually the largest and most brightly colored parts of a flower, and they function mainly to attract pollinators. Inside the corolla is a whorl of **stamens**, the reproductive organs that produce the plant's male gametophytes. In most flowers, stamens consist of a thin filament with an **anther** at the tip ❹. A typical anther contains four elongated pouches called pollen sacs. Meiosis of diploid cells in each sac produces haploid, walled spores. These spores develop into **pollen grains**, which are immature male gametophytes. The flower's innermost whorl consists of modified leaves folded and fused into a **carpel**, the reproductive organ that produces the plant's female gametophytes ❺. Carpels are sometimes called pistils, and some flowers have several of them. The upper region of a carpel is a sticky or hairy **stigma** that is specialized to receive pollen grains. Typically, the stigma sits on top of a slender stalk called a style. The lower, swollen region of a carpel is the **ovary**, which contains one or more ovules. An **ovule** is a bulge of tissue in which a haploid female gametophyte forms.

anther Part of the stamen that produces pollen.
carpel Floral reproductive organ that consists of an ovary, stigma, and often a style.
ovary In flowering plants, the enlarged base of a carpel, inside which one or more ovules form.
ovule Of a seed-bearing plant, a structure inside an ovary in which a haploid female gametophyte forms.
pollen grain Immature male gametophyte of a seed plant.
pollinator An organism that moves pollen from one plant to another.
stamen Floral reproductive organ that consists of an anther and, often, a filament.
stigma Upper part of a carpel; adapted to receive pollen.

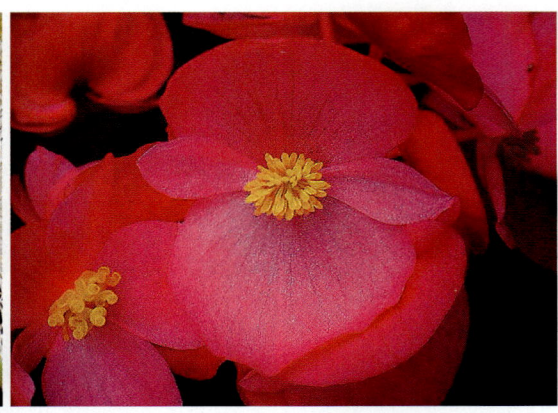

A The irregular blossom of lady's slipper (an orchid).

B Elongated inflorescence of hyacinth.

C A daisy is a composite of many individual flowers.

D Incomplete flowers of eucalyptus have no petals.

E Imperfect flowers: in some begonias, stamen-less female blossoms (*left*) form on the same plants as carpel-less male blossoms (*right*).

FIGURE 28.4 Examples of structural variation in flowers.

Credits: (a) © Salawin Chanthapan/ Shutterstock; (b–d) © Brian Johnston; (e) © Lauren Doyle.

❭ Diversity of Floral Structure

Master gene mutations (Section 7.7) give rise to dramatic variations in flower structure, and we see many such variations in the range of diversity of flowering plants. For example, the number of floral parts differs between monocots and eudicots (Section 27.2). Flowers may form as solitary blossoms (Figure 28.4**A**) or in clusters called inflorescences (Figure 28.4**B**). A composite flower is an inflorescence of many flowers ("florets") grouped as a single head (Figure 28.4**C**). Cherry blossoms (shown in Figure 28.2 and 28.3) and other regular flowers are symmetric around their center axis: If the flower were cut like a pie, the pieces would be roughly identical. Irregular flowers are not radially symmetric (the flower in Figure 28.4**A** is an example).

A cherry blossom has all four sets of modified leaves (sepals, petals, stamens, and carpels), so it is called a "complete" flower. An "incomplete" flower lacks one or more of these structures (Figure 28.4**D**). Cherry blossoms are also "perfect" flowers, because they have both stamens and carpels. Perfect flowers can be cross-pollinated (fertilized by pollen from other plants), or they can self-pollinate. Self-pollination can be adaptive in situations where plants are widely spaced, such as in newly colonized areas. However, offspring of self-pollinated flowers or plants are generally less vigorous than those of cross-pollinated plants. Accordingly, floral adaptations of many plant species encourage or even require cross-pollination. For example, a perfect flower may release pollen only after its stigma is no longer receptive to pollen. Imperfect flowers that lack stamens or carpels (Figure 28.4**E**) cannot self-fertilize. In some species, stamen-bearing and carpel-bearing flowers form on different plants, so fertilization occurs only by cross-pollination.

❭ Pollinators

Sexual reproduction in flowering plants involves the transfer of pollen. Unlike most animals, plants cannot move about to find a mate, so pollen transfer depends on environmental agents. The diversity of flower form in part reflects a dependence on pollination vectors. A **pollination vector** is an agent that delivers pollen from an anther to a compatible stigma.

Grasses and many other plants are pollinated by wind, which disperses pollen randomly throughout the environment. Such plants often release pollen grains by the billions, insurance in numbers that some of their pollen will reach a receptive stigma. Other plants enlist the help of animal pollinators (Figure 28.5).

nectar Sweet fluid exuded by some flowers; attracts animal pollinators.

pollination vector Any agent that moves pollen grains from one plant to another; e.g., the wind or a bee.

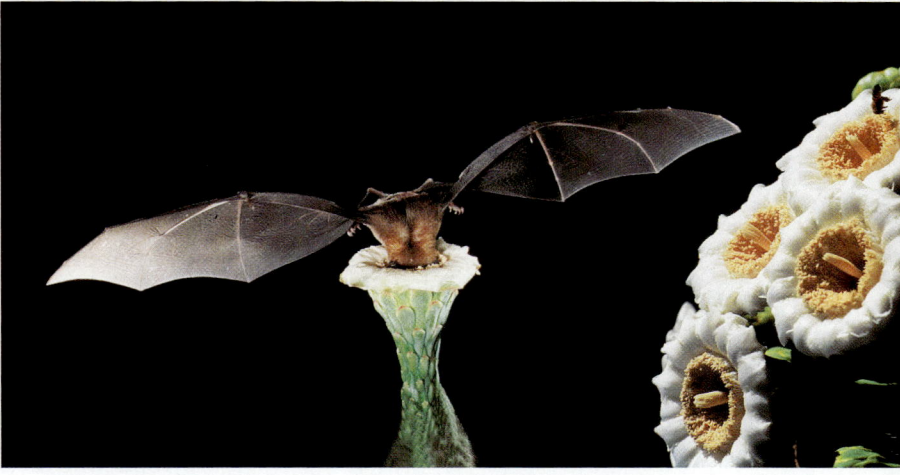

A Insects. Blueberry bees are efficient pollinators of a variety of plants, including this barberry.

B Birds. Pollen accumulates on the feathers of a little wattlebird as it sips nectar from a torch lily.

C Mammals. The large, fragrant flowers of giant saguaro cactus open at night, when they are visited by bats, and close the next day. They produce a sweet nectar.

FIGURE 28.5 Examples of animal pollinators.

Credits: (a) Photo by Jack Dykinga, USDA, ARS; (b) © Jordan Tan/ Shutterstock.com; (c) left, Merlin D. Tuttle, Bat Conservation International; right, John Alcock, Arizona State University.

An animal that is attracted to a particular flower often picks up pollen on a visit, then inadvertently transfers it to another flower on a later visit. The more specific the attraction, the more efficient the transfer of pollen among plants of the same species. Given the selective advantage for flower traits that attract specific pollinators, it is not surprising that about 90 percent of flowering plants have coevolved animal pollinators.

A flower's structure, color, and fragrance are evolutionary adaptations for pollination by particular vectors. For example, birds rely more on vision than smell, so bird-pollinated flowers typically have a brilliant color and no odor. Bees have an excellent sense of smell and they can see UV light, so bee-pollinated flowers tend to be fragrant, with a bull's-eye pattern of UV-reflecting pigments in their petals (Figure 28.6). Night-flying bats have a good sense of smell but poor vision, so bat-pollinated flowers are large and light-colored, with a strong nighttime fragrance (Figure 28.5**C**). Not all flowers smell sweet; odors like dung or rotting flesh beckon beetles and flies. Wind-pollinated flowers are typically small, pale, and have no scent.

An animal pollinator visits a flower in search of a reward, which may be **nectar** (a sweet fluid exuded by flowers), nutritious pollen, or even sex. Nectar is the only food for most adult butterflies, and it is the food of choice for hummingbirds. Honeybees convert nectar to honey, which helps feed the bees and their larvae through the winter, and their main source of protein is pollen. Beetles eat pollen directly; bees mix it with honey to make bee bread that they feed to their larvae. Pollen is the only reward for a pollinator of roses, poppies, and some other flowers that do not produce nectar. Some plants produce a tasty but infertile pollen in addition to normal, fertile pollen.

Some flower specializations prevent pollination by all but one type of pollinator. For example, nectar or pollen at the bottom of a long floral tube may be accessible only to an insect with a matching feeding device. In some plants, stamens adapted to brush against a pollinator's body or lob pollen onto it will function only when triggered by that pollinator (Figure 12.15 shows an example). Such relationships are to both species' mutual advantage: A flower that captivates the attention of an animal has a pollinator that spends its time seeking (and pollinating) only those flowers. Pollinators also benefit when they receive an exclusive supply of the flower's reward.

A Dandelion   **B** Evening primrose

FIGURE 28.6 UV-reflecting patterns in flower petals. The *top* row shows flowers as we see them. The *bottom* row shows the same flowers photographed with a special filter that allows us to see reflected UV light, here represented by the color *red*. The bull's-eye patterns guide bees, which can see UV light, directly to the flower's reproductive parts.

Bjorn Rorslett/ Photo Researchers, Inc.

> **A New Generation Begins** Figure 28.7 shows how a typical flowering plant (cherry) reproduces. The production of female gametes begins when a mass of tissue—the ovule—starts growing on the inner wall of an ovary **1**. One cell in the middle of the mass undergoes meiosis and cytoplasmic division, forming four haploid megaspores **2**. Three of the four megaspores typically disintegrate. The remaining megaspore undergoes three rounds of mitosis without cytoplasmic division, thus producing a single cell with eight haploid nuclei **3**. The cytoplasm of this cell divides unevenly, forming a seven-celled embryo sac that constitutes the female gametophyte **4**. One of the cells, the endosperm mother cell, has two nuclei ($n + n$). Another cell is the egg.

The production of male gametes begins as masses of diploid, spore-producing cells form by mitosis inside anthers. Walls typically develop around the masses, so four pollen sacs form **5**. Each cell in the pollen sacs undergoes meiosis and cytoplasmic division to form four haploid microspores **6**. Mitosis and differentiation of a microspore produce a pollen grain.

A pollen grain consists of two cells, one inside the cytoplasm of the other, enclosed by a durable coat **7**. The coat will protect the cells inside on their journey to meet an egg. After the pollen grains form, they enter **dormancy**, a period of suspended metabolism, before being released from the anther when the pollen sacs split open **8**.

A pollen grain that lands on a receptive stigma **germinates**, which means it resumes metabolic activity after a period of dormancy. When a pollen grain germinates, one of its two cells develops into a tubular outgrowth called a pollen tube **9**. The other cell undergoes mitosis and cytoplasmic division, producing two sperm cells (the male gametes) within the pollen tube.

A pollen tube together with its contents of male gametes constitutes the mature male gametophyte. The pollen tube grows from its tip down through the tissues of the style and ovary toward the ovule, carrying with it the two sperm cells. Chemical signals secreted by the female gametophyte guide the pollen tube's growth to the embryo sac within the ovule. Many pollen tubes may grow down into a carpel, but usually only one penetrates the embryo sac.

When a pollen tube penetrates an embryo sac, it releases the two sperm cells **10**. Flowering plants undergo **double fertilization**, in which one of the sperm cells from the pollen tube fuses with (fertilizes) the egg and forms a diploid zygote; the other sperm cell fuses with the endosperm mother cell, forming a triploid ($3n$) cell. This cell gives rise to triploid **endosperm**, a nutritious tissue that forms only in seeds of flowering plants. When the seed sprouts, endosperm sustains rapid growth of the seedling until its new leaves begin photosynthesis.

Take-Home Message

How does sexual reproduction occur in flowering plants?

- Flowers are the reproductive structures of angiosperm sporophytes. Variations in flower structure reflect their dependence on pollination vectors.

- Stamens typically consist of a filament with an anther at the tip. Pollen grains form in anthers.

- Carpels typically consist of stigma, style, and ovary. A female gametophyte with an egg and an endosperm mother cell forms in an ovule inside the ovary.

- A pollen grain that germinates on a stigma develops into the male gametophyte—a pollen tube and two sperm cells. The pollen tube grows into the carpel, enters an ovule, and releases the sperm cells.

- Double fertilization occurs when one of the sperm cells fuses with the egg, and the other fuses with the endosperm mother cell.

dormancy Period of temporarily suspended metabolism.
double fertilization Mode of fertilization in flowering plants in which one sperm cell fuses with the egg, and a second sperm cell fuses with the endosperm mother cell.
endosperm Nutritive tissue in the seeds of flowering plants.
germinate To resume metabolic activity after dormancy.

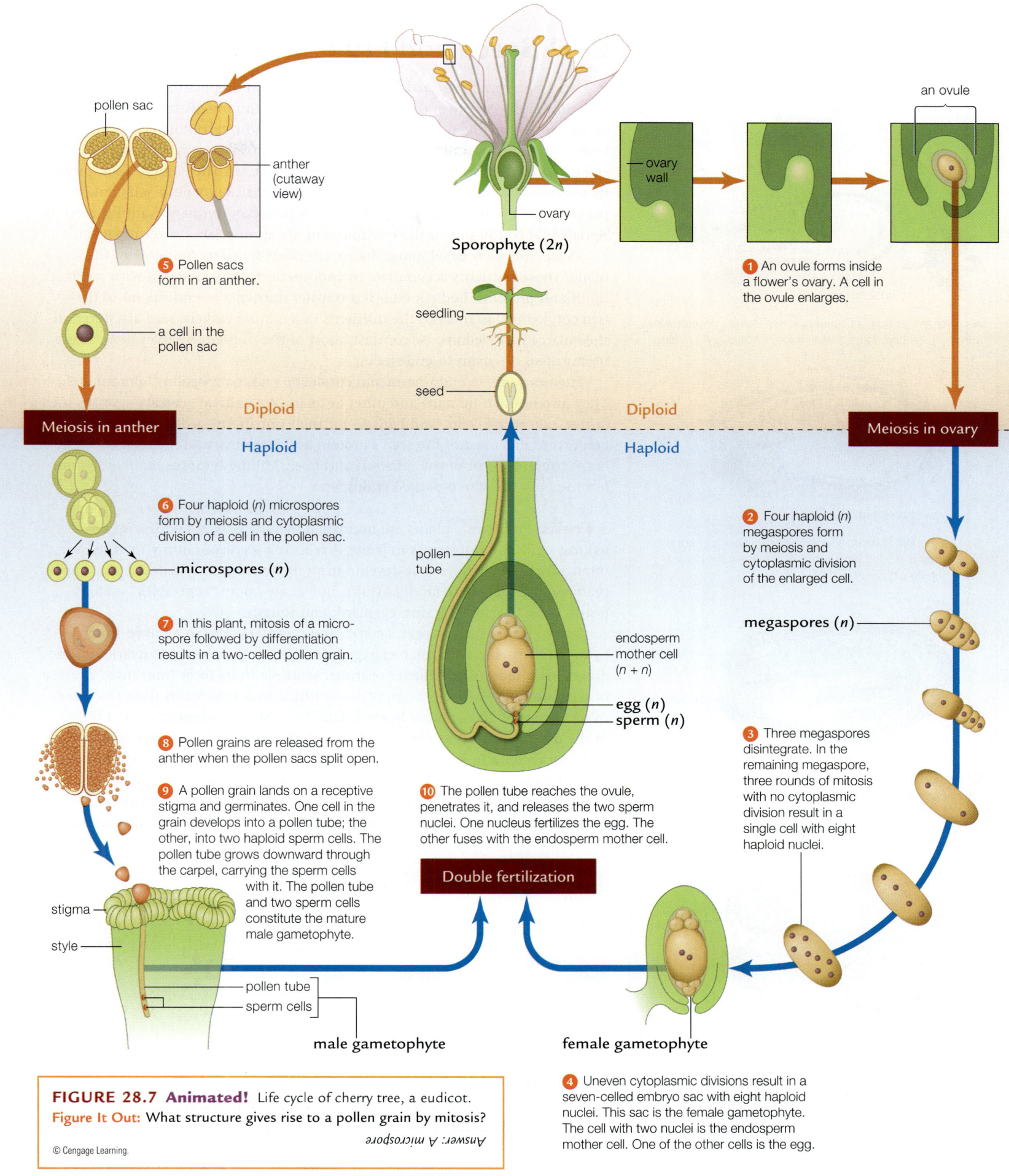

pollen sac

anther (cutaway view)

⑤ Pollen sacs form in an anther.

a cell in the pollen sac

an ovule

ovary wall

ovary

Sporophyte (2n)

① An ovule forms inside a flower's ovary. A cell in the ovule enlarges.

seedling

seed

Diploid

Meiosis in anther

Diploid

Meiosis in ovary

Haploid

Haploid

⑥ Four haploid (*n*) microspores form by meiosis and cytoplasmic division of a cell in the pollen sac.

microspores (*n*)

pollen tube

⑦ In this plant, mitosis of a microspore followed by differentiation results in a two-celled pollen grain.

endosperm mother cell (*n* + *n*)

egg (*n*)
sperm (*n*)

② Four haploid (*n*) megaspores form by meiosis and cytoplasmic division of the enlarged cell.

megaspores (*n*)

⑧ Pollen grains are released from the anther when the pollen sacs split open.

⑨ A pollen grain lands on a receptive stigma and germinates. One cell in the grain develops into a pollen tube; the other, into two haploid sperm cells. The pollen tube grows downward through the carpel, carrying the sperm cells with it. The pollen tube and two sperm cells constitute the mature male gametophyte.

⑩ The pollen tube reaches the ovule, penetrates it, and releases the two sperm nuclei. One nucleus fertilizes the egg. The other fuses with the endosperm mother cell.

③ Three megaspores disintegrate. In the remaining megaspore, three rounds of mitosis with no cytoplasmic division result in a single cell with eight haploid nuclei.

Double fertilization

stigma

style

pollen tube
sperm cells

male gametophyte

female gametophyte

④ Uneven cytoplasmic divisions result in a seven-celled embryo sac with eight haploid nuclei. This sac is the female gametophyte. The cell with two nuclei is the endosperm mother cell. One of the other cells is the egg.

FIGURE 28.7 Animated! Life cycle of cherry tree, a eudicot.
Figure It Out: What structure gives rise to a pollen grain by mitosis?

Answer: A microspore

© Cengage Learning.

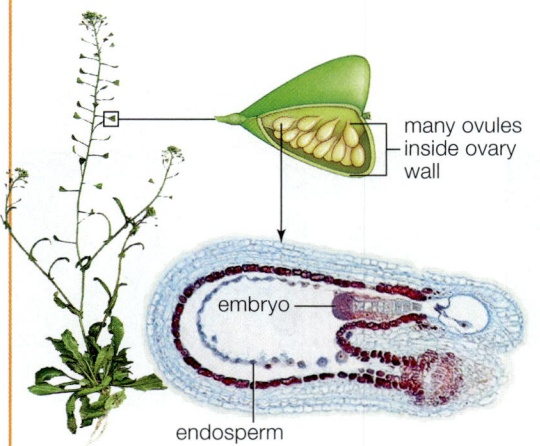

A After fertilization, a *Capsella* flower's ovary develops into a fruit. An embryo forms inside each of the ovary's ovules.

B The embryo is heart-shaped when its two cotyledons start forming. Endosperm tissue expands as the parent plant transfers nutrients into it.

C In eudicots like *Capsella*, nutrients are transferred from endosperm into two cotyledons as the embryo matures. The developing embryo becomes shaped like a torpedo when the enlarging cotyledons bend.

D A tough seed coat forms around the mature embryo and its two enlarged cotyledons.

FIGURE 28.8 Animated! Embryonic development of shepherd's purse (*Capsella*), a eudicot.

Credits: (a,c,d) Michael Clayton, University of Wisconsin; (b) © David T. Webb; (a, art) © Cengage Learning.

28.3 Seeds and Fruit

> The Embryo Sporophyte Forms In flowering plants, double fertilization produces a zygote and a triploid (3*n*) cell; both immediately begin mitotic divisions. The zygote develops into an embryo, and the triploid cell develops into endosperm (Figure 28.8). As the embryo approaches maturity, the outer cell layers of the ovule develop into a tough seed coat. The embryo sporophyte, its reserves of food, and the seed coat are now a mature ovule, a self-contained package called a **seed**. The seed may enter a period of dormancy until it receives signals that conditions in the environment are appropriate for germination.

As an embryo is developing, the parent plant transfers nutrients to the ovule. These nutrients accumulate in endosperm mainly as starch with some lipids and proteins. Eudicot embryos transfer nutrients in endosperm to their two cotyledons, so most of the nutrients in a mature eudicot seed are stored in the enlarged cotyledons. By contrast, most of the nutrient reserves in a mature monocot seed remain in endosperm.

The nutrients in endosperm and cotyledons nourish seedling sporophytes. They also nourish humans and other animals. We cultivate cereals—grasses such as rice, wheat, rye, oats, and barley—for their nutritious seeds. The embryo (the germ) contains most of the seed's protein and vitamins, and the seed coat (the bran) contains most of the minerals and fiber. Milling removes bran and germ, leaving only the starch-packed endosperm.

> Fruits Embryonic plants are nourished by nutrients in endosperm and cotyledons, but not by nutrients in fruits. A **fruit** is a seed-containing mature ovary, often with fleshy tissues that develop from the ovary wall (Figure 28.9A). Apples, oranges, and grapes are familiar fruits, but so are many "vegetables" such as beans, peas, tomatoes, grains, eggplant, and squash.

Simple fruits, such as peas, acorns, and *Capsella* pods, are derived from one ovary. Strawberries and other aggregate fruits form from separate ovaries of one flower; they mature as a cluster of fruits. Multiple fruits form from fused ovaries of separate flowers. The pineapple is a multiple fruit that forms from the fused ovary tissues of 200 or more flowers. True fruits such as cherries consist only of the ovary wall and its contents. Other floral parts, such as petals, sepals, stamens, or receptacle, expand along with the ovary in accessory fruits. Most of the flesh of an apple is an enlarged receptacle (Figure 28.9B).

Fruits may be dry or juicy (fleshy). Acorns and grains such as corn are dry fruits, as are the fruits of sunflowers, maples, and strawberries. Strawberries are

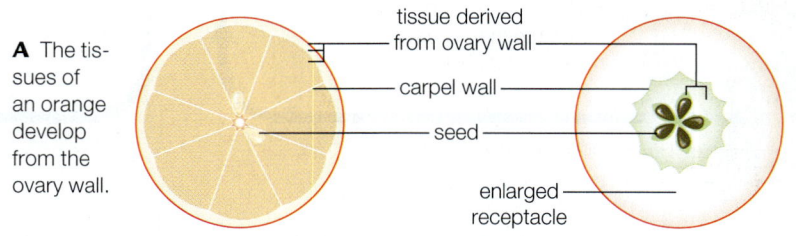

A The tissues of an orange develop from the ovary wall.

tissue derived from ovary wall

carpel wall

seed

enlarged receptacle

B The flesh of an apple is an enlarged receptacle.

FIGURE 28.9 Parts of a fruit develop from parts of a flower.

Figure It Out: How many carpels were there in the flower that gave rise to the orange in **A**?

Answer: Eight

© Cengage Learning.

A Technically, a strawberry is not a berry. The flower's carpels turn inside out as the fruits form, and the receptacle expands and becomes red, juicy flesh. The hard "seeds" on the surface are individual dry fruits.

B Blackberries, raspberries, and the fruits of other *Rubus* species are not berries, either. Each one is an aggregate of dozens of small, individual fleshy fruits.

FIGURE 28.10 Examples of dry and fleshy fruits.

Credits: (a) left, © Nicolas Raymond/ Shutterstock; right, © Roman Sigaev / Shutterstock; (b) © saiko3p/ Shutterstock.

A Wind-dispersed fruits. *Left*, dry outgrowths of the ovary wall of a maple fruit form "wings" that catch the wind and spin the seeds away from the parent tree. *Right*, wind that lifts the hairy modified sepals of a dandelion fruit may carry the attached seed miles away from the parent plant.

B Water-dispersed fruits. *Left*, the fruits of sedges native to American marshlands have seeds encased in a bladderlike envelope that floats. *Right*, buoyant fruits of the coconut palm have tough, waterproof husks. They can float for thousands of miles in seawater.

not berries and their fruits are not juicy. A strawberry's red flesh is an accessory to the many individual dry fruits on its surface (Figure 28.10**A**). Cherries, almonds, and olives are fleshy fruits, as are individual fruits of blackberries and other *Rubus* species (Figure 28.10**B**). Grapes, tomatoes, and citrus fruits are true berries—fleshy fruits produced from one ovary—as are pumpkins, watermelons, and cucumbers.

> Adaptations for Dispersal The function of a fruit is to protect and disperse seeds. Just as flower structure is adapted to certain pollination vectors, so are fruits adapted to certain dispersal vectors: environmental agents such as water or wind, or mobile organisms such as birds.

Adaptations to a specific dispersal mechanism are reflected in a fruit's form. Fruits dispersed by the wind tend to be lightweight with breeze-catching specializations (Figure 28.11**A**). Tufted fruits such as those of thistle, cattail, dandelion, and milkweed may be blown as far as 10 kilometers (6 miles) from the parent plant. Fruits dispersed by water can float, and they have water-repellent outer layers (Figure 28.11**B**). The fruits of many plants have specializations that facilitate dispersal by animals (Figure 28.11**C**). Some have hooks or spines that stick to the feathers, feet, fur, or clothing of more mobile species. Colorful, fleshy, or fragrant fruits attract birds and mammals that disperse seeds. The animal may eat the fruit and discard the seeds, or eat the seeds along with the fruit. Abrasion of the seed coat by teeth or by digestive enzymes in an animal's gut can help the seed germinate after it departs in feces.

C Animal-dispersed fruits. *Left*, curved spines attach cocklebur fruits to the fur of animals (and clothing of humans) that brush past it. *Right*, the red, fleshy fruits of hawthorn plants are an important food source for cedar waxwings, which disperse the fruits' seeds in their feces.

FIGURE 28.11 Examples of adaptations that aid fruit dispersal.

Credits: (a) left, R. Carr; right, © JupiterImages Corporation; (b) left, © T. M. Jones; right, Alex James Bramwell/ Shutterstock; (c) left, © Robert H. Mohlenbrock © USDA-NRCS PLANTS Database; right, © Vishnevskiy Vasily/ Shutterstock.

Take-Home Message

How do seeds and fruits develop?

- After double fertilization, the zygote develops into an embryo, and the endosperm becomes enriched with nutrients. A mature ovule—a seed—consists of an embryo sporophyte, nutrients in endosperm or cotyledons, and a seed coat.

- A fruit is a mature ovary, with or without accessory tissues that develop from other parts of the flower.

- A fruit protects and disperses seeds. Fruit specializations are adaptations to particular dispersal vectors.

fruit Mature ovary of a flowering plant, often with accessory parts; encloses a seed or seeds.

seed Embryo sporophyte of a seed plant packaged with nutritive tissue inside a protective coat.

28.4 Early Development

An embryonic plant complete with shoot and root apical meristems forms as part of a seed (Figure 28.12). The embryonic shoot and root tips are separated by a section of embryonic stem. As the seed matures, the embryo may dry out and enter a period of dormancy. A dormant embryo can rest in its protective seed coat for many years before it resumes metabolic activity and germinates.

❯ Breaking Dormancy Seed dormancy is a climate-specific adaptation that allows germination to occur when conditions in the environment are most likely to support the growth of a seedling. For example, seeds of many annual plants native to cold winter regions are dispersed in autumn. If the seeds germinated immediately, the tender seedlings would not survive the coming winter. Instead, the seeds remain dormant until spring, when milder temperatures and longer day length favor the growth of seedling sporophytes. By contrast, the weather in regions near the equator does not vary by season. Seeds of most plants native to such regions do not enter dormancy; they can germinate as soon as they become mature.

Other than the presence of water, the triggers for germination differ by species. Some seed coats are so dense that they must be abraded or broken (by being chewed, for example) before water can even enter the seed. Seeds of many cool-climate plants require exposure to freezing temperatures; those of some lettuce species, to bright light. In seeds of some species native to regions that have periodic wildfires, germination is inhibited by light and enhanced by smoke; in others, germination does not occur unless the seeds have been previously burned. Such requirements are evolutionary adaptations to life in a particular environment. All maximize a seedling's chance of survival.

FIGURE 28.12 Anatomy of a seed.
Left, corn, a monocot; *right*, pea, a eudicot.

As dormancy ends, cell divisions resume mainly at apical meristems of the embryonic shoot and root. These two structures are separated by a section of embryonic stem. In monocot grasses such as corn, the embryonic shoot is protected by a sheathlike coleoptile.

Credits: left, © Mike Clayton/ University of Wisconsin; right, © Jubal Harshaw/ Shutterstock.

seed coat
endosperm
cotyledon(s)
coleoptile
embryonic shoot tip
embryonic stem
embryonic root tip

coleoptile
hypocotyl
embryonic root

A As a corn grain (seed) germinates, its embryonic root and coleoptile emerge from the seed coat.

FIGURE 28.13 Animated!
Early growth of corn, a typical monocot.

Credits: (a, b right, c middle) © Cengage Learning; (b) left, Barry L. Runk/ Grant Heilman, Inc.; (c) Nigel Cattlin/Visuals Unlimited.

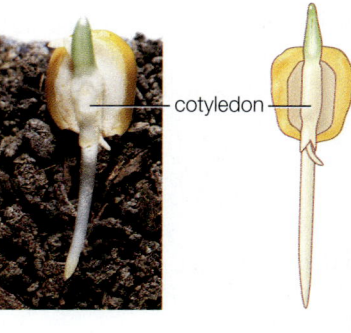

cotyledon

B The embryonic root develops into a primary root. The coleoptile grows upward, opening a channel through the soil to the surface. Once the coleoptile reaches the surface, it stops growing.

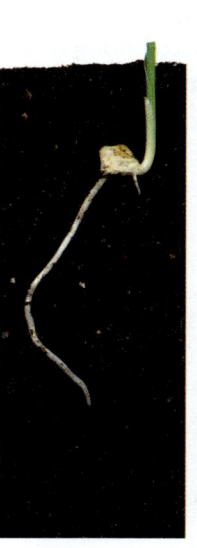

primary shoot

coleoptile

C The embryonic shoot develops into a primary shoot that emerges from the coleoptile and begins photosynthesis.

FIGURE 28.14 Animated! Early growth of the common bean, a typical eudicot.

1 The seed coat splits and the embryonic root tip emerges.

2 The embryonic stem emerges from the seed and bends in the shape of a hook.

3 As the stem lengthens, it drags the two cotyledons upward toward the surface of the soil.

4 Exposure to light makes the stem straighten.

5 Primary leaves emerge from between the cotyledons and begin photosynthesis.

6 Cotyledons typically undergo a period of photosynthesis, then wither and fall off the stem.

Nigel Cattlin/Visuals Unlimited.

The process of germination begins when water seeps into a seed. The water activates hydrolysis enzymes (Section 2.6) in endosperm, and these enzymes begin to break down stored starches into sugars. The seed's tissues swell with water, its coat ruptures, and oxygen diffuses into internal tissues. Meristem cells in the embryo use the sugars and the oxygen for aerobic respiration as they begin to divide rapidly, and the embryonic plant begins to grow. Germination ends when the first part of the embryo breaks out of the seed coat.

〉 After Germination The pattern of early growth that occurs after germination varies. For example, in corn and other monocots, a rigid sheath called a coleoptile protects the primary (first) shoot as it grows straight upward through soil. The embryo's single cotyledon, which typically stays beneath the soil, functions mainly to transfer endosperm breakdown products to the embryo (Figure 28.13). In eudicot seedlings, the primary shoot is not protected by a coleoptile (Figure 28.14). In a typical eudicot pattern of development, the embryonic root emerges from the seed **1**, followed by the embryonic stem. The stem bends into a hook shape as it lengthens **2**. The bent stem pulls the cotyledons upward through the soil until it reaches the surface **3**. There, exposure to light causes the stem to straighten **4**. Primary leaves emerge from between the cotyledons and begin photosynthesis **5**. The cotyledons typically undergo a period of photosynthesis before shriveling **6**. Eventually, the cotyledons fall off the lengthening stem, and the plant's new leaves produce all of its food.

Germination is the first step in plant development. As a sporophyte grows and matures, its tissues and organs develop in characteristic patterns. Leaves form, stems and roots lengthen and thicken, flowering occurs at a certain time of year, and so on. As you will see in Sections 28.7 and 28.8, these patterns are outcomes of gene expression and environmental influences.

Take-Home Message

What happens during early plant development?

- Seed dormancy and germination requirements are adaptations to life in a particular climate.

- After a seed germinates, nutrients stored in endosperm (in monocots) or cotyledons (in eudicots) support the seedling's growth until new leaves begin photosynthesis. A coleoptile protects the primary shoot of many monocots.

A Potatoes are tubers that grow on stolons. New plants sprout from their nodes, which we call "eyes."

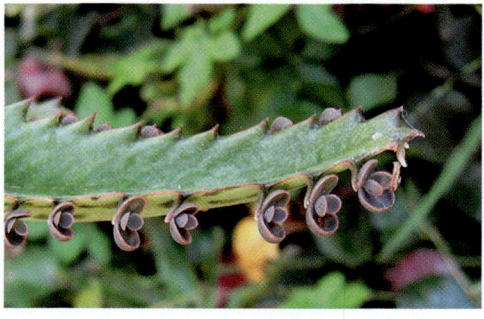

B New plants form at nodes on the leaflike cladodes of many succulents such as this devil's backbone plant.

FIGURE 28.15 Examples of asexual reproduction in plants.

Credits: (a) Image © Daniel Gale/ Shutterstock.com; (b) Lisa Starr.

28.5 Asexual Reproduction in Flowering Plants

Most flowering plants can reproduce asexually, as when new roots and shoots grow from extensions or pieces of a parent plant. Each new plant is a clone, a genetic replica of its parent. For example, new roots and shoots can sprout from nodes on modified stems (Figure 28.15). Entire forests of quaking aspen trees are actually stands of clones that grew from root suckers, which are shoots that sprout from shallow, cordlike lateral roots. Suckers sprout after aboveground parts of the aspens are damaged or removed. One stand of quaking aspens in Utah consists of about 47,000 shoots and stretches for 107 acres.

Such clones are as close as any organism gets to being immortal. The oldest known plant is a clone: the one and only population of King's holly, which consists of several hundred stems along a river gully in Tasmania. Radiometric dating of the plant's fossilized leaf litter show that the clone is at least 43,600 years old—predating the last ice age!

> Agricultural Applications For thousands of years, we humans have been taking advantage of the natural capacity of plants to reproduce asexually. For example, almost all houseplants, woody ornamentals, and orchard trees are clones that have been grown from stem fragments (cuttings) of a parent plant. Propagating some plants from cuttings may be as simple as jamming a broken stem into the soil. This method uses the plant's natural ability to form new roots and shoots from nodes. Other plants must be grafted. Grafting means inducing a cutting to fuse with the tissues of another plant. Often, the stem of a desired plant is spliced onto the roots of another.

Propagating a plant from cuttings ensures that offspring will have the same desirable traits as the parent plant. Consider familiar orchard apples, which are descendants of a wild species native to central Asia (Figure 28.16**A**). Domesticated apple varieties are grafted because species in this genus do not breed true for fruit traits—color, flavor, size, sweetness, and texture (Figure 28.16**B**). Even trees grown from seeds of the same apple produce fruits that vary dramatically.

Most apple trees grown from seed produce unpalatable fruit. In the early 1800s, John Chapman (known as Johnny Appleseed) grew millions of apple trees from seed in the midwestern United States. He sold the trees to homesteading settlers, who made hard cider from the otherwise inedible apples. About one of every hundred trees produced sweet, tasty fruit; its lucky owner would graft the tree and patent it. An estimated 16,000 varieties of edible apples were discovered during this era, but few of them remain: only 15 varieties now account for 90 percent of the apples sold in U.S. grocery stores (Figure 28.16**C**). All are clones of Chapman's original trees, and all are still propagated by grafting.

Grafting is also used to increase the hardiness of a desirable plant. In 1862, the plant louse *Phylloxera* was accidentally introduced into France via imported American grapevines. European grapevines had little resistance to this tiny insect, which attacks and kills the root systems of the vines. By 1900, *Phylloxera* had destroyed two-thirds of the vineyards in Europe, thus devastating the winemaking industry for decades. Today, French vintners routinely graft their prized grapevines onto the roots of *Phylloxera*-resistant American vines.

With **tissue culture propagation**, individual cells (typically from meristem) are coaxed to divide and form embryos in a laboratory. This technique can yield millions of genetically identical offpsring from a single parent plant. It is frequently used in research aimed at improving food crops, and also to propagate rare or hybrid ornamental plants such as orchids.

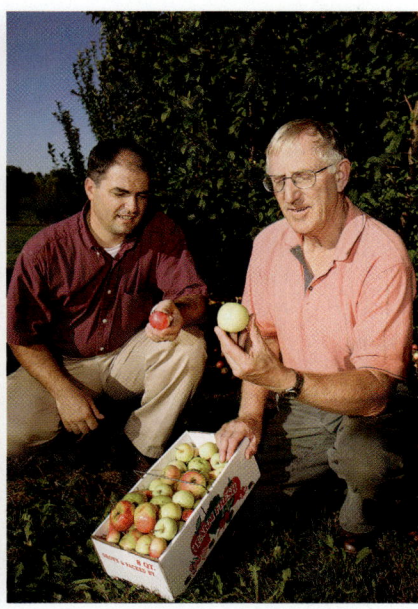

A Wild apples still grow in some regions of their native Kazakhstan, but most of the wild forests have been replanted with commercial groves of domesticated European varieties. The remaining wild forests are becoming fragmented by expanding human developments.

B Apple trees do not breed true for fruit traits. These representative fruits from 21 wild apple trees vary in color, flavor, size, sweetness, and texture.

C Gennaro Fazio (*left*) and Phil Forsline (*right*) are working to increase the genetic diversity of U.S. apples by cross-breeding them with disease-resistant wild varieties.

FIGURE 28.16 Apple agriculture.

Credits: (a) Ethan C. Roland, AppleSeed Permaculture, appleseedpermaculture.com; (b,c) Photo by Peggy Greb, USDA, ARS.

In some flowering plant species, fruits form even in the absence of fertilization. Fruits that develop from unfertilized flowers may have underdeveloped seeds or none at all. Plants with mutations that cause ovules or embryos to abort during development bear seedless fruit; such plants are necessarily sterile and must be propagated by grafting. Seedless grapes and navel oranges are like this. All commercially produced bananas are seedless because the plants that bear them are triploid (3*n*). During meiosis, the three chromosomes cannot be divided equally between the two spindle poles, so meiosis fails. Gametes do not form and neither do seeds. Grafting bananas and other monocots is notoriously difficult (if not impossible), so seedless banana plants are propagated by cutting off and replanting adventitious shoots that sprout from their corms.

Polyploid plants are common in nature (Section 12.5), but it may take a long time for one to arise spontaneously. Plant breeders often use a microtubule poison called colchicine to artificially increase the frequency of polyploidy. Tetraploid (4*n*) plants produced by colchicine treatment are then crossed with diploid (2*n*) individuals. The resulting offspring are triploid and sterile: They make seedless fruit on their own or after pollination (but not fertilization) by a diploid plant. Seedless watermelons are produced this way.

Take-Home Message

How do plants reproduce asexually?

- Many plants propagate themselves when new shoots grow from a parent plant or pieces of it. Offspring of such asexual reproduction are clones.
- Humans propagate plants asexually for agricultural or research purposes, for example by grafting or tissue culture.

tissue culture propagation Laboratory method in which individual plant cells (typically from meristem) are induced to form embryos.

Table 28.1	Major Plant Hormones
Hormone	Effect
Abscisic acid (ABA)	
	Closes stomata in times of stress
	Inhibits shoot growth
	Inhibits seed germination
	Involved in environmental stress responses
	Involved in chloroplast movement
Auxin	
	Coordinates effects of other plant hormones during development of shoots and roots
	Role in apical dominance
	Stimulates cell expansion
	Role in abscission
	Mediates tropisms
	Stimulates division of meristem cell
Cytokinin	
	Stimulates differentiation of cells in root apical meristem
	Stimulates division of cells in shoot apical meristem
	Inhibits formation of lateral roots
Ethylene	
	Involved in breaking dormancy
	Stimulates flower organ development
	Stimulates fruit ripening
	Stimulates abscission
	Involved in stress responses
Gibberellin	
	Stimulates cell division, elongation in stems
	Mobilizes food reserves in germinating seeds
	Stimulates flowering in some plants

28.6 Plant Hormones

In typical animals, most development occurs before adulthood. By contrast, most development in typical plants occurs in the adult body. Plants, unlike animals, cannot move about to avoid unfavorable conditions. Instead, each plant adapts to a stationary lifestyle by adjusting its development in response to environmental cues, including temperature, gravity, night length, the availability of water and nutrients, and even the presence of pathogens or herbivores.

Cells in different tissues and in different parts of a plant coordinate their activities by communicating with one another. This communication involves **plant hormones**, extracellular signaling molecules that exert their effects at very low concentrations (Table 28.1). All plant cells have the ability to make and release hormones, but each hormone is produced only in certain regions of the plant body at any given time. The release may occur in a region of the body that is far from cells in a targeted tissue, such as when cells in an actively growing root release cytokinins that keep shoot tips actively growing too. Some plant hormones have localized effects, such as when ethylene released from a cell in a ripening fruit affects the releasing cell as well as other cells in the same tissue.

A cell detects a hormonal signal via the receptor proteins it bears. Binding of a plant hormone to its receptor initiates a response in the cell that typically involves a change in gene expression. The cell's response varies depending on cell type, receptor, and the concentration of the hormone. In addition, one plant hormone can enhance or oppose another's effects on the same cell. Some hormones dampen or enhance the expression of others (Section 7.7), and most influence their own expression. This complex interplay gives rise to localized patterns of gene expression that can vary even from one cell layer to the next.

Most molecules that act as hormones in plants do not act as hormones in animals, and vice versa. For example, plants make phytoestrogens that are structurally similar to estrogens, which are female sex hormones in mammals (Section 26.3). The two types of compounds have similar hormonal effects in mammals, but phytoestrogens have nothing to do with sexual reproduction in plants. Rather, they function as part of plant defense mechanisms against fungal attack. The difference does not necessarily stem from differences in the structure of the hormones, but rather in the way the receptors for these compounds work in cells that bear them.

> Auxin Auxin is a plant hormone that was first discovered for its ability to promote growth (Figure 28.17). It was later discovered to have a critical role in all aspects of plant development, starting with the first division of the zygote.

FIGURE 28.17 Animated! Experiments showing that a coleoptile lengthens in response to auxin produced in its tip. © Cengage Learning.

A A coleoptile stops growing after its tip has been removed.

B A block of agar that absorbs auxin from a cut tip can stimulate a de-tipped coleoptile to resume growth.

C If an auxin-containing agar block is placed to one side of a cut tip, the coleoptile will continue to grow, but it will bend as it lengthens.

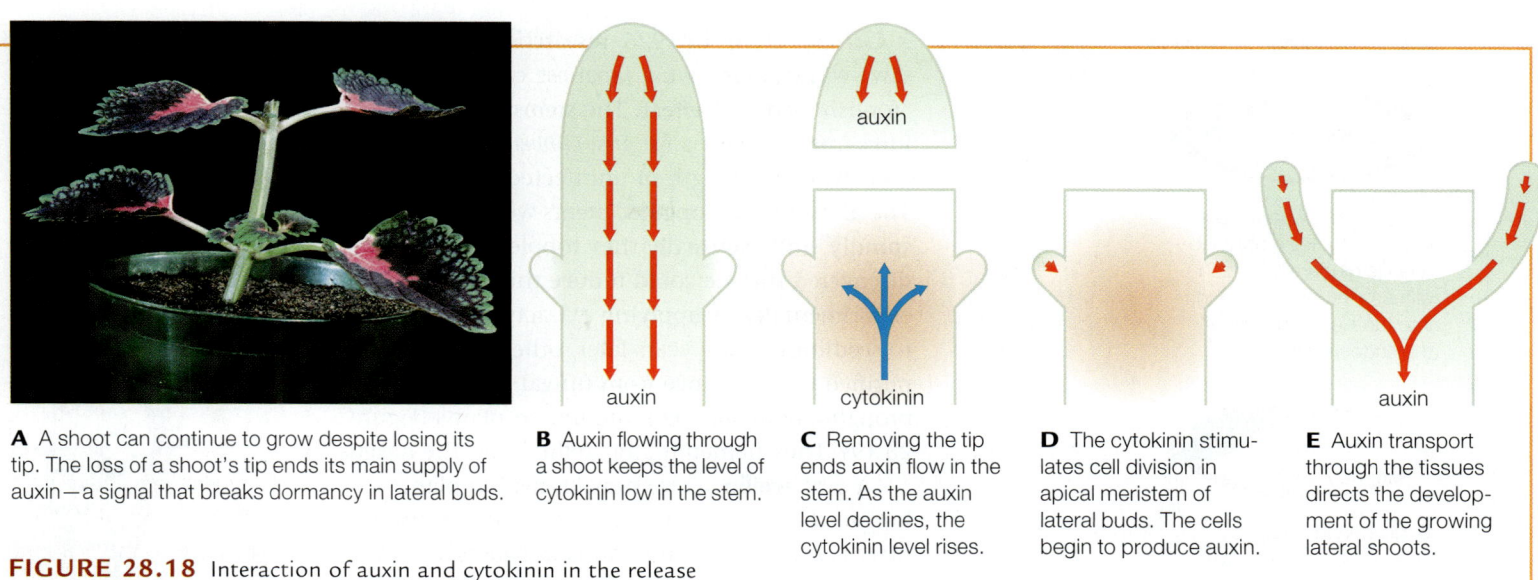

A A shoot can continue to grow despite losing its tip. The loss of a shoot's tip ends its main supply of auxin—a signal that breaks dormancy in lateral buds.

B Auxin flowing through a shoot keeps the level of cytokinin low in the stem.

C Removing the tip ends auxin flow in the stem. As the auxin level declines, the cytokinin level rises.

D The cytokinin stimulates cell division in apical meristem of lateral buds. The cells begin to produce auxin.

E Auxin transport through the tissues directs the development of the growing lateral shoots.

FIGURE 28.18 Interaction of auxin and cytokinin in the release of apical dominance.

Credits: (a) Jerome Wexler/ Visuals Unlimited; (b–e) © Cengage Learning.

Auxin is involved in embryonic development, formation of plant parts (primary leaves, shoot tips, stems, and roots), differentiation of vascular tissues, formation of lateral roots (and adventitious roots in some species), and shaping the plant body in response to environmental stimuli.

Auxin affects cells directly, and it also influences the levels of other plant hormones. It is present in almost all plant tissues, but is unevenly distributed through them. The hormone is made mainly in shoot apical meristems and in young leaves, and transported elsewhere. Auxin produced in shoot tips is loaded into phloem and travels to roots, where it is unloaded into root cells. A different mechanism that distributes auxin in a directional fashion dominates its movement over shorter distances. This mechanism is unique among plant hormones, and it is important because it establishes auxin concentration gradients across tissues and organs. These gradients have a central role in coordinating the actions of other hormones.

❯ Cytokinin A **cytokinin** is a plant hormone that affects meristem cell division and differentiation, stimulates development of lateral buds, and inhibits development of lateral roots. Auxin and cytokinin often influence the same cells, and in different contexts can enhance or oppose one another's effects. For example, in root apical meristem, auxin supports division of undifferentiated meristem cells, and cytokinin signals the cells to differentiate. In this context, the two hormones work in opposition to maintain the balance of differentiating and undifferentiated cells. By contrast, both hormones support cell division and prevent differentiation of cells in shoot apical meristem.

When a shoot is lengthening at the tip, its lateral buds are typically dormant, an effect called **apical dominance**. Apical dominance arises because auxin is produced by apical meristem in the shoot's tip. The auxin is transported down through cells of the stem, where it encourages the cells to lengthen. The presence of auxin in the stem also keeps the cytokinin level low. If the tip breaks off, the shoot's source of auxin disappears, so the cytokinin level rises in the stem (Figure 28.18**A,B**). The cytokinin moves into lateral buds and stimulates cell divisions of apical meristem there (Figure 28.18**C**). These newly active meristem cells produce auxin, which is transported down the stem away from the bud tips. This movement directs development of lateral shoots (Figure 28.18**D**).

apical dominance Effect in which a lengthening shoot tip inhibits the growth of lateral buds.

auxin Plant hormone with a central role in coordinating responses to other hormones in all stages of growth and development.

cytokinin Plant hormone that promotes cell division in shoot apical meristem and cell differentiation in root apical meristem. Often interacts antagonistically with auxin.

plant hormone Extracellular signaling molecule of plants that exerts its effect at very low concentration.

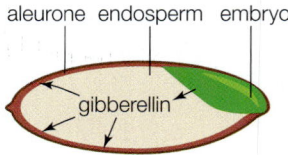

A Absorbed water causes cells of a barley embryo to release gibberellin, which diffuses through the seed into the aleurone layer of the endosperm.

B Gibberellin triggers cells of the aleurone layer to produce amylase. This enzyme diffuses into the starch-packed endosperm.

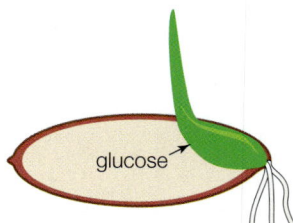

C The amylase breaks down starch into glucose. The glucose diffuses into the embryo and is used for aerobic respiration. Energy released by the reactions of aerobic respiration fuels meristem cell divisions in the embryo.

FIGURE 28.20
Gibberellin is involved in germination, as illustrated in a barley seed.

© Cengage Learning.

〉 Gibberellin In 1926, researcher E. Kurosawa was studying what Japanese call *bakane*, the "foolish seedling" effect. The stems of rice seedlings infected with a fungus, *Gibberella fujikuroi*, grew twice the length of uninfected seedlings. The abnormally elongated stems were weak and spindly, and eventually they toppled. Kurosawa discovered that he could induce the lengthening experimentally by applying extracts of the fungus to seedlings. Many years later, other researchers purified the substance from fungal extracts that brought about stem lengthening in plants (Figure 28.19). They named it gibberellin, after the fungus.

A **gibberellin**, as we now know, is a plant hormone that promotes growth in all flowering plants, among other functions. It causes a stem to lengthen between the nodes by inducing cell division and elongation. Along with auxin, gibberellin stimulates expansion along the long axis of a plant organ. Thus, it increases the length of the organ, and of the plant itself. The short stature of Mendel's dwarf pea plants (Section 9.3) is the result of a mutation that reduces the rate of gibberellin synthesis. Dwarf phenotype in plants is also caused by a defective gibberellin receptor.

Gibberellin is also involved in seed germination. During germination of a barley seed, for example, absorbed water causes cells of the embryo to release gibberellin (Figure 28.20). The hormone diffuses into the aleurone, a protein-rich layer of cells surrounding endosperm. In these cells, the presence of gibberellin turns on expression of the gene for amylase, a hydrolysis enzyme that breaks the bonds between starch's glucose monomers (Section 2.7). The cells start producing amylase and releasing it into the endosperm's starchy interior, where it breaks down stored starch molecules. The embryo then begins to use the released glucose for aerobic respiration, which fuels rapid cell divisions of meristem cells in the embryonic shoot and root tips.

FIGURE 28.19 Stem-lengthening effect of gibberellin. The three tall cabbage plants were treated with gibberellin. The two short plants in front of the ladder were not treated.

Sylvan H. Wittwer/ Visuals Unlimited.

〉 Abscisic Acid **Abscisic acid** (ABA) is a plant hormone that was named because its discoverers thought it mainly mediated **abscission**, the process by which a plant sheds leaves or other parts. However, ABA was later discovered to have a much greater role in plant stress responses than in abscission. A plant cannot run away from stressful situations: It either adapts to adverse conditions or dies. Temperature extremes, lack of water, and other environmental stresses trigger an increase in ABA synthesis, release, and transport, so that the level of this hormone rises in the plant's tissues. In turn, ABA induces expression of genes that help the plant survive the adverse conditions. For example, ABA is part of a response that causes a plant's stomata to close when water is scarce, thus preventing water loss by transpiration (Section 27.6). It also has an important role in embryo maturation, seed and pollen germination, and fruit ripening; and like cytokinin it suppresses lateral root formation.

ABA that accumulates in a seed as it forms, matures, and dries out prevents the seed from germinating too early. It exerts this effect mainly by inhibiting expression of genes involved in cell wall loosening and expansion—both critical

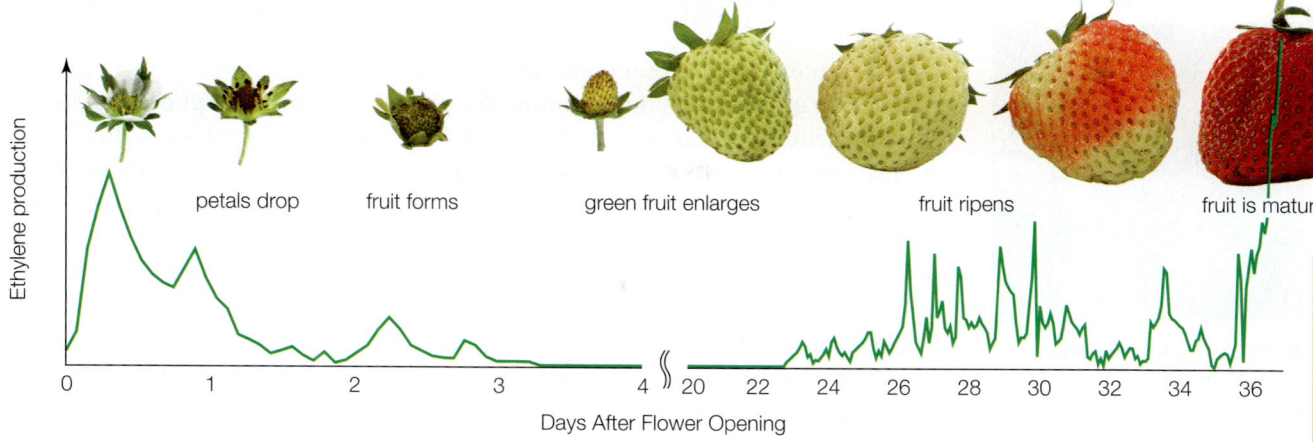

petals drop fruit forms green fruit enlarges fruit ripens fruit is mature

Ethylene production

Days After Flower Opening

FIGURE 28.21
Ethylene production during strawberry formation and ripening. Daily oscillations are due to normal cycles of gene expression (see Section 28.8).

Credits: from left, (1) © Madlen/ Shutterstock; (2) © Westend61/ SuperStock; (3–4) © Anest/ Shutterstock; (5–8) © Alena Brozova/ Shutterstock; Graph data courtesy of Dr. Frans J.M. Harren and Dr. Simona M. Cristescu, Radboud University Nijmegen, The Netherlands.

processes for growth of an embryonic plant. ABA also inhibits gibberellin synthesis. Thus, a seed cannot germinate until its ABA level declines.

〉 Ethylene **Ethylene** is a gas that is soluble in water, and it freely crosses lipid bilayers. It acts as a plant hormone in regulatory pathways that govern a wide range of metabolic and developmental processes, including germination, growth, abscission, ripening, and stress responses.

The enzymes involved in ethylene synthesis occur in multiple versions whose expression is governed by two feedback loops. A negative feedback loop maintains a low level of ethylene that helps fine-tune ongoing processes such as growth and cell expansion. A positive feedback loop produces large amounts of ethylene that trigger intermittent processes such as germination, plant defense responses, abscission, and fruit ripening. For example, the flesh of a strawberry ripens after its cells produce a burst of ethylene in a positive feedback loop (Figure 28.21). Among other effects, the burst enhances transcription of genes whose products break down tough cell walls, convert starches to sugars, and produce aromatic molecules. The softening and increased palatability attract animals that can disperse the strawberry's seeds.

Because ethylene is a gas, it can diffuse from one fruit to initiate ripening in another. Synthetic ethylene is widely used to artificially ripen fruit. Hard, unripe fruit can be transported long distances with less damage than soft, ripe fruit. Upon arrival at its final destination, the fruit is exposed to ethylene gas, which jump-starts the positive feedback ripening loop.

Take-Home Message

What coordinates growth and development in plants?

- Plant hormones have multiple effects and often work together, synergistically or in opposition, in all aspects of plant growth, development, and function.

- Auxin promotes growth and directs development by coordinating the activities of other plant hormones.

- Cytokinin and auxin act together to balance cell division and differentiation in apical meristem in shoot and root tips.

- Among other effects, gibberellin stimulates cell division and elongation in stems, which causes stems to lengthen between nodes.

- Abscisic acid is important in stress responses and stomata function.

- Ethylene produced in positive feedback loops is involved in intermittent processes such as fruit ripening.

abscisic acid Plant hormone involved in stomata function and stress responses; inhibits germination.
abscission Process by which plant parts are shed.
ethylene Gaseous plant hormone that participates in germination, abscission, ripening, and stress responses.
gibberellin Plant hormone that induces stem elongation; also helps seeds break dormancy.

statoliths

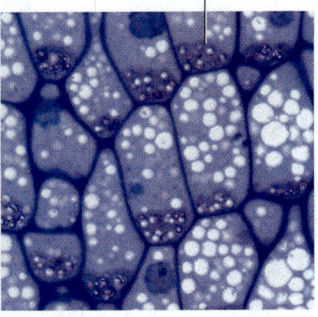

A Regardless of how a corn seed is oriented in soil, the seedling's primary root always grows down, and its primary shoot always grows up.

B Gravity-sensing in plants involves statoliths, organelles packed with heavy starch grains. This micrograph shows statoliths settled on the bottom of cells making up a root cap.

FIGURE 28.22 Gravitropism.

Credits: (a) Michael Clayton, University of Wisconsin; (b) Micrograph courtesy of Randy Moore from How Roots Respond to Gravity, M. L. Evans, R. Moore, and K. Hasenstein, Scientific American, December 1986.

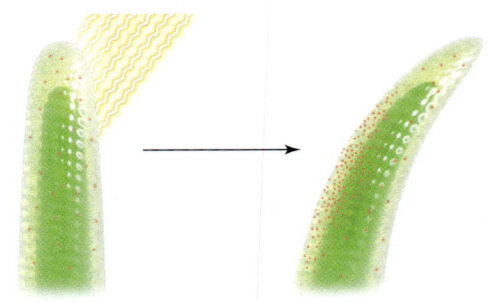

A Sunlight strikes only one side of a coleoptile. Auxin flow is directed toward the shaded side, so cells on that side lengthen more.

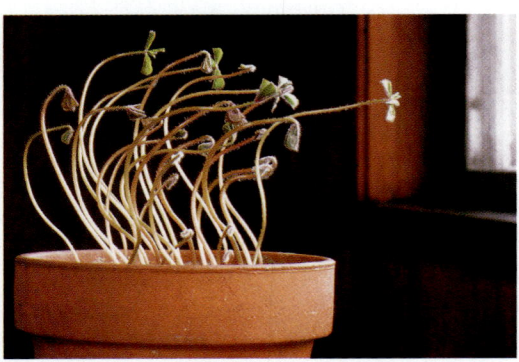

B This shamrock plant is adjusting its growth in response to a directional light source.

FIGURE 28.23 Phototropism. Auxin-mediated differences in cell elongation between two sides of a coleoptile induce bending toward light.

Credits: (a) © Cengage Learning; (b) © Cathlyn Melloan/ Stone/ Getty Images.

28.7 Adjusting the Direction of Growth

Plants respond to environmental stimuli by adjusting their growth. In this way, each plant optimizes its opportunities for photosynthesis, absorption of water and nutrients, reproduction, and other factors that affect its fitness. A directional adjustment of growth in response to an environmental stimulus is a **tropism**. Tropisms are often mediated by hormones, as the following examples illustrate.

> **Gravity** Even if a seedling is turned upside down just after germination, its primary root will grow downward and its shoot will grow upward (Figure 28.22). A growth response to gravity is called **gravitropism**. A root or shoot "bends" in response to a trigger such as gravity because of auxin concentration gradients. In shoots, auxin enhances cell elongation, so auxin that accumulates on one side of a shoot causes the shoot to bend away from that side. Auxin has the opposite effect in roots: Auxin that accumulates on one side of a growing root causes the root to bend toward that side.

Gravity-sensing in many organisms is based on organelles called statoliths. Plant statoliths are stuffed with dense grains of starch. They occur in root cap cells and also in specialized cells at the edge of vascular tissues in the stem. Starch grains are heavier than cytoplasm, so statoliths tend to sink to the lowest region of the cell, wherever that is. The shift causes the flow of auxin through the root or shoot to be directed toward the down-facing side.

> **Light** Light streaming in from one direction causes a stem to curve toward its source. This response, which is called **phototropism**, maximizes the amount of light intercepted by photosynthetic cells. Phototropism in shoots occurs in response to blue light, which is absorbed by nonphotosynthetic pigments called phototropins. These pigments also help stomata open in response to sunlight (Section 27.6). In a shoot tip or coleoptile, light-energized phototropins inhibit the expression of an auxin transport inhibitor—effectively lifting the brakes on auxin transport. Thus, a stem exposed to a directional light source ends up with auxin flowing toward its shaded side, so cells on that side elongate more than cells on the illuminated side (Figure 28.23**A**). The difference causes the entire structure to bend toward the light as it lengthens (Figure 28.23**B**).

> **Contact** Twining tendrils (*right*) result from unequal growth rates of cells on opposite sides of a shoot. A similar mechanism causes roots to grow away from contact, so they "feel" their way around rocks and other impassable objects in the soil. A change in the direction of growth in response to contact is called thigmotropism. The mechanism that gives rise to thigmotropism is not well understood, but an immediate increase in the cytoplasmic concentration of calcium ions accompanies contact, and this is likely to play a role. Both plant cells and animal cells have transport proteins that flood the cytoplasm with calcium ions when the plasma membrane gets distorted or otherwise mechanically perturbed.

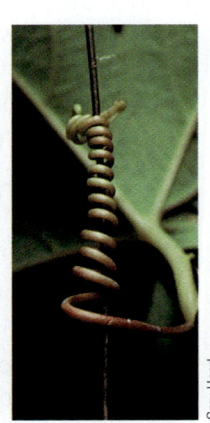

Gary Head.

Take-Home Message

How do plants respond to environmental cues?

- Plants adjust the direction and rate of growth in response to environmental stimuli such as gravity, light, contact, and mechanical stress.

1 A.M.   6 A.M.   Noon   3 P.M.   10 P.M.   Midnight

FIGURE 28.24 Animated! A circadian rhythm: leaf movements by a young bean plant. Physiologist Frank Salisbury kept this plant in darkness for twenty-four hours. Despite the lack of light cues, the leaves kept on folding and unfolding at sunrise (6 A.M.) and sunset (6 P.M.).

Frank B. Salisbury.

28.8 Responses to Recurring Change

❯ Daily Change A **circadian rhythm** is a cycle of biological activity that repeats every twenty-four hours or so (circadian means "about a day"). For example, a bean plant holds its leaves horizontally during the day but folds them close to its stem at night. Bean plants exposed to constant light or constant darkness for a few days will continue to move their leaves in and out of the "sleep" position at the time of sunrise and sunset (Figure 28.24). Similar mechanisms cause flowers of some plants to open only at certain times of day. For example, the flowers of plants pollinated mainly by night-flying bats unfurl, secrete nectar, and release fragrance only at night. Periodically closing flowers protects the delicate reproductive parts when the likelihood of pollination is typically lowest.

Circadian rhythms are driven by interconnected feedback loops involving transcription factors that regulate their own expression. These loops give rise to cycles in the levels of more than 30 percent of a plant cell's mRNAs (Figure 28.25). Transcriptional control in circadian loops often arises by methylation and demethylation of histones (Section 7.7). One recently discovered protein that methylates histones functions in circadian rhythms in both humans and plants.

Cyclic shifts in gene expression underlie cyclic shifts in metabolism, for example between daytime starch-building reactions and nighttime starch-consuming reactions. Components of photosystems, ATP synthases, and other proteins used in photosynthesis are among the many gene products produced during the day and broken down at night. By contrast, molecules that are needed at night are typically broken down during the day.

At least six types of pigments provide light input into these internal circadian "clocks," but those called phytochromes and cryptochromes are the best known. A phytochrome absorbs red light; light with a wavelength of 660 nanometers (red) causes the structure of a phytochrome to change from an inactive to an active form. Light with a wavelength of 730 nanometers (far-red, which predominates in shade) changes the molecule back to its inactive form. Cryptochromes absorb blue and UV light.

❯ Seasonal Change Except at the equator, the length of daylight varies with the season. Days are longer in summer than in winter, and the difference increases with latitude. Plants respond to these seasonal changes in light availability with seasonally appropriate behaviors such as entering or breaking dormancy. **Photoperiodism** refers to an organism's response to changes in the length of day relative to night.

Flowering is a photoperiodic response in many species. Such species are termed long-day or short-day plants depending on which season they flower,

FIGURE 28.25 Circadian cycles of gene expression for some genes involved in temperature stress responses. Plants were kept in constant light; *gray* bars show nighttime hours.

Molecules that help the plants tolerate cold nights and hot days are produced just before they are actually needed. The *blue* line shows the averaged expression of 46 cold tolerance genes; the *red* line, averaged expression of 30 warm tolerance genes.

Figure It Out: At what time of day is expression of cold tolerance genes lowest?

© Cengage Learning.

Answer: Early morning

circadian rhythm A biological activity repeated about every 24 hours.

gravitropism Plant growth in a direction influenced by gravity.

photoperiodism Biological response to seasonal changes in the relative lengths of day and night.

phototropism Plant growth in a direction influenced by light.

tropism In plants, directional growth response to an environmental stimulus.

Time (hours)

A A flash of red light at 660 nm interrupting a long night causes plants to respond as if the night were short. Long-day plants flower; short-day plants do not.

Time (hours)

B A flash of far-red light at 730 nm cancels the effect of a red light flash. Short-day plants flower; long-day plants do not.

FIGURE 28.26 Animated! Experiments showing that long- or short-day plants flower in response to night length.

Each horizontal bar represents 24 hours. Blue bars indicate night length; yellow bars, day length.

Figure It Out: Which type of pigment detected the light flashes in these experiments?   *Answer: A phytochrome*

© Cengage Learning.

A Ripening chestnuts emit ethylene that stimulates abscission of the fruits and nearby leaves.

B Horse chestnut trees take their name from abscission zones that leave horseshoe-shaped scars in this species.

FIGURE 28.27 Abscission as part of the normal life cycle of a deciduous plant.

Credits: (a) © Randy Hjelsand/ iStockphoto.com; (b) © Adrian Chalkley.

even though it was later discovered that the main trigger for flowering is the length of night, not the length of day (Figure 28.26). Irises, oats, clover, and other long-day plants flower only when the hours of darkness fall below a critical value, typically in summer. Chrysanthemums, strawberries, and other short-day plants flower only when the hours of darkness are greater than some critical value. Flowering is not photoperiodic in sunflowers, tomatoes, roses, and other day-neutral plants.

Length of night is not the only cue for flowering. Some plants flower in spring only if they have been exposed to a prolonged period of cold during the preceding winter, a process called **vernalization**. Researchers suspect that plants perceive temperature via their plasma membrane, which varies in lipid composition and calcium ion permeability depending on temperature. Regardless of the mechanism of detection, a "cold" signal influences gene expression in these plants.

Yearly cycles of abscission and dormancy are photoperiodic responses. Plants that periodically drop their leaves before a period of dormancy are typically native to regions that are too dry or too cold for optimal growth during part of the year. For example, deciduous trees of the northeastern United States begin to lose their leaves in autumn, around September. The trees remain dormant during the months of harsh winter weather that would otherwise damage leaves and buds. Growth resumes in spring (April), when milder conditions return. On the opposite side of the world, many tree species native to tropical monsoon forests of south Asia lose their leaves during the summer dry season, which is between November and May. Although the region receives a lot of rain annually, almost none of it falls during this period of the year. Dormancy offers the trees a way to survive the extended droughtlike conditions. New growth reappears at the beginning of June, just in time to be supported by the ample water of monsoon rains.

Abscission of fruits and leaves is mediated by ethylene. Let's use a horse chestnut tree native to southeastern Europe as an example. In this species, most root and shoot growth occurs between spring and early summer, from April to July. By midsummer, the tree is producing fruits and seeds. In August, the growing season is coming to a close, and the tree is routing nutrients to stems and roots for storage during the forthcoming period of dormancy. The ripe fruits (Figure 28.27**A**) and seeds release ethylene that diffuses into nearby twigs, petioles, and fruit stalks. The ethylene is a signal for cells in these zones to produce enzymes that digest their own walls. The cells bulge as their walls soften, and separate from one another as the extracellular matrix that cements them together dissolves. Tissue in the abscission zone weakens, and the structure above it drops. A scar often remains where the structure had been attached (Figure 28.27**B**).

Take-Home Message

How do plants sense and respond to recurring environmental change?

- Plants respond to recurring cues from the environment with recurring cycles of activity. Pigments provide input into circadian cycles.

- The main environmental cue for flowering is the length of night relative to the length of day, which varies by the season in most places. Low winter temperatures stimulate spring flowering in some plant species.

- The timing of seasonal abscission and dormancy is an evolutionary adaptation to recurring cycles of environmental conditions in a particular region.

vernalization Stimulation of flowering in spring by prolonged exposure to low temperature in winter.

FIGURE 28.28 Bees and other pollinators are often exposed to systemic pesticides in nectar and pollen.

© Cengage Learning 2010.

28.9 Plight of the Honeybee (revisited)

We still do not know exactly what causes colony collapse disorder. Many scientists think that CCD may not be a specific disease, but rather a characteristic of honeybee colonies suffering from a combination of environmental stresses that include parasites and diseases. Viruses such as Israeli acute paralysis virus are carried by mites that parasitize bees, and in fact parasitic mites and Israeli acute paralysis virus have been detected in many affected hives. Bees in almost all affected colonies are infected with a virus (invertebrate iridescent virus) and *Nosema*, a parasitic fungus (Section 14.9). *Nosema* is carried by phorid flies that parasitize the bees.

Some researchers suspect that synthetic pesticides called neonicotinoids are an important factor in CCD. Neonicotinoids mimic the neurotoxic (nerve-killing) effect of nicotine, a natural insecticide made by plants such as tobacco. Scientists at the U.S. Department of Agriculture's Bee Research Laboratory showed that neonicotinoids increase bees' vulnerability to *Nosema* infection even in amounts too tiny to detect in the insects themselves.

Neonicotinoids are now the most widely used insecticides in the United States, and they are still being used in other countries—about 2.5 million acres of croplands in total. The chemicals are systemic, which means they are taken up by all tissues of a plant treated with them, including the nectar and pollen that honeybees collect from flowers as they forage (Figure 28.28). Even dewdrops that form on neonicotinoid-treated plants apparently contain enough of the insecticide to affect bees.

WHERE YOU ARE GOING . . .

This chapter concludes our study of biology, but you will hear about the topics covered here again and again. We face a daunting task: sustaining a growing population without destroying natural processes on which all life depends. Your understanding of these processes and the way that humans affect them will help you form your own opinions about how we as a society should meet this challenge.

Summary

Section 28.1 Colony collapse disorder (CCD) is killing honeybees. Declines in populations of bees and other **pollinators** negatively affect plant populations as well as other animal species that depend on the plants, including humans. Widely used neonicotinoid pesticides that increase bee susceptibility to infection may contribute to CCD.

Section 28.2 Flowers consist of whorls of modified leaves. Petals surround **stamens** and **carpels**. A carpel consists of a **stigma**, style, and an **ovary** in which one or more **ovules** develop. A stamen consists of an **anther** on a thin filament. A flower's shape, pattern, color, and fragrance typically reflect an evolutionary relationship with a particular **pollination vector**. Coevolved animal pollinators receive **nectar**, pollen, or another reward for visiting a flower.

Cells in pollen sacs of anthers undergo meiosis and cytoplasmic division, then mitosis and differentiation. The resulting **pollen grains** are released after a period of **dormancy**.

A cell in an ovule undergoes meiosis and cytoplasmic division, then mitosis and cytoplasmic division. The resulting female gametophyte contains an egg and an endosperm mother cell.

A **pollen grain** arrives on a receptive stigma at pollination. Upon **germination**, a pollen grain develops into a pollen tube that contains two sperm cells. The tube grows through the carpel to the egg and releases the sperm cells. In **double fertilization**, one of the sperm cells fertilizes the egg, forming a zygote; the other fuses with the endosperm mother cell and gives rise to triploid **endosperm**.

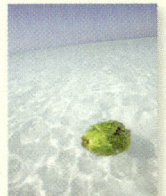

Section 28.3 As a zygote develops into an embryo, endosperm collects nutrients from the parent plant, and the ovule's protective layers develop into a seed coat. A **seed** is a mature ovule: an embryo sporophyte and endosperm enclosed within a seed coat. Nutrients stored in endosperm or cotyledons make seeds a nutritious food source. As an embryo sporophyte develops, the ovary wall and sometimes other tissues mature into a **fruit**. Fruit specializations are adaptations to seed dispersal by specific vectors such as wind, water, or animals.

Section 28.4 Mature seeds often undergo a period of dormancy that does not end until species-specific environmental cues trigger germination. Germination ends when the first part of the embryo emerges from the seed coat; other patterns of early development vary.

Section 28.5 Many types of flowering plants can produce clonal offspring by reproducing asexually. Some are propagated commercially by grafting; the common laboratory technique of **tissue culture propagation** is used to propagate some valuable ornamentals.

Section 28.6 Plants continue to develop through their lifetime. **Plant hormones** can promote or arrest development in regions of a plant by stimulating or inhibiting cell division, differentiation, or enlargement.

Some have roles in reproduction or responses to stress. Hormones typically work together, either synergistically or in opposition, and many have different effects in different regions of the plant.

Auxin coordinates the action of other hormones. **Apical dominance** is a result of auxin transport through cells of a growing shoot tip. **Cytokinin** acts together with auxin, often antagonistically, to balance growth with development in shoot and root tips. **Gibberellin** lengthens stems between nodes, and it is required for seed germination. **Abscisic acid** (ABA) plays a part in **abscission**, but a greater role in seed germination and stress responses. Bursts of **ethylene** produced in positive feedback loops provoke special processes such as abscission and fruit ripening.

Section 28.7 A **tropism** is an adjustment in the direction of plant growth in response to environmental cues. In **gravitropism**, roots grow down and stems grow up in response to gravity. In **phototropisms**, stems and leaves bend toward or away from light. Blue light is the trigger for phototropic responses. In some plants, the direction of growth changes in response to contact.

Section 28.8 **Circadian rhythms** are driven by gene expression feedback loops. Nonphotosynthetic pigments provide light input into circadian clocks. **Photoperiodisms** are responses to change in the length of day relative to night. In many species, senescence includes seasonal cycles of abscission and dormancy that are adaptations to seasonal changes in environmental conditions. Some plants require prolonged exposure to cold before they can flower, a process called **vernalization**.

Self-Quiz   Answers in Appendix I

1. The arrival of pollen grains on a receptive stigma is called _____ .

2. A pollinator may receive _____ when it visits a flower of a coevolved plant (choose all that apply).
   - a. pollen
   - b. nectar
   - c. pesticides
   - d. fruit

3. The _____ of a flower contains one or more ovaries in which eggs develop, fertilization occurs, and seeds mature.

4. In plants, the structures that produce male gametophytes are called _____ ; those that produce female gametophytes are called _____ .
   - a. pollen grains; flowers
   - b. stamens; carpels
   - c. anthers; stigma
   - d. megaspores; microspores

5. Seeds are mature _____ ; fruits are mature _____ .
   - a. ovaries; ovules
   - b. ovules; stamens
   - c. ovules; ovaries
   - d. stamens; ovaries

6. Cotyledons develop as part of _____ .
   - a. carpels
   - b. accessory fruits
   - c. embryo sporophytes
   - d. flowers

7. In some species, exposure to _____ is a trigger for seed germination.
   - a. light
   - b. cold
   - c. smoke
   - d. all can be triggers

8. The three main parts of a mature eudicot seed are the _____ .
   a. pollen grain, egg, and seed coat
   b. embryo, endosperm, and seed coat
   c. megaspores, microspores, and ovule
   d. embryo, cotyledons, and seed coat

9. A new plant forms from a stem that broke off of the parent plant and fell to the ground. This is an example of _____ .
   a. tissue culture propagation
   b. exocytosis
   c. asexual reproduction
   d. nodal cloning

10. Banana plants produce seedless fruit because they are _____ .
    a. triploid
    b. monocots
    c. propagated by grafting
    d. treated with colchicine

11. Plant hormones _____ .
    a. often have multiple, overlapping effects
    b. are active in developing plant embryos
    c. are active in adult plants
    d. all of the above

12. In some plants, flowering is a _____ response.
    a. phototropic
    b. gravitropic
    c. photoperiodic
    d. thigmotropic

13. Which of the following statements is false?
    a. Auxin and gibberellin promote stem elongation.
    b. Cytokinin promotes cell division in shoot tips.
    c. Abscisic acid promotes water loss and dormancy.
    d. Ethylene promotes fruit ripening and abscission.

14. Match the response with its main trigger.
    _____ phototropism   a. mechanical stimulation
    _____ gravitropism   b. blue light
    _____ thigmotropism   c. a long period of cold
    _____ photoperiodism   d. gravity
    _____ vernalization   e. night length

15. Match the observation with the hormone.
    _____ ethylene   a. Your cabbage plants bolt (they form elongated flowering stalks).
    _____ cytokinin
    _____ auxin   b. The potted plant in your room is leaning toward the window.
    _____ gibberellin
    _____ abscisic acid   c. The last of your apples is getting really mushy.
                          d. The seeds of your roommate's marijuana plant do not germinate no matter what he does to them.
                          e. Lateral buds are sprouting.

Critical Thinking

1. Which of the oat coleoptiles shown on the *left* will still bend toward a light source?

2. Is the seedling shown on the *right* a monocot or eudicot? How can you tell?

© Cengage Learning.

© Jubal Harshaw/ Shutterstock.

Digging Into Data

Who's the Pollinator?

Massonia depressa is a low-growing succulent plant native to the desert of South Africa. The dull-colored flowers of this monocot develop at ground level, have tiny petals, emit a yeasty aroma, and produce a thick, jellylike nectar. These features led researchers to suspect that desert rodents such as gerbils pollinate this plant.

To test their hypothesis, the researchers trapped rodents in areas where *M. depressa* grows and checked them for pollen. They also put some plants in wire cages that excluded mammals, but not insects, to see whether fruits and seeds would form in the absence of rodents. The results are shown in Figure 28.29.

1. How many of the 13 captured rodents showed some evidence of pollen from *M. depressa*?

2. Would this evidence alone be sufficient to conclude that rodents are the main pollinators for this plant?

3. How did the average number of seeds produced by caged plants compare with that of control plants?

4. Do these data support the hypothesis that rodents are required for pollination of *M. depressa*? Why or why not?

A The dull, petal-less, ground-level flowers of *Massonia depressa* are accessible to rodents such as this gerbil, who push their heads through the stamens to reach the nectar. Note the pollen on the gerbil's snout.

Type of rodent	Number caught	# with pollen on snout	# with pollen in feces
Namaqua rock rat	4	3	2
Cape spiny mouse	3	2	2
Hairy-footed gerbil	4	2	4
Cape short-eared gerbil	1	0	1
African pygmy mouse	1	0	0

B Evidence of visits to *M. depressa* by rodents.

	Mammals allowed access to plants	Mammals excluded from plants
Percent of plants that set fruit	30.4	4.3
Average number of fruits per plant	1.39	0.47
Average number of seeds per plant	20.0	1.95

C Fruit and seed production of *M. depressa* with and without visits by mammals. Mammals were excluded from plants by wire cages with openings large enough for insects to pass through. 23 plants were tested in each group.

FIGURE 28.29 Testing pollination of *M. depressa* by rodents.

Credits: (a) © Steven D. Johnson; (b,c) © Cengage Learning.

Appendix I

Chapter 1

1.	b	*1.2*
2.	c	*1.2*
3.	d	*1.3*
4.	a	*1.3*
5.	c	*1.3*
6.	c	*1.4*
7.	c	*1.3*
8.	d	*1.3*
9.	a, d, e	*1.2, 1.4*
10.	a, b	*1.4*
11.	b	*1.5*
12.	b	*1.5*
13.	b	*1.5, 1.8*
14.	b	*1.7*
15.	c	*1.2*
	f	*1.7*
	b	*1.5*
	d	*1.8*
	e	*1.6*
	a	*1.5*
	g	*1.2*

Chapter 2

1.	d	*2.2*
2.	hydrogen ion	*2.2*
3.	a	*2.3*
4.	c, b, a	*2.3*
5.	c	*2.4*
6.	a	*2.5*
7.	e	*2.7, 2.10*
8.	c	*2.8*
9.	e	*2.8*
10.	d	*2.9, 2.10*
11.	d	*2.9*
12.	a	*2.10*
13.	c	*2.4*
	b	*2.2*
	d	*2.4*
	a	*2.2*
	f	*2.4*
	e	*2.2*
14.	c	*2.8*
	e	*2.7*
	f	*2.8*
	d	*2.10*
	a	*2.9*
	b	*2.10*
15.	f	*2.9*
	a	*2.8*
	b	*2.8*
	c	*2.10*
	d	*2.7*
	e	*2.10*
	h	*2.8*
	g	*2.7*

Chapter 3

1.	c	*3.2*
2.	c	*3.2*
3.	c	*3.2*
4.	False	*3.5*
5.	c	*3.2, 3.4*
6.	b	*3.3*
7.	a	*3.3*
8.	c	*3.4, 3.5*
9.	c	*3.3*
10.	a	*3.5*
11.	a	*3.6*
12.	c, b, d, a	*3.5*
13.	d	*3.6*
14.	a	*3.6*
15.	c	*3.5*
	g	*3.5*
	e	*3.4, 3.5*
	d	*3.5*
	a	*3.6*
	h	*3.4, 3.6*
	f	*3.3*
	b	*3.6*

Chapter 4

1.	c	*4.2*
2.	b	*4.2*
3.	d	*4.3*
4.	c	*4.3, 4.4*
5.	d	*4.4*
6.	a	*4.4*
7.	d	*4.4*
8.	b	*4.5, 4.6*
9.	more, less	*4.5*
10.	c	*4.6*
11.	b	*4.6*
12.	a	*4.5*
13.	b	*4.5*
14.	d	*4.6*
15.	c	*4.3*
	e	*4.6*
	f	*4.2*
	b	*4.3*
	a	*4.4*
	g	*4.5*
	h	*4.6*
	i	*4.4*

Chapter 5

1.	b	*5.1*
2.	a	*5.1, 5.2*
3.	b	*5.3*
4.	d	*5.4*
5.	c	*5.4*
6.	False	*5.6*
7.	oxygen	*5.6*
8.	d	*5.6*
9.	b	*5.5*
10.	b	*5.6*
11.	d	*5.7*
12.	c	*5.6*
13.	f	*5.7*
14.	e	*5.4, 5.6*
15.	d	*5.8*
16.	c	*5.6*
	a	*5.7*
	d	*5.6*
	f	*5.2*
	g	*5.1*
	e	*5.5*
	b	*5.2*

Chapter 6

1.	b	*6.2*
2.	c	*6.2*
3.	b	*6.3, 6.4*
4.	c	*6.3*
5.	a	*6.4*
6.	d	*6.4*
7.	d	*6.4*
8.	d	*6.4*
9.	d	*6.5*
10.	e	*6.3*
	c	*6.1*
	d	*6.2*
	a	*6.4*
	f	*6.4*
	b	*6.4*

Chapter 7

1.	c	*7.2*
2.	c	*7.2*
3.	a	*7.2*
4.	b	*7.2, 7.4*
5.	a	*7.3, 7.5*
6.	c	*7.5*
7.	15	*7.4*
8.	c	*7.4*
9.	e	*7.6*
10.	b	*7.7*
11.	d	*7.7*
12.	b	*7.7*
13.	c	*7.7*
14.	True	*7.7*
15.	c	*7.7*
	d	*7.2*
	e	*7.4, 7.5*
	g	*7.2, 7.4, 7.5*
	f	*7.4*
	a	*7.7*
	b	*7.6*
16.	c, a, d, b	*7.3, 7.5*

Chapter 8

1.	e	*8.2*
2.	a	*8.3*
3.	b	*8.3*
4.	a	*8.2, 8.3, 8.7*
5.	d	*8.3, 8.6*
6.	c	*8.3*
7.	a	*8.3*
8.	b	*8.3*
9.	b	*8.6*
10.	c	*8.6*
11.	d	*8.2, 8.7*
12.	b	*8.7*
13.	b	*8.7*
14.	d	*8.7*
15.	b, d, a, c	*8.3*
16.	c	*8.4*
	f	*8.3*
	a	*8.5*
	g	*8.4*
	b	*8.4*
	e	*8.5*
	d	*8.3*
	h	*8.7*
	i	*8.7*

Chapter 9

1.	b	*9.2*
2.	a	*9.2*
3.	c	*9.3*
4.	d	*9.3*
5.	c	*9.3*
6.	b	*9.4*
7.	c	*9.5*
8.	b	*9.6*
9.	X from mom, Y from dad	*9.7*
10.	b	*9.8*
11.	False	*9.6*
12.	b	*9.7*
13.	c	*9.5*
14.	b	*9.3*
	d	*9.3*
	a	*9.2*
	c	*9.2*
15.	b	*9.8*
	a	*9.6*
	e	*9.6*
	c	*9.8*
	d	*9.5*
	g	*9.2*
	f	*9.5*

Chapter 10

1.	c	*10.2*
2.	a	*10.2*
3.	b	*10.2*
4.	c	*10.2*
5.	d	*10.2, 10.3*
6.	b	*10.3*
7.	a	*10.2*
8.	b	*10.3*
9.	b	*10.5*
	d	*10.2*
	e	*10.5*
	f	*10.4*
10.	d	*10.4*
11.	b	*10.5*
12.	e, a, d, b, c	*10.2, 10.3*
13.	c	*10.3*
	g	*10.4*
	d	*10.2*
	e	*10.5*
	b	*10.1*
	a	*10.4*
	f	*10.4*

Chapter 11

1.	b	*11.2*
2.	c	*11.2, 11.6*
3.	d	*11.3*
4.	b	*11.3*
5.	b	*11.4*
6.	e	*11.4*
7.	d	*11.4*
8.	Gondwana	*11.5*
9.	65.5	*11.2, 11.5*
10.	morphological divergence	*11.6*
11.	d	*11.6*
12.	b	*11.6*
13.	e	*11.6*
14.	j	*11.3*
	i	*11.2, 11.4*
	e	*11.3*
	f	*11.4*
	c	*11.3*
	b	*11.3*
	g	*11.6*
	d	*11.4*
	h	*11.6*
	a	*11.6*

Chapter 12

1.	a	*12.2*
2.	populations	*12.2*
3.	c	*12.2*
4.	a, c, b	*12.3*
5.	d	*12.4*
6.	e	*12.4*
7.	b	*12.4*
8.	f	*12.3, 12.4*
9.	a	*12.5*
10.	d	*12.4, 12.5*
11.	d	*12.7*
12.	c	*12.7*
13.	c	*12.7*
14.	b	*12.7*
15.	c	*12.4*
	e	*12.4*
	g	*12.6*
	b	*12.4*
	h	*12.7*
	f	*12.6*
	a	*12.6*
	d	*12.7*

Chapter 13

1.	oxygen	*13.3*
2.	b	*13.2*
3.	d	*13.3*
4.	b	*13.3*
5.	c	*13.5*
6.	b	*13.6*
7.	True	*13.6*
8.	choanoflagellates	*13.6*
9.	b	*13.6*
10.	c	*13.6*
11.	c	*13.6*
12.	c	*13.6*
13.	b	*13.6*
14.	d	*13.4*
15.	g	*13.6*
	d	*13.4*
	c	*13.5*
	e	*13.6*
	i	*13.5*
	f	*13.6*
	a	*13.6*
	j	*13.6*
	b	*13.6*
	h	*13.3*

Chapter 14

1.	a	*14.6, 14.7*
2.	c	*14.6, 14.7*
3.	b	*14.3*
4.	c	*14.4*
5.	a	*14.4, 14.6, 14.7*
6.	d	*14.6*
7.	b	*14.5*
8.	c	*14.6*
	d	*14.2*
	f	*14.4*
	e	*14.3*
	a	*14.2*
	b	*14.2*
	g	*14.7*
9.	c	*14.8*
10.	a	*14.8*
11.	c	*14.8*
12.	d	*14.8*
13.	b	*14.9*
14.	d	*14.8*
15.	g, f, b, c, a, d, e	all *14.8*

Chapter 15

1.	True	15.2
2.	b	15.2
3.	c	15.2
4.	b	15.2
5.	c	15.3
6.	d	15.3
7.	Notochord, dorsal nerve cord, pharynx with gill slits, tail that extends beyond anus. Pharynx with gill slits.	15.4
8.	False	15.5
9.	b	15.4
10.	f	15.6
11.	False	15.7
12.	d	15.6
13.	b	15.3
	g	15.3
	l	15.3
	j	15.3
	e	15.3
	c	15.3
	d	15.3
	f	15.3
	k	15.5
	a	15.5
	h	15.6
	i	15.6
14.	1-b	15.2
	2-a	15.2
	3-f	15.4
	4-c	15.5
	5-d	15.6
	6-e	15.7

Chapter 16

1.	a	16.2
2.	c	16.1
3.	b	16.3
4.	a	16.3
5.	b	16.3
6.	a	16.3
7.	b	16.4
8.	a	16.3
9.	a	16.5
10.	d	16.5
11.	False	16.3
12.	b	16.5
13.	c	16.3
14.	d	16.3
	c	16.3
	e	16.3
	a	16.5
	f	16.3
	b	16.4

Chapter 17

1	d	17.2
2.	b, c, a, d, e	17.3
3.	b	17.3
4.	c	17.3
5.	a	17.4
6.	d	17.5
	a	17.3
	c	17.5
	b	17.5
7.	a	17.5
8.	b, a, b, c	17.6
9.	c	17.6
10.	b	17.6
11.	a	17.6
12.	b	17.6
13.	d	17.3, 17.7
14.	d	17.6
15.	b	17.4

Chapter 18

1.	a	18.2
2.	d	18.2
3.	b	18.3
4.	d	18.3
5.	b	18.3
6.	c	18.3
7.	a	18.4
8.	a	18.4
9.	e	18.3
	d	18.3
	b	18.3
	c	18.3
	a	18.4
	f	18.3
	g	18.4
	h	18.3
10.	b	18.5
11.	c	18.5
12.	b	18.5
13.	d	18.5
14.	d	18.5
15.	c	18.5

Chapter 19

1.	a	19.3
2.	b	19.3
3.	a	19.3
4.	a	19.3
5.	a	19.3
6.	b	19.3
7.	c	19.3
8.	b	19.3
9.	c	19.3
10.	a	19.3
11.	d	19.3
12.	a	19.3
13.	d	19.4
14.	d	19.5
15.	b, h, a, f, e, c, g, i, j, d	all 19.3

Chapter 20

1.	a	20.2
2.	d	20.2
3.	d	20.2
4.	b	20.4
5.	a	20.2
6.	a	20.2
7.	a	20.4
8.	d	20.4
9.	c	20.4
10.	b	20.4
11.	d	20.4
12.	a	20.4
13.	d	20.2
14.	b, e, c, d, a	all 20.2
15.	d	20.4
	f	20.1
	g	20.4
	i	20.2
	c	20.2
	e	20.5
	a	20.5
	k	20.3
	j	20.3
	b	20.4
	h	20.4

Chapter 21

1.	b	21.2
2.	b	21.4
3.	a	21.5
4.	c	21.4
5.	c	21.5
6.	d	21.8
7.	a	21.2
8.	a	21.4
9.	c	21.8
10.	d	21.8
11.	a	21.8
12.	a	21.8
13.	g	21.6
	h	21.6
	e	21.6
	b	21.8
	d	21.6
	a	21.1
	c	21.8
	f	21.8
14.	b	21.5
	e	21.8
	i	21.8
	h	21.6
	c	21.4
	g	21.8
	d	21.8
	f	21.4
	a	21.3
15.	2, 1, 3, 4, 5	21.8

Chapter 22

1.	e	22.2, 22.3
2.	g	22.4
3.	c	22.4, 22.6
4.	d	22.4, 22.6
5.	d	22.2
6.	Inborn; general; unchanging; and immediate are all correct	22.2, 22.4
7.	d	22.5
8.	self/nonself recognition; specificity; diversity; and memory are all correct	22.6
9.	a	22.5
10.	b	22.6
11.	e	22.6
12.	b	22.6
13.	c	22.7
14.	b	22.2, 22.5, 2.6
	e	22.5
	c	22.6
	d	22.6
	a	22.6
15.	f	22.7
	e	22.6
	d	22.7
	c	22.4
	b	22.7
	a	22.2, 22.5

Chapter 23

1.	d	23.2
2.	b	23.3
3.	c	23.3
4.	b	23.3
5.	c	23.3
6.	f, b, a, d, e, c	23.3
7.	d	23.4
8.	c	23.3
9.	b	23.4
10.	d	23.5
11.	a	23.6
12.	a	23.6
13.	b	23.6
14.	c	23.6
15.	c	23.5
	a	23.5
	b	23.5
	e	23.6
	d	23.6

Chapter 24

1.	a	24.2
2.	c	24.2
3.	a	24.2
4.	b	24.4
5.	a	24.2
6.	c	24.4
7.	reflex	24.5
8.	c	24.5
9.	b	24.5
10.	c	24.4
11.	c	24.6
12.	b	24.6
13.	d	24.6
14.	c	24.6
15.	j	24.6
	c	24.6
	i	24.5
	b	24.5
	h	24.5
	f	24.6
	e	24.2
	g	24.2
	d	24.5
	a	24.6

Chapter 25

1.	a	25.2
2.	b	25.3
3.	c	25.2
4.	c, b, a, d	25.3
5.	c	25.3
6.	d	25.4
7.	b	25.5
8.	b	25.4
9.	b	25.5
10.	b	25.4
11.	False	25.7
12.	b	25.4
13.	d	25.6
14.	c	25.3
15.	d	25.6
	f	25.4
	e	25.7
	c	25.4
	g	25.5
	h	25.3
	b	25.3
	a	25.3

Chapter 26

1.	a	26.2
2.	a	26.2
3.	b	26.2
4.	a	26.3
5.	d	26.4
6.	c	26.4
7.	c	26.3
8.	c	26.3
9.	b	26.6
10.	3, 2, 1, 6, 4, 5	26.6
11.	e	26.5
12.	d	26.7
13.	a	26.7
14.	c	26.3
	h	26.3
	a	26.6
	g	26.7
	d	26.3
	e	26.3
	b	26.3
	f	26.7

Chapter 27

1.	a	27.4
2.	b	27.2
3.	c	27.2, 27.3
4.	a, b	27.2, 27.6
5.	b	27.2
6.	b	27.3
7.	b	27.4
8.	b	27.5
9.	a	27.5
10.	b	27.5
11.	c	27.6
12.	c	27.6
13.	b	27.6
14.	d	27.6
15.	a	27.6
16.	b	27.4
	d	27.4
	e	27.2
	c	27.3
	f	27.3
	a	27.4

Chapter 28

1.	pollination	28.2
2.	a, b, c	28.2, 28.9
3.	carpel	28.2, 28.3
4.	b	28.2
5.	c	28.3
6.	c	28.3, 28.4
7.	d	28.4
8.	d	28.3
9.	c	28.5
10.	a	28.5
11.	d	28.6
12.	c	28.8
13.	c	28.6
14.	b	28.7
	d	28.7
	a	28.7
	e	28.8
	c	28.8
15.	c, e, b, a, d	28.6

Appendix II

Periodic Table of the Elements

Group

Noble Gases (18)

Atomic number → 11
Symbol → Na
Atomic mass → 22.99

Atomic masses are based on carbon-12. Numbers in parentheses are mass numbers of most stable or best known isotopes of radioactive elements.

Transition Elements

Inner Transition Elements

Lanthanide Series 6

| | 58 Ce 140.1 | 59 Pr 140.9 | 60 Nd 144.2 | 61 Pm (145) | 62 Sm 150.4 | 63 Eu 152.0 | 64 Gd 157.3 | 65 Tb 158.9 | 66 Dy 162.5 | 67 Ho 164.9 | 68 Er 167.3 | 69 Tm 168.9 | 70 Yb 173.0 | 71 Lu 175.0 |

Actinide Series 7

| | 90 Th 232.0 | 91 Pa 231.0 | 92 U 238.0 | 93 Np 237.0 | 94 Pu (244) | 95 Am (243) | 96 Cm (247) | 97 Bk (247) | 98 Cf (251) | 99 Es (252) | 100 Fm (257) | 101 Md (258) | 102 No (259) | 103 Lr (260) |

Appendix III

Annotations to A Journal Article

This journal article reports on the movements of a female wolf during the summer of 2002 in northwestern Canada. It also reports on a scientific process of inquiry, observation and interpretation to learn where, how and why the wolf traveled as she did. In some ways, this article reflects the story of "how to do science" told in section 1.6 of this textbook. These notes are intended to help you read and understand how scientists work and how they report on their work.

---

(1) ARCTIC

(2) VOL. 57, NO. 2 (JUNE 2004) P. 196–203

(3) Long Foraging Movement of a Denning Tundra Wolf

(4) Paul F. Frame,[1,2] David S. Hik,[1] H. Dean Cluff,[3] and Paul C. Paquet[4]

(5) (Received 3 September 2003; accepted in revised form 16 January 2004)

---

(6) ABSTRACT. Wolves (*Canis lupus*) on the Canadian barrens are intimately linked to migrating herds of barren-ground caribou (*Rangifer tarandus*). We deployed a Global Positioning System (GPS) radio collar on an adult female wolf to record her movements in response to changing caribou densities near her den during the summer. This wolf and two other females were observed nursing a group of 11 pups. She traveled a minimum of 341 km during a 14-day excursion. The straight-line distance from the den to the farthest location was 103 km, and the overall minimum rate of travel was 3.1 km/h. The distance between the wolf and the radio-collared caribou decreased from 242 km one week before the excursion to 8 km four days into the excursion. We discuss several possible explanations for the long foraging bout.

(7) *Key words:* wolf, GPS tracking, movements, *Canis lupus*, foraging, caribou, Northwest Territories

(8) RÉSUMÉ. Les loups (*Canis lupus*) dans la toundra canadienne sont étroitement liés aux hardes de caribous des toundras (*Rangifer tarandus*). On a équipé une louve adulte d'un collier émetteur muni d'un système de positionnement mondial (GPS) afin d'enregistrer ses déplacements en réponse au changement de densité du caribou près de sa tanière durant l'été. On a observé cette louve ainsi que deux autres en train d'allaiter un groupe de 11 louveteaux. Elle a parcouru un minimum de 341 km durant une sortie de 14 jours. La distance en ligne droite de la tanière à l'endroit le plus éloigné était de 103 km, et la vitesse minimum durant tout le voyage était de 3,1 km/h. La distance entre la louve et le caribou muni du collier émetteur a diminué de 242 km une semaine avant la sortie à 8 km quatre jours après la sortie. On commente diverses explications possibles pour ce long épisode de recherche de nourriture.

Mots clés: loup, repérage GPS, déplacements, *Canis lupus*, recherche de nourriture, caribou, Territoires du Nord-Ouest

Traduit pour la revue *Arctic* par Nésida Loyer.

---

(9) Introduction

Wolves (*Canis lupus*) that den on the central barrens of mainland Canada follow the seasonal movements of their main prey, migratory barren-ground caribou (*Rangifer tarandus*) (Kuyt, 1962; Kelsall, 1968; Walton et al., 2001). However, most wolves do not den near caribou calving grounds, but select sites farther south, closer to the tree line (Heard and Williams, 1992). Most caribou migrate beyond primary wolf denning areas by mid-June and do not return until mid-to-late July (Heard et al., 1996; Gunn et al., 2001). Conse-quently, caribou density near dens is low for part of the summer.

During this period of spatial separation from the main caribou herds, wolves must either search near the homesite for scarce caribou or alternative prey (or both), travel to where prey are abundant, or use a combination of these strategies. **(10)**

Walton et al. (2001) postulated that the travel of tundra wolves outside their normal summer ranges is a response to low caribou availability rather than a pre-dispersal exploration like that observed in terri torial wolves (Fritts and Mech, 1981; Messier, 1985). The authors postulated this because most such travel was directed toward caribou calving grounds. We report details of such a long-distance excursion by a breeding female tundra wolf wearing a GPS radio collar. We discuss the relationship of the excursion to movements of satellite-collared caribou (Gunn et al., 2001), supporting the hypothesis that tundra wolves make directional, rapid, long-distance movements in response to seasonal prey availability. **(11)**

[1] Department of Biological Sciences, University of Alberta, Edmonton, Alberta T6G 2E9, Canada
[2] Corresponding author: pframe@ualberta.ca
[3] Department of Resources, Wildlife, and Economic Development, North Slave Region, Government of the Northwest Territories, P.O. Box 2668, 3803 Bretzlaff Dr., Yellowknife, Northwest Territories X1A 2P9, Canada; Dean_Cluff@gov.nt.ca
[4] Faculty of Environmental Design, University of Calgary, Calgary, Alberta T2N 1N4, Canada; current address: P.O. Box 150, Meacham, Saskatchewan S0K 2V0, Canada

196

---

1 Title of the journal, which reports on science taking place in Arctic regions.

2 Volume number, issue number and date of the journal, and page numbers of the article.

3 Title of the article: a concise but specific description of the subject of study—one episode of long-range travel by a wolf hunting for food on the Arctic tundra.

4 Authors of the article: scientists working at the institutions listed in the footnotes below. Note #2 indicates that P. F. Frame is the *corresponding author*—the person to contact with questions or comments. His email address is provided.

5 Date on which a draft of the article was received by the journal editor, followed by date one which a revised draft was accepted for publication. Between these dates, the article was reviewed and critiqued by other scientists, a process called peer review. The authors revised the article to make it clearer, according to those reviews.

6 ABSTRACT: A brief description of the study containing all basic elements of this report. First sentence summarizes the *background* material. Second sentence encapsulates the *methods* used. The rest of the paragraph sums up the *results*. Authors introduce the main *subject* of the study—a female wolf (#388) with pups in a den—and refer to later *discussion* of possible explanations for her behavior.

7 Key words are listed to help researchers using computer databases. Searching the databases using these key words will yield a list of studies related to this one.

8 RÉSUMÉ: The French translation of the abstract and key words. Many researchers in this field are French Canadian. Some journals provide such translations in French or in other languages.

9 INTRODUCTION: Gives the background for this wolf study. This paragraph tells of known or suspected wolf behavior that is important for this study. Note that (a) major species mentioned are always accompanied by scientific names, and (b) statements of fact or *postulations* (claims or assumptions about what is likely to be true) are followed by references to studies that established those facts or supported the postulations.

10 This paragraph focuses directly on the wolf behaviors that were studied here.

11 This paragraph starts with a statement of the *hypothesis* being tested, one that originated in other studies and is supported by this one. The hypothesis is restated more succinctly in the last sentence of this paragraph. This is the *inquiry* part of the scientific process—asking questions and suggesting possible answers.

12 This map shows the study area and depicts wolf and caribou locations and movements during one summer. Some of this information is explained below.

13 STUDY AREA: This section sets the stage for the study, locating it precisely with latitude and longitude coordinates and describing the area (illustrated by the map in Figure 1).

14 Here begins the story of how prey (caribou) and predators (wolves) interact on the tundra. Authors describe movements of these nomadic animals throughout the year.

15 We focus on the denning season (summer) and learn how wolves locate their dens and travel according to the movements of caribou herds.

Figure 1. Map showing the movements of satellite radio-collared caribou with respect to female wolf 388's summer range and long foraging movement, in summer 2002.

13 **Study Area**

Our study took place in the northern boreal forest–low Arctic tundra transition zone (63° 30′ N, 110° 00′ W; Figure 1; Timoney et al., 1992). Permafrost in the area changes from discontinuous to continuous (Harris, 1986). Patches of spruce (*Picea mariana, P. glauca*) occur in the southern portion and give way to open tundra to the northeast. Eskers, kames, and other glacial deposits are scattered throughout the study area. Standing water and exposed bedrock are characteristic of the area.

14 *Details of the Caribou-Wolf System*

The Bathurst caribou herd uses this study area. Most caribou cows have begun migrating by late April, reaching calving grounds by June (Gunn et al., 2001;

Figure 1). Calving peaks by 15 June (Gunn et al., 2001), and calves begin to travel with the herd by one week of age (Kelsall, 1968). The movement patterns of bulls are less known, but bulls frequent areas near calving grounds by mid-June (Heard et al., 1996; Gunn et al., 2001). In summer, Bathurst caribou cows generally travel south from their calving grounds and then, parallel to the tree line, to the northwest. The rut usually takes place at the tree line in October (Gunn et al., 2001). The winter range of the Bathurst herd varies among years, ranging through the taiga and along the tree line from south of Great Bear Lake to southeast of Great Slave Lake. Some caribou spend the winter on the tundra (Gunn et al., 2001; Thorpe et al., 2001).

15 In winter, wolves that prey on Bathurst caribou do not behave territorially. Instead, they follow the herd throughout its winter range (Walton et al., 2001; Musiani, 2003). However, during denning (May–

Table 1. Daily distances from wolf 388 and the den to the nearest radio-collared caribou during a long excursion in summer 2002.

Date (2002)	Mean distance from caribou to wolf (km)	Daily distance from closest caribou to den
12 July	242	241
13 July	210	209
14 July	200	199
15 July	186	180
16 July	163	162
17 July	151	148
18 July	144	137
19 July[1]	126	124
20 July	103	130
21 July	73	130
22 July	40	110
23 July[2]	9	104
29 July[3]	16	43
30 July	32	43
31 July	28	44
1 August	29	46
2 August[4]	54	52
3 August	53	53
4 August	74	74
5 August	75	75
6 August	74	75
7 August	72	75
8 August	76	75
9 August	79	79

[1] Excursion starts.
[2] Wolf closest to collared caribou.
[3] Previous five days' caribou locations not available.
[4] Excursion ends.

August, parturition late May to mid-June), wolf movements are limited by the need to return food to the den. To maximize access to migrating caribou, many wolves select den sites closer to the tree line than to caribou calving grounds (Heard and Williams, 1992). Because of caribou movement patterns, tundra denning wolves are separated from the main caribou herds by several hundred kilometers at some time during summer (Williams, 1990:19; Figure 1; Table 1).

 Muskoxen do not occur in the study area (Fournier and Gunn, 1998), and there are few moose there (H.D. Cluff, pers. obs.). Therefore, alternative prey for wolves includes waterfowl, other ground-nesting birds, their eggs, rodents, and hares (Kuyt, 1972; Williams, 1990:16; H.D. Cluff and P.F. Frame, unpubl. data). During 56 hours of den observations, we saw no ground squirrels or hares, only birds. It appears that the abundance of alternative prey was relatively low in 2002.

Methods

Wolf Monitoring

We captured female wolf 388 near her den on 22 June 2002, using a helicopter net-gun (Walton et al., 2001). She was fitted with a releasable GPS radio collar (Merrill et al., 1998) programmed to acquire locations at 30-minute intervals. The collar was electronically released (e.g., Mech and Gese, 1992) on 20 August 2002. From 27 June to 3 July 2002, we observed 388's den with a 78 mm spotting scope at a distance of 390 m.

Caribou Monitoring

In spring of 2002, ten female caribou were captured by helicopter net-gun and fitted with satellite radio collars, bringing the total number of collared Bathurst cows to 19. Eight of these spent the summer of 2002 south of Queen Maud Gulf, well east of normal Bathurst caribou range. Therefore, we used 11 caribou for this analysis. The collars provided one location per day during our study, except for five days from 24 to 28 July. Locations of satellite collars were obtained from Service Argos, Inc. (Landover, Maryland).

Data Analysis

Location data were analyzed by ArcView GIS software (Environmental Systems Research Institute Inc., Redlands, California). We calculated the average distance from the nearest collared caribou to the wolf and the den for each day of the study.

Wolf foraging bouts were calculated from the time 388 exited a buffer zone (500 m radius around the den) until she re-entered it. We considered her to be traveling when two consecutive locations were spatially separated by more than 100 m. Minimum distance traveled was the sum of distances between each location and the next during the excursion.

We compared pre- and post-excursion data using Analysis of Variance (ANOVA; Zar, 1999). We first tested for homogeneity of variances with Levene's test (Brown and Forsythe, 1974). No transformations of these data were required.

Results

Wolf Monitoring

Pre-Excursion Period: Wolf 388 was lactating when captured on 22 June. We observed her and two other females nursing a group of 11 pups between 27 June and 3 July. During our observations, the pack consisted of at least four adults (3 females and 1 male) and 11 pups. On 30 June, three pups were moved to a location 310 m from the other eight and cared for by an uncollared female. The male was not seen at the den after the evening of 30 June.

Before the excursion, telemetry indicated 18 foraging bouts. The mean distance traveled during these bouts was 25.29 km (± 4.5 SE, range 3.1–82.5 km). Mean greatest distance from the den on foraging

16 Other variables are considered—prey other than caribou and their relative abundance in 2002.

17 METHODS: There is no one scientific method. Procedures for each and every study must be explained carefully.

18 Authors explain when and how they tracked caribou and wolves, including tools used and the exact procedures followed.

19 This important subsection explains what data were calculated (average distance ...) and how, including the software used and where it came from. (The calculations are listed in Table 1.) Note that the behavior measured (traveling) is carefully defined.

20 RESULTS: The heart of the report and the *observation* part of the scientific process. This section is organized parallel to the Methods section.

21 This subsection is broken down by periods of observation. Pre-excursion period covers the time between 388's capture and the start of her long-distance travel. The investigators used visual observations as well as telemetry (measurements taken using the global positioning system (GPS)) to gather data. They looked at how 388 cared for her pups, interacted with other adults, and moved about the den area.

22 The key in the lower right-hand corner of the map shows areas (shaded) within which the wolves and caribou moved, and the dotted trail of 388 during her excursion. From the results depicted on this map, the investigators tried to determine when and where 388 might have encountered caribou and how their locations affected her traveling behavior.

23 The wolf's excursion (her long trip away from the den area) is the focus of this study. These paragraphs present detailed measurements of daily movements during her two-week trip—how far she traveled, how far she was from collared caribou, her time spent traveling and resting, and her rate of speed. Authors use the phrase "minimum distance traveled" to acknowledge they couldn't track every step but were measuring samples of her movements. They knew that she went at least as far as they measured. This shows how scientists try to be exact when reporting results. Results of this study are depicted graphically in the map in Figure 2.

Figure 2. Details of a long foraging movement by female wolf 388 between 19 July and 2 August 2002. Also shown are locations and movements of three satellite radio-collared caribou from 23 July to 21 August 2002. On 23 July, the wolf was 8 km from a collared caribou. The farthest point from the den (103 km distant) was recorded on 27 July. Arrows indicate direction of travel.

bouts was 7.1 km (± 0.9 SE, range 1.7–17.0 km). The average duration of foraging bouts for the period was 20.9 h (± 4.5 SE, range 1–71 h).

The average daily distance between the wolf and the nearest collared caribou decreased from 242 km on 12 July, one week before the excursion period, to 126 km on 19 July, the day the excursion began (Table 1).

23 Excursion Period: On 19 July at 2203, after spending 14 h at the den, 388 began moving to the northeast and did not return for 336 h (14 d; Figure 2). Whether she traveled alone or with other wolves is unknown. During the excursion, 476 (71%) of 672 possible locations were recorded. The wolf crossed the southeast end of Lac Capot Blanc on a small land bridge, where she paused for 4.5 h after traveling for 19.5 h (37.5 km). Following this rest, she traveled for 9 h (26.3 km) onto a peninsula in Reid Lake, where she spent 2 h before backtracking and stopping for 8 h just off the peninsula. Her next period of travel lasted 16.5 h (32.7 km), terminating in a pause of 9.5 h just 3.8 km from a concentration of locations at the far end of her excursion, where we presume she encountered caribou. The mean duration of these three movement periods was 15.7 h (± 2.5 SE), and that of the pauses, 7.3 h (± 1.5). The wolf required 72.5 h (3.0 d) to travel a minimum of 95 km from her den to this area near caribou (Figure 2). She remained there (35.5 km2) for 151.5 h (6.3 d) and then moved south to Lake of the Enemy, where she stayed (31.9 km^2) for 74 h (3.1 d) before returning to her den. Her greatest distance from the den, 103 km, was recorded 174.5 h (7.3 d) after the excursion began, at 0433 on 27 July. She was 8 km from a collared caribou on 23 July, four days after the excursion began (Table 1).

The return trip began at 0403 on 2 August, 318 h (13.2 d) after leaving the den. She followed a relatively direct path for 18 h back to the den, a distance of 75 km.

The minimum distance traveled during the excursion was 339 km. The estimated overall minimum travel rate was 3.1 km/h, 2.6 km/h away from the den and 4.2 km/h on the return trip.

24 **Post-Excursion Period:** We saw three pups when recovering the collar on 20 August, but others may have been hiding in vegetation.

Telemetry recorded 13 foraging bouts in the post-excursion period. The mean distance traveled during these bouts was 18.3 km (+ 2.7 SE, range 1.2–47.7 km), and mean greatest distance from the den was 7.1 km (+ 0.7 SE, range 1.1–11.0 km). The mean duration of these post-excursion foraging bouts was 10.9 h (+ 2.4 SE, range 1–33 h).

When 388 reached her den on 2 August, the distance to the nearest collared caribou was 54 km. On 9 August, one week after she returned, the distance was 79 km (Table 1).

Pre- and Post-Excursion Comparison

25 We found no differences in the mean distance of foraging bouts before and after the excursion period (F = 1.5, df = 1, 29, p = 0.24). Likewise, the mean greatest distance from the den was similar pre- and post-excursion (F = 0.004, df = 1, 29, p = 0.95). However, the mean duration of 388's foraging bouts decreased by 10.0 h after her long excursion (F = 3.1, df = 1, 29, p = 0.09).

26 *Caribou Monitoring*

Summer Movements: On 10 July, 5 of 11 collared caribou were dispersed over a distance of 10 km, 140 km south of their calving grounds (Figure 1). On the same day, three caribou were still on the calving grounds, two were between the calving grounds and the leaders, and one was missing. One week later (17 July), the leading radio-collared cows were 100 km farther south (Figure 1). Two were within 5 km of each other in front of the rest, who were more dispersed. All radio-collared cows had left the calving grounds by this time. On 23 July, the leading radio-collared caribou had moved 35 km farther south, and all of them were more widely dispersed. The two cows closest to the leader were 26 km and 33 km away, with 37 km between them. On the next location (29 July), the most southerly caribou were 60 km

farther south. All of the caribou were now in the areas where they remained for the duration of the study (Figure 2).

A Minimum Convex Polygon (Mohr and Stumpf, 1966) around all caribou locations acquired during the study encompassed 85 119 km².

Relative to the Wolf Den: The distance from the nearest collared caribou to the den decreased from 241 km one week before the excursion to 124 km the day it began. The nearest a collared caribou came to the den was 43 km away, on 29 and 30 July. During the study, four collared caribou were located within 100 km of the den. Each of these four was closest to the wolf on at least one day during the period reported.

28 Discussion

Prey Abundance

Caribou are the single most important prey of tundra wolves (Clark, 1971; Kuyt, 1972; Stephenson and James, 1982; Williams, 1990). Caribou range over vast areas, and for part of the summer, they are scarce or absent in wolf home ranges (Heard et al., 1996). Both the long distance between radio-collared caribou and the den the week before the excursion and the increased time spent foraging by wolf 388 indicate that caribou availability near the den was low. Observations of the pups' being left alone for up to 18 h, presumably while adults were searching for food, provide additional support for low caribou availability locally. Mean foraging bout duration decreased by 10.0 h after the excursion, when collared caribou were closer to the den, suggesting an increase in caribou availability nearby.

Foraging Excursion

One aspect of central place foraging theory (CPFT) deals with the optimality of returning different-sized food loads from varying distances to dependents at a central place (i.e., the den) (Orians and Pearson, 1979). Carlson (1985) tested CPFT and found that the predator usually consumed prey captured far from the central place, while feeding prey captured nearby to dependants. Wolf 388 spent 7.2 days in one area near caribou before moving to a location 23 km back towards the den, where she spent an additional 3.1 days, likely hunting caribou. She began her return trip from this closer location, traveling directly to the den. While away, she may have made one or more successful kills and spent time meeting her own energetic needs before returning to the den. Alternatively, it may have taken several attempts to make a kill,

24 Post-excursion measurements of 388's movements were made to compare with those of the pre-excursion period. In order to compare, scientists often use *means*, or averages, of a series of measurements— mean distances, mean duration, etc.

25 In the comparison, authors used statistical calculations (F and df) to determine that the differences between pre- and post-excursion measurements were *statistically insignificant*, or close enough to be considered essentially the same or similar.

26 As with wolf 388, the investigators measured the movements of caribou during the study period. The areas within which the caribou moved are shown in Figure 2 by shaded polygons mentioned in the second paragraph of this subsection.

27 This subsection summarizes how distances separating predators and prey varied during the study period.

28 DISCUSSION: This section is the *interpretation* part of the scientific process.

29 This subsection reviews observations from other studies and suggests that this study fits with patterns of those observations.

30 Authors discuss a prevailing *theory* (CBFT) which might explain why a wolf would travel far to meet her own energy needs while taking food caught closer to the den back to her pups. The results of this study seem to fit that pattern.

31 Here our authors note other possible explanations for wolves' excursions presented by other investigators, but this study does not seem to support those ideas.

32 Authors discuss possible reasons for why 388 traveled directly to where caribou were located. They take what they learned from earlier studies and apply it to this case, suggesting that the lay of the land played a role. Note that their description paints a clear picture of the landscape.

33 Authors suggest that 388 may have learned in traveling during previous summers where the caribou were. The last two sentences suggest ideas for future studies.

34 Or maybe 388 followed the scent of the caribou. Authors acknowledge difficulties of proving this, but they suggest another area where future studies might be done.

35 Authors suggest that results of this study support previous studies about how fast wolves travel to and from the den. In the last sentence, they speculate on how these observed patterns would fit into the theory of evolution.

36 Authors also speculate on the fate of 388's pups while she was traveling. This leads to . . .

---

which she then fed on before beginning her return trip. We do not know if she returned food to the pups, but such behavior would be supported by CPFT.

31 Other workers have reported wolves' making long round trips and referred to them as "extraterritorial" or "pre-dispersal" forays (Fritts and Mech, 1981; Messier, 1985; Ballard et al., 1997; Merrill and Mech, 2000). These movements are most often made by young wolves (1–3 years old), in areas where annual territories are maintained and prey are relatively sedentary (Fritts and Mech, 1981; Messier, 1985). The long excursion of 388 differs in that tundra wolves do not maintain annual territories (Walton et al., 2001), and the main prey migrate over vast areas (Gunn et al., 2001).

Another difference between 388's excursion and those reported earlier is that she is a mature, breeding female. No study of territorial wolves has reported reproductive adults making extraterritorial movements in summer (Fritts and Mech, 1981; Messier, 1985; Ballard et al., 1997; Merrill and Mech, 2001). However, Walton et al. (2001) also report that breeding female tundra wolves made excursions.

Direction of Movement

32 Possible explanations for the relatively direct route 388 took to the caribou include landscape influence and experience. Considering the timing of 388's trip and the locations of caribou, had the wolf moved northwest, she might have missed the caribou entirely, or the encounter might have been delayed.

A reasonable possibility is that the land directed 388's route. The barrens are crisscrossed with trails worn into the tundra over centuries by hundreds of thousands of caribou and other animals (Kelsall, 1968; Thorpe et al., 2001). At river crossings, lakes, or narrow peninsulas, trails converge and funnel towards and away from caribou calving grounds and summer range. Wolves use trails for travel (Paquet et al., 1996; Mech and Boitani, 2003; P. Frame, pers. observation). Thus, the landscape may direct an animal's movements and lead it to where cues, such as the odor of caribou on the wind or scent marks of other wolves, may lead it to caribou.

33 Another possibility is that 388 knew where to find caribou in summer. Sexually immature tundra wolves sometimes follow caribou to calving grounds (D. Heard, unpubl. data). Possibly, 388 had made such journeys in previous years and killed caribou. If this were the case, then in times of local prey scarcity she might travel to areas where she had hunted successfully before. Continued monitoring of tundra wolves may answer questions about how their food needs are met in times of low caribou abundance near dens.

34 Caribou often form large groups while moving south to the tree line (Kelsall, 1968). After a large aggregation of caribou moves through an area, its scent can linger for weeks (Thorpe et al., 2001:104). It is conceivable that 388 detected caribou scent on the wind, which was blowing from the northeast on 19–21 July (Environment Canada, 2003), at the same time her excursion began. Many factors, such as odor strength and wind direction and strength, make systematic study of scent detection in wolves difficult under field conditions (Harrington and Asa, 2003). However, humans are able to smell odors such as forest fires or oil refineries more than 100 km away. The olfactory capabilities of dogs, which are similar to wolves, are thought to be 100 to 1 million times that of humans (Harrington and Asa, 2003). Therefore, it is reasonable to think that under the right wind conditions, the scent of many caribou traveling together could be detected by wolves from great distances, thus triggering a long foraging bout.

Rate of Travel

35 Mech (1994) reported the rate of travel of Arctic wolves on barren ground was 8.7 km/h during regular travel and 10.0 km/h when returning to the den, a difference of 1.3 km/h. These rates are based on direct observation and exclude periods when wolves moved slowly or not at all. Our calculated travel rates are assumed to include periods of slow movement or no movement. However, the pattern we report is similar to that reported by Mech (1994), in that homeward travel was faster than regular travel by 1.6 km/h. The faster rate on return may be explained by the need to return food to the den. Pup survival can increase with the number of adults in a pack available to deliver food to pups (Harrington et al., 1983). Therefore, an increased rate of travel on homeward trips could improve a wolf's reproductive fitness by getting food to pups more quickly.

Fate of 388's Pups

36 Wolf 388 was caring for pups during den observations. The pups were estimated to be six weeks old, and were seen ranging as far as 800 m from the den. They received some regurgitated food from two of the females, but were unattended for long periods. The excursion started 16 days after our observations, and it is improbable that the pups could have traveled the distance that 388 moved. If the pups died, this would have removed parental responsibility, allowing the long movement.

Our observations and the locations of radio-collared caribou indicate that prey became scarce in the area of the den as summer progressed. Wolf 388 may have abandoned her pups to seek food for herself. However, she returned to the den after the excursion, where she was seen near pups. In fact, she foraged in a similar pattern before and after the excursion, suggesting that she again was providing for pups after her return to the den.

37 A more likely possibility is that one or both of the other lactating females cared for the pups during 388's absence. The three females at this den were not seen with the pups at the same time. However, two weeks earlier, at a different den, we observed three females cooperatively caring for a group of six pups. At that den, the three lactating females were observed providing food for each other and trading places while nursing pups. Such a situation at the den of 388 could have created conditions that allowed one or more of the lactating females to range far from the den for a period, returning to her parental duties afterwards. However, the pups would have been weaned by eight weeks of age (Packard et al., 1992), so nonlactating adults could also have cared for them, as often happens in wolf packs (Packard et al., 1992; Mech et al., 1999).

Cooperative rearing of multiple litters by a pack could create opportunities for long-distance foraging movements by some reproductive wolves during summer periods of local food scarcity. We have recorded multiple lactating females at one or more tundra wolf dens per year since 1997. This reproductive strategy may be an adaptation to temporally and **38** spatially unpredictable food resources. All of these possibilities require further study, but emphasize both the adaptability of wolves living on the barrens and their dependence on caribou.

Long-range wolf movement in response to caribou **39** availability has been suggested by other researchers (Kuyt, 1972; Walton et al., 2001) and traditional ecological knowledge (Thorpe et al., 2001). Our report demonstrates the rapid and extreme response of wolves to caribou distribution and movements in summer. Increased human activity on the tundra (mining, road building, pipelines, ecotourism) may influence caribou movement patterns and change the interactions between wolves and caribou in the region. Continued monitoring of both species will help us to assess whether the association is being affected adversely by anthropogenic change.

40 Acknowledgements

This research was supported by the Department of Resources, Wildlife, and Economic Development, Government of the Northwest Territories; the Department of Biological Sciences at the University of Alberta; the Natural Sciences and Engineering Research Council of Canada; the Department of Indian and Northern Affairs Canada; the Canadian Circumpolar Institute; and DeBeers Canada, Ltd. Lorna Ruechel assisted with den observations. A. Gunn provided caribou location data. We thank Dave Mech for the use of GPS collars. M. Nelson, A. Gunn, and three anonymous reviewers made helpful comments on earlier drafts of the manuscript. This work was done under Wildlife Research Permit – WL002948 issued by the Government of the Northwest Territories, Department of Resources, Wildlife, and Economic Development.

41 References

BALLARD, W.B., AYRES, L.A., KRAUSMAN, P.R., REED, D.J., and FANCY, S.G. 1997. Ecology of wolves in relation to a migratory caribou herd in northwest Alaska. Wildlife Monographs 135. 47 p.

BROWN, M.B., and FORSYTHE, A.B. 1974. Robust tests for the equality of variances. Journal of the American Statistical Association 69:364–367.

CARLSON, A. 1985. Central place foraging in the red-backed shrike (*Lanius collurio* L.): Allocation of prey between forager and sedentary consumer. Animal Behaviour 33:664–666.

CLARK, K.R.F. 1971. Food habits and behavior of the tundra wolf on central Baffin Island. Ph.D. Thesis, University of Toronto, Ontario, Canada.

ENVIRONMENT CANADA. 2003. National climate data information archive. Available online: http://www.climate.weatheroffice.ec.gc.ca/Welcome_e.html

FOURNIER, B., and GUNN, A. 1998. Musk ox numbers and distribution in the NWT, 1997. File Report No. 121. Yellowknife: Department of Resources, Wildlife, and Economic Development, Government of the Northwest Territories. 55 p.

FRITTS, S.H., and MECH, L.D. 1981. Dynamics, movements, and feeding ecology of a newly protected wolf population in northwestern Minnesota. Wildlife Monographs 80. 79 p.

GUNN, A., DRAGON, J., and BOULANGER, J. 2001. Seasonal movements of satellite-collared caribou from the Bathurst herd. Final Report to the West Kitikmeot Slave Study Society, Yellowknife, NWT. 80 p. Available online: http://www.wkss.nt.ca/HTML/08_ProjectsReports/PDF/Seasonal MovementsFinal.pdf

HARRINGTON, F.H., and ASA, C.S. 2003. Wolf communication. In: Mech, L.D., and Boitani, L., eds. Wolves: Behavior, ecology, and conservation. Chicago: University of Chicago Press. 66–103.

HARRINGTON, F.H., MECH, L.D., and FRITTS, S.H. 1983. Pack size and wolf pup survival: Their relationship under varying ecological conditions. Behavioral Ecology and Sociobiology 13:19–26.

HARRIS, S.A. 1986. Permafrost distribution, zonation and stability along the eastern ranges of the cordillera of North America. Arctic 39(1):29–38.

HEARD, D.C., and WILLIAMS, T.M. 1992. Distribution of wolf dens on migratory caribou ranges in the Northwest

37 Discussion of cooperative rearing of pups and, in turn, to speculation on how this study and what is known about cooperative rearing might fit into the animal's strategies for survival of the species. Again, the authors approach the broader theory of evolution and how it might explain some of their results.

38 And again, they suggest that this study points to several areas where further study will shed some light.

39 In conclusion, the authors suggest that their study supports the hypothesis being tested here. And they touch on the implications of increased human activity on the tundra predicted by their results.

40 ACKNOWLEDGEMENTS: Authors note the support of institutions, companies and individuals. They thank their reviewers ad list permits under which their research was carried on.

41 REFERENCES: List of all studies cited in the report. This may seem tedious, but is a vitally important part of scientific reporting. It is a record of the sources of information on which this study is based. It provides readers with a wealth of resources for further reading on this topic. Much of it will form the foundation of future scientific studies like this one.

Territories, Canada. Canadian Journal of Zoology 70:1504–1510.

HEARD, D.C., WILLIAMS, T.M., and MELTON, D.A. 1996. The relationship between food intake and predation risk in migratory caribou and implication to caribou and wolf population dynamics. Rangifer Special Issue No. 2:37–44.

KELSALL, J.P. 1968. The migratory barren-ground caribou of Canada. Canadian Wildlife Service Monograph Series 3. Ottawa: Queen's Printer. 340 p.

KUYT, E. 1962. Movements of young wolves in the Northwest Territories of Canada. Journal of Mammalogy 43:270–271.

———. 1972. Food habits and ecology of wolves on barren-ground caribou range in the Northwest Territories. Canadian Wildlife Service Report Series 21. Ottawa: Information Canada. 36 p.

MECH, L.D. 1994. Regular and homeward travel speeds of Arctic wolves. Journal of Mammalogy 75:741–742.

MECH, L.D., and BOITANI, L. 2003. Wolf social ecology. In: Mech, L.D., and Boitani, L., eds. Wolves: Behavior, ecology, and conservation. Chicago: University of Chicago Press. 1–34.

MECH, L.D., and GESE, E.M. 1992. Field testing the Wildlink capture collar on wolves. Wildlife Society Bulletin 20:249–256.

MECH, L.D., WOLFE, P., and PACKARD, J.M. 1999. Regurgitative food transfer among wild wolves. Canadian Journal of Zoology 77:1192–1195.

MERRILL, S.B., and MECH, L.D. 2000. Details of extensive movements by Minnesota wolves (Canis lupus). American Midland Naturalist 144:428–433.

MERRILL, S.B., ADAMS, L.G., NELSON, M.E., and MECH, L.D. 1998. Testing releasable GPS radiocollars on wolves and white-tailed deer. Wildlife Society Bulletin 26:830–835.

MESSIER, F. 1985. Solitary living and extraterritorial movements of wolves in relation to social status and prey abundance. Canadian Journal of Zoology 63:239–245.

MOHR, C.O., and STUMPF, W.A. 1966. Comparison of methods for calculating areas of animal activity. Journal of Wildlife Management 30:293–304.

MUSIANI, M. 2003. Conservation biology and management of wolves and wolf-human conflicts in western North America. Ph.D. Thesis, University of Calgary, Calgary, Alberta, Canada.

ORIANS, G.H., and PEARSON, N.E. 1979. On the theory of central place foraging. In: Mitchell, R.D., and Stairs, G.F., eds. Analysis of ecological systems. Columbus: Ohio State University Press. 154–177.

PACKARD, J.M., MECH, L.D., and REAM, R.R. 1992. Weaning in an arctic wolf pack: Behavioral mechanisms. Canadian Journal of Zoology 70:1269–1275.

PAQUET, P.C., WIERZCHOWSKI, J., and CALLAGHAN, C. 1996. Summary report on the effects of human activity on gray wolves in the Bow River Valley, Banff National Park, Alberta. In: Green, J., Pacas, C., Bayley, S., and Cornwell, L., eds. A cumulative effects assessment and futures outlook for the Banff Bow Valley. Prepared for the Banff Bow Valley Study. Ottawa: Department of Canadian Heritage.

STEPHENSON, R.O., and JAMES, D. 1982. Wolf movements and food habits in northwest Alaska. In: Harrington, F.H., and Paquet, P.C., eds. Wolves of the world. New Jersey: Noyes Publications. 223–237.

THORPE, N., EYEGETOK, S., HAKONGAK, N., and QITIRMIUT ELDERS. 2001. The Tuktu and Nogak Project: A caribou chronicle. Final Report to the West Kitikmeot/Slave Study Society, Ikaluktuuttiak, NWT. 160 p.

TIMONEY, K.P., LA ROI, G.H., ZOLTAI, S.C., and ROBINSON, A.L. 1992. The high subarctic forest-tundra of northwestern Canada: Position, width, and vegetation gradients in relation to climate. Arctic 45(1):1–9.

WALTON, L.R., CLUFF, H.D., PAQUET, P.C., and RAMSAY, M.A. 2001. Movement patterns of barren-ground wolves in the central Canadian Arctic. Journal of Mammalogy 82:867–876.

WILLIAMS, T.M. 1990. Summer diet and behavior of wolves denning on barren-ground caribou range in the Northwest Territories, Canada. M.Sc. Thesis, University of Alberta, Edmonton, Alberta, Canada.

ZAR, J.H. 1999. Biostatistical analysis. 4th ed. New Jersey: Prentice Hall. 663 p.

Appendix IV

A Plain English Map of the Human Chromosomes

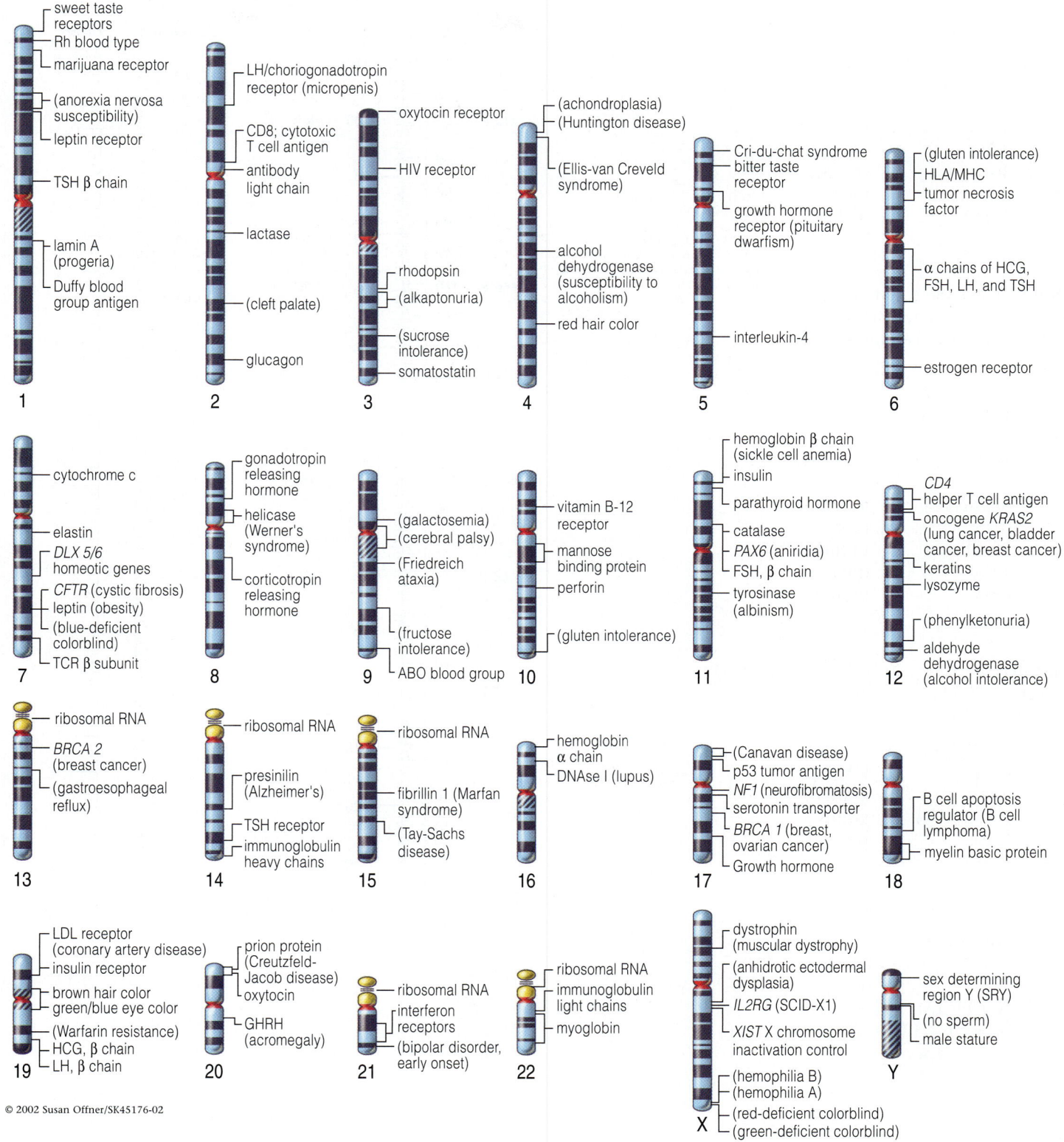

© 2002 Susan Offner/SK45176-02

Haploid set of human chromosomes. The banding patterns characteristic of each type of chromosome appear after staining with a reagent called Giemsa. The locations of some of the 20,065 known genes (as of November, 2005) are indicated. Also shown are locations that, when mutated, cause some of the genetic diseases discussed in the text.

Appendix V

Length

1 kilometer (km) = 0.62 miles (mi)
1 meter (m) = 39.37 inches (in)
1 centimeter (cm) = 0.39 inches

To convert	multiply by	to obtain
inches	2.25	centimeters
feet	30.48	centimeters
centimeters	0.39	inches
millimeters	0.039	inches

Area

1 square kilometer = 0.386 square miles
1 square meter = 1.196 square yards
1 square centimeter = 0.155 square inches

Volume

1 cubic meter = 35.31 cubic feet
1 liter = 1.06 quarts
1 milliliter = 0.034 fluid ounces = 1/5 teaspoon

To convert	multiply by	to obtain
quarts	0.95	liters
fluid ounces	28.41	milliliters
liters	1.06	quarts
milliliters	0.03	fluid ounces

Weight

1 metric ton (mt) = 2,205 pounds (lb) = 1.1 tons (t)
1 kilogram (kg) = 2.205 pounds (lb)
1 gram (g) = 0.035 ounces (oz)

To convert	multiply by	to obtain
pounds	0.454	kilograms
pounds	454	grams
ounces	28.35	grams
kilograms	2.205	pounds
grams	0.035	ounces

Temperature

Celcius (°C) to Fahrenheit (°F) :
$$°F = 1.8 \, (°C) + 32$$

Fahrenheit (°F) to Celsius:
$$°C = \frac{(°F - 32)}{1.8}$$

	°C	°F
Water boils	100	212
Human body temperature	37	98.6
Water freezes	0	32

Glossary

abscisic acid Plant hormone involved in stomata function and stress responses; inhibits germination. **572**

abscission Process by which plant parts are shed. **572**

absorption *See* nutrient absorption.

acid Substance that releases hydrogen ions in water. **32**

acid rain Rain containing sulfuric and/or nitric acid; forms after pollutants mix with water vapor in the atmosphere. **367**

actin Cysoskeletal protein; main component of thin filaments in muscle fibers. **399**

action potential Brief reversal of the charge difference across a neuron membrane. **472**

activation energy Minimum amount of energy required to start a reaction. **66**

active site Of an enzyme, pocket in which substrates bind and a reaction occurs. **68**

active transport Energy-requiring mechanism in which a transport protein pumps a solute across a cell membrane against its concentration gradient. **74**

adaptation (adaptive trait) A heritable trait that enhances an individual's fitness in a particular environment. **194**

adaptive immunity In vertebrates, set of immune defenses that can be tailored to specific pathogens encountered by an organism during its lifetime. **429**

adaptive radiation Macroevolutionary pattern in which a burst of genetic divergences from a lineage gives rise to many new species. **227**

adhesion protein Plasma membrane protein that helps cells stick to one another and (in animals) to extracellular matrix. **51**

adipose tissue Connective tissue consisting of fat-storing cells. **382**

adrenal cortex Outer portion of adrenal gland; secretes aldosterone and cortisol. **504**

adrenal gland Endocrine gland located above kidney; secretes hormones with roles in urine formation and stress responses. **504**

adrenal medulla Inner portion of the adrenal gland; secretes epinephrine and norepinephrine. **504**

aerobic Involving oxygen or occurring in its presence. **89**

aerobic respiration Oxygen-requiring metabolic pathway that breaks down carbohydrates to produce ATP. **89**

age structure Of a population, distribution of individuals among various age groups. **327**

AIDS Acquired immune deficiency syndrome. A collection of diseases that develops after a virus (HIV) weakens the immune system. **443**

alcoholic fermentation Anaerobic carbohydrate breakdown pathway that produces ATP, CO_2, and ethanol. **92**

algal bloom Population explosion of single-celled aquatic organisms such as dinoflagellates. **255**

allele frequency Abundance of a particular allele among members of a population. **214**

alleles Forms of a gene with slightly different DNA sequences; may encode slightly different versions of the gene's product. **141**

allergen A normally harmless substance that provokes an immune response in some people. **441**

allergy Sensitivity to an allergen. **441**

allopatric speciation Speciation pattern in which a physical barrier arises and separates members of a population, thus ending gene flow between them. **224**

alveoli In the lung, tiny, thin-walled air sacs where gas exchange occurs. **420**

amino acid Small organic compound that is a subunit of proteins. Consists of a carboxyl group, an amine group, and a characteristic side group (R), all typically bonded to the same carbon atom. **38**

amnion Extraembryonic membrane that encloses an amniote embryo and the amniotic fluid. **526**

amniote Vertebrate that produces amniote eggs; a reptile, bird, or mammal. **304**

amniote egg Egg with four membranes that allows an embryo to develop away from water. **304**

amoeba Solitary heterotrophic protist that feeds and moves by extending pseudopods. **258**

amphibian Vertebrate with a three-chambered heart and scaleless skin that typically develops in water, then lives on land. For example a salamander or frog. **303**

anaerobic Occurring in the absence of oxygen. **89**

analogous structures Similar body structures that evolved separately in different lineages. **204**

anaphase Stage of mitosis during which sister chromatids separate and move toward opposite sides of the cell. **137**

aneuploidy A chromosome abnormality in which there are too many or too few copies of a particular chromosome. **166**

angiosperm Seed plant that produces flowers and fruits. **274**

animal A eukaryotic heterotroph that is made up of unwalled cells and develops through a series of stages. Most ingest food, reproduce sexually, and move. **9, 287**

animal hormone Signaling molecule secreted by an endocrine gland or cell. **494**

annelid Segmented worm with a coelom, complete digestive system, and closed circulatory system. **292**

antenna Of some arthropods, sensory structure on the head that detects touch and odors. **295**

anther Of flowering plants, the part of the stamen that produces pollen. **274, 559**

anthropoids Primate lineage that includes monkeys, apes, and humans. **308**

antibody Y-shaped antigen receptor made only by B cells. **435**

antibody-mediated immune response Immune response in which antibodies are produced in response to antigen. **438**

anticodon Set of three nucleotides in a tRNA; base-pairs with an mRNA codon. **119**

antidiuretic hormone Hormone produced by the pituitary gland; increases water reabsorption by the kidney. **463**

antigen A molecule or particle that the immune system recognizes as nonself. Triggers an immune response. **429**

anus Opening through which wastes leave a complete digestive tract. **451**

apical dominance In plants, effect in which a lengthening shoot tip inhibits the growth of lateral buds. **571**

apical meristem In plants, meristem in the tip of a shoot or root; gives rise to primary growth. **546**

apicomplexan Parasitic protist that enters and lives inside the cells of its host. **255**

apoptosis Programmed cell death. **514**

appendix Tubular projection from the first part of the large intestine. **456**

aquifer Porous rock layer that holds some groundwater. **347**

arachnids Land-dwelling arthropods with four pairs of walking legs and no antennae; for example, a spider, scorpion, or tick. **296**

archaea Prokaryotes most closely related to eukaryotes; many live in extreme environments. **8, 248**

arteriole Blood vessel that delivers blood from an artery to capillaries; site of adjustments to blood distribution. **414**

artery Large-diameter blood vessel that carries blood away from the heart. **410**

arthropod Invertebrate with jointed legs and a hard exoskeleton that it periodically molts. **295**

artificial selection Process whereby humans alter traits of a domestic species by selective breeding. **194**

asexual reproduction Reproductive mode by which offspring arise from a single parent only. **133, 513**

atherosclerosis Narrowing of an artery's interior due to lipid deposition and inflammation. **416**

atom Fundamental particle that is a building block of all matter. Consists of varying numbers of protons, neutrons, and electrons. **5**

atomic number Number of protons in the atomic nucleus; determines the element. **25**

ATP Adenosine triphosphate. Nucleotide that consists of an adenine base, a ribose sugar, three phosphate groups. Functions as a subunit of RNA and as a coenzyme in many reactions. Important energy carrier in cells. **41**

atrium Heart chamber that pumps blood into a ventricle. **410**

australopiths Informal name for a lineage of hominins that lived in Africa between 4 million and 1.2 million years ago and includes some likely human ancestors. **309**

autoimmune response Immune response that targets one's own tissues. **442**

autonomic nervous system Set of nerves that relay signals to and from internal organs and to glands. **478**

autosome Of eukaryotic cells, a chromosome that is not a sex chromosome. **102**

autotroph Producer; an organism that makes its own food using carbon from inorganic molecules such as CO_2, and energy from the environment. **249**

auxin Plant hormone with a central role in coordinating responses to other hormones in all stages of growth and development. **570**

axon Cytoplasmic extension of a neuron that transmits electrical signals along its length and releases chemical signals at its endings. **471**

B cell B lymphocyte. Lymphocyte that can make antibodies. **430**

B cell receptor Antigen receptor on the surface of a B cell; an antibody that has not been released from the B cell's plasma membrane. **435**

bacteria Most diverse and well-known lineage of prokaryotes. **8, 248**

bacteriophage Virus that infects bacteria. **244**

bark In woody plants, all living and dead tissue that lies outside of the vascular cambium. **547**

Barr body Inactivated, condensed X chromosome in a cell of a female mammal. The other X chromosome is active. **126**

base Substance that accepts hydrogen ions in water. **32**

base-pair substitution Mutation in which a single base pair changes. **122**

bell curve Bell-shaped curve; typically results from graphing frequency versus distribution for a trait that varies continuously. **160**

bilateral symmetry Having right and left halves with similar parts, and a front and back that differ. **288**

bile Mix of salts, pigments, and cholesterol that is produced by the liver, stored in the gallbladder, and functions in fat digestion in the small intestine. **454**

binary fission Method of asexual reproduction in which a prokaryote divides into two identical descendant cells. **248**

biodiversity Scope of variation among organisms. Of a region, the genetic diversity within its species, variety of species, and variety of ecosystems. **8, 370**

biofilm Community of microorganisms living within a shared mass of slime. **53**

biogeochemical cycle A nutrient moves among environmental reservoirs and into and out of food webs. **347**

biogeography Study of patterns in the geographic distribution of species and communities. **191**

biological magnification A chemical pollutant becomes increasingly concentrated as it moves through a food chain. **367**

biology The scientific study of life. **4**

bioluminescent Able to use ATP to produce light. **254**

biome Any of Earth's major land ecosystems, characterized by climate and main vegetation and found in several regions. **358**

biosphere All regions of Earth where organisms live. **5, 356**

biotic potential Maximum possible population growth under optimal conditions. **319**

bipedalism Habitually walking upright. **309**

bird Modern amniote with feathers. **305**

bivalve Mollusk with a hinged two-part shell. **293**

blastula Hollow ball of cells that forms early in animal development. **514**

blood Fluid connective tissue with cells that form inside bones. **383**

blood–brain barrier Protective mechanism that prevents unwanted substances from entering cerebrospinal fluid. **479**

blood pressure Pressure exerted by blood against the walls of blood vessels. **414**

body mass index Formula used to determine a healthy weight based on height. **459**

bone tissue Connective tissue with cells surrounded by a mineral-hardened matrix of their own secretions. **383**

bony fish Jawed fish with a skeleton composed mainly of bone. **302**

boreal forest Biome dominated by conifers that can withstand the cold winters at high northern latitudes. **359**

bottleneck Drastic reduction in population size as a result of severe selection pressure. **221**

bronchiole Small airway leading to alveoli. **420**

bronchus Airway connecting the trachea to a lung. **420**

brood parasite An animal that tricks another species into raising its young, for example a cowbird. **340**

brown alga Multicelled, photosynthetic protist with brown accessory pigments; for example, a kelp **256**

bryophyte A that does not have vascular tissue; for example, a moss. **265**

buffer Set of chemicals that can keep the pH of a solution stable by alternately donating and accepting ions that contribute to pH. **32**

C3 plant Type of plant that uses only the Calvin–Benson cycle to fix carbon. **88**

C4 plant Type of plant that minimizes photorespiration by fixing carbon twice, in two cell types, using a C4 pathway in addition to the Calvin–Benson cycle. **88**

Calvin–Benson cycle Light-independent reactions of photosynthesis; cyclic carbon-fixing pathway in which sugars form. **87**

camouflage Body shape, pattern, or behavior that causes a plant or animal to blend into its surroundings. **339**

CAM plant Type of plant that minimizes photorespiration by fixing carbon twice, at different times of day. **88**

cancer Disease that occurs when a malignant neoplasm physically and functionally disrupts body tissues. **140**

capillaries Smallest-diameter blood vessels; site of exchanges between blood and other tissues. **409**

carbohydrate Molecule that consists primarily of carbon, hydrogen, and oxygen atoms in a 1:2:1 ratio. Complex kinds (e.g., cellulose, starch, glycogen) are polymers of simple kinds (sugars). **34**

carbon cycle Movement of carbon among rocks, water, the atmosphere, and living organisms. **349**

carbon fixation Process by which carbon from an inorganic source such as carbon dioxide gets incorporated into an organic molecule. **87**

cardiac cycle Sequence of contraction and relaxation of heart chambers that occurs with each heartbeat. **412**

cardiac muscle tissue Striated, involuntary muscle of the heart wall. **383**

cardiac pacemaker Group of heart cells (SA node) that emits rhythmic signals calling for muscle contraction. **413**

carpel Floral reproductive organ that consists of an ovule-containing ovary, a pollen-receiving stigma, and often a style. **274, 559**

carrying capacity Maximum number of individuals of a species that a specific environment can sustain. **320**

cartilage Connective tissue with cells surrounded by a rubbery matrix of their own secretions. **382**

cartilaginous fish Jawed fish with paired fins and a skeleton made of cartilage; for example, a shark. **301**

cartilaginous joint Joint where cartilage holds bones together and provides cushioning, as between vertebrae. **397**

cell Smallest unit of life; at minimum, consists of plasma membrane, cytoplasm, and DNA. **5**

cell cycle Series of events from the time a cell forms until its cytoplasm divides. **134**

cell junction Structure that connects a cell to another cell or to extracellular matrix in tissues; e.g., tight junction, adhering junction, or gap junction (of animals); plasmodesmata (of plants). **58**

cell-mediated immune response Immune response involving cytotoxic T cells and NK cells that destroy infected or cancerous body cells. **438**

cell theory Theory that all organisms consist of one or more cells, which are the basic unit of life; all cells come from division of preexisting cells; and all cells pass hereditary material to offspring. **47**

cell wall Semirigid but permeable structure that surrounds the plasma membrane of some cell types. **52**

cellular slime mold Heterotrophic protist that usually lives as a single-celled, amoeba-like predator. When conditions are unfavorable, cells form a cohesive group and produce a fruiting body. **258**

cellulose Tough, insoluble carbohydrate that is the major structural material in plants. **34**

central nervous system Brain and spinal cord. **476**

centromere Of a duplicated eukaryotic chromosome, constricted region where sister chromatids attach to one another. **101**

cephalopod Predatory mollusk with a closed circulatory system; moves by jet propulsion; for example a squid or octopus. **293**

cerebellum Hindbrain region that coordinates voluntary movements. **481**

cerebrospinal fluid Fluid that surrounds the brain and spinal cord. **479**

cerebrum Forebrain region that controls higher functions. **481**

cervix Narrowed region of uterus that connects to the vagina. **518**

chaparral Biome with cool wet winters, hot dry summers; dominant plants are shrubs with small, leathery leaves. **360**

character Quantifiable, heritable characteristic or trait. **229**

charge Electrical property; opposite charges attract, and like charges repel. **25**

chemical bond An attractive force that arises between two atoms when their electrons interact. *See* covalent bond, ionic bond. **28**

chemical synapse Region where a neuron's axon terminal transmits signals to another cell. **474**

chemoreceptor Sensory receptor that responds to the binding of a particular chemical. **483**

chlorophyll *a* Main photosynthetic pigment in plants. **83**

chloroplast Organelle of photosynthesis in the cells of plants and many protists. Has two outer membranes enclosing semifluid stroma. Light-dependent reactions occur at its inner thylakoid membrane; light-independent reactions, in the stroma. **56**

choanoflagellates Heterotrophic protists with a collared flagellum; protist group most closely related to animals. **258**

chordates Animal phylum characterized by four embryonic traits: a notochord, dorsal nerve cord, pharyngeal gill slits, and a tail that extends beyond the anus. Includes the lancelets, tunicates, and all vertebrates. **299**

chorion Outermost extraembryonic membrane of amniotes; major component of the placenta in placental mammals. **527**

chromosome A structure that consists of DNA and associated proteins; carries part or all of a cell's genetic information. **101**

chromosome number The sum of all chromosomes in a cell of a given species. **101**

chyme Mix of food and gastric fluid. **453**

ciliate Single-celled, heterotrophic protist with many cilia. **254**

cilia Singular, cilium. Short, movable structures that project from the plasma membrane of some eukaryotic cells. **57**

circadian rhythm A biological activity repeated about every 24 hours. **575**

clade A group whose members share one or more defining derived traits. **229**

cladistics Making hypotheses about evolutionary relationships among clades. **229**

cladogram Evolutionary tree diagram that shows a network of evolutionary relationships among clades. **230**

cleavage Mitotic division of cells in an early embryo. **514**

cleavage furrow In a dividing animal cell, the indentation where cytoplasmic division will occur. **138**

climate Average weather conditions in a region over a long time period. **357**

cloaca Of some vertebrates, body opening that releases urinary and digestive waste, and functions in reproduction. **303**

clone Genetically identical copy of an organism. **100**

cloning vector A DNA molecule that can accept foreign DNA and be replicated inside a host cell. **175**

closed circulatory system Circulatory system in which blood flows through a continuous network of vessels. **409**

cnidarian Radially symmetrical invertebrate with two tissue layers and tentacles that have stinging cells; for example a jelly or sea anemone. **290**

cochlea Coiled structure in the inner ear that holds the sound-detecting organ of Corti. **488**

codominance Effect in which two alleles are both fully expressed in heterozygous individuals and neither is dominant over the other. **156**

codon In mRNA, a sequence of three nucleotides that codes for an amino acid or stop signal during translation. **118**

coelom A body cavity completely lined by tissue derived from mesoderm. **289**

coenzyme An organic cofactor. **69**

coevolution Joint evolution of two closely interacting species; each species is a selective agent that shifts the range of variation in the other. **228, 336**

cofactor A metal ion or organic molecule that associates with an enzyme and is necessary for its function. **69**

cohesion Property of a substance that arises from the tendency of its molecules to resist separating from one another. **31**

cohesion–tension theory Explanation of how water moves through vascular plants: Transpiration creates a tension that pulls a cohesive column of water through xylem, from roots to shoots. **551**

cohort Group of individuals born during the same interval. **322**

collenchyma Simple plant tissue composed of living cells with unevenly thickened walls. **539**

colon Longest portion of the large intestine. **456**

colonial theory of animal origins Well-accepted hypothesis that animals evolved from a colonial protist. **287**

commensalism Species interaction that benefits one species and neither helps nor harms the other. **336**

community All populations of all species in a given area. **5, 334**

compact bone Dense bone with concentric layers of matrix. **397**

companion cell In phloem, parenchyma cell that loads sugars into an associated sieve tube member and provides it with metabolic support. **540**

comparative morphology The scientific study of anatomical patterns in body plans. **192**

competitive exclusion When two species compete for a limiting resource, one drives the other to local extinction. **337**

complement A set of proteins that circulate in inactive form in blood, and when activated play a role in immune responses. **429**

complete digestive tract Tubular gut with two openings. **288, 451**

compound Molecule that has atoms of more than one element. **29**

compound eye Of some arthropods, an eye that consists of many individual units, each with its own lens. **295**

concentration The amount of molecules or ions per unit volume of a solution. **32**

cone cell Photoreceptor that provides sharp vision and allows detection of color. **487**

conifer Cone-bearing gymnosperm such as a pine. **272**

conjugation Mechanism of gene exchange in which one prokaryotic cell transfers a plasmid to another. **248**

connective tissue Animal tissue with an extensive extracellular matrix; provides structural and functional support. **381**

conservation biology Field of applied biology that surveys biodiversity and seeks ways to maintain and utilize it. **370**

consumer Organism that gets energy and nutrients by feeding on tissues, wastes, or remains of other organisms; a heterotroph. **6, 344**

continuous variation A range of small differences in a shared trait. **160**

contraception Methods used to prevent pregnancy. **523**

contractile vacuole In freshwater protists, an organelle that collects and expels excess water that enters the cell by osmosis. **253**

control group In an experiment, a group of individuals who are not exposed to the variable under study, as compared with an experimental group. **13**

coral reef In tropical sunlit seas, a formation composed of the secretions of coral polyps; serves as home to many other species. **362**

cork Component of periderm; waterproofs, insulates, and protects the surfaces of woody stems and roots. **547**

cork cambium Lateral meristem that gives rise to periderm. **547**

cornea Outermost layer at the front of a vertebrate eye; bends light. **485**

corpus luteum Hormone-secreting structure that forms from follicle cells after ovulation. **519**

cortisol Adrenal cortex hormone that influences metabolism and immunity; stress increases its rate of secretion. **504**

cotyledon Seed leaf; part of a flowering plant embryo. Monocots have one; eudicots, two. **541**

covalent bond Type of chemical bond in which two atoms share a pair of electrons. May be polar or nonpolar. **28**

critical thinking Judging information before accepting it. **12**

crossing over Process in which homologous chromosomes exchange corresponding segments during meiosis. **144**

crustaceans Group of mostly marine arthropods with two pairs of antennae; for example, a shrimp, crab, lobster, or barnacle. **296**

cuticle Secreted covering at a body surface. **58, 265**

cytokinin Plant hormone that promotes cell division in shoot apical meristem and cell differentiation in root apical meristem. Often interacts antagonistically with auxin. **571**

cytoplasm Semifluid substance enclosed by a cell's plasma membrane. **47**

cytoskeleton Network of interconnected protein filaments that support, organize, and move eukaryotic cells and their internal structures. *See* microtubules, microfilaments, intermediate filaments. **56**

data Values or other factual information obtained from experiments or surveys. **13**

decomposer Consumer that feeds on organic wastes and remains, breaking them down into their inorganic subunits. **249, 344**

deforestation Removal of all trees from a forested area. **366**

deletion Mutation in which one or more nucleotides are lost from DNA. **122**

demographics Statistics that describe a population. **317**

demographic transition model Model describing the changes in human birth and death rates that occur as a region becomes industrialized. **328**

denature Regarding a biological molecule, to become so altered in shape that some or all function is lost. **39**

dendrite Neuron's signal-receiving cytoplasmic extension. **471**

dendritic cell Phagocytic white blood cell concentrated in body tissues that contact the external environment. Important antigen presenting cell in adaptive immune responses. **430**

dense connective tissue Connective tissue with many fibroblasts and fibers arranged a random or a regular arrangement. **382**

density-dependent limiting factor Factor whose negative effect on growth is felt most in dense populations; for example, infectious disease or competition for food. **320**

density-independent limiting factor Factor that limits growth in populations regardless of their density; for example a natural disaster or harsh weather. **321**

dental plaque On teeth, a thick biofilm composed of bacteria, their extracellular products, and saliva proteins. **434**

derived trait A character present in a clade but not in the clade's ancestors. **229**

dermal tissues Tissues that cover and protect the plant body. E.g., epidermis, periderm. **539**

dermis Deep layer of skin that consists of connective tissue with nerves and blood vessels running through it. **385**

desert Biome with little precipitation; its inhabitants are adapted to withstand drought. **360**

desertification Conversion of grassland or woodlands to desertlike conditions. **366**

detritivores Consumers that feed on small bits of organic material (detritus). **344**

development Multistep process by which the first cell of a new individual becomes a multicelled adult. **7**

diaphragm Dome-shaped smooth muscle at base of thoracic cavity; contracts during inhalation. **421**

diastolic pressure Blood pressure when ventricles are relaxed. **414**

diatom Single-celled photosynthetic protist with brown accessory pigments and a two-part silica shell. **256**

differentiation The process by which cells become specialized during development; occurs as different cell lineages in an embryo begin to express different subsets of their genes. **108, 514**

diffusion Spontaneous spreading of molecules or atoms through a liquid or gas. **71**

digestion Breakdown of food into smaller bits and then into its component molecules. **451**

dihybrid cross Cross between two individuals identically heterozygous for two genes; for example $AaBb \times AaBb$. **154**

dinoflagellate Single-celled, aquatic protist typically with cellulose plates and two flagella; may be heterotrophic or photosynthetic. **254**

diploid Having two of each type of chromosome characteristic of the species ($2n$). **102**

directional selection Mode of natural selection in which a phenotype at one end of a range of variation is favored. **215**

disease vector Organism that carries a pathogen from one host to the next. **245**

disruptive selection Mode of natural selection that favors forms of a trait at the extremes of a range of variation; intermediate forms are selected against. **218**

DNA Deoxyribonucleic acid. Nucleic acid that carries hereditary information about traits; consists of two chains of nucleotides (adenine, guanine, thymine, and cytosine) twisted into a double helix. **7, 41**

DNA cloning Set of procedures that uses living cells to make many identical copies of a DNA fragment. **175**

DNA fingerprinting *See* DNA profiling.

DNA library Collection of cells that host different fragments of foreign DNA, often representing an organism's entire genome. **176**

DNA polymerase DNA replication enzyme. Assembles a new strand of DNA based on the sequence of a DNA template. **105**

DNA profiling Identifying an individual by analyzing the unique parts of his or her DNA. **179**

DNA replication Process by which a cell copies its DNA before it divides. **105**

DNA sequence The order of nucleotides in a strand of DNA. **105**

dominant Refers to an allele that masks the effect of a recessive allele in heterozygous individuals. **152**

dormancy Period of temporarily suspended metabolism. **562**

double fertilization In flowering plants, one sperm fertilizes the egg, forming the zygote, and another fertilizes a diploid cell, forming what will become endosperm. **275, 562**

eardrum Membrane of middle ear that vibrates in response to sound waves. **488**

echinoderms Invertebrates with a water–vascular system and an endoskeleton made of hardened plates and spines; for example a sea star. **298**

ECM *See* extracellular matrix.

ecological footprint Area of Earth's surface required to sustainably support a particular level of development and consumption. **328**

ecological restoration Actively altering an area in an effort to restore or recreate a functional ecosystem. **371**

ecological succession A gradual change in a community in which one array of species replaces another. **341**

ecology The study of interactions among organisms, and between organisms and their environment. **316**

ecosystem A community interacting with its environment through a one-way flow of energy and cycling of materials. **5, 344**

ectoderm Outermost tissue layer of an animal embryo. **514**

ectotherm Animal that regulates its temperature by altering the rate at which it gains heat from or lose heat to the environment; commonly called "cold-blooded." **304**

effector cell Antigen-sensitized B cell or T cell that forms in an immune response and acts immediately. **438**

egg Female gamete. **144**

ejaculation Smooth muscle contractions propel semen through the penis and out of the body. **517**

electron Negatively charged subatomic particle. **25**

electron transfer chain Array of enzymes and other molecules in a cell membrane that accept and give up electrons in sequence, thus releasing the energy of the electrons in small, usable steps. **70**

electron transfer phosphorylation Process in which the flow of electrons through electron transfer chains sets up a hydrogen ion gradient that drives ATP formation. **87**

electrophoresis Laboratory technique that separates DNA fragments by size. **178**

element A pure substance that consists only of atoms with the same number of protons. **25**

elimination Expulsion of unabsorbed material from the digestive tract. **451**

endangered species A species that faces extinction in all or part of its range. **364**

endemic species A species that evolved in one place and is found nowhere else. **364**

endocrine gland Ductless gland that secretes hormones. **381**

endocrine system Hormone-producing glands and secretory cells of a vertebrate body. **495**

endocytosis Process by which a cell takes in extracellular substances in bulk; a portion of the cell's plasma membrane balloons inward and pinches off as a vesicle. **76**

endoderm Innermost tissue layer of an animal embryo. **514**

endodermis In plant roots, a layer of cells that separates vascular cylinder from cortex; prevents soil water from seeping through cell walls into xylem. **544**

endoplasmic reticulum (**ER**) Membrane-enclosed organelle that is a continuous system of sacs and tubes extending from the nuclear envelope. Smooth ER makes lipids and breaks down carbohydrates and fatty acids; rough ER modifies polypeptides made by ribosomes on its surface. **55**

endorphins Natural pain-reliever molecules. **484**

endoskeleton Internal skeleton; hard internal parts that muscles attach to and move. **298, 395**

endosperm Nutritive tissue in the seeds of flowering plants. **275, 562**

endosymbiosis One species lives inside another. **241**

endotherm Animal that regulates its temperature by adjusting its production of metabolic heat; commonly called "warm-blooded." **304**

energy The capacity to do work. **65**

energy pyramid Diagram that illustrates the energy flow in an ecosystem. **346**

enzyme Protein or RNA that speeds up a reaction without being changed by it. **34**

epidemic Disease outbreak that occurs in a limited region. **247**

epidermis Outermost tissue layer of an animal or a young plant. **384, 539**

epigenetic Refers to DNA modifications that do not involve a change in DNA sequence, e.g., DNA methylation. **126**

epiglottis Tissue flap that folds down to prevent food from entering the airway during swallowing. **420**

epiphyte Plant that grows on the trunk or branches of another plant but does not harm it. **269**

epistasis Polygenic inheritance, in which a trait is influenced by the products of multiple genes. **157**

epithelial tissue Sheetlike animal tissue that covers outer body surfaces and lines internal tubes and cavities. **380**

equilibrial life history Life history favored in stable environments; individuals grow large, then invest heavily in each of a few offspring. **323**

erection Penis becomes erect as a result of increased fluid pressure inside its spongy tissue. **521**

esophagus Muscular tube that connects the pharynx (throat) to the stomach. **452**

essential amino acid Amino acid that the body cannot make and must obtain from food. **458**

essential fatty acids Fatty acids that the body cannot make and must obtain from the diet. **457**

estrogens Sex hormones produced by a female's ovaries. **505**

estuary A highly productive ecosystem where nutrient-rich water from a river mixes with seawater. **362**

ethylene Gaseous plant hormone that participates in germination, abscission, ripening, and stress responses. **573**

eudicots Largest lineage of angiosperms; includes herbaceous plants, woody trees, and cacti. **275**

eugenics Idea of deliberately improving the genetic qualities of the human race. **184**

eukaryote Organism whose cells characteristically have a nucleus; a protist, fungus, plant, or animal. **8**

evaporation Transition of a liquid to a vapor. **31**

evolution Change in a line of descent. **193**

exaptation Evolutionary adaptation of an existing structure for a completely new purpose. **227**

exocrine gland Gland that secretes milk, sweat, saliva, or some other substance through a duct. **381**

exocytosis Process by which a cell expels a vesicle's contents to extracellular fluid. **76**

exon Nucleotide sequence that is not removed from an RNA during post-transcriptional modification. **117**

exoskeleton External skeleton; hard external parts that muscles attach to and move. **295, 395**

exotic species Species that has been introduced to a new habitat and become established there. **343**

experiment Test designed to support or falsify a prediction. Often involves experimental and control groups. **13**

experimental group In an experiment, a group of individuals who are exposed to a variable under study, as compared with a control group. **13**

exponential model of population growth Model for population growth when resources are unlimited. The per capita growth rate remains constant as population size increases. **319**

extinct Refers to a species that no longer has living members. **227**

extracellular matrix (**ECM**) Complex mixture of cell secretions; its composition varies by cell type. E.g., basement membrane of epithelial tissue. **58**

extreme halophile Organism that lives where the salt concentration is high. **250**

extreme thermophile Organism that lives where the temperature is very high. **250**

fat Lipid that consists of a glycerol molecule with one, two, or three fatty acid tails. *See* saturated fat, unsaturated fat. **36**

fatty acid Organic compound that consists of a chain of carbon atoms with an acidic carboxyl group at one end. Carbon chain of saturated types has single bonds only; that of unsaturated types has one or more double bonds. **36**

feces Mixture of unabsorbed food material and cellular waste that is stored in the rectum until expelled from the digestive tract. **456**

feedback inhibition Mechanism by which a change that results from some activity decreases or stops the activity. **70**

fermentation An anaerobic pathway by which cells harvest energy from carbohydrates. E.g., alcoholic fermentation, lactate fermentation. **92**

ferns Most diverse lineage of seedless vascular plants. **269**

fertilization Egg and sperm combine and form a zygote. **514**

fetus Developing human between 9 weeks after fertilization and birth. **528**

fever An internally induced rise in core body temperature above the normal set point as a response to infection or tissue damage. **433**

fibroblast Main cell type in soft connective tissue; secretes collagen and other components of extracellular matrix. **381**

fibrous joint Joint where dense connective tissue holds bones firmly in place. **397**

filtration Blood pressure forces water and small solutes, but not blood cells or proteins, out across the walls of capillaries. **462**

first law of thermodynamics Energy cannot be created or destroyed. **65**

fish Aquatic vertebrate belonging to the oldest and most diverse vertebrate group. **301**

fitness Degree of adaptation to an environment, as measured by an individual's relative genetic contribution to future generations. **194**

fixed Refers to an allele for which all members of a population are homozygous. **221**

flagellated protozoans Lineages of mostly heterotrophic protists having one or more flagella; for example a trypanosome. **252**

flagellum Long, slender cellular structure used for movement. **53**

flatworm Bilaterally symmetrical invertebrate with a saclike gastrovascular cavity and no coelom; for example, a planarian or tapeworm. **291**

flower Reproductive structure of angiosperms; generally consists of sepals, petals, carpel, stamens. **274**

fluid mosaic Model of a cell membrane as a two-dimensional fluid of mixed composition. **50**

food chain Sequence of steps by which energy moves from one trophic level to the next. **344**

food web System of cross-connecting food chains. **344**

foraminiferan Heterotrophic protist that secretes a calcium carbonate shell. **253**

fossil Physical evidence of an organism that lived in the ancient past. **192**

founder effect After a small group of individuals found a new population, allele frequencies in the new population differ from those in the original population. **221**

fovea Of a vertebrate eye, region of the retina with the densest concentration of photoreceptors. **486**

free radical Atom with an unpaired electron; most are highly reactive and can damage biological molecules. **27**

fruit Mature ovary of a flowering plant, often with accessory parts; encloses a seed or seeds. **274, 564**

fungus Single-celled or multicelled eukaryotic consumer with cell walls of chitin; feeds by extracellular digestion and absorption. For example, a yeast, mold, or mushroom. **9, 277**

gallbladder Organ that stores bile from the liver and squirts it into the small intestine. **454**

gamete Mature, haploid reproductive cell. **144**

gametophyte Haploid gamete-forming body that forms during a plant life cycle. **265**

ganglion Cluster of nerve cell bodies. **476**

gastric fluid Fluid secreted by the stomach lining; contains enzymes, acid, and mucus. **453**

gastropod Mollusk that moves about on its enlarged foot. **292**

gastrovascular cavity Of flatworms and cnidarians, a saclike gut that also functions in respiration. **288, 451**

gastrula Three-layered body formed when an animal blastula undergoes gastrulation during early development. **514**

gastrulation Animal developmental process by which cell movements produce a three-layered gastrula. **514**

gene Unit of hereditary information; DNA sequence that encodes an RNA or protein product. **115**

gene expression Process by which the information in a gene guides assembly of an RNA or protein product. **115**

gene flow The movement of alleles into and out of a population. **222**

gene pool All the alleles of all the genes in a population; a pool of genetic resources. **214**

gene therapy Treating a genetic defect or disorder by transferring a normal or modified gene into the affected individual. **183**

genetic code Complete set of sixty-four mRNA codons. **118**

genetic drift Change in allele frequencies in a population due to chance alone. **221**

genetic engineering Process by which deliberate changes are introduced into an individual's genome. **180**

genetically modified organism (GMO) Organism whose genome has been modified by genetic engineering. **181**

genome An organism's complete set of genetic material. **176**

genomics The study of genomes. **179**

genotype The particular set of alleles carried by an individual's chromosomes. **152**

genus plural **genera** A group of species that share a unique set of traits. **10**

geologic time scale Chronology of Earth's history. **202**

germinate To resume metabolic activity after dormancy. **562**

gibberellin Plant hormone that induces stem elongation; also helps seeds break dormancy. **572**

gills Of aquatic animals, folds or body extensions that increase the surface area available for exchange of gases with the surrounding water. **419**

gland *See* endocrine gland and exocrine gland.

global climate change Wide-ranging changes in rainfall patterns, average temperature, and other climate factors that result from rising concentrations of greenhouse gases. **350**

glottis Opening formed when the vocal cords relax; air flows through it into the trachea. **420**

glucagon Pancreatic hormone that causes liver cells to break down glycogen and release glucose, thus raising the blood glucose level. **502**

glycolysis Set of reactions in which glucose or another sugar is broken down to two pyruvate for a net yield of two ATP. First part of aerobic respiration and fermentation pathways. **90**

Golgi body Membrane-enclosed organelle that modifies polypeptides and lipids, then packages the finished products into vesicles. **55**

gonads Ovaries and testes; organs that produce gametes and secrete sex hormones. **505**

Gondwana Supercontinent that existed before Pangea, more than 500 million years ago. **201**

gravitropism Plant growth in a direction influenced by gravity. **574**

green alga Single-celled, colonial, or multicelled photosynthetic protist; the group most closely related to land plants. **257**

greenhouse effect Warming of Earth's lower atmosphere and surface as a result of heat trapped by greenhouse gases. **350**

greenhouse gas Atmospheric gas that helps keep heat from escaping into space and thus warms the Earth. **350**

ground tissue Tissue that makes up the bulk of the plant body; all plant tissue other than dermal and vascular tissue. **540**

groundwater Water between soil particles and in aquifers. **347**

growth In multicelled species, an increase in the number, size, and volume of cells. **7**

growth hormone (GH) Anterior pituitary hormone that promotes growth and development throughout the body. **498**

guard cell One of a pair of cells that define a stoma across the epidermis of a leaf or stem. **550**

gymnosperm Seed plant that produces "naked" seeds (seeds not encased within a fruit). **272**

habitat The type of place in which a species lives. **335**

half-life Characteristic time it takes for half of a quantity of a radioisotope to decay. **198**

haploid Having one of each type of chromosome characteristic of the species. **143**

heart Muscular organ that pumps fluid through a body. **409**

herbivory An animal feeds on a plant, which may or may not die as a result. **339**

hermaphrodite Individual animal that makes both eggs and sperm. **290, 513**

heterotroph Consumer; an organism that obtains carbon from organic compounds assembled by other organisms. **249**

heterozygous Having two different alleles of a gene. **152**

histone Type of protein that structurally organizes eukaryotic chromosomes. **101**

HIV (human immunodeficiency virus) Enveloped RNA virus that causes AIDS. **246**

homeostasis Process in which cells and multicelled organisms keep their internal conditions within tolerable ranges by sensing and responding to change. **7, 379**

hominin Humans and extinct humanlike species. **309**

Homo erectus Human species that dispersed out of Africa. **309**

Homo habilis Earliest named human species. **309**

Homo neanderthalensis Neanderthals. Closest extinct relatives of modern humans; had large brain, stocky body. **310**

Homo sapiens Modern humans; evolved in Africa, then expanded their range worldwide. **310**

homologous chromosomes Paired chromosomes with the same length, shape, and set of genes. In sexual reproducers, one member of a homologous pair is paternal and the other is maternal. **135**

homologous structures Body parts or structures that are similar in different lineages because they evolved in a common ancestor. **204**

homozygous Having identical alleles of a gene. **152**

humans Members of the genus *Homo*. **309**

human chorionic gonadotropin (HCG) Hormone produced by the chorion and later by the placenta; encourages maintenance of the uterus, thus sustaining a pregnancy. **527**

humus Decaying organic matter in soil. **549**

hydrogen bond Attraction between a covalently bonded hydrogen atom and another atom taking part in a separate covalent bond. Collectively, they impart special properties to liquid water and stabilize the structure of biological molecules. **30**

hydrophilic Describes a substance that dissolves easily in water. **30**

hydrophobic Describes a substance that resists dissolving in water. **30**

hydrostatic skeleton Fluid-filled chamber or chambers that muscles act on, redistributing the fluid. **395**

hydrothermal vent Place where hot, mineral-rich water streams out from an underwater opening in Earth's crust. **238, 363**

hypertension Chronically high blood pressure. **417**

hypertonic Describes a fluid that has a high overall solute concentration relative to another fluid. **72**

hypha A single filament in a fungal mycelium. **277**

hypothalamus Forebrain region that controls processes related to homeostasis and has endocrine functions. **481, 498**

hypothesis Testable explanation of a natural phenomenon. **12**

hypotonic Describes a fluid that has a low overall solute concentration relative to another fluid. **72**

immunity The body's ability to resist and fight infections. **429**

immunization Any procedure designed to promote immunity to a specific disease. **444**

inbreeding Mating among close relatives. **221**

incomplete dominance Effect in which one allele is not fully dominant over another, so the heterozygous phenotype is between the two homozygous phenotypes. **157**

indicator species A species that is particularly sensitive to environmental changes and can be monitored to assess whether an ecosystem is threatened. **370**

inflammation A local response to tissue damage or infection; characterized by redness, warmth, swelling, and pain. **432**

ingestion Taking food into the digestive system. **451**

inheritance Transmission of DNA to offspring. **8**

innate immunity In all multicelled organisms, set of immediate, general defenses against infection. **429**

insects Land-dwelling arthropods with compound eyes, a pair of antennae, six legs, and—in the most diverse groups—wings. **297**

insertion Mutation in which one or more nucleotides become inserted into DNA. **122**

insulin Pancreatic hormone that causes cells to take up glucose from the blood, thus lowering blood glucose level. **502**

intercostal muscles Muscles between the ribs; help alter the size of the thoracic cavity during breathing. **421**

intermediate filament Stable cytoskeletal element that structurally supports cells and tissues. **56**

interneurons Neurons that receive signals from, and send signals to, other neurons; they integrate information. **471**

interphase In a eukaryotic cell cycle, the interval between mitotic divisions when a cell grows, roughly doubles the number of its cytoplasmic components, and replicates its DNA. **134**

interspecific competition Two species compete for a limited resource and both are harmed by the interaction. **337**

interstitial fluid Fluid between cells of a multicelled body. **409**

intervertebral disk Cartilage disk between two vertebrae. **396**

intron Nucleotide sequence that intervenes between exons and is excised during post-transcriptional modification. **117**

invertebrate Animal without a backbone. **289**

***in vitro* fertilization** Procedure in which eggs are removed from a woman's ovary and fertilized outside her body. **512**

ion Atom that carries a charge because it has an unequal number of protons and electrons. **27**

ionic bond Type of chemical bond in which a strong mutual attraction links ions of opposite charge. **29**

iris Ring of smooth muscle with pupil at its center; adjusts how much light enters the eye. **485**

iron–sulfur world hypothesis Hypothesis that life began in rocks rich in iron sulfide near deep-sea hydrothermal vents. **238**

isotonic Describes two fluids with identical solute concentrations. **72**

isotopes Forms of an element that differ in the number of neutrons their atoms carry. **25**

jawless fish Fish with a skeleton of cartilage, and no jaws or paired fins; for example, a lamprey. **301**

joint Region where bones come together. **397**

karyotype Image of an individual's complement of chromosomes arranged by size, length, shape, and centromere location. **102**

key innovation An evolutionary adaptation that gives its bearer the opportunity to exploit a particular environment more efficiently or in a new way. **227**

keystone species A species that has a disproportionately large effect on community structure. **342**

kidney Organ of the vertebrate urinary system that filters blood, adjusts its composition, and forms urine. **300, 461**

kidney dialysis Procedure used to cleanse blood and restore proper solute concentrations in a person with impaired kidney function. **464**

knockout An experiment in which a gene is deliberately inactivated in a living organism. **125**

Krebs cycle Cyclic pathway that helps break down pyruvate to carbon dioxide during aerobic respiration. **90**

labor Expulsion of a placental mammal from its mother's uterus by muscle contractions. **532**

lactate fermentation Anaerobic carbohydrate breakdown pathway that produces ATP and lactate. **92**

lancelet Invertebrate chordate with a fishlike shape; retains the defining chordate traits into adulthood. **299**

large intestine Organ that concentrates and stores waste, then delivers it to the anus for excretion. **456**

larva Preadult stage in some animal life cycles. **290**

larynx Short airway containing the vocal cords (voice box). **420**

lateral meristem In plants, a cylindrical sheet of meristem that gives rise to secondary growth; vascular cambium or cork cambium. **546**

law of nature Generalization describing a natural phenomenon that is consistent but has an incomplete scientific explanation. **18**

lens Structure that focuses light on an eye's photoreceptors. **485**

lethal mutation Mutation that alters phenotype so drastically that it causes death. **214**

lichen Composite organism consisting of a fungus and a single-celled photosynthetic alga or bacterium. **280**

life history Set of traits related to growth, survival, and reproduction such as life span, age-specific mortality, age at first reproduction, and number of breeding events. **322**

ligament Strap of dense connective tissue that holds bones together at a joint. **397**

lignin Compound that stiffens walls of some cells (including xylem) in vascular plants. **265**

lineage Line of descent. **193**

lipid A fat, steroid, or wax. **36**

lipid bilayer A double layer of lipids (mainly phospholipids) arranged tail-to-tail; structural foundation of all cell membranes. **37**

liver Organ that produces bile, stores glycogen, and detoxifies many substances. **454**

lobe-finned fish Bony fish with bony supports in its fins. **302**

logistic model of population growth Model for growth of a population limited by density-dependent factors; numbers increase exponentially at first, then the growth rate slows and population size levels off at carrying capacity. **320**

loose connective tissue Connective tissue with relatively few fibroblasts and fibers scattered in its matrix. **382**

lung Internal saclike organ inside which blood exchanges gases with the air; serves as the respiratory surface in most land vertebrates and some fish. **300, 419**

lymph Fluid that has left capillaries and entered lymph vessels. **415**

lysosome Enzyme-filled vesicle that breaks down cellular wastes and debris. **55**

lysozyme Antibacterial enzyme that occurs in body secretions such as mucus. **431**

macroevolution Large-scale evolutionary patterns and trends; e.g., adaptive radiation. **226**

macrophage Phagocytic white blood cell that patrols tissues and tissue fluids. **430**

mammal Vertebrate that nourishes its young with milk from mammary glands. **306**

mantle Skirtlike extension of tissue in mollusks; covers the mantle cavity and secretes the shell in species that have a shell. **292**

marsupial Mammal in which young complete development in a pouch on the mother's body. **306**

mass extinction Simultaneous loss of many lineages from Earth. **190**

mass number Of an isotope, the total number of protons and neutrons in the atomic nucleus. **25**

master gene Gene encoding a product that affects the expression of many other genes. **124**

mechanoreceptor Sensory receptor that responds to pressure, position, or acceleration. **483**

medusa Bell-shaped, free-swimming cnidarian body form. **290**

megaspore In seed plants, a haploid cell that gives rise to a female gametophyte. **271**

meiosis Nuclear division process that halves the chromosome number. Basis of sexual reproduction. **133**

melatonin Hormone produced by the pineal gland; affects onset of puberty, sleep–wake cycles. **506**

membrane potential *See* resting potential, action potential.

memory cell Long-lived, antigen-sensitized B or T cell that can act in a secondary immune response. **438**

meninges Membranes that surround and protect the brain and spinal cord. **479**

menopause Of a human female, the end of fertility and of menstrual cycles. **521**

menstrual cycle Approximately monthly cycle in which the uterine lining thickens in preparation for pregnancy, then is shed if pregnancy does not occur. **520**

menstruation Flow of blood and bits of shed uterine lining out through the vagina. **520**

meristem In plants, zone of undifferentiated cells that can divide rapidly. **546**

mesoderm Middle tissue layer of a three-layered animal embryo. **514**

messenger RNA (mRNA) Type of RNA that has a protein-building message. **115**

metabolic pathway Series of enzyme-mediated reactions by which cells build, remodel, or break down an organic molecule. **70**

metabolism All the enzyme-mediated chemical reactions by which cells acquire and use energy as they build and break down organic molecules. **34**

metamorphosis Dramatic remodeling of body form during the transition from larva to adult. **295**

metaphase Stage of mitosis at which all chromosomes are aligned in the middle of the cell. **137**

metastasis The process in which cancer cells spread from one part of the body to another. **140**

methanogen Organism that produces methane gas as a metabolic by-product. **250**

MHC markers Self-proteins on the surface of human body cells. **435**

microevolution Change in an allele's frequency in a population or species. **214**

microfilament Cytoskeletal element that reinforces cell membranes and tissues; fiber of actin subunits. **56**

microspore In seed plants, a haploid cell that gives rise to a male gametophyte (pollen grain). **271**

microtubule Cytoskeletal element involved in movement; hollow filament of tubulin subunits. **56**

microvilli Thin projections that increase the surface area of cells such as those lining the small intestine. **454**

mimicry Two or more species come to resemble one another. **338**

mineral Inorganic substance that is required in small amounts for normal metabolism. **459**

mitochondrion Membrane-enclosed organelle that produces ATP by aerobic respiration. **55**

mitosis Nuclear division mechanism that maintains the chromosome number. Basis of body growth, tissue repair and replacement in multicelled eukaryotes; also asexual reproduction in some plants, animals, fungi, and protists. **133**

model Analogous system used to test an object or event that cannot be tested directly. **13**

molecule Two or more atoms joined by chemical bonds. **5**

mollusk Invertebrate with a reduced coelom and a mantle. **292**

monocot Member of a lineage of angiosperms that includes grasses, orchids, and palms. **275**

monohybrid cross Cross between two individuals identically heterozygous for one gene; for example $Aa \times Aa$. **153**

monomer Molecule that is a subunit of polymers. **33**

monophyletic group An ancestor in which a derived trait evolved, together with all of its descendants. **229**

monotreme Egg-laying mammal. **306**

morphological convergence Evolutionary pattern in which similar body parts evolve separately in different lineages. **204**

morphological divergence Evolutionary pattern in which a body part of an ancestor changes in its descendants. **204**

mosses Most diverse group of bryophytes (nonvascular plants). Low-growing plants with flagellated sperm; disperse by producing spores. **267**

motor neurons Neurons that control muscles and glands. **471**

motor protein Type of energy-using protein that interacts with cytoskeletal elements to move the cell's parts or the whole cell. **56**

mRNA *See* messenger RNA.

mutation Permanent change in DNA sequence. **106**

mutualism Species interaction that benefits both species. **280, 336**

mycelium Mass of threadlike filaments (hyphae) that make up the body of a multicelled fungus. **277**

mycorrhiza Fungus–plant root partnership. **280, 549**

myelin Material made by some neuroglia, insulates axons. **471**

myofibrils Within a muscle fiber, threadlike, cross-banded components that consist of sarcomeres arranged end to end. **399**

myosin Motor protein that makes up the thick filaments of muscle fibers. **399**

natural selection The differential survival and reproduction of individuals of a population based on differences in their shared, heritable traits. Driven by environmental pressures. **194**

nectar Sweet fluid exuded by some flowers; attracts animal pollinators. **561**

negative feedback A change causes a response that reverses the change. **388**

neoplasm An accumulation of abnormally dividing cells. **139**

nephron One kidney tubule and the associated capillaries; filters blood and forms urine. **462**

nerve Axons bundled in connective tissue. **476**

nerve net Mesh of interacting neurons with no central control, as in the nervous system of jellies. **476**

nervous tissue Animal tissue composed of neurons and supporting cells (neuroglia); detects stimuli and controls responses to them. **384**

neuroglia Cells that support neurons. **471**

neuron Main type of cell in nervous tissue; transmits electrical signals along its plasma membrane and communicates with other cells by secreting chemical signals. **384, 471**

neurotransmitter Chemical signal released by a neuron's axon terminals. **474**

neutron Uncharged subatomic particle in the atomic nucleus. **25**

niche The role of a species in its community. **335**

nitrogen cycle Movement of nitrogen among the atmosphere, soil, and water, and into and out of food webs. **348**

nitrogen fixation Process of combining nitrogen gas with hydrogen to form ammonia. **250, 348**

node In plants, a region of stem where leaves form. **542**

nondisjunction Failure of chromosomes to separate properly during nuclear division. **166**

normal flora Collection of microorganisms that normally live in or on a healthy animal or person. **251, 430**

notochord Stiff rod of connective tissue that runs the length of the body in chordate larvae or embryos; in vertebrates develops into part of the vertebral column. **299**

nuclear envelope A double membrane that constitutes the outer boundary of the nucleus. Pores in the membrane control which substances can cross. **54**

nucleic acid Single- or double-stranded chain of nucleotides joined by sugar–phosphate bonds; DNA or RNA. **41**

nucleic acid hybridization Base-pairing between DNA or RNA from different sources. **176**

nucleotide Small organic compound that is a subunit of nucleic acids. Consists of a five-carbon sugar, nitrogen-containing base, and one or more phosphate groups. E.g., adenine, guanine, cytosine, thymine, uracil. **41**

nucleus Of an atom, core region occupied by protons and neutrons. **25** In eukaryotic cells, organelle that holds the DNA; has two membranes. **8, 47**

nutrient Substance that an organism needs for growth and survival but cannot make for itself. **6**

nutrient absorption Movement of nutrient molecules from a digestive cavity into the body proper. **451**

oncogene Gene that helps transform a normal cell into a tumor cell. **139**

oocyte Immature egg. **519**

open circulatory system Circulatory system in which blood leaves vessels and flows among tissues before returning to a heart. **409**

opportunistic life history Life history favored in unpredictable environments; individuals reproduce while young and invest little in each of many offspring. **323**

organ Structural unit that is composed of two or more tissues and adapted to carry out a particular task. **379**

organ of Corti Sound-detecting organ in the cochlea of the inner ear. **488**

organ system Organs that interact closely in some task. **379**

organelle Structure that carries out a special metabolic function inside a cell; e.g., a mitochondrion. **47**

organic Describes a molecule that consists mainly of carbon and hydrogen atoms. **33**

organism Individual that consists of one or more cells. **5**

orgasm Sexual climax during which neurotransmitter release causes smooth muscle contractions accompanied by a feeling of release and pleasure. **522**

osmosis Diffusion of water across a selectively permeable membrane; occurs when the fluids on either side of the membrane are not isotonic. **72**

ovarian follicle Immature egg and surrounding cells. **519**

ovary In flowering plants, the enlarged base of a carpel, which holds one or more ovules. In animals, the egg-producing female gonad. **274, 516, 559**

oviduct Ciliated tube connecting an ovary to the uterus. **518**

ovulation Release of a secondary oocyte from an ovary. **519**

ovule Of seed plants, chamber inside which megaspores form and develop into female gametophytes; after fertilization becomes a seed. **271, 559**

ovum Mature animal egg. **522**

ozone layer Atmospheric layer with a high concentration of ozone that prevents much UV radiation from reaching Earth's surface. **240, 367**

pancreas Organ that secretes digestive enzymes into the small intestine and hormones into the blood. **454**

pandemic Disease outbreak with cases worldwide. **247**

Pangea Supercontinent that formed about 270 million years ago. **200**

parasite A species that withdraws nutrients from another species (its host), usually without killing it. **339**

parasitoid An insect that lays eggs in another insect, and whose young devour their host from the inside. **340**

parasympathetic neurons Neurons of the autonomic system that encourage housekeeping tasks. **478**

parathyroid glands Four small endocrine glands in the neck whose hormone product increases the level of calcium in blood. **501**

parenchyma Simple plant tissue made up of living cells; main component of ground tissue. **539**

parthenogenesis Individuals develop from unfertilized eggs. **513**

passive transport Mechanism in which a concentration gradient drives the movement of a solute across a cell membrane through a transport protein. Requires no energy input. **74**

pathogen Disease-causing agent, such as bacteria that cause syphilis or a virus that causes influenza. **236**

PCR Polymerase chain reaction. Laboratory method that rapidly generates many copies of a targeted DNA fragment. **176**

pedigree Chart showing the pattern of inheritance of a trait in a family. **161**

pellicle Layer of proteins that gives shape to many unwalled, single-celled protists. **252**

peptide bond A bond between the amine group of one amino acid and the carboxyl group of another. Joins amino acids in proteins. **38**

per capita growth rate The number of individuals added during some interval divided by the initial population size. **318**

periderm In plants, dermal tissue that replaces epidermis during secondary growth of woody stems and roots. **547**

peripheral nervous system Nerves that extend through the body and carry signals to and from the central nervous system. **476**

permafrost Layer of permanently frozen soil in the Arctic. **361**

peroxisome Enzyme-filled vesicle that breaks down amino acids, fatty acids, and toxic substances. **55**

pH Measure of the amount of hydrogen ions in a fluid. Decreases with increasing acidity. **32**

phagocytosis Endocytic pathway by which a cell engulfs large particles such as microbes. **76**

pharynx In vertebrates, the throat; opens to airways and digestive tract. **420**

phenotype An individual's observable traits. **152**

pheromones Signaling molecules that affect another member of the same species. **484**

phloem Complex vascular tissue of plants; distributes sugars through its sieve tubes. Consists of sieve-tube members and companion cells. **265, 540**

phospholipid A lipid with a phosphate in its hydrophilic head, and two nonpolar fatty acid tails. Major component of cell membranes. **37**

phosphorus cycle Movement of phosphorus among rocks, water, soil, and living organisms. **348**

phosphorylation A phosphate-group transfer. **69**

photoperiodism Biological response to seasonal changes in the relative lengths of day and night. **575**

photoreceptor Light-sensitive sensory receptor. **483**

photosynthesis Metabolic pathway by which most autotrophs use light energy to make sugars from carbon dioxide and water. Converts light energy into chemical bond energy. **6, 82**

phototropism Plant growth in a direction influenced by light. **575**

phylogeny Evolutionary history of a species or group of species. **229**

pigment An organic molecule that selectively absorbs light of certain wavelengths. Reflected light imparts a characteristic color. E.g., chlorophyll. **83**

pilus Protein filament that projects from the surface of some bacterial cells. **53**

pineal gland Endocrine gland in the forebrain that secretes melatonin; secretion declines when the eye is exposed to light. **506**

pioneer species Species that can colonize a new habitat. **341**

pituitary gland Pea-sized endocrine gland in the forebrain that interacts closely with the adjacent hypothalamus. **498**

placenta Of placental mammals, an organ composed of maternal and embryonic tissues that allows the exchange of substances between a developing individual and its mother. **527**

placental mammal Mammal in which developing young are nourished within their mother's body by way of a placenta. **306**

plankton Community of mostly microscopic drifting or swimming organisms. **253**

plant A multicelled, typically photosynthetic eukaryote; develops from an embryo that forms on the parent and is nourished by it. **9, 265**

plant hormone Extracellular signaling molecule of plants that exerts its effect at very low concentration. **570**

plasma Protein-rich fluid portion of blood. **414**

plasma membrane A cell's outermost membrane. **47**

plasmid Of many prokaryotes, a small ring of nonchromosomal DNA replicated independently of the chromosome. **248**

plasmodial slime mold Heterotrophic protist that moves and feeds as a multinucleated mass; forms a fruiting body when conditions are unfavorable. **258**

plate tectonics Theory that Earth's outer layer of rock is cracked into plates, the slow movement of which rafts continents to new locations over geological time. **201**

platelet Cell fragment that functions in blood clotting. **414**

pleiotropic Refers to a gene whose product influences multiple traits. **158**

pluripotent Capable of developing into any type of cell in a multicelled body. **378**

polarity Any separation of charge into distinct positive and negative regions. **28**

pollen grain Immature male gametophyte of a seed plant. **266, 559**

pollen sac Of seed plants, chamber in which microspores form and develop into male gametophytes (pollen grains). **271**

pollination Delivery of pollen to female part of a plant. **271**

pollination vector Any agent that moves pollen grains from one plant to another; e.g., the wind or a bee. **560**

pollinator Animal that moves pollen from one plant to another, thus facilitating pollination. **275, 558**

pollutant A natural or man-made substance that is released into the environment in greater than natural amounts and that damages the health of organisms. **367**

polymer Molecule that consists of multiple monomers. **34**

polyp Tubular, typically sessile, cnidarian body form. **290**

polyploid Having three or more of each type of chromosome characteristic of the species. **166**

population A group of organisms of the same species who live in a specific location and breed with one another more often than they breed with members of other populations. **5, 213, 316**

population density Number of members of a population in a given area. **317**

population distribution The way in which members of a population are dispersed in their environment. **317**

population size Number of individuals in a population. **317**

prairie Temperate grassland biome of North America. Its grasses and other nonwoody plants are adapted to recover after grazing and the occasional fire. **360**

predation One species (the predator) captures, kills, and feeds on another (its prey). **338**

prediction Statement, based on a hypothesis, about a condition that should exist if the hypothesis is correct. **13**

pressure flow theory Explanation of how organic compounds move through vascular plants: Fluid flow through sieve tubes is driven by a difference in turgor between source and sink regions. **553**

primary growth In plants, lengthening of young shoots and roots; originates at apical meristems. **546**

primary production The energy captured by an ecosystem's producers. **346**

primary succession Ecological succession occurs in an area where there was previously no soil. **341**

primates Mammalian order characterized by grasping hands with nail-tipped digits; forward-facing eyes, and great shoulder mobility; includes lemurs, tarsiers, monkeys, apes, and humans. **307**

primer Short, single strand of DNA or RNA that base-pairs with a targeted DNA sequence. **105**

prion Infectious protein. **40**

probability The chance that a particular outcome of an event will occur; depends on the total number of outcomes possible. **17**

probe Short fragment of DNA designed to hybridize with a nucleotide sequence of interest and labeled with a tracer. **176**

producer Organism that makes its own food using energy and nonbiological raw materials from the environment. **6, 344**

product A molecule that is produced by a reaction. **66**

progesterone Sex hormone produced by a female's ovaries. **505**

prokaryote Single-celled organism in which the DNA resides in the cytoplasm; a bacterium or archaean. **8**

promoter In DNA, a nucleotide sequence that serves as a binding site for RNA polymerase. **117**

prophase Stage of mitosis during which chromosomes condense and become attached to a newly forming spindle. **137**

prostate gland Exocrine gland that encircles a male's urethra; its secretions contribute to semen. **517**

protein Organic compound that consists of one or more chains of amino acids (polypeptides). **38**

protist A eukaryote that is not a fungus, plant, or animal. **8, 252**

protocells Membranous sac that contains interacting organic molecules; hypothesized to have formed prior to the earliest life forms. **238**

proton Positively charged subatomic particle that occurs in the nucleus of all atoms. Determines the element. **25**

pseudopod A temporary protrusion that helps some eukaryotic cells move and engulf prey. **57**

puberty Period when reproductive organs mature and begin to function. **505**

pulmonary circuit Circuit through which blood flows from the heart to the lungs and back. **410**

Punnett square Diagram used to predict the genetic and phenotypic outcome of a cross. **152**

pupil Opening through which light enters the eye. **485**

pyruvate Three-carbon end product of glycolysis. **89**

radial symmetry Having parts arranged around a central axis, like spokes around a wheel. **288**

radioactive decay Process by which atoms of a radioisotope emit energy and subatomic particles when their nucleus spontaneously disintegrates. **25**

radioisotope Isotope with an unstable nucleus. **25**

radiometric dating Method of estimating the age of a rock or fossil by measuring the content and proportions of a radioisotope and its daughter elements. **198**

radula Tonguelike, chitin-hardened feeding organ of many mollusks. **292**

rain shadow Dry region on the downwind side of a coastal mountain range. **357**

ray-finned fish Bony fish with fins supported by thin rays derived from skin. **302**

reabsorption Water and solutes enter capillaries. **462**

reactant A molecule that enters a reaction and is changed by participating in it. **66**

reaction Process of molecular change, in which reactants become products. **35**

receptor protein Plasma membrane protein that triggers a change in cellular activities when it binds to a particular substance outside of the cell. **51**

recessive Refers to an allele with an effect that is masked by a dominant allele in heterozygous individuals. **152**

recognition protein Plasma membrane protein that identifies a cell as belonging to self (one's own body or species). **51**

recombinant DNA A DNA molecule that contains genetic material from more than one organism. **175**

rectum Portion of the large intestine that stores feces until they are expelled. **456**

red alga Single-celled or multicelled photosynthetic protist with red accessory pigment. **257**

red blood cell Hemoglobin-filled blood cell that functions in oxygen transport. **414**

red marrow Bone marrow that makes blood cells. **397**

reflex Automatic response to a stimulus. **480**

reproduction Process by which parents produce offspring. *See* sexual reproduction, asexual reproduction. **7**

reproductive cloning Technology that produces genetically identical individuals. **108**

reproductive isolation The end of gene flow between populations. **223**

reptile Amniote lineage that includes lizards, snakes, turtles, crocodilians, and birds. **304**

resource partitioning Use of different portions of a limited resource; allows species with similar needs to coexist. **338**

respiration Physiological process by which gases enter and leave an animal body. **418**

respiratory cycle One inhalation and one exhalation. **421**

respiratory surface Moist surface across which gases are exchanged between animal cells and their environment. **418**

resting potential Membrane potential of a neuron at rest. **472**

restriction enzyme Type of enzyme that cuts DNA at a specific nucleotide sequence. **175**

retina Layer of eye that contains photoreceptors. **485**

rhizome Stem that grows horizontally along or just below the ground. **269**

ribosomal RNA (rRNA) Type of RNA that, together with structural proteins, composes ribosomes. Has enzymatic activity. **115**

ribosome Organelle of protein synthesis. An intact ribosome has two subunits, each composed of rRNA and proteins. **52**

RNA Ribonucleic acid. Nucleic acid with roles in gene expression; consists of a single-stranded chain of nucleotides (adenine, guanine, cytosine, and uracil). *See* messenger RNA, transfer RNA, ribosomal RNA. **41**

RNA polymerase Enzyme that carries out transcription. **116**

RNA world hypothesis Hypothesis that RNA served as the first material of inheritance. **238**

rod cell Photoreceptor most active in dim light; provides coarse perception of image and detects motion. **487**

root hairs Fine, absorptive extensions of an epidermal cell in young roots. **544**

root nodules Of some plant roots, swellings that contain nitrogen-fixing bacteria. **549**

roundworm Unsegmented worm with a pseudocoelom and a cuticle that is molted as the animal grows. **294**

rRNA *See* ribosomal RNA.

salt Ionic compound that releases ions other than H^+ and OH^- when it dissolves in water. **30**

sampling error Difference between results derived from testing an entire group of events or individuals, and results derived from testing a subset of the group. **17**

sarcomere Contractile unit of skeletal and cardiac muscle. **399**

saturated fat Fat that consists mainly of triglycerides with three saturated fatty acid tails. **36**

savanna Tropical biome dominated by grasses and other plants adapted to grazing, as well as a scattering of shrubs. **360**

scales Hard, flattened elements that cover the skin of reptiles and some fishes. **301**

science Systematic study of the observable world. **12**

scientific method Systematically making, testing, and evaluating hypotheses about nature. **13**

scientific theory Hypothesis that has not been disproven after many years of rigorous testing. **18**

sclerenchyma Simple plant tissue that is dead at maturity; its lignin-reinforced cell walls structurally support plant parts. **539**

scrotum Pouch that holds the testes. **516**

seamount An undersea mountain. **363**

second law of thermodynamics Energy tends to disperse spontaneously. **65**

second messenger Molecule that forms inside a cell when a hormone binds at the cell surface; sets in motion reactions that alter enzyme activity inside the cell. **496**

secondary growth In plants, thickening of older stems and roots; originates at lateral meristems. **271, 546**

secondary succession Ecological succession occurs in an area where a community previously existed and soil remains. **341**

seed Embryo sporophyte of a seed plant packaged with nutritive tissue inside a protective coat. **266, 564**

semen Sperm mixed with fluid secreted by exocrine glands. **517**

semiconservative replication Describes the process of DNA replication, which produces two copies of a DNA molecule: one strand of each copy is new, and the other is a strand of the original DNA. **106**

seminal vesicles Exocrine glands that secrete fluid into the vasa deferentia; the main contributors to semen volume. **517**

sensory neurons Neurons that detect specific stimuli. **471**

sequencing Method of determining the order of nucleotides in a DNA molecule. **178**

sex chromosome Of eukaryotes, a member of a pair of chromosomes that differs between males and females. **102**

sexual reproduction Reproductive mode by which offspring arise from two parents and inherit genes from both. **133, 513**

sexual selection Mode of natural selection in which some individuals outreproduce others because they are better at securing mates. **219**

shell model Model of electron distribution in an atom. **27**

short tandem repeats In chromosomal DNA, sequences of 4 or 5 nucleotides repeated multiple times in a row. Used in DNA profiling. **180**

single-nucleotide polymorphism (SNP) One-nucleotide DNA sequence variation carried by a measurable percentage of a population. **174**

sink Region of a plant where sugars are being used or stored. **552**

sister chromatid Of a duplicated eukaryotic chromosome, one of the two attached molecules of DNA. **101**

sister groups The two lineages that emerge from a node on a cladogram. **230**

skeletal muscle tissue Striated, voluntary muscle that interacts with bone to move body parts. **383**

sliding-filament model Explanation of how interactions among actin and myosin filaments shorten a sarcomere and bring about muscle contraction. **400**

small intestine Longest portion of the digestive tract, and the site of most digestion and absorption. **454**

smooth muscle tissue Involuntary muscle that lines blood vessels and hollow organs; not striated. **383**

solute A dissolved substance. **30**

solution Homogeneous mixture of solute and solvent. **30**

solvent Liquid that can dissolve other substances. **30**

somatic cell nuclear transfer (SCNT) Reproductive cloning method in which the DNA from a body cell is transferred into an unfertilized, enucleated egg. **108**

somatic nervous system Set of nerves that control skeletal muscle and relay signals from joints and skin to the central nervous system. **478**

somatic sensations Sensations that arise when sensory neurons in skin and near joints or muscle are activated. **483**

sorus Cluster of spore-forming chambers on a fern frond. **269**

source Region of a plant where sugars are being produced or released from storage. **552**

speciation Evolutionary process in which new species arise. **223**

species Unique type of organism. Of sexual reproducers, often defined as one or more groups of individuals that can potentially interbreed, produce fertile offspring, and do not interbreed with other groups. **10**

species diversity The number of species and their relative abundance. **335**

sperm Male gamete. **144**

sphincter Ring of muscle that controls passage through a tubular organ or body opening. **453**

spinal cord Portion of central nervous system that runs through the vertebral column and connects peripheral nerves with the brain. **480**

spindle Apparatus that moves chromosomes during nuclear division; consists of dynamically assembled and disassembled microtubules. **137**

sponge Porous aquatic invertebrate that filters food from the water; has no tissues or organs. **290**

spongy bone Lightweight bone with many internal spaces. **397**

sporophyte Diploid spore-forming body that forms in a plant life cycle. **265**

stabilizing selection Mode of natural selection in which an intermediate form of a trait is favored over extreme forms. **217**

stamen Floral reproductive organ that consists of an anther and, often, a filament. **274, 559**

stasis Evolutionary pattern in which a lineage persists with little or no change over evolutionary time. **226**

statistically significant Refers to a result that is statistically unlikely to have occurred by chance alone. **17**

stem cell A cell that can divide and create more stem cells or differentiate to become a specialized cell type. **378**

steroid A type of lipid with four carbon rings and no fatty acid tails. **37**

steroid hormone Hormone derived from cholesterol. **496**

stigma Upper, pollen-receiving part of a carpel. **274, 559**

stomach Muscular organ that mixes food with gastric fluid secreted by its lining. **453**

stomata Singular, stoma. Closable gaps formed by guard cells in plant epidermis. When open, they allow water vapor and gases to diffuse into and out of the plant's tissues. **88, 265**

stroma The cytoplasm-like fluid between the thylakoid membrane and the two outer membranes of a chloroplast. Site of light-independent reactions of photosynthesis. **85**

stromatolites Dome-shaped structures composed of layers of prokaryotic cells and sediments; form in shallow seas. **240**

style Elongated portion of a carpel that holds the stigma above the ovary. **274**

substrate Of an enzyme, a reactant that is specifically acted upon by the enzyme. **68**

surface-to-volume ratio A relationship in which the volume of an object increases with the cube of the diameter, and the surface area increases with the square. Limits cell size. **47**

survivorship curve Graph showing the decline in numbers of a cohort over time. **322**

symbiosis One species lives on or inside another in a commensal, mutualistic, or parasitic relationship. **336**

sympathetic neurons Neurons of the autonomic system that prepare the body for danger or excitement. **478**

sympatric speciation Pattern in which speciation occurs in the absence of a physical barrier. **225**

synovial joint Joint such as the knee that is lubricated by fluid and allows movement of bones around the joint. **397**

systemic circuit Circuit through which blood flows from the heart to the body tissues and back. **410**

systolic pressure Blood pressure when the ventricles are contracting. **414**

T cell T lymphocyte. Lymphocyte central to adaptive immunity. E.g., helper T cell, cytotoxic T cell. **430**

T cell receptor (TCR) Antigen receptor on the surface of a T cell; basis of self/nonself discrimination in immune responses. **435**

T lymphocyte *See* T cell.

taiga *See* boreal forest.

taxon plural **taxa** Group of organisms that share a unique set of traits. **11**

taxonomy The science of naming and classifying species. **10**

telophase Stage of mitosis during which chromosomes arrive at opposite ends of the cell and decondense, and two new nuclei form. **137**

temperate deciduous forest Biome dominated by broadleaf trees that grow during warm summers, then drop their leaves and become dormant during cold winters. **359**

temperature Measure of molecular motion. **31**

tendon Strap of dense connective tissue that connects a skeletal muscle to bone. **398**

testes Sperm-producing organs. **516**

testosterone Sex hormone produced by a male's testes. **505**

tetrapod Vertebrate with four limbs. **300**

theory *See* scientific theory.

thermoreceptor Temperature-sensitive sensory receptor. **483**

threatened species A species likely to become endangered in the near future. **364**

threshold potential Membrane potential at which voltage-gated sodium channels in a neuron axon open, resulting in an action potential. **472**

thylakoid membrane Highly folded inner membrane of a chloroplast. Molecules embedded in this membrane carry out the light-dependent reactions of photosynthesis. **85**

thyroid gland Endocrine gland at the base of the neck; produces thyroid hormone, which increases the metabolic rate and is essential for normal brain development. **500**

tissue One or more types of cells that are organized in a specific pattern and that carry out a particular task. **288, 379**

tissue culture propagation Laboratory method in which individual plant cells (typically from meristem) are induced to form embryos. **568**

total fertility rate Average number of children born to females of a population over the course of their lifetimes. **327**

tracer A molecule with a detectable component; researchers can track it after delivery into a body or other system. **26**

trachea Largest airway; carries air to the bronchi; commonly called the windpipe. **420**

tracheal system Tubes that deliver air from body surface to tissues of insects. **419**

transcription RNA synthesis; process by which an RNA is assembled from nucleotides using a DNA template. **115**

transcription factor Protein that influences transcription by binding to DNA. **124**

transfer RNA (tRNA) Type of RNA that delivers amino acids to a ribosome during translation. **115**

transgenic Refers to a genetically modified organism that carries a gene from a different species. **181**

translation Process by which a polypeptide chain is assembled from amino acids in the order specified by an mRNA. **115**

translocation Movement of organic molecules through phloem. **552**

transpiration Evaporation of water from aboveground plant parts. **551**

transport protein Protein that allows specific ions or molecules to cross a membrane. Those that function in active transport require an energy input, as from ATP; passive transport proteins do not. **51**

triglyceride A fat that has three fatty acid tails. **36**

tRNA *See* transfer RNA.

trophic level Position of an organism in a food chain. **344**

tropical rain forest Multilayered forest that occurs where warm temperatures and continual rains allow plant growth year-round. Most productive and species-rich biome. **358**

tropism In plants, directional growth response to an environmental stimulus. **574**

tubular secretion Substances are moved out of capillaries and into kidney tubules. **462**

tumor A neoplasm that forms a lump. **139**

tundra Northernmost biome, dominated by low plants that grow over a layer of permafrost. **360**

tunicates Invertebrate chordates that have a fish-like shape as larvae, then undergo metamorphosis to become adults that secrete a baglike "tunic"; lose their defining chordate traits during the transition to adulthood. **299**

turgor Pressure that a fluid exerts against a membrane, wall, or other structure that contains it. **73**

unsaturated fat Fat that consists mainly of triglycerides with one or more unsaturated fatty acid tails. **36**

ureter Tube that carries urine from a kidney to the bladder. **461**

urethra Tube through which urine from the bladder flows out of the body. **461**

urinary bladder Hollow, muscular organ that stores urine. **461**

urinary system Organ system that filters blood, and forms, stores, and expels urine. **461**

urine Mix of water and soluble wastes formed and excreted by the urinary system. **461**

uterus Muscular chamber where offspring develop; womb. **518**

vaccine A preparation introduced into the body in order to elicit immunity to a specific antigen. **444**

vacuole A fluid-filled, empty-looking organelle that isolates or disposes of waste, debris, or toxic materials. **55**

vagina Female organ of intercourse and birth canal. **518**

variable In an experiment, a characteristic or event that differs among individuals or over time. **13**

vascular bundle In plants, a multistranded cord of xylem and phloem in a stem or leaf. **542**

vascular cambium Lateral meristem that produces secondary xylem and phloem. **547**

vascular cylinder In plant roots, a central column of vascular tissue. **543**

vascular plant A plant that has xylem and phloem. **265**

vascular tissues Tissues that distribute water and nutrients through a plant body. *See* xylem, phloem. **540**

vas deferens Main duct that conveys mature sperm toward the urethra. **517**

vein Large-diameter vessel that returns blood to the heart. **411**

ventricle Heart chamber that pumps blood out of the heart and into an artery. **410**

venule Blood vessel that connects a capillary to a vein. **415**

vernalization Stimulation of flowering in spring by prolonged exposure to low temperature in winter. **576**

vertebrae Bones of the backbone (vertebral column). **396**

vertebral column Backbone. **300, 395**

vertebrate Animal with a backbone. **289, 301**

vesicle Small, membrane-enclosed organelle; different kinds store, transport, or break down their contents; e.g., a peroxisome or lysosome. **55**

vestibular apparatus In vertebrates, an organ of equilibrium in the inner ear. **488**

villi Multicelled projections that increase the surface area of the lining of the small intestine. **454**

viral envelope A layer of cell membrane derived from the host cell in which an enveloped virus was produced. **244**

viral reassortment Two viruses of the same type infect an individual at the same time and swap genes. **247**

virus A noncellular infectious particle with a protein coat and a genome of RNA or DNA; replicates only in living cells. **244**

visceral sensations Sensations that arise when sensory neurons in soft internal organs are activated. **483**

vitamin Organic substance required in small amounts for normal metabolism. **459**

warning coloration Distinctive color or pattern that makes a well-defended prey species easy to recognize. **338**

water cycle Water moves from its main reservoir—the ocean—into the air, falls as rain and snow, and flows back to the ocean. **347**

water mold Heterotrophic protist that forms a mesh of nutrient-absorbing filaments. **256**

water–vascular system Of echinoderms, a system of fluid-filled tubes and tube feet that function in locomotion. **298**

wavelength Distance between the crests of two successive waves of light. **83**

wax Complex, water-repellent mixture of lipids with long fatty acid tails bonded to long-chain alcohols or carbon rings. **37**

white blood cell Blood cell that helps defend the body against pathogens and cleans up cellular debris. **414**

wood Lignin-stiffened secondary xylem of some seed plants. **271, 547**

xenotransplantation Transplantation of an organ from one species into another. **183**

xylem Complex vascular tissue of plants; distributes water through its tracheids and vessel members. **265, 540**

yellow marrow Bone marrow that is mostly fat; fills cavity in most long bones. **397**

yolk Nutritious material in many animal eggs. **514**

zygote First cell of a new individual; a diploid cell that forms by the fusion of two haploid gametes. **144, 514**

Index

The letter *f* designates figure; *t* designates table; bold designates key terms; ■ indicates human health applications; ■ indicates environmental applications.